CONCEPTS in BIOINFORMATICS and GENOMICS

CONCEPTS in BIOINFORMATICS and GENOMICS

JAMIL MOMAND
California State University, Los Angeles

ALISON McCURDY
California State University, Los Angeles

CONTRIBUTORS

Silvia Heubach
California State University, Los Angeles

Nancy Warter-Perez
California State University, Los Angeles

New York   Oxford
OXFORD UNIVERSITY PRESS

Oxford University Press is a department of the University of Oxford.
It furthers the University's objective of excellence in research,
scholarship, and education by publishing worldwide.

Oxford   New York
Auckland   Cape Town   Dar es Salaam   Hong Kong   Karachi
Kuala Lumpur   Madrid   Melbourne   Mexico City   Nairobi
New Delhi   Shanghai   Taipei   Toronto

With offices in
Argentina   Austria   Brazil   Chile   Czech Republic   France   Greece
Guatemala   Hungary   Italy   Japan   Poland   Portugal   Singapore
South Korea   Switzerland   Thailand   Turkey   Ukraine   Vietnam

Published by Oxford University Press
198 Madison Avenue, New York, New York 10016
http://www.oup.com

Library of Congress Cataloging-in-Publication Data

Names: Momand, Jamil, author. | McCurdy, Alison, author.
Title: Concepts in bioinformatics and genomics / Jamil Momand, California
    State University, Los Angeles, Alison McCurdy, California State
    University, Los Angeles ; contributors, Silvia Heubach, California State
    University, Los Angeles, Nancy Warter-Perez, California State University,
    Los Angeles.
Description: New York : Oxford University Press, 2016. | Includes
    bibliographical references and index.
Identifiers: LCCN 2015048200| ISBN 9780199936991 (alk. paper) | ISBN
    9780199936984 (online material) | ISBN 9780190610531 (professor material)
    | ISBN 9780190610548 (professor material)
Subjects: LCSH: Bioinformatics--Textbooks. | Genomics--Textbooks.
Classification: LCC QH324.2 .M66 2016 | DDC 570.285--dc23

Printing number: 9 8 7 6 5 4 3 2 1

Printed in the United States of America
on acid-free paper

To our colleague Ray Garcia,
whose devotion to students will never be forgotten.

BRIEF CONTENTS

TABLE OF CONTENTS

PREFACE

In 2001, in a massive collaborative effort of scientists, working from a multitude of disciplines, including biology, biochemistry, chemistry, genetics, engineering, and computer science, one of the tremendous feats in the history of science was accomplished: the sequencing of the human genome.

Today, we can imagine a not-too-distant future in which our personal genomes are entirely known to us. We will download our genetic data and know by our sequences whether we are susceptible to particular diseases such as diabetes, cancer, and stroke. We'll modify our behaviors and mitigate these risks—our lives will change. For some of us, a poor genetic profile will affect our outlook on life, or the economics of our lives. How will medicine adapt to common knowledge of the genome? We do not quite know yet what this world looks like, but some of its weightiest questions are already being asked and debated—and studied by a rapidly expanding field of genomics and bioinformatics research. These are questions about the modern world, the modern person, and the future of biological science.

Welcome to the world of bioinformatics.

THE APPROACH

Concepts in Bioinformatics and Genomics takes a conceptual approach to its subject, balancing biology, mathematics, and programming, while highlighting relevant real-world applications. Topics are developed from the fundamentals up, like in an introductory textbook. This is a comprehensive book for students enrolled in their first course in bioinformatics. A compelling case study gene, the *TP53* gene, a human tumor suppressor with strong clinical applications, runs throughout, engaging students with a continuously relevant example. The textbook thoroughly describes basic principles of probability as they lead up to the concept of Expect value (*E*-value) and its use in sequence alignment programs. *Concepts in Bioinformatics and Genomics* also describes, from a mathematical perspective, the development of the hidden Markov model and how it can be used to align sequences in multiple sequence alignment programs. Finally, it introduces students to programming exercises directly related to bioinformatics problems. Thought-provoking exercises stretch the students' imaginations and learning, giving them a deeper understanding of software programs, molecular biology, basic probability, and program-coding methodology underpinning the discipline. The material covered in this book provides students with the fundamental tools necessary to analyze biological data.

ORGANIZATION

Introduction to Bioinformatics: Chapters 1–5

CHAPTER 1 is an overview of molecular biology. It will provide the essential biology vocabulary for understanding bioinformatics. Chapter 2 introduces GenBank, the database that stores the vast amounts of DNA and RNA sequence data crucial for bioinformatics research.

CHAPTER 3 discusses molecular evolution, which explains the diversity of sequences and how mutations get passed to progeny. Chapter 4 delves into the derivation of amino acid substitution matrices, the basis of sequence comparison programs, which help us connect molecular evolution to protein structure and function. Chapter 5 discusses amino acid substitution matrices and pairwise sequence comparison programs. Here, we begin to get into the nuts and bolts of algorithms that use data from evolution and protein domain conservation to infer whether two genes are homologs.

Biology: Chapters 6–10

CHAPTER 6 further develops the topic of pairwise sequence comparison by describing the Basic Local Alignment Search Tool (BLAST) and discusses multiple sequence alignment programs with an emphasis on the first popular program of this class—ClustalW. Chapter 7 is devoted to protein structure prediction programs. This chapter provides strong foundational knowledge of protein structures and the Protein Data Bank. Chapter 8 introduces phylogenetics with a discussion of DNA, protein sequence information, and the construction of phylogenetic trees. Chapter 9 presents genomics analysis with an emphasis on next-generation sequencing (NGS), and annotation of bacterial genomes. Chapter 10 is all about gene expression. Approximately half of this chapter is devoted to methods to measure transcript levels with an emphasis on microarrays and RNA-seq. The other half is devoted to proteomics, where we describe how mass spectrometry is used to identify proteins isolated from 2D-gels.

Mathematics: Chapters 11–12

CHAPTER 11 introduces you to probability, a requisite component of bioinformatics research, with an emphasis on counting methods, dependence, Bayesian inference, and random variables. In Chapter 12 the subject of a continuous random variable, introduced in the previous chapter, will be further developed into a discussion of the extreme value distribution and its use in analyzing the significance of an alignment. We conclude the chapter with stochastic processes, specifically Markov chains and hidden Markov models, as well as a mathematical derivation of the Jukes-Cantor model.

Programming: Chapters 13–14

CHAPTER 13 focuses on Python, a popular bioinformatics programming language. The Kyte-Doolittle Hydropathy sliding window program (one of the first popular bioinformatics programs) is used to illustrate Python fundamentals and to introduce you to the program design process. Chapter 14 follows this design process and steps you through the development of a pairwise sequence alignment tool.

FOR PROFESSORS

Approach and Rationale

The bioinformatics discipline has matured to the point where there is general agreement on the software programs and databases that are standards in the field. The algorithms that form the foundations of these software programs will not significantly change within the next three to four years. Similarly, databases that are bulwarks of the field will not vanish in the foreseeable future. Understanding the rationale for the basis of these bioinformatics tools is critical for students pursuing molecular life science or bioinformatics careers.

Flexible Organization

Overall, biology, mathematics, and computer science are presented in an order that systematically develops a student's understanding of the area. To highlight relevant connections between the three, we include cross-references in the main text and in footnotes. Those who wish to teach the course with the biology-heavy chapters in the beginning may consider presenting the chapters in the order listed in the table of contents. In this order, the biology-heavy chapters (Chapters 1 through 10) are followed by two mathematics-heavy chapters (Chapters 11 and 12) and two computer science-heavy chapters (Chapters 13 and 14).

If instructors wish to integrate computer programming early into the course, they may want to consider presenting the chapters in the following order: 1–5, 13, 14, and 6–12. Chapters 1 through 5 provide the biological rationale for pairwise sequence alignment and Chapters 13 and 14 provide the computer programming background so that students can create their own software tools to align sequences. The programming concepts in Chapters 13 and 14 reinforce the biological principles covered in Chapters 1 through 5. To provide students with more time to learn the Python programming basics, instructors may wish to intersperse topics from Chapters 13 and 14 among topics covered in Chapters 1 through 5. After covering Chapters 1 through 5, 13 and 14, material from the more biology-heavy chapters (Chapters 6–10) and the mathematics-heavy chapters (Chapters 11–12) can be covered.

Some bioinformatics and genomics courses are taught in a format consisting of a lecture section and a separate computer lab section. If this is the case, the lecture section can focus on Chapters 1 through 12, the lab section on Chapters 13 and 14. The lab section may allow more time for students to work through small coding assignments that together provide a foundation for a more extensive programming project (described in Chapter 14) to be completed by the end of the lab course. Another way of dividing the material between lecture and lab sections is to focus the lecture on the biology-heavy chapters (Chapters 1–10) and include Chapters 11–14 in the lab.

If instructors would like to integrate mathematics earlier in the course they may consider covering Chapters 11 and 12 just prior to Chapter 6. The introductory basic probability segment of Chapter 11, followed by the explicit derivation of extreme value distribution in Chapter 12, provide a strong foundation for the discussion of E-value, an important component of the BLAST

SUGGESTED ALTERNATIVE PRESENTATIONS OF TEXTBOOK

PRESENTATION ORDER OF FIRST FIVE CHAPTERS	ALTERNATE CHAPTER PRESENTATION ORDER	TYPE OF INTEGRATION
1, 2, 3, 4, 5,	6, 7, 8, 9, 10, 11, 12, 13, 14	Biology-heavy chapters first with cross-references to mathematics- and computer science-heavy chapters.
1, 2, 3, 4, 5,	13, 14, 6, 7, 8, 9, 10, 11, 12	Biology-foundation chapters first with computer science-heavy chapters more integrated.
1, 2, 3, 4, 5,	6, 7, 8, 9, 10, 11, 12-lecture 13, 14-lab	Biology-heavy and mathematics-heavy lecture section with a lab focused on computer science.
1, 2, 3, 4, 5,	6, 7, 8, 9, 10, 11, 12–lecture 11, 12, 13, 14–lab	Biology-heavy lecture section with a lab focused on mathematics and computer science.
1, 2, 3, 4, 5,	11, 12, 6, 7, 8, 9, 10, 13, 14	Biology-foundation chapters first with mathematics-heavy chapters more integrated.

program discussed in Chapter 6. The segment of Chapter 12 that introduces hidden Markov models will strengthen the students' understanding of multiple sequence alignment discussed in Chapter 6.

The table above shows our suggestions for alternative sequences of the textbook chapters that can be tailored to your particular needs.

THE FEATURES

Balance of Biology, Mathematics, and Programming

Concepts in Biochemistry and Genomics strikes a balance of topics for all students, no matter their background. Biology students will appreciate the reinforcement of the molecular life science topics and the gradual introduction to basic probability and programming concepts. Basic probability and programming use examples in biology to help biology students see the relevance of these concepts to molecular life science. Mathematics is expertly interwoven with bioinformatics concepts. Students with a background in computer programming will appreciate the basic biology primer in the first chapter. For students who already know how to program in another language, this textbook offers the opportunity to learn the fundamentals of a new language, Python.

Genomics

Genomics is a field that studies the entire sequenced genomes of organisms. Bioinformatics programs and databases are highly applicable to genomics because of the critical need to analyze and store a large amount of sequence data. Without bioinformatics, we cannot fully assess the genomics data we have collected. Chapters that emphasize genomics are Chapter 8 ("Phylogenetics"),

Chapter 9 ("Genomics") and Chapter 10 ("Transcript and Protein Expression Analysis").

Case studies of *TP53*, the Tumor Suppressor Gene

The *TP53* tumor suppressor is mutated in virtually all cancer types, and there is wide interest in using this knowledge to develop better cancer therapies. In Chapter 1, we discuss how p53 was discovered as a protein bound to a monkey virus oncoprotein, and in the last chapter, we show students how to create sequence alignment programs that quantify the similarities between p53 and its paralogs, p63 and p73. By the end of this textbook, students and instructors will have a deep understanding of the molecular biology of this gene and how bioinformatics can be used to further research progress in the fight against cancer.

Scientist Spotlight

Scientists who made significant contributions to the bioinformatics field are highlighted in "Scientist Spotlight" boxed sections. The scientists who created the first widely applicable amino acid substitution matrices (Margaret Dayhoff), the first global sequence alignment program (Christian Wunsch), the first local sequence alignment program (Michael Waterman), and the first program that successfully predicted protein membrane spanning regions (Russell Doolittle)—these are just a few of the brilliant discoveries and minds featured.

A Closer Look

From the *TP53* gene to DNA fingerprinting and the Neanderthal genome, this boxed material examines in detail some of the most important elements of *Concepts in Bioinformatics and Genomics*. Replete with figures, photographs, and excerpts from published texts,

"A Closer Look" provides the background and clarity needed to fully grasp the relevance of bioinformatics.

Thought Questions

Interspersed throughout the text, "Thought Questions" ask the important conceptual questions and prompt students to problem-solve and apply their knowledge on the fly. These questions provide students opportunities to self-test and better engage with their reading. Answers are found at the end of the chapter.

End-of-Chapter Exercises

Additionally, a robust list of end-of-chapter exercises encourages students to apply their bioinformatics knowledge holistically. Exercises are qualitative *and* quantitative, specific and comprehensive.

Glossary Terms

Glossary terms are highlighted and defined the first time they appear in the text. Concise explanations of the terms are also provided in the glossary section at the end of the book.

SUPPORT PACKAGE

Oxford University Press offers a comprehensive ancillary package for instructors and students using *Concepts in Bioinformatics and Genomics*.

For Students

Companion website (**www.oup.com/us/momand**): Resources and links to bioinformatics software, tools, and databases are available on the companion website. These are stable resources, such as Dotter, BLAST, GenBank, and many more, that have matured with the discipline into the essential tools for the bioinformatician. The companion site also provides downloadable programming tools that are necessary for students to complete the programming projects and end-of-chapter exercises.

For Instructors

The Ancillary Resource Center (ARC), located at **www.oup-arc.com/momand**, contains the following teaching tools:

- **Digital Image Library** includes electronic files in PowerPoint format of every illustration, photo, graph, figure caption, and table from the text—both labeled and unlabeled versions.

- **Answers to End-of-Chapter Questions** includes detailed solutions to all of the many exercises provided at the end of each chapter.

- **Editable Lecture Notes** in PowerPoint format for each chapter help make preparing lectures faster and easier than ever. Each chapter's presentation includes a succinct outline of key concepts and incorporates the graphics from the chapter.

ACKNOWLEDGMENTS

When writing a textbook for the first time it is difficult to foresee the amount of time it will require. For this project we resorted to carving out time between academic terms, during summers, on weekends, and late at night. Time usually spent with our families was, instead, dedicated to *Concepts in Bioinformatics and Genomics*. We would not have persevered without the encouragement and sacrifices of our spouses and families. For this we owe them tremendous gratitude.

We would also like to express our appreciation to Jason Noe, senior editor at Oxford University Press, who five years ago listened to our pitch and read our proposal to create a new textbook. His combination of patience, encouragement, and optimism moved the project along to its favorable conclusion. Working with Jason were assistant editor Andrew Heaton and editorial assistant Ben Olcott, who were very responsive to our long lists of questions. Jason also selected Dragonfly Media Group, whose team of Craig Durant, Caitlin Duckwall, and Rob Duckwall did a superb job refining our rough drafts of the illustrations. In production, art director Michele Laseau, designer Renata De Oliveira, and production manager Lisa Grzan produced an accurate and beautiful printed work. We would also like to thank Patrick Lynch, editorial director; John Challice, vice president and publisher; Frank Mortimer, director of marketing; David Jurman, marketing manager; Ileana Paules-Bronet, marketing assistant; and Bill Marting, national sales manager, along with the Oxford University Press sales force for their support.

We also thank our colleagues at California State University, Los Angeles (Cal State LA), Sandra Sharp and Kirsten Fisher, who read early drafts of chapters and offered invaluable advice on how to improve them. We acknowledge the National Science Foundation and National Institutes of Health, which jointly funded a project at Cal State LA; it gave us the opportunity to train more than 100 undergraduate and graduate students in bioinformatics and genomics over the course of several years. This project, called the Southern California Bioinformatics Summer Institute, created an environment of collegiality that made us realize how to use our respective fields to complement each other to create a rigorous bioinformatics training curriculum. Finally, we thank our Cal State LA students who, over the many years of enrolling in our bioinformatics course, collectively guided us to effectively convey the multidisciplinary concepts in this rapidly evolving field.

MANUSCRIPT REVIEWERS

More than 70 reviewers were commissioned to read draft manuscript chapters. We are grateful to each one for sharing insights and suggestions, which contributed greatly to the published work. We thank Oxford University Press for locating and commissioning top reviewers. To these reviewers we give a big thank you for your insightful comments and suggestions. We hope you find much in the book that will captivate and benefit your students.

Preston Aldrich, Benedictine University

Stephane Aris-Brosou, Ottawa University

Erich Baker, Baylor University

Guy F. Barbato, Richard Stockton College

Serdar Bozdag, Marquette University

Claudio Casola, Saint Louis University

Kari L. Clase, Purdue University

Soochin Cho, Creighton University

Tin-Chun Chu, Seton Hall University

Garrett Dancik, Eastern Connecticut State University

Heather Dehlin, Carroll University, Medical College of Wisconsin

Justin DiAngelo, Hofstra University

Qunfeng Dong, University of North Texas

Derek Dube, University of Saint Joseph

Bert Ely, University of South Carolina

Matthew Escobar, California State University San Marcos

Chester S. Fornari, DePauw University

Karl Fryxell, George Mason University

Arezou A. Ghazani, Harvard Medical School

Eugenia Giannopoulou, CUNY City College

Cynthia Gibas, University of North Carolina at Charlotte

Santhosh Girirajan, Pennsylvania State University–University Park

James Godde, Monmouth College

Michael Gribskov, Purdue University

Karen Guzman, Campbell University

Jeremiah Hackett, University of Arizona

Xiyi Hang, California State University, Northridge

Barry Hoopengardner, Central Connecticut State University

Yasha Karant, California State University San Bernardino

Anuj Kumar, University of Michigan

Sathish A. P. Kumar, Coastal Carolina University

Stephen Levene, University of Texas at Dallas

Zhijun Li, University of the Sciences in Philadelphia

Li Liao, University of Delaware

David A. Lightfoot, Southern Illinois University, Carbondale

Ping Ma, University of Illinois Urbana–Champaign

Padmanabhan Mahadevan, University of Tampa

Shaun Mahony, Penn State University

Susan McDowell, Ball State University

Brett McKinney, University of Tulsa

Vida Mingo, Columbia College

Murlidharan T. Nair, Indiana University South Bend

Michael Persans, University of Texas Pan-American

Helen Piontkivska, Kent State University

Sarah Prescott, University of New Hampshire

Catherine Putonti, Loyola University Chicago Lakeshore

Srebrenka Robic, Agnes Scott College

Michael Rosenberg, Arizona State University Tempe

Eric Rouchka, University of Louisville

Elizabeth Ryder, Worcester Polytechnic Institute

Eva Sapi, University of New Haven

Scott C. Schaefer, Lenoir-Rhyne University

Amarda Shehu, George Mason University

Kim Simons, Emporia State University

Malathi Srivatsan, Arkansas State University

Aurelien Tartar, Nova Southeastern University

Bryan Thines, Claremont Colleges

Vladimir Uversky, University of South Florida

Virginia Oberholzer Vandergon, California State University Northridge

Sabrina Walthall, Mercer University

Xiaofei Wang, Tennessee State University

David Weisman, University of Massachusetts, Boston

Amy Wiles, Mercer University

Zhenyu Xuan, University of Texas at Dallas

Mai Zahran, CUNY New York City College of Technology

Jamil Momand

Alison McCurdy

Silvia Heubach

Nancy Warter-Perez

ABOUT THE AUTHORS

JAMIL MOMAND has been a professor of biochemistry at Cal State LA since 1999. In 2000, he and Dr. Nancy Warter-Perez (contributing author and professor of electrical and computer engineering) created a course that introduces bioinformatics to students. The philosophy of the course is to appeal to students with a background in either the mathematics/computer science area or the molecular life science area. The two professors designed the course to help students be intelligent users of existing databases and software programs as well as to help them become developers of new bioinformatics software programs. The experience of teaching this course for many years was harnessed to write Chapters 1–6 and Chapters 13–14 in *Concepts in Bioinformatics and Genomics*. Dr. Momand also created the Bioinformatics and Computational Biology Minor Program at Cal State LA, and he is the student advisor for students in this program. He received his Ph.D. in biochemistry at UCLA studying the metabolism of aged proteins in Dr. Steven Clarke's laboratory. Dr. Momand was awarded a postdoctoral fellowship at Princeton University in molecular biology, where, in the laboratory of Dr. Arnold J. Levine, he demonstrated that the MDM2 proto-oncoprotein forms a complex with the p53 tumor suppressor protein. At the City of Hope National Medical Center Dr. Momand continued to study p53 and MDM2. He showed that p53 is susceptible to cysteine residue oxidation and that the oxidation destroys p53's ability to bind to DNA. Dr. Momand's research remains focused on the molecular mechanisms of cell growth control and cancer. One line of investigation his laboratory is currently pursuing is understanding how cancer cells become resistant to chemotherapy agents. Dr. Momand received the Cal State LA Outstanding Professor Award for the 2014–2015 academic year. His hobbies include camping, hiking, jogging, and ultimate Frisbee.

ALISON MCCURDY is a professor of chemistry at Cal State LA. After earning her B.S. in biological chemistry from University of Chicago in 1988 and her Ph.D. in chemistry at California Institute of Technology in 1995, she was a Camille and Henry Dreyfus Postdoctoral Fellow. She has been enjoying teaching chemistry—general, organic, bioorganic, and biochemistry—courses first at Harvey Mudd College and currently at Cal State LA. She strives to improve her pedagogy as an ongoing process and is active in securing funding for curriculum development in science courses. She has been heavily involved in the development of innovative curriculum and educational policies at all levels of the university. Dr. McCurdy collaborates and publishes on externally funded research projects with her research students, applying the techniques of chemistry to tackle challenging biological questions. Most recently, her laboratory students have been working with her to develop an organic chemical photoswitchable tool that may be used to help understand aspects of oscillatory calcium signaling. She was the recipient of the 2009 Cal State LA Distinguished Woman Award. Outside of work, she enjoys good meals, good books, and long, hilly bicycle rides with her husband Kerry.

SILVIA HEUBACH is a professor of mathematics at Cal State LA. Since she joined the faculty in 1994, she has been very active in interdisciplinary curricular development. She taught the mathematics component in the bioinformatics summer institute jointly run by Dr. Jamil Momand and Dr. Nancy Warter-Perez, which led to her involvement in this book project. She was also the principal investigator on a National Science Foundation grant to develop a general education modeling course for non-mathematics majors. In this course, students use the computer-algebra system *Mathematica* to explore the modeling process without the advanced background needed to create and analyze mathematical models. More recently, she was the driving force for the redesign of the mathematics sequence for life sciences majors, a major part of a project funded by the National Institutes of

Health to improve quantitative skills of life-science majors at Cal State LA. The newly created courses now form the required mathematics course sequence and have been taught since fall 2011. Dr. Heubach holds a master's degree in mathematics and economics from the University of Ulm, Germany, and both master's and Ph.D. degrees in applied mathematics from the University of Southern California. Her current research areas are enumerative combinatorics and combinatorial games, and she is the coauthor of a graduate text titled *Combinatorics of Compositions and Words*. She also served as the chair of her department for four years and as a vice-chair of the Southern California/Nevada Section of the Mathematical Association of America. Her accomplishments have been recognized by the 2013 Cal State LA Distinguished Woman Award and the 1999 Cal State LA Outstanding Professor Award. She received several research travel grants and was several times an invited researcher at the University of Haifa, Israel. Dr. Heubach enjoys watching foreign movies, dancing, hiking, camping, traveling to interesting places, and spending time with her husband and family at large.

NANCY WARTER-PEREZ is a professor of electrical and computer engineering at Cal State LA. Since joining Cal State LA in 1993, she has been dedicated to improving STEM education and outreach, particularly for underrepresented minorities, and has been actively involved in academic governance, currently serving as the Academic Senate chair. In 1994 Dr. Warter-Perez established the Cal State LA Compiler Research Group to study advanced compiler techniques for high-performance processors. She has developed and taught a broad range of computer engineering courses and since 2000 has co-developed curricula for training biologists and computer scientists in the field of bioinformatics. She has published widely on collaborative project based learning, an active learning strategy designed to help students persist and succeed in engineering. As director of the IMPACT LA NSF GK-12 Program from 2008 to 2014, she has worked to promote STEM by teaching graduate students how to communicate their research to a broad audience and by providing a wide range of opportunities for local middle and high school students to explore science and engineering careers. Dr. Warter-Perez holds a B.S. in electrical engineering from Cornell University and an M.S. and Ph.D. in electrical engineering from the University of Illinois at Urbana–Champaign. She is the recipient of the 2013 Cal State LA Outstanding Professor Award, the 2010 Cal State LA Distinguished Woman Award, and the NSF Young Investigator Award. She is also an avid supporter and cheerleader of her husband and four children and is always up for a game of ultimate Frisbee.

INTRODUCTION

In our time and for the foreseeable future, biology data is being collected at an astounding rate. Whether the data is a new DNA sequence from an exotic animal or the 3D structure of a novel protein, the information is invariably stored in a database—some of which is accessible through the Internet. In parallel, software tools to analyze this data have been developed steadily. It is essential for those who wish to perform research in life sciences to be aware of the databases and software programs commonly used by scientists. Bioinformatics is the field that has emerged from this information explosion.

Before delving into bioinformatics, let's review its formal definition. **Bioinformatics** is the research, development, or application of computational tools and approaches for expanding the use of biological, medical, behavioral or health data, including those tools and approaches to acquire, store, organize, archive, analyze or visualize such data. The origin of the term "bioinformatics" is obscure. In the late 1970s, a bioinformatics group was established at the University of Utrecht. Paulien Hogeweg published the first paper from this group on a program that graphically represents the predictions of different models for simulating real systems. Although she did not explicitly use the term "bioinformatics" in her paper, because of her affiliation with the bioinformatics group, it was the first time the term appeared in print. Perhaps independently, Jean-Michel Claverie used the French word "bioinformatique" in the 1980s and translated it to English during one of his lectures. In today's usage, bioinformatics overlaps with a number of other interdisciplinary subjects including, but not limited to, systems biology, computational biology, and genomics. Although nothing in the formal definition of bioinformatics limits it to molecular data, the bioinformatics field expanded rapidly in this area when the human genome sequence was published in 2001. Many software programs and databases have been developed to analyze molecular data; it is from this perspective that this textbook is written.

Because bioinformatics depends on the collection and availability of biological data, the question that emerges is why there is so much interest in the analysis and storage of this data. A few examples of the application of biological data to our society may help answer this broad question. In 1985, Alec Jeffreys discovered that it was possible to distinguish two individuals on the basis of their "DNA fingerprints." He found regions in the human genome that are highly variable and are called minisatellites. The lengths of the minisatellites could be used to identify individuals, much like traditional fingerprints. Fast forward several decades, and the United States passed a law that requires individuals convicted of felonies to submit their DNA to a national database (the Combined DNA Index System) for DNA fingerprint (or profile) analysis. Today, the database of DNA fingerprints from felons provides critical evidence to the criminal justice system.

Another example of the application of bioinformatics to our society is the discovery of how related humans are to Neanderthals (also written as Neandertals). Neanderthals belong to a species of hominids that, for a time, cohabited the earth with humans (*Homo sapiens*) up until 30,000 years ago, after which Neanderthals died out. Fossil data had long suggested that Neanderthals constituted a species distinct from humans, meaning they could not mate. Recently, the genome of Neanderthals was sequenced from bones and compared to the human genome using bioinformatics software tools. Although these tools were essential to this study, the feasibility of the sequencing and assembling the entire Neanderthal genome is largely attributed to the advances in the field of genomics. Genome comparison reveals that humans and Neanderthals shared a common ancestor 500,000 years ago. Interestingly, the comparison data shows that humans and Neanderthals had some degree of genetic mixing between 50,000 and 100,000 years ago. In fact, humans in Europe and Asia carry 1 to 4% Neanderthal DNA! These are just a few examples of the transformative power of bioinformatics and genomics.

In academia and in the workplace, it is generally accepted that those who practice bioinformatics can be placed in one of two camps: **users** or **developers.** Users access databases and software programs to answer biological problems. For example, a researcher may want to know if a new drug might have toxic effects on the body. If it is known precisely how the drug binds to its intended protein target, then one approach is to search for similar regions on other potential unintended protein targets. The programs and databases required to identify similar regions are readily available through the Internet. On the other hand, developers create new programs and databases that add to the repertoire of tools that increase our knowledge of molecular life sciences. For example, a group of developers at the National Center for Biotechnology Information (NCBI) created a software program called PSI-BLAST, which allows mining of databases for similar genes in widely different organisms. PSI-BLAST has been extensively used to determine the function of newly discovered genes. This book will cover areas essential for the user as well as the developer. The user will gain insight into how bioinformatics algorithms and databases operate; the developer will gain insight into the biological problems that are solved through bioinformatics tools.

Another field related to bioinformatics is **genomics,** which is the study of whole genomes from organisms. Bioinformatics tools are required for the scientific discoveries made in genomics. We will devote three chapters to specialized topics within the genomics field: Phylogenetics (Chapter 8), Genomics (Chapter 9), and Transcript and Protein Expression Analysis (Chapter 10). These topics are crucial because it is likely that the cost of sequencing the human genome will fall to the point where it is affordable for most people to have their own genomes sequenced. If used properly, personal genomes could be used in targeted preventative care and incentivize people to maintain a healthy lifestyle. The future of bioinformatics and its application to genomics will undoubtedly be surprising and exciting.

REFERENCES

Altschul, S. F., T. L. Madden, A. A. Schäffer, J. Zhang, Z. Zhang, W. Miller, and D. J. Lipman. 1997. "Gapped BLAST and PSI-BLAST: A New Generation of Protein Database Search Programs." *Nucleic Acids Research* 25: 3389–3402.

Bioinformatics Definition Committee. 2000. "NIH Working Definition of Bioinformatics and Computational Biology, July 17, 2000." http:// www.bisti.nih.gov/docs/compubiodef.pdf.

Claverie, J. M. and C. Notredame. 2007. *Bioinformatics for Dummies*, 2nd ed. Hoboken, NJ: Wiley Publishing.

Green, R. E, J. Krause, S. E. Ptak, M. T. Ronan, J. F. Simons, L. Du, M. Egholm, J. M. Rothberg, M. Paunovic, and S. Pääbo. 2006. "Analysis of One Million Base Pairs of Neanderthal DNA." *Nature* 444: 330–336.

Green, R. E. et al. 2010. "A Draft Sequence of the Neandertal Genome." *Science* 328: 710–722.

Hogeweg, P. and B. Hesper. 1978. "Interactive Instruction on Population Interactions." *Computers and Biology in Medicine* 8: 319–327.

Jeffreys, A. J., V. Wilson, and S. L. Thein. 1985. "Individual-Specific 'Fingerprints' of Human DNA." *Nature* 316: 76–79.

REVIEW OF MOLECULAR BIOLOGY

AFTER STUDYING THIS CHAPTER, YOU WILL:

■ Understand the relationship between genes, transcripts, proteins, and some functions carried out by proteins.

■ Know that genes are composed of DNA, transcripts are composed of RNA, and proteins are composed of amino acid residues.

■ Be familiar with the one-letter code for nucleotides and amino acids.

■ Appreciate the chemical nature of nucleotides and proteins.

■ Understand the term "sequence alignment."

■ Be familiar with the chemical properties of amino acid side chains.

■ Know that there are four levels of protein structure, each with a higher degree of complexity.

■ Understand that there exists a universal genetic code that creates a correspondence between an order of nucleotides in DNA to an order of amino acid residues in proteins.

■ Appreciate that DNA alterations can lead to protein alterations that affect protein function.

■ Understand how the first experiment demonstrating the relationship between a mutation and a disease was carried out.

■ Be familiar with the *TP53* tumor suppressor gene and its protein product p53.

1.1 INTRODUCTION

In this chapter, we review basic concepts in molecular biology necessary for comprehending bioinformatics. You will then be introduced to an important molecule, p53 protein (sometimes referred to as "p53," for short), that plays a significant role in preventing cancer. This molecule, as well as others, will be used to demonstrate the utility and power of bioinformatics research throughout this book.

1.2 GENES AND DNA

The genome is the DNA (deoxyribonucleic acid) that is found in an organism. Each cell in the organism contains a complete genome.[1] Part of the genome is devoted to making proteins. The segments of the genome that encode proteins are called genes.[2] Each protein has at least one function that it carries out for the cell, and, ultimately, for the organism. Figure 1-1 shows the hierarchy of genome, genes, proteins, and functions. Proteins carry out the many functions or tasks that collectively allow the organism to live and reproduce. A single function could be repairing damaged DNA or it could be a step in the breakdown of glucose for the generation of energy. We are just beginning to appreciate the fact that some proteins have more than one

---

[1] There are a few exceptions. For example, in some mammals, red blood cells, also known as erythrocytes, lose their DNA once they reach maturity, whereupon they fail to divide further and have a very limited lifespan.

[2] A small minority of genes express functional RNAs such as transfer RNA, ribosomal RNA, small nuclear RNA, and other types of RNA. These RNAs do not produce proteins but they perform important functions in the cell.

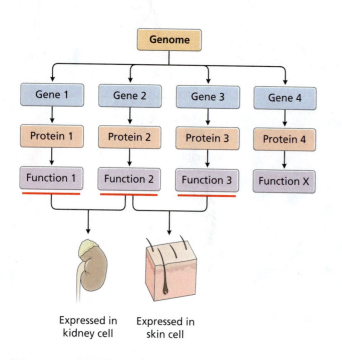

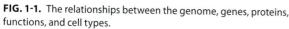

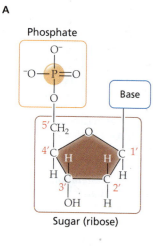

Expressed in kidney cell

Expressed in skin cell

FIG. 1-1. The relationships between the genome, genes, proteins, functions, and cell types.

function, but we will not explore this complexity further here. Note that although they have the same genome, not all cells express the same proteins. These differences in protein expression occur because different cell types have different specialized functions, and it is the repertoire of expressed proteins that gives each cell its specialized function. There is a subset of proteins unique to every cell type and there is a subset of proteins common to all cells. Proteins unique to a cell type are called specialized proteins, and proteins common to all cells are called housekeeping proteins. Figure 1-1 shows that a kidney cell produces protein 1 and protein 2, but not protein 3. A skin cell produces protein 2 and protein 3, but not protein 1. Proteins 1 and 3 are therefore specialized proteins and protein 2 is a housekeeping protein.

There are estimated to be 20,000–25,000 genes in the human genome, but they constitute only approximately 2% of the total DNA found in each cell.[3] The non-gene part of the genome remains a mystery and its exact function is an active area of research. All DNA, whether it is constituted of genes or not, is composed of paired nucleotides (sometimes called base pairs). It is important to know the structures and abbreviations associated with nucleotides. These abbreviations are crucial in bioinformatics, for they allow efficient storage and manipulation of the information contained in these fairly complex molecular structures. As you will see later in this chapter, abbreviations have been developed for other biomolecules as well.

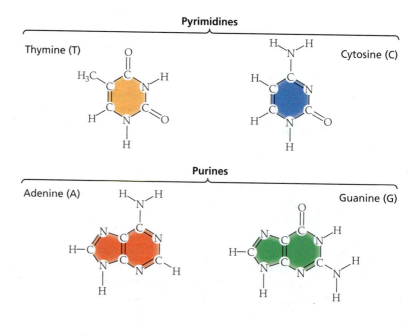

FIG. 1-2. Structure of a nucleotide within DNA.
A. The yellow-bordered atoms form the phosphate, the brown-bordered atoms form the sugar (deoxyribose), and the base (*bordered in blue*) can be any one of four types. **B.** The four types of bases are thymine, cytosine, adenine, and guanine.

[3] Here, we are referring to the genome that resides in the cell nucleus, sometimes referred to as the nuclear genome. Outside the cell nucleus there are much smaller genomes that reside in organelles called mitochondria (found in all eukaryotes) and chloroplasts (found in eukaryotes that undergo photosynthesis). Compared to the nuclear genome, these smaller genomes produce relatively few proteins.

Nucleotides are composed of three parts: a phosphate, a sugar, and a base. The parts of nucleotides that distinguish them from one another are called bases, and there are four bases found in DNA. Figure 1-2 shows the chemical structures of DNA nucleotides.

DNA is composed of two strands of nucleotides that are joined together. Between the strands of DNA, thymine associates specifically with adenine and cytosine associates specifically with guanine. The two strands are complementary if all of their bases can associate with each other. Figure 1-3A shows the chemical structure

protein One or more chains of amino acid residues where each chain is a minimum of 50 residues long. Proteins perform most of the biochemical functions in the organism.

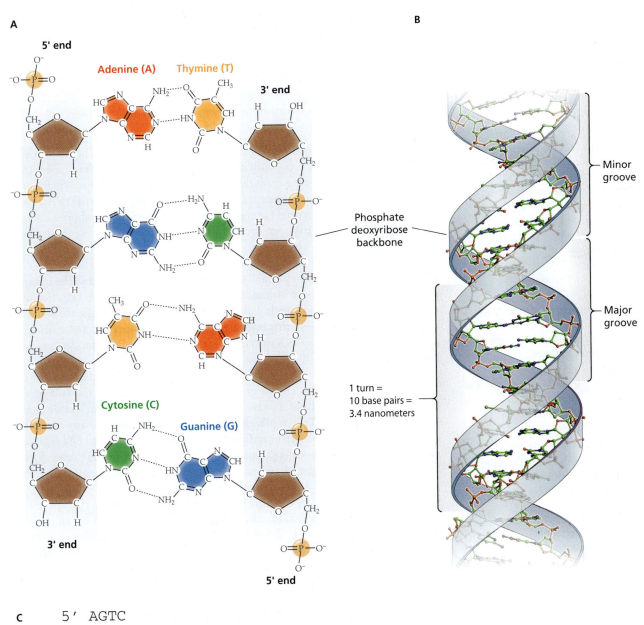

FIG. 1-3. Structure of DNA.
A. Close-up view of the bases and phosphate-sugar backbone. The two strands of DNA are joined together by base pairing where adenine associates with thymine and guanine associates with cytosine through hydrogen bonds (*signified by dots*). **B.** This image shows more clearly the double helix structure of DNA. The base pairs appear as rungs, and phosphate groups form the sides of the ladder. The positions of the atoms are exactly as they are in nature. **C.** The one-letter abbreviations used by bioinformaticists to write DNA sequences. Often this sequence is presented simply as AGTC because the lower strand is complementary and its sequence is inferred from the top strand.

TABLE 1-1. SINGLE-LETTER ABBREVIATIONS USED FOR DNA NUCLEOTIDE SEQUENCES

ONE-LETTER ABBREVIATION	NUCLEOTIDE NAME	BASE NAME	CATEGORY
A	Adenosine monophosphate	Adenine	Purine
C	Cytidine monophosphate	Cytosine	Pyrimidine
G	Guanosine monophosphate	Guanine	Purine
T	Thymidine monophosphate	Thymine	Pyrimidine
N	Any nucleotide	Any base	NA
R	A or G	A or G	Purine
Y	C or T	C or T	Pyrimidine
– or *	—	—	Gap

genome The DNA found in the organism. For some cells a separate genome exists in some organelles. In such cases, we distinguish the two genomes by saying there is a nuclear genome and an organelle genome (mitochondrial genome, chloroplast genome, etc.). In rare instances the genome can be composed of RNA, but this is limited to RNA viruses.

gene A segment of the genome that produces a protein or a functional RNA, such as transfer RNA, small nuclear RNA, ribosomal RNA, and so on.

nucleotide A molecule composed of a nitrogenous base, a sugar, and a minimum of one phosphate.

complementary strand The DNA strand that binds to the written strand through hydrogen bonds formed by pairs of nucleotide bases.

sequence alignment Optimized pairwise matching of nucleotide or amino acid sequences.

of DNA where two complementary strands associate with each other through hydrogen bonding. Note that the ends of each strand are labeled with numbers (5′ or 3′). These numbers originate from the naming conventions of the sugars found in DNA. Figure 1-3B shows the chemical structure of a longer stretch of DNA, revealing its double helical structure. Figure 1-3C shows the single-letter abbreviations of the bases of the DNA that correspond to the DNA depicted in Figure 1-3A. By convention, nucleotides are written as a sequence of letters from left to right with the left nucleotide being the 5′ nucleotide and the right nucleotide being the 3′ nucleotide. The complementary strand is listed below the first strand in the opposite orientation. Because the two strands of DNA are always complementary, the lower strand is often not explicitly included when writing a sequence. Thus, the sequence can be written as AGTC. In this book we will adopt the convention of writing the sequence of one strand of DNA with the 5′ nucleotide on the far left to represent the double strand DNA.

Table 1-1 lists the single-letter abbreviations used for each nucleotide. The two larger bases (see Figure 1-2) as a group are called purines, and the two smaller bases are called pyrimidines. In rare instances, when the DNA nucleotide cannot be identified experimentally, a placeholder, N, is used to represent any nucleotide when writing the sequence. In bioinformatics, it is common to try to find regions of similarity in DNA sequences that may reveal structural, functional, or evolutionary relationships. To find such regions one typically lines up the sequences to find sections that are identical. This process is called sequence alignment. If two DNA sequences are compared by aligning them, one may need to place a gap in one sequence to achieve an optimal match, or alignment (see Chapter 5 for an in-depth discussion of gaps). A gap is denoted as a single dash or an asterisk. Figure 1-4 shows the alignment of two nucleotide sequences where gaps are introduced to create an optimal alignment.

1.3 RNA: THE INTERMEDIARY

We are now able to broach the subject of RNA (ribonucleic acid), which, as the name suggests, is similar to DNA. RNA is the molecule that is intermediary between DNA and protein. RNA is transcribed from DNA and looks very similar to DNA, with three exceptions. First, RNA is usually a single strand; second, it uses the uracil (U) base instead of the thymine (T) base; and finally, it has a hydroxyl

Human $\longrightarrow$ CTGCCATGGAGGAGCCGCAGTCAGA
Frog $\longrightarrow$ CTGCCATG- - -GAACCTTCTTCTGA

FIG. 1-4. Segments of two gene sequences that are optimally aligned. The top gene sequence is from human, and the bottom is from *Xenopus silurana* (frog). The dashes in the frog sequence were inserted by the computer software program ClustalW to achieve optimal alignment between the two sequences.

group (−OH) on the 2′ carbon atom of the sugar. Figure 1-5 shows the structure of the RNA nucleotide and the uracil base. RNA nucleotides are linked to each other into a single strand not unlike a single strand of DNA. However, due to these small chemical deviations from DNA, RNA adopts a variety of configurations. One important function of RNA is to transform the gene information within DNA into protein.

Now might be a good time to introduce the central dogma of molecular biology (Figure 1-6). Developed in the 1960s, the central dogma explains how information encoded in DNA relates to information encoded in RNA and proteins. According to the central dogma, DNA replicates itself, RNA is transcribed from DNA, and protein is translated from RNA. In rare instances, RNA can be reverse transcribed into DNA.

It should be mentioned that RNA molecules are not always mere intermediaries that convey information from DNA in order to create proteins. RNA molecules can also be found in important structures of the cell, especially in the translation machinery, and can carry out a few chemical reactions. The structures of RNA are diverse and are often found in tight complexes with proteins. More recently, RNA was discovered to control the process of transcription and translation through RNA interference, often abbreviated RNAi.

1.4 AMINO ACIDS: THE BUILDING BLOCKS OF PROTEINS

By far the most abundant biological molecules on earth are proteins. Whenever you look at someone's face, you are actually looking at proteins. Proteins are made up of 20 amino acids. The amino acids are linked together in a linear fashion, and each amino acid has a unique side chain. Figure 1-7A shows a picture of two amino acids combining together to form a dipeptide (a peptide with two amino acids bound to

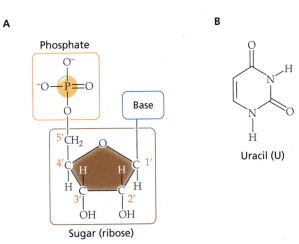

FIG. 1-5. Structure of a nucleotide within RNA. A. The green atoms form the phosphate, the black atoms form the sugar (ribose), and the base (*in blue*) can be any one of four types: uracil, cytosine, adenine, and guanine. **B.** The structure of uracil.

RNA (ribonucleic acid) A chain of nucleotides where each nucleotide is one of four bases, adenine, guanine, cytosine, uracil, and a ribose sugar and a single phosphate within each nucleotide. RNA is transcribed from DNA and performs many functions including the coding of proteins.

central dogma A term that explains the relationship between DNA, RNA, and proteins. Briefly, DNA replicates and serves as the template for its transcription into RNA. RNA serves as the template for its translation into protein. RNA can also be reverse transcribed into DNA.

transcription The process of polymerizing nucleotides to produce RNA from DNA.

translation The process of polymerizing amino acids in an order dictated by messenger RNA.

amino acid A molecule composed of an alpha carbon, a carboxyl group, an amino group, and a side chain. Amino acids can be polymerized to form proteins with distinct molecular functions. Each side chain gives a unique chemical property to its amino acid.

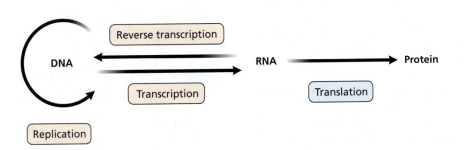

FIG. 1-6. Central dogma of molecular life science. The curved arrow surrounding the DNA signifies that DNA is capable of replicating itself.

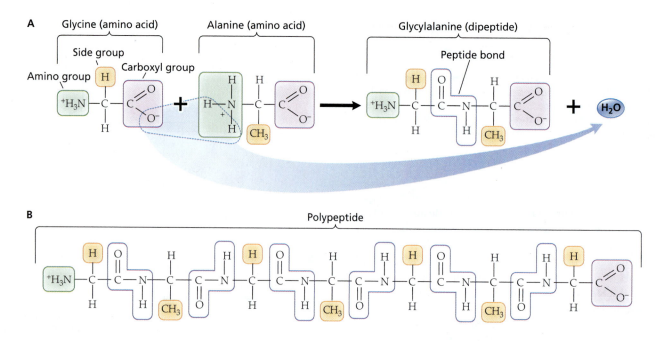

FIG. 1-7. Amino acids, the building blocks of proteins. **A.** Two amino acids combined together to form a dipeptide. **B.** Seven amino acids combined together to form a peptide.

each other).[4] When the two amino acids such as glycine and alanine combine, water is removed and the resulting bond between the amino acids is called a peptide bond. Once peptides are more than 50 amino acids long, they are called polypeptides, which are also known as proteins. Polypeptides (or proteins) can contain up to several hundred amino acids. Figure 1-7B shows seven amino acids combined together to form a peptide. By convention, the sequence of amino acids within a protein or peptide is written starting on the left with the first amino acid that contains a free amino group (sometimes called the amino terminus or N-terminus). The last amino acid in the protein contains the free acid group (sometimes called the carboxyl terminus or C-terminus).

It is incredible that the diversity of life that one sees can be attributed to just 20 amino acids. The proteins composed of these amino acids, in different arrangements and with different lengths, facilitate the multitude of chemical reactions that constitutes life. The side chains of the amino acids give proteins their particular characteristics or structures. Just as the bases are what distinguish nucleotides from one another, the side chains are what distinguish amino acids from one another. Scientists have divided the amino acids into different classes based on their relative ability to dissolve in aqueous (water) solutions. Figure 1-8 shows three classes of amino acids. In the hydrophobic class are those that are least able to dissolve in water. In the hydrophilic class are those that readily dissolve in water. The third class contains those amino acids that dissolve in water only slightly. Importantly, each amino acid has both a three-letter abbreviation and a single-letter abbreviation. The single-letter abbreviations (often called single-letter codes) are used extensively by bioinformaticians.

Hydrophobicity is a simple way of classifying amino acids, and there are other, more detailed classification schemes. Hydrophobic amino acids can be divided into those with aromatic and aliphatic side chains, both of which are composed of only carbon and hydrogen. Aromatic side chains are planar, and aliphatic side chains have kinks in them. Hydrophilic amino acids can be divided further into those with side chains that are, at physiological pH, electrically charged and uncharged. The amino

---

[4] When amino acids are chemically bonded to each other in proteins they are named "amino acid residues," or simply "residues."

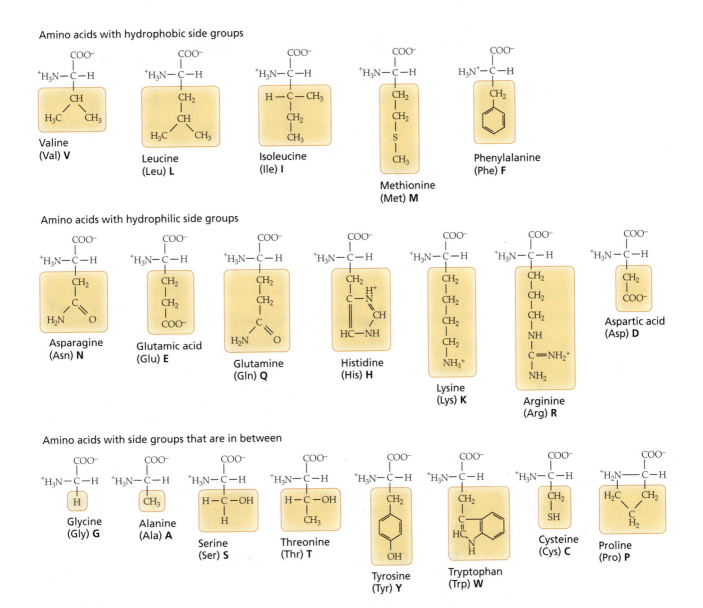

FIG. 1-8. **Structures, names, three-letter abbreviations, and single-letter codes for 20 amino acids.** In this figure, amino acids are classified on the basis of hydrophobicity. Note that there are other classification schemes as well (for example, see Table 13-7).

acids with positively charged side chains are basic, and those with negatively charged side chains are acidic. Amino acids with side chains that contain hydroxyl groups ($-OH$), sulfydryl groups (SH), carboxylic acid groups ($-CO_2H$), amide groups ($-CONH_2$), or amino groups ($-NH_2$) are polar because they can easily associate with water molecules through hydrogen bonds. Amino acid side chains can also be classified as large or small.[5]

In addition to the abbreviations for the 20 amino acids, there are seven more rarely used abbreviations. Amino acid abbreviations typically used in bioinformatics are shown Figure 1-8; a few additional unusual abbreviations appear in Table 1-2. B is used when it is difficult to distinguish between Asn and Asp experimentally. Many years ago, when proteins were first sequenced, the harsh chemical conditions required for protein sequencing prevented chemists from distinguishing these two

[5] See Table 13-7 for a more detailed amino acid classification scheme.

TABLE 1-2. ABBREVIATIONS USED FOR AMBIGUOUS AND RARE AMINO ACIDS

ONE-LETTER ABBREVIATION	THREE-LETTER ABBREVIATION	MEANING
B	Asn or Asp	Asparagine or aspartic acid
J	Xle	Isoleucine or leucine
O	Pyr	Pyrrolysine
U	Sec	Selenocysteine
Z	Gln or Glu	Glutamine or glutamic acid
X	Xaa	Any amino acid
– or *	—	No corresponding residue (gap)

amino acids. The same difficulties occurred when experimentally identifying Gln and Glu, and the abbreviation Z is used to represent these residues when they are indistinguishable. Because the overwhelming majority of amino acid sequences are now derived indirectly from DNA and not directly from protein sequencing, both B and Z are becoming obsolete. J is used when it is difficult to distinguish Ile from Leu. Mass spectrometry, a relatively new technology that is now used to sequence proteins, is unable to readily distinguish between Ile and Leu. O (Pyr) and U (Sec) are amino acids pyrrolysine and selenocysteine. These rare amino acids are sometimes called the 21st and 22nd amino acids. Pyr and Sec are coded by stop codons (see the genetic code section below) read by the translation machinery in a specific context. X is used when the amino acid is not known at a particular location. Finally, just as in DNA sequence alignments, protein sequences may be compared by aligning them. In protein alignments, a dash (–) or asterisk (*) is used to denote a gap that is inserted to optimize the alignment.

1.5　LEVELS OF PROTEIN STRUCTURE

Proteins have complicated structures that are necessary for them to perform their varied functions. Scientists have divided these structures into four levels, ranging from primary to quaternary. Figure 1-9 shows a chart listing the four types of protein structure. The primary structure is merely the order of bonded amino acids in a protein. For example, MAGTAK is a protein with the sequence Met-Ala-Gly-Thr-Ala-Lys, where methionine is at the amino terminus and lysine is at the carboxyl terminus. The secondary structure is the first type of folding the protein undergoes. There are three basic types of secondary structures: alpha helix, beta sheet, and turn (sometimes referred to as coil or loop). Relatively accurate structure prediction software programs have been developed that can successfully predict the secondary structure of a protein when the sequence is known. We will discuss these computer programs in Chapter 7. Once the protein begins to fold back onto itself, it forms a tertiary structure. There are many types of tertiary structures found in proteins, and predicting the tertiary structure from a primary sequence is a challenge.

When more than one polypeptide (chain of amino acids) associates specifically with one another, they form a quaternary structure. Sometimes, tertiary and quaternary structures are maintained, in part, by pairs of cysteine amino acids. Cysteines possess side chains that can form strong bonds to each other. These are called disulfide bonds. Figure 1-10 shows two cysteines forming a disulfide bond in a polypeptide. Insulin, a natural hormone responsible for uptake of sugar into tissues from blood, contains three disulfide bonds.

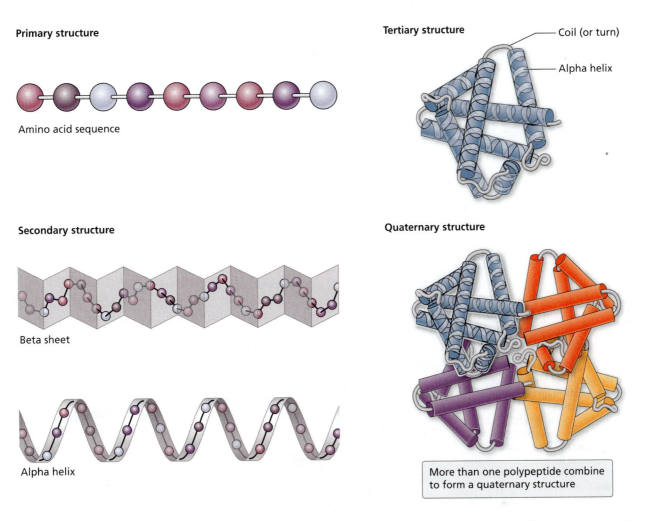

FIG. 1-9. Levels of protein structure. Primary structure is the sequence of amino acids. Secondary structure is the initial fold of proteins (alpha helix, beta sheet composed of beta strands, coil [not shown]). Tertiary structure is the folding of the protein on itself and often results in the active form of the protein. Quaternary structure occurs when more than one polypeptide associates specifically with one another.

A good example of a protein with quaternary structure is hemoglobin. This protein, abundant in red blood cells, is responsible for delivering oxygen within the blood from the lungs to the tissues. Hemoglobin contains four separate polypeptides, and it was one of the first proteins whose structure was determined by X-ray crystallography. Hemoglobin's ability to pick up and deliver the right amount of oxygen under different physiological circumstances depends on its quaternary structure. Not all proteins form quaternary structures, because in many cases one chain is all that is necessary for the proteins to carry out their functions.

1.6 THE GENETIC CODE

We now come to a topic that integrates the relationship between DNA, RNA, and protein—the genetic code. For scientists, cracking the genetic code was a breakthrough similar in magnitude to finding the Rosetta Stone more than two centuries ago. In 1798, Napoleon Bonaparte, a general from France, conquered Egypt and established the Institut de l'Egypt in Cairo. The Institut had several scientists available to catalog the antiquities that were being unearthed by the French. In 1799, during the construction of Fort Julien in the city of Rashid, an engineer discovered a stone with letters engraved on it. Scientists at the Institut immediately knew that the stone

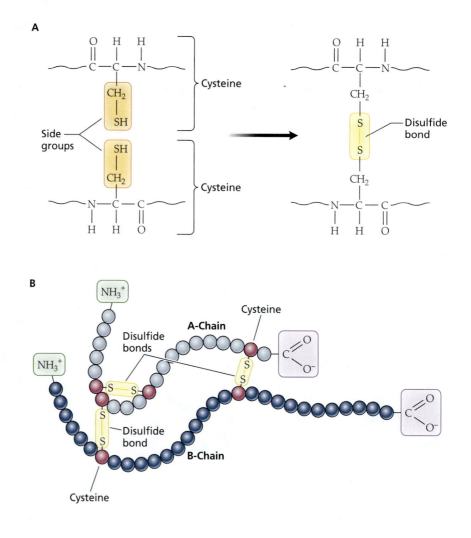

FIG. 1-10. Disulfide bonds between cysteines help maintain tertiary and quaternary structures.
A. Two polypeptides with cysteines positioned so that their thiol groups are in close proximity. Oxidation causes a disulfide bond to form. **B.** Insulin, an important hormone, contains three disulfide bonds that maintain quaternary structure. Chain A and chain B are two polypeptides.

was important. The engraving, dated to 196 BC, describes repealing taxes and gives instructions for raising statues in temples. The engraved decree, from the ruler Ptolemy V, was written in three languages: hieroglyphic, demotic, and Greek. Over the next 25 years, through the principal efforts of Jean-François Champollion, the demotic and hieroglyphic scripts were translated into French. The Rosetta Stone was incredibly important because it enabled archeologists to translate other ancient Egyptian texts into French and English. In a sense, the Rosetta Stone of molecular life science is the genetic code.

The genetic code was discovered through the hard work and the brilliant experiments of scientists in the 1960s. This code allows one to predict with nearly 100% accuracy the amino acid sequence of the protein if the DNA sequence is known. Astonishingly, the genetic code is universal—all organisms use the same code (with few minor exceptions). Thus, a DNA sequence that codes for insulin in humans can be used to produce human insulin in a bacterium known as *Escherichia coli* (*E. coli*). Both humans and bacteria use the same code to translate DNA sequences into amino acid sequences. Let us now turn our attention to the details of the genetic code.

In the genetic code, three nucleotides in a row (called a codon) code for one amino acid. Figure 1-11 shows a table depicting the genetic code. The letters in the left column are the first nucleotides of the codons. The letters in the top row are the second nucleotides of the codons, and the letters in the right column are the third nucleotides of the codons (Figure 1-11). For example, going from 5′ to 3′, the nucleotides ATG code for methionine. Methionine is usually the first amino acid coded in a protein-coding gene. This first methionine is coded by the start codon.

First position	Second position								Third position
	T		**C**		**A**		**G**		
	Code	Amino acid	Code	Amino acid	Code	Amino acid	Code	Amino acid	
T	T T T	phe	T C T	ser	T A T	tyr	T G T	cys	T
	T T C	phe	T C C	ser	T A C	tyr	T G C	cys	C
	T T A	leu	T C A	ser	**T A A**	**STOP**	T G A	**STOP**	A
	T T G	leu	T C G	ser	**T A G**	**STOP**	T G G	trp	G
C	C T T	leu	C C T	pro	C A T	his	C G T	arg	T
	C T C	leu	C C C	pro	C A C	his	C G C	arg	C
	C T A	leu	C C A	pro	C A A	gln	C G A	arg	A
	C T G	leu	C C G	pro	C A G	gln	C G G	arg	G
A	A T T	ile	A C T	thr	A A T	asn	A G T	ser	T
	A T C	ile	A C C	thr	A A C	asn	A G C	ser	C
	A T A	ile	A C A	thr	A A A	lys	A G A	arg	A
	A T G	met	A C G	thr	A A G	lys	A G G	arg	G
G	G T T	val	G C T	ala	G A T	asp	G G T	gly	T
	G T C	val	G C C	ala	G A C	asp	G G C	gly	C
	G T A	val	G C A	ala	G A A	glu	G G A	gly	A
	G T G	val	G C G	ala	G A G	glu	G G G	gly	G

FIG. 1-11. The genetic code.

There are 64 codons in total. In many cases, the same amino acid is coded by more than one codon. This phenomenon is called degeneracy. For example, serine is specified by six degenerate codons. Three codons, called stop codons, code for a stop: TAA, TAG, and TGA. Any one of these stop codons signal the cell to stop assembling amino acids into the polypeptide chain. In DNA, the codon immediately before the stop codon codes for the amino acid at the carboxyl terminus. Although one can convert DNA codons directly into amino acids on paper, in the cell, things are a bit more complicated. DNA is transcribed into RNA, and it is the RNA that is read by the protein translation machinery to assemble amino acids into polypeptides (remember the central dogma).

If you were given the DNA sequence of a small peptide made by a cell, say ATGTCTTCCTACAGAGGTTAA, and asked to determine its protein sequence, how would you do it? The sequence can first be broken up into the codons, ATG TCT TCC TAC AGA GGT TAA. Then, with the use of the genetic code table, you can assign an amino acid to each codon: MSSYRG. Remember, the last codon in the DNA codes for a stop. Protein sequences usually start with methionine, so it is a safe bet that our conversion on paper gave us the correct sequence of amino acids. However, if we had only a partial DNA sequence (one that did not include the start codon), then we do not know whether the sequence of nucleotides starts in the middle of a codon, and therefore we do not know if we should read the codons starting from the very first (leftmost) nucleotide. Furthermore, we would not know whether the given sequence describes the *complementary* strand of the DNA (the strand that we usually do not write down). Computer programs developed to translate DNA sequences into protein sequences cover all these possibilities. They give six possible protein sequences for a single DNA sequence: three for the strand that is shown (each reading begins at a different nucleotide within the first three nucleotides of the sequence) and three for the complementary strand. As long as the translation produces

codon Three nucleotides that code for an amino acid or a signal to terminate protein translation.

Relative sizes of matter

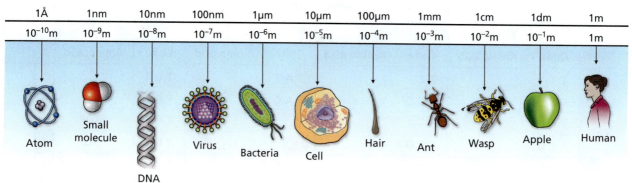

FIG. 1-12. Relative sizes of matter.

a protein sequence without stops, the nucleotide sequence is called an open reading frame (ORF). You can see for yourself whether we properly translated the DNA sequence above by using the translation tool found at the ExPASy website (http://web.expasy.org/translate/). The translation tool will show all six possible translations of the DNA sequence.

1.7 RELATIVE SIZES OF MATTER

Up to now we have discussed the molecules of life at the structural and chemical level. Much of what is new in bioinformatics centers on DNA, RNA, and protein, but it is important to keep a perspective of these molecules in terms of their relative sizes and their sizes relative to other common forms of matter. In Figure 1-12, starting with the size of an atom (a hydrogen atom, H, for example), we will step through the sizes of selected matter until we end with that of the human. Along the way, we will pause to flesh out some characteristics of matter relevant to bioinformatics and genomics.

We begin with the hydrogen atom, which is approximately angstrom (Å) or 10^{-10} meter (m) in diameter. In gas form, hydrogen exists as a diatomic molecule (H_2) and is the most abundant gas in the universe. A small molecule such as water (H_2O) has a diameter of 3.86 Å and the volume of approximately 30 cubic Å or 30 Å^3. A DNA molecule has a width of 20 Å and a length that varies from hundreds to billions of Å, depending on the organism. For eukaryotes, organisms with a cell nucleus, DNA is divided into chromosomes with lengths that are more manageable to manipulate. Humans have 23 sets of chromosomes (46 chromosomes in total), and houseflies (*Drosophila melanogaster*) have four sets of chromosomes. Figure 1-13 shows a chromosome composed of two chromatids that contain almost identical sequences of DNA. The chromosomes are held together near the middle by a centromere and the four tips of the chromosomes are called telomeres. Chromosomes consist of DNA and protein. Among their many functions, proteins help to activate and inactivate transcription and keep the DNA tightly compacted to fit within the confines of the cell nucleus. One type of protein that keeps DNA tightly wound and controls transcription is histone. The DNA found in one cell is considered the genome.

Viruses contain DNA (or RNA) and have a protein coat. Some viruses, such as human immunodeficiency virus 1 (HIV1) are wrapped in a coat made of lipid and protein. The sizes of most viruses fall with the 50 to 3000 Å range. With a light microscope one can barely detect free-living organisms called bacteria and archaea. They are free living, because unlike viruses, they can live outside of another living organism. Bacteria and archaea typically fall within a size range of 10,000 to 100,000 Å. Both are considered prokaryotes, although they are sufficiently distinct to be considered separate domains in the living world. The third domain is that of the eukaryotes, to which humans belong.

The cell is the basic unit of life. Organisms can be unicellular (consist of one cell) or multicellular (consist of many cells). Most eukaryotes are multicellular, where the

chromosome A segment of the genome tightly wound and combined with protein. The segments range in size. In diploid organisms, such as humans, there are two sets of almost identical segments (with the exception of the XY set in males). After genome replication each replicated segment is called a chromatid (a pair of identical chromatids are called sister chromatids). Chromatids are held together by a protein structure called a centromere near the center of the segment. The ends of the chromatids are called telomeres.

organelle Membrane-bound entities in cells, which perform specialized functions in eukaryotic cells. Examples include mitochondria, chloroplast, Golgi apparatus, nucleus, and nucleolus.

mutation An alteration in DNA sequence that produces a sequence that is different from normal DNA and is passed on to daughter cells.

neutral mutation A mutation that does not alter the ability of the organism to produce viable progeny (i.e., does not alter the fitness of the organism).

cells have specialized functions. Cells contain many subcellular components or organelles (see Figure 1-13). One organelle is the nucleus where the organism's genome resides and where the genome is transcribed into RNA. This freshly transcribed RNA, called the primary transcript, is exported from the nucleus to the cytoplasm where it is translated into protein. Other important organelles include the mitochondria, where chemical energy in the form of ATP is generated; the chloroplasts, where in plants sunlight is converted into ATP and carbon dioxide is incorporated into sugar; and the Golgi apparatus, where proteins are sorted into vesicles for transport to different parts of the cells or to the exterior of the cell. There are several other organelles that carry out critical functions for the cell.

Hair has a width of 100 microns (100×10^{-6} m) and consists of many protein polypeptides that are folded together in a quaternary structure. Ants, wasps, apples, and humans range in size from 1 millimeter (1×10^{-3} m) to 3 m, and are complex and multicellular. Each contains many cells that are grouped into specialized functions. In humans, tissues and organs consist of millions of cells that work together to perform a specialized function. As we continue to explore concepts in bioinformatics and genomics it is a good idea to keep relative sizes in mind.

1.8 DNA ALTERATIONS

Mutations are alterations in the DNA that produce a sequence that is different from normal DNA. Mutations are rare and are detected at a particular DNA location in less than 1% of the organisms in a species population.[6] They can result from errors in DNA replication or from DNA damage that has not been properly repaired by the cell. Mutations can cause disease, especially if they occur within genes. If a mutation gives benefit to the organism, the organism may thrive and reproduce and help the organism's offspring compete with other organisms for resources. This is called positive selection. On the other hand, if the mutation causes the organism to be sickly, the offspring will not survive and the mutation will not be observed after a generation. This is called negative selection. Most mutations, however, neither help nor harm the organism, primarily because mutations rarely occur in genes (remember that only 2% of the human genome codes for genes). Furthermore, even if mutations do occur in genes, many do not alter the amino acid sequence of proteins due to the degeneracy of the genetic code. Mutations that do not alter the fitness of the organism are called neutral mutations.

[6] Note that less than 1% frequency of occurrence at a particular DNA location is a practical definition of mutation. When a DNA alteration occurs at more than 1% frequency, the alteration is considered a polymorphism.

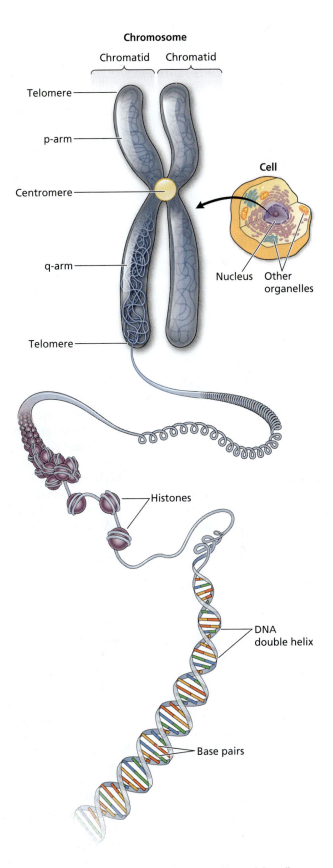

FIG. 1-13. DNA, histones, chromosomes, organelles, and the cell: how they fit together.

copy number variation (CNV) A DNA segment that is present at a number that is different from the number found in a reference genome with a usual copy number. For humans and other diploid organisms the usual copy number of the reference genome is two.

The most devastating mutations are those in which an extra chromosome is placed into every cell of the organism (called trisomy) and those that cause a loss of a chromosome in every cell of the organism (called monosomy). Trisomy and monosomy result from nondisjunction during meiosis—a process where chromosomes fail to properly segregate during the formation of sperm or egg cells. Most nondisjunctions are not compatible with life and result in miscarriages. Another type of severe mutation is the loss of a chromosome arm (a p-arm or q-arm), sometimes called a large deletion, which results in the loss of several genes. Alternatively, a mutation can result in gene amplification, which results in several more copies of the gene than normal. DNA amplification occurs in response to environmental signal or malignant transformation. In another type a mutation, a large segment of a chromosome can abnormally duplicate itself, causing gene duplication. In many cases of gene duplication, it is actually several genes that are abnormally duplicated, which can sometimes be detected as an elongation of the chromosome. In general, the alteration of the number of copies of a gene, whether more or less than normal, is called copy number variation, abbreviated CNV.

There is also a type of severe mutation, called translocation, where there is exchange of portions of arms of chromosomes. Translocations can result in abnormal expression of genes near the region where the two chromosome portions join (this junction where the two chromosomes join is called the breakpoint). Large deletions, gene amplification, abnormal gene duplications, and translocations are usually not tolerated by normal cells, but are frequent events in cancer cells (we will discuss these in further detail in Chapter 3).

Some mutations are more readily tolerated by the organism, but tolerance depends on the specific genes that sustain the mutations. An insertion/deletion mutation (called an indel mutation) is where one or more nucleotides are inserted or deleted. If an indel mutation occurs in a gene, it frequently causes a change in the amino acid sequence. Let's take a closer look at deletion mutations. If the sequence ATGTCTTCCTACAGAGGTTAA had a deletion of the 13th nucleotide, A, the codons resulting from this deletion, ATG TCT TCC TAC GAG GTT AA, would produce the amino acid sequence MSSYEV.[7] This sequence differs from the original sequence, MSSYRG, and it does not have a proper stop codon. When a deletion causes a change in the amino acid sequence after the deletion point, it is called a frameshift; it is often the case that the protein translation machinery adds amino acids to the growing polypeptide chain that were not originally intended for the protein. If the protein is essential for life, the deletion mutation can be life threatening. Equally devastating is an insertion mutation. Here, one or more nucleotides are inserted into the gene causing a gross alteration in the protein sequence.

More subtle are point mutations. A point mutation occurs when one different nucleotide is found in place of the normal (also known as wild-type) nucleotide. In some cases the point mutation will cause no change in the amino acid sequence. This lucky circumstance occurs when the point mutation changes the codon into one that codes for the same amino acid. In this situation, the mutation is called a neutral or silent mutation. For example, Ser is coded by six codons so there is a good chance that a point mutation in a Ser codon will result in no change in the amino acid. However, if the point mutation alters the codon so that it codes for a different amino acid, the mutation is called a missense mutation. A change in one amino acid out of an entire protein may not seem like a significant change, but, as we will see in the next section, even a single amino acid change in a protein can have severe consequences.

---

[7] This is also sometimes called a frameshift mutation because, after the indel mutation, the codons "read" by the translation machinery are different than the nonmutated gene.

1.9 A CASE STUDY: SICKLE CELL ANEMIA

What Are the Symptoms of Sickle Cell Anemia?

We are now ready to apply the knowledge we have gained to study an important disease that afflicts thousands of people, particularly those who live in the malaria belt in sub-Saharan Africa. The disease is sickle cell anemia. Patients with the disease exhibit a number of symptoms including pain episodes, strokes, increased infections, leg ulcers, bone damage, jaundice, early gallstones, lung blockage, kidney damage, loss of water in urine, painful erections in men (priapism), blood blockage in spleen or liver, lower red blood cell counts (anemia), and delayed growth. These symptoms can all be traced to the fact that tissues are not receiving enough oxygen. When viewed under the microscope, the carrier of oxygen, the red blood cell, normally appears discoid. This discoid appearance is similar to that of a hockey puck, except that the center is concave. The sickle cell, on the other hand, is a red blood cell that is misshapen and looks like a sickle (Figure 1-14). The sickle cells clump together and tend to stick to the blood capillary walls, causing an interruption in blood flow. When blood flow is interrupted by sickle cells, tissues become damaged due to lack of oxygen.

sickle cell anemia Disease caused by a mutation that results in an abnormal sickle shape of the erythrocyte. The abnormal erythrocytes stick to endothelial cells of the blood capillaries and prevent blood flow.

paper chromatography Experimental method that separates molecules on the basis of charge and hydrophobicity.

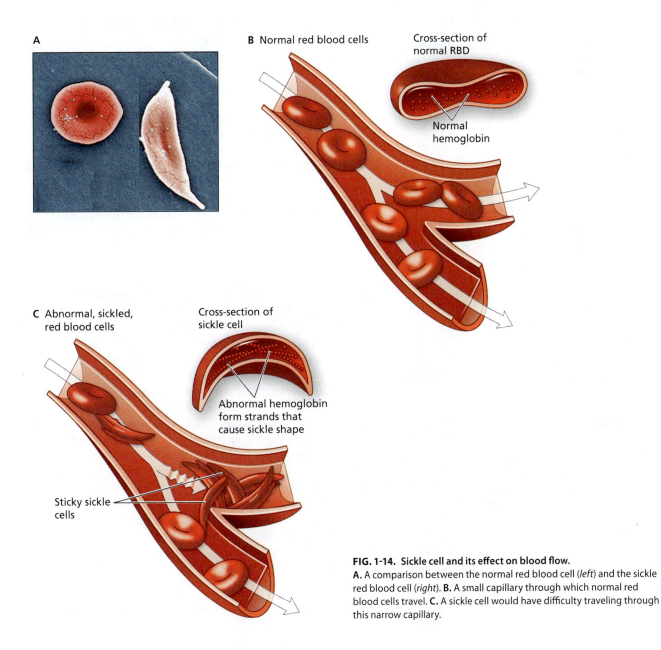

A. Normal red blood cells
Cross-section of normal RBD
Normal hemoglobin

B. Normal red blood cells

C. Abnormal, sickled, red blood cells
Cross-section of sickle cell
Abnormal hemoglobin form strands that cause sickle shape
Sticky sickle cells

FIG. 1-14. Sickle cell and its effect on blood flow.
A. A comparison between the normal red blood cell (*left*) and the sickle red blood cell (*right*). **B.** A small capillary through which normal red blood cells travel. **C.** A sickle cell would have difficulty traveling through this narrow capillary.

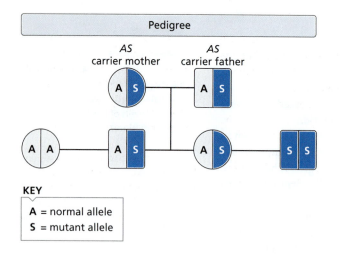

KEY

A = normal allele
S = mutant allele

FIG. 1-15. A pedigree showing the inheritance pattern of sickle cell anemia. A is the normal allele of the gene and S is the mutant allele. Both the mother (*top left circle*) and father (*top right square*) are carriers of the disease. On average, one quarter of the children will be normal (*female on left*), one half of the children will be carriers (the male and female with shapes that are divided into red and white), and one quarter of the children have the sickle cell disease (*male child on the right*).

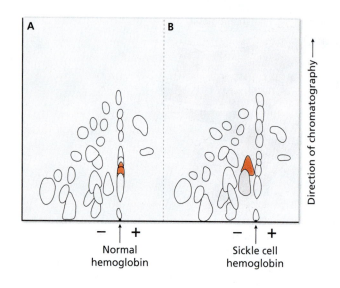

FIG. 1-16. Paper chromatography separation of hemoglobin peptides.
A. Peptides from normal hemoglobin.
B. Peptides from sickle cell hemoglobin. The peptide in red has a different migration pattern on the two chromatography plates. Arrowheads near bottom indicate where peptides were spotted prior to separation.

Sickle cell anemia is a familial disease, so patients inherit the disease from their parents. It is considered a recessive disorder because one must receive two mutated copies of the DNA in order to display symptoms of the disease.[8] It is possible to be a carrier of the disease without showing symptoms. The carriers are heterozygous because they carry one good copy of the gene and one bad (mutated) copy; they are said to have the sickle cell trait. A pedigree is a good way to keep track of the disease from generation to generation. Figure 1-15 shows a pedigree in a sickle cell disease family. Here, the mother (*top left circle*) and the father (*top right square*) both carry one normal copy (A) and one mutant copy (S) of the gene that causes sickle cell anemia. The child will exhibit the disease symptoms if s/he receives a mutant copy from each parent. For example, the son on the right bottom row in the pedigree will have sickle cell anemia.

Sickle Cell Anemia Is the First Disease Linked to a Specific Mutation

In 1949, Linus Pauling and his colleagues came up with an ingenious experiment to show that sickle cell disease was caused by an amino acid change in the protein hemoglobin. The experiment was to determine the electrical charge of hemoglobin from different people. Hemoglobin is the protein that binds to oxygen within the red blood cell. He found that hemoglobin from sickle cell anemia patients was more positively charged than hemoglobin from normal individuals. He suggested that this charge difference was the reason that hemoglobin was abnormal in sickle cell patients and that this charge difference could be due to a single amino acid change. Later, in 1956, Vernon Ingram devised an experiment to visualize this difference more clearly.

To detect the origin of the charge difference, Ingram clipped hemoglobin into small pieces with an enzyme called a protease. The peptide products of hemoglobin were separated from each other by paper chromatography. In paper chromatography, a mixture of peptides is placed in the lower right corner of the paper. Peptides are separated by net charge in the X direction by applying a voltage across the paper. Once they are separated by their charge, the peptides are separated by hydrophobicity in the Y direction by allowing a solvent to run up the paper. As a result, each peptide migrates to a particular position on the paper depending on its amino acid composition. Incredibly, Ingram found that only one peptide from sickle cell hemoglobin migrated differently from that particular peptide from normal hemoglobin. Figure 1-16 is an illustration of the peptide spots on the paper. One year later it was found that the peptide from the normal individual had a Glu and that this Glu was replaced by Val in the sickle cell hemoglobin. This change was the result of a missense point mutation in the gene that expresses hemoglobin. This stunning finding was the first demonstration of a single amino acid change causing a disease and was thought by many to have ushered in the field of molecular diagnostics.

---

[8] A dominant disorder is where one needs only one mutated copy of the gene to show the full effect of the disease.

1.10   INTRODUCTION TO p53

Sickle cell anemia was the first disease linked to an amino acid change, and therefore to a mutation in DNA.[9] This discovery led to the demonstration that other diseases were caused by mutations in DNA. Another gene linked to disease is *p53* (also known as *TP53*). Just a note about writing convention is needed. We will adhere to the U.S. National Library of Medicine's gene-naming conventions and use italics to denote the gene and transcripts; and use normal font to denote the protein product of the gene. *TP53* is mutated in approximately 50% of human cancers. It is known as a tumor suppressor gene because it normally suppresses tumors. In such cancers, one copy of *TP53* often contains a point mutation that codes for a protein with a single amino acid substitution (missense mutation). A second copy of *TP53* is completely deleted from the cells (deletion mutation). Without any normal *TP53*, cells grow uncontrollably.

One may ask, what causes mutations? Environmental pollutants, carcinogens in cigarettes, harmful radiation, and even diet can promote mutations. In rare cases, not unlike sickle cell anemia, cancer is an inherited disease (see Box 1-1). In 1969 two physicians, Frederick Li and Joseph Fraumeni Jr., described a rare cancer that seemed to run in families. The cancer could be in any tissue, but it was often a sarcoma, a cancer of the connective tissue. A patient with Li-Fraumeni syndrome has an 80% chance of dying from cancer, and often the cancer is diagnosed before the patient reaches twenty years of age. Several years after the cancer was first described, when it became clear that *TP53* was a tumor suppressor gene, normal tissues from patients with Li-Fraumeni syndrome were tested for mutations in *TP53*. Sure enough, every tested tissue of patients with this syndrome had one mutated copy of *TP53* (the mutation caused a loss of function of the p53 protein, usually a missense mutation in the DNA-binding domain of the protein). The second copy of *TP53* was normal (called wild-type). However, cancer tissues of these patients had two mutated copies of *TP53*—one a missense and the other a deletion. These patients are prone to cancers because all of their cells start out with only one copy of wild-type *TP53*. On the other hand, the overwhelming majority of people start out with two copies of wild-type *TP53* in all of their cells.

The protein product of the *TP53* gene was discovered in 1979 by four research groups working independently. The discovery was fascinating and illustrates how scientists used new methods to make seminal discoveries. The scientists used cells infected with a virus, called the Simian Vacuoling 40 Virus (SV40), which was discovered in African Green monkeys (*Cercopithecus aethiops*) in 1960. When rodents were inoculated with SV40, the rodents contracted cancers. Investigators wanted to uncover the cancer-causing agent in SV40. To do this, they obtained proteins from the rodent cancers and created antibodies against these proteins. The majority of the antibodies reacted against one protein coded by the SV40 virus DNA. That protein was large T antigen—which is necessary for the virus to cause cancer.

Interestingly, a few of the antibodies recognized a second protein from the cancer cells. The second protein was not encoded by the virus DNA but rather by the DNA of the rodent cells. The second protein appeared to have a molecular weight of 53,000 daltons (or 53 kilodaltons) and it was clear that this protein formed a tight association with large T antigen. Because not much was known about the function of this protein, it was called p53—for protein 53 kilodaltons (a kilodalton is the mass of approximately 9 amino acids). Much later, it was discovered that large T antigen caused cancer largely due to its ability to associate with p53 and *inhibit* p53 from performing its normal function. The fascinating story of how p53 eventually became known as a central defender against cancer will be explained serially in each chapter of this book as we proceed to learn more about bioinformatics and genomics.

p53 A tumor suppressor protein (canonical length in humans is 393 residues) responsible for initiating cell cycle arrest, apoptosis, or DNA repair in response to cell stress (DNA damage, ribosome denaturation, oncogene activation, etc.).

TP53 The gene that codes for p53, located on chromosome location 17p13.1, approximately in the base pair range 7,571,720–7,590,868. The gene codes for multiple alternatively spliced forms of mRNA.

---

[9] This particular sickle cell mutation is a point mutation that results in a T to A nucleotide transversion (T→A) within codon 6 of the β globin gene. Codon 6 is converted from CTC to CAC. This DNA point mutation results in a Glu→Val change.

A Rare Inherited Cancer Is Caused by Mutated *TP53*

BRIDE WITH RARE CANCER GENE OPTS FOR MASTECTOMY AFTER WEDDING

Russell Jenkins

Joan Crossland is approaching her wedding day with more nervous anxiety than most because only a day after she returns from honeymoon she will have to undergo a double mastectomy to save her life.

She has inherited a rare genetic disorder, Li-Fraumeni syndrome, which means that she is susceptible to the same kind of cancers that claimed the lives of her mother and brother.

Miss Crossland, 29, from Ashton-under-Lyne, Greater Manchester, says the radical surgery offers the best chance of breaking the deadly cycle. Like other sufferers, she contracted a cancerous growth at an early age. She survived an adrenal tumour as a child and had hoped that she was stronger than the syndrome that had blighted her family.

She has undergone regular screenings since the age of 19 because of her family's medical history but was devastated to notice an incipient lump on her breast several months ago. This month she and her fiancé, Mat Swift, 27, were told that it was an aggressive type of breast cancer.

Miss Crossland said, "I wandered around in a daze. I cried a lot until I got to the stage where I thought 'yes, this is real, I have got to deal with it.' Mat's initial reaction was anger. He wanted someone to blame, but when he calmed down he said, 'Right, we know what it is, now how do we fix it?'

"I am not vain but I have always considered my breasts to be one of my best features. Mat reassured me, though. He said, 'You are much more to me than a pair of breasts.'" Li-Fraumeni, which is named after two American physicians, is an extremely rare inherited predisposition to cancer.

One of the symptoms of the disorder is that cancer can strike at a young age. Miss Crossland's mother, Kathleen, died aged 28 and her brother, David, died of leukaemia aged 2.

"I was dreading telling my father, Leonard," said Miss Crossland. "Of course, he had hoped that the day would never come. When it did, he cried but he realises that technology has advanced dramatically since Mum died, and that unlike Mum I quickly identified a lump and sought medical advice straight away."

Miss Crossland is now preparing for the prospect of chemotherapy. "I might get away with hormone therapy," she said. "But if it does have to be chemotherapy and I have my head shaved, then so be it." Her syndrome will also have implications for any decision to start a family.

Meanwhile, she is continuing to plan for her wedding on August 11. She said, "Before the diagnosis I was worrying about table settings. Now I do not care. So long as everyone turns up and has a good time that is all that matters. I have to look at the bigger picture now."

SUMMARY

This chapter is a brief overview of the molecules that make life possible. The structure of DNA, discovered in 1953, is used as a template to create or transcribe RNA. The three nucleotide codons within RNAs are then translated into the amino acid sequences of proteins using a universal genetic code. Proteins have sufficient structural complexity to guide the development of simple and complex organisms. Much of bioinformatics is directed at using computers to compare and contrast nucleotide and amino acid sequences and predicting structures of proteins and RNAs. The variability of sequences is often caused by mutations that became stable in genomes either through positive or neutral selection. The discovery by Linus Pauling and his colleagues of the specific amino acid change that causes sickle cell anemia ushered in an era of molecular diagnostics—an area that is still expanding. Related to the fact that mutations cause disease is the realization that mutations also drive selection processes that, in turn, promote evolution. One gene that is frequently mutated in cancers is *TP53*, a fascinating gene that has been dubbed the guardian of the genome. As we explore different bioinformatics and genomics topics we will come to understand how *TP53* earned this curious moniker.

EXERCISES

1. **A good resource for information on disease-related molecules is the Online Mendelian Inheritance of Man (OMIM). OMIM is linked to the National Center for Biotechnology Information (NCBI) website. To optimally use this website it is important to become familiar with chromosomes. Chromosomes are large segments of DNA bound to protein found in the nucleus of eukaryotic cells. For many organisms, their genomes are distributed as chromosomes. Humans have 23 sets of chromosomes (46 total chromosomes). Each chromosome is divided into two major areas. The short arm (p-arm) is located above the centromere, and the long arm (q-arm) is located below the centromere. Telomeres are located at the very ends of the chromosome (Figure 1-17).**

   To determine the location of a gene on a chromosome it is important to note that chromosomes can be stained with special dyes that create a series of bands on the chromosomes. If a gene is located at 8q21.3, that would mean chromosome 8, q-arm, band 21.3. The higher the band number, the farther away from the centromere the gene is located.

   a. Give the chromosomal location (in terms of banding position) of the gene that causes sickle cell anemia.

   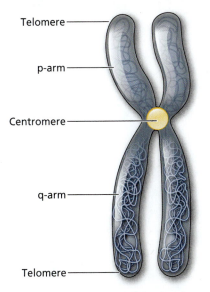

   FIG. 1-17. Human chromosome showing the centromere, telomeres, p-arm, and q-arm.

   b. Give the name of the gene that causes sickle cell anemia.
   c. Find out the nucleotide change and amino acid change that leads to sickle cell anemia.
   d. Explain the change in distribution of spots in the peptide map (Figure 1-16) based on properties of the side chain of the mutant amino acid in the disease.
   e. If sickle cell anemia is so devastating, why has it persisted in the population for such a long time? Give a molecular explanation.

2. **One important exercise a bioinformatician performs is to compare amino acid sequences. One reason to make comparisons is to determine the parts of the proteins that are critical for function. These regions are generally conserved within proteins that perform the same duties. Conserved regions are those that have nearly the same amino acid sequences. Proteins that perform the same duties are called homologs and can be found in different species. For example, p53 from humans and p53 from frogs perform the same functions. There are some regions within these proteins that will be similar in both humans and frogs. We call these regions conserved sequences. A multiple sequence alignment allows the bioinformatician to readily line up amino acid sequences of related proteins. The conserved regions are identified in the alignment. For this exercise, perform a multiple sequence alignment of three homologs of cytochrome C. The three homologs are from human, yeast, and dog, and the accession numbers for these sequences are AAA35732, NP_001183974, and 1YCC. The first two sequences can be found in the protein database at the NCBI. The last sequence can be found in the structure database at the Protein Data Bank. Print out your multiple sequence alignment result and attach a short paragraph explaining how the alignment gives you a clue as to which parts of the cytochrome C protein you would hypothesize are most important to its function. (The function is the same in all three organisms.) For those of you unfamiliar with NCBI, here are specific instructions:**

   • Go to the NCBI website.
   • Use the dropdown menu to search in the "Protein" database.

- Enter the accession number and click "Search."
- Change the format to FASTA.
- Copy the FASTA output into a sequence alignment window of a website that hosts a sequence alignment program. Make sure the header (the line with the > symbol) is placed on top of the sequence within the window.
- Repeat for each FASTA output of the remaining two proteins.
- Once all three sequences are pasted into sequence alignment input window, run the program.

REFERENCES

DeLeo, A. B., G. Jay, E. Appella, G. C. Dubois, L. W. Law, and L. J. Old. 1979. "Detection of a Transformation-Related Antigen in Chemically Induced Sarcomas and Other Transformed Cells of the Mouse." *Proceedings of the National Academy of Science USA* 76: 2420–2424.

Elgar, G., and T. Vavouri. 2008. "Tuning in to the Signals: Noncoding Sequence Conservation in Vertebrate Genomes." *Trends in Genetics* 24: 344–352.

Fire, A. Z. 2007. "Gene Silencing by Double Stranded RNA." *Cell Death and Differentiation* 14: 1998–2012.

Ingram, V. M. 1956. "A Specific Chemical Difference Between Globins of Normal and Sickle-Cell Anaemia Haemoglobins." *Nature* 178: 792–794.

Ingram, V. M. 1957. "Gene Mutations in Human Haemoglobin: The Chemical Difference Between Normal and Sickle Haemoglobin." *Nature* 180: 326–328.

International Human Genome Sequencing Consortium. 2004. "Finishing the Euchromatic Sequence of the Human Genome." *Nature* 431: 931–945.

Kress, M., E. May, R. Cassingena, and P. May. 1979. "Simian Virus 40-Transformed Cells Express New Species of Proteins Precipitable by Anti-simian Virus 40 Tumor Serum." *Journal of Virology* 31: 472–483.

Lane, D. P., and L. V. Crawford. 1979. "T Antigen Is Bound to a Host Protein in SV40-Transformed Cells." *Nature* 278: 261–263.

Lane, D. P. 1992. "p53, Guardian of the Genome." *Nature* 358: 15–16.

Levine, A.J. 2009. "The Common Mechanisms of Transformation by the Small DNA Tumor Viruses: the Inactivation of Tumor Suppressor Gene Products: p53." *Virology* 384: 285–293.

Li, F. P., and J. R. Fraumeni Jr. 1969. "Soft-Tissue Sarcomas, Breast Cancer and Other Neoplasms: A Familial Syndrome?" *Annals of Internal Medicine* 71: 747–752.

Linzer, D. I., and A. J. Levine. 1979. "Characterization of a 54K Dalton Cellular SV40 Tumor Antigen Present in SV40-Transformed Cells and Uninfected Embryonal Carcinoma Cells." *Cell* 17: 43–52.

Pauling, L., I. Harvey, R. S. Singer, and I. Wells. 1949. "Sickle Cell Anemia, a Molecular Disease." *Science* 110: 543–548.

INFORMATION ORGANIZATION AND SEQUENCE DATABASES

2.1 INTRODUCTION

In Chapter 1, you learned the vocabulary of molecular life science—a necessary prerequisite for bioinformatics. We also reviewed DNA, RNA, and proteins in the context of structure and the central dogma. Finally, you learned that single-base mutations in the genome can lead to devastating diseases. In this chapter, you will learn about the major public databases that are repositories for sequence data. We describe in detail one of these databases, GenBank, the grandparent of all nucleic acid sequence databases. GenBank started in 1983 as a joint venture between Walter Goad's group at the Los Alamos National Laboratory and Baranek and Newman, Inc. (Box 2-1), and is now the largest repository of nucleotide sequences. GenBank is the oldest continuously running database containing nucleic acid sequence data. It accepts DNA and RNA sequence data directly from scientists who are sequencing in their laboratories. Most journals require researchers to deposit new sequence data into GenBank prior to acceptance of manuscripts for publication. The data in GenBank has been used as source material for several other databases. One of these is Reference Sequence (RefSeq), a database that contains natural (wild-type) sequences—sequences that exist naturally in organisms and contain no mutations. Another database that gets information from GenBank is Protein Knowledge Database (UniProtKB), a

Refseq = database for WT sequences, but what is truly WT?

Walter Goad
GenBank Founder

BOX 2-1 | **SCIENTIST SPOTLIGHT**

Born September 5, 1925, in Marlowe, Georgia, Walter Goad explored the world beyond the rural south at an early age and in 1942 he landed a job in Schenectady, New York, as a radio station engineer. At his employer's urging he enrolled in Union College, where he received his bachelor's degree in physics.

In 1950, while a graduate student at Duke University, Goad joined the staff of the Theoretical (T) Division at Los Alamos, where, except for sabbaticals at the University of Colorado Medical Center and the Medical Research Council (MRC) Laboratory of Molecular Biology in Cambridge, England, he would spend his entire scientific career. He received his doctorate in theoretical physics from Duke University in 1954 for studies in cosmic ray physics under Lothar Nordheim. From 1950 to 1965, Goad was a member of the team that developed the first and subsequent generations of thermonuclear weapons.

In the 1960s, Goad turned his attention to questions in molecular biology. Goad's vision and leadership, coupled with his knowledge of computers, mathematics, the physical sciences, and biology, resulted in the creation of the first nucleic acid database, GenBank. GenBank, in turn, would become a cornerstone in the revolutionary fields of bioinformatics and genomics.

In 1970 and 1971, Goad spent a year with Francis Crick at the MRC Laboratory of Molecular Biology. Upon his return he devoted his full scientific efforts to biology, providing theoretical support to various experimental biology programs at Los Alamos. In 1974 George Bell created the Theoretical Biology and Biophysics (T-10) Group, which Goad joined. With the advent of methods to obtain the exact nucleotide sequences of genes, it became clear that sequence data would accumulate at a great rate. In 1979 a meeting was organized at Rockefeller University to discuss how these data could be managed and exploited. From Los Alamos, Michael Waterman (see Box 12-1) and Temple Smith attended that meeting and upon their return talked to Goad, who had been thinking about sequence data and how to analyze these data with computers. After the meeting, Goad's group began collecting sequences on the computer and writing software for the analysis of his pilot sequence database.

Starting in 1979, Goad headed the Los Alamos effort to create a national data bank and analysis center for nucleic acid sequences. Goad began to collaborate with other groups. In particular he contacted Margaret Dayhoff (see Box 4-1), who was at the National Biomedical Research Foundation in Washington, DC. Dayhoff had already started a protein sequence database and was just beginning to collect data on nucleic acid sequences. Goad submitted a proposal to the National Institutes of Health to extend his data collection efforts. This was a joint proposal with the company Baranek and Newman, Inc.

In 1982, Goad's efforts were rewarded when the National Institutes of Health funded the proposal for the creation of GenBank, a national nucleic acid sequence databank. By the end of 1983, more than 2,000 sequences (about two million base pairs) were annotated and stored in GenBank. Shortly thereafter, the European Molecular Biology Laboratory in Heidelberg and GenomeNet in Japan each started their own nucleic acid databases.

REFERENCES

Goad, W. B., 1983. "GenBank." *Los Alamos Science* (Fall): 53–61.
Hanson, T. 2000. "Walter Goad, GenBank Founder, Dies." *Los Alamos National Laboratory Daily Newsbulletin*, November 21.

well-annotated database focused on the protein products of the genes found in GenBank. We will take a tour of RefSeq and UniProtKB to gain insight into some of the databases central to bioinformatics. Interwoven in our discussion of GenBank is an analysis of the basic organization of a gene or, in other words, how the sequence segments are arranged. Furthermore, most gene transcripts in complicated organisms such as humans undergo a phenomenon known as alternative splicing. Alternative splicing is a method of adding more variability to messenger RNAs and their protein products. The nomenclature associated with gene organization and alternative splicing will deepen your understanding of bioinformatics and genomics and their applications.

GenBank An annotated database that contains nucleotide sequences derived from DNA or RNA sources. Each record in the database consists of three sections: header, feature keys section, and nucleotide sequence.

alternative splicing A process that produces different messenger RNAs (mRNAs) from a single transcript. The different mRNAs are called alternatively spliced variants.

2.2 PUBLIC DATABASES

GenBank is a primary or archival database[1] that contains nucleotide sequences from DNA and from DNA that is copied from mRNA. The person who submits the sequences also annotates the DNAs. Annotations describe the characteristics and locations of these characteristics in the submitted sequences. If the submitted DNA contains protein coding regions, those are usually annotated. GenBank is maintained by the National Center for Biotechnology Information (NCBI). When a researcher performs an experiment and obtains sequence information of a gene, the typical first act is to see whether GenBank already contains an identical or similar sequence. If the sequence in GenBank is identical or similar, then the researcher can obtain information from the GenBank record annotations to offer insight into the function of the newly sequenced gene (see Box 2-2). The researcher may also decide to submit their sequence to GenBank. Authors of submitted sequences have full authority over the content of the records they create.

GenBank and two other public nucleic acid databases are stored on servers located in three cities (Figure 2-1). GenBank's server is maintained by the NCBI and is located in Bethesda, Maryland, United States. The European Nucleotide Archive (ENA) database is maintained by the European Molecular Biology Laboratory-European Bioinformatics Institute (EMBL-EBI) server, located in Hinxton, United Kingdom. The DNA Data Bank of Japan (DDBJ) is maintained by GenomeNet server in Kyoto. The three public nucleic acid databases form the International Nucleotide Sequence Database Collaboration (INSDC). When a DNA sequence is submitted to one database, the data is shared with the other two databases within one day. We will discuss the organizational structure of NCBI.

One of the programs hosted by the NCBI is Entrez. Entrez is a search engine that allows you to search several databases simultaneously for certain subjects. We will discuss only a few of these databases here. Medline is a database that contains science literature and is often accessed through the PubMed portal. PubMed allows you to obtain published biology and medicine article abstracts. Full articles are obtained through a linked database named PubMed Central or through libraries. The Online Mendelian Inheritance of Man (OMIM) is a database that describes human genes and disorders associated with genes. OMIM is often a convenient database to begin

[1] GenBank contains primary sequence data. The National Center for Biotechnology Information typically does not curate the data in this database, leaving that to the submitter. There may be multiple submissions of a gene sequence from different submitters.

BOX 2-2 | **A CLOSER LOOK**

GenBank Is Critical to the Discovery of the MDM2 Oncoprotein, an Inhibitor of p53

MDM2 is the murine double minute-2 protein, a protein that forms a complex with p53 and prevents p53 from performing its tumor suppressor activities. The discovery of MDM2 as a p53 inhibitor illustrates the importance of GenBank to discovery in science.

A protein with an apparent mass of 90 kilodaltons (kDa) was discovered to bind to p53 in mammalian cells (Hinds et al. 1990). To identify the 90 kDa protein, an affinity column was created (Momand et al. 1992). The affinity column was made with an antibody that recognizes p53. The experimental approach was to grow mammalian cells in dishes, break open the cells, and pour the cell contents through the affinity column (see figure to the right). Because the 90 kDa protein was bound to p53, the 90 kDa protein was captured by the affinity column as well. The p53 protein and its associated 90 kDa protein were released from the column by adding a peptide that dissociated the p53 from the column. The released p53/90 kDa protein complex was separated by gel electrophoresis. The 90 kDa protein was treated with a protease called trypsin to digest the protein into peptides. Three of these peptides were sequenced. The primary structures of the three peptides, determined by Edman degradation sequencing, were: PLLLK, AKLESSDQAEEGLDVPDGK, and VAQMLLSQESDDYSQPSTS.

The researchers wanted to use the peptide sequences to identify the 90 kDa protein by comparing them to other known protein sequences in GenBank. They used a software program called tBLASTn to search GenBank for a match to the two long peptide sequences above (see Chapter 6 for a discussion of BLAST programs). tBLASTn uses the genetic code to translate all the DNA sequences in GenBank to amino acid sequences. When this was first attempted no matches were found. However, when the researchers attempted the same search a month later, GenBank contained a match! The match was to a gene that had just been deposited to GenBank by Donna George (University of Pennsylvania).

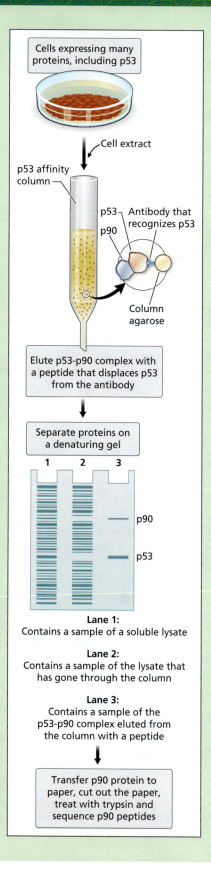

Cells expressing many proteins, including p53

Cell extract

p53 affinity column

p53
p90

Antibody that recognizes p53

Column agarose

Elute p53-p90 complex with a peptide that displaces p53 from the antibody

Separate proteins on a denaturing gel

1　2　3

p90

p53

Lane 1:
Contains a sample of a soluble lysate

Lane 2:
Contains a sample of the lysate that has gone through the column

Lane 3:
Contains a sample of the p53-p90 complex eluted from the column with a peptide

Transfer p90 protein to paper, cut out the paper, treat with trypsin and sequence p90 peptides

The GenBank annotations of George's gene indicated that it coded for a protein that promoted tumor growth. The match between the gene and the 90 kDa protein showed that the gene protein product, MDM2, and the 90 kDa protein were one and the same. Furthermore, it suggested that MDM2 causes cancer by binding to and inhibiting p53 tumor suppressor activity. Later, it was found that MDM2 modifies p53 with ubiquitin, which ultimately targets p53 for degradation. This snippet of research history shows the importance of GenBank in scientific discovery.

REFERENCES

Fakharzadeh, S. S., S. P. Trusko, and D. L. George. 1991. "Tumorigenic Potential Associated with Enhanced Expression of a Gene That Is Amplified in a Mouse Tumor Cell Line." *EMBO Journal* 10: 1565–1569.

Hinds, P. W., C. A. Finlay, R. S. Quartin, S. J. Baker, E. R. Fearon, B. Vogelstein, and A. J. Levine. 1990. "Mutant p53 DNA Clones from Human Colon Carcinomas Cooperate with ras in Transforming Primary Rat Cells: A Comparison of the 'Hot Spot' Mutant Phenotypes." *Cell Growth & Differentiation* 1: 571–580.

Momand, J., G. P. Zambetti, D. C. Olsen, D. George, and A. J. Levine. 1992. "The *mdm-2* Oncogene Product Forms a Complex with the p53 Protein and Inhibits p53-Mediated Transactivation." *Cell* 69: 1237–1245.

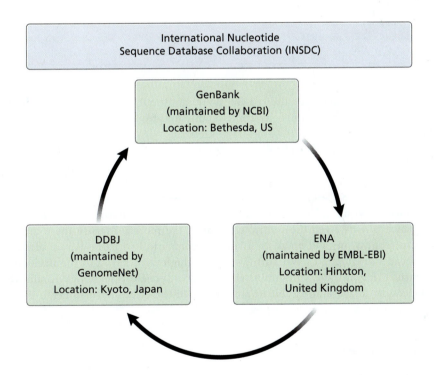

FIG. 2-1. Three public nucleotide databases form the International Nucleotide Sequence Database Collaboration (INSDC). Sequence data is shared between the databases within one day.

searching for genes associated with human diseases. The Protein database contains amino acid sequences and the Taxonomy database presents the organism names associated with sequences in GenBank and Protein database.

GenBank is an annotated collection of nucleotide sequences. Each record in the database contains a single contiguous stretch of DNA or DNA obtained from RNA. The sequences in GenBank that were obtained from RNA were actually experimentally copied from RNA through a process known as reverse transcription. The product of reverse transcription is complementary DNA (cDNA). So it is actually cDNA

organism division

refer this record

accession #

3 sections of a GENBANK record

1. Header = ID, organism, DNA/RNA,

2. FEATURE KEYS = info on segments of sequence

3. sequence = end of record denoted by "//"

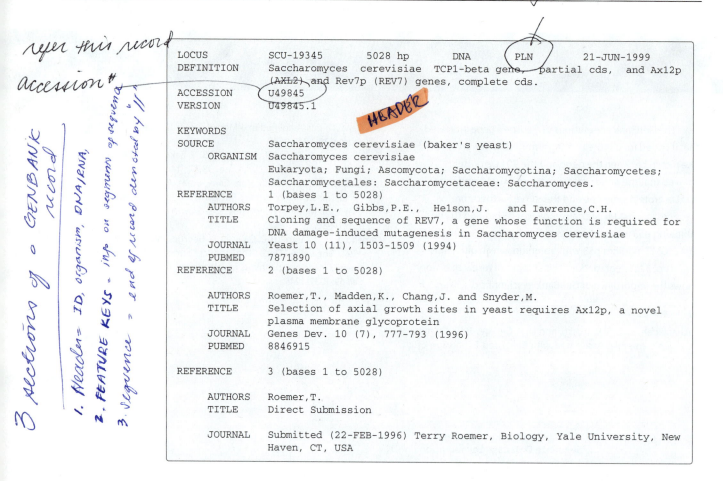

```
LOCUS       SCU-19345       5028 hp        DNA         PLN        21-JUN-1999
DEFINITION  Saccharomyces  cerevisiae  TCP1-beta gene,  partial cds,  and Ax12p
            (AXL2) and Rev7p (REV7) genes, complete cds.
ACCESSION   U49845
VERSION     U49845.1

KEYWORDS
SOURCE      Saccharomyces cerevisiae (baker's yeast)
    ORGANISM  Saccharomyces cerevisiae
            Eukaryota; Fungi; Ascomycota; Saccharomycotina; Saccharomycetes;
            Saccharomycetales: Saccharomycetaceae: Saccharomyces.
REFERENCE   1 (bases 1 to 5028)
    AUTHORS   Torpey,L.E.,  Gibbs,P.E.,  Helson,J.  and Iawrence,C.H.
    TITLE     Cloning and sequence of REV7, a gene whose function is required for
            DNA damage-induced mutagenesis in Saccharomyces cerevisiae
    JOURNAL   Yeast 10 (11), 1503-1509 (1994)
    PUBMED    7871890
REFERENCE   2 (bases 1 to 5028)

    AUTHORS   Roemer,T., Madden,K., Chang,J. and Snyder,M.
    TITLE     Selection of axial growth sites in yeast requires Ax12p, a novel
            plasma membrane glycoprotein
    JOURNAL   Genes Dev. 10 (7), 777-793 (1996)
    PUBMED    8846915

REFERENCE   3 (bases 1 to 5028)

    AUTHORS   Roemer,T.
    TITLE     Direct Submission

    JOURNAL   Submitted (22-FEB-1996) Terry Roemer, Biology, Yale University, New
            Haven, CT, USA
```

HEADER

FIG. 2-2. Header section of a GenBank record.

cDNA Complementary DNA synthesized from RNA using reverse transcriptase. The final product consists of two strands.

that is sequenced and submitted to GenBank, not the RNA. GenBank records are usually generated from direct electronic submissions from investigators responsible for sequencing the DNA.

A GenBank record can be divided into three sections. The first is a header that provides information about the entire record. The second section contains feature keys. Feature keys are associated with descriptions that annotate segments of the sequence. The third section is the nucleotide sequence. The sequence in the third section ends with a "//" symbol, which denotes the end of the record. The nucleotide sequence is considered the central element of the record. Because these records can be lengthy, we will break up a single record into two parts. Part 1 contains the header (Figure 2-2), and part 2 contains the feature keys and the nucleotide sequence (Figure 2-3). To examine the full record, one can search GenBank (the Nucleotide database at the NCBI website) with the Accession number U49845. Now we are ready to analyze one record from GenBank.

2.3 THE HEADER

The header contains fields that describe the entire GenBank record. As shown in Figure 2-2, these fields are located in a column on the left. On the top line, the first field we encounter is Locus.[2] The locus is associated with a number SCU49845, the number of base pairs (5028), the type of nucleic acid (DNA), the organism division (PLN—plant, fungal, and algal), and the date of the last modification to the record

---

[2] Just a note about our writing conventions in this chapter. Underlined terms are identical to words in the database record. Italicized words fall into two classes: (1) those that are of significance to bioinformatics and genomics and may be new to the reader; (2) those that describe a biological species.

[handwritten annotations: "nucleotide sequence for CDS that is incomplete on left", "nucleotide range", "STOP codon will not be a part of the CDS."]

```
FEATURES         Location/Qualifiers
     source      1..5028
                 /organism="Saccharomyces cerevisiae"
                 /db_xref="taxon:4932"
                 /chromosome="IX"
                 /map:"9"
     CDS         <..206
                 /codon_start=3
                 /product="TCP1-beta"
                 /protein_id="AAA98665.1"
                 /db_xref="GI:1293614"
                 /translation="SSIYNGISTSGLDLNNGTIADMRQLGIVESYKLKRAVVSSASEA
                 AEVLLRVDNIIRARPRTANRQHM"
     gene        687..3158
                 /gene="AXL2"
     CDS         687..3158
                 /gene="AXL2"
                 /note="plasma membrane glycoprotein"
                 /codon_start=1
                 /function="required for axial budding pattern of S.
                 cerevisiae"
                 /product="Ax12p"
                 /protein_id="AAA98666.1"
                 /db_xref="GI:1293615"
                 /translation="MTQLQISLLLTATISLLHVVATPYEAYPIGKQYPPVARVNESF
                 TFQISNDTYKSSVDKTAQITYNCFDLPSWLSFDSSSRTFSGEPSSDLLSDANTTILP..."
     gene        complement(3300..4037)
                 /gene="REV7"
     CDS         complement(3300..4037)
                 /gene="REV7"
                 /codon_start=1
                 /product="Rev7p"
                 /protein_id="AAA98667.1"
                 /db_xref="GI:1293616"
                 /translation="MNRWVEKWLRVYLKCYINLILFYRNVYPPQSFDYTTYQSFNLPQ
                 FVPINRHPALIDYIEELILDVLSKLTHVYRFSICIINKKNDLCIEKYVLDFSE..."
ORIGIN
        1 gatcctccat atacaacggt atctccacct caggtttaga tctcaacaac ...
     4021 ...tctacccatc tattcataaa gctgacgcaa cgattactat tttttttttc ...
     4981 ...tgccatgact cagattctaa ttttaagcta ttcaatttct ctttgatc
//
```

[handwritten annotations: "FEATURE KEYS = important segments", "another database for cross reference", "accession number in protein database", "CDS is incomplete in 5' end", "no '<' sign means CDS is complete", "Methionine b/c CDS is complete", "the gene is on the complement strand", "SEQUENCE", "3'", "+ strand", "5' of + strand", "5' of ⊖ strand", "just a sequence - multiple genes in this locus"]

FIG. 2-3. The feature keys and the nucleotide sequence of a GenBank record (Accession number U49845). Due to space constraints, the entire Axl2p and Rev7p protein products are not displayed. A large segment of the nucleotide sequence (nts 51-4020 and 4071-4980) is not displayed. The 4035th, 4036th, and 4037th nucleotides are displayed in red.

CDS The protein coding sequence within a nucleotide sequence.

[handwritten: "cds = protein coding sequence nucleic acid sequence that codes for protein = EXONS"]

(21-Jun-1999). The type of nucleic acid is RNA if the RNA was converted to cDNA and deposited in GenBank. The first two letters of the locus are the first letters of the species and genus, respectively, followed by the Accession number. The accession number is a number randomly assigned to the record. The field Definition describes the organism species from which the DNA sequence was obtained and the known genes within the record. In Figure 2-2 the field Definition shows that the species is a yeast named *Saccharomyces cerevisiae* and that the genes in this sequence are TCP1-beta, Ax12p, and Rev7p. The DNA in this record contains a partial cds of the TCP1-beta gene and the complete cds's of the other two genes. The letters cds, sometimes displayed in capital letters CDS, refer to protein coding sequence (we will use "CDS" to describe this sequence). In other words, the CDS is the nucleotide sequence that codes for amino acids that constitute the protein product. The Ax12p and Rev7p genes in this record are complete because the CDS's (from the start to stop codons) are in the record. The TCP-1-beta gene is partial because its CDS is truncated. We will discuss the CDS in more detail when we explore the features section.

The field below Definition is Accession. As described above, accession refers to a unique identifying number for this record that never changes. The field Version signifies whether there is an update to the record. The record in Figure 2-1 is version 1 because there is a "1" after the decimal point (U49845.1). The field Source describes the organism from which the DNA was derived, and the field <u>Organism</u> describes the taxonomy of the organism. The field <u>Reference</u> lists the DNA segment that is described in a publication, and the remainder of the reference field gives selected literature citations associated with the DNA segment. The literature citations are listed beginning with the oldest citation and ending with the most recent citation prior to submission of the sequence to GenBank.

2.4 THE FEATURE KEYS

Now that we are finished with the header section, we will explore the feature keys and the nucleotide sequence (Figure 2-3). Feature keys give information about specific parts of the nucleotide sequence. Feature keys are displayed in a two-column format. In the first column are feature key names, and in the second column are nucleotide locations followed by qualifiers. Nucleotide locations and qualifiers are separated by the "/" symbol. In this record there are three feature keys: <u>source</u>, <u>CDS</u>, and <u>gene</u>. The source describes the range of the complete sequence—in this case 1–5028. As mentioned earlier, CDS describes the location of the protein-coding sequence within that range. The gene feature key describes the locations of complete genes. There may be more than one qualifier associated with a particular location.

Let's take a look at the qualifiers associated with the source feature key in Figure 2-3. The first term is a category description of the qualifier. In this case, the qualifier is <u>organism</u>. This is followed by an equal sign and the name of the qualifier in quotes (=“Saccharomyces cerevisiae”). The only time the qualifier is not in quotes is when it is a number. After the last quotation mark, other qualifiers may follow. Many qualifiers are database cross-references. In Figure 2-3, the second qualifier, db_xref, directs you to the Taxonomy database entry number 4932.

The CDS Feature Key and Gene Structure

A critical gene feature is the CDS, the protein coding sequence. When this term is a feature key, it is capitalized. To have a deeper understanding of the CDS it is necessary to understand gene structure. Gene structure describes the arrangement of DNA regions that are transcribed into RNA. Together, these regions constitute the gene. For simple organisms, such as bacteria, the gene structure is straightforward. The gene consists of three regions: the upstream untranslated region (UTR), the CDS, and the downstream untranslated region (Figure 2-4). The upstream DNA region is closer to the 5′ phosphate end of the DNA (often called the 5′ UTR) and the downstream DNA region is closer to the 3′ end (often called the 3′ UTR) (see Figure 1-3). The transcribed RNA includes the 5′ UTR and 3′ UTR. The start of the untranslated region marks the beginning of the gene. As implied by its name, the untranslated region does not code for protein. Only the CDS codes for protein and the stop codon. The gene, therefore, is the DNA that codes for the entire RNA transcript from its 5′ end to its 3′ end. Even farther upstream to the gene is the promoter, a region of DNA where the enzyme that synthesizes RNA initially sits prior to transcription. Recall that the DNA is double stranded. The DNA strand that is identical to its transcribed RNA sequence (with the exception of T's and U's) is called the coding strand. The DNA strand that is complementary to the coding strand is called the template strand. Sometimes the DNA strand shown in the GenBank record is the coding strand, and sometimes it is the complementary strand (more on this later). In bacteria, the CDS of the gene is contiguous. However, in other organisms, the CDS may be interrupted by introns, segments of DNA that do not code for protein (see "Reference Sequence" section below).

Now we are in a position to discuss the CDS feature key in the GenBank record (Figure 2-3). The number range listed in the second column refers to the range of nucleotide numbers within a stretch of DNA that codes for the amino acid sequence plus the stop codon. The first three nucleotides that start the CDS usually begin with

UTR Untranslated region at either of the two extreme ends of the gene. This is a segment of RNA that is not translated into protein and does not contain the stop codon. The 5′ UTR is a UTR that is at the 5′ end of the gene, and the 3′ UTR is located at the 3′ end of the gene.

coding strand The DNA strand that is identical in sequence to the RNA transcript with the exception that DNA has T's and the RNA has U's.

template strand In replication and in transcription, the strand that is being used as the basis for synthesis of the new nucleotide strand. The template strand is complementary to the newly synthesized strand.

CDS

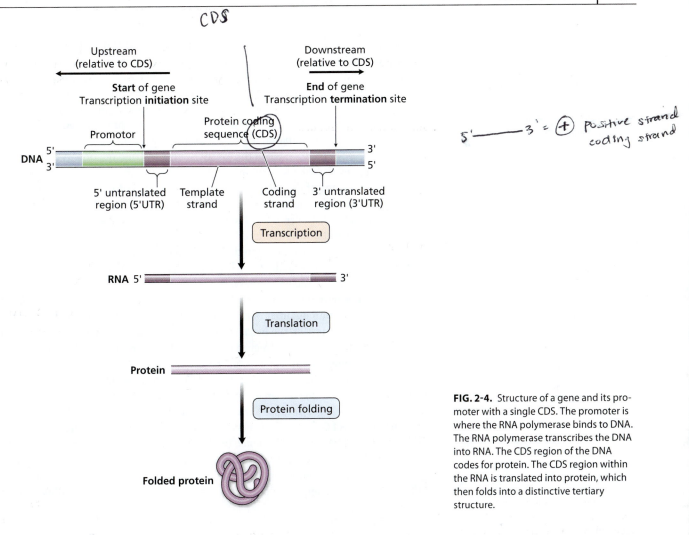

FIG. 2-4. Structure of a gene and its promoter with a single CDS. The promoter is where the RNA polymerase binds to DNA. The RNA polymerase transcribes the DNA into RNA. The CDS region of the DNA codes for protein. The CDS region within the RNA is translated into protein, which then folds into a distinctive tertiary structure.

ATG, which codes for the amino acid Met. Sometimes the nucleotide sequence submitted in the GenBank record is partial: a complete protein sequence is not in the record. When the CDS is partial, the nucleotide number range will show a "<" symbol at the beginning or a ">" symbol at the end. The "<" symbol at the beginning of the range indicates that the beginning of the protein sequence coded by the CDS is not in the record. Similarly, the ">" at the end of the range means that the end of the protein sequence coded by the CDS is not in the record.

In the example shown in Figure 2-3, the first CDS range is <1. .206, which means that the beginning of the sequence that codes for the protein is not in the record. If you scroll down a few lines to the qualifier translation, you can see that the first amino acid is Ser and not Met. This record does not contain coding information for the amino acids that are upstream of Ser. Another qualifier associated with this CDS is "codon_start=3." This means that the first nucleotide in the Ser codon begins with the third nucleotide from the left under the ORIGIN (tcc). The upstream segment of the gene is missing from this record. The last nucleotide in the range, 206, is the last nucleotide in the stop codon. The name of the gene product is TCP1-beta. Because this segment of DNA does not contain the complete CDS of TCP1-beta, there is no gene feature key listed.

The Gene Feature Key and FASTA Format

The next feature key shown in Figure 2-3 is gene. The gene is named "AXL2," and its location is 687..3158, or nucleotides 687–3158 of the record's sequence. The complete CDS of AXL2 is contained in the record. The next feature key is CDS.

```
nt 4035 ──┐    ┌── nt 4037
        5'cat 3'—(displayed strand shown in fig. 2-3)
        3'gta 5'—(complementary strand)
         Met ——(coded amino acid at beginning of sequence)
```

FIG. 2-5. Small region of REV7 displayed strand with its complementary strand and coded amino acid.

The CDS corresponds to AXL2 and has the same nucleotide range as its corresponding gene. One might ask, why is the gene nucleotide range not wider than the CDS? Where are the untranslated upstream and downstream regions? One must recall that GenBank records are annotated by the investigators who submit the sequence. In this particular case, the investigator may not have had knowledge of the extent of untranslated regions. When this segment of the yeast genome was submitted the researcher may have used software to translate the regions of the DNA segment into protein and thus declared those regions to be the CDS regions. Software tools for translation of DNA into protein are relatively reliable. However, it is difficult to predict, using software tools, the extent of the upstream and downstream untranslated regions. Without experimental data, the investigator may have annotated this GenBank record to show that the CDS region and gene region are identical, implying that the CDS region is the *minimum* region that the gene could encompass.

To continue with our analysis of the AXL2 gene, note that there is no "<" or ">" symbol in the location qualifier of the CDS. The translated gene product shows that the first amino acid is Met, the usual starting amino acid for proteins, so this CDS is complete. Another gene in the record is "Rev7p." Its nucleotide range is 3300. .4037, but the word "complement" appears just before the range. Complement signifies that the coding strand of the DNA is located on the complementary strand (the strand that complements the one shown under ORIGIN).[3] Again, the CDS and gene share the same nucleotide range. Because the coding strand is on the complementary strand, the start codon ATG (codes for Met) is on the complementary strand and in reverse orientation (3'GTA5'). The sequence on the strand shown in red font in Figure 2-3 under the ORIGIN is CAT, which base pairs to 3'GTA5'. The three nucleotide sequence CAT and its complement are shown in Figure 2-5. The 4035th, 4036th, and 4037th bases in the sequence correspond to c, a, and t, respectively, illustrated in Figure 2-5.

It should be noted that the translated sequences of Ax12p and Rev7p have been shortened in Figure 2-3 to conserve space. The complete record in GenBank can be found in the GenBank database by searching the nucleotide database with the accession number U49845 at the NCBI website. The third and last section in this GenBank record is the nucleotide sequence of the entire deposited DNA sequence located just below the ORIGIN. The end of the record is marked with the "//" symbol.

Thought Question 2-1

Redraw Figure 2-5 to show the displayed and complementary strands that correspond to the first *three* translated amino acids of TCP1-beta in this record. Which strand in this figure is the template strand?

The default display mode for GenBank is shown in Figure 2-3. This display is good for ascertaining annotated information about the nucleotide sequence in the record. However, this mode is not useful if one needs to copy the sequence and paste it into another program, such as Basic Local Alignment Search Tool (BLAST).

If your objective is to analyze the sequence with a software program such as BLAST, you must change the display mode to FASTA. There is a hyperlink in the NCBI host website that will switch the format to FASTA. In FASTA format, also known as Pearson format, a header line is followed by the sequence. The header line

---

[3] Note that the strand shown under ORIGIN is always named the plus strand. This is the strand deposited into GenBank. The strand that complements the strand under ORIGIN is the minus strand.

is denoted by a ">" sign on the far left. You can copy the FASTA output and paste it into the BLAST software program window and run the program.[4] Most sequence analysis programs accept FASTA-formatted sequences as input. An example of a FASTA-formatted sequence is:

```
>gi|23491728|dbj|AB082923.1| Homo sapiens mRNA for p53, complete cds
CGTGCTTTCCACGACGGTGACACGCTTCCCTGGATTGGCCAGACTGCCTTCCGGGTCACTGCCATGGAGG
AGCCGCAGTCAGATCCTAGCGTCGAGCCCCCTCTGAGTCAGGAAACATTTTCAGACCTATGGAAACT
```

Note that this record was truncated after three lines.

2.5 LIMITATIONS OF GENBANK

A limitation of GenBank is that there are many records with identical or almost identical sequences. This redundancy makes it difficult for the user to decide which sequences are wild-type (natural, nonmutated) sequences and which sequences may contain sequencing errors or mutations. For example, there are at least 46 GenBank records that contain all or part of the human *TP53* tumor suppressor gene. Because *TP53* sequences are often derived from DNA in cancer tissue, the majority of these sequences contain mutations. If the record is carefully annotated, it will state that the DNA sequence contains a mutation and give its location. On the other hand, it may be possible that the record is not carefully annotated, or that the sequence was deposited before it was clear whether the *TP53* contained a mutation.

Another limitation of GenBank is that it is not immediately clear whether the gene sequences in the record are complete. One might think the full-length gene is obtained, but in reality only the beginning (or end) segment of the gene may be in the record. Rather than assume that the full-length sequence is present, it is prudent to carefully scrutinize the annotations in the record. Reading the literature referenced in the annotations will often reveal whether the entire sequence of the gene is contained within the record. These limitations should not detract from the important impact GenBank continues to make. GenBank and other databases that use the data stored in GenBank have become indispensable resources for researchers. One of these other databases is Reference Sequence (RefSeq).

FASTA format Also known as Pearson format, formatted data in which the first line begins with ">" to signify that it is the header line. The following lines contain a nucleotide or amino acid sequence.

2.6 REFERENCE SEQUENCE (REFSEQ)

Because several versions of a gene may be submitted to GenBank, it became imperative to develop a database that contains only wild-type sequences. The RefSeq database contains only wild-type sequences of DNA, RNA, and proteins. The RefSeq database information is derived from GenBank records[5] and is deposited, annotated, updated, and reviewed by the staff at the NCBI. RefSeq is a secondary database (sometimes called curatorial or curated database) because it contains information from a primary database. The staff at NCBI actually accumulates information from several databases and the literature to ensure that the records are correctly annotated. Interestingly, more than one RefSeq record can be generated from a single stretch of DNA. Each record in RefSeq represents a unique naturally occurring molecule, whether it is DNA, RNA, or protein. The RefSeq database shows whether more than one RNA molecule can be transcribed from a single DNA region.

RefSeq A secondary database derived from GenBank that contains wild-type sequences.

What is wild-type?

---

[4] More recent web versions of the NCBI-sponsored GenBank records contain links that allow the user to directly perform a BLAST search of a record without manually converting to FASTA format.

[5] In fact, RefSeq obtains its sequence information from the International Nucleotide Sequence Database Collaboration (INSDC), of which GenBank is one contributor.

Alternative Splicing

To understand how more than one RNA molecule, or transcript, can be derived from one DNA region, recall from Chapter 1 our discussion of the central dogma. RNA is transcribed from DNA and RNA is translated into protein. In bacteria (more generally known as prokaryotes), the RNA transcribed from DNA undergoes very few changes prior to translation. This RNA is known as messenger RNA (mRNA). Recall that eukaryotes, such as humans and plants, contain a structure inside their cells known as a nucleus. The nucleus contains the DNA that codes for the vast majority of proteins inside the cell.[6] In contrast to prokaryotes, RNA initially transcribed from DNA in eukaryotic cells undergoes extensive processing prior to translation. In eukaryotes, the RNA initially transcribed from DNA is called the primary transcript. After processing, the RNA derived from the primary transcript is called messenger RNA (mRNA).

What exactly is meant by processing of primary transcripts? In eukaryotes, segments of nucleotides in the primary transcript are removed and the surviving segments are joined together. The removal of nucleotides and rejoining of RNA segments is called splicing. Exons are segments of DNA that are transcribed into the segments of a primary transcript that survive the splicing process and end up in the mRNA. Introns are DNA segments that are transcribed into the primary transcript, but are spliced out prior to the creation of mRNA. Figure 2-6A shows how splicing of the primary transcript generates the mRNA. Here, four exons in DNA give rise to mRNA that is transcribed into a single protein.

Genes from eukaryotes can contain several exons. For example, the human *TP53* gene is composed of 11 exons and 10 introns. Splicing gives rise to mRNAs that are considerably shorter than the primary transcript. The *TP53* primary transcript is 19,149 bases in length. However, the *TP53* mRNA, transcript variant 1 is only 2,591 bases in length.[7]

Now we can address the question as to how more than one mRNA can be derived from a single gene. One way that this occurs is through *alternative splicing* of the primary transcript. In alternative splicing, the maturation process from primary transcript to mRNA is not consistent. Along with the intron sequences, one or more exon sequences may be removed from the primary transcript, giving rise to mRNAs called alternatively spliced variants. Some of the variants can be translated into different proteins. Figure 2-6B shows two alternatively spliced mRNA variants created from a primary transcript. Each alternatively spliced mRNA variant derived from a single gene is assigned a unique RefSeq record.

Another mechanism for producing different mRNAs from a single DNA segment is to begin RNA synthesis at multiple transcription initiation sites. RNA polymerase, the enzyme that transcribes the RNA from the DNA template, initiates transcription on a region of DNA called the promoter. Usually, the promoter is located upstream of the first exon (see Figure 2-4). Sometimes a gene can have more than one promoter—one located in the usual position and another located farther downstream. Each promoter produces a unique primary transcript that is processed into a unique mRNA that can code for a unique protein. Human p53 has at least 15 known transcript variants—some of which are produced from alternative splicing,

exon A segment of a gene that is transcribed as part of the initial transcript. The initial transcript undergoes splicing to keep the RNA encoded by the exon.

intron A segment of a gene that is transcribed as part of the initial transcript. The initial transcript undergoes splicing to remove the RNA encoded by the intron.

---

[6] We say vast majority because a few organelles such as mitochondria and chloroplasts have their own genomes and express a few proteins that contribute to the cell's protein repertoire. It is likely that mitochondria and chloroplasts were once independent organisms that were engulfed by larger organisms, which led to a symbiotic relationship.

[7] Sequence lengths are based on data derived from Human Genome Assembly number GRCh38. The lengths of the primary transcript and mRNA transcript 1 may change slightly depending on which version of the human genome sequence is being searched. The RefSeq database reports a consensus view of the human genome and its gene transcripts. But due to individual gene length and RNA length variations, the reported lengths of the primary transcript and its alternatively spliced forms are average values that fluctuate depending on the genome assembly version. Variability is usually found in the 5′ and 3′ UTRs.

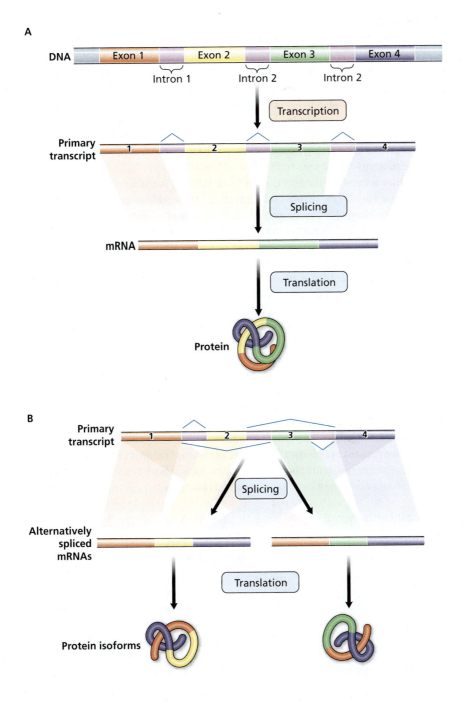

FIG. 2-6. Splicing of RNA primary transcript.
A. The process of splicing is shown beginning with transcription of a DNA gene into an RNA primary transcript. The gene has four exons (*colored rectangles*) and three introns. The splicing pattern is indicated by the blue lines connected to the primary transcript. After splicing, the intron sequences of the primary transcript are removed and the mRNA is created. The mRNA is translated into protein. Here, each exon contributes to the protein product. **B.** Alternative splicing can give rise to two mRNAs. The path of splice form A is shown in light blue lines above the primary transcript. The path of splice form B is shown in light blue lines below the primary transcript. Splice form A mRNA is derived from exon sequences 1, 2, and 4. Splice form B mRNA is derived from exon sequences 1, 3, and 4. Two protein isoforms are translated from two alternatively spliced mRNAs.

and some of which are produced from alternate transcription initiation sites. The most abundant mRNA is transcript variant 1, which codes for a protein of 393 amino acids. Interestingly, in a small study of 3,500 genes it was shown that 93% of the genes produce *more than one* mRNA. RefSeq will undoubtedly continue to expand as we gather more information on wild-type molecules.

2.7 PRIMARY AND SECONDARY DATABASES

We have discussed GenBank and RefSeq databases. As mentioned previously, GenBank is a primary database and RefSeq is a secondary database. Also, recall that other primary nucleotide databases are ENA and DDBJ. In addition, another primary database is the RCSB Protein Data Bank (PDB). It contains molecular

Protein Data Bank A database that contains Cartesian coordinates of atoms of biomolecules. The majority of the coordinates are derived from X-ray crystallography experiments. Molecular viewers can be used to display the atoms on a computer screen.

structure data on proteins, DNA, and RNA. The data is the atom identity (carbon, nitrogen, etc.), its XYZ Cartesian coordinates in space, and B-factor (uncertainty of atom position).[8] Sequences of the biomolecules are also stored in the PDB. Another primary database is ProSite. Each record in ProSite contains a consensus pattern of amino acids sequences that form a common three-dimensional structure. The consensus pattern can also constitute a site where a post-translational modification can occur—called a functional site. These consensus patterns are stored in the ProSite database and can be used to preliminarily predict the structures and functional sites of new proteins. Ultimately, experiments in the laboratory will confirm (or not confirm) the proteins' predicted structures and functional sites.

Another secondary database, aside from RefSeq, is the non-redundant nucleotide database (nt). In nt, records with identical nucleotide sequences from primary databases are merged into one record. The data in nt is obtained from GenBank, ENA, DDBJ, and PDB. Another example of a secondary database is UniProt Knowledge Base (UniProtKB). The information in this protein sequence database comes from the translated nucleotide databases of GenBank, ENA, and DDBJ. Recall from Box 2-2 that the identity of the 90 kDa protein was obtained by comparing the translated GenBank sequences to the peptides generated from the 90 kDa protein. If UniProtKB contained the MDM2 gene sequence deposited in GenBank, it would have translated the MDM2 CDS into amino acid sequence. To use the UniProtKB to identify the 90 kDa protein, BLAST could be used to directly compare the experimentally derived peptides sequences to protein sequences in UniProtKB.

UniProt Knowledge Base

UniProtKB (formerly Swiss-Prot) was begun by Amos Bairoch at the University of Geneva. UniProtKB is a database kept current through human curators and the software program TrEMBL. TrEMBL translates new DNA sequences and annotates the translated sequences automatically. If there are several records in GenBank for a particular gene of interest, it may be prudent to also look at both UniProtKB and RefSeq. Both databases contain information on natural proteins.

TP53 record in UniProtKB

The *TP53* gene record in UniProtKB is laid out in a style that is easy for the user to understand. The protein p53 record, P04637, is divided into the following headings: names and origin, protein attributes, general annotation (comments), ontologies, binary interactions, alternative products, sequence annotation (features), sequences, references, web resources, cross-references, entry information, and relevant documents. Protein attributes include the number of amino acids, the completeness of the sequence, and evidence for the existence of the protein. That evidence could be inferred from computer-generated translation of the DNA, or it could be based on experimental determination. In the case of p53, there is direct evidence, beginning with its discovery in 1979, that the protein does exist. The general annotation section gives a summary of the *TP53* gene and protein functions. Ontologies describe biological processes associated with p53. For example, in the disease ontology, the record includes the Li-Fraumeni syndrome (introduced in Chapter 1). This is a rare familial disease in which a mutant form of p53 is passed from parents to offspring. The outcome of the disease is usually a cancer contracted by the time the patient reaches early adulthood (see Box 1-1). Another heading of the p53 UniProtKB record is "Binary interactions." Binary interactions refer to known biological molecules that bind to p53 within the cell. The p53 record lists more than 55 proteins that bind to p53 and gives links to the UniProtKB records of the binding proteins. As expected, one of the binding proteins is MDM2.

MDM2 A ubiquitin ligase that places multiple ubiquitin units onto p53, which marks p53 for destruction. MDM2 also binds to the transactivation domain of p53 and escorts p53 from the nucleus to the cytoplasm.

Li-Fraumeni syndrome A rare familial cancer exhibiting autosomal dominant inheritance and early onset of tumors, multiple tumors within an individual, and multiple affected family members. The most common cancer types are soft tissue sarcomas, osteosarcomas, breast cancer, brain tumors, leukemia, and adrenocortical carcinoma. Most Li-Fraumeni syndrome patients inherit one mutant *TP53* allele. The second *TP53* allele is commonly observed to be mutated in the cancer tissue, but not normal tissue.

---

[8]See Chapter 7 for a more extensive discussion of RCSB Protein Data Bank.

The sequence annotation (features) section of the p53 UniProtKB record describes information pertaining to specific segments of the protein sequence. One of the features in the sequence annotation section is <u>Region</u> and the qualifier is <u>DNA binding</u>. The amino acid sequence that corresponds to this region spans amino acids 102–292. This region is critical for p53 function because p53 must bind to DNA in order to mediate its tumor suppressor activity. p53 is a transcription factor. It binds to DNA and assists RNA polymerase to transcribe specific genes, including those that repair damaged DNA, prevent cells from dividing, and cause programmed cell death. Interestingly, the vast majority of mutations in the p53 gene found in human cancers are located in the region of p53 gene exons that code for the DNA binding region. The mutations code for p53 proteins devoid of DNA binding capability, thus knocking out p53's important function as a tumor suppressor.

SUMMARY

We began our journey into information organization with a discussion of GenBank, the longest running molecular biology database. GenBank records are divided into three sections: header, feature keys, and nucleotide sequence. While GenBank is a primary database, another database, RefSeq, is a secondary database that derives its data from GenBank and other primary nucleotide sequence databases. RefSeq stores sequence information of wild-type molecules. More than one transcript can be derived from a single gene. Such variant transcripts are the products of alternative splicing and multiple transcription start sites. There is a unique RefSeq record for each RNA variant. Another secondary database is UniProtKB. UniProtKB is protein centered and is rich with annotations that describe the functions of the protein.

EXERCISES

1. **For this exercise, you will need to access GenBank by going to the NCBI website and using the drop-down menu to search "Nucleotide." Note that the definition of the coding strand is the strand of DNA within the gene that is identical to the transcript (for genetic code, see Fig. 1.11). On the other hand, the template strand is the strand that is complementary to the coding strand.**

   a. Use the following accession number to access the nucleotide sequence in GenBank: CU329670

   b. Go to the FEATURES section of the record.

   c. Link to the CDS to gain access to the first 5662 nucleotides of the sequence.

   d. Name the protein product of the CDS.

   e. Write the first four amino acids (starting from the N-terminus).

   f. Write the nucleotide sequence of the coding strand that corresponds to these amino acids.

   g. Write the nucleotide sequence of the template strand that corresponds to these amino acids.

   h. Using the sequence shown in the record, give the nucleotide number range that corresponds to these amino acids.

2. **Genes in eukaryotes are often organized into exons and introns, which require splicing to produce an mRNA that can be translated. The gene** organization is the order of the DNA segments that comprise the gene starting with the promoter, the first exon, the first intron, the second exon, and so on. The interspersed introns can make gene identification difficult in eukaryotes—particularly in higher eukaryotes with many introns and alternative spliced mRNAs. Prediction of many genes and their organization has been based on similarity searches between genomic sequence and known protein amino acid sequences and genomic sequence and the corresponding full-length cDNAs. cDNAs are reverse-transcribed mRNAs and therefore generally do not contain intron sequences. cDNAs (i.e., copied DNA) can be considered mRNAs. A comparison of a genomic sequence (with introns) to its corresponding cDNAs will reveal where introns begin and end. GenBank will contain the genomic sequence and the cDNA sequence. To find out the structure of the gene (i.e., the arrangement of the exons and introns) we simply need to perform a sequence comparison between the genomic sequence and the cDNA sequence. Shown below is a genomic sequence from the species *C. elegans*. The Basic Local Alignment Sequence Tool (BLAST) can be used to elucidate part of the gene organization (arrangement of exons and introns) of a

genomic sequence. BLAST can be used to compare genomic DNA sequence with all RNA sequences (i.e., cDNA sequences) in GenBank. The top hit of the output will be a sequence comparison between your sequence (the query sequence) and the most similar sequence in the database (subject sequence). Subsequent hits will display sequence comparisons between the query sequence and subject sequences that are increasingly less similar. If all hits have 100% identity, use the hit with the most extensive percent coverage to report on. Use the nucleotide BLAST tool and appropriate databases to construct a schematic diagram that shows the arrangement of introns and exons in the genomic sequence. Remember that the species source of genomic sequence is *Caenorhabditis elegans.*

```
ATTTTTAAAAATGTACAAAATCAAACGCCCTACAA
ATCATGTGTGTGAAGAAGAATAATAACTAACATAT
CTATTTATATTTACCGAATAAATATATATTCATCAAT
TAACCTGAAGAACAAACGAATTCGGCTACAGGC
GTCGATCAGTCTCGAATCTAGTAACAACAAGAGAG
CAATACGAAAACCGGTAAATCAATAGGGGGAAGCG
AAACAGTAGGTACAAATTGGAGGGGAAGCACCAAT
ACATTAGGTGGGGGGTACGACTTGAAAAATGAGCT
GATTTTCGAATAGTTAAAGCGATGATCGTGTCCGA
AAAACAGTTCATTTTTCAAGACAACATTGAGACTG
GGAGTACGGGGAAGCTCATTTACGGTGAGAGGAA
TTGGTGAGATCTTTAGAATATGCTTAAGGAGTTGGG
GTGGCTGGAGAAGTTCCTGTAGCCTCCGTGCCGG
GATTCGATGGAGAAGTCGTTGCGGCTGGTCCCTTTT
CCTTCACTGGTGCTGGATCCTTGGCTGGAAGACAT
ATGCGTGGCTTGACAGTCGATGAGGTGCGAGCCGA
CGAGTCCTTGTGAACTTCGTATCTGGAAATATTTTA
CTTAGATAGCAAATACTAAAATTGTAAAATTACC
TCAAAATCTCAGTATCCGGAATGCTCAATTTCTGCT
TCAAAACCTGTCCGATGCGAAGATTGACATCATC
GCGAGTAGCATCACGAGTCCACAAGGAAACCTTGT
CACCCTTTTGACGAACATTCACGACAGCTCCGCAG
ATGTAGTCTCCGTACTCGTCGAATTGCTCTCCAA
CAATAGCCATCAACAGCTCCAACCAGTAGTGATCGA
GCAATTGCGTTCTTCTCTGAAGCTTCTATGATTCAT
TGAATAAAATATATTTCTCAAAACGTACTTGCTT
ATCGACAACAACCAACCAACGTCCACCTTGAACGTT
GTTGACGTCCTCCCACATTGGCTTGATTCCTTCC
TTGAACAAGTAATAATCGGATCCCCAGTTCAATCCT
CCGGCAGACTGAATGTGATTGTACAGCGACCAGA
AGTCCTCGACAGTGTCGAAAAGTGAAACCATCTGGA
AAAAATCGATAAAAGACGTATTTAAAAATCTTCT
```

```
ACCTTCAGACAATCCTCCCATTCCTTGTTACGGTCA
GCTTTCAAGTACCAGAGAGCCCAGCGATTCTGGA
GGGGGTGTCTGGTGAGAAGCTCTGGAGGAACTGAAGC
ATCGGACGCATTCACATCGCCGGAAGCTGACAA
TGCTTTGTTTTCCGCTACGGATGTGCTCATTTAGC
TGAAAATAGGTAATATTATATACGATTAGAGCTCG
GAAAACGATAAAATAGAGAAGAGTATGAATTTGGTT
CAAATAACTCGGATTTTATAGGAAATTTTGTTTT
ACTGCACATTTTCGGCTAGTTTCCAAGCTTTTTAGA
TTTTTCAAGTGTAATTGGTAACATCGGGCACAAT
AAATTGATATTAAAGCTTGGAAAACAATAA
```

In addition to construction of the schematic diagram answer the following:

a. Give the name and accession numbers of each distinct mRNA produced from this gene.

b. Give the names and accession numbers of the protein product(s).

c. Note the numbering of the sequences in the alignments. Does the database genomic sequence progress in the same direction as the database mRNA? In other words is it the same orientation (see below):

1 . . . . 114 = query

61 . . . . 98 = subject

or opposite orientation (below):

1 . . . . 114 = query

98 . . . . 61 = subject

d. Consider the alignment of the query sequence and the subject sequence. What does the orientation of the sequences relative to each other tell you about the sequence that was used as the query sequence?

e. Give the amino acid sequences separately translated from each exon sequence of the longest transcript.

f. How many alternative splice variants are associated with this genomic sequence? List their accession numbers.

g. Give the chromosome position numbers that denote the start and end of the *TP53* gene. The position number is the base number on chromosome 17. Calculate the length of the primary transcript. Give the lengths, in base pairs, of each exon and intron that is used for the transcription of *TP53* into mRNA isoform a. Cite the sources you used to gather your information.

ANSWER TO THOUGHT QUESTION

2-1.

```
5'gatcctccata3' (displayed strand)
3'ctaggaggtat5' (comp. strand)
   SerSerIle
```

Bottom strand is template strand. Top strand (displayed strand) is coding strand.

(Note: the start of the sequence begins at frame 3; the DNA coding for amino terminus of this protein is not in this record.)

REFERENCES

Claverie, J.-M., and C. Notredame. 2007. *Bioinformatics for Dummies*. Hoboken, NJ: Wiley.

Fakharzadeh, S. S., S. P. Trusko, and D. L. George. 1991. "Tumorigenic Potential Associated with Enhanced Expression of a Gene That Is Amplified in a Mouse Tumor Cell Line." *EMBO Journal* 10: 1565–1569.

Goad, W. B. 1983. "GenBank." *Los Alamos Science* (Fall): 53–61.

Hanson, T. 2000. "Walter Goad, GenBank Founder, Dies." *Los Alamos National Laboratory Daily Newsbulletin*, November 21.

Hinds, P. W., C. A. Finlay, R. S. Quartin, S. J. Baker, E. R. Fearon, B. Vogelstein, and A. J. Levine. 1990. "Mutant p53 DNA Clones from Human Colon Carcinomas Cooperate with ras in Transforming Primary Rat Cells: A Comparison of the 'Hot Spot' Mutant Phenotypes." *Cell Growth & Differentiation* 1: 571–580.

Lamb, P., and L. Crawford. 1986. "Characterization of the Human p53 Gene." *Molecular and Cell Biology* 6: 1379–1385.

Momand, J., G. P. Zambetti, D. C. Olsen, D. George, and A. J. Levine. 1992. "The *mdm-2* Oncogene Product Forms a Complex with the p53 Protein and Inhibits p53-Mediated Transactivation." *Cell* 69: 1237–1245.

Mortazavi, A., B. A. Williams, K. McCue, L. Schaeffer, and B. Wold. 2008. "Mapping and Quantifying Mammalian Transcriptomes by RNA-Seq." *Nature Methods* 5: 621–628.

NCBI. 2016. "Sample GenBank Record." Accessed January 5. http://www.ncbi.nlm.nih.gov/genbank/samplerecord/.

Pevsner, J. 2009. *Bioinformatics and Functional Genomics*, 2nd ed. Hoboken, NJ: Wiley-Liss.

AFTER STUDYING THIS CHAPTER, YOU WILL:

- **Understand that mutations are the drivers of evolution.**

- **Understand that, from an evolutionary perspective, mutations may be negative, neutral, or positive.**

- **Describe the types of mutations, including point mutation, indel, repeat expansion, partial gene duplication, and whole gene duplication.**

- **Compare and contrast mutations acquired during cancer progression to mutations acquired during evolution.**

- **Distinguish the terms homolog, ortholog, and paralog and how they arise during evolution.**

- **Appreciate that conserved regions within homologs infer functional, structural, or evolutionary relationships.**

- **Describe how exon shuffling can account for the modular nature of proteins.**

- **Describe the difference between vertical and horizontal gene transfer.**

- **Know that germ cell mutations, as opposed to somatic cell mutations, can be transferred to offspring.**

- **Understand that the DNA binding domain within *TP53* gene is often mutated in human cancers.**

MOLECULAR EVOLUTION

3.1 INTRODUCTION

In Chapter 2 you learned a bit about sequence information stored in databases. One database, GenBank, contains records with DNA or RNA sequences, annotations of the sequences, and cross-references to other databases. You learned some of the specific vocabulary bioinformaticians use in these annotations. The annotations are used as a starting point to explore basic molecular life science concepts such as gene structure, alternative splicing, and translation. You are probably aware that sequences, whether composed of nucleotides or amino acid residues, are often compared with one another. Such comparisons reveal, at the molecular level, how organisms are related to one another. Sequence differences in proteins that carry out the same function in their respective species are due to mutations that have been accepted in the molecules millions of years ago. Mutations can also change the function of proteins and change the level of protein expression. Such changes contribute to the evolution of new species. Of course mutations can have harmful effects. Mutations are necessary for cancer initiation and progression. At the molecular level, just as in the process of evolution, cancer mutations cause changes to protein function and expression level. In this chapter, we will use cancer mutations to explain the process of molecular evolution. As you learn about the many mechanisms that produce cancer mutations, keep in mind that the same mechanisms occur in the process of evolution.

Bioinformaticians routinely perform DNA sequence and protein sequence alignments. DNA sequence alignments are often useful for discovering genetic alterations associated with traits and diseases. For example, comparison of genes from a cancer tissue to the wild-type genes can reveal whether the genes from cancer tissues have acquired mutations. Protein sequence alignments are useful for inferring function, structure, and evolutionary history of proteins. Protein sequences are richer in diversity than nucleotide sequences because there are 20 possibilities at each amino acid position in a protein, while there are only four nucleotide possibilities at each nucleotide position in DNA. Due to this richness, proteins can have a wider variety of structures than DNA or RNA.

The diversity of protein structures gives them the ability to perform a multitude of functions. For example, they catalyze the vast majority of chemical reactions in the organism, they create the shape of the organism, they replicate DNA, they transcribe genes into mRNA, and they translate mRNA into proteins. In short, they are the workhorses of the organism. Mutations within the coding regions of genes can produce altered proteins with the same or even enhanced ability to perform their functions through a process called molecular evolution. If a gene mutation enables the organism to be more capable of producing progeny, then that mutation, which causes a change in protein sequence, is accepted into the genome. The mutation process, coupled with natural selection over millions of years,[1] produces proteins that perform the same function in different species. Interestingly, those proteins often differ slightly in their sequences.

The most convenient way to determine whether a protein of interest is similar to other proteins is to scan annotated protein sequence databases. In practice, we can do this by uploading a sequence to a sequence alignment program such as BLAST, clicking a button, and examining the many aligned sequences in the output of the program. The ease with which we perform a BLAST search might lead us to treat the process as a "black box" that we neglect to understand. However, there are several questions a bioinformatician should be able to answer about this process before attempting to draw meaningful conclusions from the output. First, what types of information do we get from sequence alignment? Second, why do certain regions of aligned sequences contain a larger fraction of residues that are identical between two or more sequences? Third, what is the basis for scoring sequence alignments? Fourth, what is the difference between identity and similarity in the BLAST output? In this chapter and the following three chapters we will answer these questions.

---

[1] Natural selection is the process by which traits become either more or less common over a long period of time. The traits have a beneficial or harmful effect on the production of progeny and the progeny's survival. Aside from the traits carried by organisms in the species, a component of natural selection is the random occurrence of environmental effects. For example, there may be a fraction of a species population that exhibit a particular trait. That fraction resides in an area separate from other members of the species. An earthquake may destroy the fraction of the species that resides in the separate area. The surviving species have been selected to survive and reproduce. Natural selection is a driving force of evolution.

3.2 CONSERVED REGIONS IN PROTEINS

conserved region A segment of a sequence that, after multiple alignments with other ortholog or paralog sequences, has a higher fraction of identical or similar aligned residues than other segments of a sequence.

catalytic site Region within an enzyme that binds to a substrate, assists in the conversion of the substrate into the product, and releases the product.

When sequences of proteins that perform the same function in different species are aligned, it is usually the case that some regions contain a higher fraction of identical amino acids than others.[2] Such regions within aligned protein sequences are called conserved regions. These regions are conserved due to one or more of the following: (1) common functional relatedness, (2) structural relatedness, and (3) ancestral gene relatedness. Although these three properties are often intertwined, for the sake of clarity let's discuss them one at a time.

Proteins that perform the same function, or nearly the same function, in different species often have regions with nearly identical sequences. One function of certain proteins is to speed up the rate of chemical reactions. Proteins that speed up or catalyze chemical reactions are called enzymes and in these chemical reactions, substrates are converted into products. Typically, the part of the enzyme that catalyzes the reaction is composed of three to five amino acids, a region called the catalytic site of the enzyme that interacts with the substrate. Enzymes from different species that catalyze the same chemical reaction are likely to have identical or nearly identical amino acids in their catalytic sites. This is a conserved region of the proteins. It is conserved because, although the species that harbor these enzymes may have separated (diverged) from a common ancestor species gene millions of years ago, the catalytic sites have maintained (or conserved) the same amino acids.

Regions outside of the catalytic sites may have amino acid differences. These outside regions are important for positioning the catalytic site amino acids in their correct orientations, but there is some flexibility in the types of amino acids that can serve this role. We call these outside regions nonconserved. Bear in mind that natural selection and neutral mutations over millions of years allow us to distinguish conserved regions and nonconserved regions. Figure 3-1 shows a three-dimensional structure of a protein called trypsin (from a cow), with three regions (colored red, gold, and blue) that come together to form the catalytic site. The three colored regions are also depicted below the protein structure in the schematic diagram showing the alignment of this cow trypsin sequence with two other proteins that perform the same function in other species. The amino acids in the colored regions are conserved in three species.

While catalytic sites in enzymes are often conserved, there are other regions of functional importance that may also be conserved. Some are critical for binding to large molecules, such as DNA, RNA, or proteins in the cell. For DNA-binding proteins, the conserved region is often composed of amino acids that interact with DNA. The protein p53 has three fundamental regions required for its tumor suppressor activity: a transactivation region (near the amino terminus), a DNA-binding region (the middle part of the protein), and an oligomerization region (near the carboxyl terminus). When p53 sequences from different species are aligned, it turns out that all three regions have some identical amino acids; however, the DNA-binding domain is more invariant than the others. Even species that are only distantly related to humans still retain those identical amino acids. In other words, the least variation in amino acids is found in the DNA-binding domain of p53 among diverse species, and it turns out that the DNA-binding region is the most defining characteristic used to identify p53 proteins. Knowing that this region is most conserved helps us identify p53 in species that diverged from humans hundreds of millions of years ago.

[2] Organisms within a single species can mate and reproduce. Organisms from two different species are unable to produce progeny that, in turn, can continue to generate more progeny. Dogs and cats belong to different species because they are unable to produce progeny. Male donkeys and female horses can produce progeny, called mules. However, because mules are unable to produce more progeny, donkeys and horses appear to belong to different species. Sometimes male horses and female donkeys produce progeny (called hinnies) that can, on rare occasions, mate and have offspring. The latter example shows the complexity of deciding whether some organisms belong to the same or different species.

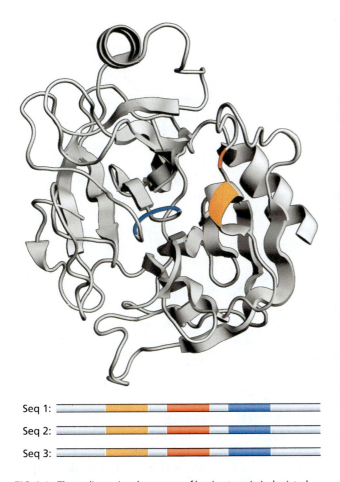

Seq 1:
Seq 2:
Seq 3:

FIG. 3-1. Three-dimensional structure of bovine trypsin is depicted with the conserved amino acids that comprise the catalytic site in color. The protein backbone residues of the catalytic amino acids histidine 57 (*gold*), aspartate 102 (*red*), aspartate 194 (*blue*), and serine 195 (also part of the blue region) are shown. The numbers listed after the amino acids mark their locations in the primary sequence starting from the amino terminus. Below the structure is a schematic diagram showing the primary sequences of trypsin from three species. The primary sequences show the approximate locations of the conserved catalytic amino acids.

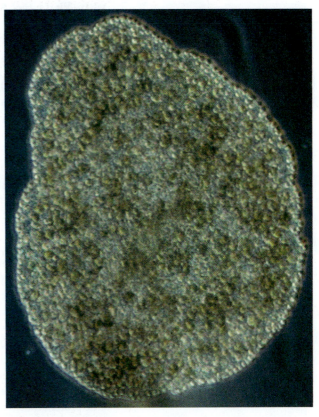

FIG. 3-2. *Trichoplax adhaerens* (placozoa) is the simplest known animal, with the smallest known animal genome (50 million base pairs). It contains a gene that is similar to *TP53*.

For example, a gene coding for p53 is found in placozoa (*Trichoplax adhaerens*) (Figure 3-2). Placozoa are relatively simple organisms that live on the ocean floor. These animals measure only a millimeter in length, and it is estimated that the last common ancestor shared between placozoa and humans lived at least *700 million* years ago. It is conjectured that placozoa p53 is responsible for a process known as programmed cell death (apoptosis) during the development of the organism.

Let's explore the DNA binding function of p53 a bit more. Human p53 binds to the following DNA sequence (also known as p53-response element or p53 DNA binding site):

```
5'-PuPuPuC(A/T)(T/A)GPyPyPy-(0-13  nucleotides)-PuPuPuC(A/T)
(T/A)GPyPyPy-3'³
```

---

³ Using the International Union for Pure and Applied Chemistry (IUPAC) nomenclature, the p53-response element sequence is RRR(A/T)(T/A)GYYY-(0-13 nucleotides)-RRRC(A/T)(T/A)GYYY.

where Pu is either A or G, Py is either C or T, (A/T) is either A or T, and (T/A) is either T or A.

In this representation of the p53-response element,[4] when (A/T) near the 5′ end is an A, then the next nucleotide is a T. When (A/T) is a T, then the next nucleotide is an A. The first and last 10 nucleotide sections are called "half-sites." The term in the sequence "0-13 nucleotides" means that any sequence of DNA from 0 through 13 nucleotides can separate the two half-sites. This is called the spacer.

The length of spacer affects the presentation of the two half-sites to the p53 protein. Because there are about 10 nucleotides per turn of the double helix, a spacer of 10 nucleotides presents the two half-sites on the same side of the double helix. The two half-sites can then easily engage the p53 protein. The spacer is not directly involved in binding to p53, so its nucleotide composition is not important.

In general, many response elements have such a half-site arrangement, which reflects the fact that proteins often bind to DNA in the form of an oligomer. When the double helix presents chemically identical or similar half-sites on the same face, in fact it attracts a protein oligomer. The tetrameric form of p53 (i.e., four identical p53 polypeptides bound together) binds to the p53-response element.

Recall that only one of the two DNA strands in the p53-response element is shown. Because the p53 DNA binding region is conserved in many species, it is thought that p53 targets the same DNA sequence in nonhuman species (though this has not been rigorously tested in all nonhuman species). What is incredible is the fact that the p53 DNA binding region changed little in the approximately 700 million years since the time humans and placozoa last shared a common ancestor.

As mentioned earlier, regions within proteins that have been found in different species may be conserved because they each interact with the same protein sequence. One protein that binds to p53 is MDM2. MDM2 binds to another conserved region of p53 called the transactivation domain (one of the three conserved regions within p53). Recall from Chapter 2 that MDM2 binds to and inhibits p53 tumor suppressor activity. The amino acid sequence within the human p53 transactivation domain that binds to MDM2 is TFSDLWKLL, which is conserved in some species. Although this p53 sequence is not as conserved as the DNA binding domain, it is identical in chimpanzee, orangutan, and dog p53.

disulfide bond A covalent bond created between two sulfur atoms. Some cysteine amino acid residues form disulfide bonds to stabilize the tertiary and quaternary structures of proteins.

Let's take a moment to review what we have learned thus far in this section. We have discussed that proteins may have conserved regions because they catalyze the same reaction (the example of trypsins) or bind to the same specific biological molecules (the examples of p53 binding to the p53-response element and p53 binding to MDM2 protein). In other words, these proteins share the same function. Another reason regions may be conserved is that they have structural relatedness. Some conserved amino acids are critical for preservation of the structures or shapes of proteins. For example, proteins that have the same structures often contain cysteines at the same locations within the 3D shapes of proteins. In extracellular proteins, the side chains of cysteines frequently form disulfide bonds at these key positions, which are necessary for maintaining protein structure. These disulfides are extremely strong bonds and keep the cysteines physically close to each other to help maintain tertiary and quaternary structures (see Chapter 1). In intracellular proteins, cysteine side chains often bind to metal ions, which serve a variety of functions including maintenance of tertiary structure and catalysis of enzymatic reactions. It should be noted that because these two properties are intertwined, it is often difficult to tell whether regions are conserved due to a particular function or due to a particular structure.[5] Function depends on protein structure, and protein structure usually connotes a specific function.

---

[4] Actually, this is a generalized view of the p53-response element. This representation of the p53-response element is called a consensus p53-response element because any one of several particular p53-response elements could conform to this consensus sequence. A particular p53-response element that drives a gene (such as the *p21* gene) in a human cell would have one sequence that conforms to this consensus sequence.

[5] If 3D structures are available, it is sometimes possible to assign regions to a functional or structural role.

Finally, a third reason two sequences may be conserved is that they evolved from a common ancestral gene. An ancestral gene is the precursor gene from an extinct species that gave rise to at least one gene in a living species.[6] During natural selection, the sequences do not diverge to the point where the sequences appear completely unrelated. The two proteins expressed from these genes retain sequences that have some identical or nearly identical amino acids due to the fact that they share a common ancestral gene. One may ask if it is ever possible to ascertain whether aligned sequences show high percent identity solely due to an ancestral gene relationship, but not a functional relationship. The answer is yes. As we will see later in this chapter, two proteins, named His4 and HisF, catalyze different reactions in the biosynthetic pathway of histidine in the bacterial strain *Thermotoga maritima*. These paralogs appear to be derived from a duplication event in a common ancestor gene.

3.3 MOLECULAR EVOLUTION

We have discussed the sequence similarity and its relationship to function, structure, and ancestral history, so we know that proteins that perform the same function in different species sometimes have slightly different sequences. When one compares orangutan p53 to human p53, one finds that there are nine amino acid differences out of 393 amino acids. When a comparison is made between human p53 and p53 from other more dissimilar species, one finds that the number of amino acid differences increases. The observation that proteins with the same function do not have identical sequences in all species raises the question, what causes these differences? To answer this important question, it is necessary to understand how changes in DNA sequences can occur and how those changes can be maintained in descendant cells.

Transformation of Normal Cells to Cancer Cells

A useful model to show how such changes can take place is the transformation of normal cells into cancer cells. This transformation is a limited type of natural selection that takes place in the body of multicellular organisms. The cancer cells and their progeny outcompete normal cells for resources and continue to replicate— so, in a sense, the cancer cell progeny are more fit than normal cell progeny. Cancer cells undergo higher rates of mutation to achieve this phenotype, and these mutations have been studied in detail. These same types of mutations, albeit at a much lower frequency, drive the evolution of species, creating the diversity of life we see today.

From molecular analyses of cancers we know that protein sequences can change when DNA is damaged or when there is an error in DNA replication. If the cell fails to repair the damage or correct the error, a mutation occurs. A mutation is a change in the DNA sequence of the genome of a cell that is passed on to daughter cells. As we learned in Chapter 1, if a mutation occurs in the protein coding part of the DNA the protein sequence may be altered and affect its function. In the case of p53, in Chapter 2 we learned that a mutation that leads to a single amino acid replacement can alter p53 so it cannot bind to DNA. The inability of p53 to bind to DNA incapacitates p53's tumor suppressor function and allows cells with DNA damage to survive and proliferate. Extensive analysis of the *TP53* gene in thousands of cancer tissues from patients has shown that p53 is frequently mutated in the DNA binding region (Figure 3-3). This raises important questions: *Are the mutations in the DNA binding*

---

[6] It should be noted that we typically cannot determine the sequences of ancestral genes directly through experiment. Their sequences can only be *inferred* by studying their homologs in existing species. The first scoring matrices that quantify amino acid similarities were developed by Margaret Dayhoff. She used inferred ancestral genes to develop her scoring matrices, called PAM matrices. See Chapter 4 for more details on PAM matrices.

FIG. 3-3. Histogram of p53 mutation frequency in 8,355 cancer tissues. Frequency of p53 mutations observed in cancer tissue samples is plotted on the y-axis. The p53 amino acid number is plotted on the x-axis. Amino acids 175, 248, and 273 are "hotspots" because they are frequently replaced by other amino acids in cancers due to mutation. The hotspots are amino acids that bind to the p53-response element or are required for p53 to remain properly folded. These mutations render p53 incapable of binding to the p53-response element. Splice site mutations not shown.

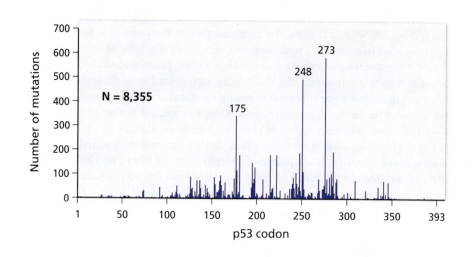

domain observed in cancers because those regions in p53 are particularly susceptible to DNA damage and replication errors?[7] Or, alternatively, are all regions in p53 equally susceptible to mutation, but only those that result in damaged cell survival are the ones observed in cancers? To be honest, there is some debate on this issue, but notwithstanding this debate, it is known that the vast majority of other genes in cancer tissues are not mutated. It is possible that mutations in these other genes are lethal to the cell. The cancer cells that survive the mutation process display mutations limited to particular genes that give cancer cells a growth advantage over normal cells.

The transformation of a normal cell into a cancer cell shows, in a microcosm, how a mutation can give daughter cells a growth advantage over neighboring cells. DNA damage to a nucleotide base within *TP53*, coupled with failure to repair the damage, causes misreading of the parental DNA during replication. The replicated DNA contains an incorrect nucleotide because the damaged base of the template DNA is misread. Similarly, the replication machinery could incorporate an incorrect nucleotide into the *TP53* during DNA replication even if there were no DNA damage. In either case, after replication the mutated genome ends up in the daughter cell.

Here is probably a good place to discuss what, exactly, the *TP53* protein product p53 actually does. Why is it that when the p53 protein is mutated, cancers often result? In normal replicating cells, p53 levels are quite low because p53 is quickly degraded after it is synthesized. Once the cell is stressed the p53 protein becomes phosphorylated and its degradation rate is drastically lowered. The level of p53 protein increases, and the protein accumulates in the nucleus. At sufficiently high levels, p53 binds to DNA sequences called p53-response elements in the genome. Upon binding to the p53-response elements, the transcription rate of certain genes near these response elements increases (this is explained in more detail in Chapter 10). These genes control DNA repair, cell cycle arrest, and programmed cell death.[8] Many types of cell stress activate p53, but we will explore just one here for illustrative purposes. If the stress is severe, for example, DNA damage due to harmful radiation, p53 will activate genes that cause the host cell to commit cell suicide (apoptosis). Apoptosis rids the organism of cells that could become cancerous had the damaged DNA persisted. If p53 itself becomes nonfunctional due to a mutation, the cells will not undergo apoptosis and radiation damaged DNA could be converted into a mutation upon cellular division. If the mutations occur in genes that control cell

---

[7] One study has shown that cigarette smoke preferentially forms benzo[a]pyrene adducts with the *TP53* gene at bases that are frequently found to be mutated in lung cancers in smokers. See Denissenko, Pao, Tang, and Pfeifer (1996).

[8] Actually, the list of gene categories that p53 controls is fairly lengthy.

proliferation it is possible that cells will divide uncontrollably. When p53's molecular function became known it inspired David Lane, a pioneer in p53 research, to name p53 the "guardian of the genome."

Let's analyze the consequence of a cell having a mutated *TP53* a bit further. Because the daughter cell with the mutated *TP53* is compromised in its ability to properly respond to cell stress, the DNA of the daughter cell is susceptible to more mutations.[9] If further mutations occur in other tumor suppressor genes, the cell will not properly respond to growth inhibitory signals. Furthermore, if mutations activate genes that normally promote growth (called proto-oncogenes) the cell will divide more frequently. The upshot is that the cells with mutations in tumor suppressor and proto-oncogenes have a growth advantage over normal cells. This scenario demonstrates that it is possible for genes to mutate and that this change brings about a growth advantage for cells.

Are Mutations Inherited?

The transformation of normal cells to cancer cells shows that mutations in specific genes can give cells a growth advantage. To drive evolution, mutations must be passed on to the next generation of organisms. In complex organisms, there are two cell types: somatic and germ. Somatic cells constitute the majority of cells in a complex multicellular organism (liver cells, skin cells, kidney cells, etc.), but the DNA from these cells is not passed on to the next generation. In contrast, germ cells (for example, egg cells and sperm) contain DNA that can be passed on to the next generation. The vast majority (~99%) of cancers arise from somatic cell mutations. Recall that humans are diploid organisms; they have two sets of genes (called alleles) in every somatic cell. One allele comes from the mother and the other comes from the father. The *TP53* alleles from the mother and the father are both deactivated by mutation in the somatic cancer cells of the patient. The deactivation must occur at some point after the egg and sperm combine and the organism begins to develop. It could occur during the development of the embryo. It could occur during childhood, or it could occur after the organism has reached adulthood. In the case of *TP53*, the cancer cells usually contain one *TP53* allele deleted (the DNA encoding p53 is missing) and one *TP53* allele with a point mutation (a single nucleotide change) that gives rise to an amino acid substitution (see Figure 3-3). Thus, there is no normal (wild-type) p53 in somatic cancer cells. On the other hand, noncancerous normal cells have two normal *TP53* alleles.

Rarely, a cancer *does* arise from a mutation in germ cell DNA. Germ cells contribute their DNA to the progeny of the organism. In the case of *TP53*, there is one syndrome, called Li-Fraumeni, where mutant *TP53* is inherited (see Box 1-1). In this syndrome, the mutant *TP53* is part of either the egg or sperm genome. Patients with Li-Fraumeni syndrome typically inherit a *TP53* allele that contains a point mutation from one parent. The second inherited *TP53* allele is wild-type. This means that when a Li-Fraumeni baby is born, *all* of its cells have one mutant *TP53* and one wild-type *TP53*. As the baby grows, somehow, the remaining wild-type *TP53* gene gets deleted in some cells and those particular cells are primed to become cancerous.[10]

Up to now we have discussed mutations mainly in the very specific context of cancer. Mutations in certain genes (tumor suppressor genes and proto-oncogenes, for example) give cells a growth advantage over normal cells. Most mutations occur

tumor suppressor gene A gene whose normal function is to suppress cancer formation. The gene's normal function is to restrain cell growth or cause apoptosis. In cancers, tumor suppressors are inactivated.

proto-oncogene A gene whose normal function is to promote cell growth or cell survival. This is the wild-type version of the oncogene.

allele Alternative sequence variant that occurs at a particular locus in a species.

---

[9] Note that humans and other diploid mammals have two copies of *TP53*. In cancers, both copies of *TP53* are usually mutated. These mutations usually occur in separate events. Daughter cells with one mutant copy may create many generations of cells prior to mutation of the second copy of *TP53*.

[10] The average age of cancer onset in Li-Fraumeni patients varies by gender. In women it is 29 years, and in men it is 40 years. Li-Fraumeni syndrome was considered a clinical disease in 1969, and in 1990 it was discovered that *TP53* is commonly mutated in patients with this disease. We now know that germ-line mutations in *TP53* are observed in approximately 70% of patients with Li-Fraumeni syndrome.

in somatic cells, but in rare cases mutations can occur in germ cells. This raises a broader question: can mutations cause new species to arise? According to Charles Darwin's theory of natural selection, the answer is yes.

Natural Selection

Charles Darwin (1809–1882) was a British naturalist who spent the majority of his life studying organisms in their natural habitats. In 1859 he published his book *On the Origin of Species*, in which he posits that natural selection is a mechanism by which new species could come into existence.[11] In the introduction to his book, Darwin explains:

> As many more individuals of each species are born than can possibly survive; and as, consequently, there is a frequently recurring struggle for existence, it follows that any being, if it vary however slightly in any manner profitable to itself, under the complex and sometimes varying conditions of life, will have a better chance of surviving, and thus be *naturally selected*. From the strong principle of inheritance, any selected variety will tend to propagate its new and modified form. (emphasis added)

natural selection A process by which biological traits become more or less common in a population over a period of time. The traits have a beneficial or harmful effect on the production of progeny and the progeny's survival.

Of course Darwin did not have knowledge of the genes that underlie the process of natural selection. From our vantage point in history we can begin to understand the mechanism and repercussions of natural selection by examining gene and protein sequences. By aligning the p53 protein sequences from humans and placozoa (Figure 3-4), one can immediately surmise that there are particular regions that are more identical than others. One may hypothesize that the *TP53* genes from these organisms could be derived from the same ancestor (known as the ancestral gene). The regions of two *TP53* genes that are nearly identical (conserved) must be essential for life, or have given a survival advantage for placozoa and humans over other organisms during natural selection. The nonconserved regions (those that are not highlighted in Figure 3-4) are the result of mutations in the descendants of the ancestral gene that gave rise to amino acid differences in the *TP53* genes from placozoa and humans. The differences may slightly alter the function of the p53 protein in these two organisms, or, on the other hand, the differences could have no effect if they are the result of neutral mutations (see below). In humans, p53 is not required for development, but is essential for tumor suppression. However, the role of p53 in placozoa development and tumor suppression is presently unknown.

neutral mutation A mutation that does not alter the ability of the organism to produce viable progeny (i.e., does not alter the fitness of the organism).

There are several criteria necessary for evolution by natural selection to occur: (1) variation in a trait important to fitness/reproduction (this is called phenotypic variation), (2) heritability in that trait, (3) competition for resources, the outcome of which is influenced by that trait. At the organism level, if mutations in germ cells give the progeny a survival advantage over other progeny then those mutations will be maintained in the genome of those progeny and future progeny. Mutations can lead to changes in protein structure and function, thus altering the physical properties of the organism. If mutations in germ cells do not alter the survival advantage of the progeny they may be maintained in the genome for a time. Such mutations are called neutral mutations.

Interestingly, researchers have tested whether specific gene replacements from one species into another affect the phenotype of the recipient species. A gene that codes for the development of the mouse eye (*Sey*) was used to replace the fly *ey* gene in the fly. With the mouse *Sey* gene, the fly eye developed normally and the fly was

---

[11] Alfred Wallace (1832–1912) arrived at the principle of natural selection at nearly the same time as Darwin. Wallace's theory, captured in the paper titled "On the Tendency of Species to Form Varieties; and on the Perpetuation of Varieties and Species by Natural Means of Selection," which Darwin presented to the Linnean Society of London on July 1, 1858 (with proper credit given to Wallace). Wallace's work appeared in print one month later in the *Journal of the Proceedings of the Linnean Society of London. Zoology* 3 (August 20): 46–50.

```
human      1    MEEPQSDPSVEPPLSQETF----SDLWKLL-------------PENNVLS        33
                .|.||.|||..:|     |...|:|:            .I...:.
placozoa   1    -------MSDEPTLSQLSFSQELSSSWQLMIDEITQGKFNTNEDEGTAIY        43

human      34   PLPSQAMDDLMLSPDDIEQWFTEDPGPDEAPRMP-EAAPPVAPAP-----        77
                ....|..||..|....:.|:.:.:.....:.::||  |.|....|:|
placozoa   44   SYSEQNPDDRYLMRPNEPQYISAGYPDGQVGQLPREFAVNQIPSPRTFSD        93

human      78   ---------------AAPTPAAPAPAPSWPLSSSVPSQKTYQGSYGFRLG        112
                .|.....:...:.|......|:||...|.|.::||..:.
placozoa   94   NVSSSADKAREAYYGQAVNGVSAETSPPLKRDPSLPSNAEYIGNFGFDIA        143

human      113  F-LHSGTAKSVTCTYSPALNKMFCQLAKTCPV------------------        143
                . .:....|:...|||..|.|.:.::....|:
placozoa   144  IDQNDNPTKATNNTYSTHLKKLFIKMECLFPIHITIERMDYTFKIAYGSL        193

human      144  -------QLWVDSTPPPGTRVRAMAIYKQSQEMTEVVRRCPHHERCSDSD        186
                       ||.:...||..:.:||..:|.:.|.:.|.||||:|  ...|..
placozoa   194  ATRRNCNQLIIPGEPPANSYIRAYVMYTKPQDVYEPVRRCPNH-ALRDQG        242

human      187  GLAPPQHLIRVEGNLRVEYLDDRNTFRHSVVVPYEPPEVGSDCITIHYNY        236
                ......|::|.|.. |.||.:| .:.||||.|||...:.|..:|.|..
placozoa   243  KYESSDHILRCESQ-RAEYYED-TSGRHSVRVPYTAPAVGELRSTLLYQF        290

human      237  MCNSSCMGGMNRRPILTIITLEDSSGNLLGRNSFEVRVCACPGRDRRTEE        286
                ||.|||.|.:|||||..:|||.: |:|||...|||||||||| |:.|
placozoa   291  MCFSSCSGSINRRPIELVITLENGT-NVLGRKKVEVRVCACPGRD-RSNE        338

human      287  ENLRKKGEPHHELPP------------------GSTKRALPNNTSSSPQ        317
                |....|.|..|:.||             ..:||.:.:..|:.
placozoa   339  ERAAMKSEKEHKQPPNKKLKTSKTVSREVTGVISNESKRIMERSVESTS-        387

human      318  PKKKPLDGEYFTLQIRGRERFEMFRELNEALE----LKDAQAG--KEPGG        361
                :.:.||:.:|||:.:  :..:::|:||    |.|||.  |..|.
placozoa   388  ------NDUVFTITVRGRKNYEILAKMSESLEVLDKLSDAQINEIKSHGT        431

human      362  -----SRAHSSHLKSKKGQSTSRHKKLMFKTEGPDSD-------------        393
                .|.::.:|...:.:...........|..|..
placozoa   432  LTAPLERTNTEELVRRQSRNLDTLQNAVTTKENSDGADLNLSISRWLSNI        481

human      394  --------------------------------------------------        393

placozoa   482  NMEKYTQEFIKHGFKVCGHLANVSYSDMKKIIKNMEDCKKISAYLLESNF        531

human      394  --------------------------------------------------        393

placozoa   532  SSGNEEDIPCSQIGNSFRASQMSMNSTASQELDITRFTLRQTITL-----        576
```

FIG. 3-4. Alignment of human p53 and putative placozoa p53. The human p53 sequence has 393 amino acids, and the placozoa p53 has 576 amino acids. The DNA binding region of human p53 is highlighted in yellow. Vertical lines between amino acids denote identical amino acids, two dots denote very similar amino acids, and one dot denotes dissimilar amino acids shared between the two sequences.

able to see![12] Only 29% of the amino acids are identical in the protein products of the mouse gene and the fly gene, which must be critical for proper functioning. Thus, only specific locations of mutations within the genome are critical to organism development. In fact, to create a new species, a mutation within the coding region of the

---

[12] The *ey* gene codes for a protein in *Drosophila melanogaster*, called eyeless (GenBank accession number NP_524628.2), that is a transcription factor, and, similar to p53, it activates the transcription of genes, except the genes it activates leads to eye development and not tumor suppression. The *Sey* gene (sometimes known as *Paired Box 6* gene or *Pax6*) codes for a protein in mouse, called Pax-6 (GenBank accession number NP_001231127.1). It too is a transcription factor that leads to eye development in the mouse. When the protein sequences of eyeless and Pax-6 are aligned only 29% of their amino acids are identical. Although it is incredible that the *Sey* gene can substitute for the *ey* gene in flies, multigenerational studies have not been performed to determine whether there are survival disadvantages for the flies that have this substitution.

gene may be less important than a mutation that alters the timing of expression of genes during development. Such mutations occur in gene promoters and elsewhere in the genome, and are likely to be the strongest drivers of change in phenotype of the organism. We will discuss this issue in Chapter 8, but for now, let's take a tour of different classes of mutations.

Mechanisms of Mutation

At this point in our discussion it is useful to define the vocabulary that will help us discuss molecular evolution at a deeper level. Just to review, a mutation is a change in the DNA that can be passed on to daughter cells. Mutations occur when the cell is unable to repair damaged DNA or to correct errors during DNA replication. Once the damaged or changed DNA replicates and the cell divides to create two daughter cells, the alteration in the DNA sequence is called a mutation. The daughter cells contain the mutated genome. Only germ cell mutations can lead to changes in the progeny of the organism. It is quite possible that mutations occur in a region of the genome that does not code for protein or RNA. In this case, the mutation may be neutral (i.e., does no harm or is good for the organism).

A neutral mutation can occur even in the coding region of a gene and have no effect on the organism. For example, if a sequence changes from TCT to TCC in a particular gene, this would have no effect on the protein structure, because both sequences encode a serine in the protein. A neutral mutation can even change the amino acid sequence of a protein. If the amino acid residue change does not affect the protein function, then this too is a neutral mutation. Mutations in the coding region of a gene that leave proteins with the same amino acid sequence are called synonymous mutations (also known as silent mutations or silent substitutions).[13] On the other hand, mutations that lead to any amino acid alteration in the protein are called nonsynonymous mutations.

At a basic level, a synonymous mutation is a neutral mutation.[14] However, it is also possible that a nonsynonymous mutation is a neutral mutation. In this case, an amino acid change in a part of the protein that is not critical may not have an apparent effect on protein function. This raises the question, how does one decide whether a mutation is neutral? One measure of this quality is the *fitness* of an organism. Fitness is defined as the ability of the organism to survive to the age where it reproduces and creates viable offspring. By this definition, a 90-year-old woman who is in excellent health but never had children is considered less fit than a woman who dies at 40 and leaves behind children. Having children ensures that one's DNA survives to the next generation. A mutation that increases the propensity of an organism to leave behind viable progeny is called an advantageous mutation. A mutation that decreases the propensity of an organism to leave behind viable progeny is called a deleterious mutation. A mutation that has no effect on fitness is called a neutral mutation.

Throughout the more than three billion years in which life has existed on Earth, there has been a selection process (natural selection) where only the organisms (and their genomes) that are the fittest survive. The organisms have sustained advantageous, deleterious, and neutral mutations. Mutations that are advantageous or neutral will survive in future generations, but those that are deleterious (no matter how small) will eventually not survive in future generations. Once the mutation is observed in more than 1% of the population of a species it is no longer considered a

synonymous mutation A mutation in the coding region of a gene that does not alter the protein sequence. This is also known as a silent mutation.

nonsynonymous mutation A mutation in the coding region of a gene that alters the protein sequence.

fitness The ability of the organism to survive to the age where it reproduces and creates viable offspring.

deleterious mutation A mutation that decreases the propensity of an organism to create viable progeny.

advantageous mutation A mutation that increases the propensity of an organism to create viable progeny

polymorphism A DNA sequence alteration observed in more than 1% of the population of the species. The most common polymorphism is a single nucleotide polymorphism (SNP).

---

[13] Degenerate codons are codons that differ in their nucleotide sequences, but code for the same amino acids. For example, serine is coded by six degenerate codons.

[14] Actually, some degenerate codon sequences are much less common than others and there are less abundant tRNAs that recognize them. An uncommon codon can actually lower the rate of translation, causing a drop in the protein level. If that drop causes a change in the organism's ability to survive, then mutations that change codon sequences that do not result in different amino acids are not neutral mutations.

mutation. Instead, it is called a *polymorphism*.[15] The 1% cutoff ignores the presence of extremely rare polymorphisms. This is a practical definition of a polymorphism because it is difficult to distinguish a polymorphism from a mutation if the frequency of the polymorphism is less than 1%.

At this point it is useful to define other terms related to sequence variation. A locus is the name for a specific position on a chromosome. Until recently, a locus was typically the position of a gene that coded for a protein, but now, with many entire genomes sequenced, a broader definition of a locus is the position of a sequence that is experimentally detectable. Alleles are alternative sequence variants that occur at the same locus in a species. A polymorphic locus (or polymorphism) is one where there is more than one allele. Now let's apply these terms to a specific gene in a diploid organism (an organism that contains two copies of genes in every somatic cell).

TP53 is a gene as well as the name of a locus. In humans, there is polymorphism in *TP53* at codon 72. One allele codes for Pro at this codon and another allele codes for Arg at this codon. In humans and other diploid organisms, there are two *TP53* genes in every somatic cell, so there are three possible *TP53* allele arrangements. In the first arrangement, both *TP53* alleles code for Pro at codon 72 (Pro/Pro). In the second, one *TP53* allele codes for Pro and the other codes for Arg (Pro/Arg) and in the third, both alleles code for Arg (Arg/Arg). Interestingly, 61% of African-Americans have Pro/Pro alleles and 21% of Caucasians have the Pro/Pro alleles. Nonhuman primates such as chimpanzee, rhesus macaque, and orangutan appear to have only Pro at position 72 suggesting that Pro was the original amino acid. Although some hypotheses have been put forward to account for the association of *TP53* allele frequency differences with race, none have been supported. At the moment, the mutation that gave rise to this polymorphism appears to be a neutral mutation, hence its prevalence. It would be interesting to know whether, millions of years from now, the Arg polymorphism will still exist in humans. If not, then we may need to revise our assessment of the Arg polymorphism and name it a deleterious mutation due to the fact that natural selection prevented future generations of humans from carrying this amino acid change.

Point mutations and indel mutations

To discuss mutations at the DNA level, recall from Chapter 1 that A and G nucleotides are purines and C and T nucleotides are pyrimidines. A mutation from one purine to another or from one pyrimidine to another is called a transition. A mutation that exchanges a purine for a pyrimidine or vice versa is called a transversion. Collectively, these are called point mutations, and they are the most frequent of all mutation types. When a point mutation changes the codon for one amino acid into that of another, it is a missense mutation. When a point mutation changes the codon for one amino acid into a stop codon, it is called a nonsense mutation. More rare than point mutations are mutations that arise from the insertion or deletion of one or more nucleotides, called indels. Repeated nucleotide sequences are very susceptible to indel mutations. An example of a repeated sequence is CAGCAGCAG. Such sequences, called simply "repeats," are susceptible to insertion because during replication the newly synthesized DNA appears to dissociate from the template strand and forms a hairpin structure (Figure 3-5). Newly synthesized DNA (daughter strand) can dissociate from the template strand and then bind to the template strand at a different location. This slipped mispairing results in a daughter strand that is longer than the template strand. The progeny cells may use the daughter strand for future generations of DNA replication. If an indel occurs in a protein coding region and it is

locus A specific sequence on a chromosome that is experimentally detectable.

[handwritten: unstable b/c of bulk]

[handwritten: transitions = same type g/purine]
[handwritten: ex. A ↔ G]
[handwritten: T ↔ C]

transition A mutation that results in the substitution of one purine for another or one pyrimidine for another.

transversion A mutation that results in the substitution of one purine for one pyrimidine or vice versa.

indel A mutation that results in the insertion or deletion of nucleotides into the genome.

[handwritten: transversion = diff # of rings]
[handwritten: ex.]
[handwritten: A ↔ T, A ↔ C]
[handwritten: G ↔ T, G ↔ C]

---

[15] The most common polymorphism is a single nucleotide polymorphism (SNP), which accounts for 90% of all polymorphisms in the human genome. A SNP occurs approximately once in every 1,000 nucleotides in the human genome.

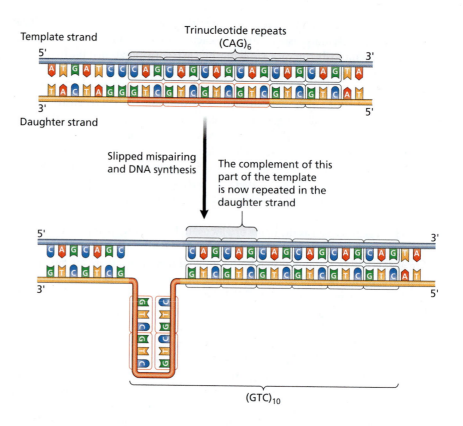

FIG. 3-5. Model for a mechanism that causes indel mutations. This type of indel mutation is known as repeat expansion. The top template strand is from the parent DNA. The bottom daughter strand is newly synthesized. The process of "slipped mispairing" creates a hairpin structure (red) consisting of repeats of GTC. The daughter strand (yellow) to the right of the hairpin forms nucleotide base pairs with the parent strand and DNA replication continues. The expanded daughter strand will cause an insertion mutation in the progeny cells.

microsatellites Segments of DNA, known as simple repeats, that have a range of 1–3 nucleotides that are repeated in tandem in blocks of up to 200 nucleotides in length. The lengths of these blocks often vary from individual to individual and are used for DNA profiling.

translocation Mutation where one part of a chromosome fuses to another chromosome.

not a multiple of three nucleotides, it will cause a frameshift mutation. The frameshift will change the amino acid sequence downstream of the mutation (Figure 3-6).

Recent evidence indicates that there is a system that recognizes these DNA hairpins and repairs them before an indel mutation occurs. Some individuals may lack the repair system, causing them to be more susceptible to indel mutations. Repeated sequences, which are particularly vulnerable to indel mutations, are called microsatellites, and their lengths often vary within the human population. When repeated DNA sequences code for proteins, the potential for indel mutations may cause devastating effects on the protein and the individual. In Huntington's disease the number of CAG repeats in the gene *huntingtin* increases as the patient ages (a phenomenon known as repeat expansion). CAG codes for Gln and when the number of Glns in the huntingtin protein increases beyond 35, the patient shows signs of brain wasting.[16]

Translocations, inversions, and gene amplifications

Although point mutations and indel mutations are relatively common, on occasion large segments of DNA can move from one part of the genome to the other. Collectively, these are called genome rearrangements. One example of this is the translocation mutation that occurs in a type of blood cancer called chronic myelogenous leukemia (CML). In CML a translocation occurs in which the *bcr* gene from chromosome 22 and the *abl* gene from chromosome 9 come together to create a hybrid gene (in this case, an oncogene named *bcr-abl*).[17] With a light microscope, the translocation

---

[16] Singer/songwriter Woody Guthrie (1912–1967), composer of the American folk song "This Land Is Your Land," suffered from Huntington's disease—a disease characterized by lesions in the brain detected through postmortem autopsy. Symptoms include hallucinations, paranoia, psychosis, unsteady gait, and unusual facial movements.

[17] This was the first DNA change associated with a cancer. It was discovered at Fox Chase Cancer Center in Philadelphia, Pennsylvania, in 1960 by Peter Nowell. The shorter of the new hybrid chromosomes is called the Philadelphia chromosome.

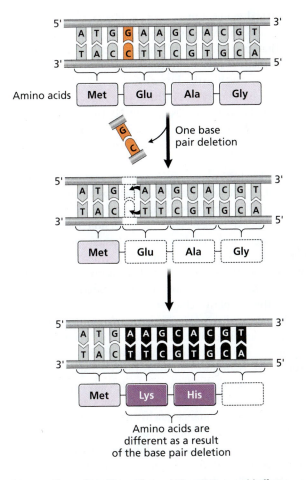

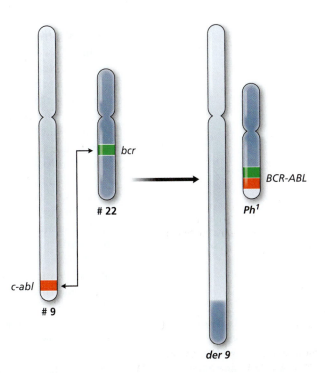

FIG. 3-6. Illustration of how a frameshift mutation could affect the amino acid sequence of the protein product. In the top panel, a GC base pair is deleted from the gene. This shifts the reading frame of the gene so that the amino acids read downstream of Met are different from the original gene.

FIG. 3-7. Reciprocal translocation between one chromosome 9 and one chromosome 22. The *c-abl* gene (*red*) is located near the telomere of the long arm (q arm) of chromosome 9. The *bcr* gene (*green*) is located near the centromere on the long arm of chromosome 22. After translocation, an extra long chromosome 9 called *der 9* and an altered chromosome 22 called *Ph*[1] (Philadelphia chromosome) are formed. The *bcr* gene from chromosome 22 fuses with the *c-abl* gene from chromosome 9 to form the *BCR-ABL* oncogene. High expression of the BCR-ABL oncoprotein from *Ph*[1] drives cell division in chronic myelogenous leukemia. (Refer to Figure 1-13 to see the nomenclature associated with different parts of a chromosome.)

can be detected as an altered chromosome in the cancer cells. Figure 3-7 shows a schematic diagram of single alleles of normal chromosome 9 and chromosome 22. After translocation, the majority of the remaining chromosome 9 plus the added DNA from chromosome 22 becomes the "der 9" chromosome. The majority of chromosome 22 plus the added DNA from chromosome 9 becomes Ph[1] (the Philadelphia chromosome). The hybrid protein BCR-ABL expressed from this mutant hybrid gene, located at the junction (called the breakpoint) within the Ph[1] chromosome, is a cancer growth promoting protein. BCR-ABL is an oncoprotein, and is key to driving abnormal cell division in CML. An important point is that the BCR-ABL protein is abnormal for two reasons: it promotes growth and, due to its new location in the genome, is expressed at higher-than-normal levels. We find that translocation mutations may not only affect the protein sequence, but may also affect the expression level of proteins.[18]

When chromosomes are compared from two species we often find evidence of translocations of large regions of genomes. These translocations appear to contribute to the phenotypic differences between the species. For example, human and

---

[18] In the Philadelphia chromosome, the translocation results in higher-than-normal levels of the BCR-ABL protein. Translocations can also disrupt gene function, constitutively activate gene transcription, and cause DNA deletion near the breakpoint.

chimpanzees have almost identical genes, but there is evidence that translocations occurred millions of years ago. In one case, chromosome 2 in humans is the result of two chromosomes that fused together[19] in an ancestor common to humans, Neanderthals, and Denisovans. Neanderthals and Denisovans appear to be two subspecies (now extinct) within the human species.[20] Chimpanzees, which diverged from humans approximately five million years ago, have retained two separate chromosomes, named chromosomes 2A and 2B, that have DNA sequences similar to two portions of human chromosome 2. All other nonhuman hominidae species, such as gorilla and orangutan, have also retained the two separate chromosomes.

Similar to translocation, another type of genome rearrangement occurs when a particular chromosome breaks off, inverts, and reattaches itself to the same chromosome. In this mutation, called a chromosome inversion, a segment of DNA (potentially containing thousands of genes) is reversed. An inversion may lead not necessarily to a loss or gain of genetic material, but merely a rearrangement of the DNA. Interestingly, human chromosomes and chimpanzee chromosomes have almost identical amounts of DNA, but there are nine large inversions and more than 1,500 small inversions that account for much of the overt genetic differences between the two species (see exercise 6 at the end of this chapter).

Under this umbrella of genome arrangement, another mutation that can affect large segments of genomes is gene amplification. In DNA amplification, large segments or small regions of a genome, sometimes containing genes, are repeated. For example, the *MDM2* proto-oncogene is amplified in many cancers, especially in sarcomas (i.e., cancers derived from non-epithelial tissue), which leads to high expression of its protein, MDM2—the natural inhibitor of p53. To be clear, there is no mutation in the MDM2 protein coding sequence itself. Instead, the mutation is the increase in the copy number of the gene beyond what is normal in the genome. Sarcomas arise because p53 is unable to overcome the inhibition caused by MDM2. Gene amplification of proto-oncogenes that code for oncoproteins prevents tumor suppressor genes from working properly or directly drive cell growth.

We discussed several types of mutations that lead to changes in DNA. These mutations stem from DNA replication errors and unrepaired DNA damage. We have used examples from human cancers to illustrate how these mutations can occur. In the context of evolution, it is necessary to accept the possibility that these mutations can occur in germ cells and, furthermore, in rare instances give a selection advantage to progeny. Evidence for mutations being passed to progeny can be found in the sequences of genomes of existing organisms.

A good example of mutation being passed to progeny that gives a survival advantage is a well-known point mutation in hemoglobin. Recall from Chapter 1 that hemoglobin S is a hemoglobin protein expressed from a mutated hemoglobin gene. The point mutation results in a single amino acid change that causes red blood cells to take on a sickle shape. Progeny born with two alleles for hemoglobin S suffer from sickle cell anemia. Progeny born with one hemoglobin S allele and one normal hemoglobin allele have the sickle cell trait. Symptoms associated with sickle cell trait are much less severe than those associated with sickle cell anemia. Interestingly, sickle cell trait individuals have high resistance to malaria, a disease transmitted through mosquitoes. In fact, their resistance is stronger than that of individuals who have two alleles of normal hemoglobin. In areas of the world where malaria is prevalent, individuals with sickle cell trait have lower mortality rates than both sickle cell anemia patients and normal individuals. Below, we will explore two more mechanisms of mutation that are also critical for natural selection.

amplification (also known as DNA amplification) DNA copy number is increased several-fold. This can occur in some parts of a chromosome in cancers. Amplification is also used to describe the process of polymerase chain reaction (PCR).

---

[19] This particular type of translocation is called a telomere-telomere fusion.

[20] We discuss the evolution of the great apes in more detail in Chapter 8.

Viruses can cause mutations

Besides unrepaired DNA replication errors and unrepaired DNA damage, viral infection can mutate the DNA sequence of the genome. In 1970, David Baltimore at the Massachusetts Institute of Technology and Howard Temin at the University of Wisconsin independently discovered an enzyme in viruses that reverse transcribes RNA into DNA. They named the enzyme reverse transcriptase, and the virus that harbors this enzyme is called a retrovirus. Figure 3-8 depicts the life cycle of the retrovirus. Once the retrovirus gets inside the host cell, its RNA genome is reverse transcribed in a single stranded DNA complementary sequence. The reverse transcriptase then turns around and makes a complement of the DNA strand it just synthesized. The now double stranded DNA (called viral DNA) integrates into the genome of the host cell. Incredibly, the integrated viral DNA gets transcribed into viral RNA by the host's transcription machinery. The viral RNA has two functions.

retrovirus A virus containing two copies of a RNA genome and reverse transcriptase enzyme. In part of the virus life cycle its genome is reverse transcribed into double strand viral DNA and this incorporates into the host cell genome.

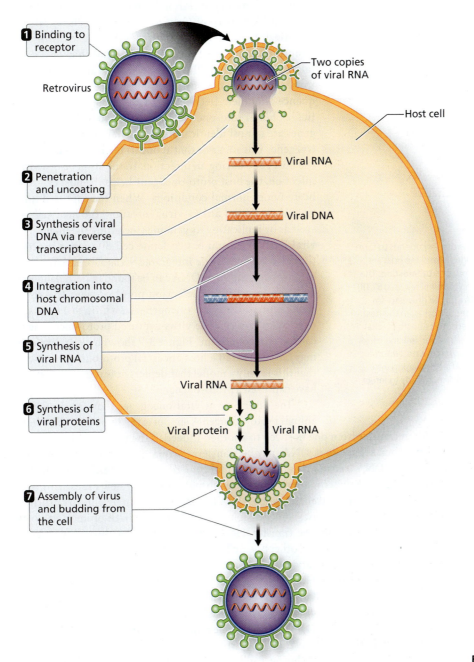

1 Binding to receptor

Retrovirus

Two copies of viral RNA

Host cell

2 Penetration and uncoating

Viral RNA

3 Synthesis of viral DNA via reverse transcriptase

Viral DNA

4 Integration into host chromosomal DNA

5 Synthesis of viral RNA

Viral RNA

6 Synthesis of viral proteins

Viral protein   Viral RNA

7 Assembly of virus and budding from the cell

FIG. 3-8. Life cycle of the retrovirus.

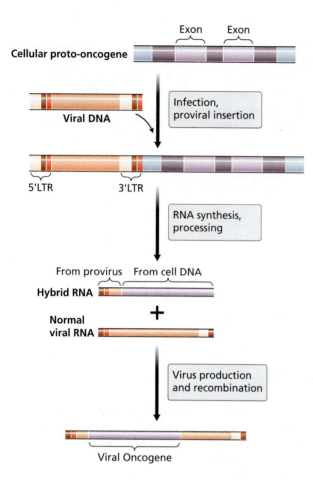

FIG. 3-9. Model for transduction of cellular proto-oncogenes to form retroviral oncogenes. Exons of a proto-oncogene from the host are located downstream of a retroviral DNA introduced by infection. The strong promoter of the virus (from the 5′ long terminal repeats; 5′LTR) is responsible for transcribing the virus DNA and the downstream host DNA. RNA splicing removes the intron sequences from the host DNA. The processed virus-host hybrid RNA recombines with normal viral RNA during virus production. After recombination (mixing of genetic material) with normal viral RNA the new retrovirus genome has captured the cellular gene from the host. After packaging into its protein lipid coat (not shown) the new retrovirus is ready to infect other cells in the host.

oncogene A mutated form of a gene whose normal function is to promote cell growth or cell survival. The mutation may cause abnormally high levels of oncogene product or abnormally high levels of oncogene function.

horizontal gene transfer (also known as lateral gene transfer) Transfer of DNA from one cell to another or from one organism to another without sexual or asexual reproduction. An example is when a virus captures a host gene from one cell and transfers the host gene to another cell.

First, it is translated into proteins that comprise the viral coat and the viral enzymes for a new infection, and second, it becomes the genome for future viruses. The viral RNA is then packaged into new retrovirus particles assembled from viral coat and viral enzymes. The retrovirus particles leave the infected cell (killing the cell in the process) and infect other cells of the organism.

Scientists working with retroviruses considered the possibility that the transcription process may, on rare occasions, transcribe some host DNA along with the integrated viral DNA. If the host DNA codes for protein, then the retrovirus virus particles could capture RNA that is comprised of a combination of host and viral genes. Strong circumstantial evidence supporting this possibility came from the laboratory of Harold Varmus and Michael Bishop (both at University of California, San Francisco). These investigators found that retroviruses that cause cancers in nonprimate animals (typically rodents and chickens) contain viral oncogenes. The viral oncogenes are very similar to normal proto-oncogenes in animal cells. In fact, sequence alignment between the viral oncogene and a particular host proto-oncogene demonstrated that the two genes are almost identical and likely have the same ancestral gene. This suggests that at one time, the virus captured the host gene during an infection. In the absence of a viral infection, the host proto-oncogene helps the host cell replicate under normal conditions. When the captured host gene is expressed after a virus infection, the host cell responds by replicating its genome (including the integrated viral DNA) to begin or accelerate cell division, a process that also replicates the integrated virus DNA. After host cell division, the virus DNA can be found in both daughter cells. From a natural selection point of view, the virus that captured the proto-oncogene has a growth advantage over other viruses. The virus-captured proto-oncogene is called a viral oncogene. Figure 3-9 shows how a retrovirus may have captured a host ancestral proto-oncogene. Note that this is only a model that depicts how retroviruses *may* have captured a proto-oncogene.

The discovery of viral oncogenes that are similar to animal proto-oncogenes supports the hypothesis that host DNA can transfer from one host cell to another. Indeed, if the virus infects other organisms, then segments of DNA may be transferred from one organism to another. This type of DNA transfer is called horizontal gene transfer or lateral gene transfer. Horizontal gene transfer is another way organisms can acquire new genes. This is in contrast to vertical gene transfer, where DNA transfer goes from parents to offspring. Upon capture by retroviruses, it is possible that other non-growth-promoting genes may undergo horizontal gene transfer. Horizontal gene transfer through viruses is common for prokaryotes (and other one-celled organisms) and plays a large role in the production of genetic diversity of prokaryotic cells. This genetic diversity is key to natural selection in bacteria. Horizontal gene transfer as a mechanism for creating genetic diversity in multicellular eukaryotes (such as humans) may be more limited because it usually occurs only in somatic cells. The egg and the sperm are usually spared infection by viruses. However, given the millions of years that multicellular life has existed on Earth, it is likely that retroviruses had a major role in transferring genes into and out of germ cells during evolution. Substantial DNA mutations can be brought about by viruses, especially in one-celled organisms.

Transposons can cause mutations in eukaryotes

In addition to viral infections, there is one more major mutational mechanism that has the potential to greatly alter the genome: transposase-mediated DNA transfer. A transposase is an enzyme that moves a segment of DNA from one location to another in the genome. The segment that is moved is called a transposon.[21] In turns out that the majority of the DNA in the human genome consists of transposons and deactivated transposons. There are two mechanisms of DNA mutations caused by transposons: direct transfer and indirect transfer.

Direct transfer occurs when the transposon is cut out from one part of the genome and pasted into another part of the genome (first discovered in maize by Barbara McClintock at Cold Spring Harbor; see Box 3-1). Direct transfer is carried out by class II transposons. The human genome has approximately 300,000 transposons (also known as transposable elements) integrated into its sequence. None of them appear to be currently active, but it has been suggested that direct transfer may have been responsible for gene duplications and inversions in chromosomes. Sequence analysis of transposons shows that at one time they coded for a few genes—one of them being transposase. Interestingly, human enzymes RAG1 and RAG2 have retained the ability to cut and integrate DNA on a limited scale. The RAG1 and RAG2 gene sequences are similar to the transposase sequences found in class II transposases. The ability of our immune system to respond to many foreign substances relies on creating many different combinations of DNA sequences in our immunoglobulin genes, which code for a tremendous variety of protein sequences in our immunoglobulin molecules. RAG1 and RAG2 are responsible for creating these combinations.[22]

Mutations can also be created by class I transposons, which use an indirect transfer method. Here, a copy of the transposon is made and inserted into another section of the genome (there is no DNA removal). Because class I transposons require an RNA intermediate, the resulting DNA product is called a retrotransposon.[23] It is likely that retrotransposons are remnants of retrovirus DNAs generated by infections of our germ cells millions of years ago.

Retrotransposons in humans start with a segment of DNA and typically fall into two categories: long interspersed nuclear elements (LINEs) and short interspersed nuclear elements (SINEs). LINEs are approximately 6–7 kb in length, and SINEs are 300 bp (bp = base pair) in length. Figure 3-10 shows a schematic diagram of a LINE. Typically, the LINE DNA is transcribed into RNA by the cell's RNA polymerase. The RNA is then reverse transcribed into cDNA by a reverse transcriptase coded by the LINE ORF2 and the cDNA is inserted into another part of the genome by the DNA endonuclease coded by LINE ORF1. In humans, there are nearly 900,000 LINEs that constitute 17% of the human genome. Most LINEs belong to the LINE-1 (L1) group.[24] However, only approximately 50 L1's remain active in retrotransposition. Two genes in active L1's are ORF1, which codes for an RNA binding protein,

vertical gene transfer Transfer of DNA through sexual or asexual reproduction (i.e., from parent to offspring or parent cell to daughter cell).

transposon (also known as transposable elements, mobile elements, and mobile genetic elements) Segment of DNA moved from one location to another in the genome. There are two mechanisms for this movement: direct transfer and indirect transfer.

retrotransposon A transposon that requires an RNA intermediate. Also known as class I transposon.

---

[21] Transposons are also called transposable elements, mobile elements, or mobile genetic elements. Sometimes scientists do not distinguish between active transposable elements (those still capable of moving) and inactive transposable elements (those that did, at one time, move but no longer do so). They will refer to active and inactive transposons as transposable elements. The act of moving from one genome location to another is called transposition.

[22] RAG1 and RAG2 are the products of recombination activating gene-1 and recombination activating gene-2. They are responsible for creating antibody diversity in B-cells and T-cell receptor diversity in T-cells. In the lab, mice have been generated that have no RAG1 or RAG2 genes. These mice have no immune system.

[23] When retrotransposons move from one location to another in the genome, the process is called retrotransposition.

[24] Line-1 (L1) elements are approximately 6 kb long and possess a 5' UTR, within which resides an RNA polymerase II promoter. Full-length L1 elements encode two open reading frames that produce a reverse transcriptase, endonuclease, and an RNA binding protein. L1 elements have a 3' UTR and an oligo-dA tail, and are flanked by direct repeat sequences.

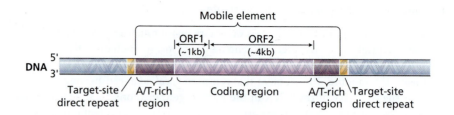

FIG. 3-10. General structure of an L1 LINE element, a common eukaryotic retrotransposon. The flanking target-site direct repeats appear to be generated from the genomic sequence during insertion of the mobile element. The A/T-rich region on the left end is the promoter region to which the RNA polymerase binds. The majority of L1 LINE elements have mutations that result in the production of multiple stop codons that prevent the expression of proteins. Although the full-length L1 element is ≈ 6 kb long, variable amounts of the left end are absent at more than 90% of the sites where this mobile element is found. The shorter open reading frame (ORF1) encodes an RNA binding protein. The longer ORF2 encodes a reverse transcriptase and endonuclease. The A/T-rich region at the right end is thought to be critical for retrotransposition.

and ORF2, which codes for a protein with two enzymatic activities: reverse transcription and DNA endonuclease activity. It has been hypothesized that retrotransposition may have assisted in increasing the diversity of proteins, and is a major driver in evolution (see *Modular proteins and protein evolution* section below).

3.4 ANCESTRAL GENES AND PROTEIN EVOLUTION

identity (also known as percent identity) In two optimally aligned sequences, the number of identical residues divided by the number of residues plus gaps in the alignment multiplied by.100.

Until now we have discussed mechanisms of how mutations contribute to genome changes, which places us in a position to begin to interpret protein sequence comparisons. Before we delve into this important area we will discuss some sequence comparison terminology. Identity is a quantity that describes how much two aligned sequences are alike *in strictest terms*. We can demonstrate this by assigning a score of 1 when two residues in aligned sequences are identical and a score of 0 when the two residues in aligned sequences are not identical. When one compares the five residue sequence ACDEF to the five residue sequence AVDEF one can calculate that the two sequences share 80% identity. The equation for identity is as follows:[25]

$$\frac{(\text{number of identical residues})}{(\text{number of residues and gaps in the alignment})} \times 100$$

similarity (also known as percent similarity) In two optimally aligned sequences, the number of similar residues divided by the number of residues plus gaps in the alignment multiplied by 100.

similarity scoring by BLOSUM, PAM, or others

In our case, $4/5 \times 100$ equals 80% identity. Similarity is a quantity that describes the *degree* to which two residues in aligned sequences are alike. A similarity score of 1 could be assigned for two residues that are identical, such as Leu and Leu. For two residues with chemically similar side chains, such as Leu and Ile, we may assign a similarity score between 0 and 1—perhaps 0.7. For two residues with dissimilar side chains, such as Leu and Tyr, we may give a score of 0. In these examples of similarities we assigned scores in a somewhat arbitrary fashion, but there are rigorous methods for quantifying similarities. (In Chapter 4 we will discuss how sophisticated scoring systems have been developed to quantify amino acid similarities.) Similarity is calculated in the same manner as percent identity in two aligned sequences:

$$\frac{(\text{number of similar residues})}{(\text{number of residues and gaps in the alignment})} \times 100$$

[25] The definition of identity can vary depending on the software program used. We will use the definition discussed here and will describe exceptions to this definition as they arise when we discuss a particular software program. A discussion of how gaps are placed in sequences to create optimal alignments is presented in Chapter 5.

BOX 3-1   **SCIENTIST SPOTLIGHT**

Barbara McClintock

BARBARA MCCLINTOCK earned her Ph.D. in botany from the College of Agriculture at Cornell University in 1927. She showed that the genetics of maize (corn) could be studied by examining its chromosomes through the microscope. She identified and characterized individual maize chromosomes by their lengths, shapes, and banding patterns. She became a cytogeneticist—one who relates physical characteristics of the organism to the characteristics of the chromosomes. In 1931, McClintock and her student Harriet Creighton published a paper demonstrating that gene exchange in maize germ cells correlates to an exchange of chromosomal material. The paper was hailed as a cornerstone in cytogenetics. In 1941, McClintock joined Cold Spring Harbor, a genetics research center on Long Island in New York. McClintock became a member of the National Academy of Sciences in 1944 and was president of the Genetics Society in 1945.

While McClintock was a respected cytogeneticist, the 1940s marked the beginning of a scientific revolution in biology that shifted the focus of the scientific community away from cytogenetics. In the 1940s through the 1960s, the advancement in the analysis of molecules opened the door for many biologists to study genetics at the molecular level. Scientists used simple organisms—viruses and bacteria—to work out the central dogma: DNA is transcribed into RNA and RNA is translated into protein. Yet McClintock continued to study maize, a relatively complex organism. In 1944, McClintock found genes she called "controlling elements" that could cause parts of DNA to move from one location to another in the maize genome. In 1947 she described one of these controlling elements, *Activator*. She found that *Activator* and other controlling elements led to a phenomenon called variegation—the altered pigmentation of maize kernels. We now know that *Activator* is a 4.6 kb (kb = kilobase pair) segment of DNA that cuts itself out of one part of the genome and inserts itself into another part of the genome. *Activator* codes for one protein, called transposase, which is essential for this "cut-and-paste" mechanism in plants. The capacity to change the position of a segment of DNA, called transposition, was greeted with skepticism by the research community. Transposition of DNA was a challenge to the central dogma, because it meant that something besides DNA was controlling information flow to proteins. Something besides DNA was causing *Activator* to cut itself out of one part of the genome and insert itself into another. McClintock described transposition as a response to genome shock, but did not know the molecular details of how transposition occurred.

Although McClintock had an excellent reputation as an investigator, few scientists understood her work. She was labeled as "mad" and "obscure." Then, in the 1960s, molecular biologists discovered that parts of the bacterial genome appeared to "jump around." These were called "jumping genes," "transposons," and "insertion elements." Later it was found that mutations in the fly *Drosophila* could be caused by transpositions of P elements. The development of immunity genes in B-cells and T-cells in mammals is due to transposition. When it became clear that transposition is a commonplace occurrence in the majority of, if not all, organisms, McClintock was awarded the Nobel Prize in Physiology or Medicine in 1983 for her early discovery of "mobile genetic elements." To date, she is the only woman to have won an unshared Nobel Prize in this category.

REFERENCE

Keller, E. F. 1983. *A Feeling for the Organism*. New York: W. H. Freeman and Company.

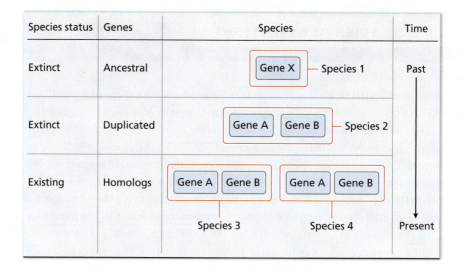

FIG. 3-11. A model of how orthologs and paralogs arise in evolution.

[handwritten annotations: HOMOLOG = similar genes by evolutionary descent; smit; loose; PARALOG = duplicated gene; ORTHOLOG = a diverged gene w/ a new distinct function from another ortholog; = may be diff functions]

homolog A gene related to other genes by evolutionary descent from a common ancestral DNA sequence. A homolog may be an ortholog or a paralog.

ortholog One gene of a set of genes that descended from a single gene in a common ancestor. The set of genes diverged from one another due to the evolution of a new species.

paralog One gene of a set of genes that that underwent a duplication event in a common ancestor. The set of genes diverged from one another due to evolution of a new species.

Two aligned residues are considered similar if the scoring system gives a similarity value greater than zero. Many sequence alignment programs place a single dot between residues that are dissimilar, a colon between residues that are similar, and a vertical line between residues that are identical.[26] A similarity score can be obtained by summing the similarity values. So upon comparing the sequence WLLI to LLLL we can produce a similarity score and similarity as follows:

```
W  L  L  I
·  |  |  :
L  L  L  L
```

Similarity values: 0   1   1   .7

Similarity score: 0 + 1 + 1 + .7 = 2.7

Similarity: 3/4 × 100 = 75%

When two sequences have substantial identity or similarity it is believed they are related. They were likely derived from the same ancestral gene, and they are called homologs. There are two types of homologs—orthologs and paralogs. Orthologs are genes from different species that are descended from a common ancestral sequence and diverged from one another due to the evolution of new species. Orthologs often carry out the same function in different species. Paralogs are genes derived from a gene duplication event of one original gene. Paralogs may be found in the same species or different species and may or may not carry out identical functions.

Figure 3-11 shows a model of how orthologs and paralogs can arise. The ancestral gene, Gene X in Species 1, was passed to Species 2 and duplicated into Gene A and Gene B. Gene A and Gene B are related by sequence but, after mutation and selection, diverged in sequence within Species 2. Species 3 and Species 4 evolved from the common ancestor, Species 2. Genes A and B are passed to both Species 3 and Species 4. Species 1 that harbored Gene X is now extinct, and Species 2 that harbored Genes A and B is now extinct. The two living species, Species 3 and Species 4, retain Genes A and B but the genes have undergone more mutations. We shall consider relationships of genes in existing species. Orthologs are genes in different species that evolved from a common gene. Gene A in Species 3 and Gene A in Species 4 are

---

[26] In some sequence alignment programs, a space between aligned residues signifies that they are dissimilar whereas a single dot signifies some similarity, and a colon signifies close similarity.

```
human    1   IVGGYNCEENSVPYQ-----VSLNSGYHFCGGSLINEQWVVSAGHCYKSR   45
             ||||.:......:|||        :|....:||||.|:.||.|.:.:||||.:-.
shrimp   1   IVGGTDATPGELPYQLSFQDISFGFAWHFCGASIYNENWAICAGHCVQGE   50

                                                 ^
human   46   -------IQVRLGEHNIEVLEGNEQFINAAKIIRHPQYDRKTLNNDIMLI   88
             :||..||.|..:|..||.||..:.:|||:|..|:..|::||.|.:
shrimp  51   DMNNPDYLQVVAGELNQDVDEGTEQTVILSKIIQHEDYNGFTISNDISLL   100

human   B9   KLSSRAVINARVSTISLPTAPPATGTKCLISGWGNTASSGADYPDELQCI   138
             |||.....|..|..|.:|....|.....|::||||  |.|.|..|..||.:
shrimp  101  KLSQPLSFNDNVRAIDIPAQGHAASGDCIVSGWG-TTSEGGSTPSVLQKV   149

                                                      ^^
human   139  DAPVLSQAKCEASY-PGKITSNMFCVGFLEGGKDSCQGDSGGPVVCNG--  185
             ..|.:|:|..:|..:  ...|...:|.|.|..||||||||||||||:.|.:..
shrimp  150  TVPIVSDDECRDAYGQSDIEDSMICAGVPEGGKDSCQGDSGGPLACSDTA  199

human   186  --QLQGVVSWGDGCAQKNKPGVYTKVYNYVKWIKNTIAANS-         224
             .|.|:||||.|||:...||||.:|..:|.|||                  II:
shrimp  200  STYLAGIVSWGYGCARPGYPGVYAEVSYHVDWIK----ANAV         237
```

handwritten margin note: Basic Local Alignment in BLAST uses neighborhood words of 3 or so nucleotides

orthologs. Paralogs evolved from a common gene that duplicated, where each gene represent one of the duplicated genes. The following pairs of genes are paralogs: Gene A and Gene B in Species 3; Gene A and Gene B in Species 4; Gene A in Species 3 and Gene B in Species 4; and Gene B in Species 3 and Gene A in Species 4. All genes shown in Species 3 and Species 4 belong to a single gene family. Members of a single gene family are homologs.

Examples of orthologs are the *TP53* genes from humans and mice. The proteins from these genes share 84% identity and both suppress tumors through the same mechanism. Examples of paralogs are human *TP53* and human *TP63*. The proteins coded by these genes share 42% identity and perform different functions (the product of *TP63*, p63, is necessary for proper limb, facial-cranial, and skin development). One should not come away from this discussion with the impression that orthologs always share higher identity than paralogs. Two orthologs from very divergent species (humans and placozoa, for example) can share a lower identity than two paralogs in the same species. This is especially true if the two species shared a common ancestor prior to the gene duplication event.

When ortholog sequences from two distantly related species are aligned, there are regions that contain a mix of identical and similar residues. Shown in Figure 3-12 is an alignment of human and shrimp trypsin. As mentioned before, trypsin is an enzyme that degrades other proteins. More specifically, trypsin is secreted into the small intestine where it catalyzes the degradation of food proteins into peptides. The peptides will be further degraded by other enzymes into amino acids to nourish the organism. The alignment in Figure 3-12 is annotated to highlight residues in the catalytic site (denoted with carat signs: ^). The thick horizontal bars show where cysteine side chains form disulfide bonds to stabilize the structure of the enzymes. The catalytic site amino acids and many of the cysteines are conserved. These two trypsin enzymes are homologs (and orthologs) that share 42% identity.

The next example shows two proteins with significant identity that have different functions. These paralogs appear to come from the same ancestor gene and, after a gene duplication event, diverged in function. One protein is His4 and the other is HisF.[27] They perform different functions in the biosynthesis of histidine in the bacterial strain *Thermotoga maritima*. Figure 3-13 shows a sequence alignment of these two paralogs.

FIG. 3-12. Two trypsin orthologs showing conserved regions. Human trypsin (Swiss Prot accession number P07477) and *Litopenaeus vannamei* (shrimp) trypsin (Swiss Prot accession number Q27761) were aligned with the Needleman-Wunsch global alignment program (http://www.ebi.ac.uk/emboss/align/) using default parameters. (See Chapter 5 for more on this program.) Disulfide bridges (——) show connections between cysteines. Amino acids involved in catalysis are denoted by ^. Aligned amino acids with vertical lines between them are identical, aligned amino acids with colons between them are similar, aligned amino acids with single dots between them are not similar. Note that amino acids near the N-termini of the two proteins were removed to optimize alignment. The two sequences share 42% identity.

---

[27] His4 is the enzyme phosphoribosylformimino-5-aminoimidazolecarboxamide ribonucleotide (ProFAR) isomerase. HisF is the enzyme imidazole glycerol phosphate synthase. His4 catalyzes the fourth step in histidine biosynthesis and HisF catalyzes the fifth step in histidine biosynthesis.

```
HIS4_THEMA      1   ML---VVPAIDLFRGKVARMIKGRKENTIFYEKDPVELVEKLIEEGFTLI   47
                    ||    ::..:|:    |..|::||..........:.....||||:...:.|.|...:
HISF_THEMA      1   MLAKRIIACLDV---KDGRVVKGTNFENLRDSGDPVELGKFYSEIGIDEL   47

HIS4_THEMA     48   HVVDLSNAIENSGENLPVLEKLSEFAEHIQI----GGGIRSLDYAEKLRK   93
                    ..:|::.::|.....|.::||:    ||.|.|    ||||....:.|.:|..
HISF_THEMA     48   VFLDITASVEKRKTMLELVEKV---AEQIDIPFTVGGGIHDFETASELIL   94

HIS4_THEMA     94   LGYRRQIVSSKVLEDPSFLKSLRE----------IDVEPV---FSLDTRG   130
                    .|..:.::::..:|:||.:..:.:    ||.:.|    |..:.|..
HISF_THEMA     95   RGADKVSINTAAVENPSLITQIAQTFGSQAVVVAIDAKRVDGEFMVFTYS   144

HIS4_THEMA    131   GR----VAFKGWLAEEEIDPVSLLKRLKEYGLEEIVHTEIEKDGTLQEHD   176
                    |:    :..:.|:.|.|    ||    |..||:.|.|::|||...:|
HISF_THEMA    145   GKKNTGILLRDWVVEVE-------KR----GAGEILLTSIDRDGTKSGYD   183

HIS4_THEMA    177   FSLTKKIAIEAEVKVLAAGGISSENSLKTAQKVHTETNGLLKGVIVGRA-   225
                    ..:.::.:.....:.:||.||    |:|........|.|.....|
HISF_THEMA    184   TEMIRFVRPLTTLPIIASGG---------AGKMEHFLEAFLAGADAALAA   224

HIS4_THEMA    226   ----FLEGILTVEVMKRYAR            241
                    |.|    :.|..:|.|.:
HISF_THEMA    225   SVFHFRE--IDVRELKEYLKKHGVNVRLEGL   253
```

FIG. 3-13. Two enzymes from the histidine biochemical pathway of *Thermotoga maritima* perform different functions but retain significant sequence identity. His4 Swiss Prot accession number is Q9X0C7, and the HisF Swiss Prot accession number is Q9X0C6. The identity is 66/270 * 100 or 24% identity. It is likely that the genes coding these enzymes are paralogs.

synteny The similar arrangement of genes in the genomes of two species that share a common ancestor.

global alignment Optimal pairing of two sequences over the entire lengths of the sequences.

local alignment Optimal pairing of two subsequences within two sequences such that the similarity score remains above a set threshold.

As we have just demonstrated, we can use sequence alignment software programs to detect homologs. However, we also need to consider the fact that some amino acids will align by chance. If we have two long sequences of random letters and place one sequence under the other, undoubtedly some letters will match. Bioinformaticians need to establish a threshold for deciding whether aligned sequences are homologs, which is typically 18–25% identity across a significant proportion of the aligned sequences. When the percent identity falls within this range we still need to be cautious before claiming that the two sequences are homologs. A statistical analysis of aligned sequences is required to conclude that two sequences are homologs.[28]

If you are attempting to determine whether two genes are orthologs, the gene structure contains useful information. In eukaryotes, the exon lengths of orthologs across different species should be nearly constant. Furthermore, the order of genes within a set of genes on the chromosome of one organism should match the order of genes within a set of genes on the chromosome of another organism. This property, called synteny, can help one decide whether two genes are orthologs. Paralogs usually do not have the same degree of conservation of gene structure, nor do they display the same synteny as do orthologs.

3.5 MODULAR PROTEINS AND PROTEIN EVOLUTION

The sequences in Figures 3-12 and 3-13 were aligned from the N-terminus ends to the C-terminus ends. Alignment over entire lengths of the proteins is called global alignment and is very effective for proteins that maintain some conserved amino acids over their entire sequences after divergence from a common ancestral gene. However, many proteins have only smaller segments that are conserved. For these proteins, it is best to consider aligning shorter sequences within the proteins. Programs that detect alignments of short sequences are called local alignment programs. The rationale for local alignment stems from the idea that protein contain short

---

[28] Statistical analysis of the similarity score, to be discussed in Chapter 6 (with mathematical derivation in Chapter 12), quantifies the probability that the similarity score could have been achieved by chance. This probability helps bioinformaticians decide whether a given alignment constitutes evidence for homology.

segments that provide specific functions. These segments can be repeated in a single protein or can be found in different proteins. In fact, many proteins are modular, and as a result one segment of the protein can be removed and placed onto a second protein. The second protein would then have an additional function carried out by the added segment. Let's explore how modular proteins may have evolved.

Recall from Chapter 2 that genes from bacteria and eukaryotes differ in one important way. In bacteria, the coding sequence for a protein is found in one continuous segment of DNA. In eukaryotes, the coding sequence for a protein is usually found in separate segments called exons. Proteins from eukaryotic cells are usually derived from more than one exon. Exons are interesting because they are involved in a proposed phenomenon called exon shuffling, first suggested by Walter Gilbert (Harvard University). Exon shuffling is a proposed mechanism to account for the fact that many proteins are composed of modular components. No one has witnessed exon shuffling, but evidence from the genome sequence suggests it has occurred at some point during evolution. Exon shuffling occurs when an exon from one gene is copied and inserted into another gene. This is called intergenic shuffling. By natural selection, the inserted exon sequence may change and diverge from the original exon sequence. On the other hand, exons may be duplicated within the same gene, which is called intragenic shuffling.

For illustrative purposes, let's explore one potential mechanism of exon shuffling. Recall that LINEs retrotransposons are segments of DNA that can be copied and moved, through an RNA intermediate, from one genome location to another. How can LINEs shuffle exons during indirect transposon transfer? A LINE can shuffle an exon by initially inserting itself into an intron located upstream of a functioning exon. In Figure 3-14, a hypothetical gene, Gene A, is shown with three exons, E1, E2, and E3. A LINE is inserted in the intron just upstream of E3. The LINE has its own transcription promoter, and when it is transcribed, both the LINE and E3 are transcribed into a long single RNA. The RNA is reverse transcribed by the LINE reverse transcriptase into cDNA, and the cDNA is inserted into a second gene, Gene B.

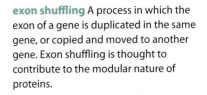

exon shuffling A process in which the exon of a gene is duplicated in the same gene, or copied and moved to another gene. Exon shuffling is thought to contribute to the modular nature of proteins.

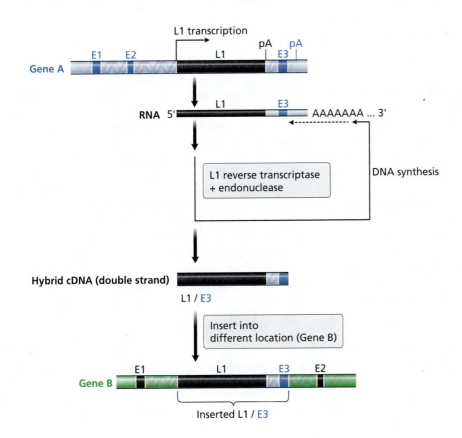

FIG. 3-14. Exon shuffling between genes can be mediated by transposable elements. The LINE1 (L1) sequence family contains members that actively transpose in the human genome. L1 elements have weak poly(A) signals and so transcription can continue past such a signal until another nearby poly(A) signal is reached, as in the case of Gene A (in blue font) at top. The resulting RNA copy can contain a transcript not just of L1 sequences, but also of a downstream exon (in this case E3). The L1 reverse transcriptase complex can then act on the extended poly (A) sequence to produce a cDNA copy that contains both L1 and E3 sequences. Subsequent transposition into a new chromosomal location may lead to insertion of exon 3 into a different gene (Gene B). In Gene B, the blue-colored exon, E3, originated from Gene A.

Gene B now has an extra exon that is identical to the E3 from Gene A. That is, E3 has shuffled. Because exons code for protein and exons can shuffle, exon shuffling may be the origin of *modular* proteins.

Proteins that are modular contain segments that are structurally and/or functionally independent of one another. These segments are called motifs if they contain fewer than 40 amino acids and they are called domains if they contain more than 40 amino acids. A motif or a domain can have an independent function that works without the rest of the protein. Let's take the example of the p53 tumor suppressor. As previously mentioned, the protein has three separate regions, or domains. The entire protein has 393 amino acids, and the first 42 amino acids are required for activating transcription. This domain, called the transactivation domain, interacts with transcription machinery associated with RNA polymerase II so that RNA polymerase II is more likely to transcribe genes. Another domain of p53 stretches from amino acid 100 to 300. Called the DNA binding domain, it is responsible for recognizing and binding to the p53-responsive element. The third domain encompasses amino acids 307–355. It is called the oligomerization domain and is responsible for creating the quaternary structure of p53. Like hemoglobin (see Chapter 1), p53 is a tetramer. Four identical p53 polypeptides bind to each other through the oligomerization or, more specifically, tetramerization domain. Because p53 functions can be separated into structurally and functionally distinct segments it is a modular protein. Thus, one domain of p53 may be very similar to a domain of another protein, such as its paralog p63, but its other domains may not be as similar. To be able to detect similarities between proteins, a prudent approach is to attempt to detect small regions of similarity rather than similarity along the entire length of the protein. It is for this reason that local sequence alignment programs were developed. The most popular local alignment program is BLAST (Basic Local Alignment Sequence Tool), which will be described in more detail in Chapter 6.

SUMMARY

In this chapter we discussed conserved regions and nonconserved regions of proteins and gave three explanations for why certain regions are conserved: (1) functional relatedness, (2) structural relatedness, and (3) ancestral gene relatedness. These properties are intertwined, often making it difficult to distinguish them from one another. Mutations, derived from DNA damage, DNA replication errors, viruses, and transposons all contribute to genome changes. Mutations that occur in germ cells have the potential to be passed to the progeny, and fitness is a measure of the ability of passing genes to progeny. Deleterious mutations decrease fitness, neutral mutations have no effect on fitness, and advantageous mutations increase fitness. Natural selection is a mechanism by which genes are selected for passage to the next generation. Neutral mutations cause changes in the amino acid sequences of proteins noticeable when homolog sequences are aligned. Deleterious mutations serve to maintain conserved regions within homologs. Conserved regions are segments of amino acids within aligned homologs, which tend not to vary.

Orthologs are genes derived from the same ancestral gene. Paralogs are genes resulting from a gene duplication event. Together, orthologs and paralogs are called homologs—which belong to a gene family. While gene duplication events will create paralogs of entire genes, specific regions within genes can be duplicated as well. Those duplicated regions can remain part of the original gene or may be inserted into other genes.

Retrotransposons can duplicate a segment of a gene known as an exon. Duplicating exons and transferring them to other parts of the same genes or to other genes is called exon shuffling. It has been proposed that retrotransposons mediate exon shuffling. Exon shuffling is a theory that may account for the observation of multiple domains in proteins and the fact that many proteins possess domains that have similar sequences.[29] BLAST, the software program that aligns sequences, creates alignments based on the idea that proteins are modular. So a segment of one protein may be aligned to a segment of a second protein.[30] This is called local alignment. Local alignment can be contrasted with global alignment,

[29] Another proposed mechanism for the observation of multiple domains is recombination (mixing of genetic material). Sequences with as few as 5–6 base pairs in common can recombine to create proteins with multiple domains.

[30] The same principle also applies to nucleotide sequences.

another type of sequence alignment that attempts to align protein sequences from the amino terminus to the carboxyl terminus. Both types of alignment tools are useful to the bioinformatician.

Molecular evolution is a dynamic process that leads to genomic sequence changes. The environment selects for only those changes that help progeny of the species survive. An area of molecular evolution that will continue to be intensely studied by scientists will be those mutations that affect the timing of gene expression and silencing during organism development. Such mutations will likely have dramatic effects on phenotype.

EXERCISES

1. **Obtain the human p63 isoform 1 protein sequence and the human p53 isoform a protein sequence. Use the Needleman-Wunsch program (http://www.ebi.ac.uk/Tools/psa/) to align the two proteins. Which amino acid sequences within p53 are conserved in p63? (Hint: find a region longer than 50 amino acids with no gaps spanning longer than 5 amino acids.) A second paralog of p53 is p73. Which amino acid sequences within p53 isoform a are conserved in p73 isoform a? Is the particular domain common within p53, p63, and p73 associated with a specific function? Use the following parameters when you run the Needleman-Wunsch program:**

   Matrix: BLOSUM62; Gap Open Penalty: 10; Gap Extend Penalty: 0.5; End Gap Penalty: False; End Gap Open Penalty: 10; End Gap Extend Penalty: 0.5

2. **Retinoblastoma is a rare childhood cancer of the retina. The retina is located in the back of the inside of the eye and is responsible for detecting light. Alfred Knudson, at the University of Texas at Houston, observed that retinoblastomas fall into two classes. In Class I, the retinoblastoma is first diagnosed in children at the mean age of three and the tumors are often found in both eyes. In Class II, the retinoblastoma is diagnosed at the mean age of five, and a single tumor is detected in one eye. Given what you know about germ cell mutations and somatic cell mutations, give a plausible explanation for the observation of two classes of retinoblastoma.**

3. **Peyton Rous (1879–1970) was a relatively young man when, in 1911, he discovered a virus that causes sarcomas in chickens. The virus was named Rous sarcoma virus (RSV). Later, it was found that RSV is a retrovirus that contains an oncogene, *v-src*, in its RNA genome. Perform pairwise global alignments with the Needleman-Wunsch algorithm between v-src protein and the following proteins: chicken c-src, human c-src, and mouse c-src proteins. The "c-" prefix is short for cellular. Sometimes proto-oncogenes are distinguished from viral oncogenes with the prefix "c-" and "v-" respectively. Report the identities for each pairwise alignment. Given what you know about the origin of v-src, does the result match your expectations? Use the Schmidt-Ruppin A strain of RSV as your source for the v-src sequence.**

4. **In the Li-Fraumeni syndrome *TP53* is often mutated. Compare the mRNAs from a Li-Fraumeni patient to transcript variant 1 of wild-type p53 with the Needleman-Wunsch global alignment software program.**

   Matrix: DNAfull; Gap Open Penalty: 10; Gap Extend Penalty: 0.5; End Gap Penalty: False; End Gap Open Penalty: 10; End Gap Extend Penalty: 0.5

   Classify the mutations in the areas that overlap. List the nucleotide location (using the wild-type p53 nucleotide numbering system as the reference) of the mutation. Indicate whether it is a point mutation or indel, transversion or transition, missense or other type of DNA mutation. Give the amino acid substitution(s), if there is any, that occurs in the p53 protein. The p53 transcript sequence from a Li-Fraumeni patient can be obtained from GenBank with accession number BT019622.1. For a codon table, please see Chapter 1.

5. **The sequence of two different forms of a gene starting with ATG codon is shown below. Which of the base differences in the second sequence are synonymous changes, and which are non-synonymous changes?**

   Form 1: ATGTCTCATGGACCCCTTCGTTTG

   Form 1: ATGTCTCAAAGACCACATCGTCTG

   (This exercise is adapted from Hartwell et al., 2008)

6. **Synteny is the conservation of the physical order of genetic loci in the genomes of two species. For example, a segment of mouse chromosome 2 is similar to the entirety of human chromosome 20 (chromosome 20 is one of the smaller chromosomes in humans). Cytogeneticists use chemicals to stain chromosomes to produce banding patterns. The banding patterns mark where sections of genes are located. Similar banding patterns denote regions of synteny. The figure shows the banding patterns on human chromosome 5 and the chimpanzee chromosome 5. Describe the region(s) where an inversion event occurred. (Exercise adapted from Lesk, 2008.)**

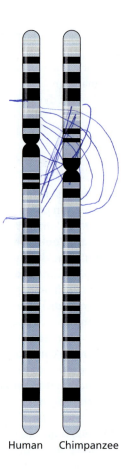

Human   Chimpanzee

REFERENCES

Altschul, S. F., W. Gish, W. Miller, E. W. Myers, and D. J. Lipman. 1990. "Basic Local Alignment Search Tool." *Journal of Molecular Biology* 215: 403–410.

Bachinski, L. L., S.-E. Olufemi, X. Zhou, C.-C. Wu, L. Yip, S. Shete, G. Lozano, C. I. Amos, L. C. Strong, and R. Krahe. 2005. "Genetic Mapping of a Third Li-Fraumeni Syndrome Predisposition Locus to Human Chromosome 1q23." *Cancer Research* 65: 427–431.

Baxevanis, A. D., and B. F. F. Ouellette, eds. 2004. *Bioinformatics: A Practical Guide to the Analysis of Genes and Proteins*, 3rd ed. New York: Wiley.

Beckman, G., R. Birgander, A. Sjalander, N. Saha, P. A. Holmberg, A. Kivela, et al. 1994. "Is p53 Polymorphism Maintained by Natural Selection?," *Human Heredity* 44: 266–270.

Denissenko, M. F., A. Pao, M. Tang, and G. P. Pfeifer. 1996. "Preferential Formation of Benzo[a]pyrene Adducts at Lung Cancer Mutational Hotspots in P53." *Science* 274: 430–432.

el-Deiry, W. S., S. E. Kern, J. A. Pietenpol, K. W. Kinzler, and B. Vogelstein. 1992. "Definition of a Consensus Binding Site for p53." *Nature Genetics* 1: 45–49.

Garrido-Ramos, M. A., ed. 2012. *Repetitive DNA*. Genome Dynamics, edited by M. Schmid, vol. 7. Basel, Switzerland: Karger Publishers.

Gonzalez, K. D., C. H. Buzin, K. A. Noltner, D. Gu, W. Li, D. Malkin, and S. S. Sommer. 2009. "High Frequency of de Novo Mutations in Li-Fraumeni Syndrome." *Journal of Medical Genetics* 46: 689–693.

Griffiths, A. J. F., S. R. Wessler, R. C. Lewontin, and S. B. Carroll. 2008. *Introduction to Genetic Analysis*, 9th ed. New York: W. H. Freeman.

Hamroun, D., S. Kato, C. Ishioka, M. Claustres, C. Béroud, and T. Soussi. 2006. "The UMD TP53 Database and Website: Update and Revisions." *Human Mutation* 27: 14–20. PubMed (PMID: 16278824).

Hartwell, L. H., L. Hood, M. L. Goldberg, A. E. Reynolds, L. M. Silver, and R. C. Veres. 2008. *Genetics: From Genes to Genomes*, 3rd ed. New York: McGraw-Hill Higher Education.

Higgs, P. G., and T. K. Attwood. 2005. *Bioinformatics and Molecular Evolution*. Malden, MA: Blackwell Publishing.

Ijdo, J. W., A. Baldini, D. C. Ward, S. T. Reeders, and R. A. Wells. 1991. "Origin of Human Chromosome 2: An Ancestral Telomere-Telomere Fusion." *Proceedings of the National Academy of Sciences of the USA* 88: 9051–9055.

Keller, E. F. 1983. *A Feeling for the Organism*. New York: W. H. Freeman and Company.

Knudson, A. G., Jr., H. W. Hethcote, and B. W. Brown. 1975. "Mutation and Childhood Cancer: A Probabilistic Model for the Incidence of Retinoblastoma." *Proceedings of the National Academy of Sciences of the USA* 72: 5116–5120.

Lane, D. P. 1992. "Cancer: p53, Guardian of the Genome." *Nature* 358, 15–16.

Lesk, A. M. 2008. *Introduction to Bioinformatics*, 3rd ed. Oxford, UK: Oxford University Press.

Li, F. P., and J. F. Fraumeni. 1969. "Rhabdomyosarcoma in Children: An Epidemiologic Study and Identification of a Familial Cancer Syndrome." *Journal of the National Cancer Institute* 43: 1364–1373.

Malkin, D., F. P. Li, L. C. Strong, J. F. Fraumeni Jr., C. E. Nelson, D. H. Kim, J. Kassel, M. A. Gryka, F. Z. Bischoff, M. A. Tainsky, and S. H. Friend. 1990. "Germ Line p53 Mutations in a Familial Syndrome of Breast Cancer, Sarcomas, and Other Neoplasms." *Science* 250: 1233–1238.

Matlashewski, G. J., S. Tuck, D. Pim, P. Lamb, J. Schneider, and L. V. Crawford. 1987. "Primary Structure Polymorphism at Amino Acid Residue 72 of Human p53." *Molecular and Cellular Biology* 7: 961–963.

Moran, J. V., R. J. DeBerardinis, and H. H. Kazazian Jr. 1999. "Exon Shuffling by L1 Retrotransposition." *Science* 283: 1530–1534.

Nobel Media AB. 2012. "All Nobel Prizes in Physiology or Medicine." Accessed February 20. http://www.nobelprize.org/nobel_prizes/medicine/laureates/.

Petitjean, A., E. Mathe, S. Kato, C. Ishioka, S. V. Tavtigian, P. Hainaut, and M. Olivier. 2007. "Impact of Mutant p53 Functional Properties on TP53 Mutation Patterns and Tumor Phenotype: Lessons from Recent Developments in the IARC TP53 Database (Version R15)." *Human Mutation* 6: 622–629.

Pettersen, E.F., T. D. Goddard, C. C. Huang, G. S. Couch, D. M. Greenblatt, E. C. Meng, and T. E. Ferrin. 2004. "UCSF Chimera—A Visualization System for Exploratory Research and Analysis." *Journal of Computational Chemistry* 105: 1605–1612.

Vogel, C., M. Bashton, N. D. Kerrison, C. Chothia, and S. A. Teichmann. 2004. "Structure, Function and Evolution of Multidomain Proteins." *Current Opinion in Structural Biology* 14: 208–216.

Weston, A., and J. H. Godbold. 1997. "Polymorphisms of H-ras-1 and p53 in Breast Cancer and Lung Cancer." *Environmental Health Perspectives* 105: 919–926.

Yang, A., M. Kaghad, D. Caput, and F. McKeon. 2002. "On the Shoulders of Giants: p63, p73 and the Rise of p53." *Trends in Genetics* 18: 90–95.

Yang, A., M. Kaghad, Y. Wang, E. Gillett, M. D. Fleming, V. Dotsch, N. C. Andrews, D. Caput, and F. McKeon. 1998. "p63, a p53 Homolog at 3q27-29, Encodes Multiple Products with Transactivating, Death-Inducing, and Dominant-Negative Activities." *Molecular Cell* 2: 305–316.

AFTER STUDYING THIS CHAPTER, YOU WILL:

- Learn the notation conventions for mathematical matrices.

- Understand that natural selection is the foundation for the development of the PAM substitution matrices.

- Understand that protein domain conservation is the foundation for the development of the BLOSUM substitution matrices.

- Understand that the frequency of occurrence of amino acids is considered in the creation of PAM and BLOSUM substitution matrices.

- Understand that amino acids with distinctive structures and essential roles in protein function are rarely replaced in nature.

- Calculate percent identity, identity score, similarity, and similarity score given a sequence alignment and substitution matrix.

- Understand that sequence clustering is used to create different BLOSUM substitution matrices.

SUBSTITUTION MATRICES

4.1 INTRODUCTION

In Chapter 3 we discussed natural selection as a driving force for creating new species—a process called speciation. A new species will often have protein sequences that differ from the sequences of its ancestral species because proteins acquire advantageous, neutral, or even mildly deleterious amino acid changes. Such changes are observed in living species that arose from a common ancestral gene. A bioinformatician might need to establish whether two proteins can be traced to a common ancestral gene to learn about their shared and nonshared functions. Sequence alignment of proteins will give a strong indication of their function, structure, and evolutionary history. As we saw in the last chapter, two proteins can be tentatively classified as homologs by aligning their sequences and measuring their percent identity. For example, the proteins p53, p63, and p73 are homologs that share a similar DNA binding domain, but other domains differ.

In Chapter 3 we also introduced the concept of similarity within aligned amino acids. How does our understanding of the relatedness between aligned sequences change if we consider both identity and similarity? In other words, let us suppose that Ile and Val are in the same position within two aligned sequences. Ile and Val obviously have side chains that are nearly the same, so we would say that this aligned pair strengthens our conclusion that the two proteins are homologs. In this chapter, you will learn about the methods used to quantify this and other amino acid pair similarities. Values associated with similarities are used in amino acid substitution matrices, which are necessary tools in many pairwise and multiple sequence alignment programs.

In 1978, Margaret Dayhoff, while working at the National Biomedical Research Foundation (NBRF) in Silver Spring, Maryland, led a group of scientists that used evolutionary theory to quantify the likelihood of amino acid changes in homologs. Her group developed a series of amino acid substitution matrices, called the PAM substitution matrices, based on the evolutionary distances of protein sequences diverged from common ancestral proteins.[1] Approximately 14 years later, using a different approach to capture amino acid substitutions, David Jones, William Taylor, and Janet Thornton (at University College Gower Street and the National Institute for Medical Research, both in the United Kingdom) created a software program that generates substitution matrices—named the PET substitution matrices. In 1992, a group headed by Steven Benner (at the Institute for Scientific Computation, Swiss Federal Institute of Technology, Zurich, Switzerland) created a series of substitution matrices, called the Gonnet substitution matrices, through a reiterative process of evolutionary distance measurements, similar to the technique developed by Dayhoff. Also in 1992, a series of substitution matrices was developed by Steven and Jorja Henikoff at the Fred Hutchinson Cancer Center in Seattle, Washington. Called the BLOSUM substitution matrices, the Henikoffs used data from computer-generated sequence alignments of motifs and domains. Frequencies of aligned amino acid pairs were quantified and converted into the BLOSUM substitution matrix series. The methods used to derive the PAM, PET, and Gonnet substitution matrices were somewhat similar, although that of the BLOSUM matrix was somewhat different. Because the PAM and BLOSUM series are typically used in bioinformatics software programs, we will focus much of our discussion here on their development.

4.2 IDENTITY SUBSTITUTION MATRIX

One way to determine whether two sequences are homologs is to consider the identity score. We can align two short sequences by eye and compute an identity score by assigning a value of 1 for match and 0 for mismatch for the aligned amino acids. The identity score is the sum of the values in the matched amino acids in the aligned sequences. If the identity score is high it is likely that the two sequences are homologs. To create an identity substitution matrix (also known as a binary matrix), values for pairs of aligned amino acids are placed into a matrix. For our discussion of this matrix, it is useful to review the notation conventions for matrices.

identity score Sum of matched residues in two optimally aligned sequences.

A matrix is a rectangular array of numbers, symbols, or expressions. The individual items in a matrix are called its *elements* or *entries*. An example of a 2 × 3 matrix (2 rows, 3 columns) with six elements is

elements (also known as entries) Individual items in a matrix.

$$\begin{bmatrix} 1 & 9 & -13 \\ 20 & 5 & -6 \end{bmatrix}$$

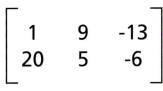

Matrices are read as ↔ by ↕ rows columns

like codon tables, and like paper 8½ × 11 ↔ ↕

---

[1] Evolutionary distance (also known as genetic distance) is the estimated number of substitutions that have occurred since two species shared a common ancestor genome.

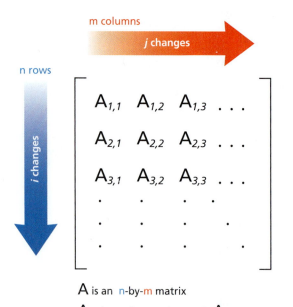

A is an n-by-m matrix

$A_{i,j}$ is an element in matrix A

FIG. 4-1. Notation conventions for a matrix.

A

	L	I	K	Q	T	V
L	1	0	0	0	0	0
I	0	1	0	0	0	0
K	0	0	1	0	0	0
Q	0	0	0	1	0	0
T	0	0	0	0	1	0
V	0	0	0	0	0	1

B

	L	I	K	Q	T	V
L	4	2	-2	-2	-1	1
I	2	4	-3	-3	-1	3
K	-2	-3	5	1	-1	-2
Q	-2	-3	1	5	-1	-2
T	-1	-1	-1	-1	5	0
V	1	3	-2	-2	0	4

FIG. 4-2. Two 6 × 6 substitution matrices.
A. The elements corresponding to the amino acid pairs are located in the grid. The value, or score, for the first element in the substitution matrix, $A_{L,L}$ is 1. **B.** A different kind of substitution matrix used to score alignments. These types of substitution matrices are more useful than the identity substitution matrix when there is little shared identity in a pair of sequences. These particular values are from the BLOSUM62 substitution matrix.

Specific elements in a matrix are often denoted by a variable with two subscripts. For instance, in the 2 × 3 matrix, $A_{2,1}$ refers to the element at the second row and first column of a matrix **A** and has a value of 20.

Figure 4-1 shows another matrix we name **A**. It is an *n*-by-*m* matrix, where *n* is the number of rows and *m* is the number of columns. The number of elements equals the product of *n* and *m*. The generic term for an element in this matrix is $A_{i,j}$ where *i* refers to a specific row and *j* refers to a specific column. The first element in this matrix is specified by $A_{1,1}$, where *i* = 1, *j* = 1.

In a substitution matrix **A**, each row (*i*) corresponds to one of 20 amino acids and each column (*j*) corresponds to one of 20 amino acids in the same order, giving us 400 elements total. The element $A_{i,j}$ specifies a score associated with the pair of amino acids *i* and *j*. For each amino acid pair in aligned sequences, the substitution matrix is consulted and the score corresponding to the amino acid pair is retrieved. A type of substitution matrix called an amino acid identity substitution matrix can be used to calculate an identity score for an alignment. This identity substitution matrix gives a score of 1 to pairs of identical amino acids and a score of 0 to pairs of nonidentical amino acids. The identity score is then the sum of all scores for the aligned amino acids. An example of an identity substitution matrix consisting of just six amino acids is shown in Figure 4-2A. We can use our eye to align the following sequences and the identity substitution matrix to score them as follows:

Sequence 1: TLKKVQKT
Sequence 2: TLKKIQKQ

The alignment will be:

```
TLKKVQKT
|||| ||
TLKKIQKQ
```

The identity score (also known as the alignment score)[2] is 6 because six amino acid pairs have positive values from the identity substitution matrix. The percent identity

---

[2] Alignment score is a general term that can apply to an identity score or a similarity score.

is the number of pairs with positive values divided by the total number of amino acid residues (including gaps, if any) in the alignment multiplied by 100:

$$6/8 \times 100 = 75\%$$

An identity substitution matrix is convenient for quantifying aligned sequences from homologs that are nearly identical, but not practical for homologs that have evolved to such an extent that there is very little shared identity. In such cases it is better to rely on one of the several other types of amino acid substitution matrices (such as one shown in Figure 4-2B that will be described in more detail later in the chapter) to assess whether two sequences are homologs. Sequence alignment software programs Needleman-Wunsch, BLAST, Smith-Waterman, CLUSTALW, and in fact all such programs rely on substitution matrices to optimally align sequences. The software programs produce similarity scores from the substitution matrices, which quantify the degree of similarity between two aligned sequences. As we shall see, one approach to developing substitution matrices is to consider natural selection.

similarity score In two optimally aligned sequences, the sum of scores of residue matches and mismatches. Mismatches include residue mismatches and residues aligned with gaps.

4.3 AMINO ACID SUBSTITUTION SYSTEM BASED ON NATURAL SELECTION

Natural selection determines which amino acid residue substitutions are acceptable in homologous proteins. In order for a substitution to be tolerated, the physical and chemical properties of the replacement residue should be nearly the same as those of the original residue. A substitution that results in a radically different residue could alter the protein structure such that its function is compromised. PAM and BLOSUM substitution matrices utilize the natural selection approach. Both consider the relative frequency and probability that one residue is substituted by another in aligned sequences. Frequencies are derived from the number of instances in which residues are substituted in aligned sequences of a reference set of known homologous proteins, and probabilities are derived from the natural abundance of each residue in these sequences.

Margaret Dayhoff and her colleagues developed the first popular set of amino acid substitution matrices (see Box 4-1) named PAM (an acronym for accepted point mutation; the letter order was altered, perhaps to make the acronym easier to pronounce). Conceptually, she wanted to develop a substitution system based on the principle of point mutations accepted by natural selection. However, instead of accepted point mutation, it is more accurate to use the phrase "accepted amino acid substitution" because only amino acid substitutions could be observed in her data (DNA sequencing had not yet been developed).[3] Accepted amino acid substitutions means that in sequences that are very similar, say in a human globin and in a chimpanzee globin, the amino acid differences at equivalent positions in the sequences have been accepted by natural selection.

The amino acid substitution theory used in the development of the PAM substitution matrices is based on (1) neighbor independence, (2) position independence, and (3) history independence. Neighbor independence means that one assumes that each amino acid residue substitutes randomly and independently of other amino acid residues nearby. Position independence assumes that the probability of substituting one amino acid residue for another depends only on those two residues and not their positions in the sequence. For example, a change from Cys to Ser near

PAM substitution matrix A general term for a set of amino acid substitution matrices based on the principle of missense mutations accepted by natural selection. These were the first amino acid substitution matrices that were widely used to calculate similarity scores in aligned sequences.

---

[3] Due to Dayhoff's emphasis on comparing proteins that were closely related, substitutions resulting from single DNA base mutations are much more likely to be detected than those resulting from successive two base substitutions. Also, amino acid substitutions resulting from DNA transitions would be more likely to be detected than those from DNA transversions.

Margaret Belle Dayhoff

BOX 4-1 | **SCIENTIST SPOTLIGHT**

MARGARET BELLE (Oakley) DAYHOFF (March 11, 1925–February 5, 1983) was trained as a physical chemist and became a pioneer in bioinformatics. Dayhoff was a professor at Georgetown University Medical Center and headed a research team at the National Biomedical Research Foundation (NBRF) where she applied mathematics and computational methods to analyze protein sequence data. In 1961 she created the first computer software program, named Comprotein, to assemble long protein sequences from shorter (partial) sequences that overlap, which eventually led to the publication of the first protein sequence database in 1965. With her colleague Richard Eck, she created the first computer-generated reconstruction of a phylogenetic tree from molecular sequences, using a maximum parsimony method. She earned a Ph.D. from Columbia University, Department of Chemistry, where, in her dissertation, she devised computational methods to calculate molecular resonance energies of organic compounds. She conducted postdoctoral research at the Rockefeller Institute (now Rockefeller University) and the University of Maryland, and joined the then newly established private nonprofit NBRF in 1959. She was the associate director of the NBRF for 21 years. She was the first woman to hold office in the Biophysical Society, first as secretary and eventually as president. She served on the editorial boards of three scientific journals: *DNA*, *Computers in Biology and Medicine*, and the *Journal of Molecular Biology*.

Her contributions to bioinformatics can be summarized in three major feats:

1. She originated one of the first sets of amino acid substitution matrices, point accepted mutations (PAMs).
2. She developed the one-letter code for amino acids in an attempt to reduce the size of the data files used to describe amino acid sequences in an era of punch-card computing.
3. Under her editorship, she, along with her colleagues at the NBRF, compiled the first comprehensive collection of macromolecular sequences in the *Atlas of Protein Sequence and Structure*, published from 1965 to 1978.

The first publication of the *Atlas*, which contained 70 protein sequences, was the forerunner of the Protein Information Resource (PIR) database. In 1980, Dayhoff also established a nucleic acid sequence database, predating GenBank, that was made available to scientists through the telephone line via modem. PIR and GenBank (founded by Walter Goad; see Box 2-1) became the two leading databases for protein and nucleotide sequence information, respectively. Perhaps presciently, she described the *Atlas* in the following manner: "We sift over our fingers the first grains of this great outpouring of information and say to ourselves that the world be helped by it. The *Atlas* is one small link in the chain from biochemistry and mathematics to sociology and medicine."

Margaret Dayhoff held the vision that sequence information from proteins and DNA would, with the advent of computers, lead a revolution in molecular life science research. The major research question for which she searched for answers focused on the use of sequence information to understand the nature of evolution. She used computers to make major progress in this broad area.

REFERENCES

Hunt, L. T. 1983. "Margaret O. Dayhoff 1925–1983." *DNA* 2: 97–98.

NCBI. 2012. "Bruno J. Strasser Speaking at the Genbank 25th Anniversary." YouTube. Accessed April 3. http://www.youtube.com/watch?v=VRnY5HP3wjM.

Readers711. 2012. "Margaret Dayhoff." Scribd. Accessed April 3. http://www.scribd.com/doc/50146164/margaret-dayhoff.

the amino terminus of the sequence has the same probability of occurrence as a change from Cys to Ser near the carboxyl terminus. Finally, history independence means that a change from one residue to another depends only on the present state and not past states.[4] For example, a series of changes may occur where Cys is changed to Ser and then that Ser is changed to Thr. The probability of Cys to Thr change does not depend on the fact that Ser was an intermediate state. To decrease the chance of a double amino acid residue change occurring, PAM substitution matrices are derived from aligned sequences that have high identity—in other words, closely related sequences. The reasoning for this restriction is that for closely related sequences (i.e., they diverged from a common ancestral gene relatively recently), it would be unlikely that DNA mutations occurred so frequently as to result in a double amino acid residue change in the same position. This is important because, ideally, we want to determine the likelihood of *single* amino acid substitutions for all amino acid pairs in this evolutionary model.

4.4 DEVELOPMENT OF THE MATRIX OF "ACCEPTED" AMINO ACID SUBSTITUTIONS

The first step to developing a PAM substitution matrix is to create a matrix of accepted amino acid substitutions. Amino acid substitutions are documented in closely related proteins and their inferred ancestral sequences. Altogether Dayhoff recorded 1,572 substitutions. To show how to create such a matrix we will start with a simple example. Figure 4-3 shows a phylogenetic tree that compares four short sequences (at the bottom of the tree), each from a different living species. At the two nodes above these sequences are ancestral sequences inferred from the observed sequences. Amino acid substitutions are noted with double-headed arrows. To create the matrix, substitutions were counted between the known sequences, between known sequences and inferred ancestral sequences, and between inferred ancestral sequences. In the phylogenetic tree in Figure 4-3 let's compare the ancestral proteins to the observed proteins, and we will concentrate on the substitutions affecting residue A. On the left node, residue A in the ancestral sequence changed to D, and on the right node residue A in the ancestral sequence changed to C.

---

[4] History independence is a major tenet of the Markov chain theory of Andrey Markov (St. Petersburg University, Russia), a famous Russian mathematician who lived from 1856 to 1922. In Markov chain theory, the probability of change depends only on the current state and not past states. (See Chapter 12 for more details on Markov chain theory.)

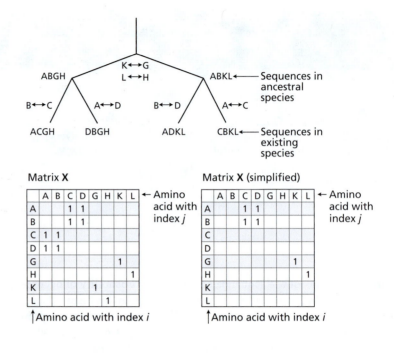

FIG. 4-3. Building a matrix of accepted amino acid substitutions. A phylogenetic tree with four observed sequences in existing species is shown. The inferred ancestral sequences are located at the nodes of the tree, and amino acid substitutions are indicated with double-headed arrows. In the lower left portion of the figure is a symmetrical matrix of accepted amino acid substitutions derived from the phylogenetic tree. In the lower right portion of the figure is a simplified version of the symmetrical matrix, where only the upper right values are shown.

Thought Question 4-1

In Figure 4-3, why do the ancestral sequences at the two nodes above the existing species start with residue A and not with D in the left sequence, and start with residue A and not with C in the right sequence?

In the bottom left in Figure 4-3 is a matrix, named Matrix **X**, of accepted substitutions based on our simple phylogenetic tree where the number of accepted substitutions relating to amino acid A are presented in positions $\mathbf{X}_{A,C}$, $\mathbf{X}_{A,D}$, $\mathbf{X}_{C,A}$, and $\mathbf{X}_{D,A}$. In addition, for this particular phylogenetic tree, Matrix **X** shows all other numbers of accepted substitutions compiled for changes between existing species sequences and inferred ancestral sequences, as well as between inferred ancestral sequences. It is assumed that the likelihood of replacing amino acid i with amino acid j is the same as that of replacing amino acid j with amino acid i, which results in a symmetrical matrix of accepted substitutions. To save space and to make it easier to read, a symmetrical matrix is often presented with elements only in the upper right section filled (see Matrix **X** [simplified] in Figure 4-3).[5]

Figure 4-3 shows a simplified example of the matrix of substitutions, but in the matrix of substitutions that Dayhoff created, the numbers of substituted amino acids from closely related sequences in 71 evolutionary trees were tabulated. Sequences from existing species that shared at least 85% identity were used, and sequences from observed species and inferred ancestral sequences were even greater than 85%. One of the reasons Dayhoff used nearly identical sequences was that she could easily infer the ancestral sequences from existing sequences.

Figure 4-4 indicates the numbers of substitutions in the Dayhoff matrix of accepted amino acid substitutions, matrix **A**. The numbers of substitutions in matrix **A** are multiplied by 10 so that no fractional numbers are displayed in Figure 4-4. How can a fractional number be obtained in the first place? When the ancestral sequences are ambiguous, Dayhoff allowed for the possibility that more than one amino acid be located at a single position. Where there is this ambiguity, the change from the ancestral sequence to the living sequence is counted as a *fractional change*. For example,

---

[5] Some authors choose to show the elements in the lower left portion of a symmetrical matrix.

	A	R	N	D	C	Q	E	G	H	I	L	K	M	F	P	S	T	W	Y	V
A	—	30	109	154	33	93	266	579	21	66	95	57	29	20	345	772	590	0	20	365
R		—	17	0	10	120	0	10	103	20	17	477	17	7	57	137	20	27	3	20
N			—	532	0	50	94	156	226	36	37	322	0	7	27	432	169	3	36	13
D				—	0	76	831	162	43	13	0	85	0	0	10	98	57	0	0	17
C					—	0	0	10	10	17	0	0	0	0	10	117	10	0	30	33
Q						—	422	30	243	8	75	147	20	0	93	47	37	0	0	27
E							—	112	23	35	15	104	7	0	40	85	31	0	10	37
G								—	3	0	17	50	7	17	49	450	50	0	0	97
H									—	3	40	23	0	20	50	26	14	3	40	30
I										—	253	43	57	90	7	20	129	0	13	661
L											—	39	207	167	43	32	52	13	23	303
K												—	90	0	43	168	200	0	10	17
M													—	17	4	20	28	0	0	77
F														—	7	40	10	10	260	10
P															—	269	73	0	0	50
S																—	696	17	22	43
T																	—	0	23	186
W																		—	6	0
Y																			—	17
V																				—

FIG. 4-4. Numbers of accepted amino acid substitutions (multiplied by 10) indicated in matrix **A**. A total of 1,572 substitutions are recorded in matrix **A**.

if there are five equally possible amino acids at a single position in the ancestral sequences, then each of the five amino acids contributes a value of 0.2 to the ancestral amino acid (see $A_{Q,E}$ in Figure 4-4).

One of the drawbacks to Dayhoff's substitution matrix is that, due to the scarcity of data at the time, some amino acid substitutions are never observed in her set of aligned sequences. Of the 190 amino acid substitutions that were possible, 35 were never observed.[6] This meant that Dayhoff was forced to put zeros into her matrix for these substitutions (for example, the Trp (W) to Cys (C) change was never observed in her data). As a matter of fact, Trp and Cys (C) changes are rarely observed in aligned sequences from two homologs. Trp has a side chain that is quite different from all other amino acids so that a suitable substitution with another amino acid is rare. Also, the side chain of Cys is very reactive and is often used to maintain tertiary and quaternary structure by forming disulfides with other Cys residues. Cys residues also bind to diverse metals that help maintain structure as well as serve catalytic roles in the active sites of enzymes. For these reasons, one can imagine that changing Trp or Cys likely results in irreparable harm to the protein, perhaps causing the organism to die early in development.

To address the issue of the absence of some amino acid changes in Dayhoff's substitution matrix, 14 years later Jones, Taylor, and Thornton used a larger data set of protein sequences that led to the documentation of 59,190 substitutions. By this time, there were thousands more sequences available. Like Dayhoff, Jones and colleagues used sequences that were closely related, which they captured from the Swiss-Prot database by searching for sequences that shared ≥85% identity (they did not use inferred ancestral sequences). Given the extensive number of sequences available to them, they observed at least two substitutions for every amino acid (Figure 4-5). Interestingly, even with more data, amino acid substitutions tabulated in the Dayhoff matrix did not significantly change, and the few that did are summarized at the end of this chapter (see Table 4-3).

---

[6] The number of possible amino acid substitutions equals 20 × 20 (for all possible amino acid pairs) minus 20 (to remove self-pairing such as A to A for example) divided by 2 (because the matrix is symmetric).

	A	R	N	D	C	Q	E	G	H	I	L	K	M	F	P	S	T	W	Y	V
A	—	247	216	386	105	208	600	1183	46	173	257	200	100	51	901	2413	2440	11	41	1766
R		—	116	48	125	750	119	614	446	76	206	2348	61	16	217	413	230	109	46	69
N			—	1433	32	157	180	291	466	130	63	758	39	15	31	1738	693	2	114	55
D				—	13	130	2914	577	144	37	34	102	27	8	39	244	151	5	89	127
C					—	9	8	98	40	19	36	7	23	66	15	353	66	38	164	99
Q						—	1027	84	635	20	314	858	52	9	395	182	149	12	40	58
E							—	610	41	43	65	754	30	13	71	156	142	12	15	226
G								—	41	25	56	142	27	18	93	1131	164	69	15	276
H									—	26	134	85	21	50	157	138	76	5	514	22
I										—	1324	75	704	196	31	172	930	12	616	3938
L											—	94	974	1093	578	436	172	82	84	1261
K												—	103	7	77	228	398	9	20	58
M													—	49	23	54	343	8	17	559
F														—	36	309	39	37	850	189
P															—	1138	412	6	22	84
S																—	2258	36	164	219
T																	—	8	45	526
W																		—	41	27
Y																			—	42
V																				—

FIG. 4-5. Numbers of accepted amino acid substitutions using a total of 59,190 substitutions.

4.5 RELATIVE MUTABILITY CALCULATIONS

The matrix of amino acid substitutions is one step in the development of the PAM substitution matrix. Another step is to consider the *probability* that each amino acid will change in a given small evolutionary interval. According to Dayhoff, a small evolutionary interval is the amount of time required for two sequences to diverge from a common ancestor to the point where the two sequences share ≥85% identity. The probability of change is called relative mutability. To calculate relative mutability, Dayhoff used the same sequence data from existing species as was used for the construction of her matrix of accepted amino acid substitutions. Relative mutability is calculated by tallying the number of changes an amino acid undergoes in the aligned sequences and dividing that number by the frequency of the amino acid occurrences in the data set. Figure 4-6 illustrates how relative mutability is calculated in a hypothetical case. Here, two sequences, ADAE and ADCE, are aligned. For each amino acid, the number of changes and frequency of occurrence in the two sequences is tabulated. Relative mutability is calculated by dividing the number of changes by the frequency of occurrence. In this hypothetical case, C has a highest relative mutability, and D (as well as E) the lowest.

From Dayhoff's sequence alignment data, the relative mutability, m_j, of each amino acid is calculated. In the next section we will show how the relative mutabilities and the matrix of amino acid substitutions are used to construct the PAM1 mutation probability matrix. The PAM1 mutation probability matrix is the precursor to the PAM1 log-odds substitution matrix.

Aligned sequences	A D A E			
	A D C E			
Amino acids	A	C	D	E
Changes	1	1	0	0
Frequency of occurrences	3	1	2	2
Relative mutability	0.33	1	0	0

FIG. 4-6. Hypothetical example to show how relative mutability is calculated when the sequence ADAE is aligned with sequence ADCE.

4.6 DEVELOPMENT OF THE PAM1 MUTATION PROBABILITY MATRIX

In general, a mutation probability matrix describes the probabilities for an amino acid to change into another amino acid, for each possible pair. We obtain such a mutation probability matrix (also known as transition matrix) from the amino acid substitution matrix. More specifically, the PAM1 mutation probability matrix is denoted as matrix $\mathbf{M}$.[7] Here, the *probability of amino acid substitutions* is calculated over an evolutionary distance of an average of 1 substitution for every 100 amino acids. There are two calculations required for $\mathbf{M}$, both of which we will carefully go over. One calculation is for the $\mathbf{M}_{i,j}$ elements for $i \neq j$, which give the probability that amino acid i is changed to another amino acid j. The second calculation is for the elements $\mathbf{M}_{i,i}$, which gives the probability that amino acid i does not change. These elements are located in the longest diagonal starting at the upper left corner of the matrix, called the main diagonal.

For matrix $\mathbf{M}$, the $\mathbf{M}_{i,j}$ elements have the following values:

$$\mathbf{M}_{i,j} = \lambda m_i \mathbf{A}_{i,j} / \left(\sum_j \mathbf{A}_{i,j} \right) (\text{for } i \neq j) \tag{4.1}$$

where

λ is the proportionality constant (to be discussed below)

m_i is the relative mutability of amino acid i

$\mathbf{A}_{i,j}$ is an element of the amino acid substitution matrix (see Fig. 4-4)

$\sum_j \mathbf{A}_{i,j}$ is the sum of the elements in row i of matrix $\mathbf{A}$

$i \neq j$ (amino acid i and amino acid j are not the same)

The following equation applies to the condition where the original amino acid remains unchanged:

$$\mathbf{M}_{i,i} = 1 - \lambda m_i \tag{4.2}$$

These are the values in the main diagonal in the matrix.

The proportionality constant λ is the same in both equations, and is chosen so that the average probability of amino acid substitution is 1%. In other words, the λ value is 0.01, which means, conversely, that the value for $\mathbf{M}_{i,i}$ is 0.99. Dayhoff defined the matrix $\mathbf{M}$ as the PAM1 mutational probability matrix (PAM1, for short) where PAM1 represents a matrix in which an average of 1% ($\lambda = 0.01$) of the amino acids have changed during evolution. A PAM1 *distance* refers to the amount of evolutionary time necessary for an amino acid sequence to undergo a change of 1%.

Figure 4-7 is a scaled version of PAM1 mutational probability matrix $\mathbf{M}$ where all element values are multiplied by 10,000 to make the matrix easier to read. For each row in PAM1, the sum of all elements is 1. The *average* value for all $\mathbf{M}_{i,i}$'s in the longest diagonal is 0.99—in other words, an average of 99% of the amino acids remains unchanged. This is consistent with the fact that PAM1 is created in a way such that 1% of the amino acids, on average, are replaced.

4.7 DETERMINATION OF THE RELATIVE FREQUENCIES OF AMINO ACIDS

PAM1 is constructed in a way such that an average of 1% of the amino acids is substituted. In the previous section this was shown by summing the $\mathbf{M}_{i,i}$'s in the main diagonal and then dividing the sum by 20 to give an average $\mathbf{M}_{i,i}$ of 0.99. The average $\mathbf{M}_{i,i}$ of 0.99 represents the probability that the amino acids remain unchanged. We can

main diagonal Elements located in the longest diagonal of a matrix, starting at the upper left corner.

PAM1 mutational probability matrix A matrix where an average of 1% of the amino acids have changed during evolution.

---

[7] Dayhoff used the i index to denote the replacement amino acid and the j index to denote the original amino acid. To be consistent with more recent indices and algorithms, we have adopted the convention that the i index is the original amino acid and the j index is the replacement amino acid. Furthermore, $\mathbf{M}$ is formally defined as a transition matrix where the state is the amino acid at a specific position.

SUBSTITUTED AA

ORIGINAL AA

	A	R	N	D	C	Q	E	G	H	I	L	K	M	F	P	S	T	W	Y	V
A	9867	1	4	6	1	3	10	21	1	2	3	2	1	1	13	28	22	0	1	13
R	2	9913	1	0	1	9	0	1	8	2	1	37	1	1	5	11	2	2	0	2
N	9	1	9822	42	0	4	7	12	18	3	3	25	0	1	2	34	13	0	3	1
D	10	0	36	9859	0	5	56	11	3	1	0	6	0	0	1	7	4	0	0	1
C	3	1	0	0	9973	0	0	1	1	2	0	0	0	0	1	11	1	0	3	3
Q	8	10	4	6	0	9876	35	3	20	1	6	12	2	0	8	4	3	0	0	2
E	17	0	6	53	0	27	9865	7	1	2	1	7	0	0	3	6	2	0	1	2
G	21	0	6	6	0	1	4	9935	0	0	1	2	0	1	2	16	2	0	0	3
H	2	10	21	4	1	23	2	1	9912	0	4	2	0	2	5	2	1	0	4	3
I	6	3	3	1	1	1	4	0	0	9872	22	4	5	8	1	2	11	0	1	57
L	4	1	1	0	0	3	1	1	1	9	9947	1	8	6	2	1	2	0	1	11
K	2	19	13	3	0	6	4	2	1	2	2	9926	4	0	2	7	8	0	0	1
M	6	4	0	0	0	4	1	1	0	12	45	20	9874	4	1	4	6	0	0	17
F	2	1	1	0	0	0	0	1	2	7	13	0	1	9946	1	3	1	1	21	1
P	22	4	2	1	1	6	3	3	3	0	3	3	0	0	9926	17	5	0	0	3
S	35	6	20	5	5	2	4	21	1	1	1	8	1	2	12	9840	32	1	1	2
T	32	1	9	3	1	2	2	3	1	7	3	11	2	1	4	38	9871	0	1	10
W	0	8	1	0	0	0	0	0	1	0	4	0	0	3	0	5	0	9976	2	0
Y	2	0	4	0	3	0	1	0	4	1	2	1	0	28	0	2	2	1	9945	2
V	18	1	1	1	2	1	2	5	1	33	15	1	4	0	2	2	9	0	1	9901

FIG. 4-7. Scaled PAM1 mutational probability matrix, **M**. In the left column are original amino acids, and on the top row are the replacement amino acids. Elements corresponding to $M_{i,i}$ are shown in red. Elements of **M** were multiplied by 10,000 to make the matrix easier to read. The average of the scaled $M_{i,i}$ elements is 9,900, and the average of the sums of rows is 10,000.

obtain the same probability by considering the relative frequencies of amino acids in the data set and $M_{i,i}$. Let f_i be the relative frequency of amino acid i. The frequency is the number of times an amino acid is observed in the existing sequences. The relative frequency is the frequency divided by the total number of amino acids in the data. If all amino acids are represented equally in the data, then f_i would be 0.05 and the sum of all the relative frequencies would be 20 × 0.05, which gives a value of 1. Actually, adding up all of the f_i's regardless of whether the f_i's are equal will give a value of 1. In nature, not all amino acids are represented equally—for example, Trp and Tyr occur infrequently and the amino acids Ser and Ala are common. Table 4-1 shows the relative frequencies of amino acids calculated from Dayhoff's data.

TABLE 4-1. RELATIVE FREQUENCIES OF AMINO ACIDS

AMINO ACID	RELATIVE FREQUENCY	AMINO ACID	RELATIVE FREQUENCY
G	0.089	R	0.041
A	0.087	N	0.040
L	0.085	F	0.040
K	0.081	Q	0.038
S	0.070	I	0.037
V	0.065	H	0.034
T	0.058	C	0.033
P	0.051	Y	0.030
E	0.050	M	0.015
D	0.047	W	0.010

From M. O. Dayhoff, R. M. Schwartz, and B. C. Orcutt, "A Model of Evolutionary Change in Proteins," in *Atlas of Protein Sequence and Structure*, Vol. 5, Suppl. 3 (Washington, DC: National Biomedical Research Foundation, 1978).

< 0 = less likely to occur than random, "you really don't wanna change isoforms"

0 = totally random substitution

> 0 = occurs more likely than random "you're fine w/ switching to this"

Gly is the most abundant amino acid in the data set that Dayhoff used, and Trp is the least abundant. To determine the number of amino acids that remain unchanged in a group of 100 amino acids in PAM1 we can use the following equation:

$$100 \times \sum_i f_i \mathbf{M}_{i,j} \tag{4.3}$$

where f_i = relative frequency of amino acid i.

This value is 99, which is expected because λ is set to 0.01 in equations 4.1 and 4.2. Thus, there is an average of 1 change for every 100 amino acids when one uses the PAM1.

In the next section we will learn how to use the relative frequencies to improve the scoring of amino acid similarities using a PAM substitution matrix.

4.8 CONVERSION OF THE PAM1 MUTATIONAL PROBABILITY MATRIX TO THE PAM1 LOG-ODDS SUBSTITUTION MATRIX

At first glance, one may want to use PAM1 directly for scoring amino acid similarities. However, PAM1 does not take into consideration the possibility that two amino acids could be aligned by chance. To illustrate, because Gly is much more abundant than Trp, the chance of Gly being the replacement amino acid for any amino acid is much higher at all positions in the protein sequence. So we must consider the relative frequencies of both the initial amino acid, f_i, and the replacement amino acid, f_j (from Table 4-1). In an aligned pair of sequences, the fraction of sites at which the first sequence has amino acid i and the second sequence has amino acid j at a particular location is $f_i \mathbf{M}_{i,j}$. To account for the possibility that such a change could occur by chance, you can imagine that the amino acids in each sequence are shuffled. The relative frequencies of amino acids remain the same in the shuffled sequences. In the shuffled sequences, the fraction of sites at which the first sequence is i and the second sequence is j at a particular location is $f_i f_j$. We define $\mathbf{R}_{i,j}$ as an element in the matrix $\mathbf{R}$ consisting of the ratio of $f_i \mathbf{M}_{i,j}$ to $f_i f_j$:

$$\mathbf{R}_{i,j} = f_i \mathbf{M}_{i,j}/f_i f_j \text{ or } \mathbf{R}_{i,j} = \mathbf{M}_{i,j}/f_j \tag{4.4}$$

where $\mathbf{R}_{i,j}$ is the ratio of the number of times that an i is aligned with a j in two protein sequences evolving at an evolutionary rate of 1% ($\lambda = 0.01$) to the number of times that an i would be aligned with a j in random protein sequences. If $\mathbf{R}_{i,j} > 1$, then amino acid i and j are more likely to be aligned than they would be by chance.

The matrix $\mathbf{R}$ is symmetrical (the values above the main diagonal are equal to the values below the diagonal). This makes sense because the probability that i and j are aligned with each other does not depend on which of the two sequences we choose to contain the i amino acid and which we choose to contain the j amino acid.

We are now in a position to determine the likelihood of a pair of aligned amino acids appearing in sequences evolving according to the PAM1 model relative to the likelihood of the pair of aligned amino acids in random sequences:

relative likelihood =

$$\prod_{k=1}^{L} R_{a_k, b_k} \tag{4.5}$$

where

a_k and b_k are amino acids at the kth site in two aligned sequences

R_{a_k,b_k} is the $\mathbf{R}_{i,j}$ element for those amino acids

L is the length of the two sequences

The relative likelihood is the product of those values from $k = 1$ through $k = L$.

Alignment algorithms, to be discussed in Chapters 5 and 6, attempt to maximize the relative likelihood that amino acids align due to sequence conservation during evolution. However, such algorithms more easily tabulate sums rather than products, thus it is more convenient to create a log-odds substitution matrix so that values of the elements can be summed. Dayhoff converted the PAM1 mutational probability matrix into the PAM1 log-odds substitution matrix, $\mathbf{S}$. $\mathbf{S}_{i,j}$, an element in the log-odds substitution matrix $\mathbf{S}$, can be described as follows:

$$\mathbf{S}_{i,j} = 10\log_{10}(\mathbf{R}_{i,j}) \tag{4.6}$$

where $\mathbf{S}_{i,j}$ is the log-odds score for amino acid i changing to amino acid j.

The value 10 was included in equation 4.6 to ensure that no decimals would be present in the matrix. Because we will frequently refer to this and other log-odds substitution matrices, we will shorten the name to simply "substitution matrices." To understand the PAM1 substitution matrix in more detail, let's use information in the scaled PAM1 mutational probability matrix (Figure 4-7) to derive a corresponding value in the PAM1 substitution matrix $\mathbf{S}$. We will use, as an example, Ala changing to Gly. From the scaled PAM1 mutational probability matrix in Figure 4-7 we obtain a value of 21 for the Ala to Gly change. Because the values in this matrix were multiplied by 10,000, the $\mathbf{M}_{i,j}$ for this value is 21/10,000 or 0.0021. The relative frequency, f_j, of Gly is 0.089 (from Table 4-1). Thus, the $\mathbf{R}_{i,j}$ value for this replacement is 0.0021/0.089, or 0.024. The value in the PAM1 substitution matrix is:

$$\mathbf{S}_{i,j} = 10\log_{10}(0.024) = -16 \tag{4.7}$$

A negative score indicates very little similarity between Ala and Gly. Actually, it means that Ala and Gly would be expected to align $10^{-1.6}$ (i.e., 25/1000), as frequently in homologous sequences as random chance would predict. In the PAM1 substitution matrix, negative values mean that the pair is expected to occur less frequently than random chance, zero means that the pair is expected to occur at a frequency equal to random chance, and positive means that the pair is expected to occur more frequently than random chance. If you were to continue to calculate the $\mathbf{S}_{i,j}$'s of all of the amino acid pairs using the PAM1 mutational probability matrix $\mathbf{M}$, you would find that all amino acid replacements would give negative values in the PAM1 substitution matrix $\mathbf{S}$. When amino acid pairs are the same (i.e., where all of the $\mathbf{M}_{i,i}$ values are high), the $\mathbf{S}_{i,i}$ values are positive. The PAM1 substitution matrix is similar to the identity matrix discussed earlier in this chapter in that the only positive values are for amino acids that do not change. Importantly, other substitution matrices that have practical applications can be derived from the PAM1 mutational probability matrix.

4.9 CONVERSION OF THE PAM1 MUTATIONAL PROBABILITY MATRIX TO OTHER PAM SUBSTITUTION MATRICES

The PAM1 substitution matrix is not practical for comparing sequences that are from potential homologs because all amino acid substitutions are heavily penalized. There is a need to create PAM substitution matrices that give similar amino acid substitutions positive scores and dissimilar amino acid substitutions negative scores. To create more practical substitution matrices, Dayhoff converted the PAM1 mutational probability matrix into other mutational probability matrices where $\lambda > 0.01$.

TABLE 4-2. CORRESPONDENCE BETWEEN THE NUMBER OF TIMES PAM1 MUTATIONAL PROBABILITY MATRIX IS MULTIPLIED BY ITSELF AND THE OBSERVED PERCENT AMINO ACID DIFFERENCE

PAM MUTATIONAL PROBABILITY MATRICES[1]	OBSERVED PERCENT AMINO ACID DIFFERENCE[2]
1...........................	1
5...........................	5
11...........................	10
17...........................	15
30...........................	25
56...........................	40
80...........................	50
112...........................	60
159...........................	70
246...........................	80

[1] The value equals n, where n is the exponent in the term (PAM1)^n.
[2] Observed percent amino acid differences were calculated by the following: $100*(1 - \Sigma f_i \mathbf{M}_{ii})$ where f_i is the relative frequency of amino acid i and $\mathbf{M}_{ii}$ is the mutational probability of amino acid i remaining unchanged, obtained from PAM mutational probability matrices.

The probability that amino acid i will change to amino acid j increases if λ increases. You can multiply the PAM1 mutational probability matrix by itself tens or hundreds of times, which leads to mutational probability matrices in which the observed percent amino acid differences increases. To denote these derived mutational probability matrices, the number following the acronym PAM describes the number of times the PAM1 mutational probability matrix is multiplied by itself. For example PAM2 = (PAM1)^2. Table 4-2 shows a comparison between the exponent and the observed percent amino acid differences.

The PAM mutational probability matrices in Table 4-2 can be converted to log-odds substitution matrices with equations 4.4 and 4.6. These are called PAMx substitution matrices, where x is the number of times PAM1 was multiplied by itself. A popular matrix utilized to determine whether two sequences are homologs is the PAM250 substitution matrix (shown in Figure 4-8). Positive scores in the PAM250 substitution matrix (shaded pink) represent substitutions that are more likely to occur than by chance.

4.10 PRACTICAL USES FOR PAM SUBSTITUTION MATRICES

PAM substitution matrices are useful when comparing two sequences to determine whether they are homologous. You simply line up two sequences and add the scores at each position. For example, when you compare the sequences AFRRSGN and AFLLTGN, you can use the PAM250 substitution matrix to compute a similarity score of 13:

Sequence 1:	A F R R S G N
	| | . . : | |
Sequence 2:	A F L L T G N
Score:	2 9 -3 -3 1 5 2
Similarity score:	13
Similarity:	71%

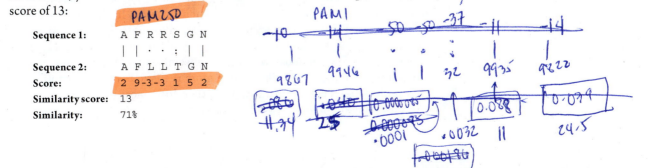
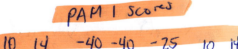

	A	R	N	D	C	Q	E	G	H	I	L	K	M	F	P	S	T	W	Y	V	B	Z	X
A	2	-2	0	0	-2	0	0	1	-1	-1	-2	-1	-1	-3	1	1	1	-6	-3	0	0	0	0
R		6	0	-1	-4	1	-1	-3	2	-2	-3	3	0	-4	0	0	-1	2	-4	-2	-1	0	-1
N			2	2	-4	1	1	0	2	-2	-3	1	-2	-3	0	1	0	-4	-2	-2	2	1	0
D				4	-5	2	3	1	1	-2	-4	0	-3	-6	-1	0	0	-7	-4	-2	3	3	-1
C					12	-5	-5	-3	-3	-2	-6	-5	-5	-4	-3	0	-2	-8	0	-2	-4	-5	-3
Q						4	2	-1	3	-2	-2	1	-1	-5	0	-1	-1	-5	-4	-2	1	3	1
E							4	0	1	-2	-3	0	-2	-5	-1	0	0	-7	-4	-2	3	3	-1
G								5	-2	-3	-4	-2	-3	-5	0	1	0	-7	-5	-1	0	0	-1
H									6	-2	-2	0	-2	-2	0	-1	-1	-3	0	-2	1	2	-1
I										5	2	-2	2	1	-2	-1	0	-5	-1	4	-2	-2	-1
L											6	-3	4	2	-3	-3	-2	-2	-1	2	-3	-3	-1
K												5	0	-5	-1	0	0	-3	-4	-2	1	0	-1
M													6	0	-2	-2	-1	-4	-2	2	-2	-2	-1
F														9	-5	-3	-3	0	7	-1	-4	-5	-2
P															6	1	0	-6	-5	-1	-1	0	-1
S																2	1	-2	-3	-1	0	0	0
T																	3	-5	-3	0	0	-1	0
W																		17	0	-6	-5	-6	-4
Y																			10	-2	-3	-4	-2
V																				4	-2	-2	-1
B																					3	2	-1
Z																						3	-1
X																							-1

FIG. 4-8. The PAM250 log-odds substitution matrix. Positive scores are shaded pink. A score of -10 means that the substitution would occur at 1/10th as frequently in the two sequences as random chance would predict. A score of +2 means that the replacement would occur 1.6 as frequently as random chance would predict. A score of 0 means that the frequency of substitution in the two sequences is equal to the frequency of that which random chance would predict. Recall the equation for the $S_{i,j}$, is $S_{i,j} = 10\log_{10}(R_{i,j})$.

Similarity is the number of pairs with a positive score (denoted with either the "|" or ":" symbol) divided by the total number of amino acids and gaps in alignment multiplied by 100. Pairs with negative scores are denoted by the "." symbol. In our example, the similarity is 5 divided by 7 multiplied by 100, which equals 71%. There are several PAM substitution matrices available, but typically you start with the PAM250 substitution matrix to align sequences with unknown sequence similarity. In the rare case where you know the average overall percent difference in all sequences in two species (called divergence), then using a PAM substitution matrix that matches this difference may be of some benefit (Table 4-2). For example, when comparing chimpanzee proteins with human proteins you might use the PAM5 substitution matrix because the divergence between these organisms is approximately 5%. The PAM5 substitution matrix is much more stringent than the PAM250 substitution matrix, thus requiring more identities to achieve a positive similarity score. If you are comparing two sequences that are extremely divergent, the PAM250 substitution matrix is most useful. Typically, the PAM250 substitution matrix is the practical upper limit for determining whether two sequences are homologous. It is also useful to bear in mind that it is difficult to draw the conclusion that two sequences are homologs if they have less than 18–25% identity.

Fourteen years after Dayhoff published the method she used to derive the PAM250 substitution matrix, Jones, Wilson, and Thornton used an automated procedure to collect thousands of sequences that shared ≥85% identity to tally amino acid substitutions (no ancestral sequences were inferred). They used amino acid substitution data, mutational probabilities, and amino acid frequencies to create a set of amino acid substitution matrices in the same way that Dayhoff did (see Figure 4-5). They derived a log-odds substitution matrix named PET91 (Figure 4-9), which is, in a sense, an updated version of the PAM250 substitution matrix. For the most part, PET91 resembles PAM250. The scores for substitutions of Trp (W) are likely to be more realistic because Dayhoff's data lacked information on Trp substitutions. Other differences between PET91 and PAM250 are highlighted in Table 4-3 (see section 4.12).

	A	R	N	D	C	Q	E	G	H	I	L	K	M	F	P	S	T	W	Y	V
A	2	-1	0	0	-1	-1	-1	-1	-2	0	-1	-1	-1	-3	1	1	2	-4	-3	1
R		5	0	-1	-1	2	0	0	2	-3	-3	4	-2	-4	-1	-1	-1	0	-2	-3
N			3	2	-1	0	1	0	1	-2	-3	1	-2	-3	-1	1	1	-5	-1	-2
D				5	-3	1	4	1	0	-3	-4	0	-3	-5	-2	0	-1	-5	-2	-2
C					11	-3	-4	-1	0	-2	-3	-3	-2	0	-2	1	-1	1	2	-2
Q						5	2	-1	2	-3	-2	2	-2	-4	0	-1	-1	-3	-2	-3
E							5	0	0	-3	-4	1	-3	-5	-2	-1	-1	-5	-4	-2
G								5	-2	-3	-4	-1	-3	-5	-1	1	-1	-2	-4	-2
H									6	-3	-2	1	-2	0	0	-1	-1	-3	4	-3
I										4	2	-3	3	0	-2	-1	1	-4	-2	4
L											5	-3	3	2	0	-2	-1	-2	-1	2
K												5	-2	-5	-2	-1	-1	-3	-3	-3
M													6	0	-2	-1	0	-3	-2	2
F														8	-3	-2	-2	-1	5	0
P															6	1	1	-4	-3	-1
S																2	1	-3	-1	-1
Y																	2	-4	-3	0
W																		15	0	-3
Y																			9	-3
V																				4

FIG. 4-9. The PET91 log-odds substitution matrix of Jones, Taylor, and Thornton (1992) that corresponds to PAM250. These values were calculated from data presented in Figure 4-5. The main diagonal values are from $S_{i,i}$ elements that correspond to amino acids that do not change. All other values correspond to substituted amino acids, $S_{i,j}$. Elements with positive values are shaded pink. Values for ambiguous amino acids B, Z, and X were not calculated in the creation of this matrix.

TABLE 4-3. IMPORTANT FEATURES OF PAM250, PET91, AND BLOSUM45 SUBSTITUTION MATRICES

	PAM250	PET91	BLOSUM45
Most conserved amino acids	1. W 2. C 3. Y 4. F	1. W 2. C 3. Y 4. F	1. W 2. C 3. H 4. P
Least conserved amino acids	1. A,S,N 2. T 3. D,E,Q,V	1. A,S,T 2. N 3. I,V	1. S,V 2. A,I,L,K,T,V
Highly scored substitutions	1. F↔Y 2. I↔V, L↔M 3. K↔R, H↔Q, D↔E	1. F↔Y 2. I↔V, K↔R, D↔E, H↔Y 3. I↔M, L↔M	1. F↔Y, I↔V, K↔R 2. D↔E, E↔Q, H↔Y, I↔L, I↔M, L↔M, S↔T, W↔Y

Note: PAM250 and PET91 are derived using the nearly the same methods, but approximately 40 times more data was used to create the PET91 matrix. BLOSUM45 was derived from multiple aligned sequences in conserved domains and motifs.

4.11 BLOSUM SUBSTITUTION MATRIX

The PAM substitution matrices constituted the first system that utilized an evolutionary model to quantify similarities of amino acids in sequences. A different series of substitution matrices created by Steve and Jorga Henikoff used aligned short multiple sequences from conserved regions of proteins and the information from aligned sequences to create BLOSUM substitution matrices (<u>bl</u>ocks <u>su</u>bstitution <u>m</u>atrix).

BLOSUM substitution matrix
A general term of a set of amino acid substitution matrices derived from amino acids changes observed in multiply aligned sequences found in motifs and domains.

MOTIFS = short (<40 AA) functional / structural ~~and~~ sequences

DOMAINS = LONGER MOTIFS

A = how many times the AA = apparent occurrence ÷ ~~frequency~~ (frequency – 1) aligns

```
ID      P53SUPPRESSR; BLOCK
036006              ( 324) EYFTLKIRGRARFEMFQELNEALEL
P53_RABIT|Q95330    ( 324) EYFILKIRGRERFEMFRELNEALEL
P53_CERAE|P13481    ( 326) EYFTLQIRGRERFEMFRELNEALEL
P53_MACFA|P56423    ( 326) EYFTLQIRGRERFEMFRELNEALEL
P53_MACMU|P56424    ( 326) EYFTLQIRGRERFEMFRELNEALEL
P53_CANFA|Q29537    ( 314) EYFTLQIRGRERYEMFRELNEALEL
P53_FELCA|P41685    ( 319) EYFTLQIRGRERFEMFRELNEALEL
P53_HUMAN|P04637    ( 326) EYFTLQIRGRERFEMFRELNEALEL
Q16848              ( 326) EYFTLQIRGRERFEMFRELNEALEL
P53_RAT|P10361      ( 324) EYFTLKIRGRERFEMFRELNEALEL
                       .      .      .
                       .      .      .
                       .      .      .
```

FIG. 4-10. BLOCK record of sequences from the p53 tumor suppressor protein. ID is a description of the sequence. Letters in blue are accession numbers (from UniProt database) and p53 species descriptors, if available. The numbers in parentheses denote the starting amino acid from the N-terminus of the sequence.

```
Position #:   12345678
         A:   SQDTFQTL
         B:   SQDTFQDL
         C:   SQEFAELT
         D:   SQETFSLL
         E:   SQETFSDL
         F:   SQETFSDL
         G:   SQETFSDL
```

FIG. 4-11. Sequences A through G are used to illustrate how BLOSUM substitution matrices were developed.

	A	D	E	F	L	Q	S	T
A			6					
D		14	10		8			1
E		10	20			2	4	
F	6			30				6
L		8		32				8
Q			2			44	8	
S			4			8	54	
T		1		6	8			30

FIG. 4-12. Matrix Ã is a matrix of aligned amino acid pairs derived from information in the sequence alignment in Fig. 4-11.

BLOSUM = based on identical function in DIFFERENT proteins

PAM = within same family of proteins

Blocks are aligned sequences that use the ProSite database as their starting point. The ProSite database contains aligned sequences from motifs and domains. Recall that motifs are short functional or structural modules of proteins consisting of fewer than 40 amino acids and domains are simply long motifs. To create the ProSite database, sequences that are similar in different proteins and perform the same function or exhibit the same structure are aligned. Scientists chosen for their expertise in particular protein structure/function areas were invited to contribute aligned sequences with annotations that describe the sequences in ProSite. The Henikoffs added more sequences to the ProSite sequences in an automated fashion. Aligned sequences compose a block and, in total, 2,106 blocks were created. An example of a block is shown in Figure 4-10.

To show how the BLOSUM substitution matrix is created, we will use a simple sequence alignment of seven sequences where each sequence has eight amino acids (Figure 4-11). Matrix Ã, the matrix of aligned amino acid pairs (also known as the amino acid replacement matrix), is created by tabulating the number of times each amino acid is aligned with each of the other amino acids. These frequencies are recorded as elements, $\tilde{A}_{i,j}$. We will use two positions of aligned amino acids to illustrate how frequencies are calculated. Let's start with the first position, which only contains S's. At this position, we have the amino acid S in a total number of sequences (k) of 7. Each S-containing sequence has a match with $k – 1$ sequences—in this case, $k – 1 = 6$. Using the multiplication principle, there are $k*(k – 1)$ matches or $7*(6) = 42$ matches.[8] The value of 42 contributes to the $\tilde{A}_{S,S}$ element. There are other positions in the aligned sequences that also contribute to the $\tilde{A}_{S,S}$ element to give a total value of 54.

Now let's determine the values that contribute to $\tilde{A}_{i,j}$ from the fifth position in the aligned sequences. In the fifth position there are 6 F's and one A, so these amino acids will contribute to elements $\tilde{A}_{F,F}$, $\tilde{A}_{A,F}$, and $\tilde{A}_{F,A}$. First we will calculate the contribution to $\tilde{A}_{F,F}$. Using the multiplication principle above where matches are calculated as $k*(k – 1)$ we get $6*(5)$, or 30, which contributes to $\tilde{A}_{F,F}$. What about the value for $\tilde{A}_{A,F}$? Because these amino acids do not match, we simply use the mathematical expression $x*y$ where x is the number of sequences with amino acid A and y is the number of sequences with amino acid F. At this position, we find that the number of times amino acid A can match with F is $1*6 = 6$, which contributes to $\tilde{A}_{A,F}$. Because there are no other positions where A can match with F, the value 6 is placed in matrix Ã (blue-shaded element in Figure 4-12). To calculate the contribution to $\tilde{A}_{F,A}$ at this position, we again use expression $x*y$ to get $6*1 = 6$. Because there are no other positions where F can match with A, the value 6 is placed in matrix Ã (yellow-shaded element in Figure 4-12). Using these two mathematical expressions for matches or mismatches, we can calculate all of the values to fill matrix Ã (Figure 4-12).

Thought Question 4-2

Can you use the information of the sequence alignment in Figure 4-11 to show how you would get a value of 44 for $\tilde{A}_{Q,Q}$ in the matrix shown in Figure 4-12?

---

[8] This multiplication principle is further explained in Chapter 11.

Next, matrix **Q** is created, which contains elements $Q_{i,j}$, the fractions of alignments of amino acid i with amino acid j:

$$\mathbf{Q}_{i,j} = \tilde{\mathbf{A}}_{i,j}/\tilde{\mathbf{A}}_{total} \qquad (4.8)$$

where $\tilde{\mathbf{A}}_{total}$ is $\sum_i \sum_j \tilde{\mathbf{A}}_{i,j}$, the sum of all the elements in the matrix $\tilde{\mathbf{A}}$.

From the $\mathbf{Q}_{i,j}$ elements we obtain the relative frequencies, $\hat{\mathbf{R}}_{i,j}$ of the amino acid pairs compared to what we would expect for randomly shuffled sequences with the same amino acid frequencies:

$$\mathbf{R}_{i,j} = \mathbf{Q}_{i,j}/f_i f_j \qquad (4.9)$$

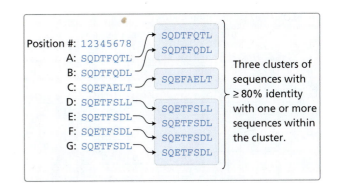

FIG. 4-13. Clustering to create a BLOSUM80 substitution matrix.

where f_i and f_j are relative frequencies of amino acids i and j in the data set. Note that these values differ slightly from the relative frequencies found in Table 4-1 used for the PAM matrices because the number of sequences in the blocks database was much larger than what was available to Dayhoff.

The quantity of the BLOSUM $\hat{\mathbf{R}}_{i,j}$ is comparable to the $\mathbf{R}_{i,j}$ calculated for the PAM model (*equation 4.4*). The log-odds score, $\hat{\mathbf{S}}_{i,j}$ for the BLOSUM log-odds substitution matrix, $\hat{\mathbf{S}}$, is calculated as follows:

$$\hat{\mathbf{S}}_{i,j} = 2\log_2 \hat{\mathbf{R}}_{i,j} \qquad (4.10)$$

Matrix $\hat{\mathbf{S}}$ is the BLOSUM log-odds substitution matrix (we will call this simply the BLOSUM substitution matrix), but in fact there are many BLOSUM substitution matrices that differ by the type of groupings of aligned sequences (called clustering). Clustering influences the calculation of $\tilde{\mathbf{A}}_{i,j}$ values and is discussed briefly below.

To derive practical BLOSUM substitution matrices the block sequences are grouped into clusters such that each cluster within a block contains sequences with a percent identity greater than or equal to a specified cutoff value (for example, 80% identity). When the numbers of amino acid pairs are calculated to fill matrix $\tilde{\mathbf{A}}$, pairs in the same cluster are *not* counted. Instead, pairs are counted *between* clusters. All sequences in the same cluster have a combined weight of 1. In other words, if there are two sequences in a single cluster, each would count half the weight of a single sequence. This treatment alters the $\tilde{\mathbf{A}}_{i,j}$ elements and, ultimately, the log-odds substitution matrix $\hat{\mathbf{S}}$. In effect, clustering and weighting reduces the contribution of redundant sequences to the log-odds substitution matrix.

The log-odds substitution matrix that results from clustering and weighting is BLOSUMx where x is the cutoff value for clustering. We can take, for example, the sequences in Fig. 4-11 and choose a cutoff value of 80% to generate three clusters: sequences AB, sequence C, and sequences DEFG (Figure 4-13).[9]

Each cluster contains sequences with more than 80% identity to at least one other sequence in the cluster. Sequences A and B each contribute 50% weight to their cluster. Sequence C contributes 100% weight to its cluster, and sequences D, E, F, and G each contribute 25% weight to their cluster. We have shown that for clusters containing higher numbers of sequences, the weight contribution of each sequence to the matrix $\tilde{\mathbf{A}}$ is reduced. The more unique sequences are given more weight.

---

[9] For example, in the case of the cluster containing sequences A and B, each sequence must exhibit ≥ 80% identity to *at least one other sequence* in the cluster. Sequence A is 87.5% identical to sequence B. Neither of these sequences can belong to other clusters because they are less than 80% identical to each of the sequences in the remaining two clusters.

A BLOSUM62 substitution matrix

B BLOSUM45 substitution matrix

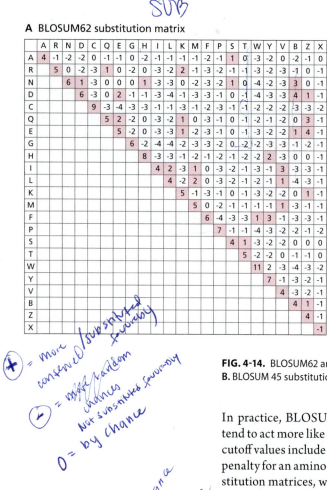

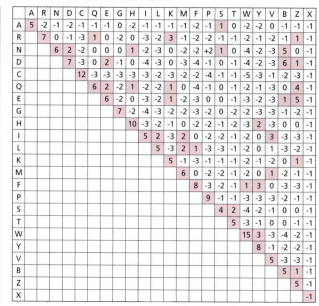

FIG. 4-14. BLOSUM62 and BLOSUM45 substitution matrices. **A.** BLOSUM62 substitution matrix. **B.** BLOSUM 45 substitution matrix. Note that these are symmetric matrices.

In practice, BLOSUM substitution matrices with high cutoff values for clustering tend to act more like identity matrices, and BLOSUM substitution matrices with low cutoff values include more unique sequences into the clusters, thereby giving less of a penalty for an amino acid mismatch. These trends contrast those seen with PAM substitution matrices, where higher numbers are indicative of greater evolutionary distances (i.e., nonidentical amino acids are tolerated). The BLOSUM62 matrix is, for now, the most widely used default substitution matrix in sequence alignment computer software programs (Figure 4-14A).

Let's review the methods used for development of the PAM and BLOSUM substitution matrices. First, sequences used to create the PAM substitution matrices possessed high identity whereas sequences used to create the BLOSUM substitution matrices contained some sequences with low percent identity. Second, to generate the PAM substitution matrices, phylogenetic trees were created from pairs of sequences and accepted amino acid substitutions were inferred from the trees. The BLOSUM substitution matrices were built on counting amino acid pairs from multiple aligned sequences in conserved motifs and domains.

If the alignments in the blocks database are correct, then it is probable that those sequences had a common ancestral gene, but no phylogenetic trees were used. Proteins that share only motifs or domains are often more diverged than the proteins used to create the PAM matrix. Both PAM substitution matrices and BLOSUM substitution matrices are useful for detecting sequences that are conserved in proteins and have diverged from a common ancestral gene.

4.12 PHYSICO-CHEMICAL PROPERTIES OF AMINO ACIDS CORRELATE TO VALUES IN MATRICES

At the beginning of this chapter, it was suggested that some amino acid pairs in aligned sequences should be scored highly because their side chains looked nearly identical. We discussed two different approaches used by bioinformaticians to quantify the degree of similarity between two amino acid side chains. To appreciate the quantification of amino acid similarities, you may find it helpful to review the amino acid structures (see Figure 1-8) as we compare substitution matrices often used by sequence alignment software programs—PAM250 (Figure 4-8), PET91 (Figure 4-9), BLOSUM45 (Figure 4-14B).[10]

Here are some examples of high value non-identical amino acid pairs (Table 4-3):

1. F↔Y. In PAM250, PET91, and BLOSUM45 substitution matrices, the values are 7, 5, and 3, respectively. These are among the highest values in their respective matrices. The side chains of F and Y are both aromatic and differ only by the presence of a hydroxyl (–OH) group on the tyrosine.

2. D↔E. In PAM250, PET91, and BLOSUM45 substitution matrices, the values are 3, 4, and 2, respectively. The side chains of D and E have a carboxylate (–COO⁻) group and differ only because E has an extra methylene (–CH₂–) group.

3. K↔R. In PAM250, PET91, and BLOSUM45 substitution matrices, the values are 3, 4, and 3, respectively. The side chains of K and R have basic groups and they are both long.

4. I↔V. In PAM250, PET91, and BLOSUM45 substitution matrices, the values are 4, 2, and 3, respectively. V and I should have a high similarity value, as they are both hydrophobic and their side chains differ only by the extra methylene (–CH₂–) group in I.

As one would expect, pairs of amino acid side chains that possess significantly different properties tend to have negative values. Hydrophobic amino acids paired with hydrophilic ones tend to have negative values. Large, bulky amino acids paired with those that are small also tend to have negative values. As mentioned earlier in this chapter, there are two amino acids that have very distinctive side chain structures. One, W, has a side chain such that substitution by other amino acids almost invariably gives a negative value. Changes at a W position occur slowly on an evolutionary timescale. Recall that Dayhoff rarely found amino acid substitutions for W in her data. Due to the unique properties of W, an amino acid substitution at its position likely results in a protein that is nonfunctional and is ultimately fatal to the organism. Another amino acid, C, also has a unique side chain in that it forms disulfide bonds with other C's. As one would expect, the substitution matrices show that replacement of W or C with other amino acids gives either a negative value or zero. The substitution matrices reinforce the principle that natural selection and structure strongly depend on amino acid side chain properties. A more comprehensive comparison of PAM250, PET91, and BLOSUM45 substitution matrices is shown in Table 4-3.

Similarities in amino acid side chains are allowed in homologous sequences. The more similar the side chains, the higher the value in the substitution matrix. Another influence on amino acid substitution data may be the ease of mutating one codon into another. It would be expected that a single point mutation in a codon would be more common than multiple point mutations in a codon. Analysis of PET91 (Figure 4-9) reveals that all similar amino acid pairs (i.e., those amino acids that have positive scores for substitutions) differ by single nucleotides in their codons, suggesting that part of the driving force for accepted amino acid substitutions in homologous proteins is the ease of converting from one codon to another. This must be balanced by the fact that a few dissimilar amino acid pairs also differ by single nucleotides, indicating that the ease by which one codon can be converted into another is not the only consideration in natural selection. From the perspective

---

[10] These matrices are comparable in terms of the amount of average mutual information per amino acid pair, also known as the relative entropy. For more information on relative entropy, see Jones, Taylor, and Thornton (1992).

of physicochemical properties, a somewhat dissimilar amino acid pair is H↔Y. Yet, this substitution is assigned a +4 score in PET91 matrix and a +2 score in the BLOSUM45 matrix (Figure 4-14B). Interestingly, the two codons that code for H can each be converted to two codons that code for Y with just a single nucleotide change in the DNA (C→T), suggesting that the ease of DNA mutation may contribute, at least in part, to the amino acid substitutions observed in related sequences. In sum, it appears that there are two natural selection forces that drive amino acid changes: one is the selection against amino acids that are radically different in structure, and the other is the ease by which codons can be mutated.

4.13 PRACTICAL USAGE

When you compare PAM substitution matrices to BLOSUM substitution matrices, both generally perform well when attempting to search sequence databases for proteins that are homologs to a protein of interest. BLOSUM substitution matrices appear to be better performers when detecting conserved regions in sequences. This is logical given that BLOSUM substitution matrices were derived from aligned sequences found in motifs and domains. In alignment software programs such as BLAST, which are designed to detect local regions of similarity, BLOSUM is often the preferred substitution matrix. What is interesting is that PAM substitution matrices still perform quite well despite the small amount of data underlying them. The most likely reasons for this are the care used in constructing the alignments and phylogenetic trees used in counting amino acid substitutions, and the fact that they are based on a simple model of evolution. The BLOSUM62 substitution matrix is generally the best for detecting the majority of weak alignment similarities. BLOSUM45 is best for detecting weak alignments within long sequences. An estimate of equivalent PAM and BLOSUM matrices, based on calculated relative entropies, is as follows:[11]

PAM100 ==> BLOSUM90

PAM120 ==> BLOSUM80

PAM160 ==> BLOSUM60

PAM200 ==> BLOSUM52

PAM250 ==> BLOSUM45

If you are attempting to align sequences that are not too evolutionarily distant, it is best to use a low PAM numbered matrix or high BLOSUM numbered matrix. If you are attempting to align sequences that are evolutionarily distant, it is best to use a high PAM numbered matrix or low BLOSUM numbered matrix. Another factor that influences the outcome of sequence alignments is the treatment of indel mutations that result in loss or gain of one or more amino acid(s) in a sequence. When sequences are aligned, a gap must be inserted to account for the indel mutation to maintain optimal alignment. Values associated with gaps are negative because indel mutations usually cause a radical change in the amino acid sequence. Such values are called gap penalties, and they can heavily influence the similarity score (see next chapter).

---

[11] Equivalent substitution matrices were calculated from relative entropies by Henikoff and Henikoff (1992).

SUMMARY

Substitution matrices are critical for detecting protein homologs. They also help to rank which homologs are more closely related than others. Two commonly used substitution matrices are PAM and BLOSUM substitution matrices. PAM substitution matrices are derived from PAM mutational probability matrices. The PAM mutational probability matrices are mathematically derived from the PAM1 mutational probability matrix, which, in turn, is created from an evolutionary model where the overall frequency of amino acid substitutions is scaled to 1%. Higher percentages of amino acid substitutions are created by PAM1 mutational probability matrix multiplications. As expected, amino acid substitutions with nearly the same side chain compositions result in positive similarity scores, and substitutions with radically different side chains give negative scores. It is likely that such amino acid substitutions are often deleterious to the organism. Another force that drives amino acid substitutions appears to be the ease of converting from one codon to another. A single mutation in the codon is more likely than a double mutation.

The BLOSUM substitution matrices are derived from short aligned sequences within motifs and domains. Aligned amino acid pairs are counted within the aligned sequences. Clusters of sequences are created with percent identity cutoff values. Within each cluster, individual sequences are assessed lower weights per sequence, resulting in more weight given to divergent sequences. BLOSUM substitution matrices with different identity cutoff values are created through this clustering and weighting process. BLOSUM62 (where the identity cutoff for clustering is 62%) is the default substitution matrix for many sequence alignment software programs. This matrix is especially proficient at detecting short sequence similarities in protein sequences. In the next chapter we will discuss the logic and development of pairwise sequence alignment programs, which use substitution matrices to carry out their functions.

EXERCISES

1. Give two reasons why Dayhoff used closely related sequences to create her matrix of amino acid substitutions.

2. Use the value for $S_{W,W}$ in the PAM250 log-odds substitution matrix to calculate the value of $M_{W,W}$ in the PAM250 mutation probability matrix.

3. Download the PAM250 and PAM30 substitution matrices and place them side by side on your computer screen. Describe the differences between the two matrices. Explain why you see these differences.

4. Show how to calculate the value for $A_{L,I}$ in the amino acid replacement matrix A shown in Figure 4-12. Use the data in Figure 4-11 for this calculation.

5. Download BLOSUM30 and BLOSUM80 substitution matrices and place them side by side on your computer screen. What are the differences between the two matrices? Why do you see these differences?

6. Which of the following statements concerning the BLOSUM62 substitution matrix is correct?

   a. Ala is aligned with Arg more often than expected by chance.
   b. Ala is never substituted by Cys.

   c. Tryptophan is substituted less frequently than any other amino acid.

7. What can we say about the alignment of Asp with Glu in two sequences when considering the BLOSUM62 substitution matrix with respect to alignment by chance?

8. Given the following two sequences, create a pairwise alignment by hand. Do not place any gaps in the sequences to optimize alignment. Use an identity matrix and BLOSUM62 substitution matrix to create the two sets of pairwise alignments. Report the identity, identity score, similarity, and similarity score.

   Sequence 1: ATPLM

   Sequence 2: ATKIM

9. Although it is not used in practice for determining similarities, describe what a PAM0 substitution matrix would look like.

10. The PAM1 mutational probability matrix is not a symmetrical matrix (for example, the value of element $M_{A,R}$ does not equal to the value of element $M_{R,A}$). Why?

11. Use the Needleman-Wunsch software program to perform two pairwise sequence alignments. For one alignment, use the BLOSUM30

substitution matrix, and for the other, use BLOSUM90 substitution matrix. Report the identity and similarity score for each alignment. Explain why these values differ depending on the substitution matrix. Use human p53 and *Xenopus laevis* p53 as inputs for your pairwise sequence alignments.

Here are parameters to use for the alignments when running the software program:

Matrix: Gap Open Penalty: 10; Gap Extend Penalty: 0.5; End Gap Penalty: False; End Gap Open Penalty: 10; End Gap Extend Penalty: 0.5

ANSWERS TO THOUGHT QUESTIONS

4-1. Comparison of existing species sequences shows that the two of the sequences start with A and only one starts with D or C. Therefore, you can infer that the ancestral sequences start with an A.

4-2. There are two positions where Q can match with Q: position 2 and position 6. We can calculate the contribution to $\tilde{A}_{Q,Q}$ for each position separately and then sum those contributions to arrive at a final value for $\tilde{A}_{Q,Q}$. Use the mathematical expression $k*(k-1)$.

Position 2: $7*6 = 42$
Position 6: $2*1 = 2$
Final value for $\tilde{A}_{Q,Q} = 42 + 2 = 44$

REFERENCES

Dayhoff, M. O., R. M. Schwartz, and B. C. Orcutt. 1978. "A Model of Evolutionary Change in Proteins." In *Atlas of Protein Sequence and Structure*, Vol. 5, Suppl. 3. Washington, DC: National Biomedical Research Foundation.

European Bioinformatics Institute. 2012. "Help about Matrices Tutorial." Accessed April 7. http://www.ebi.ac.uk/2can/tutorials/matrices.html.

Gonnet, G. H., M. A. Cohen, and S. A. Benner. 1992. "Exhaustive Matching of the Entire Protein Sequence Database." *Science* 256: 1443–1445.

Henikoff, S., and J. G. Henikoff. 1992. "Amino Acid Substitution Matrices from Protein Blocks." *Proceedings of the National Academy of Sciences of the USA* 89: 10915–10919.

Henikoff, S., and J. G. Henikoff. 1994. "Position-Based Sequence Weights." *Journal of Molecular Biology* 243: 574–578.

Higgs, P. G., and T. K. Attwood. 2005. *Bioinformatics and Molecular Evolution*. Malden, MA: Blackwell Publishing.

Hunt, L. T. 1983. "Margaret O. Dayhoff 1925–1983." *DNA* 2: 97–98.

Jones, D. T., W. R. Taylor, and J. M. Thornton. 1992. "The Rapid Generation of Mutation Data Matrices from Protein Sequences." *Computer Applications in the Biosciences* 8: 275–282.

NCBI. 2012. "Bruno J. Strasser Speaking at the Genbank 25th Anniversary." YouTube. Accessed April 3. http://www.youtube.com/watch?v=VRnY5HP3wjM.

Pevsner, J. 2009. *Bioinformatics and Functional Genomics*. Hoboken, NJ: John Wiley and Sons.

Readers711. 2012. "Margaret Dayhoff." Scribd. Accessed April 3. http://www.scribd.com/doc/50146164/margaret-dayhoff.

Zvelebil, M., and J. O. Baum. 2008. *Understanding Bioinformatics*. New York: Garland Science, Taylor & Francis Group.

PAIRWISE SEQUENCE ALIGNMENT

5.1 INTRODUCTION

In Chapter 3 we discussed sequence alignments that can be used to identify homologs and provide insight to the function, structure, and evolutionary history of proteins and genes. In Chapter 4 we reviewed substitution matrices, which are essential for determining the degree to which sequences are related. In this chapter you will learn the details of how pairwise sequence alignment is performed by computer programs originally written in the 1970s and 1980s and gradually improved. Pairwise sequence alignment is the process by which two biological sequences are matched to show optimal similarity. Such programs guarantee that the user receives the alignment with the maximum similarity score. Statistical analysis of the similarity score, to be discussed in Chapters 6 and 11, quantifies the probability that the similarity score could have been achieved by chance. This probability helps the user decide whether a given alignment is sufficient evidence for homology. The three essential outputs of alignment software programs are (1) the sequence alignment, (2) the similarity score, and (3) the statistical analysis.

To approach the pairwise sequence alignment programs, we will first discuss the sliding window. The sliding window accumulates information or data about the properties of a segment of amino acid residues in a window of specific length within a long polypeptide.

The data collected in the window is quantified and plotted on a graph. The window then shifts to the right and the process is repeated until the window reaches the end of the sequence to be studied. The sliding window can be used to analyze both nucleotide and amino acid sequences. We will discuss how the sliding window program is used to create dot plots. A dot plot enables the user to easily visualize similar regions in aligned sequences. One program that creates the dot plot is Dotter, which utilizes a substitution matrix and a sliding window to create a visual graph of sequence alignment. We will then delve into the nuts and bolts of two pairwise sequence programs: the Needleman-Wunsch global alignment program and the Smith-Waterman local alignment program. We will also discuss two updated versions of the Needleman-Wunsch global alignment programs. These programs provide strong foundations that undergird the bioinformatics field.

5.2 SLIDING WINDOW

sliding window program A portion of data within a larger data set is captured. The captured data is held within a window. A sliding window program performs a calculation on data in the window and then the window moves incrementally. At each increment, the program repeats the calculation. The output from each calculation may be displayed in a graphical format.

In bioinformatics, the sliding window is defined as captured data from a larger data set. A calculation is performed on the data in the window, and the window moves incrementally through the data set where it repeats the calculation. Some of the earliest bioinformatics computer programs employed the sliding window concept to gather information. In 1978, Peter Chou and Gerald Fasman (Brandeis University, Waltham, Massachusetts) used rules developed from protein structure data analysis to create a sliding window program that predicts protein secondary structure. The program was very popular in the 1980s and, although only ~60% accurate, helped launch the field of protein structure prediction. A sliding window program can also identify segments of a nucleotide sequence that contain a high proportion of guanine (G) and cytosine (C). Recall from Chapter 1 that GC-rich areas in DNA exhibit more hydrogen bonds between their base pairs than AT-rich areas, which helps to make it more difficult to separate the two strands of a GC-rich DNA double helix. Genomes with high proportions of GC-rich areas are found in organisms that live in areas where there are high temperatures (for example, hydrothermal vents in the ocean floor and natural hot springs). Some GC-rich areas in genomes are modified by enzymes called DNA methyltransferases that methylate cytosines.[1] These GC-rich areas, known as GC islands (or CpG islands) in eukaryotes, are sometimes found in promoters upstream of genes. Methylation represses transcription of genes downstream of the promoters. One can develop a simple sliding window program to visualize GC-rich areas in long stretches of DNA. Interestingly, in cancer cells the pattern of methylation in the human genome is significantly altered such that some tumor suppressor genes are transcribed less frequently and some proto-oncogenes are transcribed more frequently.

As a simple approach to determining GC-rich areas in a long sequence, you can use a sliding window to focus on a small segment of the sequence and the percentage

larger window sizes
reduces ~~that~~ background
noise /
false matches

---

[1] DNA methyltransferases add methyl ($-CH_3$) groups onto the 5-position of the cytosines in DNA. In eukaryotes, methylation creates a DNA structural change that represses transcription in the area near the methylation event. DNA methylation is part of a wide series of genome-altering, nonmutational events known collectively as epigenetics.

of G plus C within the window is calculated. That number is plotted on an *x-y* coordinate plane where *x* is the nucleotide sequence number and *y* is the percent of G plus C in the window. The window then slides to the right 1 nucleotide and the calculation is repeated. The major difficulty for extrapolating biologically useful information with sliding window programs is deciding on the size of the window (Figure 5-1). If you choose a window that is too wide, you may detect general trends in the data set but miss important details. If you use a window that is too narrow, a high level of detail is obtained—making it difficult to find general trends.

Sliding window programs can also be used to predict protein structures. One such program, called the Kyte-Doolittle program, is useful in the prediction of secondary structures of membrane proteins—a notoriously difficult class of proteins to determine the structure of experimentally. The Kyte-Doolittle program predicts regions of a protein embedded in a membrane and regions exposed to the cytoplasm or extracellular milieu. The program is also used to predict those segments of a globular protein that are on the surface and those that are located inside the protein. Experimental work showed that the Kyte-Doolittle program correctly predicted which segments of the bacteriorhodopsin membrane protein were embedded in the membrane. To create their sliding window program, Jack Kyte and Russell Doolittle (University of California, San Diego) developed a hydropathy scale (sometimes referred to as hydrophobicity scale). The scale assigns the degrees to which amino acids are hydrophobic (water hating).

Kyte and Doolittle used two methods to develop their scale (Table 5-1). In the first method, proteins with known tertiary and quaternary structures were studied.

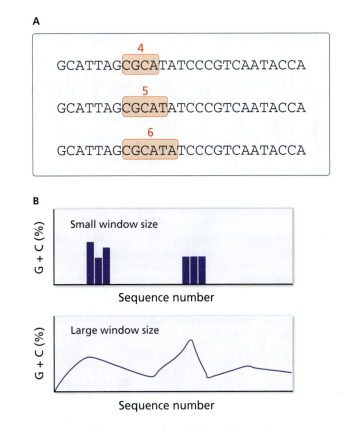

FIG. 5-1. A sliding window program gathers information about the properties of nucleotides or amino acids in a segment of a sequence. A. Three identical sequences are shown with window sizes set to 4, 5, and 6. **B.** Percent (G + C) plotted as a function of sequence number is depicted. In the top schematic plot, the percent (G + C) is shown when a small window is selected. In the bottom schematic plot, the percent (G + C) is shown when a large window is selected.

TABLE 5-1. HYDROPATHY SCALE USED FOR CONSTRUCTING THE HYDROPATHY PLOTS OF KYTE AND DOOLITTLE (1982). THE HIGHER THE VALUE, THE MORE HYDROPHOBIC THE AMINO ACID

AMINO ACID	HYDROPATHY	AMINO ACID	HYDROPATHY
A	1.8	M	1.9
C	2.5	N	−3.5
D	−3.5	P	−1.6
E	−3.5	Q	−3.5
F	2.8	R	−4.5
G	−0.4	S	−0.8
H	−3.2	T	−0.7
I	4.5	V	4.2
K	−3.9	W	−0.9
L	3.8	Y	−1.3

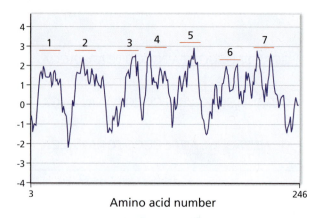

FIG. 5-2. Hydropathy plot of bacteriorhodopsin. Kyte and Doolittle used a sliding window program (with a window length of seven) to predict the membrane-spanning regions of bacteriorhodopsin from the amino acid sequence. The seven known membrane-spanning regions are numbered 1 to 7 in red on the plot. Note that this particular software program averaged the hydropathy values in the window (http://www.vivo.colostate.edu/molkit/hydropathy/index.html). The original program by Kyte and Doolittle summed the hydropathy values.

If the amino acid side chains were buried on the inside of the proteins, the amino acids were considered more hydrophobic, and if the amino acid side chains were exposed on the surface of the proteins, they were considered hydrophilic. In the second method, Kyte and Doolittle used experimentally determined measurements of the amount of energy required to transfer amino acids from the water phase to the vapor phase (phase transition). The higher the energy level required to achieve phase transition, the less hydrophobic the amino acid. The data collected from these two approaches, combined with some intuition, was used to create the hydropathy scale.

With the hydropathy scale in hand, the sliding window program was created (see Chapter 13 for detailed information on how to create the Kyte-Doolittle sliding window program). A plot is created such that the x-axis is the amino acid sequence number and the y-axis is the average hydropathy for a chosen window length. An amino acid sequence is entered and a window length is chosen by the user. The sequence is segmented by the window, and the hydropathy values from the scale are averaged in the window. In the plot, the first averaged hydropathy value is depicted. The window slides to the right one amino acid in the sequence, and the process is repeated. Figure 5-2 shows a hydropathy plot that predicts the hydrophobic regions of bacteriorhodopsin from its amino acid sequence. In the plot are regions that are relatively more hydrophobic (generally, above a hydropathy value of 1.6). Kyte and Doolittle adjusted the window size so that it gave a plot with approximately seven transmembrane segments—consistent with experimental data available to them at the time. The window size they used was seven amino acids.

Dot Plots

Another example of an application of a sliding window program is the dot plot. A dot plot is a convenient method of displaying similarity regions of two sequences. First you create an $n \times m$ matrix where n is the number of amino acids in the first sequence and m is the number of amino acids in the second sequence. You place the first sequence on top of the first row along the x-axis and the second sequence to the left of the first column along the y-axis.[2] When the letters within the matrix match, a dot is placed in the cell found at the intersection of the two letters. To simplify this exercise we will use nucleotide sequences. In Figure 5-3A, a dot plot of two nucleotide sequences is shown. The dots denoting identical nucleotides form a diagonal line that stretches from the upper left corner to the lower right corner. The diagonal line without breaks shows that the two sequences are identical.

One challenge with dot plots is the background noise, which appear as dots located in "off-main diagonal" locations in the matrix. You can calculate the percent of background noise from the number of times a nucleotide is expected to randomly match another nucleotide. The percent chance of a nucleotide to match randomly at a single position is $1/4 \times 100$, or 25% (where 4 is the number of nucleotides). This high percentage of background noise is due to the fact that the window size is only 1, but if the window size is increased to 3, the background noise is reduced considerably (Figure 5-3B). Now there must be three ordered positions that match to create a dot in the matrix. The percent chance of a random match with a window size of 3 is

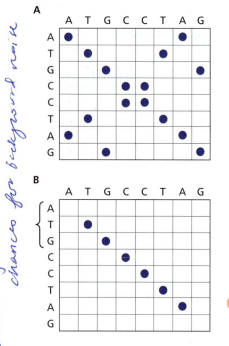

FIG. 5-3. Dot plots of nucleotide sequences.
A. Two nucleotide sequences aligned with a window size of one. **B.** The window size increased to three (bracket depicts the window size).

increasing window size decreases chances for background noise

---

[2] This order of sequence assignment to the matrix is unconventional in bioinformatics. Typically, the first sequence runs alongside the first column (n rows) and the second sequence is displayed over the first row (m columns).

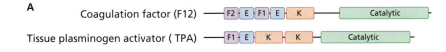

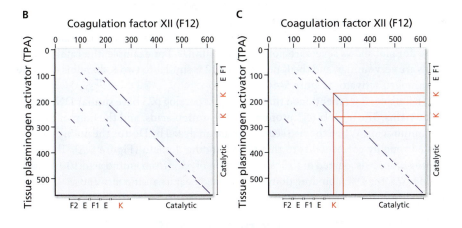

FIG. 5-4. Dot plots of two related proteins.
A. Schematic depiction of coagulation factor XII (F12) and tissue plasminogen activator (TPA) showing locations of domains. **B.** Dot plot of sequence alignment. Note that diagonals represent regions of similarity. **C.** Dot plot of sequence alignment highlighting Kringle domain (K) alignments.

calculated as the probability of occurrence of all three independent events: $1/4 \times 1/4 \times 1/4 \times 100$, or 1.56%. *lower than in a small program*

The Dotter Program

Dot plots are useful graphics that show where within two sequences there are identical residues. More sophisticated dot plots can be made by the computer software program Dotter. Dotter is a graphical dot plot program, developed by Erik L.L. Sonnhammer (at the Sanger Center in Cambridge, United Kingdom) and Richard Durbin (at the MRC Laboratory of Molecular Biology in Cambridge), that creates dot plots from the alignments of two protein sequences. Every residue in one sequence is compared to every residue in the other sequence. Just like our simple dot plots shown in Figure 5-3, in regions where the two sequences are similar, a series of dots will run diagonally across the dot plot from the upper left to the lower right, called the main diagonal. When a sequence is compared against itself, the main diagonal scores maximally, because it is a 100% perfect self-match. Pairwise similarity scores are averaged in a sliding window that slides through all possible sequence alignments. The scores within the window are averaged, and the higher the average score, the darker the dot in the plot. The pairwise scores are usually obtained from PAM or BLOSUM substitution matrices.

Figure 5-4A shows a schematic diagram of two proteins with similar sequences. One protein is coagulation factor XII (F12) and the other is tissue plasminogen activator (TPA). They perform vital functions in blood clotting, and both are modular proteins consisting of several domains. Coagulation factor XII is composed of two epidermal growth factor domains (each abbreviated as E) and two fibronectin domains, F1 and F2, that are similar to one another.[3] Coagulation factor XII also has a Kringle (K) domain, a segment that contains loops stabilized by three disulfide bonds. The E, F1, F2, and K domains are also found in tissue plasminogen activator, but their arrangement along its primary sequence is altered. In both proteins, these domains are located upstream (toward the N-terminus) of their respective catalytic domains.

hydropathy plot A plot created from an output of a sliding window program that shows the hydrophobic areas of proteins.

dot plot A plot created by placing dots in a matrix to create a main diagonal when two sequences are similar.

Dotter A sliding window program that compares the similarity of two proteins by producing a dot plot.

↓ *synteny*
= conservation of the order of genes in chromosomes

---

[3] The E domain has a sequence similar to another protein called epidermal growth factor, which is responsible for driving epidermal cell proliferation. The two domains F1 and F2 are commonly found in fibronectin, a protein required for cell adhesion, growth, and differentiation.

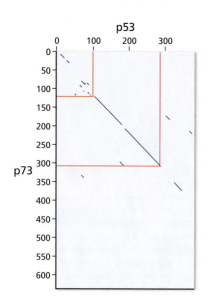

p53

p73

FIG. 5-5. Dot plot comparing p53 to p73 sequences. The red lines bracket the segment of the main diagonal that corresponds to the similar DNA binding domains of the two proteins. Note that p53 is nearly 200 residues shorter than p73.

global alignment Optimal pairing of two sequences over the entire lengths of the sequences.

match = 1
mis = 0
gap = 0

Alignment A

Sequence 1: ADCDN-R-CKCRWP
Sequence 2: AWC-NDRQCLCR-P
Score: 10101010101101
Total score 1: 8

Alignment B

Sequence 1: ADC-DNR-CKCRWP
Sequence 2: AWCND-RQCLCR-P
Score: 10101010101101
Total score 1: 8

FIG. 5-6. Two sequences aligned by eye. Two alignments are possible, both of which return a score of 8 with the simple substitution matrix where match = 1, mismatch = 0, gap = 0.

When the protein sequences are compared using Dotter, a dot plot is created that shows that the two proteins have several similar domains (Figure 5-4B). The dot plot reveals that it is highly probable that the two share the same ancestral gene. Long diagonal lines signify large areas of similarity, and the darker the line, the higher the similarity. Shorter diagonal lines that are parallel show that some domains are repeated along the primary sequence. For instructional purposes, the schematic diagrams of the proteins are displayed along the two axes of the plot. Careful examination of the dot plot shows several regions that are similar. For example, the catalytic domains are very similar, and the K domain in F12 is similar to two K domains in tissue plasminogen activator (Figure 5-4C).

In Chapter 3 we mentioned that p53 and its paralog p73 have similar DNA binding domains. The human p53 protein has 393 amino acids, and the human p73 has 588 amino acids. When the two sequences are analyzed by Dotter, the main diagonal created roughly corresponds to this DNA binding domain (Figure 5-5). The DNA binding domain is located in the central region of p53 (from amino acid 100 to 300). Within p73, the DNA binding domain is located near its amino terminus.

5.3 THE NEEDLEMAN-WUNSCH GLOBAL ALIGNMENT PROGRAM

Now that we are familiar with sliding window programs, we are in a position to understand how sequence alignment programs operate. These sequence alignment programs find an optimal alignment by, if necessary, placing gaps in the sequences and delivering a similarity score. The Needleman-Wunsch global alignment program, the first software program to align two sequences, was developed by Saul Needleman and Christian D. Wunsch (Northwestern University, Evanston, Illinois) (Box 5-1). The Needleman-Wunsch (N-W) global alignment program is called a *global* alignment program because it aligns pairs of sequences across their *entire* lengths.

To demonstrate how the program aligns and scores pairs of sequences, we will use a simple substitution matrix. An amino acid match will receive a score of 1, a mismatch will receive a score of 0, and a gap will receive a score of 0. Let's first attempt to align the hypothetical sequences "ADCDNRCKCRWP" (Sequence 1) and "AWCNDRQCLCRP" (Sequence 2) by eye. The expected alignments with the least number of gaps show that there are two possible outcomes (Figure 5-6). Both outcomes are valid because they both return a score of 8, the highest score possible. We will show how the N-W program creates these alignments and gives the same score. The N-W program uses a three-stage approach named initialization, matrix fill, and traceback (traceback is also called backtracking).

Initialization and Matrix Fill

In the initialization phase, Sequence 1 is placed to the left of the first column and Sequence 2 is placed above the top row of matrix **M**, the scoring matrix. For this scoring matrix, we will use two sequences and values from an amino acid substitution matrix as inputs. For now, our substitution matrix is very simple, 1 for match, 0 for mismatch, 0 for gap. In the initialization phase, **M** is filled with 1s for each $M_{i,j}$ element where there is a match (Figure 5-7A).[4]

In the matrix fill phase, values are tabulated for each element beginning with the element in the bottom right corner. Proceeding from right to left and from the bottom row to the top row **M** is filled. Zeros are placed in the $M_{i,j}$ elements when there are amino acid mismatches in the bottom row. Proceeding to the next row up from

---

[4] Needleman and Wunsch's alignment program predated the modern concept of the amino acid substitution matrix, but the authors acknowledged that there is a hierarchy of amino acid similarity. They created a rough substitution matrix based on the number of nucleotide changes required for a codon to change its specificity from one amino acid to another. Modern global alignment programs use BLOSUM and PAM substitution matrices.

A

	A	W	C	N	D	R	Q	C	L	C	R	P
A	1											
D					1							
C			1					1		1		
D					1							
N				1								
R						1					1	
C			1					1		1		
K												
C			1					1		1		
R						(1)					1	
W		1										
P												1

B

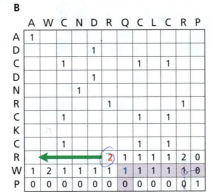

	A	W	C	N	D	R	Q	C	L	C	R	P
A	1											
D					1							
C			1					1		1		
D					1							
N				1								
R						1					1	
C			1					1		1		
K												
C			1					1		1		
R						(2)	1	1	1	1	2	0
W	1	2	1	1	1	1	1	1	1	1	1	0
P	0	0	0	0	0	0	0	0	0	0	0	1

C

	A	W	C	N	D	R	Q	C	L	C	R	P
A	8	7	6	6	5	4	4	3	3	2	1	0
D	7	7	6	6	6	4	4	3	3	2	1	0
C	6	6	7	6	5	4	4	4	3	3	1	0
D	6	6	6	5	6	4	4	3	3	2	1	0
N	5	5	5	6	5	4	4	3	3	2	2	0
R	4	4	4	4	4	5	4	3	3	2	2	0
C	3	3	4	3	3	3	3	4	3	3	1	0
K	3	3	3	3	3	3	3	3	3	2	1	0
C	2	2	3	2	2	2	2	3	2	3	1	0
R	2	1	1	1	1	2	1	1	1	1	2	0
W	1	2	1	1	1	1	1	1	1	1	1	0
P	0	0	0	0	0	0	0	0	0	0	0	1

FIG. 5-7. Needleman-Wunsch global alignment program.
A. Initialization. Values for matches are placed in appropriate elements in **M**. In this matrix, *i* corresponds to a specific row and *j* corresponds to a specific column where there are a total of *n* rows and *m* columns. **B.** Matrix fill. In the matrix fill stage we begin to fill the matrix starting with the element in the bottom right corner. Here, the matrix fill is shown partially completed with the bottom two rows filled and the third row partially filled. Mismatch elements in the bottom row are filled with zeros. To complete the next row up, the values of the subrow/subcolumn are scanned for the highest value. This highest value is added to the $M_{i,j}$ element of the second row. Each element in the row is calculated in the same fashion. In the third row from the bottom, the value for the element corresponding to a match between R and R (in red font) is the sum of the initial fill value (1) and the maximum value in the subrow and subcolumn (purple-shaded). If several elements in the subrow or subcolumn have maximum values, the value of the element nearest to the R-R element is chosen (in blue font). The broad arrow shows the direction of the matrix fill in the row. **C.** Final matrix fill and traceback of optimal paths. Beginning with the value in the top row/left column, the traceback follows the path of the predecessor elements in the matrix. The first predecessor element is shaded yellow. The traceback shows a bifurcation at the element corresponding to a match between C and C (value 7). When the traceback follows the top path, a column is skipped, which creates a gap between C and D in the left vertical sequence (Sequence 1). When the traceback follows the bottom path, a row is skipped which creates a gap between C and N in the horizontal sequence (Sequence 2).

the bottom, we add the value of $M_{i,j}$ to the maximum value in the row and column diagonally juxtaposed to $M_{i,j}$. The diagonally juxtaposed row and column are called subrow and subcolumn (we will name this region subrow/subcolumn). As an example, let's calculate the value for the element corresponding to a match between R and R in the third row from the bottom (Figure 5-7B). We will call this element the target element. The value for the target element is the sum of its initial fill value (1) and the maximum value located in the subrow/subcolumn (1). Thus, the final value target element is 1 + 1 = 2. This value is in red font in Figure 5-7B. The element in the subrow/subcolumn that contributed its value is called the predecessor element. The predecessor element must have the maximum value and have a distance that is closest to the target element. Before moving onto the next target element, the program records the location of the predecessor element. The predecessor element is in blue font in Figure 5-7B. After initialization and matrix fill, the resulting matrix represents all possible alignments between two sequences, including gaps. Each possible alignment is a pathway through the matrix. We will next describe how to use the relative location of the target elements and predecessor elements to generate an alignment.

If a predecessor element is diagonal to the target cell, there is no gap in the alignment of the sequences. If the predecessor element is not diagonal to the target cell, then a gap will be placed in one of the sequences for alignment. Arrows can be used to show the relationships between the predecessor elements and target elements. If an arrow from a target element pointing to a predecessor element does not skip a row or a column, there is no gap in the sequence alignment. If the arrow from a target element pointing to a predecessor element skips a row, a gap will be placed in Sequence 2, the horizontal sequence. If the arrow from a target element pointing to a predecessor element skips a column, a gap will be recorded in Sequence 1.

Figure 5-7C shows the scoring matrix **M** after completion of the matrix fill phase with arrows pointing from some target elements to their respective predecessor elements. The arrows are critical because they indicate those elements that contribute

Christian D. Wunsch

BOX 5-1 | **SCIENTIST SPOTLIGHT**

CHRISTIAN D. WUNSCH was enrolled in both graduate school and medical school at Northwestern University in the late 1960s. He had become fascinated with molecular orbital calculations and had an idea to analyze the interactions between serotonin analog substrates and the enzyme monoamine oxidase (MAO). He thought it would be great if he could relate the kinetic properties of MAO to the molecular orbital calculations he would compute for the serotonin analogs. To prepare himself for his Ph.D. project, he enrolled in courses to learn about quantum chemistry, linear algebra, and matrix algebra. He began to become proficient at writing Fortran subroutines to solve problems in quantum mechanics.

Wunsch happened to attend a department lecture by Saul Needleman, a professor of biochemistry at Northwestern. In his lecture, Needleman stated that it would be important to find the best match between two amino acid sequences. The problem captivated Wunsch because he realized he could cast it as a matrix problem. Once the amino acid similarity values were calculated in the matrix, he could trace a path through the matrix that minimized the number of gaps. The shortest path (or minimum sequence length) through the matrix would start at the top left corner of the matrix and end at the bottom right corner, and the path could be used to print a sequence alignment (see Figure 5-7).

It took Wunsch about a month of dedicated effort to come up with this solution. However, in order to test his algorithm he needed to purchase some computing time, which was quite expensive in those days. He approached Needleman to discuss the algorithm and ask for some computing time. Needleman agreed to the request, and, once Wunsch demonstrated that the program was robust, together they published their findings in *the Journal of Molecular Biology* in 1970.

Wunsch dropped his original thesis project of calculating molecular orbitals and refined what would come to be known as the Needleman-Wunsch global alignment program. Wunsch received his medical degree and doctorate in 1970 and entered a residency program in pathology clinical chemistry at Brigham and Young Hospital. He accepted a position at the Jackson Memorial Hospital Department of Pathology in the University of Miami and was instrumental in creating and managing the computing center at the hospital until 2002. He continues to do research in clinical chemistry.

Needleman left Northwestern University to chair the biochemistry department of Roosevelt University. He became the coordinator of scientific affairs at Abbott Laboratories and the director of clinical affairs at Schering-Plough.

values that create the shortest path through the matrix. The arrows start from the element in the bottom right corner of the matrix and proceed to the upper left corner of the matrix. The target element at the match between R and R ($M_{10,11}$) has a predecessor element at $M_{12,12}$ that creates an arrow that skips a row (see green arrow in Figure 5-7C). The sequence alignment for the small part of the sequence that follows the path of the green arrow is:

Sequence 1: RWP
Sequence 2: R-P

Rule: skipping a row results in a gap in Sequence 2 in the alignment.

If the arrow to the predecessor element skips a column, a gap will be created in Sequence 1, the vertical sequence. The target element located at the match between R and R ($\mathbf{M}_{6,6}$) has an arrow to its predecessor element at $\mathbf{M}_{7,8}$ resulting in a skipped column (see blue arrow in Figure 5-7C). The sequence alignment for the small part of the sequence that follows the blue arrow will be:

Sequence 1: R-C
Sequence 2: RQC

Rule: skipping a column results in a gap in Sequence 1 in the alignment.

Another scenario involving gap treatment is where two target elements will have the same predecessor cell. The target elements corresponding to D and D ($\mathbf{M}_{4,5}$) and N and N ($\mathbf{M}_{5,4}$) both use the predecessor element corresponding to R and R ($\mathbf{M}_{6,6}$). When this scenario is encountered it means that there are two possible sequence alignments:

Alignment 1 (row is skipped):
Sequence 1: DNR
Sequence 2: D-R

Alignment 2 (column is skipped):
Sequence 1: N-R
Sequence 2: NDR

The element values and predecessor element information that result from initiation and matrix filling will be used in the traceback phase, described below, to identify the alignments (pathways through the matrix) with the highest similarity score, and therefore an optimized global alignment that includes gaps.

Traceback

After the matrix fill phase, the maximum value within the top row or left column is the maximum score. In our example, the maximum score is 8, located at $\mathbf{M}_{1,1}$. Starting with the maximum score, you can create a sequence alignment by accounting for each predecessor element one by one until the element in the bottom row or last column is reached; this is called traceback or backtracking. The first predecessor element is highlighted in yellow in Figure 5-7C. Although Needleman and Wunsch did not print out a sequence alignment, it is easy to do so starting with the N-terminus amino acids (i.e., the first amino acids) of both sequences. Later, Peter Sellers, Temple Smith, Michael Waterman, and Osamu Gotoh significantly modified the Needleman-Wunsch program to be more efficient and account for gap penalties. For our sample sequences, there are two alignments (Figure 5-7C). If the traceback follows the top path the alignment is:

Sequence 1: ADC-DNR-CKCRWP
              | | | | | | || |
Sequence 2: AWCND-RQCLCR-P
Score: 8

If the traceback follows the bottom path the alignment is:

Sequence 1: ADCDN-R-CKCRWP
              | | | | | | ||| |
Sequence 2: AWC-NDRQCLCR-P
Score: 8

In the above example, a simplified scoring system was used for illustrative purposes (match = 1, mismatch or gap = 0). In practice, scores are from a substitution matrix such as PAM250 or a BLOSUM62. When these substitution matrices are used in conjunction with long sequences it is highly unusual to have more than one optimal alignment.

Gap Penalties

As illustrated in the N-W global alignment program, gaps are sometimes necessary to create optimal alignments. The treatment of gaps is critical because it can alter the similarity score. At the molecular level, gaps may be acceptable to optimally align sequences if the location of the gap corresponds to a segment of a protein that is not necessary to maintain tertiary structure. A gap in a protein sequence may correspond to a segment of a similar protein that is dispensable for maintenance of tertiary structure. In other words, a break in the sequence in these segments should not greatly affect a protein's function. Sometimes, these "breakable" segments are found in loops—segments that loop out of the protein and then return back to the protein. The amino acid composition and length of loops may not be conserved in orthologs from highly divergent species. Based on the mechanism of mutagenesis (see below), bioinformaticians generally recognize two types of gap penalties. One is a gap opening penalty, and the other is a gap extension penalty. The penalty for opening a gap is greater than the gap extension penalty.

The theory behind the greater penalty for gap opening is that it would require a mutation event to cause an insertion or deletion (indel mutation). Gap extension is not penalized as heavily as gap opening because "extension" is inclusive of the many cases in which more than one base is inserted or deleted in a single event. Recall in Chapter 3 that indel mutations can be caused by repeat expansion during DNA replication. Once natural selection has accepted a repeat expansion event, the length of the expansion should not be a large issue as long as the organism survives. In other words, if the range of expansion was one to three codons, there should not be a much higher penalty for the three codon expansion over the one codon expansion if nature has accepted it, because it is likely the result of the same mutation event.

In Chapter 4, we discussed scientific principles behind the derivation of the PAM and BLOSUM substitution matrices. For PAM substitution matrices, a strong evolutionary theory was used, and for BLOSUM substitution matrices sequence comparison data was used. The quantification of gap penalties has less of a scientific foundation.

gap penalty A value that is subtracted from the similarity score in alignment programs. The penalty may be linear or affine. A linear gap penalty means that the same value is subtracted for each gap added to optimize sequence alignment. An affine gap penalty, which is more commonly used in sequence alignment programs, is composed of two components: a gap opening penalty and a linear gap penalty.

We can define $W(L)$ as the penalty for a gap of length L amino acids. A simple way of calculating a gap penalty is:

$$W(L) = g*L \tag{5.1}$$

where g is the cost of a single gap.

In equation 5.1, there is a linear relationship between gap length and the penalty. This simplistic relationship does not take into account the idea that once a gap is opened (gap opening penalty) there should be a smaller penalty for gap extension. The most widely used gap penalty equation that takes this distinction into account is the affine gap penalty function. For this function, we can define the penalty for opening a new gap as g_{open}, and the penalty for extending the gap, for each subsequent increment, as g_{ext}. The affine gap penalty function is then,

$$W(L) = g_{open} + g_{ext}*L \tag{5.2}$$

From the arguments presented above, it is typical that $g_{ext} < g_{open}$. The affine gap penalty function is described in more detail in Chapter 14. For now, we can develop a general alignment scoring system for two aligned sequences. First, let us define a similarity score $S(a,b)$ for every aligned pair of amino acids a, b in two aligned sequences. Then the total similarity score would be defined by the following:

$$\text{Total similarity score} = \sum_{\substack{aligned \\ pairs}} S(a,b) - \sum_{gaps} W(L) \qquad (5.3)$$

Equation 5.3 is helpful for conceptualizing the components necessary to align sequences. However, we need a general scoring system that compares sequences and calculates similarity scores that drive the alignment. Needleman and Wunsch's global alignment program was a good start toward this goal. Other scientists improved the program.

5.4 MODIFIED NEEDLEMAN-WUNSCH GLOBAL ALIGNMENT (N-Wmod) PROGRAM WITH LINEAR GAP PENALTY

An efficient method of aligning sequences and scoring them is presented here. We call this method the modified Needleman-Wunsch global alignment (N-Wmod) program because its method of computation is different than the original.

Suppose we have two sequences with lengths n and m. Sequence 1 has length n and Sequence 2 has length m. The amino acids in Sequence 1 will be denoted $seq1_i$, where $1 \leq i \leq n$. The amino acids in Sequence 2 will be denoted $seq2_j$, where $1 \leq j \leq m$. The matrix $\mathbf{M}$ is created with $n + 1$ rows and $m + 1$ columns. For each cell in the matrix, we want to know the maximum possible score for an alignment ending at that point. To do this, a scoring system is created where $\mathbf{M}_{i,j}$ is the maximum value of three possible calculations for an element in $\mathbf{M}$. The maximum value is always placed into each location of the matrix. The formal definition of $\mathbf{M}_{i,j}$ is:

$\mathbf{M}_{i,j} = \text{MAXIMUM}\,[$

   $\mathbf{M}_{i-1,j-1} + \mathbf{S}_{seq1i, seq2j}$ (match or mismatch), diagonal move

   $\mathbf{M}_{i-1,j} + w$ (gap in Sequence 2), vertical move

   $\mathbf{M}_{i,j-1} + w$ (gap in Sequence 1), horizontal move

   $]$ $\hspace{6cm}$ (5.4)

where $\mathbf{M}_{i-1,j-1}$ is the value of the element diagonally juxtaposed to $\mathbf{M}_{i,j}$ and the $i-1, j-1$ element is up and to the left of $\mathbf{M}_{i,j}$.

where $\mathbf{S}_{seq1i, seq2j}$ is an element in the substitution matrix $\mathbf{S}$. It is the score for the match or mismatch for pairing amino acid $seq1_i$ and $seq2_j$,[5]

where $\mathbf{M}_{i-1,j}$ is the value of the element above $\mathbf{M}_{i,j}$ in the matrix,

where w is the value for the linear gap penalty, and

where $\mathbf{M}_{i,j-1}$ is the value of the element to the left of $\mathbf{M}_{i,j}$ in the matrix.

Equation 5.4 shows three options for $\mathbf{M}_{i,j}$. If the first option is the maximum score, then $\mathbf{M}_{i,j}$ equals the value of $\mathbf{M}_{i-1,j-1}$ added to the value of the amino acid pair match or mismatch. A diagonal arrow is recorded starting from $\mathbf{M}_{i,j}$ and pointing

---

[5] In the actual algorithm, the entire substitution matrix (either PAM or BLOSUM) is used as the matrix **S**. We will create a smaller **S** as a subset of the BLOSUM62 matrix to make it easier to follow this discussion.

A

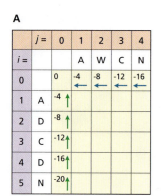

$j =$	0	1	2	3	4
$i =$		A	W	C	N
0	0	-4 ←	-8 ←	-12 ←	-16 ←
1 A	-4 ↑				
2 D	-8 ↑				
3 C	-12 ↑				
4 D	-16 ↑				
5 N	-20 ↑				

B

	A	W	C	N
A	4	-3	0	-2
D	-2	-4	-3	1
C	0	-2	9	-3
D	-2	-4	-3	1
N	-2	-4	-3	6

C

$j =$	0	1	2	3	4
$i =$		A	W	C	N
0	0	-4 ←	-8 ←	-12 ←	-16 ←
1 A	-4 ↑	4	0		
2 D	-8 ↑				
3 C	-12 ↑				
4 D	-16 ↑				
5 N	-20 ↑				

D

$j =$	0	1	2	3	4
$i =$		A	W	C	N
0	0	-4 ←	-8 ←	-12 ←	-16 ←
1 A	-4 ↑	4	0 ←	-4 ←	-8 ←
2 D	-8 ↑	0 ↑	0	-3 ↑	-3 ↑
3 C	-12 ↑	-4 ↑	-2 ↑	9	0 ←
4 D	-16 ↑	-8 ↑	-6 ↑	5 ↑	10
5 N	-20 ↑	-12 ↑	-10 ↑	1 ↑	11

E

Sequence 1: ADCDN

| · | |

Sequence 2: AWC−N

Similarity score: 11

FIG. 5-8. N-Wmod of Sequence 1, ADCDN, with Sequence 2, AWCN.
A. Matrix **M** consists of $m + 1$ rows and $n + 1$ columns, shaded yellow, where m is the number of amino acids in Sequence 1 and n is the number of amino acids in Sequence 2. Columns and rows corresponding to sequences (*white*) and i (*purple*) and j (*pink*) values are added for instructional purposes. Values for $M_{i,0}$ and $M_{0,j}$ have been added to **M** with a gap penalty value set to -4. **B.** Scores for amino acid matches and mismatches, $S_{seq1i, seq2j}$ are for the amino acids in Sequences 1 and 2. The amino acid match and mismatch scores are from the BLOSUM62 substitution matrix. **C.** Partial matrix fill where values for $M_{1,1}$ and $M_{1,2}$ have been added in the row $i = 1$. Arrows signify the three calculations performed to determine each of the maximum values for elements $M_{1,1}$ and $M_{1,2}$. Rectangle-bordered arrows point to the predecessor elements from which the maximum values for $M_{1,1}$ and $M_{1,2}$ were derived. **D.** The matrix fill is complete. Arrows point to the predecessor elements for all elements in the **M**. Circled arrows mark the path for traceback and sequence alignment. The alignment score (also known as the similarity score) is 11 and is located in the bottom right corner of **M**. **E.** The alignment of Sequence 1 with Sequence 2 printed from the traceback.

to $M_{i-1, j-1}$. If the second option is the maximum score, then $M_{i,j}$ equals the value of $M_{i-1, j}$ added to the gap penalty value. A vertical arrow is recorded starting from $M_{i,j}$ and pointing to $M_{i-1, j}$. If the third option is the maximum score, then $M_{i,j}$ equals the value of $M_{i, j-1}$ added to the gap penalty value. A horizontal arrow is recorded starting from $M_{i,j}$ and pointing to $M_{i, j-1}$. Equation 5.4 says that the value for $M_{i,j}$ in the matrix requires that we know the elements of the three neighbors located diagonally, vertically, and horizontally to $M_{i,j}$.

The equation is used repeatedly until all elements in the matrix are calculated. The diagonal, vertical, and horizontal tag associated with each $M_{i,j}$ is recorded and will be used to create the traceback path and sequence alignment.

N-Wmod Initialization

In the N-Wmod program, we will eventually fill the scoring matrix proceeding from the top left to the bottom right. To begin initialization, we need to know the element values in the first row (i) and first column (j). Recall that matrix **M** is created with $n + 1$ rows and $m + 1$ columns. We define the first row as $i = 0$ and the first column as $j = 0$. We define the following equations:

$$M_{i,0} = w*i \tag{5.5}$$

$$M_{0,j} = w*j \tag{5.6}$$

where $M_{i,0}$ refers to elements in the first column and $M_{0,j}$ refers to elements in the first row,

where $w*i$ is the linear gap penalty multiplied by the index, i, and

where $w*j$ is the linear gap penalty multiplied by the index, j.

Now let's align the two following sequences:

Sequence 1: ADCDN (**vertical sequence**)

Sequence 2: AWCN (**horizontal sequence**)

In the first column, Sequence 1 amino acids align with none of the amino acids in Sequence 2 (Figure 5-8A). In the first row, Sequence 2 amino acids align with none of the amino acids in Sequence 1. The value of $M_{0,0}$ is 0 because $w*i = 0$ and $w*j = 0$. The value of each successive element in $M_{i,0}$ and $M_{0,j}$ is equal to the gap penalty multiplied by the value of i or j. If we choose a gap penalty of -4 ($w = -4$), then each successive vertical element in $M_{i,0}$ will be equal to $w*i$ ($-4, -8, -12$, etc.) for $0 \leq i \leq n$. This is the same as saying that when three amino acids (A, D, C) of Sequence 1 are aligned with gaps, the score will be -12. Similarly, each successive horizontal element in $M_{0,j}$ will be equal to $w*j$ for $0 \leq j \leq m$. We will keep track of the predecessor elements with vertical and horizontal arrows that point to the elements from where the $M_{i,0}$ values and $M_{0,j}$ values are derived (Figure 5-8A).

N-Wmod Matrix Fill

In the matrix fill phase, we will calculate the values of $M_{i,j}$. These elements will depend on scores for matches or mismatches for amino acid pairs. We will use scores from the BLOSUM62 substitution matrix (see Chapter 4) and store them in substitution matrix S with elements $S_{seq1i, seq2j}$ for each possible amino acid pair in Sequence 1 and Sequence 2 (Figure 5-8B). We are now ready to complete the matrix fill for M. For the element $M_{1,1}$ we must consider three options. For option 1, $M_{0,0} + S_{A,A}$, the value of $M_{0,0}$ is 0 and the value of $S_{A,A}$ is 4. The total value for option 1 is $0 + 4 = 4$. For option 2, $M_{0,1} + w$, the value of $M_{0,1}$ is -4 and the value of w is -4. The total value for option 2 is $-4 + -4 = -8$. For option 3, $M_{1,0} + w$, the value of $M_{1,0}$ is -4 and the value of w is -4. The total value for option 3 is $-4 + -4 = -8$. Because option 1 gave the maximum value we choose this option for the value of $M_{1,1}$ (Figure 5-8C). We will record option 1 by placing a rectangle around the diagonal arrow that points to the predecessor of $M_{1,1}$. The diagonal arrow keeps track of the predecessor element that contributed to $M_{1,1}$. Let's proceed to the next element in the same row, $M_{1,2}$. The three options for this element are as follows:

$$M_{0,1} + S_{A,W} = -4 + -3 = -7$$

$$M_{0,2} + w = -8 + -4 = -12$$

$$M_{1,1} + w = 4 + -4 = 0$$

Because option 3 gives the maximum score, we choose a horizontal value for our calculation and place a rectangle around the horizontal arrow. We repeat this process for all elements in the matrix, being careful to keep track of predecessor elements (Figure 5-8D).

N-Wmod Traceback

Once matrix M is filled, we locate the element corresponding to the bottom right corner of the matrix, $M_{n,m}$. This element is $M_{5,4}$ and the similarity score (also called the alignment score) is equal to the value of this element, 11. Starting from this element, we traceback to element $M_{0,0}$. The traceback follows the path of the circled arrows (i.e., pointing to the predecessor elements) and, like the Needleman-Wunsch (N-W) program, denotes the optimal path. The sequence alignment is printed beginning with the N-terminus amino acids at $M_{1,1}$ and proceeding to the C-terminus amino acids at $M_{5,4}$. If a diagonal arrow starts at an element, then the corresponding amino acid pair (i, j) from Sequence 1 and Sequence 2 is printed. If a vertical arrow starts at an element, then the amino acid in Sequence 1 aligns with a gap. If a horizontal arrow starts at an element, then the amino acid in Sequence 2 aligns with a gap (Figure 5-8E).

This similarity score was produced using a linear gap penalty of -4 and a BLOSUM62 substitution matrix for amino acid matches and mismatches. If we had used a linear gap penalty of -6 we would have obtained the same alignment, but the similarity score would have been 9. Also, note that there may be more than one path that leads to the similarity score located at M. In other words, an element in M with a maximum value could have been obtained from two predecessor cells (for example, diagonal and vertical). In such cases, the N-Wmod program is usually designed to choose only one predecessor cell, and only one traceback path would be obtained. However, more advanced programs can be written to account for multiple traceback paths and print all alignments that give the same similarity score.

5.5   ENDS-FREE GLOBAL ALIGNMENT

The N-Wmod program works well for sequences that are similar throughout their entire lengths. One drawback is that the ends of the sequences, if they do not have identical lengths, can significantly lower the similarity score because they are given a

A

	A	W	C	N
T	0	-2	-1	0
H	-2	-2	-3	1
A	4	-3	0	-2
D	-2	-4	-3	1
C	0	-2	9	-3
D	-2	-4	-3	1
N	-2	-4	-3	6

B

C

Sequence 1: THADCDN

         | . | :

Sequence 2: --AWCN-

Similarity score: 10

FIG. 5-9. Ends-free global alignment of Sequence 1, THADCDN, with Sequence 2, AWCN.
A. Scores for amino acid matches and mismatches, $S_{seq1i, seq2j}$, for the pairs in Sequences 1 and 2. These scores are from the BLOSUM62 substitution matrix. The linear gap penalty is -6 per increment. **B.** The matrix fill is complete. Arrows point to the predecessor elements for all elements in the matrix. Circled arrows mark the path for traceback and sequence alignment. The similarity score is the maximum score in the bottom row or last column. **C.** The alignment of Sequence 1 with Sequence 2 printed from the traceback starting at $M_{0,0}$. The program must be adjusted to include the unpaired C-terminal amino acids to be aligned with gaps without penalty. In this case, the amino acid N at the C-terminus in Sequence 1 is unpaired.

local alignment Optimal pairing of two subsequences within two sequences such that the similarity score remains above a set threshold.

gap penalty for each amino acid. If two sequences have significantly different lengths, the similarity score could be quite low even though they are identical over a large segment of the alignment. The N-Wmod program can be modified to create what is known as an ends-free global alignment program (also known as semiglobal alignment program). In this modification, the $M_{i,0}$ and $M_{0,j}$ values are set to 0 for $0 \leq i \leq n$ and $0 \leq j \leq m$. This means that there is no penalty when the sequences overlap with no amino acids at the N-terminus or C-terminus. The key is that the traceback starts at the highest value in the bottom row or in the last column. We will use the following sequences to illustrate this:

Sequence 1: THADCDN

Sequence 2: AWCN

We will use match and mismatch scores from BLOSUM62 (Figure 5-9A) and a linear penalty -6 ($w = -6$) to create matrix **M**. The traceback begins at the maximum value in the bottom row or the last column. In our example, the traceback begins at $M_{6,4}$ where the similarity score is 10. Then the traceback follows the path depicted in Figure 5-9B until the element $M_{0,0}$ is reached.

5.6 LOCAL ALIGNMENT PROGRAM WITH LINEAR GAP PENALTY

Global alignment programs align two sequences from the N-terminus to the C-terminus. We reviewed the Needleman-Wunsch program, the N-Wmod program and the ends-free global alignment program. For proteins that are very similar over their entire sequences, these are useful programs, but suppose this is not the case. Instead, they may be similar over a single domain. Such similarities are thought to arise through duplication and exon shuffling, and are indicative of the modularity of proteins (see Chapter 3). Global alignment programs may not detect small, local segments that are similar. Therefore, local alignment programs have been developed to "catch" such similar segments in large sequences and ignore parts of sequences that are not similar.

In 1981, Temple Smith (Northern Michigan University) and Michael Waterman (Los Alamos Scientific Laboratory) developed an elegant local alignment program. To achieve local alignment with the Smith-Waterman program, we need to make three changes to the N-Wmod program. First, we modify equation 5.4 to include a fourth option, 0, for the calculation of $M_{i,j}$. This means that no $M_{i,j}$ element will be less than 0.[6] Second, traceback starts from the highest value of $M_{i,j}$ anywhere within the matrix and continues through predecessor elements until 0 is encountered. Once 0 is encountered, the traceback stops, which marks the end of the sequence alignment.[7] Third, the elements in the top row ($i = 0$) and leftmost column ($j = 0$) are all set to 0 for $0 \leq i \leq n$ and $0 \leq j \leq m$. We have already discussed this third change in our discussion of ends-free global alignment.

---

[6] The addition of $M_{i,j} = 0$ as one of the possible definitions of $M_{i,j}$ gives the option of creating a short (local) pairwise alignment within two larger sequences because the alignment stops once a predecessor element with a value of 0 is obtained.

[7] It is possible to create more than one alignment between two sequences. You can set a threshold value for starting traceback. Then the program will start tracebacks at elements in the matrix with values above the threshold value.

A

	H	E	A	G	A	W	G	H	E	E
P	0	0	-1	-2	-1	-3	-2	0	0	0
A	-2	-1	5	0	5	-2	0	-2	-1	-1
W	-3	-3	-2	-2	-2	15	-2	-3	-3	-3
H	10	0	-2	-2	-2	-3	-2	10	0	0
E	0	6	-1	-2	-1	-3	-2	0	6	6
A	-2	-1	5	0	5	-2	0	-2	-1	-1
E	0	6	-1	-2	-1	-3	-2	0	6	6

B

		H	E	A	G	A	W	G	H	E	E
	0	0	0	0	0	0	0	0	0	0	0
P	0	0	0	0	0	0	0	0	0	0	0
A	0	0	0	5	0	5	0	0	0	0	0
W	0	0	0	0	3	0	20	12	4	0	0
H	0	10	2	0	0	1	12	18	22	14	6
E	0	2	16	8	0	0	4	10	18	28	20
A	0	0	8	21	13	5	0	4	10	20	27
E	0	0	6	13	19	12	4	0	4	16	26

E

Sequence 1: AW-HE

       || ||

Sequence 2: AWGHN

Similarity score: 28

  % similarity: 4/5 x 100 = 80

FIG. 5-10. Local alignment of Sequence 1, PAWHEAE, with Sequence 2, HEAGAWGHEE.
A. Scores for amino acid matches and mismatches, $S_{seq1i,\,seq2j}$, for the pairs in Sequences 1 and 2. These scores are from the BLOSUM45 substitution matrix. We will use a linear gap penalty of -8 for this alignment. **B.** The matrix fill is complete. Circled arrows point to predecessor elements for the traceback in the matrix beginning with the highest value, 28 (shaded green), and ending with 0 (shaded pink). The traceback is used for printing the sequence alignment. **C.** The alignment of Sequence 1 with Sequence 2 printed from the traceback, the similarity score, and the percent similarity.

For the local alignment program, the definition of the element $\mathbf{M}_{i,j}$ in matrix $\mathbf{M}$ is:

$$\mathbf{M}_{i,j} = \text{MAXIMUM} \Big[$$

$\mathbf{M}_{i-1,j-1} + \mathbf{S}_{seq1i,\,seq2j}$ (match or mismatch), diagonal move

$\mathbf{M}_{i-1,j} + w$ (gap in Sequence 2), vertical move

$\mathbf{M}_{i,j-1} + w$ (gap in Sequence 1), horizontal move

0

$$\Big] \qquad\qquad (5.7)$$

We will use the following the sequences to demonstrate local alignment:

Sequence 1: PAWHEAE

Sequence 2: HEAGAWGHEE

We will use a linear gap penalty of -8 ($w = -8$) and amino acid match and mismatch scores from the BLOSUM45 substitution matrix (Figure 5-10A). After matrix fill, we find that the maximum value in $\mathbf{M}$ is 28 (Figure 5-10B). This is the similarity score of the best local alignment. Traceback starts at the element with the maximum value and is followed through predecessor elements until an element with the value 0 is reached. The sequence alignment is printed starting from the amino acids corresponding to the element that points to the element with a value of 0 through the traceback path until it reaches the amino acid pair corresponding to the element with the maximum value. In this case, the sequence alignment would be printed out starting from $\mathbf{M}_{2,5}$ and ending at $\mathbf{M}_{5,9}$ (Figure 5-10C). The percent similarity is calculated by dividing the amino acid pairs in the alignment with positive scores by the number of amino acids and gaps in the alignment and multiplying by 100.

SUMMARY

This chapter discussed the intersection between biological data and a few of the programs designed to quantify and display this data. Sliding window programs quantify data associated with nucleotides and amino acids and display that data in graphical format. The percent G plus C and Kyte-Doolittle programs are simple, yet powerful, methods for visualizing molecular biology phenomena. We underscored the importance of experimenting with different window sizes, which are used to reduce background or increase detail. Sliding window programs may be used to generate a dot plot—a visualization tool that helps the bioinformatician readily detect areas of similarity between two sequences. The output of the dot plot depends on the threshold setting for display of the dots, the window size, and the substitution matrix (see exercise 4 at the end of this chapter).

Dot plots display the alignment of two sequences in a plot that resembles a matrix, with one sequence displayed on the vertical axis and the other on the horizontal axis. These two sequences could then be aligned using a scoring matrix. The first scoring matrix was created by Needleman and Wunsch, who used it to calculate a similarity score of two amino acid sequences. Further development of the scoring matrices led to powerful global and local alignment computer programs that, today, constitute a major foundation in the field of bioinformatics.

Chapter 5 is a jumping off point for one of three chapters: Chapter 6, Chapter 11, or Chapter 13. In Chapter 6, we will discuss how efficient computer programs such as BLAST perform local alignments between a query sequence and a large database. We will also discuss how a sequence can be efficiently aligned with several sequences (multiple sequence alignment). Multiple sequence alignments can help you detect regions of sequences that are functionally, structurally, and evolutionarily conserved. In Chapters 11 and 12, we delve into basic probability and discuss the foundations of the *E*-value and hidden Markov models. In Chapter 13, we get down to the nuts and bolts of computer programming. We will present you with the tools of the computer language Python and assist you to create your own Kyte-Doolittle computer program. The programming exercises will build your skill level so that you can eventually create your own sequence alignment program (see Chapter 14).

EXERCISES

1. Create a %(G+C) plot of the sequence GAACTCAT-ACGAATTCACGTCAGCCCATCGTGCCACGT. On the *y* axis will be %(G+C) and on the *x* axis will be the nucleotide sequence number. Use a sliding window of 3 nucleotides and slide the window 1 nucleotide at a time. Calculate the %G+C as a function of nucleotide sequence number. You may use a spreadsheet program to create the plot. Change the sliding window to 5 nucleotides and create a second plot. Overlap the two plots. Explain any differences in the two plots.

2. Given the following sequence, PLSQETFSDLWKLL-PENNVLSP, use the Kyte/Doolittle Hydropathy scale and a window size of 7 to construct a hydropathy plot (calculate average hydrophobicities in the window). You may use an online spreadsheet program for this exercise.

3. Download the amino acid sequence of bacteriorhodopsin from GenBank (accession number 1O0A_A). Use a spreadsheet program to plot the hydropathy of the protein as a function of sequence number with a window size of 7. Compare your plot to the plot in Figure 5-2.

4. Download the Dotter software program from the textbook website to perform the following exercises.

   a. Plot wild-type human p53 against itself. You will find one main diagonal line. In addition, you will detect a series of faint shorter lines near the main diagonal in the 70–93 amino acid region. What do the faint diagonal lines signify? How does the program generate these lines?

   b. Obtain five homologs of p53 from species that are evolutionarily distant from human. Plot each of the homologs against human p53. Use the cursor to find the starting amino acid and ending amino acid of the domains that are similar in each plot. Always use the human p53 sequence numbers as a reference for reporting similar domains. Report the shortest domains that are shared between human p53 and all other species (the minimum conserved regions). From the literature or a reliable annotated database, describe the functions of the conserved regions.

   c. Use Dotter to compare human p53 with squid (*Loligo forbesii*) p53 protein sequences. Change the substitution matrix from default (BLOSUM62) to BLOSUM35. Is there a difference in the plot appearance? Explain why. Now compare the plot obtained with BLOSUM62 to one obtained with BLOSUM100. Is there a difference in the plot appearance? Explain why. You may download the substitution matrices from the textbook website.

d. Use Dotter to compare human p53 with mouse p53 (use the longest isoforms of these proteins). Use a window size of 5 for one plot and a window size of 25 for another plot. Is there a difference in the plot appearance? Explain why. → window size

e. What does the GreyRamp tool do in Dotter?

5. **Use the N-Wmod method to align ATATGC and ATATGA. Use the following matrix for this alignment: Match = 1, Mismatch = 0, Gap = 0. Report the score.**

6. **Use the N-Wmod method to align APVEEDFI and PPVQQDHT. Use the BLOSUM62 substitution matrix as a source for match and mismatch scores and a linear gap penalty of −6. Report the similarity score and percent similarity.**

7. **Use the ends-free method to align the sequences APVEEDFI and PTVQQDHT. Give the similarity score. Use the BLOSUM62 substitution matrix and a linear gap penalty of −3.**

8. **Use the local alignment method to align the following sequences:**

Sequence 1: VEPPLSQE

Sequence 2: ELPLC

Use a linear gap penalty of −3 and the BLOSUM62 substitution matrix as a source for match and mismatch values.

REFERENCES

Baxenavis, A. D., and B. F. F. Ouellette, eds. 2004. *Bioinformatics: A Practical Guide to the Analysis of Genes and Proteins*, 3rd ed. New York: Wiley.

Baylin, S. B., J. G. Herman, J. R. Graff, P. M. Vertino, and J. P. Issa. 1998. "Alterations in DNA Methylation: A Fundamental Aspect of Neoplasia." *Advances in Cancer Research* 72: 141–196.

Gotoh, O. 1982. "An Improved Algorithm for Matching Biological Sequences." *Journal of Molecular Biology.*" 162: 705–708.

Higgs, P. G., and T. K. Attwood. 2005. *Bioinformatics and Molecular Evolution*. Malden, MA: Blackwell Publishing.

Kyte, J., and R. F. Doolittle. 1982. "A Simple Method for Displaying the Hydropathic Character of a Protein." *Journal of Molecular Biology* 157: 105–132.

Needleman, S. B., and C. D. Wunsch. 1970. "A General Method Applicable to the Search for Similarities in the Amino Acid Sequence of Proteins." *Journal of Molecular Biology* 48: 443–453.

Pevsner, J. 2009. *Bioinformatics and Functional Genomics.* Hoboken, NJ: John Wiley and Sons.

Sellers, P.H. 1974. "On the theory and computation of evolutionary distances." *SIAM Journal on Applied Mathematics* 26: 787–793.

Smith, T. F., and M. S. Waterman. 1981. "Identification of Common Molecular Subsequences." *Journal of Molecular Biology* 147: 195–197.

Sonnhammer, E. L., and R. Durbin. 1995. "A Dot-Matrix Program with Dynamic Threshold Control Suited for Genomic DNA and Protein Sequence Analysis." *Gene* 167: GC1–10.

Zvelebil, M., and J. O. Baum. 2008. *Understanding Bioinformatics*. New York: Garland Science, Taylor & Francis Group.

BASIC LOCAL ALIGNMENT SEARCH TOOL (BLAST) AND MULTIPLE SEQUENCE ALIGNMENT

LEARNING OUTCOMES

AFTER STUDYING THIS CHAPTER, YOU WILL:

- **Understand the logic of how BLAST aligns a query sequence with a subject sequence from a database.**

- **Understand that the *E*-value is a measure of the false positive rate of a similarity score.**

- **Know the major BLAST programs available through NCBI.**

- **Know that the PSI-BLAST program can be used to identify distantly related sequences through the use of an iteration of BLAST runs that utilize position specific substitution matrices.**

- **Know that multiple sequence alignment programs can be used to identify conserved regions within a gene or protein.**

- **Understand the logic of how ClustalW aligns multiple sequences.**

- **Become familiar with other multiple sequence alignment programs.**

6.1   INTRODUCTION

The Needleman-Wunsch and Smith-Waterman programs discussed in Chapter 5 are useful for aligning two sequences and for searching a small sequence database for sequences that are similar to a short query sequence. In the 1990s, advances in DNA sequencing technology led to a significant expansion of the number of sequences deposited in databases. This data explosion required the development of "shortcut" programs (also known as heuristics) that reduce the computing time necessary to perform large database searches with long query sequences. The most popular shortcut program is the Basic Local Alignment Search Tool (BLAST), which is available from the National Center for Biotechnology Information. First introduced in 1990, BLAST captures database sequences that meet a threshold for very short sequences within the query. Once the threshold is met, the alignment between the query and the database sequence is extended until the score drops below a "dropoff" threshold. Two significant advantages of BLAST over other sequence alignment programs are its increased speed and its use of a statistical measurement, the *E*-value, to assess the significance of the similarity score. Because BLAST is a widely used program, we will delve into the details of how it works in this chapter.

Another important technique crucial to many bioinformatics projects is alignment of more than two sequences, called multiple sequence alignment. Multiple sequence alignment is the alignment of three or more protein or nucleic acid sequences. The rationale for creating multiple sequence alignments is actually rooted in the evolutionary process. In a simplistic view, the evolutionary relationship of organisms on the phenotypic level has its underpinnings in the genomes. For example, humans look more similar to chimpanzees than to frogs, leading you to expect that human and chimpanzee DNA would be more similar than human and frog DNA. Although this is certainly true, you will find quite a bit of variability in the degree of similarity among different ortholog (see Chapter 3) gene sets. Even within a particular set of orthologs, certain regions within genes are more conserved (i.e., more similar across orthologs) than others. In other words, these conserved regions have not significantly changed over millions of years of evolution.

Regions are often conserved because they code for parts of proteins that create structures necessary for protein function. Conserved regions often give insight into the functions of proteins, and these regions may be useful targets for drugs if the functions are involved in human disease. These considerations have led to an effort to tease out conserved regions within proteins, and multiple sequence alignment is the best way to identify such regions. A multiple sequence alignment of p53 orthologs from vertebrates shows that the region spanning amino acid residues 100 through 300 is clearly more conserved than other regions of the protein. It turns out that this region is necessary for binding to DNA, a function required for its tumor suppressor function. As mentioned in Chapter 3, it is this DNA binding domain coded by the *TP53* gene that is often mutated in human cancers. One popular multiple sequence alignment program is Cluster Alignment Weighted (ClustalW). As a matter of fact, the paper describing ClustalW has been cited more than 31,000 times, making it one of the most cited papers in bioinformatics. You will spend some time getting acquainted with this classic bioinformatics software program, and then learn about some newer multiple sequence alignment programs.

BLAST (Basic Local Alignment Sequence Tool) A popular heuristic program that rapidly compares a query sequence to a subject sequence.

multiple sequence alignment (MSA) Alignment of more than two protein or nucleotide sequences. MSA is useful for detection of conserved regions of proteins that may be of functional and structural significance.

ClustalW A popular multiple sequence alignment program that creates a rooted tree from a distance matrix. The rooted tree, called a guide tree, is used to determine the order of progressive pairwise alignments. The sequence at the tip of the longest branch is added last to the alignment. Through clustering, an intermediate unrooted tree is created. Sequences are associated with different weights depending on the length of the branches that connect them to the guide tree.

6.2 THE BLAST PROGRAM

Often, the first analysis a scientist conducts with a newly sequenced gene is to "BLAST it." The gene's coding sequence (CDS) is pasted into the query box in the software program and the scientist performs a BLAST analysis against the non-redundant sequence database or the UniProtKP (both databases are described in Chapter 2). These are secondary databases that contain all CDS regions translated from GenBank plus protein sequences from protein databases. Within a few minutes, the scientist receives an output of accession numbers, in decreasing rank of similarity, along with their associated sequences aligned with the query sequence. Figure 6-1 shows an example of a BLAST output using human p53 as the query sequence that was run against the UniProtKP, where the database was restricted to proteins from

A Distribution of 17 blast hits on the query sequence

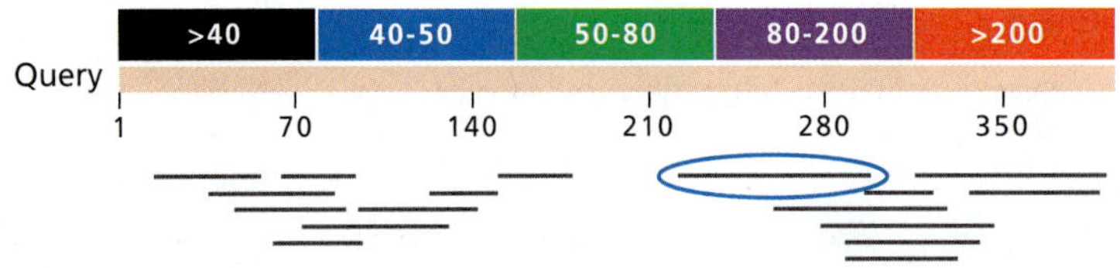

	>40	40-50	50-80	80-200	>200

Query

```
1        70      140      210      280      350
```

(handwritten annotations: "% of two sequences that aligned", "# of hits due to chance (↓ = better)")

B Sequences producing significant alignments

Description	Max score	Total score	Query cover	E value	Ident	Accession
RecName: Full=Transcription factor cep-1; AltName: Full=C elegans p53-like protein 1	34.3	34.3	19%	0.020	31%	Q20646_2
RecName: Full=Zinc finger protein lin-13: AltName: Full=Abnormal cell lineage protein 13	28.1	28.1	19%	1.6	32%	Q11101_2
RecName: Full=Chromatin remodeling complex ATPase chain isw-1; AltName: Full=Nucleosome remodeling factor subunit isw-1	28.1	28.1	10%	1.8	36%	P41877_2
RecName: Full=Uncharacterized protein C27D6_1	26.9	26.9	21%	4.2	25%	Q10030_1

C

RecName: Full=Transcription factor cep-1; AltName: Full=C elegans p53-like protein 1
Sequence ID: spIQ2064.2ICEP1_CAEEL Length:644 Number of matches:1
Range 1:337 to 427 GenPept Graphics

Score	Expect	Method	Identities	Positives	Gaps
34.3 bits (77)	0.020	Compositional matrix adjust.	28/91 (31%)	43/91 (47%)	14/91 (15%)

```
Query    224  EVGSDCTTIHY----------NYMCNSSCMGGMNRRPILTI-ITLEDSSGN-LLGRNSFE  271
              E  GS T IY          +MC   C+   +RR + + + L+D +GN +L       +
Subject  337  EKGSTFTLIMYPGAVQANFDIIFMCQEKCLDLDDRRKTMCLAVFLDDENGNEILHAYIKQ  396

Query    272  VRVCACPGRDRRT--EEENLRKKGEPHHELP  300
              VR+ A P RD +   E E+ ++K    ELP
Subject  397  VRIVAYPRRDWKNFCEREDAKQKDFRFPELP  427
```

FIG. 6-1. BLAST output of human p53 query alignment (NP_000537.3 accession number) to the UniProtKP/Swiss-Prot database (database restricted to the species *C. elegans*).
A. Cartoon showing the alignment score color key, with a higher score being a higher quality match and the length of the human p53 query (393 residues). The black lines, signifying low quality matches, underneath the query represent the alignments of all 17 hits from the database. The circled black line corresponds to the top hit. Note that all 17 hits produce scores of less≈than 40 bits (see Chapter 12 for an explanation of the bit score). **B.** The second part of the output gives information on the top four hits (this was truncated to save space). The top hit is a transcription factor named cep-1, a known homolog of human p53. The score corresponding to the top hit is 34.3 bits (here named max score). The total score is the sum of the max scores for two or more segments within the query aligned with one subject. The alignment of human p53 and cep-1 consists of 19% of the 393 residues of human p53. The *E*-value (or Expect value) is 0.020, and the percent identity shared within the aligned segments is 31%. The accession number of cep-1 within the UniProtKP/Swiss-Prot database is Q20646.2. **C.** The third part of the BLAST output shows, in addition to the information displayed in the second part, the raw score (77) and the method of compositional matrix adjustment, which takes into account the amino acid residue frequency difference between humans and *C. elegans*. The Identities column shows the number of identical amino acids shared within the alignment divided by the number of amino acids and gaps in the paired alignment. The Positives column consists of the sum of the number of similar and identical amino acids divided by the number of amino acids in the paired alignment. The Gaps column shows the number of amino acids that align against a gap divided by the number of amino acids in the paired alignment. The BLOSUM62 substitution matrix was used to compute similarity scores. An Affine Gap Penalty is used to score the alignment: the gap opening penalty is −11, and the gap extension penalty is −1.

the worm *C. elegans*.[1] In the output, the hits near the top are more similar to the query sequence than those near the bottom. The top hit is cep-1, a protein known to be the p53 ortholog in *C. elegans*.[2] How does BLAST perform this feat in just a few minutes? How do we know that the top ranked aligned protein is truly biologically related to the query sequence? Can BLAST be used for other types of analyses? As we describe BLAST, these questions will be answered.

Our discussion of BLAST will be largely limited to protein sequences, but bear in mind BLAST can also be used to perform database searches with nucleotide sequences (this version of BLAST is called blastn). When BLAST compares a query sequence to database sequences, it generates high-scoring segment pairs (HSPs). They are high-scoring because the alignment produces similarity scores that are above an alignment threshold *A* either preset by the BLAST program or set by the user. The "segment pair" refers to the two stretches of amino acids that are aligned. The length of aligned sequences will be extended as long as the extension improves the similarity score *s*. Once *s* is maximally improved in a local area, it may be reported to the user if its value is above the alignment threshold *A*.[3] The significance of the similarity score for a segment pair is computed by calculating its *E*-value. The *E*-value is used as a measure to assess whether similarity score *s* could have been obtained by random chance (see equation 6.1 later in this chapter).[4] The lower the *E*-value, the lower the chance that the similarity score *s* was obtained by random chance. The alignment threshold *A* for HSPs is actually set by a cutoff *E*-value. For protein BLAST (called BLASTp) offered by the National Center for Biotechnology Information, the default cutoff *E*-value is 10, but this value can be altered by the user.

E-value For a given similarity score *s*, the expected number of HSPs with at least score *s*, assuming random sequences. Used as a measure to assess whether similarity score *s* could have been obtained by random chance.

high-scoring segment pair Two amino acid or nucleotide sequences that, upon alignment, produce a similarity score *s* above some alignment threshold *A*.

Four Phases in the BLAST Program

There are four major phases in the BLAST program: (1) compiling a list of high-scoring words (short sequences of defined length), (2) scanning the database for hits, (3) extending the hits, and (4) trimming the segment pair to the maximum score. Figure 6-2 summarizes the four phases of BLAST. In the high-scoring word compilation phase, each amino acid in the query sequence is taken in the context of its surrounding amino acids to create *query* words. In BLAST, query words with a length of three amino acids are created from the query sequence.[5] There are $n - w + 1$ words in the query sequence, where n = the length of the query sequence and w = length of the word. In a 10-amino acid query there are 8 query words, each with a length of 3 amino acids. In the query sequence RCPHHERCSD, these 8 query words are RCP, CPH, PHH, HHE, HER, ERC, RCS, and CSD.

The query words are compared to other amino acid words of length 3 in a table containing all other possible words. That means the table contains 20^3 words, or 8,000 words.[6] A similarity score *s* is obtained by comparing one query word with all the words in the table. A score for each match is obtained by consulting the BLOSUM62 substitution matrix (see Chapter 4 for discussion of BLOSUM62 substitution matrix). If the score is above a certain word threshold (*T*), then that word and its associated score are placed into a list of acceptable, or "neighborhood" words. These words are acceptable because they represent the query word plus conservatively substituted variations of the query words. This list of acceptable words

trimming Shortening of aligned sequences of extended hits until the maximum similarity score is obtained.

query words Short segments of a query sequence used to create a hit in BLAST.

= windows

neighborhood words A list of words that are similar to or identical to the query words. The words in the list have, upon alignment to the query words, similarity scores above a word threshold set by the user or the BLAST program.

---

[1] The BLAST program implemented by the NCBI allows the option to restrict the database so that sequences of only one species are searched. For this particular search, the *E*-value cutoff for reporting results to the user was set to 20.

[2] Derry, Putzke, and Rothman (2001).

[3] To be reported to the user, the *E*-value of the alignment must fall within a top range set by the user. So, if the user sets the program to report back the top 50 hits and this particular hit has an *E*-value that falls within this range, the user will receive output.

[4] A mathematical derivation of *E*-value is presented in Chapter 12.

[5] The length of the words can be altered by the user if desired.

[6] See Exercise 11.4.

Phase 1: Compile a list of high-scoring words at or above word threshold T (T=17).

Query sequence: human p53: ...RCPHHERCSD...

Query words derived from query sequence: RCP,CPH,PHH,HHE,HER,ERC,RCS,CSD

List of acceptable words above word threshold T for the query word RCP:

Word	Scores from BLOSUM62 substitution matrix	Total score
RCP	5 + 9 + 7	21
KCP	2 + 9 + 7	18
QCP	1 + 9 + 7	17
ECP	0 + 9 + 7	16
-	-	-
-	-	-
-	-	-

Note: dotted line is the word threshold where T=17

Phase 2: Scan the database for short segments that match the list of acceptable words with scores above or equal to word threshold T.

Phase 3: Extend the hits (see arrows below) and terminate when the tabulated score exhibits significant decay.

```
                ←———    ———→
Query    EVVRRCPHHERCSD
         EVVRRCPHHER S+
Sbjct    EVVRRCPHHERSSE      (Chinese hamster p53 009185)
```

Note: exact matches are shown in the middle sequence. The absence of a character in the middle sequence indicates that the amino acid pair has a non-positive score. A "+" in the middle sequence indicates an amino acid pair with a positive similarity score. The Sbjct sequence is the subject sequence which was detected from the database as a hit.

Phase 4: The sequence alignment is trimmed back until the maximum score of the segment pair is retrieved. The segment pair is a high scoring pair (HSP). The sequence alignment of the HSP and its score s may then be reported to the user.

FIG. 6-2. Four phases of BLAST sequence alignment algorithm.

alignment score threshold
A similarity score used by the BLAST program to determine whether that score is to be reported to the user.

significant decay A decrease in the similarity score created by extending the hit further than the maximum similarity score.

maximum similarity score The highest similarity score created by extending the hit by the BLAST program.

contains all words associated with the query words with s scores above the word threshold T.

The sequence database is scanned to detect exact matches between the list of acceptable words and words in the sequence database. If an amino acid word of length 3 in the database matches an acceptable word, then a "hit" is registered. Two nonoverlapping "hits" must be found within a certain distance of each other for the alignment to be considered further (note that Figure 6-2 shows only one hit). At this point, the hit is extended in both directions, and the alignment is scored as it is extended. As shown in Figure 6-3, the extension continues as matches outweigh mismatches, and the cumulative score remains above the word threshold T. Another threshold, called the alignment score threshold (A), is preset by the BLAST program. If extension of the alignment gives a similarity score that is higher than A, the similarity score may be reported to the user.[7] In the web-based version of BLAST from NCBI, the default number of hits above A reported to the user is 100; however, this number may be adjusted.

Of course, extending the alignment too far will often result in too many mismatches and the insertion of gaps, causing the similarity score to fall. If the similarity score drops significantly relative to the maximum similarity score, the extension will terminate. This drop is called significant decay and is triggered when the score for a segment pair falls a certain distance below the best score obtained from shorter

---

[7] See footnote 3 regarding the output report to user.

extensions (Figure 6-3). Once the extension is terminated, the alignment is trimmed back until the maximum score is reached. The maximum similarity score along with the sequence alignment associated with the score is reported to the user as a high-scoring segment pair (HSP). Other HSPs with similarity scores and associated sequence alignments may be reported to the user as long as the scores are higher than the threshold *A*.

How Does BLAST Account for Gaps?

When you use BLAST there will be situations in which BLAST must insert gaps into either the query or the subject sequence to create an optimal alignment. However, if the gaps are too wide, BLAST will terminate extensions of hits due to significant decay. To demonstrate how BLAST takes gaps into account when hits are extended, a dot plot is useful. Recall the Dotter program discussed in Chapter 5. The Dotter program aligns two sequences, one displayed horizontally on top of the dot plot and the other displayed vertically on the left of the dot plot. The diagonals in the plot area represent matches between the horizontal and vertical sequences. Let's use the dot plot format to help us visualize the steps of BLAST sequence alignment. The final step of BLAST alignment is called local gapped alignment.

Figure 6-4A shows an empty dot plot, the query sequence on the vertical axis, and the protein database to be searched on the horizontal axis. Assume that the individual records in the database are sewn together to create one long unbroken chain of amino acids. Figure 6-4B shows four three-letter query words created from the first 6 amino acids of the query sequence on the vertical axis. Two of the three letters in the query words overlap each other. In addition, there are neighborhood words (i.e., words that score above the word threshold *T*) that are used to find matches between the query sequence and the database, but for simplicity these neighborhood words are not shown. Short diagonals displayed in the dot plot indicate the locations where hits are created between the database and the query words. Figure 6-4C shows the extension and termination of the hits. The diagonals extend because the scores remain above the word threshold *T*. The termination occurs due to significant decay. The sequence alignments are trimmed (not shown). If, after trimming, the end of one diagonal is close to the beginning of a second diagonal, it is reasonable to extend the alignment to cover the two diagonals and fill in the junction between the two diagonals with gaps. This is called a local gapped alignment.

Figure 6-4D demonstrates how BLAST creates local gapped alignments. Here, the blue zones delineate areas where gaps can be used to join two diagonals. The gaps are depicted as red squares in the figure. If two diagonals fall into a single blue zone, the sequence alignment can be extended with gaps to join the diagonals. However, if two diagonals are separated by a wide gap, the diagonals are not joined and, instead, the two diagonals will be reported to the user as separate alignments.

How Is the Hit Deemed to Be Statistically Significant?

The score reported to you is called the similarity score, which is the maximum similarity score above the alignment threshold *A*. The maximum similarity score depends on the length of the aligned sequences, the number of gaps inserted by the

A

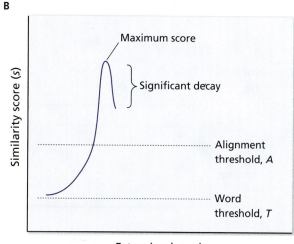

```
PPGTRVRAMAIYKQSQHMTEVVRRCPHHERCSD-SDGLAPPQHLIRVEGN
P G+ +RA A+YK+S+H+ EVV+RCPHHER  +  +  APP HL+RVEGN
PRGSILRATAVYKKSEHVAEVVKRCPHHERSVEPGEDAAPPSHLMRVEGN
```

B

FIG. 6-3. The relationship between extension length and similarity score, *S*.
A. The sequence alignment shows a portion of the BLAST output for the query sequence of human p53 protein used to probe the *Xenopus laevis* (frog) protein sequence database. The query sequence acceptable word RCP matches a record within the sequence database, and a hit is generated. The hit is extended upstream and downstream of the matched word. **B.** Let's just look at the extension downstream of the matched word. As the hit is extended, the similarity score (*s*) increases, and when the score surpasses the alignment threshold, *A*, the hit may be reported to the user. Upon further extension of the alignment, the score may rapidly decrease (significant decay), causing the extension to terminate. The sequence alignment is trimmed back to the point where the score is maximum. The maximum score, its associated sequence alignment, and the record number of the subject sequence are reported to the user. The same process occurs when the sequence alignment is extended upstream.

local gapped alignment A type of BLAST program that bridges two pairs of aligned sequences that are separated by relatively few gaps.

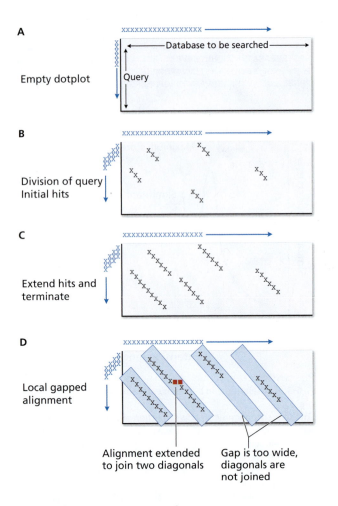

A Empty dotplot

B Division of query Initial hits

C Extend hits and terminate

D Local gapped alignment

Alignment extended to join two diagonals

Gap is too wide, diagonals are not joined

FIG. 6-4. The steps to a BLAST search.
A. The query sequence is placed on the vertical axis and the subject database sequences are placed on the horizontal axis. **B.** Overlapping query words are created from the query sequence. When the query words match the database sequence, the dot plot shows a diagonal corresponding to the location of the hit. **C.** Hits are extended until the cumulative score undergoes significant decay. **D.** Two diagonals that share a blue region may be joined by inserting gaps (*red square*). This produces longer sequence alignments. The joining of two diagonals is called local gapped alignment.

E-value = # of high scoring segment pairs w/ enough similarity score by chance

↓ E-value =

algorithm required to align the sequences, and the number and types of amino acid matches or mismatches that are aligned (as defined by the BLOSUM substitution matrix). You might be tempted to use the similarity score to conclude that an alignment is biologically significant. The potential pitfall here is that a high similarity score could be obtained by chance. If the query sequence has only three amino acids and a database containing 10,000 amino acids is randomly ordered, there is a good chance that one exact match will be due to chance.[8]

BLAST takes into consideration the possibility that the query sequence aligns with a subject sequence by chance. BLAST helps you determine whether an alignment is statistically significant by calculating the likelihood that the alignment could have occurred by chance: the lower this likelihood, the higher the probability that the alignment is significant. BLAST uses the *E*-value to measure the false positive rate. The *E*-value is the number of HSPs having a score equal to or greater than the similarity score *s* by chance alone. When you set an *E*-value cutoff prior to a BLAST run, the program will perform a calculation to determine the corresponding alignment threshold *A*. The equation for *E* is:

$$E = Kmne^{-\lambda s} \tag{6.1}$$

where
E = *E*-value

s = similarity score

m = length of query sequence

n = length of database

λ = a parameter that scales the scoring system[9]

K = a scaling factor for the search space

An *E*-value = 1 means that one sequence match with a score *s* or better could have been obtained by chance.

An *E*-value closer to zero is desired because such a value lowers the likelihood of obtaining the same score as the HSP by chance. This means the match is more statistically (and likely to be biologically) significant. Because *s* is a mathematical exponent, *E* is heavily dependent on *s*; the higher the score *s*, the lower the *E*-value. *E* is also dependent on query length (*m*) and database length (*n*), but less so. According to the equation above, when either *m* or *n* increases, the *E*-value increases. However, you must take into account the effect of *m* and *n* on *s*.

The database length (i.e., the total number of sequences to be searched) can be chosen by you so that the smallest database can be used for your BLAST search.

---

[8] For a three-amino-acid sequence there are 20 × 20 × 20 (or 8,000) permutations. This means that, on average, one three-amino-acid sequence in a database of 8,000 randomly ordered amino acids will match. As the database becomes larger, the chance of a match becomes greater. See Chapter 11 for more insights into permutations.

[9] The equation for *E*-value introduced here is also given in Chapter 12 as equation 12.10. A more detailed description of parameter λ and scaling factor *K*, from a probabilistic perspective, is also discussed in Chapter 12.

This is desired because then there is less likelihood of obtaining a hit due to chance. For example, if you are only interested in identifying a human homolog then it behooves you to perform a BLAST search on a database restricted to human proteins.

Thought Question 6-1

A particular homolog to a query resides in two databases, UniProtKP and PDB. After performing BLAST against the UniProtKP you obtain an *E*-value of 1.0 for an HSP with the homolog. After performing BLAST against the PDB database you obtain an *E*-value of 0.0625 for an HSP with the same homolog. What is the relative length of the two databases?[10]

Why Is the BLAST Program Faster Than the Smith-Waterman Program?

The Smith-Waterman program finds the highest similarity score for a pair of sequences; however, it is slow when searching the large sequence databases that are accessible through NCBI (see Chapter 14 for further discussion on computation time of pairwise alignment programs). As stated previously, BLAST is a shortcut program or heuristic (i.e., a program that sacrifices precision and accuracy for speed) that attempts to maintain the sensitivity of the Smith-Waterman program. We can use "$O(N)$" notation (also known as "big O" notation) to discuss what makes the BLAST program generally faster than the Smith-Waterman program when large databases are searched. $O(N)$ notation describes the order of growth of a program—in other words, the program's worst-case performance changes as the size of the data set the program operates on increases. Performance usually refers to speed of operation, which is what we mean here; but on occasion performance refers to computer memory space, if specified. O is the order, and N is the amount of data in the data set (in our case N is the number of amino acids). The larger the order of growth, the longer the time it takes to process the information. In a simple case, when the order of growth is $O(N)$, the program's performance is proportional to the size of the data set being processed. There is a linear relationship between the size of the data set and the time required to run the program.

In the case of Smith-Waterman, the order of growth is $O(N*M)$ because there are two data sets that need to be processed—one is the query, which we can assign to N, and the other is the database that is searched, M. A calculation is required for potential matches between each amino acid in the query and each amino acid in the database.

Determining the order of growth of the BLAST program is a bit more complicated. BLAST is faster than Smith-Waterman because it does not perform a calculation on each potential amino acid match. From the query sequence BLAST creates a table of words, each of which is associated with a score. To begin the process of creating a high-scoring sequence pair (HSP), BLAST searches for matches between those scored words above threshold T and the database. To determine the order of growth, we need to consider the different phases of the BLAST program. In phase 1, there is the time required to create the list of words which results in an order of growth of $O(N)$. In phase 2, the time required to scan the database with the words is $O(M)$. In phase 3, the time required to extend the hit is $O(N*M)/20^w$, where w is the length of the words (typically, the length is three). Extending the hit requires calculations, but only on a subset of the database. The longer the word length, the lower value of the $O(N*M)$. In phase 4 there is trimming, but the computer time needed for this is negligible. Putting all of the phases of BLAST together we get the following order of growth: $O(N) + O(M) + O(N*M/20^w)$. At first glance the order of growth of BLAST

---

[10] Database length is defined by the sum of lengths of the amino acid sequences in all records. One could make an argument that database length or database size is dictated by the number of records in the database. However, for our purposes, unless explicitly stated otherwise, database length and database size will refer to the sum of the lengths of amino acid sequences in the database.

appears to be larger than that of Smith-Waterman. However, as *N* or *M* increases, the term $O(N*M/20^w)$ dominates this mathematical expression. In fact, the term *M* is not the entire database to be searched but, instead, only the part of the database that is extended by matches with words compiled in phase 1. A BLAST search will take longer if you lower the threshold, *T*, or if you reduce the word size, *w*.

Low Complexity Regions and Masking

low complexity region A sequence with a repeated pattern of amino acids or nucleotides.

Analysis of the *E*-values of the BLAST output is a robust method to measure the significance of HSP scores. However, you should also consider whether the query sequence contains low complexity regions. Low complexity regions are those that contain repeated patterns of amino acids. If the query sequence has repeating patterns, there is a higher probability that it will produce a hit with the database. There are many proteins that contain repeats of single, double, or triple amino acids. If the query contains repeats, it has a strong likelihood of generating an HSP from the database sequence containing these repeats, and you might obtain erroneous hits. BLAST offers you the option of excluding the low complexity regions, and this should be considered if you believe that the low complexity region contains a part of the protein that is not necessary for function.

exclude or mask region from analysis & alignment

The problem of low complexity regions is compounded when a nucleotide query is used against a nucleotide sequence database. There are only four nucleotides, so there is a higher chance of finding low complexity regions in DNA sequences. DNA sequences with known low complexities include retrotransposons, ALU regions, microsatellites, centromeric sequences, telomeric sequences, and 5′ untranslated regions of expressed sequence tags (ESTs). An example of a low complexity region in DNA is shown in Figure 6-5. You should consider using the low complexity filter if the query sequence has repeated amino acids or nucleotides that are not essential for function or structure of the biological molecule.

masking A feature of BLAST that allows the user to exclude a region of the query sequence from contributing to the alignment. It is used when there is a region in the query sequence that is very common to many sequences in the database. The user is more interested in regions of similarity found between the unmasked areas of the query and the database.

A strategy similar to low complexity filtering is masking. Masking is an option that allows you to exclude a region of a query sequence from contributing to the alignment. When would you use masking? Masking is used when you want to exclude a region of a protein that is common to a wide variety of proteins. Some domains of proteins are very common. For example, DNA binding proteins often contain zinc finger domains. Performing a BLAST search with a query containing a zinc finger domain will result in numerous hits of proteins with zinc finger domains. These proteins may not be highly related to the query in any domain other than in the zinc finger. To exclude hits to all zinc finger-containing proteins, you can mask the zinc finger domain in the query. With masking, the hits will be due to sequence alignments with regions other than zinc finger domains.

Usefulness of BLAST

BLAST, when properly used, can be a quick way to obtain extensive knowledge of your protein or gene of interest. BLAST comes in several forms that manipulate your query sequence or the database to give you several alignment options. Here is a list of the common types of BLAST programs available:

BLASTp: compares an amino acid query sequence against a protein sequence database.

BLASTn: compares a nucleotide query sequence against a nucleotide sequence database.

GGGTGCAGGAATTCGGCACGAGTCTCTCTCTCTCTCTCTCTCTCTCTCTCT
CTCTCTCTCTCTCTCTCTCTCTCTCTCTCTCTCTCTCTCTCTC

FIG. 6-5. An example of a low complexity region in a cDNA sequence. GenBank record T27311.

BLASTx: compares a nucleotide query sequence translated in all six reading frames against a protein sequence database.

tBLASTn: compares a protein query sequence against a nucleotide sequence database dynamically translated in all six reading frames.

tBLASTx: compares the six-frame translations of a nucleotide query sequence against the six-frame translations of a nucleotide sequence database.

PSI-BLAST: compares a protein sequence to a protein database in a manner that allows detection of evolutionarily distant homologs (see next section).

DELTA-BLAST: compares a protein sequence to a database of conserved protein domains in a manner that allows detection of evolutionarily distant homologs.

BLAST2: sometimes abbreviated as bl2seq, compares two user-defined protein or nucleotide sequences (or sets of sequences). In the online version of BLAST available from NCBI, blast2 is embedded in all of the BLAST programs listed above. Once you are in a BLAST program (for example, BLASTp), BLAST2 is activated by checking a box next to the option that states "Align two or more sequences."

Table 6-1 gives some useful tips for using BLAST to obtain information about your query. Experimental approaches are generally required to confirm results from BLAST searches.

PSI-BLAST

A bioinformaticist can easily use BLAST to align a query sequence with sequences stored in publicly available databases. A BLAST run can sometimes result in the discovery of protein homologs from distantly related species. To further aid in this discovery, a BLAST program named Position-Specific Iterated BLAST (PSI-BLAST) was developed. The concept behind PSI-BLAST is that you can detect distantly related sequences if a custom substitution matrix tailored to the query sequence could be used for sequence alignments. A custom substitution matrix can be imagined where strongly conserved positions in the query sequence penalize mismatched amino acids and weakly conserved positions in the query sequence allow mismatched amino acids without severe penalties. The difference between a custom substitution matrix and the amino acid substitution matrices we have already encountered, such as BLOSUM62 and PAM250, is that the former contains amino acid substitution values that are different for each position in the query. The custom substitution matrix in PSI-BLAST is called a position-specific substitution matrix (PSSM, pronounced "possum").

Let's take a look at how a PSI-BLAST run on a protein sequence works step by step. We will use the default PSI-BLAST parameters used in the online version sponsored by NCBI as we proceed through these steps. In the first step, PSI-BLAST uses protein BLAST (BLASTp) to create HSPs between the query protein sequence and subject sequences from a large database that contains sequence data from many species. The HSPs are scored with a BLOSUM62 substitution matrix, and the user receives an output ranked by E-value. In the second step, the subject sequences from the HSPs with a threshold A better than $E = 0.005$ (i.e., $E \leq 0.005$) are aligned through multiple sequence alignment (a multiple sequence alignment procedure is described later in this chapter). In step 3, the aligned sequences are used to create a PSSM. In step 4, the PSSM (sometimes called a profile) is BLASTed against the database (this is called the second iteration). In step 5 another output of E-value ranked HSPs is reported to the user. Starting from step 2, this process is repeated so that the user can perform successive iterations. With each BLAST run the PSSM will change as new sequences are captured in HSPs. In most cases, after the fifth iteration the PSSM does not change, and no new sequences are captured.

PSI-BLAST (Position-Specific Iterated BLAST) A software program that creates a position-specific substitution matrix (PSSM) from the top hits of a BLAST run. The PSSM is used as a substitution matrix for more rounds of BLAST searching. PSI-BLAST is useful for detecting distant homologs of proteins.

[Handwritten margin notes:]

PSI = Position-Specific Iteration (a Repetition)

matrix specific for your sequence of interest. the matrix represents the substitution rates for between a LARGE library of very closely related sequences

the PSSM is modified by subsequent hits w/ E ≤ 0.005.

The PSSM stops changing usually after 5 iterations.

TABLE 6-1. UTILIZING BLAST TO GET QUICK ANSWERS TO BIOINFORMATICS PROBLEMS

TASK	TRADITIONAL METHOD	BLAST METHOD
Predict protein function (1)	Perform wet-lab experiments	Perform BLASTp with the protein sequence query against a protein sequence database. If a hit that covers a large proportion of the query is returned, and that hit has a low *E*-value, then the annotation of the subject sequence may reveal the putative function.
Predict protein function (2)	Perform wet-lab experiments	If the above method does not work, perform tBLASTn with the protein sequence query against a nucleotide sequences. This function first translates all nucleotide sequences into six reading frames. The annotation of the nucleotide sequence database is often more complete than the protein databases.[1]
Predict protein structure	Perform X-ray crystallography experiment or NMR experiment on purified protein. One can also run structure prediction software.	Use BLASTp with the protein sequence query against Protein Data Bank. If a hit with a low *E*-value is returned, then the structures of the proteins could potentially be similar.
Identify genes in a newly sequenced genome	Perform microarray experiment or RNA-seq on organism's RNA. One can also run gene-prediction software.	Divide genome nucleotide sequence into two- to five-kilobase segments. Paste one segment at a time into query box in BLAST program. Use BLASTx to search a nucleotide query against the non-redundant protein sequence database.
Identify distantly related proteins	No traditional method. However, once identification is made with psi-blast you can create a recombinant version of the sequence and test to see if it performs the same function as your query sequence.[2]	Use PSI-BLAST or DELTA-BLAST (see next section). Finds distantly related sequences. After an initial BLASTp search, the query sequence is replaced with a position-specific substitution matrix. PSI-BLAST then uses the matrix to find distantly related sequences.
Identify DNA sequence (non-protein)	Experimentally screen genomic DNA library with radiolabeled or fluorescently labeled probe. Perform microarray experiment on organism's DNA.	Use BLASTn. Finds alignments between nucleotide sequences.

(handwritten annotation:) to see crystal structure

[1] This was the BLAST program used to identify MDM2 as an inhibitor of p53. See Box 2-2 for details.
[2] If the distantly related organism is amenable to genetic manipulation, you can knockout or knockdown the distantly related gene and check the phenotype of the organism. The phenotype may match that of the organism when its gene corresponding to the query sequence is genetically removed.

Key to this procedure is the PSSM—a substitution matrix that assigns a score to an alignment based on the position of the amino acid in the query sequence. To create the PSSM, a multiple sequence alignment is generated with a length equal to the query sequence length and a depth equal to the number of sequences captured in the first BLAST run (minus sequences that are ≥94% identical to the query sequence). Gaps are allowed in the aligned sequences, but not in the query sequence. At each position in the aligned sequences, 20 scores are placed, reflecting the frequency of all 20 amino acids. For example, if the majority of the aligned sequences have a valine that aligns to the first position in the query, then a valine at this position is assigned a high score in the PSSM, and scores assigned to the other 19 amino acids at this position will be lower. The multiple sequence alignment process assigns a score to each of 20 amino acids at each position in the query. Figure 6-6 shows a PSSM for a query sequence that is 18 amino acids in length.

Thought Question 6-2

Figure 6-6 shows a PSSM aligned with the query sequence. In the query sequence, a Gln is found in positions 4, 12, and 14. Can you rank the positions in terms of relative degree of Gln conservation? At position 12, what other amino acid(s) is (are) often substituted?

		A	R	N	D	C	Q	E	G	H	I	L	K	M	F	P	S	T	W	Y	V
1	V	-2	-4	-5	-5	-2	-4	-4	-5	-5	4	3	-4	0	-2	-4	-4	-2	-4	-2	4
2	G	-1	-1	1	1	-4	-1	0	5	-1	-5	-5	0	-2	-4	-3	0	0	-4	-4	-4
3	S	-2	-2	0	1	-5	2	4	-2	0	-4	-4	0	-2	-4	-1	0	-1	-5	-2	-1
4	Q	0	1	1	-1	-4	0	0	0	-1	-2	-1	2	-1	0	-1	0	-2	2	4	-2
5	G	0	-3	-2	-2	-4	-3	-3	4	-4	-4	-4	-2	-4	-5	5	-1	0	-4	-2	-1
6	S	1	0	1	0	-2	-1		0	3	0	-2	0	-2	0	-2	2	0	-4	-1	-1
7	Q	-1	0	3	0	-4	3	3	0	0	-3	-3	0	0	-4	-2	0	0	-4	-2	-2
8	L	-3	-4	-5	-6	-2	-4	-5									-4	-3	-3	-1	0
9	S	0	-3	0	-2	-3	-2	-2									7	0	-5	-4	-4
10	R	-1	-3	-2	-2	-2	-2	-2	6	-2	-1	-3	-1	-4	-3	-1	-1	-3	-1	-2	0
11	G	-1	-4	-2	-3	-5	-4	-2	7	-4	-6	-6	-4	-5	-5	-4	-1	-4	-4	-5	-5
12	Q	-3	-1	0	-1	-5	6	4	-4	-1	-4	-3	-1	4	-4	-3	-2	-3	2	-3	-2
13	K	-1	5	-2	-3	0	4	-1	-3	-2	-3	-2	4	-1	-3	-3	-2	-2	0	-3	-4
14	Q	-2	2	0	-2	-5	7	0	-3	-2	-5	-3	1	0	-5	-1	-1	-2	-4	-3	-4
15	R	-2	7	-2	-4	-3	-1	-2	-3	-2	-3	0	2	0	-3	-4	-3	-3	-5	-4	-3
16	I	0	-4	-5	-5	-3	-4	-4	-5	-5	3	3	-4	0	-2	-3	-3	-1	-5	-3	4
17	A	4	-2	-1	-2	1	-1	0	0	-3	-1	0	-1	1	-2	-3	1	0	0	-3	-1
18	I	-3	-4	-5	-5	-3	-4	-5	-6	-5	6	3	-4	1	1	-5	-4	-3	-4	-2	1

Glutamine scored differently in these two positions

FIG. 6-6. A position-specific substitution matrix (PSSM). The top row displays 20 naturally occurring amino acids. The first vertical column shows the position number of the amino acid in the query sequence. The second vertical column shows the query sequence. The second row through the last row show the amino acid scores at each sequence position. Note that Q in position 4 gets a score of 0, and a Q in position 14 gets a score of 7 (Q is highly conserved at position 14, but not conserved at position 4).

We will perform a PSI-BLAST run on human p53 protein to discover putative homologs from distantly related species. We will use BLAST version number 2.2.29 hosted at the NCBI website in this example.[11] We will use the wild human p53 accession number P04637 (from UniProtKB) as the query. In the NCBI online version of PSI-BLAST, the nr database will be searched and we will exclude all primate organism(s) from our database search using the filter. The rationale for excluding all primates is that p53 from primates, especially humans, is so frequent that they are overrepresented in the nr database. We will also change the parameter "maximum target sequences to be displayed" from 100 to 5,000. This will allow us to see subject sequences that are lower ranked (i.e., worse E values) in the output window, which are likely to be from distantly related species.

When we hit the run button, we get an output of hits that are ranked. We can repeat the run process three more times. At the end of each run, the PSI-BLAST output allows us to scroll down to new sequences that were not included in the previous output—these new sequences may be homologs from distantly related species. In the authors' fourth run (each run is called an iteration), an HSP was in the output with a subject sequence from the species *Brugia malayi*. *Brugia malayi* is one of three species of nematode (roundworm) that causes lymphatic filariasis— also known as elephantiasis (Figure 6-7). The HSP has an E-value of 1×10^{-4} (when the PSSM is utilized as the substitution matrix), a query coverage of 21% (only 21% of the 393 amino acids of query human p53 sequence align with the subject sequence), and a 16% identity over the length of the alignment. The accession number of the subject sequence is XP_001893712. Annotations in the corresponding GenBank record indicate that this segment of DNA codes for a hypothetical protein (its existence is surmised from the DNA sequence) and do not mention p53. To show that this hypothetical protein is a p53 homolog would require much more analysis of the arrangement of exons and introns in its gene and, ideally, wet-lab experiments

---

[11] It is likely that the version number of the online BLAST hosted by NCBI will be higher once you read this. In addition, you may get slightly different outputs because, as you know, sequences are constantly being added to the public databases. A different output will also result in a PSSM that will be different than the one the authors are using here. Nonetheless, the logic and the approach of this example should be instructive.

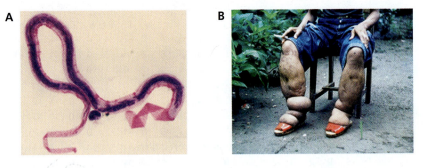

C Hypothetical protein Bm1_11215 [Brugia malayi]
Sequence ID: refjXP_001893712.1   Length: 191   Number of matches: 1

Range 1: a to 96 GenDept Graphics

Score	Expect	Method	Identities	Positives	Gaps
41.4 bits (95)	1e-04	Composition-based stats.	14/89 (16%)	30/89 (33%)	3/89 (3%)

```
Query 294  EPHHELPPGSTKRALPNNTSSSPQPKKK---PLDGEYFTLQIRGRERFEMFRELNEALEL 350
             ++ P    A+    S S + K +     D  ++L+IRGR +++    +     EL
Sbjct 8    SLGNKRPAAKHYPAVAEVVSVSAKNKPEQPWDDDDIVYSLEIRGRHLYKIVCAIVGNFEL 67

Query 351  KDAQAGKEPGGERAHSSHLKSKKGCSISR 379
             +   +      +     G  +
Sbjct 68   TRNLLKDKMQRNDREMLTVSDSSGSLSQN 96
```

FIG. 6-7. *Brugia malayi*—a roundworm (nematode) parasite endemic to India, North and South Korea, Sri Lanka, China, and Southeast Asia.
A. Giemsa-stained *B. malayi* in peripheral blood magnified. **B.** Case of elephantiasis often caused by *B. malayi* infection.
C. Result of PSI-BLAST analysis of human p53. The query sequence is human p53, and the subject sequence is hypothetical protein Bm1_11215 from *B. malayi*.

conducted on the protein. This small example gives just a taste of the power of PSI-BLAST for identifying potential homologs from species that are distantly related to humans.

Another program used to detect distant homologs is domain enhanced lookup time accelerated BLAST (DELTA-BLAST). It is similar to PSI-BLAST, but instead of creating a PSSM directly from query searches of sequence databases, DELTA-BLAST searches a database of preconstructed PSSMs. These PSSMs are constructed from a database of conserved domains called the CDD. Each PSSM represents a multiple sequence alignment of a particular conserved domain. From the search of PSSMs with the query a new PSSM is created. The new PSSM is used to search a protein sequence database for distantly related homologs. In at least one test, DELTA-BLAST outperforms PSI-BLAST. If DELTA-BLAST fails to create a PSSM above a prespecified score, the online program offered by NCBI defaults to PSI-BLAST.

BLAST has become a fixture in bioinformatics. As of this writing, the papers describing BLAST and gapped BLAST/PSI-BLAST have each been cited more than 33,000 times. The team that developed BLAST included molecular life scientists, computer scientists, mathematicians, and an interdisciplinary team headed by David Lipman, who has headed the NCBI since its inception in 1988. Under his leadership the NCBI has, among other activities, developed and hosted a suite of BLAST programs and refined PubMed—an online digital library that makes scientific literature available through the Internet. His pioneering efforts in bioinformatics have been transformative for this rapidly developing field (Box 6-1).

6.3 MULTIPLE SEQUENCE ALIGNMENT

We briefly discussed multiple sequence alignment in the previous section on PSI-BLAST. Here, we go into more detail on this basic tool used by bioinformaticians. Multiple sequence alignment is the alignment of more than two protein or nucleic acid sequences. In many ways, the program that creates multiple sequence alignments is very different from protein BLAST, but, like BLAST, it is useful for

BOX 6-1 | **SCIENTIST SPOTLIGHT**

David Lipman
NCBI Director

DAVID LIPMAN of the National Institutes of Health is an Internet pioneer who has made critical medical and scientific information available online for scientists, researchers, and the general public. Now Lipman is pushing the digital boundaries even further, employing a Google-style approach to make the voluminous government databases he helped create even more accessible and user-friendly. Lipman said the goal of his "Discovery Initiative" is not simply to provide researchers with more information on medical topics, but to "offer them links to the highest quality pieces of information so that they can perform at the highest level possible." "It's like ads on Google—if you like this article, you might want to read these four articles," said Lipman, the director of the National Center for Biotechnology Information (NCBI). Lipman envisioned, helped create, and now oversees more than 40 publicly available online medical and scientific databases within NIH, although he gives much of the credit to his team. The databases, which are interconnected for maximum research capabilities, are used daily by more than two million people. Each week the equivalent of all the text content in the Library of Congress is downloaded from these databases. They include PubMed, an online service that allows the public to search abstracts from approximately 4,600 of the world's leading biomedical journals; PubMed Central, an archive of 1.7 million full-text journal articles from biomedical journals; GenBank, the world's largest genetic sequence data repository; and PubChem, a resource that connects chemical information with biological studies.

"His vision enabled NCBI to be one of the very best public resources available," said Richard J. Roberts, a molecular biologist and winner of the Nobel Prize for Medicine. "The current state of biological research would not be where it is if NCBI did not exist." "He has truly done an extraordinary job at NCBI and continues to be imaginative and forward looking," said Roberts. The readily accessible NCBI databases are proving helpful both to researchers and to the general public in finding important medical information. Heather Joseph, executive director of the Scholarly Publishing and Academic Resources Coalition, said she never expected that, a decade after becoming familiar with Lipman's work through her job, she would use PubMed Central to help her own family.

In 2008, Joseph's five-year-old son Alex was diagnosed with Type 1 diabetes. At night, Joseph would wake her crying son for his insulin shot, without which he could go into a coma—or worse. "I just thought, there has to be a better way, but nothing was out there," Joseph said. "Then it hit me—I went to PubMed Central and found an article about a brand-new technology recently approved by the FDA that can monitor Alex's glucose through the night, which will really help our family." The expansion of online resources was significantly aided by Congress, which in 2008 mandated that all taxpayer-funded medical research and clinical trials be placed online. "If it weren't for Congress's mandate and NCBI's quickness in getting the information up, we wouldn't have found something that has profoundly helped my son," said Joseph. Lipman's work was not easy. He had to overcome some resistance within the government and scientific communities. With his team, he developed the necessary tools to allow storage, rapid searches, and

continued

continued

barrier-free access to biomedical research reports. Although Lipman's focus is on using technology to make the latest medical information available, he is working with journals across the country to preserve older research data. "What this has done is made good research from 50 years ago available online," Lipman said. As he looks back over his career, Lipman says he is impressed by how much technology has influenced his profession and the positive role it plays in the medical community. "When I started my work, I never imagined that the comprehensive data that we have now would be so readily available. It's phenomenal," Lipman said.

identifying regions in sequences that are similar. Recall that PSI-BLAST uses multiple sequence alignment to create its PSSM. For proteins, a multiple sequence alignment of protein homologs is useful because it allows the user to easily visualize regions that are similar amongst many homologs. These similar regions are called conserved regions and they are likely to maintain the same structure. Because structure dictates function, the conserved regions are likely to be critical for protein function.

There are practical reasons why you would align DNA sequences as well. Recall from Chapter 2 that a region of a gene 5′ of the coding sequence contains the promoter. Alignment of 5′ regions from ortholog genes will highlight areas within promoters that are conserved. The conserved areas likely bind to the same transcription factors. The DNA sequence specificities of many transcription factors are known, thus giving you insight into identification of potential transcription factors that control gene expression.

In another example highlighting the importance of multiple sequence alignment, you may wish to design polymerase chain reaction (PCR) primers to amplify a gene from a novel organism (whose genomic DNA has never been sequenced).[12] Without knowing the genome sequence, how do you design a PCR primer that will amplify the novel gene? If homologs of the gene have been previously sequenced, you can perform a multiple sequence alignment of the homologs. Multiple sequence alignment will reveal the segments of the gene homologs that are conserved. DNA primers that match the conserved regions can be synthesized and used for PCR amplification of the gene from the novel organism. Using this PCR design strategy, a paralog of *TP53*, called *TP63*, was identified in the squid, *Euprymna scolopes*. In the squid, p63 protein is expressed during development of an organ that produces light, called the light organ. It is hypothesized that during development p63 removes unneeded cells from the light organ through programmed cell death.

ClustalW

A popular program used to align sequences is ClustalW. It employs three steps for multiple sequence alignment: (1) creation of pairwise alignments of every sequence to be aligned; (2) creation of a guide tree based on "distances" between each pair of sequences; and (3) alignment of sequences according to the guide tree. Let's explore the details of how ClustalW operates.

---

[12] Please see Chapter 9 for a detailed explanation of polymerase chain reaction.

ClustalW is a progressive alignment program. The progressive alignment algorithm, introduced by Da-Fei Feng and Russell Doolittle in 1987 (at the University of California, San Diego), performs a pairwise global alignment of all sequences given by the user. The pair that has the highest identity is aligned first. Then other sequences, in the order of their identities, are progressively added to the first aligned pair. Feng and Doolittle also introduced the concept of "once a gap, always a gap" into their algorithm. Gap treatment for multiple sequence alignment is critical because introduction of gaps is often necessary for optimal alignment. The key is to know where, within the sequence, is an appropriate place to put a gap. Feng and Doolittle thought that gaps placed in sequences with high similarity should not be altered once additional, less similar sequences are added to the alignment. They reasoned that gaps required for optimal alignment in similar proteins likely represent locations that are *not* critical for function or structure. Natural selection has allowed amino acids to be inserted into or deleted from those locations of the proteins. When more distantly related proteins are aligned, it is better to insert a gap in their sequences at these noncritical regions rather than into other regions farther upstream or downstream. ClustalW uses progressive alignment and gap treatment in a fashion similar to the Feng and Doolittle algorithm.

In the first step of the program, ClustalW performs a pairwise alignment of all sequences to be aligned. ClustalW calculates the percent identity shared by pairs of sequences. The percent identities are then converted into difference scores (D) using the following equation:

$$D = 1 - (I) \qquad (6.2)$$

where I is the number of identities in the pairwise global alignments divided by the total number of amino acids in the aligned pair. A higher value for D means a smaller percent identity.

We will use seven sequences from the globin family to illustrate how ClustalW operates. Globins are important for carrying oxygen in multicellular organisms. Two alpha globin polypeptides and two beta globin polypeptides form hemoglobin, a tetramer found in the blood of many multicellular organisms. Figure 6-8 shows the difference scores for seven globin sequences. This is called a 7×7 distance matrix. We can see from the distance matrix that Hba-Ho and Hba-Hu have the highest percent identity of all the pairwise comparisons.

In the next step of the ClustalW program, the distance matrix is converted into a guide tree that will be used to construct the multiple sequence alignment. The important feature of a guide tree is that it reflects the relatedness of proteins through the *order* of branching and the *length* of the branches. There are actually two types of trees we will be concerned with here, an unrooted tree and a rooted tree; a guide tree is a type of rooted tree.[13]

The distance matrix is converted into a guide tree by a neighbor joining (NJ) process. First, an unrooted NJ tree is created in which each sequence starts out distributed radially as if on the spokes of a wheel. More similar sequences are joined to become neighbors until an unrooted tree is created (Figure 6-9). Low distance scores (D) from Figure 6-9 are translated into joined branches of short lengths. A node is the point where two branches join. In the unrooted NJ tree, branch lengths are

		1	2	3	4	5	6	7
Hbb-Hu	1	-						
Hbb-Ho	2	.17	-					
Hba-Hu	3	.59	.60	-				
Hba-Ho	4	.59	.59	.13	-			
Myg-Ph	5	.77	.77	.75	.75	-		
Gib-Pe	6	.81	.82	.73	.74	.80	-	
Lgb-Lu	7	.87	.86	.86	.88	.93	.90	-

FIG. 6-8. Distance matrix of seven globin sequences.

---

[13] We discuss trees in greater detail in Chapter 8.

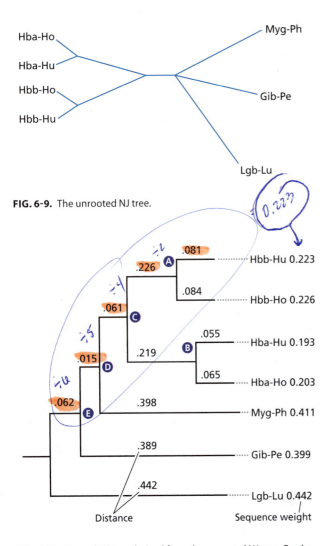

FIG. 6-9. The unrooted NJ tree.

FIG. 6-10. Rooted NJ tree derived from the unrooted NJ tree. On the left, numbers above each horizontal branch represent distances between tree nodes (lengths of vertical lines do not contribute to branch lengths). Letters A–E each represent the set of aligned sequences from branches just to the right of the nodes. The sequences will be progressively aligned by ClustalW. Numbers to the right of the named sequences are sequence weights.

proportional to the estimated distance of each sequence from an "average sequence" of all sequences to be aligned.

In the second step, the unrooted tree is converted into a rooted NJ tree. In this tree, the root is considered the ancestral sequence to the entire family of sequences to be aligned. Calculation of branch lengths and nearest neighbors in the rooted NJ tree is complex, and interested students are referred to another source for details on how to calculate these lengths and nearest neighbors.[14] A weight, sometimes called a sequence weight, is calculated from the rooted tree for each sequence to be aligned (weights are shown in Figure 6-10). The sequence that is more distant from other sequences is given the highest weight. According to Figure 6-10, the most distant sequence from all others is Lgb-Lu. The method used to create the unrooted tree is called clustering, so the multiple sequence alignment program is named "Clustal," for cluster alignment. The "W" in ClustalW stands for weight.

There are two types of information that are utilized from a rooted NJ tree. First, the lengths of the branches dictate the order of progressive alignment. Progressive alignment starts with the shortest branches and proceeds to the longest branches. Second, the lengths of the branches and the number of branch points are used to calculate sequence weights. The sequence weights are used to decide which sequences contribute more to driving the alignments. Sequences that share a branch with other sequences share the weight derived on that common branch. In the example shown in Figure 6-10, Lgb-Lu has a weight of 0.442, which is equal to the distance from the tip of the branch to the root. Calculation of sequence weights of other sequences is more involved. The sequence weight of Hbb-Hu is calculated by adding the length of the branch leading to its tip that is not shared with any other sequence (0.081) plus half the length of the branch shared with Hbb-Ho (0.226/2), plus one-fourth the length of the branch shared with Hbb-Ho, Hba-Hu, and Hba-Ho (0.061/4), plus one-fifth the length of the branch shared with Hbb-Ho, Hba-Hu, Hba-Ho, and Myg-Ph (0.015/5), plus one-sixth the length of the branch shared with Hbb-Ho, Hba-Hu, Hba-Ho, Myg-Ph, and Gib-Pe (0.062/6). The sequence weight of Hbb-Hu is 0.223. In this manner, sequences that are very similar to other sequences (such as Hba-Hu) are given *low* weights and those that are more distant (such as Lgb-Lu) are given *high* weights. The rationale behind this is that those sequences that are very similar should not be allowed to dominate the contribution to the alignments during progressive alignment.

In the third step of the ClustalW algorithm, sequences and sets of aligned sequences become progressively aligned. Initially, pairs of sequences will be aligned with each other. Then, sets of aligned sequences will be aligned with each other or with individual sequences. In Figure 6-10 the sets of aligned sequences are labeled with letters A through E. For example, A represents the aligned set of Hbb-Hu and Hbb-Ho and B represents the aligned set of Hba-Hu and Hba-Ho. After sequence

---

[14] Cristianini and Hahn (2007) do an excellent job in describing calculations of branch lengths in an NJ rooted tree.

weights have been calculated, the next step is to align the sequences starting from the tips of the tree towards the root. Sequences with the shortest branch lengths, or highest similarity, are aligned first. Here is the order of alignment for the rooted NJ tree in Figure 6-10:

1. Hba-Hu versus Hba-Ho
2. Hbb-Hu versus Hbb-Ho
3. A versus B
4. Myg-Ph versus C
5. Gib-Pe versus D
6. Lgb-Lu versus E

Just like the Smith-Waterman algorithm discussed in Chapter 5, the goal of ClustalW is to maximize the alignment score as sequences are aligned. Scores come from two sources: BLOSUM substitution matrices and the sequence weights (see Figure 6-10). We will step through the calculation of the score at a single position that will be used to align two previously aligned sets (A vs. B). In ClustalW, this calculation is repeated for all possible alignment positions between A and B.

Prior to the score calculation, the sequence weights are normalized by dividing each sequence weight by the highest sequence weight, 0.442, so this means that Lgb-Lu has a normalized weight of 1.000 and all other sequences have values less than 1.000. Sequence weights for the seven globin sequences are shown in Table 6-2.

The normalized sequence weight is used to calculate the score derived from comparison of position 7 from the 2-sequence set A to position 6 from the 2-sequence set B (Figure 6-11). To calculate this score, the BLOSUM62 substitution matrix will be used where the term $\mathbf{M}_{i,j}$ is the value of the element in substitution matrix $\mathbf{M}$ for amino acid i to amino acid j substitution score. The substitution values from the BLOSUM62 matrix are as follows:

$$\mathbf{M}_{T,V} = 0$$

$$\mathbf{M}_{T,I} = -1$$

$$\mathbf{M}_{L,V} = 1$$

$$\mathbf{M}_{L,I} = 2$$

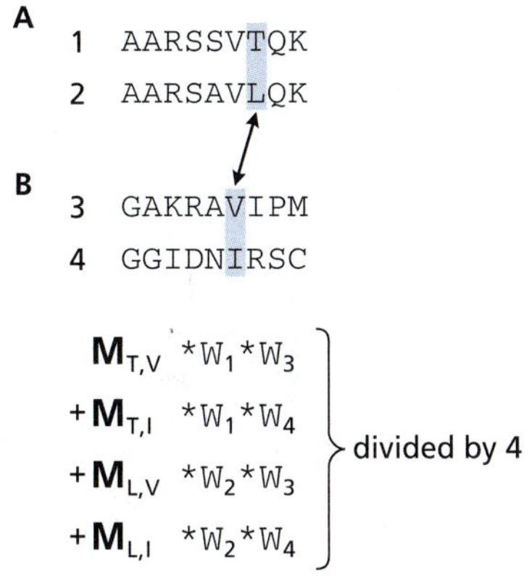

A
```
 1   AARSSVTQK
 2   AARSAVLQK
```

B
```
 3   GAKRAVIPM
 4   GGIDNIRSC
```

$$\left.\begin{array}{l} \mathbf{M}_{T,V} \;{*}W_1{*}W_3 \\ +\mathbf{M}_{T,I} \;{*}W_1{*}W_4 \\ +\mathbf{M}_{L,V} \;{*}W_2{*}W_3 \\ +\mathbf{M}_{L,I} \;{*}W_2{*}W_4 \end{array}\right\} \text{divided by 4}$$

FIG. 6-11. Scoring during progressive alignment in ClustalW. Two sets of aligned sequences are shown. Sequence 1 and 2 comprise set A, and sequence 3 and 4 comprise set B. They are truncated here to illustrate the scoring calculation. Amino acids at position 7 in set A will be scored against the amino acids at position 6 in set B. $\mathbf{M}_{i,j}$ is the BLOSUM62 substitution matrix value for amino acids i and j. W_n is the sequence weight where n is the sequence number.

TABLE 6-2. SEQUENCE WEIGHT CALCULATIONS

SEQUENCE NUMBER	SEQUENCE NAME	RAW SEQUENCE WEIGHT	NORMALIZED SEQUENCE WEIGHT
1	Hbb-Hu	0.223	0.506
2	Hbb-Ho	0.226	0.511
3	Hba-Hu	0.193	0.437
4	Hba-Ho	0.203	0.459
5	Myg-Ph	0.411	0.930
6	Gib-Pe	0.399	0.903
7	Lgb-Lu	0.442	1.000

Following the procedure shown in Figure 6-11, calculation of the score for the comparison of A and B at the outlined position is:

$$0 * 0.506 * 0.437 = 0$$

$$-1 * 0.506 * 0.459 = -.232$$

$$1 * 0.511 * 0.437 = .223$$

$$2 * 0.511 * 0.459 = .469$$

$$(0 + (-0.232) + 0.223 + 0.469)/4 = 0.460$$

In a fashion similar to the NWmod global alignment program, the calculated score is added to the scores of predecessor cells to maximize the score in an element of a scoring matrix (see Chapter 5). The procedure is repeated so that all possible amino acid comparisons between sets A and B are made, and the matrix is filled. Traceback, starting with the maximum score at the lower right side of the filled matrix, produces the best alignment.

To calculate the score in Figure 6-11, the BLOSUM62 substitution matrix was used to generate the values for $\mathbf{M}_{i,j}$. In the ClustalW program, the substitution matrix used depends on the distance scores used to create the guide tree. ClustalW switches substitution matrices as the alignments proceed progressively from the least divergent sequences to the most divergent sequences. Sequence distances are proportional to branch lengths. The substitution matrix used depends on the rooted tree branch lengths, such as those depicted in Figure 6-10. For sequences connected to a node by short branch lengths, the BLOSUM80 substitution matrix is used, and for sequences connected to a node by long branch lengths, the BLOSUM30 substitution matrix is used. The distance (D) ranges used with BLOSUM series is as follows: 0–0.20: BLOSUM80, 0.20–0.40: BLOSUM62, 0.40–0.70: BLOSUM45, 0.70–1.00: BLOSUM30. Recall from Chapter 5 that higher BLOSUM substitution matrices are used for comparing sequences that are more identical (less distant). A ClustalW (version 2.1) multiple sequence alignment of the globin sequences is shown in Figure 6-12.

Treatment of gaps

As described earlier, ClustalW has special rules for creating gaps. ClustalW follows the "once a gap, always a gap" rule first described by Feng and Doolittle. In addition to this rule, there are others that ClustalW follows. Analysis of structures of homologous proteins suggests that gap opening penalties should not be uniform. Short stretches of 5 hydrophilic residues (D, E, G, K, N, Q, P, R, or S) often form a loop or random coil in protein structures. Protein structure analysis, often performed on crystallographic data, indicates that a run of 5 of these amino acids is usually not essential for maintaining the overall structure of the protein. So creating a gap in the alignment sequence within the hydrophilic stretch, implying an insertion or deletion of amino acids in the protein, would likely not disrupt the overall protein structure. ClustalW reduces the gap opening penalty within a stretch of 5 hydrophilic amino acids. Another rule that ClustalW applies is a specific gap weight for each amino acid. A gap weight is assigned to the position immediately after each of the 20 amino acids. The gap weight is inversely proportional to the frequency with which that amino acid is located upstream of a position that must be gapped in order to align homologous proteins with solved structures. For example, the gap weight of M is 1.29 and the gap weight of G is 0.61. Given the choice between placing a gap after M or G, ClustalW will choose G. Another rule that ClustalW follows is "the rule of 8." The rule of 8 is applied because alignments of homologous proteins with solved structures show that

```
P68871 | HBB_HU      --------MVHLTPEEKSAVTALWGKVN--VDEVGGEALGRLLVVYPWTQRFFESFGDLS  50
P02062 | HBB_HO      ---------VQLSGEEKAAVLALWDKVN--EEEVGGEALGRLLVVYPWTQRFFDSFGDLS  49
P69905 | HBA_HU      ---------MVLSPADKTNVKAAWGKVGAHAGEYGAEALERMFLSFPTTKTYFPHF-DLS  50
P01958 | HBA_HO      ---------MVLSAADKTNVKAAWSKVGGHAGEYGAEALERMFLGFPTTKTYFPHF-DLS  50
P02208.2 | GLB5_PETMA MPIVDTGSVAPLSAAEKTKIRSAWAPVYSTYETSGVDILVKFFTSTPAAQEFFPKFKGLT  60
P02185 | MYG_PHYCA    ---------MVLSEGEWQLVLHVWAKVEADVAGHGQDILIRLFKSHPETLEKFDRFKHLK  51
P02240.2 | LGB2_LUPLU --------MGALTESQAALVKSSWEEFNANIPKHTHRFFILVLEIAPAAKDLFSFLKGTS  52
                             *:   :    :    *    .              :   .:   *  :   *   :     .
```

```
P68871 | HBB_HU      TPDAVMGNPKVKAHGKKVLGAFSDGLAHLDN-----LKGTFATLSELHCDKLHVDPENFR  105
P02062 | HBB_HO      NPGAVMGNPKVKAHGKKVLHSFGEGVHHLDN-----LKGTFAALSELHCDKLHVDPENFR  104
P69905 | HBA_HU      -----HGSAQVKGHGKKVADALTNAVAHVDD-----MPNALSALSDLHAHKLRVDPENFK  100
P01958 | HBA_HO      -----HGSAQVKAHGKKVGDALTLAVGHLDD-----LPGALSNLSDLHAHKLRVDPENFK  100
P02208.2 | GLB5_PETMA TADQLKKSADVRWHAERIINAVNDAVASMDDT--EKMSMKLRDLSGKHAKSFQVDPQYFK  118
P02185 | MYG_PHYCA    TEAEMKASEDLKKHGVTVLTALGAILKKKGH-----HEAELKPLAQSHATKHKIPIKYLE  106
P02240.2 | LGB2_LUPLU EVP--QNNPELQAHAGKVFKLVYEAAIQLQVTGVVVTDATLKNLGSVHVSKG-VADAHFP  109
                           .  .:: *.   .                               :   *.   *   .   :       :
```

```
P68871 | HBB_HU      LLGNVLVCVLAHHFGKEFTPPVQAAYQKVVAGVANALAHKYH------  50
P02062 | HBB_HO      LLGNVLVVVLARHFGKDFTPELQASYQKVVAGVANALAHKYH------  49
P69905 | HBA_HU      LLSHCLLVTLAAHLPAEFTPAVHASLDKFLASVSTVLTSKYR------  50
P01958 | HBA_HO      LLSHCLLSTLAVHLPNDFTPAVHASLDKFLSSVSTVLTSKYR------  50
P02208.2 | GLB5_PETMA VLAAVIADTVAAG---------DAGFEKLMSMICILLRSAY-------  60
P02185 | MYG_PHYCA    FISEAIIHVLHSRHPGDFGADAQGAMNKALELFRKDIAAKYKELGYQG  51
P02240.2 | LGB2_LUPLU VVKEAILKTIKEVVGAKWSEELNSAWTIAYDELAIVIKKEMNDAA---  52
                         .:    :   .:            ...             .   :
```

protein gaps for optimal alignment are not more frequent than once every 8 residues. Therefore, penalties for gaps increase when required at a frequency of 8 or fewer amino acids to achieve alignment. These rules help to make ClustalW a popular and effective program for aligning multiple sequences.

Practical concerns when working with ClustalW

Progressive alignment programs such as ClustalW fall into a class of programs called "greedy heuristics." In sum, ClustalW starts with pairwise alignments of sequences and generates a distance matrix. This matrix is used to generate a guide tree. ClustalW creates alignments by following the branches of the tree, starting from the shortest branches. If an error is made in the first set of alignments, this error is then propagated through the rest of the alignments and cannot be corrected later as subsequent sequences are added in. To mitigate this error, it is critical that closely related sequences be included in the set of sequences to be aligned. The alignment of the closely related sequences sets up a template for future alignments as the program progresses through the guide tree. To create this template, it is best to choose sequences that are similar to each other over their entire lengths.

Multiple sequence alignment programs are often updated to improve accuracy and speed. An improved version of ClustalW is Clustal Omega. Clustal Omega uses a modified version of a program called mBed to "emBed" each sequence in a space of *n* dimensions. Each sequence is replaced by an *n* element vector, where each element is the distance to one of *n* "reference sequences." These vectors can be clustered by UPGMA (see Chapter 8) or K-means. Alignments are then computed by the software program HHalign, which aligns sequences with two profile hidden Markov models.[15]

FIG. 6-12. Multiple sequence alignment of globin sequences using ClustalW (version 2.1). An asterisk denotes positions that have identical amino acids. A colon denotes positions that have amino acids with highly similar properties. A period denotes positions that have amino acids with weakly similar properties. Amino acid color code is as follows: red = small or hydrophobic; blue = acidic; magenta = basic (with the exception of H); green = polar + G + H.

Clustal Omega A multiple sequence alignment program that uses unweighted pair group method with arithmetic mean (UPGMA) to cluster sequences and hidden Markov models to align sequences.

---

[15] Hidden Markov models are discussed in more detail in Chapter 12.

Accuracy of multiple sequence alignment programs can be assessed by testing their ability to align sequences from protein segments with known structures. Protein segments with slightly different sequences but the same structure should be reliably aligned with robust multiple sequence alignment programs. One database with such

protein segments is BAliBASE, which contains more than 1,100 sequences. In one study Clustal Omega compared favorably to other popular multiple sequence alignment programs including T-COFFEE and MUSCLE.[16]

---

[16] Sievers et al. (2011).

SUMMARY

In this chapter, you learned how the BLAST program performs pairwise alignment between a query and a subject sequence. BLAST can quickly perform local alignments on large databases because it does not perform a Smith-Waterman analysis between each amino acid in the query and each amino acid in the database. We introduced you to the statistics used to evaluate the significance of the BLAST outputs—an essential consideration to rule out false positives. We also briefly showed you how to use the suite of different BLAST programs offered by the NCBI to answer, or at least begin to answer, specific biological questions. One interesting question is, how do you identify homologs of proteins from distantly related species? One approach is to use PSI-BLAST, a program that generates a PSSM to capture these homologs. In our discussion, we used PSI-BLAST to detect potential homologs of human p53. One putative p53 homolog we identified is Bm1_11215, a hypothetical protein from *B. malayi*—a roundworm that causes elephantiasis.

Of course, experimental tests are needed to confirm that this protein is actually expressed in the roundworm and that it is indeed a p53 homolog.

Part of the PSI-BLAST program requires a multiple sequence alignment step to create the PSSM. The PSSM is then used to score HSPs in successive BLAST runs. We expanded on the topic of multiple sequence alignment to discuss a popular program that creates alignments of more than two sequences—ClustalW. We went into detail about how ClustalW creates a distance matrix and how the distance matrix and the BLOSUM amino acid substitution matrix series are utilized to create a multiple sequence alignment. We explained how to read a ClustalW output. Recently, ClustalW has been replaced by Clustal Omega, an improved version of ClustalW that can align hundreds of sequences in a timely fashion. In the next chapter, we will explore protein structure, which has been greatly influenced by recent advances in the bioinformatics field.

EXERCISES

1. Locate the C-terminal region (approximately 215 residues) of human BRCA1 from the UniProtKB accession number P38398, isoform 1. Perform a PSI-BLAST search of the nr protein database with this query sequence. Save your search results. Now perform a second iteration. Compare your new search results to the first search. Some sequence alignments from the second search have higher HSP scores than the same sequence alignment obtained from the first search. Why? Alternatively, some sequence alignments from the second search have lower HSP scores than the same sequence alignments obtained from the first search. Why?

2. MDM2, an oncoprotein, is an inhibitor of p53 and is observed to be overexpressed in about 7% of human cancers. In sarcomas, the frequency of MDM2 overexpression is 20%. MDM2 belongs to a class of proteins known as E3 ligases. MDM2 transfers the small protein ubiquitin onto p53, which tags p53 for destruction by the 26S

proteasome. An analysis of the conserved regions of MDM2 assists scientists to uncover regions that are critical for ubiquitin transfer activity. Obtain the following protein sequences from any public database you wish and align using Clustal program. The protein sequences are: human MDM2, chimpanzee MDM2, mouse MDM2, Xenopus MDM2, and zebrafish MDM2. For your alignment use the longest sequences available. Give amino acid ranges of *two* areas that are conserved among these orthologs. Use the human MDM2 amino acid numbers to report on your ranges. According to the alignment score table, which pairwise alignment has the highest score? Perform Clustal again using only human MDM2 and zebrafish MDM2 sequences. Does the human/zebrafish alignment in this output, differ from the human/zebrafish alignment obtained in the first output? Explain.

3. There exists a paralog of MDM2 named MDM4 (also known as MDMX). MDM4 is found to be

overexpressed in nervous tissue cancers, breast cancers, and soft tissue tumors at a frequency of 10–25%. Obtain the sequence of the human paralog MDM4 and the mouse paralog MDM4 and perform multiple sequence alignment together with the original five MDM2 sequences listed in problem 2. Give amino acid ranges of the domains (in human MDM2 amino acid numbers) that are highly conserved within sequences of this entire homolog family.

4. The quagga is an animal indigenous to Africa that is now extinct. It looked partly like a donkey and partly like a zebra. In 1872, the last living quagga was photographed. Mitochondrial DNA was obtained from a museum quagga specimen and sequenced.

a. Perform a BLAST search with the quagga (*Equus burchellii quagga*) mitochondrial 12S ribosomal DNA query and find out if the quagga is more closely related to the donkey (*Equus asinus*) or the zebra (*Equus zebra*). Support your answer with data.

b. Perform a Clustal analysis of mitochondria 12S ribosomal DNA sequences from the three species. Along with the outputted multiple sequence alignment an identity matrix is created. Use the identity matrix to confirm the conclusion you made from the BLAST analysis in part a.

ANSWERS TO THOUGHT QUESTIONS

6-1. The length of the UniProtKP database is approximately 16 times longer than the length of PDB because the ratio of the E-values for a particular HSP is 16.[17]

---

[17] The equation for calculating E slightly accounts for database size with scaling factor K, but the ratio of the database lengths will not be greatly altered.

6-2.

Gln CONVERSATION RANK	Gln POSITION NUMBER
1	14
2	12
3	4

Met and Glu are often substituted for Gln at position 12.

REFERENCES

Altschul, S. F., W. Gish, W. Miller, E. W. Myers, and D. J. Lipman. 1990. "Basic Local Alignment Search Tool." *Journal of Molecular Biology* 215: 403–410.

Altschul, S. F., T. L. Madden, A. A. Schäffer, J. Zhang, Z. Zhang, W. Miller, and D. J. Lipman. 1997. "Gapped BLAST and PSI-BLAST: A New Generation of Protein Database Search Programs." *Nucleic Acids Research* 25: 3389–3402.

Boratyn, G. M., A. A. Schäffer, R. Agarwala, S. F. Altschul, D. J. Lipman, and T. L. Madden. 2012. "Domain Enhanced Lookup Time Accelerated BLAST." *Biology Direct* 7, no. 12. doi:10.1186/1745-6150-7-12.

Cristianini, N., and M. W. Hahn. 2007. *Introduction to Computational Genomics: A Computational Approach*. Cambridge: Cambridge University Press.

Derry, W. B., A. P. Putzke, and J. H. Rothman. 2001. "Caenorhabditis elegans p53: Role in Apoptosis, Meiosis, and Stress Resistance." *Science* 294: 591–595.

Feng, D., and R. F. Doolittle. 1987. "Progressive Sequence Alignment as a Prerequisite to Correct Phylogenetic Trees." *Journal of Molecular Evolution* 60: 351–360.

Goodson, M. S., W. J. Crookes-Goodson, J. R. Kimbell, and M. J. McFall-Ngai. 2006. "Characterization and Role of p53 Family Members in the Symbiont-Induced Morphogenesis of the *Euprymna scolopes* Light Organ." *Biology Bulletin* 211: 7–17.

Goujon, M., H. McWilliam, W. Li, F. Valentin, S. Squizzato, J. Paern, and R. Lopez. 2010. "A New Bioinformatics Analysis Tools Framework at EMBL-EBI." *Nucleic Acids Research* 38 (Web Server issue): W695–699.

Higgins, D. G., and P. M. Sharp. 1988. "CLUSTAL: A Package for Performing Multiple Sequence Alignment on a Microcomputer." *Gene* 73: 237–244.

Higgs, P. G., and T. K. Attwood. 2005. *Bioinformatics and Molecular Evolution*. Malden, MA: Blackwell Publishing.

Karlin, S., and S. F. Altschul. 1993. "Applications and Statistics for Multiple High-Scoring Segments in Molecular Sequences." *Proceedings of the National Academy of Sciences USA* 90: 5873–5877.

Larkin, M. A., G. Blackshields, N. P. Brown, R. Chenna, P. A. McGettigan, H. McWilliam, F. Valentin, I. M. Wallace, A. Wilm, R. Lopez, J. D. Thompson, T. J. Gibson, and D. G. Higgins. 2007. "Clustal W and Clustal X Version 2.0." *Bioinformatics* 23: 2947–2948.

McWilliam, H., W. Li, M. Uludag, S. Squizzato, Y. M. Park, N. Buso, A. P. Cowley, and R. Lopez. 2013. "Analysis Tool Web Services from the EMBL-EBI." *Nucleic Acids Research* 41 (Web Server issue): W597–600.

NCBI. 2010. "Introduction to Molecular Biology Information Resources: Slide List for This Module [BLAST]."

Accessed November 30. http://www.ncbi.nlm.nih.gov/Class/MLACourse/Modules/BLAST/slide_list.html.

Saitou, N., and M. Nei. 1987. "The Neighbor-Joining Method: A New Method for Reconstructing Phylogenetic Trees." *Molecular Biology and Evolution* 4: 406–425.

Schuler, G. D. 2001. "Sequence Alignment and Database Searching." In *Bioinformatics: A Practical Guide to the Analysis of Genes and Proteins*, 2nd ed., edited by A. D. Baxevanis and B. F. F. Ouellette, 187–214. New York: John Wiley & Sons.

Sievers, F., A. Wilm, D. Dineen, T. J. Gibson, K. Karplus, W. Li, R. Lopez, H. McWilliam, M. Remmert, J. Söding, J. D. Thompson, and D. G. Higgins. 2011. "Fast, Scalable Generation of High-Quality Protein Multiple Sequence Alignments Using Clustal Omega." *Molecular Systems Biology* 7: 539.

Tatusov, R. L., S. F. Altschul, and E. V. Koonin. 1994. "Detection of Conserved Segments in Proteins: Iterative Scanning of Sequence Databases with Alignment Blocks." *Proceedings of the National Academy of Sciences USA* 91: 12091–12095.

Thompson, J. D., D. G. Higgins, and T. J. Gibson. 1994. "CLUSTAL W: Improving the Sensitivity of Progressive Multiple Sequence Alignment Through Sequence Weighting, Position-Specific Gap Penalties and Weight Matrix Choice." *Nucleic Acids Research* 22: 4673–4680.

Yu, Y. K., J. C. Wootton, and S. F. Altschul. 2003. "The Compositional Adjustment of Amino Acid Substitution Matrices." *Proceedings of the National Academy of Sciences USA* 100: 15688–15693.

Zvelebil, M., and J. O. Baum. 2008. *Understanding Bioinformatics*. New York: Garland Science, Taylor & Francis Group.

PROTEIN STRUCTURE PREDICTION

7.1   INTRODUCTION

In Chapter 6 you learned that multiple sequence alignment programs can be used to detect conserved regions within protein sequences. These regions often have common structural features that are necessary for the proteins to carry out their biological functions. Recall from Chapter 1 that proteins can have up to four hierarchical levels of protein structure: primary, secondary, tertiary, and quaternary structure (Figure 7-1). The primary structure is the order of amino acids that, by convention, is written left to right from the N-terminus to the C-terminus. The secondary structure is the initial fold of the protein of which the three fundamental forms are α-helix, β-sheet, and loop (or turn). The tertiary structure is a composite of interacting secondary structures in a single polypeptide. The quaternary structure is a composite of two or more polypeptides that stably interact with each other.

In this chapter you will learn how secondary and tertiary structures of proteins are experimentally determined and how bioinformatics programs can be used to visualize these structures. However, the structures of many proteins, especially those associated with membranes, are difficult to elucidate through experiments. You will also learn about bioinformatics software programs that predict the structures of such proteins. Such predictions can often be verified through biochemical experiments.

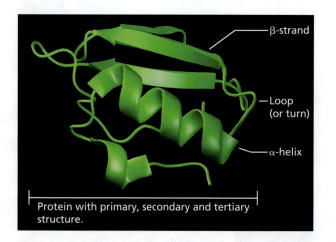

Protein with primary, secondary and tertiary structure.

FIG. 7-1. Hierarchical levels of protein structure. For simplicity, quaternary structure is not shown. A β-sheet is composed of β-strands.

Protein Data Bank A database that contains Cartesian coordinates of atoms of biomolecules. The majority of the coordinates are derived from X-ray crystallography experiments. Molecular viewers can be used to display the atoms on a computer screen.

Experimental determination of protein structures has made tremendous progress since 1960, when the first detailed structure of myoglobin was published by John Kendrew and his colleagues (at the Molecular Research Council laboratory in Cambridge, United Kingdom) using X-ray diffraction.[1] Today, the two primary methods for protein structure determination are X-ray diffraction and nuclear magnetic resonance (NMR) spectroscopy. Structure data on proteins and other biological macromolecules is stored in the Protein Data Bank (PDB). In fact, most academic journals will not publish structure data unless it has been deposited in the PDB. Approximately 80% of the files in the PDB are structures solved by X-ray crystallography, and the remainder are solved by NMR experiments.[2] Experimental structure determination is laborious and time-consuming, but the benefits are many. Structures give scientists an insight into how enzymes catalyze reactions. Structures are critical for understanding how drugs exert their effects on proteins and creating opportunities for improving drug efficacy. Structures show how proteins interact with DNA, RNA, and other proteins. It can even be said that structures illuminate the beauty of nature.

Due to the difficulty of protein structure determination and the relative ease of DNA sequencing, there is a widening gap between our knowledge of protein structures and our knowledge of protein sequences. This chasm has created an opportunity for bioinformaticists to contribute to the field of structure determination by predicting protein structures starting from sequence information.

Even prior to the development of the bioinformatics field, scientists have used protein sequence information to attempt to predict protein structures. In 1960 this challenge presented itself to the scientific community when it became clear that a linear protein sequence could correctly fold into its physiological structure without the aid of other molecules. In other words, the amino acids themselves contain instructions for a protein to fold into its correct structure. From this discovery, five main categories of structure prediction eventually developed: computational methods, amino acid residue propensity methods, artificial neural networks, homology modeling, and threading.

The earliest category of structure prediction to be established is **computational methods**. One computational method is called *ab initio*,[3] and it predicts structures by utilizing the physical and chemical properties of nuclei and electrons of the amino acid residues and the surrounding solvent. *Ab initio* methods employ quantum mechanical equations to describe the behavior of the electrons and atoms. Another computational method is molecular mechanics, which describes the interactions of atoms with each other. It uses the principles of Newtonian physics and properties of known structures to model protein structures.

A second category of structure prediction uses **amino acid propensities for secondary structures** to predict structures. When the first protein structures were experimentally solved in the 1960s, it was apparent that segments of proteins adopt one of three secondary structures: α-helix, β-sheet, or loop.[4] Today, these remain the three major classes of secondary structure, although sometimes loops are

---

[1] Kendrew et al., 1960.

[2] Structures derived from NMR experiments are often called solution structures because the proteins remain dissolved in a solvent during NMR analysis. On the other hand, the X-ray diffraction technique requires proteins to be in a crystalline form.

[3] *Ab initio* is Latin for "from the beginning."

[4] Sometimes "β-strand" is used instead of "β-sheet" because β-strands come together to form a sheet-like structure. An older term for loop is random coil, but the term random is troubling because these protein regions often have specific nonrandom structures that happen to fall outside the α-helix and β-strand categories.

subcategorized into turns and coils. Protein structure experts noticed that certain residues tend to be found in particular secondary structures. For example, proline is found in β-sheets and loops but rarely in α-helices. In the mid- to late 1970s Peter Chou and Gerald Fasman (Brandeis University, Waltham, Massachusetts) cataloged these amino acid propensities and used them to create an algorithm that predicts secondary structures of proteins. As more structures became experimentally solved, this method evolved to be fairly accurate.

A third category of structure prediction is the **artificial neural network**, or **neural network** (NN) for short. A computer neural network attempts to model computer software programs after neurons in the brain. In this model, the computer program receives many signals through an input layer of "neurons" known as units. The data from the units are transformed and are fed to a second, or output, layer of units. The second layer of units makes a decision based on the inputs from the first layer. In a computer program, a neural network can be trained to give an output of α-helix, β-strand, or loop to describe the secondary structure of an amino acid in proteins of known structures. A window of sequence (typically 13–17 amino acid residues) is fed to the input layer of units. The output layer gives the secondary structure prediction of the amino acid residue in the center of the window. Some off-center residues in the window influence the secondary structure of the center residue. Depending on its location and type, a weight is placed on the off-center residues, signifying their contribution to the prediction of the center amino acid residue structure.

Using the data from the training set of known structures, weights from the input layer are adjusted and transmitted to the output neurons. The weights are adjusted by the computer program so that correct predictions are given as output. Once the weights are set and the NN is tested for accuracy, the computer program can predict secondary structures from protein sequences with unknown structures. Like the Chou-Fasman method, information from known structures is used to create a neural network program that will predict structures solely from protein sequences. The NN method is currently the most accurate at predicting secondary structures.

A fourth category for predicting structures is **homology modeling**. With homology modeling the goal is to predict the tertiary structure of a protein sequence (query sequence) using a template. Homology modeling requires a three-dimensional template from which to derive the structure. The template comes from one or more sequences that have known structure. The key to homology modeling is that the query sequence and the template sequence share at least 50% identity. The steps to homology modeling are: (1) template search and selection, (2) building a multiple sequence alignment, (3) assignment of spatial coordinates to the query sequence, (4) model refinement, and (5) model evaluation. Upon completion of step 5, a model of the structure is built.

The fifth category that we will discuss is **threading**. Threading is useful for creating models of tertiary structures when there is low sequence identity between the query sequence and available three-dimensional templates. Threading is predicated on the fact that experimental data has shown that two sequences with low identity can, nonetheless, possess nearly the same tertiary structure. Like homology modeling, a template is required to build a model of the structure. However, the template is selected from a combination of properties. Prominent among these properties are sequence alignment between the query sequence and the potential templates, compatible folding patterns between the query and the potential templates, and predicted query sequence solvation potential (i.e., how well the amino acid side chains of the query sequence are stabilized in the local environment within the template). Threading analyzes each amino acid in the query for its capability of forming a structure compatible with potential templates found in a protein fold library. The final step to building a model is to use one template from the fold library to build a structural model of the query sequence, much like homology modeling.

Depending on the software program, these prediction methods have the capability of predicting secondary or tertiary levels of protein structure. In practice, the secondary structure prediction programs can be quite accurate and the tertiary

homology modeling A protein structure prediction method that uses a protein template of known structure to build a structural model of a sequence. This method works well when the percent sequence identity is ≥50%.

threading A protein structure prediction method that uses the CATH database to predict the structure of each residue of a protein sequence. Each residue is tested for optimal compatibility with each fold in the CATH database.

structure prediction programs less so. Quaternary structure prediction is still in its infancy. This chapter is not an exhaustive discussion of structure prediction. Instead, it gives you a taste of the current activities in this fascinating area of bioinformatics. Prior to diving into structure prediction methods, it is critical to understand the experimental methods of structure determination. Here we discuss the two predominant experimental methods, X-ray crystallography and NMR spectroscopy.

7.2 EXPERIMENTAL METHODS OF STRUCTURE DETERMINATION

X-ray Crystallography

X-ray crystallography A method that uses X-ray bombardment of crystallized molecules to create a diffraction pattern. The diffraction pattern is used to create a three-dimensional arrangement of the atoms in the molecules.

X-ray diffraction of crystals began in 1912 when the father and son team of William Henry Bragg and William Lawrence Bragg got the idea that X-ray diffraction patterns could be used to elucidate the three-dimensional arrangement of atoms in space. They discovered a new law, Bragg's law, which precisely describes the angles of scattering of X-rays upon colliding with electron clouds of atoms.[5] A film is placed near the protein crystal that is subjected to X-ray bombardment. A pattern of spots is created on film caused by X-rays deflecting from atoms. From this pattern the Braggs determined the spatial arrangement of the atoms in the crystal. They experimented on a few simple crystals—sodium chloride, zinc sulfide, diamond—to prove that their law is robust, but it would take almost another 40 years for X-ray crystallography to advance to the point of accurately describing the three-dimensional arrangement of atoms in a protein.

As we discuss the subject of structure analysis it is important to bear in mind the resolution of the structure. Resolution is a measure of the smallest detail that can be distinguished by a sensor system. The human eye has the ability to distinguish two points that are ≥ 0.2 millimeters apart. We can say that the resolution limit of the eye is 0.2 millimeters. A higher resolution means that two points can be distinguished at a shorter distance. The ability to resolve points separated by a shorter distance is increased with the aid of a microscope. In fact, the resolution power of a light microscope is limited only by the wavelength of visible light used to examine a sample. Resolution, r, is defined as:

$$r = \lambda/2$$

where λ is the wavelength of the light or electromagnetic radiation.

The shortest wavelength of visible light is a violet color with a wavelength of 4,000 Å, which makes the resolution limit of a light microscope 2,000 Å. In proteins the distance between atoms is 1–2 Å, so it is impossible to use a light microscope to detect individual atoms. However, the wavelength of X-rays (a form of electromagnetic radiation) ranges from 0.1 to 100 Å, which means that individual atoms can be distinguished with X-rays.

There are some challenges when we use X-rays for protein structure determination. Although X-rays have the correct wavelength for resolving atoms, unlike visible light X-rays cannot be focused and the atoms of proteins deflect the X-rays only weakly. Additionally, unless all the molecules are oriented in the same way, X-rays will scatter in different directions. Therefore, proteins must be arranged identically in space and be highly concentrated. To meet these requirements, the proteins must be crystallized; hence this type of analysis is called X-ray crystallography.

---

[5] The Braggs were the only father-son team to jointly win the Nobel Prize (1915). The son, William Lawrence, was 25 years old when he received the Nobel Prize, making him the youngest recipient ever to win a Nobel Prize in science or literature. William Lawrence Bragg went on to head the Cavendish laboratory (later renamed the MRC laboratory) at Cambridge University, United Kingdom. He hired scientists who used X-ray crystallography to determine the structures of proteins and help solve the structure of DNA.

X-ray crystallography, step by step

There are several steps to X-ray crystallography. Note that many proteins do not crystallize and those that do may require months to form crystals large enough for X-ray diffraction. Crystallization is not an exact science. No single method or procedure causes proteins to form high-quality crystals suitable for structure analysis, which makes crystallization the major hurdle in this field. Several companies sell kits with solutions that have different buffers and salts to "coax" proteins into forming crystals. The following is a brief summary of the steps to X-ray crystallography:

STEP 1. *Isolation and purification of the protein.* The protein must be purified for crystallization. Typically, recombinant DNA of a single domain is over-expressed in bacteria.[6] It is rare for a multidomain protein to be crystallized because the domains are typically joined by flexible regions, which are naturally unstructured and inhibit the crystallization process. In the case of p53, the DNA binding domain was first crystallized and its structure was determined. Later, the tetramerization domain and transactivation domain were separately purified and crystallized. Sometimes, amino acid residues that are sensitive to oxidation, such as cysteine and methionine, are replaced by other residues because random reaction of their side chains with oxygen molecules causes the protein molecules to be heterogeneous, which inhibits crystallization. Although there is no theoretical limit to the size of a protein that may be studied by X-ray crystallography, in practice it is typical for a protein to be no more than 30 kDa in mass for successful crystallization.

STEP 2. *Crystallization.* A protein with a mass in the 10–30 kDa range must be concentrated to approximately 10 milligrams per milliliter. The pH of the crystallization solution should be far away from the isoelectric point (pI)[7] of the protein. This ensures that the protein remains charged in solution and is therefore less likely to have interactions that unfold the protein and cause the protein to aggregate and precipitate (form a solid) without crystals forming. Such precipitates are not desired because the protein structure is destroyed (denatured—more on this subject later in this chapter). The pI of the protein can be calculated with a computer program. If, for example, the pI of a protein is 4.5, then a solution with a pH that is 7 or 8 may be used to ensure that the protein remains negatively charged in solution. Protein crystals must have a minimum dimension of 20 μm^3 before treatment with X-rays. The hanging drop method of crystallization is a common technique for crystallization (Figure 7-2).

  In this method, a hanging drop containing the protein, water, and salts is suspended from a horizontal plate. The drop is suspended above a reservoir that contains a high concentration of salt, but no protein. Because there is a higher salt concentration in the reservoir, the water in the hanging drop evaporates, causing the protein to increase in concentration and co-crystallize with the salt.

isoelectric point The pH of a solution where a dissolved molecule has a net neutral charge.

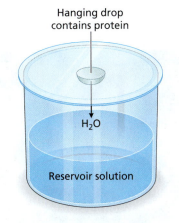

FIG. 7-2. The hanging drop method of protein crystallization. The salt concentration in the hanging drop suspended from the glass lid is half that of the salt concentration in the reservoir solution. To achieve equilibrium, the water evaporates from the more dilute hanging drop and condenses into the more concentrated reservoir solution. The lowered water concentration in the hanging drop increases the protein concentration, gradually causing it to crystallize.

---

[6] Recombinant DNA is DNA from two or more sources that have been put together. Often a DNA that codes for a human protein is spliced into a circular DNA from bacteria known as a plasmid. The plasmid has a promoter that drives the production of many mRNA transcripts from the human DNA. The plasmid is placed (transformed) into bacteria. The bacteria transcribe and translate the human DNA. Because the promoter is usually very strong, the production of protein in this manner is called overexpression. The protein is purified and used for crystallization trials. Sometimes the recombinant DNA that is overexpressed is a variant of the original protein of interest in order to increase the likelihood of getting good crystals.

[7] The isoelectric point determines the net charge of the protein at a given pH. If the pH of the solution equals the pI of the protein, the net charge on the protein is zero. The protein charge is caused by the N-terminus, C-terminus, and amino acid residues with ionizable side chains (His, Lys, Arg, Asp, Glu).

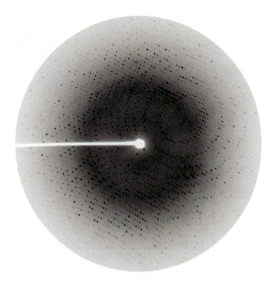

FIG. 7-3. Diffraction pattern of APS kinase.

This process can take weeks to months. Sometimes the protein is mixed with a partner molecule that it naturally binds to in the organism. The protein-partner molecule complex can help the protein crystallize and gives insight into how the two molecules interact in their natural environment. For example, the p53 DNA binding domain was co-crystallized with its DNA target (see Box 7-1).

STEP 3. *X-ray irradiation and data collection.* A single crystal is kept at −180°C or lower to decrease the rate of degradation of the crystal during irradiation. The crystal is mounted on an instrument called a goniometer, which rotates the crystal at a speed and angle that is dictated by the crystal type. The crystal type is assessed by microscopic observation. X-ray sources are often metals such as Mo, Cu, and Co. The best X-ray source is the synchrotron—a large particle accelerator that produces monochromatic X-rays of high luminosity. The X-ray beam is directed at the rotating crystal and a diffraction pattern of spots (reflections) is captured by a detector (Figure 7-3). The intensity and spacing of the spots are used to calculate the atom locations in the crystal. In Figure 7-3, the white center of the diffraction pattern is caused by the X-ray trap for the incident beam. This is where the majority of the X-rays land, and it is bleached out to prevent background darkness in the image. The narrow white line from the center to the circumference is the route of the X-ray beam from the X-ray source to the crystal. The X-ray detector does not collect scattered X-rays in this area.

STEP 4. *Diffraction pattern analysis.* X-rays are diffracted by the surface electrons of the atoms. There is a relationship between the diffraction pattern and the density of the electrons. If we assume that the electron density around the atoms represents a mathematical function, the resulting diffraction pattern is the Fourier transform of this function. Through a mathematical treatment called the inverse Fourier transform, we can obtain the electron densities of the atoms from the diffraction pattern. To build an electron density map of the molecule, two properties of the diffraction spots must be obtained: the amplitude and the phase. Teasing out these two properties from the diffraction pattern is not trivial, and other sources listed at the end of this chapter explain this subject in more detail. For our discussion, suffice it to say that molecular structures are built by fitting atoms and bonds into an electron density map (Figure 7-4). Because hydrogen atoms have very little electron density, they are usually not observed. Often there are several rounds of data collection, electron density map generation, and model building to fit the electron density map to the nonhydrogen atoms. With each round the model matches more closely with the data collected.

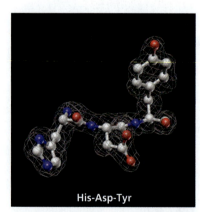

His-Asp-Tyr

FIG. 7-4. Amino acid residues fitted into an electron density map. The wire mesh represents the electron density map generated from X-ray diffraction data. The solid atoms and bonds are components of residues fitted into the electron density map. Note that hydrogen atoms are not shown because their electron densities are too small. The sequence is listed below the map.

It is generally held that as long as the structure is solved at a suitable resolution, say below 3 Å,[8] the structure is reliable, but this is not always the case. A better measure of the correctness of the structure is the R-factor. The R-factor corresponds to the difference in the data collected and the structure predicted by the model. A low R-factor indicates a good fit between the data and the structure. For proteins, an R-factor in the 0.15–0.25 range is considered satisfactory. Unfortunately, R-factors are not always reported when a new structure is submitted for publication.

It is almost always the case that some regions of the structure will be better resolved than others. Areas of high resolution include the segments buried beneath the

---

[8] Note that the diameter of a hydrogen atom is 1 Å.

BOX 7-1 | **A CLOSER LOOK**

p53 Co-crystallized with DNA Reveals Insights into Cancer

When it became clear that p53 is a DNA binding protein and its gene is frequently mutated in cancers, there was an interest in determining its structure. Early attempts to obtain crystals failed because p53 has multiple domains connected by highly flexible segments. Nikola Pavletich and his colleagues at the Memorial Sloan-Kettering Cancer Center (New York City) used limited proteolysis to show that the DNA binding domain is stable and is connected through flexible segments to other parts of the protein. After treatment with the protease, they sequenced the p53 DNA binding domain to determine its beginning and ending amino acids. They created a DNA binding domain of human p53 (amino acids 94–292) and overexpressed the protein in *E. coli*. They co-crystallized the purified protein with a synthetic DNA molecule (called the target DNA) with a sequence that binds to p53 and reported a structure with a resolution of 2.2 Å (PDB file name is 1TSR).

As stated in Chapter 3, the cancer mutations in humans cluster in the DNA binding domain of p53. Earlier studies showed that, in the majority of cases, the mutations are missense mutations that prevent p53 from binding to its DNA targets in the promoters of genes. Hotspot mutations are located at Arg175 (R175), Arg248 (R248), and Arg273 (R273). Other mutations that are frequently found in cancers are at Gly245 (G245) and Arg282 (R282). The crystal structure shows that R248 and R273 interact with the target DNA. R248 anchors p53 to the DNA minor groove with four hydrogen bonds, and R273 anchors the p53 to the phosphate backbone of one of the DNA nucleotides. These critical amino acid residues are located at or near loops that project, like fingers, into the DNA target. One can readily see that when these amino acid residues get mutated, p53 can no longer bind to DNA. What about the residues that do not interact with DNA? Why are they found in cancers as well? It turns out that

R175, G245, R249, and R248 are necessary for stabilizing the native structure of p53.

Sensitive biochemistry experiments showed that mutations of certain residues lead to p53 denaturation. R175, G245, R249, and R248 create critical interactions with internal residues that keep p53 tightly packed. A close look at the spectrum of mutations in p53 (see Chapter 3) shows that virtually all residues in the p53 DNA binding domain are mutated in cancers. It is likely that the location and type of mutation is due a combination of the types of carcinogens the *TP53* gene is exposed to as well as the mutation's effectiveness in inhibiting the p53 DNA binding activity.

There is a third consideration as well. All humans have two *TP53* alleles. In the beginning stage of conversion of a normal cell to a cancer cell, it is likely that a point mutation occurs in one allele while the other allele remains normal. For the cancer cell to survive, it would be helpful for the protein with the mutation to be dominant over the normal protein. In fact, this happens because p53 acts as a tetramer. The mutant protein can "poison" the normal protein by forming part of the p53 tetramer. When a mutation knocks

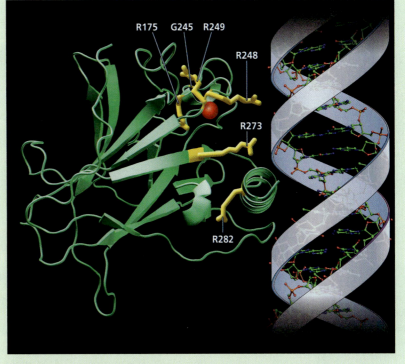

Chain B of the p53 DNA binding domain bound to its DNA target. The backbone of p53 is green. The side chains of amino acids frequently mutated in cancers are shown in yellow. The red sphere is a zinc atom held by p53.

continued

p53 Co-crystallized with DNA Reveals Insights into Cancer

out an activity (DNA binding in this case) and the mutant protein acts dominantly over its normal allele (by poisoning the tetramer in this case), it is called a dominant-negative mutation. It is likely that the dominance is not complete because in most cancer samples the second p53 allele (the normal one) is deleted. The structure of the DNA binding domain bound to DNA presents challenges for small molecule drug design. How can we design a drug that restores p53 DNA binding activity? Some researchers have proposed small molecules that stabilize the p53 native state so that denaturing mutations are less effective. Designing such molecules will likely prove difficult. It is likely that such drugs will be specific for a particular mutation—so the number of patients receiving benefit from such drugs will be small.

Recent studies by Alan Fersht and his colleagues at the Medical Research Council Laboratory of Molecular Biology in the United Kingdom suggest that p53 slides along the DNA through its C-terminal domain. Once it detects a p53 target sequence it clamps down on the DNA with its DNA binding domain. p53 interacts with a large complex of proteins that include RNA polymerase II. RNA polymerase II then transcribes p53 target genes into RNA. On the right is a schematic diagram showing the p53 domains and a structure of the p53 tetramer bound to DNA. The structure was derived from many experimental techniques.

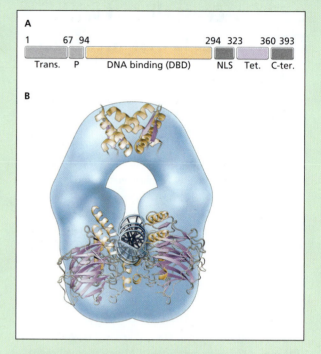

A. p53 functional domains and segments. Numbers represent the amino acid number starting from the N-terminus. **B.** Structure of p53 tetramer derived from X-ray crystallography, NMR spectroscopy, electron microscopy, and other experimental techniques. This conformation is proposed for p53 bound to its target DNA through the DNA binding domains. The yellow molecule in the middle is the target DNA. The DNA binding domain is in close proximity to the target DNA (3 to 9 o'clock). The tetramerization domain is located at 12 o'clock.

REFERENCES

Cho, Y., S. Gorina, P. D. Jeffrey, and N. P. Pavletich. 1994. "Crystal Structure of a p53 Tumor Suppressor-DNA Complex: Understanding Tumorigenic Mutations." *Science* 265: 346–355.

Melero, R., S. Rajagopalan, M. Lázaro, A. C. Joerger, T. Brandt, D. B. Veprintsev, G. Lasso, D. Gil, S. H. Scheres, J. M. Carazo, A. R. Fersht, and M. Valle. 2011. "Electron Microscopy Studies on the Quaternary Structure of p53 Reveal Different Binding Modes for p53 Tetramers in Complex with DNA." *Proceedings of the National Academy of Sciences USA* 108: 557–562.

Friend, S. 1994. "p53: A Glimpse at the Puppet Behind the Shadow Play." *Science* 265: 334–336.

R-factor (reliability factor, residual factor, R-value, R_{work}) A measure of how well the final crystal structure model predicts the observed data used to create the model. An R-factor value in the 0.15–0.25 range is considered satisfactory.

protein surface and those that contain well-ordered secondary structures such as α-helices and β-sheets. Low resolution areas may be flexible or have poor electron densities and are therefore difficult to model. Flexible regions include loops and surface amino acids.

X-ray crystallography has some limitations. It is especially difficult to crystallize membrane-bound proteins, so there is less structural information available for this important class of proteins. Also, when examining a structure obtained by crystallography, it is important to remember that it is a static structure, and therefore it cannot reveal some aspects of protein motion, or may not be able to reveal different protein conformations that contribute to its function. Finally, the protein structure obtained may not be the native structure of the protein in solution due to perturbations that result from being packed into a crystal lattice.

NMR Spectroscopy

Another experimental method for determining molecular structures is nuclear magnetic resonance (NMR) spectroscopy. NMR, based on the method discovered by Isador Rabi in 1938,[9] is the exploitation of the spin of an atomic nucleus to gain insight on its neighboring atoms' nuclei. The following common nuclei have spin: ^{1}H, ^{13}C, ^{15}N, ^{19}F, and ^{31}P. When such atoms are in a molecule that is dissolved in a liquid and placed in a magnetic field created within the NMR instrument, their nuclear spins align in one of two orientations. One alignment is in parallel to the magnetic field and the other is antiparallel. The spins in the antiparallel orientation have higher energy. In proteins, as in all molecules placed in a strong magnetic field, more nuclei will have spins in the low energy alignment. The imbalance of the parallel and antiparallel magnetic moments will create a polarization of the spins, or a net magnetization, that is in a state of equilibrium.

This equilibrium state is destabilized when the NMR instrument fires an energy pulse at the protein sample. When the magnitude of the fired energy pulse is equal to the energy difference between the parallel and antiparallel states, the nuclei momentarily reverse their spins. This is called excitation, and the energy of excitation is called the resonance energy. After the pulse, the nuclei revert to their original spin states. As the nuclei revert, they emit energy that is detected by the NMR instrument. This reversion process is called relaxation. Because energy and frequency are related to one another, scientists often refer to the resonance frequency instead of the resonance energy. The frequencies of energy required for nuclear spin excitation and relaxation in an NMR instrument fall within the range of radiowave frequencies. By convention, the resonance frequency of a nucleus is reported in terms of how much it is shifted relative to that of the resonance frequency of a reference molecule. This convention means resonance frequencies are reported as chemical shifts, in units of parts per million (ppm). Chemical shifts are recorded graphically by the NMR instrument as an NMR spectrum.

A one-dimensional NMR experiment produces a one-dimensional NMR spectrum and provides the scientist with the resonance frequency required to change a given nucleus's spin. Nuclei at different locations in the molecule can be distinguished because the resonant frequency depends on the nearby environment. For small, simple molecules, a scientist may easily interpret the resonance frequencies as well as other features of the one-dimensional NMR spectrum plus some additional data to infer the structure of the molecule.

The difficulty with using NMR to determine protein structure is that there are thousands of signals emitted from ^{1}H or ^{13}C in a sample, and these signals are usually overlapping. In the 1970s Kurt Wüthrich's group, working at Eidgenössische Technische Hochschule (ETH) in Zurich, Switzerland discovered a way to fractionate these signals into a two-dimensional format. This allowed many of the overlapping signals to be resolved on a two-dimensional NMR spectrum. By 1982 Wüthrich's group had developed several NMR techniques that were used collectively to determine the structure of the protein basic pancreatic trypsin.[10]

nuclear magnetic resonance (NMR) spectroscopy The perturbation of atom nuclei by the application of a magnetic field to gain information about the location of neighboring atoms. An NMR spectrum is produced by chemical shifts produced by nuclei resonance frequencies relative to the resonance frequency of a reference molecule.

---

[9] Isador Rabi first discovered nuclear magnetic resonance working on molecular beams (a gas of atoms, molecules, or ions traveling at equal velocities). He placed magnets in close proximity to molecular beams composed of lithium-containing molecules and was able to measure their magnetic moments. He was awarded the Nobel Prize in Physics for this work in 1944. Other scientists expanded the technique for use with liquids and solids.

[10] Kurt Wüthrich was awarded the Nobel Prize in Chemistry in 2002 for the development of nuclear magnetic resonance spectroscopy for determining the three-dimensional structure of biological macromolecules in solution.

A comparison of a complex one-dimensional ¹H-NMR spectrum of a 10.6 kDa protein and two types of better-resolved two-dimensional spectra (¹⁵N-HSQC and ¹³C-HSQC) is provided in Figure 7-5. In the one-dimensional ¹H-NMR spectrum, the resonance frequency of each proton nucleus in a sample is seen as a peak. A two-dimensional NMR experiment is possible because the excited spin state of the nuclei lasts long enough that some of the excitation may be transferred to neighboring nuclei. This transfer allows you to identify the nuclei that are located short

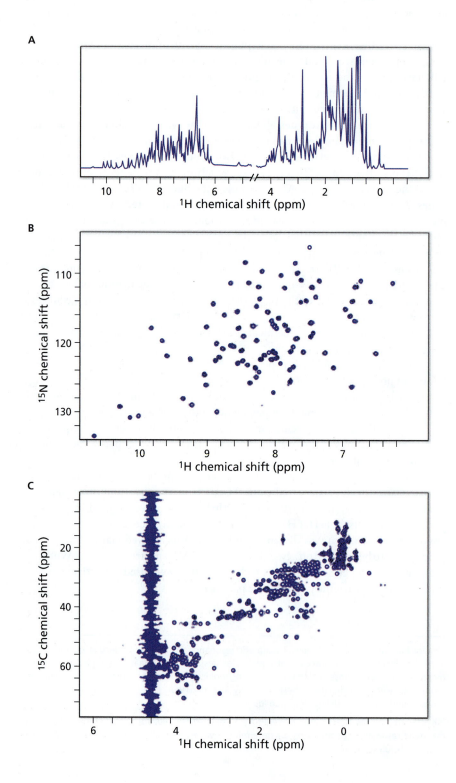

FIG. 7-5. Protein NMR spectra.
A. 1D ¹H-NMR spectrum of CtBP-THAP.
B. ¹⁵N- heteronuclear single-quantum coherence (HSQC) spectrum (2D), showing one peak for each backbone nitrogen and each side chain nitrogen. **C.** ¹³C-HSQC spectrum showing one peak for each bonded ¹H-¹³C in the sample (2D). The natural abundances of ¹⁵N and ¹³C are not high enough to run the NMR experiment shown in this figure, so recombinant proteins must be overproduced in media containing amino acids that are enriched with the isotopes ¹⁵N and ¹³C.

distances from one another. A two-dimensional NMR spectrum looks like a topographical map, in which an absorbance peak is represented by concentric circles. Two nuclei that transfer excitation may be identified by a cross-peak in the two-dimensional spectrum. Each type of amino acid residue has a unique pattern of cross-peaks; even the same amino acid residues in different parts of a protein structure (for example, an Ala near the N-terminus and an Ala near the C-terminus) may have slightly shifted patterns from one another.

Several different types of two-dimensional NMR experiments are needed to produce all the data required to infer a protein structure. Once all of the NMR data is collected, there are additional steps to complete. First, the resonance frequencies of all atoms in the structure must be determined. Next, the distances between close nuclei must be determined. Dihedral angles of bonded sets of atoms are then determined. Hydrogen bonding interactions are identified. These distances, angles, and hydrogen bonds are structural constraints that are then used by computer software to calculate a three-dimensional protein structure that is consistent with all of the NMR data. The result is a group of 20–25 reasonable protein structures.

Using NMR to infer the structure of proteins is also challenging because the protein must be soluble enough (at concentrations on the order of 50 micromolar), and it is very challenging to determine the structure of a protein that is larger than 30 kDa. The advantage of using NMR to determine a protein structure is that it may reveal more than one conformation in solution. Even disordered parts of proteins are visible in NMR, unlike in crystallography.

7.3 INFORMATION DEPOSITED INTO THE PROTEIN DATA BANK

As mentioned in the introduction to this chapter, prior to publication in a journal article the scientist must submit a file containing structure information to the PDB. Along with other information, the atom types, the X, Y, and Z coordinates, and the connectivity of the atoms are necessary. Connectivity describes how the atoms are bonded together. Although the crystal is kept at a very low temperature during X-ray diffraction, atoms still have motion, or kinetic energy, which causes disorder. There are two types of disorder. In *static disorder*, the disorder results from different copies of the molecule (from different unit cells) having different conformations. A unit cell is the simplest, smallest shape from which the overall protein structure can be constructed. A crystal is composed of repeated unit cells (Figure 7-6A). Electron densities within a unit cell are averages of densities of hundreds of copies of atoms in the crystal. In *dynamic disorder*, every copy of the molecule (from different unit cells) in the crystal has its atoms at slightly different locations due to thermal motion. The location of an atom fitted to the electron density map is actually an average location for that atom. The more ill-defined the electron density map due to disorder, the more uncertainty there is in the average location of the atom.

Both static disorder and dynamic disorder contribute to the deviation of the atom from its average location. This deviation is called the B-factor (also known as B-value, Debye-Waller factor, or temperature factor).[11] The B-factor usually ranges from 20 to 80 $Å^2$. A cross section of the structure of galactose/glucose binding protein shows a range of B-factors in this molecule (Figure 7-6B). Note that the amino acid residues on the protein surface have higher B-factors than those that are buried. This is expected because there are fewer packing forces on the surface to stabilize and constrain the atoms.

unit cell A small three-dimensional segment that contains atoms. The environment of the atoms at the vertices formed by the intersection of the unit edges must be identical. The unit can be repeated to create a three-dimensional structure of the molecule.

B-factor (B-value, Debye-Waller factor, temperature factor) A measure of the deviation of an atom from its average location. The B-factor usually ranges from 20 to 80 $Å^2$.

---

[11] B-factor is actually a measure of how spread out the atom's electron density is, which implies how large its range of motion is. It can be modeled by a simple harmonic oscillator, and therefore the B-factor is interpreted as the maximum displacement from the lowest energy equilibrium position of that oscillating atom.

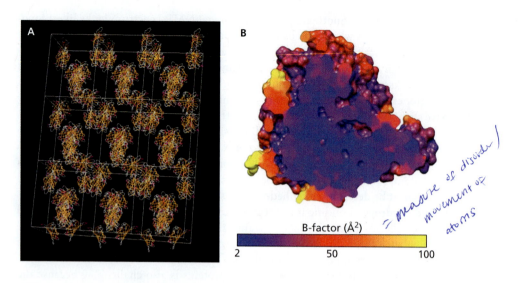

FIG. 7-6. Repeated unit cells and B-factor.
A. A depiction of nine unit cells (3 × 3) packed in a crystal with electron densities converted to protein backbone structures. **B.** Cross section of galactose/glucose binding protein with B-factor scale. Blue denotes smaller ranges of motion, and yellow denotes larger ranges of motion. Note how the B-factor is smaller in the core of the protein and gets larger at the surface of the protein.

TABLE 7-1. RELATIONSHIP BETWEEN CRYSTAL STRUCTURE RESOLUTION AND OBSERVABLE FEATURES IN THE PROTEIN MOLECULE

RESOLUTION (Å)	DESCRIPTION OF STRUCTURAL FEATURES
0.8	Shows hydrogen atoms, covalent bonds, and associated water molecule orientations
1.0–1.2	Shows individual atoms, except for hydrogen atoms
2.0–2.5	Shows backbone and amino acid side chains
3.5	Shows outline of backbone
4.0–5.0	Shows location of secondary structures and overall protein shape

Finally, the average resolution presented in the PDB file refers to the smallest separation between the planes of atoms that diffract the X-rays. Table 7-1 presents a general guide that correlates resolution to observable structural features.

A PDB file consists of several sections, and each can contain one or more record types. One record type is ATOM. Each ATOM record contains data that corresponds to an atom in the unit cell of the crystal. Figure 7-7 shows a small portion of the PDB file 1TSR that contains 14 ATOM records, which cover two amino acid residues. The 1TSR file is the human p53 DNA binding domain bound to its target DNA. The atom names shown in Figure 7-7 are N (nitrogen), CA (alpha carbon), C (carbonyl carbon), and O (carbonyl oxygen); these are considered polypeptide backbone atoms. Side chain atoms are shown as well: CB (beta carbon), CG1 (gamma carbon 1), CG2 (gamma carbon 2), and CD (delta carbon). "Residue name" is the name of the amino acid residue that contains the atoms. "Chain" refers to the polypeptide or other biomolecule chain in the crystal structure. In the 1TSR PDB file there are five chains: A, B, C, E, and F. The chains A, B, and C refer to three identical (in sequence) p53 polypeptides. These chains have slightly different conformations. Chains E and F refer to the two strands of the DNA double helix.

A

	Atom number	Atom name	Residue name	Chain	Residue number	X	Y	Z	Occupancy	B-factor
ATOM	19	N	VAL	A	97	76.757	21.863	73.092	1.00	22.30
ATOM	20	CA	VAL	A	97	76.056	22.304	71.921	1.00	23.28
ATOM	21	C	VAL	A	97	77.108	22.777	70.943	1.00	27.08
ATOM	22	O	VAL	A	97	78.219	23.021	71.371	1.00	29.96
ATOM	23	CB	VAL	A	97	75.293	23.510	72.373	1.00	29.06
ATOM	24	CG1	VAL	A	97	73.968	23.078	72.983	1.00	30.96
ATOM	25	CG2	VAL	A	97	76.150	24.173	73.445	1.00	34.56
ATOM	26	N	PRO	A	98	76.768	22.913	69.654	1.00	26.89
ATOM	27	CA	PRO	A	98	77.731	23.371	68.681	1.00	27.15
ATOM	23	C	PRO	A	98	78.225	24.779	68.961	1.00	30.32
ATOM	29	O	PRO	A	98	77.475	25.686	69.334	1.00	32.71
ATOM	33	CB	PRO	A	98	77.021	23.347	67.336	1.00	24.51
ATOM	31	CG	PRO	A	98	75.876	22.392	67.495	1.00	23.34
ATOM	32	CD	PRO	A	98	75.581	22.342	68.979	1.00	22.39

B

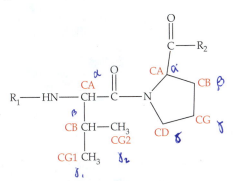

FIG. 7-7. ATOM records.
A. A portion of the ATOM records from the PDB file 1TSR. This file has been annotated to denote atom numbers, atom names, amino acid residue names, polypeptide chain, amino acid residue numbers, Cartesian coordinates of atom locations, occupancy, and B-factors. **B.** Valine and proline amino acid residues with atom nomenclature.

"Residue number" is the amino acid residue number starting from the N-terminus. In the case of 1TSR, amino acid residue numbers in Chain A range from 94 to 289. The next three columns, X, Y, and Z, refer to Cartesian coordinates of the atom in angstrom units. The occupancy column is generally a measure of the fraction of time the atom stays at that location. Sometimes this value is less than 1.00 if the atom is in a flexible region of the protein. The same atom may occupy another set of Cartesian coordinates in the same crystal due to static disorder. The B-factor column has already been discussed. There are other record types in a PDB file that give important information, but we refer the interested reader to the PDB website for further details (http://www.rcsb.org/).

7.4 MOLECULAR VIEWERS

Molecular viewers are widely available to read PDB files and display molecules. They give you a view of molecules in three dimensions. It is now standard practice to employ molecular viewers for both teaching and research purposes. Some basic rules for viewing molecules are helpful. Color is extremely important. If the color is set to CPK color format,[12] atoms are painted with what is known as

molecular viewer Software program that uses PDB files as input and displays structures from these PDB files.

---

[12] C, P, and K are the initials of three scientists who developed the paint scheme: Robert Corey, Linus Pauling, and Walter Koltun.

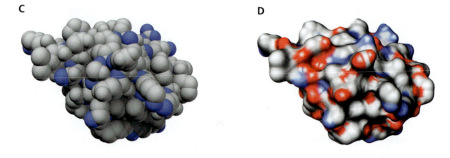

A

hydrogen (H)	white
carbon (C)	black
nitrogen (N)	dark blue
oxygen (O)	red
flourine (F), chlorine (CL)	green
bromine (BR)	dark red
iodine (I)	dark violet
noble gases (He, Ne, Ar, Xe, Kr)	cyan
phosphorus (P)	orange
sulfur (S)	yellow
boron (B), mostly transition metals	peach, salmon
alkali metals (Li, Na, K, Rb, Cs)	violet
alkali earth metals (Be, Mg, Ca, Sr, Ba, Ra)	dark green
titanium (Ti)	gray
iron (Fe)	orange
other elements	pink

B

C

D

FIG. 7-8. CPK coloring scheme of atoms and alternate views of the protein chymotrypsin inhibitor.
A. The CPK coloring scheme for atoms. **B.** The ribbon display format of chymotrypsin inhibitor (chain I). **C.** The space-filling display format. **D.** The surface display format.

CPK standard colors (Figure 7-8). A few of these are worth remembering: carbon is gray or black, nitrogen is blue, oxygen is red, hydrogen is white, and sulfur is yellow.

Thought Question 7-1

Open the PDB file 1TSR (p53 DNA binding domain) in a molecular viewer of your choice and identify the N-terminal amino acid residue on chain A. Display this amino acid in ball-and-stick format and label the carbon, nitrogen, and oxygen atoms on this residue. You do not need to attempt to display hydrogen atoms.

Molecular viewers offer several ways to view the same molecule. Three views of chymotrypsin inhibitor are shown in Figure 7-8B–D. In the ribbon display, only the trace of the backbone atoms is shown in a highly stylized schematic that describes secondary structures (Figure 7-8B). Flat arrows symbolize β-strands (the tip of the arrow always points toward the C-terminus in the linear sequence), and corkscrews symbolize α-helices. Some molecular viewers display α-helices as cylinders. Segments that are neither β-strands nor α-helices are depicted as threads. Note that CPK coloring format was not used in this ribbon display.

The space-filling display (sometimes referred to as CPK or ball display) gives the radius of each atom in the protein (Figure 7-8C). Here, the side chains of the surface amino acids dominate the view. The radius of the ball is a scaled representation of the van der Waals radius of the atom. A general rule is that atomic radii increase down a column in the periodic table of elements because the elements have more shells of electrons. This rule is balanced with another: the more electronegative the atom (i.e., located more to the right in a row of the periodic table), the smaller the atomic radius. Nitrogen (blue) and oxygen (red) have slightly higher atomic numbers than carbon (gray). However, they appear smaller than carbon because they are more electronegative than carbon.

Another, perhaps more useful, display of the protein is the surface display. This display shows the electron distribution on the surface of the protein, which may be color-coded according to electron density. Red areas are relatively electron rich (more negatively charged) and blue areas are relatively electron poor (more positively charged). Gray areas tend to be hydrophobic. Electrostatic surface coloring and coulombic surface coloring are similar methods of visualizing relative electron densities. The surface display is likely what other molecules "see" when they encounter the protein. Molecular viewers have several options that assist you to create displays and make measurements. After your first taste of viewing molecules in this format, it is easy to feel that these structures faithfully represent the natural molecules as they exist in the cell. It is worth bearing in mind that erroneous structures sometimes get deposited in the PDB. Also, proteins undergo conformation changes that are difficult to appreciate when viewing structures derived from X-ray crystallography experiments. When viewing structures it is important to consider the resolution guide (Table 7-1), the B-factor, and the R-factor.

7.5 PROTEIN FOLDING

Now that we are familiar with the major experimental methods for determining protein structures, we are in a position to explore the methods for predicting protein structures. Before we discuss prediction methods, it is important to review how proteins naturally fold. Recall that most proteins exhibit tertiary or quaternary structures. The primary structure is the sequence of the amino acids. The secondary structure is created by the initial folds that the protein makes with itself, such as helices and β-sheets. The tertiary structure is created from the interactions of secondary structures. In quaternary structures, more than one polypeptide is observed in the final protein. Protein folding is predicated on the theory that a linear polypeptide can, without assistance from other molecules, fold into its physiological form (native state),[13] first shown by Christian Anfinsen.

native state Natural state of macromolecule. It is thought that the native state is the structure of the macromolecule in nature at which it is fully functional.

Christian Anfinsen's Protein Unfolding and Refolding Experiment

In a key experiment Christian Anfinsen (Laboratory of Chemical Biology at the National Institute of Arthritis and Metabolic Diseases, now the National Institute of Diabetes and Digestive and Kidney Diseases) added urea to a sample of ribonuclease A until the protein lost its secondary and tertiary structure—a process known as denaturation. In the denatured form (or nonnative state) the protein loses its ability to catalyze its normal enzymatic activity. Anfinsen gradually removed the urea from the sample and observed that ribonuclease A regained its catalytic activity. This simple experiment showed that a protein can self-fold into its native state (Figure 7-9).

In Figure 7-9 the protein on the left illustrates one of several unfolded unstructured conformers. When unfolded, the protein is very flexible and assumes no one specific conformer for a significant length of time. After folding, the same protein is folded into a tertiary structure, which, under physiological conditions, is maintained. The folded protein is the most stable (least energetic) of all forms of the protein. During evolution, there is a selection against proteins that require higher energy conformations to maintain their activities.

Local Minimum Energy States

When Anfinsen removed the urea from ribonuclease A he was careful to do it gradually. What happens when one removes urea from the sample too quickly? We get misfolding, which leads to one of the nonnative states. The misfolded protein constitutes

misfolding Macromolecules, especially proteins, may not correctly fold to their native states due to thermal or chemical stress. Such proteins either must be degraded or refolded to their native states. Excess buildup of misfolded proteins, also known as protein aggregation, can lead to disease including Alzheimer's and Parkinson's.

---

[13] It is pertinent to point out that in a living cell protein folding takes place on a timescale of seconds to minutes whereas protein folding in the test tube takes hours. To fold proteins so quickly cells use other proteins called chaperones (also known as chaperonins) that bind to newly synthesized proteins and guide the folding process just as they emerge from the ribosomes.

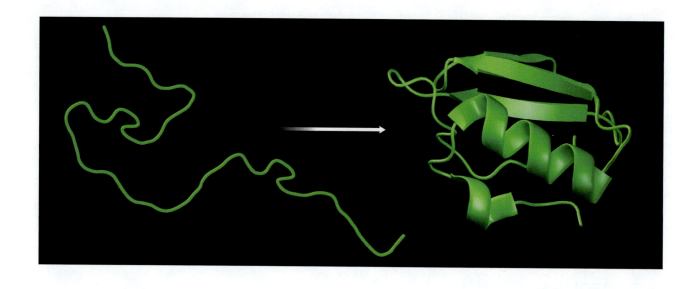

FIG. 7-9. Protein folding. On the left is an unfolded (denatured) protein. Under physiological conditions, the protein folds into a tertiary structure (the native state) shown on the right.

a local minimum energy state from which the protein is sometimes unable to escape. A helpful way to view the folding process is to think of an unfolded protein as a boulder placed at the top of a rocky hill. The top of the hill represents the highest energy form of the protein (the completely unfolded state), and the bottom of the hill represents the lowest energy form of the protein (the native state). As the boulder tumbles to the bottom of the hill it encounters rocks and bushes that may slow down or even prevent it from reaching the bottom. If the boulder does reach the bottom, it reaches its lowest energy state and the protein ends up folded into its native state. If the boulder gets "hung up" on a rock or a bush prior to reaching the bottom, it is in a local minimum energy state and is misfolded.

Another possible consequence of protein denaturation is aggregation. Once the protein is denatured it can bind nonspecifically to other protein molecules and form aggregates. Such aggregates may consist of identical polypeptides or may include a mixture of different polypeptides. The low energy states of the aggregates make it difficult for the proteins to become renatured again. An example of protein aggregation is when an egg (which is mostly protein) is placed into a frying pan and the temperature is increased. Its proteins become denatured and aggregate. Even if the cooked egg is cooled to room temperature the individual proteins are unable to release themselves from each other and renature. The egg proteins in the fried egg are in a low energy state.[14]

Energy Landscape Theory

We are now ready to apply the concepts of local and global energy minima to the energy landscape theory. By 1995 the energy landscape theory of protein folding was developed by Joseph Bryngelson (National Institutes of Health), Jose Onuchic (UC San Diego), and Peter Woynes (UC San Diego). The theory holds that proteins in the unfolded state can assume many conformations and may start to fold starting from any one of these conformations (Figure 7-10). Folding decreases the potential energy of the protein by creating new intramolecular noncovalent bonds and by removing water from hydrophobic side chains. The removal of water when hydrophobic groups associate is called the hydrophobic effect. The native state of the protein is the global minimum potential energy. As the protein folds there are local energy minima that

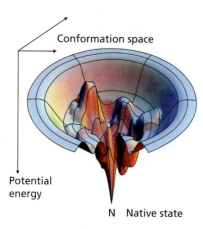

FIG. 7-10. Energy landscape theory. In the energy landscape theory one must imagine that the highest potential energy is associated with the protein located on the lip of the funnel. The protein at this energy state can assume many unfolded states, all equivalent in potential energy. As the potential energy of the protein is decreased, the protein may encounter local energy minima. If the protein assumes a local minimum it may be difficult for it to achieve the global energy minimum, N. The energy landscape theory suggests that there are many protein folding paths.

---

[14] Interestingly, when we eat the egg we recover the amino acids in the aggregates. Our digestive system has a combination of enzymes and detergents that disaggregate the proteins. The individual polypeptides without secondary or tertiary structures are degraded by proteases. The released amino acids are absorbed by the cells that line the small intestine.

the protein may encounter. At these local minima, the protein is in a misfolded state, and it is only with the addition of energy that a protein can get out of a local minimum energy state. The energy landscape theory suggests that there are many possible protein folding pathways that lead to the native state. Interestingly, in a thought experiment, Cyrus Levinthal (Columbia University) calculated, assuming the folding process is random and given the number of potential torsion angles possible in a polypeptide, that it would take an incredibly large number of trials to fold a protein correctly. Extending Levinthal's idea further, Robert Zwanzig (National Institutes of Health) estimated that by random sampling of all possible conformers, it would take approximately 10^{27} years to fold a small protein containing 101 amino acids.[15] Because proteins are observed to fold on the order of 10^{-6} s, it is likely that the protein explores only one or a few pathways before properly folding. At the moment it is a mystery as to how a protein "chooses" its folding pathway. However, this has not prevented scientists from attempting to model protein folding.

7.6 PROTEIN STRUCTURE PREDICTION METHODS

Prediction Method 1: Computational Methods

To appreciate computational methods we must define the *total energy* and the *system*. Total energy, also known as the free energy, is composed of potential energy and kinetic energy. Free energy determines the stability of the system. Energy-based predictions use the following information to predict structure: atom type, relative atom spatial location, and covalent and noncovalent bonding. Computational methods are based on two steps:

1. Calculation of the free energy of the system in one configuration.

2. Sampling many system configurations and picking out the one with the lowest energy.

We define the system as all of the protein atoms and all of the interacting solvent atoms (such as water) and other chemicals that could interact with the protein (ions and other small molecules). We define configuration as a particular spatial distribution of all system atoms at one particular time. At a particular point in time, each atom occupies a point in space. As the protein folds, several configurations are sampled. The configuration that achieves the lowest energy state corresponds to the protein native state.

configuration A particular arrangement of all atoms in space in a system at a particular time.

Ab initio methods

Given the fact that a protein can spontaneously fold into its native state, there must be a physical explanation for this process. *Ab initio* methods for protein structure prediction from amino acid sequence information rely exclusively on physical laws to predict protein folding. The physical laws that guide behaviors of nuclei and electrons are based on quantum mechanics (QM).

To give you a scope of the problem of energy calculations, it is useful to recall that physical forces acting on nuclei and electrons are governed by QM. QM governs the forces resulting from the spatial distributions of electrons around nuclei. Calculation of QM forces requires a tremendous amount of computing power, even when simplifying approximations are used. In *ab initio* protein structure prediction methods, the native state conformation of the protein requires a sampling of the entire ensemble of possible configurations of the system. All possible atom locations in the

---

[15] Zwanzig assumed that each amino acid residue could assume only three conformations. Then there would be $3^{(101-1)}$ possible conformations. This equals 5×10^{47} conformations. Considering the time it takes for a conformation change to occur, it was assumed that an amino acid could adopt a new conformation every 10^{-13} s. This means that the number of seconds required to sample all conformations is 5×10^{34}. This is equal to 1.5×10^{27} years.

system are considered, and the potential energy of each resulting configuration is calculated. The configuration with the lowest potential energy is chosen. To consider only potential energy, you need to assume that the system is at 0 Kelvin, the temperature at which all atoms are motionless. To account for kinetic energy at a temperature above 0 Kelvin, statistical mechanics is required.

Because calculation of the system configuration with the lowest potential energy is a formidable problem that is impossible to solve rigorously, scientists use different configuration-searching methods, which attempt to obtain the lowest potential energy configuration without sampling each and every possible configuration. A QM calculation for a single conformation of a small peptide may require months.

Molecular mechanics methods

Given the hurdle of calculating energies for many system configurations, scientists studying proteins often use classical mechanical forces instead of QM forces to approximate the energy of the system, because they are much easier to calculate. This approach is called molecular mechanics (MM). In MM, energy content is described by treating atoms as balls and bonds as springs. The set of mathematical equations that describes the energy content of the system is called a force field. The force field is composed of separate expressions that describe the potential energy of a single protein conformation. The potential energy results from interactions between any two atoms in the protein. One definition of total potential energy, U_{tot} is:

$$U_{tot} = U_{bond} + U_{angle} + U_{torsion} + U_{electro} + U_{vdW} \qquad (7.1)$$

where U_{bond} is the potential energy of a covalent bond, U_{angle} is the potential energy resulting from the bond angle produced by three covalently bonded atoms, $U_{torsion}$ is the potential energy resulting from rotation around the middle bond of four covalently bonded atoms, U_{elec} is the potential energy resulting from electrostatic interactions, and U_{vdW} is the potential energy resulting from atoms that are spatially near to one another (van der Waals interactions).

molecular mechanics (MM) A method of modeling structure and movement of molecules using Newtonian (or classical) physics principles. Free energy of the molecule is estimated from a force field composed of the sum of separate expressions that describe the potential energy of a single macromolecular conformation.

Let's discuss these potential energies in the context of their contributions to folding a protein (Figure 7-11A). Covalent bond energy considers the energy associated with the covalent bond. A covalent bond occurs when electrons are shared between atoms. For example, in proteins, amino acid residues are attached to each other through their backbone atoms via covalent bonds. Side chains are attached to the backbone by covalent bonds to the alpha carbon. On some occasions, side chains of amino acids can form covalent bonds directly with other side chains to help fold the protein. The classic example of this is the disulfide bond that keeps two Cys amino acids together (see Chapter 1). The stretching or contraction of covalent bonds away from their optimal interatomic distances increases potential energy. Sometimes this is referred to as stretching energy.

Angle energy refers to the energy required to widen or narrow a bond angle. Bond angles formed by three atoms attached by covalent bonds are dictated by the identity of the central atom, the number of bonded atoms, and the number of nonbonded electrons around each central atom. There is an optimal bond angle, and any deviation that is wider or narrower increases the potential energy of the structure. Sometimes this is called bending energy.

steric hindrance Electrostatic repulsion (also known as van der Waals repulsion) caused by close proximity of electron clouds of two or more atoms. Steric hindrance can restrict atom movement.

Torsion energy is dictated by the freedom of rotation about a middle bond of four covalently bonded atoms. The torsion energy necessary to achieve a stable conformation is limited by the type of covalent bonds that are formed. Atoms connected by single covalent bonds have more freedom of rotation than atoms connected by double covalent bonds.[16]

---

[16] Imagine four atoms A-B-C-D connected by single covalent bonds. When those four atoms have optimal (minimum) stretching and bending energies, they will not fall in a straight line. Instead, the

Electrostatic energy refers to the energy due to the attractive force of opposite charges (or partial charges) or a repulsive force of like charges (or partial charges) that are near one another. For example, at physiological pH the side chain of Lys is positively charged and the side chain of Asp is negatively charged. Proteins may fold into a conformation that allows Lys to electrostatically interact with Asp. This is called a charge-charge (or ion-ion) interaction. Electrostatic energy can also be calculated due to a dipole-dipole interaction. Here, a partially charged residue may interact with another partially charged residue. The hydroxyl group (–OH) in Ser can form a dipole-dipole interaction with the hydroxyl group of another Ser (Figure 7-11B). The oxygen of the side chain has a partial negative charge and the hydrogen has a partial positive charge. This particular dipole-dipole interaction is called a hydrogen bond.

The final contribution to the total energy of the system is van der Waals force. When two atoms that are not covalently bonded to each other come near to one another, the electrons in their outer shells can fluctuate. The fluctuation of the electrons can create areas of momentary positive charge on one atom and areas of momentary negative charge on another atom. Such fluctuations lead to weak attractions between the two atoms called van der Waals forces (also known as van der Waals interactions). On one hand, if the interatomic distance is too long, the temporary charges that are formed are too weak to have an attractive effect on the two nonbonded atoms. On the other hand, if the interatomic distance of the nonbonded atoms is too short, the negatively charged electron clouds of the atoms repel each other and the interaction is destabilizing. Sometimes van der Waals forces and electrostatic energy are lumped together under an energy known as the nonbonded interaction energy. Depending on the size of the atoms and groups of atoms, there can be a limitation on the rotation of groups around a middle covalent bond (see Figure 7-11A). Nonbonded interaction energies can limit the rotation because these atoms clash with one another. This is called steric hindrance, and it can have a major effect on total potential energy.

The total potential energy is the sum of these energies. MM is useful for calculating the total potential energy of protein structures. Other computational methods are used to alter the protein conformation. These methods use MM

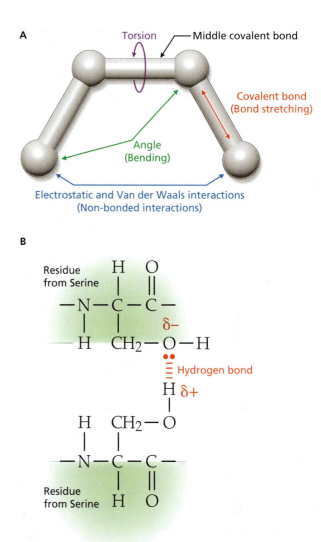

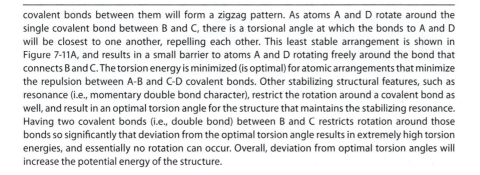

FIG. 7-11. Potential energies used for molecular mechanics. **A.** Schematic of five types of MM potential energies. **B.** A special dipole-dipole interaction called a hydrogen bond. Two Ser residues are positioned on two separate segments of a folded protein. The oxygen on the side chain of one Ser has a partial negative charge ($\delta-$), and the hydrogen on the side chain of a second Ser has a partial positive charge ($\delta+$). This dipole-dipole interaction is called a hydrogen bond.

covalent bonds between them will form a zigzag pattern. As atoms A and D rotate around the single covalent bond between B and C, there is a torsional angle at which the bonds to A and D will be closest to one another, repelling each other. This least stable arrangement is shown in Figure 7-11A, and results in a small barrier to atoms A and D rotating freely around the bond that connects B and C. The torsion energy is minimized (is optimal) for atomic arrangements that minimize the repulsion between A-B and C-D covalent bonds. Other stabilizing structural features, such as resonance (i.e., momentary double bond character), restrict the rotation around a covalent bond as well, and result in an optimal torsion angle for the structure that maintains the stabilizing resonance. Having two covalent bonds (i.e., double bond) between B and C restricts rotation around those bonds so significantly that deviation from the optimal torsion angle results in extremely high torsion energies, and essentially no rotation can occur. Overall, deviation from optimal torsion angles will increase the potential energy of the structure.

at various stages of the protein folding process to gauge the success of determining the structure of the native state of the protein.

Energy minimization through the Monte Carlo method

One method used to predict the lowest energy protein conformation is called energy minimization through the Monte Carlo method.[17] A popular implementation of the Monte Carlo method is the Metropolis procedure. Energy minimization through the Metropolis procedure starts from an initial configuration of the protein atoms and solvent atoms and follows these steps:

STEP 1. The potential energy of the initial configuration of atoms is calculated using MM methods.

STEP 2. A small change is introduced in the location of an atom in the system.

STEP 3. The potential energy of the new configuration is calculated.

STEP 4. If the new configuration has lower energy than the previous configuration, then the new configuration is adopted. Steps 2–3 are repeated. If the new configuration has a higher energy than the previous configuration, then the new configuration is rejected. Another configuration is created, and steps 2–3 are followed.

STEP 5. The above process continues (starting at step 2) until no new lower energy configurations are found.

The goal of energy minimization is to determine the global energy minimum configuration. However, this method often results in the protein conformation getting "hung up" in one of the local energy minima (see Figure 7-10). The energy hurdle that the protein must overcome in order to continue on its path to its global energy minimum is called the energy barrier.

To overcome the energy barrier and allow the system to search for new configurations, energy must be added to the atoms. This energy is added as virtual heat energy at step 4. The added heat energy increases the motions of the atoms in the system. This creates new configurations that are outside of the local minimum. This method, called simulated annealing, is used to get the misfolded protein out of the local energy minimum. Simulated annealing consists of three steps:

STEP 1. Add simulated heat energy to the system.

STEP 2. Remove simulated heat energy from the system (also known as cooling).

STEP 3. Calculate the potential energy using MM force fields.

The Monte Carlo method is a stochastic process (many system configurations are generated), and a potential energy is calculated for each configuration. The configuration with the lowest potential energy is used to predict the protein conformation.

Molecular dynamics

molecular dynamics (or molecular dynamics simulation) A method of protein structure prediction that uses heat energy and molecular mechanics to predict atom positions.

Another approach to protein structure prediction is molecular dynamics (MD). In MD, atom movements of a system are calculated as the temperature of the system is altered. First, the atom locations are determined using the lowest possible energy potential calculated from MM. The lowest possible energy occurs when the temperature of the system is at a minimum (absolute zero or 0 Kelvin). As the

---

[17] The Monte Carlo method uses random numbers to solve problems for which it is difficult to calculate an exact answer. The name was coined by Nick Metropolis (Los Alamos National Laboratory, Los Alamos, New Mexico) and is a reference to the random number generators found in the gambling casinos of Monte Carlo.

temperature increases, MD calculates the atom velocities and positions. Once the atoms move to new positions, the potential energy is calculated. After the system has reached equilibrium (where there are no major movements of atoms), the temperature is lowered. As the temperature is lowered, the positions of the atoms are tracked by calculating the potential energy. The time intervals between total potential energy calculations are very short—approximately 10^{-15} second (femtosecond) for each step. Movement of the protein and the solvent atoms can be simulated by integrating the atom positions over hundreds of steps as the temperature decreases. Typical small protein folding simulations including accounting for surrounding solvent cover the range of thousands of femtoseconds. In principle, the atoms' position information generated from the simulation can be used to fully characterize the thermodynamic state of the system. MD also provides a method to overcome the problem of protein getting trapped in a local energy minimum conformation. The scientist adds heat (called simulated heat) to the system, and then the system is allowed to cool again. Atom positions are monitored throughout this process.

Due to the calculation intensity of MD, some scientists divide protein folding simulations into two categories: a low-resolution category and a high-resolution category. In the low-resolution category, the core of the protein is built based on the consideration of only one property, the hydrophobic effect.[18] The hydrophobic effect is an energy decrease when hydrophobic groups associate, releasing water molecules. The hydrophobic effect increases the entropy (chaos) of the system by allowing water molecules more freedom of movement. Increasing entropy is another way to lower the potential energy of the system. The low-resolution structure of the protein core is driven only by the hydrophobic effect, which may not require significant computer resources. For the high-resolution category, other potential energies are used to calculate atom positions at each step during temperature change.

Laboratory experiments have shown that proteins fold on the order of 10^{-6} s. Due to the large number of calculations required, MD simulations cover only 10^{-12}–10^{-9} s of the folding pathway. Supercomputers have been able to increase the amount of computer power necessary to carry out MD simulations to a bit longer than 10^{-9} s. Furthermore, networked computers can contribute to solving protein folding problems.

Combining Computational Methods and Knowledge-Based Systems

Rosetta is a popular software program that combines computational methods with data from the PDB to fold proteins. Rosetta has been modified to enlist the help of thousands of home computers to fold proteins through gameplay. Game developers and scientists collaborate to create online protein folding programs that allow gamers to attempt to fold proteins online for points. Gamers attempt to fold the protein as the absolute value of the potential energy (calculated on the fly) is displayed.[19] One folding game is called Foldit. Feedback from the gamers' solutions to protein folding problems show which parts of Rosetta are robust and which parts need to be more carefully tuned. In some cases, top gamers are able to achieve a more accurate depiction of the native state than purely computer-generated methods (Figure 7-12).

Figure 7-12A shows a screenshot of the Foldit game as a player folds a protein. The visualizations include a steric clash (impacts nonbonded interactions potential energy)

---

[18] The hydrophobic effect can be illustrated by mixing water and oil (as you do when you shake a bottle of Italian salad dressing). During the mixing stage, water molecules surround the hydrophobic molecules in the oil. This creates a highly structured arrangement (known as a clathrate) where water molecules form cages around oil molecules. This ordered structure is not favored because all systems attempt to reach the lowest potential energy possible. The hydrophobic effect comes into play when the water and oil are allowed to separate. The oil molecules come together and form oil droplets. The water molecules are freed from the clathrate to form a more random distribution with other water molecules (they form temporary hydrogen bonds with one another). Within a few minutes the oil and water are completely separated so that the water molecules are optimally randomized.

[19] Absolute value of the potential energy is calculated and displayed. This gives a positive number, which is more intuitive for a gamer to achieve as opposed to a negative score.

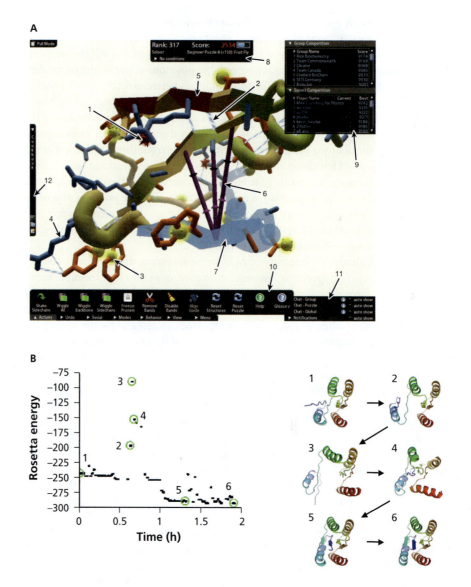

FIG. 7-12. Data from the Foldit protein folding game. **A.** Screenshot of the Foldit game as a player folds a protein. **B.** A trajectory of Rosetta structure energies as a function of time for one top gamer.

representing atoms that are too close (arrow 1); a hydrogen bond (arrow 2) (impacts nonbonded interactions); a hydrophobic side chain with a yellow blob signifying it is exposed to water (arrow 3) (impacts hydrophobic effect); a hydrophilic side chain (arrow 4); and a segment of the backbone that is red due to high residue energy possibly due to unacceptable torsional energy (arrow 5). The players can make modifications including "rubber bands" (arrow 6), which add constraints to guide the folding process and freezing (arrow 7), which prevents molecular motions. The user interface includes information about the player's current status, including score (arrow 8); a leader board (arrow 9), which shows the scores of other players and groups; toolbars for accessing tools and options (arrow 10); chat for communicating with other players (arrow 11); and a "cookbook" for making new automated tools or "recipes" (arrow 12).

Figure 7-12B shows a trajectory of Rosetta structure energies as a function of time for one top gamer. The beginning structure was misfolded and in a local minimum. The gamer, playing over the course of a two-hour period, explored high energy conformations to obtain the native state. The starting misfolded structure had a Rosetta structure energy of –243 (which correlates to total potential energy). Each point in the plot represents a solution produced by this player. A display of the structures corresponding to the circled points is shown on the right. The first structure (1) is near the starting structure. In structures 2–4, the player must explore higher energy structures to move the β-strand into its proper place, shown as a dark blue flat arrow.

In structures 5 and 6, the player refines the β-strand pairing to form a parallel β-sheet.

Two types of folding problems are presented to gamers. One is to start folding a protein from a primary structure (i.e., sequence alone) to achieve the native state. The second, shown in Figure 7-12, is to start with a misfolded protein (in a local minimum energy state) and refold the protein into the native state. Analysis of gamers' performance suggested the following. Gamers are especially good at trying different major reconstructions of misfolded proteins but have less success at folding proteins starting with only the primary structure. This might be due to the types of tools made available to the gamers. It will be exciting to see what the future holds for collaborative efforts between gamers and scientists to solve protein folding problems.

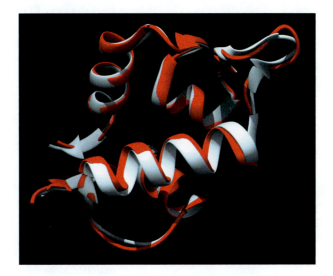

Calculation of Accuracy of Structure Predictions

At this point you might be wondering how scientists know if they have correctly predicted the native folding state of the protein. Anfinsen used enzyme catalytic activity as a primary measure of native state. Nowadays many of the proteins used for structure prediction do not have a known biochemical activity. It is standard practice to evaluate the predictive power of a methodology by comparing a predicted structure to the experimentally derived structure (either through X-ray crystallography or NMR spectroscopy). The goal, of course, is to create software programs that are highly accurate in predicting native structures. Such programs can then be used to assess the structures of proteins that are recalcitrant to experimental methods of structure determination.

Root mean square deviation is a measure of protein structure difference

The common method of comparing the experimentally derived structure to the modeled structure is the root mean square deviation (RMSD) calculation. In this method, the two structures (one predicted and the other known) are superimposed with a molecular viewer. The equation for RMSD is:

$$\text{RMSD} = (\Sigma d_i^2 / \text{n})^{1/2} \qquad (7.2)$$

where d_i is the distance (in Å) between each pair of matched atoms (one atom from each protein) and n is the number of matched pairs of atoms.

The RMSD is a measure of the differences between the compared structures, so its value increases with more structural differences. The RMSD value also greatly depends on the number of atoms being compared. Typically, one might compare certain groups of atoms within structures. For example, one may compare Cα atoms only, backbone (or main chain) atoms, or all atoms (with the exception of H atoms). Predicted structures can often have low RMSD values if only the Cα atoms are compared, but RMSD values increase considerably when side chains are compared because they can adopt different low energy side-chain orientations (rotamers).

One challenge with the RMSD approach is that there may be more than one experimentally derived structure. In theory, two experimentally derived structures should have an RMSD value of zero when compared. However, differences in the crystallography conditions or conformation perturbations due to bound substrate molecules often lead to slight structural differences (up to RMSD values of 0.5 Å when only Cα atoms are compared). Within two experimentally derived structures, the RMSD values for α-helices and β-strands are usually lower than those for loops and turns. Figure 7-13 shows a superimposition of two identical MDM2 polypeptides for an RMSD calculation. The RMSD value is calculated to be 0.713 Å (all non-H atoms are compared). The two polypeptides are identical in sequence and are

FIG. 7-13. Superimposition of two protein structures and RMSD calculation. Two identical polypeptide chains (A and B) of the MDM2 domain that binds p53 were superimposed. Using the Match command, the software program Chimera calculated an RMSD value of 0.713 Å for the 707 matched atoms. This calculation includes all nonhydrogen atoms in the polypeptides including side chains (not shown). Chain A is red, and chain B is blue. Note that the significant deviation in structure is in the loop on the right.

root mean square deviation (RMSD) A method used to quantify the difference of locations of atoms in two structures of the same atom composition. RMSD is often used to calculate the differences between a predicted protein structure and an experimentally derived protein structure.

from the same PDB file, but represent different chains. Note that the significant deviation in structure is found in a loop near the right side of the superimposed structures. When comparing a predicted structure and an experimentally derived structure, RMSD values may slightly differ due to the conditions of the experiment.

Q_3 and Sov are common measurements of secondary structure prediction accuracy

Another common measurement of structure prediction accuracy is Q_3, which is used to predict accuracy of secondary structures. Q_3 is the percentage of correctly predicted residues within a sequence of known secondary structure:

$$Q_3 = \frac{\text{number of amino acids correctly predicted}}{\text{total number of amino acids}} \times 100 \qquad (7.3)$$

Q_3 An equation used to measure the secondary structure prediction accuracy of software programs.

Sov (fractional overlap of segments) An equation used to measure the secondary structure prediction accuracy of software programs. This equation takes the order of secondary structures into account and is thought to be superior to Q_3 as a measure of structure prediction accuracy.

The Q_3 score ranges from 0% to 100%, with 100% indicating perfect prediction. Many secondary structure prediction programs have three possible outputs: α-helix, β-strand (sometimes referred to as sheet), and loop (sometimes referred to as turn or coil). In considering Q_3 as a measure of prediction accuracy, one must be cautious because sometimes a high Q_3 value can be obtained but the prediction is not useful (Figure 7-14).

As shown in Figure 7-14, the Q_3 measurement can give a false picture of structure prediction accuracy. Prediction 1 shows a high Q_3 value that reflects the experimentally derived structure. Prediction 2 shows a high Q_3 value that does not reflect the experimentally derived structure. Prediction 2 erroneously predicts two segments to be helices. An improvement would be to consider the order of secondary structures in the prediction. In this vein, the fractional overlap of segments (Sov) is a more accurate measurement of secondary structure prediction. The Sov value is sensitive to the amount of overlap between predicted and observed structures, but allows for small variation in the start and end points of the secondary structures in the predicted and experimentally derived proteins. A discussion of the Sov calculation can be found in the paper by Zemla et al. (1999) (see reference at the end of this chapter). Table 7-2 shows a comparison between Q_3 and Sov accuracy measurements.

Amino acid sequence: VLHQASGNSVILFGSDVTVPGATNAEQAR

Experimentally derived structure: HHHHHCCCCEEEECCCEEECCCCCHHHHH

Prediction1: Q_3 = 22/29 100 = 76%: CHHHCCCCEEEECCCCCEEECCCHHHHHH

Prediction2: Q_3 = 22/29 100 = 76%: HHHHHCCCCHHHHCCCHHHCCCCCHHHHH

FIG. 7-14. Shortcomings of Q_3 measurement. A sequence of 29 amino acids is shown. The experimentally derived secondary structure is below the sequence. H = helix, C = coil, and E = β-strand.

TABLE 7-2. COMPARISON OF Q_3 AND Sov

STATUS	SECONDARY STRUCTURE	Sov	Q_3
Experimentally derived	CHHHHHHHHHHC	—	—
Prediction 1	CHCHCHCHCHCC	12.5	58.3
Prediction 2	CCCHHHHHCCCC	63.2	58.3
Prediction 3	CHHHCHHHCHHC	40.6	83.3
Prediction 4	CHHCCHHHHHCC	52.3	75.0
Prediction 5	CCCHHHHHHCCC	80.6	66.7

Adapted from A. Zemla, C. Venclovas, K. Fidelis, and B. Rost, "A Modified Definition of Sov, a Segment-Based Measure for Protein Secondary Structure Prediction Assessment," *Proteins* 34 (1999): 220–223.

TABLE 7-3. CONFORMATIONAL PARAMETERS FOR α-HELICAL, β-STRAND, AND TURN AMINO ACID

AA	P(α)	AA	P(β)	AA	P(T)	AA	F(i)	F(i+1)	F(i+2)	F(i+3)
Glu	1.51	Val	1.70	Asn	1.56	Ala	0.060	0.076	0.035	0.058
Met	1.45	Ile	1.60	Gly	1.56	Arg	0.070	0.106	0.099	0.085
Ala	1.42	Tyr	1.47	Pro	1.52	Asp	0.147	0.110	0.179	0.081
Leu	1.21	Phe	1.38	Asp	1.46	Asn	0.161	0.083	0.191	0.091
Lys	1.14	Trp	1.37	Ser	1.43	Cys	0.149	0.050	0.117	0.128
Phe	1.13	Leu	1.30	Cys	1.19	Glu	0.056	0.060	0.077	0.064
Gln	1.11	Cys	1.19	Tyr	1.14	Gln	0.074	0.098	0.037	0.098
Ile	1.08	Thr	1.19	Lys	1.01	Gly	0.102	0.085	0.190	0.152
Trp	1.08	Gln	1.10	Gln	0.98	His	0.140	0.047	0.093	0.054
Val	1.06	Met	1.05	Thr	0.96	Ile	0.043	0.034	0.013	0.056
Asp	1.01	Arg	0.93	Trp	0.96	Leu	0.061	0.025	0.036	0.070
His	1.00	Asn	0.89	Arg	0.95	Lys	0.055	0.115	0.072	0.095
Arg	0.98	His	0.87	His	0.95	Met	0.068	0.082	0.014	0.055
Thr	0.83	Ala	0.83	Glu	0.74	Phe	0.059	0.041	0.065	0.065
Ser	0.77	Gly	0.75	Ala	0.66	Pro	0.102	0.301	0.034	0.068
Cys	0.70	Ser	0.75	Met	0.60	Ser	0.120	0.139	0.125	0.106
Tyr	0.69	Lys	0.74	Phe	0.60	Thr	0.086	0.108	0.065	0.079
Asn	0.67	Pro	0.55	Leu	0.59	Trp	0.077	0.013	0.064	0.167
Gly	0.57	Asp	0.54	Val	0.50	Tyr	0.082	0.065	0.114	0.125
Pro	0.57	Glu	0.37	Ile	0.47	Val	0.062	0.048	0.028	0.053

From P. Y. Chou and G. D. Fasman, "Prediction of Protein Conformation," *Annual Review of Biochemistry* 47 (1978): 251–276.

The Sov scores are relatively high when the central segment of helix is predicted (see Sov scores for predictions 2 and 5) and are relatively low when the central helical segment is interrupted by another secondary structure prediction (see predictions 3 and 4). Q_3 scores are not as sensitive to such segment breakers.

Prediction Method 2: Statistical and Knowledge-Based Methods
Chou-Fasman method

In 1974 Peter Chou and Gerald Fasman developed a popular program that predicts secondary structures. From experimentally solved protein structures, they created an algorithm for structure prediction. The program developed from this algorithm utilizes a sliding window that begins at the N-terminus of the protein sequence and ends at the C-terminus. The input is the primary structure and the output is the secondary structure prediction. As the window slides, the algorithm predicts whether residues will form α-helices, β-strands, or loops (referred to as turns by Chou and Fasman). The algorithm consults a table of conformational parameters (derived from known structures in the PDB) for the 20 amino acid residues (Table 7-3).

Residues are categorized as α-helix formers (P(α)), β-strand formers (P(β)), and turn formers (P(T)) based on their occurrence in known protein structures.

Chou-Fasman method A secondary structure prediction program created in 1974 that uses a sliding window. Information from experimentally solved crystal structures is used to predict the amino acid secondary structures from primary sequence information.

Table 7-3 ranks amino acid propensities in each of these classes. To predict turns, they considered the propensity of the residue found in turns, P(T), and the identity of the residues in the first, second, third, and fourth positions after the residue to be categorized for turn.

The steps in the Chou-Fasman algorithm are as follows:

STEP 1. Assign all of the residues in the protein the appropriate set of parameters.

STEP 2. Scan through the protein and identify regions where four out of six contiguous residues have P(α-helix) > 1.00. That region is declared an alpha-helix. Extend the helix in both directions until a set of four contiguous residues that have an average P(α-helix) < 1.00 is reached. That is declared the end of the helix. If the segment defined by this procedure is longer than five residues and the average P(α-helix) > P(β-strand) for that segment, the segment can be assigned a helix.

STEP 3. Repeat this procedure to locate all of the helical regions in the sequence.

STEP 4. Scan through the protein again and identify a region where three out of five of the residues have a value of P(β-strand) > 1.00. That region is declared a β-strand. Extend the strand in both directions until a set of four contiguous residues that have an average P(β-strand) < 1.00 is reached. That is declared the end of the β-strand. Any segment of the region located by this procedure is assigned a β-strand if the average P(β-strand) > 1.05 and the average P(β-strand) > P(α-helix) for that region.

STEP 5. Any region containing overlapping alpha-helical and beta-sheet assignments are taken to be helical if the average P(α-helix) > P(β-strand) for that region. It is a β-strand if the average P(β-strand) > P(α-helix) for that region.

STEP 6. To identify a turn at residue position number i, calculate the following value:

$$p(t) = f(i)f(i+1)f(i+2)f(i+3)$$

where the f(i+1) value for the i+1 residue is used, the f(i+2) value for the i+2 residue is used and the f(i+3) value for the i+3 residue is used. If: (1) p(t) > 0.000075; (2) the average value for P(T) > 1.00 in the tetrapeptide; and (3) the tetrapeptide obeys the inequality P(α-helix) < P(T) > P(β-strand), then a turn is predicted at that location.

The Chou-Fasman algorithm gives a Q_3 accuracy range of 60–65%, which is quite good given the lack of data on protein structure at the time of its development.

GOR method

GOR method A secondary structure prediction program that uses a sliding window. It predicts the secondary structure of each amino acid residue by considering each amino acid in a 17-residue window. The window slides in increments of one residue until the secondary structure of each amino acid is predicted.

The GOR method is based on some of the same principles as the Chou-Fasman method. It is named after its creators, Jean Garnier, David Osguthorpe, and Barry Robson (University of Manchester, United Kingdom). The GOR method uses a window of 17 residues in a protein sequence and predicts the secondary structure of the ninth residue (the central residue) in this window. The window slides through the sequence by one residue and calculates the structure of the new central residue. These steps are repeated until the end of the sequence is reached. The GOR method gathers information from all 17 residues to predict the secondary structure of the residue in the center position of the window.

We can define the residues by their positions in the window. The jth position is the location of the central residue. The residue at position j − 8 is the first residue in the window, and the residue at position j + 8 is the last residue in the window.

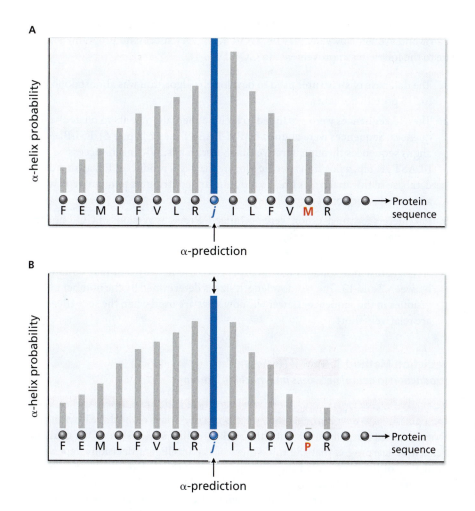

α-prediction

FIG. 7-15. Effect of directional information.
A. Met in the $j + 5$ position increases the probability that amino acid in the j position forms an α-helix. **B.** Pro in the $j + 5$ position decreases the probability that amino acid in j position forms an α-helix.

What information does GOR use for its predictions? GOR uses *self-information* and *directional information*. Self-information is similar to the Chou-Fasman data. It is the propensity that the jth amino acid residue will form an α-helix, β-strand, or loop (referred to as coil) derived from statistical data of selected sequences from the PDB. Directional information is the contribution to conformation at position j by residues in the window that are *not* at the jth position (similar to Chou-Fasman algorithm for turn prediction). These two types of information are derived from selected sequences stored in a database. These sequences are non-redundant and represent a wide variety of structures.

To illustrate the effect of directional information, consider two positions in a 17-residue window, j and $j + 5$ (Figure 7-15). We will compare two sequences that are identical with the exception of the residue at $j + 5$. In one sequence a Met is at this position, and in the other a Pro is present. The x axis shows the amino acid sequence, and the y axis shows the probability of forming an α-helix. Met is a strong α-helix former, so the effect of Met on the amino acid at the jth position is to increase its probability for forming an α-helix. Because Pro is considered an α-helix breaker, the probability of the jth amino acid forming an α-helix is decreased.

There are five successive versions of the GOR method (GOR I–V), and each version results in higher Q_3 values for test proteins. GOR IV uses single amino acid information (called singlet) and paired amino acid information (called doublet) and sums this information to come up with a secondary structure decision. Paired amino acid information means that it considers all possible pairs of amino acids in the 17-residue window. This information is derived from the PDB structure database. The average Q_3 value for GOR IV is 64%. The later version, GOR V (which also uses a 17-residue window), has improved structure prediction accuracy: average Q_3 value of 74% and average Sov value of 71%. GOR V is a very successful program, which is due to the following improvements to GOR IV:

1. The database of structures used to develop the algorithm was almost doubled to 513 protein domains.

2. The 513 sequences were expanded to include their alignments to related sequences. Sequences were aligned by PSI-BLAST (see Chapter 6). PSI-BLAST aligns sequences through iterative BLAST searches. With each successive BLAST search, a position specific scoring matrix is produced. Expansion of the database in this manner is the most significant contributor to improvement of accuracy prediction.

3. Parameters for the decision were optimized. Previous versions of GOR underestimated the α-helix state and overpredicted the loop state. These parameters were adjusted to compensate for this deficiency.

4. A resizable window length was incorporated. The window size may be adjusted between 7 and 13. The window length size is determined by the number of residues in the sequence. Lower window sizes are used when the sequences are relatively short.

Prediction Method 3: Neural Networks
Introduction to neural networks and machine learning

Artificial neural networks is a branch of a larger field called machine learning, which uses algorithms to develop behaviors based on data. A focus of machine learning is to recognize complex patterns in data and make intelligent decisions. One challenge to machine learning is that the robustness of the intelligent decisions depends on the quality of the data (called the training data set) that is fed to the computer. As long as the training data set is accurate and sufficiently varied, the machine learning program should make correct decisions when it encounters data the program has not encountered before.

PSIPRED

PSIPRED A secondary structure prediction program that uses a neural network and PSI-BLAST.

PSIPRED is the first artificial neural network-based program that achieved high accuracy in predicting secondary structure. PSIPRED was developed by David T. Jones at the University of Warwick, United Kingdom (now at University of London). It is a two-stage feed-forward artificial neural network that predicts protein secondary structure using position specific scoring matrices produced by PSI-BLAST. An artificial neural network (named simply neural network, or NN for short) is a computer science term that describes a type of program modeled after neurons in the brain. A neuron receives a signal from the outside (for example, heat sensation in the hand when the hand is placed on a hot stove) and transfers that information to a second neuron. The second neuron carries the information to a third neuron that gives the output (for example, move the hand away from the stove). In this simple example, there is a layer of input neurons, a "hidden" layer of neurons, and an output neuron. In Similarly, in an NN, one must imagine layers of neurons. The neurons in the NN, called units, can either inhibit or stimulate one another. These inhibitions and stimulations are called "weights" in computer science parlance. A high weight means that the signal is amplified, and a low weight means that a signal is repressed. Figure 7-16 is a schematic that compares an NN in the human body to an NN in a computer.

Logic of neural networks

Let's look closer at a single neuron (or unit) in a neural network (Figure 7-17A). A single neuron is a discrete decision process. In this example, the output is a

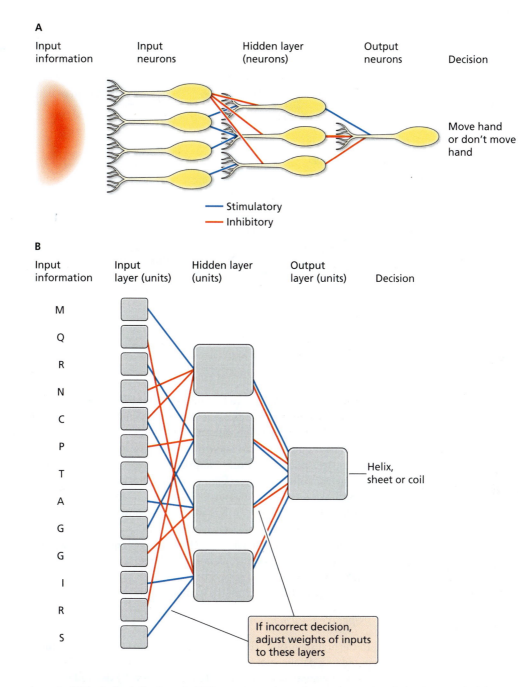

FIG. 7-16. A comparison of a NN in the body/brain and a NN in a computer program.

A. An input signal from a hot surface is received by the dendrites of the input neurons. The neurons that have red lines emanating from the bulbs are inhibitory. The neurons that have blue lines emanating from the bulb are stimulatory. There is one hidden layer of neurons. These neurons collect the information from the input neurons and integrate the signals. The hidden layer feeds information to the output layer (here, just one neuron). The output layer produces a decision depending on the number of stimulatory and inhibitory messages it receives. **B.** In the computer program with NNs, information on each amino acid is transmitted to the input layer of 13 units. In this example, there is one hidden layer of 4 units, which integrates signals from the input layer. The final layer of 1 unit makes a decision as to whether a central residue in the window is helix (α-helix), sheet (β-strand), or loop (coil). During training on a set of proteins of known structures, weights from each unit can be adjusted (blue for more weight and red for less weight) to give correct decisions.

geometric interpretation displayed as an area bounded by Cartesian coordinates. If the inputs are coordinates of a point (x, y), the neuron decides on which side of a line the point lies. The neuron has two possible outputs, 1 or 0. The output will be 1 if and only if $x + y < 2$, and the area selected is below and to the left of the line $x + y = 2$. If the above condition is not met, the output is 0, and the selected area is above and

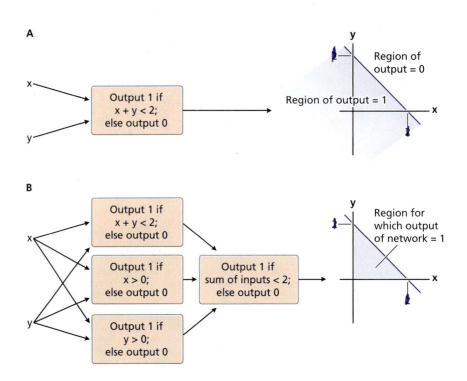

FIG. 7-17. The logic of neural networks.

to the right of the line as well as inclusive of the line $x + y = 2$. Thus, a single neuron with two inputs makes a decision as to which side of the line a point lies—a decision that results in two possible outputs.

A neural network can make more complex decisions than a single neuron. Here, we create a neural network comprised of an input layer of three neurons and an output layer of one neuron (Figure 7-17B). This neural network can select points that lie within a triangle. If the output is 1, the area inside the triangle is selected. If the output is 0, the triangle borders and the area outside of the triangle is selected.

Weights on processes may be regarded as variables. Using a set of training data, the neural network can determine the weights required for a particular outcome. To train a neural network, the network is fed inputs for which the final output is known. The output of the neural network is compared to the known output, and if the neural network output differs from the known output, the weights can be adjusted.

PSIPRED neural network

In the first stage of PSIPRED, a window of 15 residues of the query sequence is compared to a database of sequences by PSI-BLAST (Figure 7-18).[20] The database contains sequences corresponding to experimentally derived structures. PSI-BLAST is used for three iterations (see Chapter 6 for extensive coverage of PSI-BLAST). PSI-BLAST produces a position specific scoring matrix (PSSM) that is retrieved from the third iteration. The PSSM contains 15×20 elements, 15 for each residue in the query sequence and 20 elements for each score of the 20 potential residues in a single position. These scores, which range from -8 to $+8$, are converted to a standard logistical function score by the following equation:

$$\text{Standard logistical function score} = 1/(1 + e^{-x})$$

where x is the PSSM score. The standard logistical function scores range from 0 to 1.

---

[20] The sequences used by PSI-BLAST for comparison to the query sequence are stored in a database that contains hundreds of thousands of sequences that are non-redundant. This non-redundant database was previously filtered to remove low-complexity regions, transmembrane segments, and regions likely to form coiled-coil structures. These sequences were removed because structure data for them is quite varied and there are few examples of high-resolution structures.

A Raw profile from PSI-BLAST log file

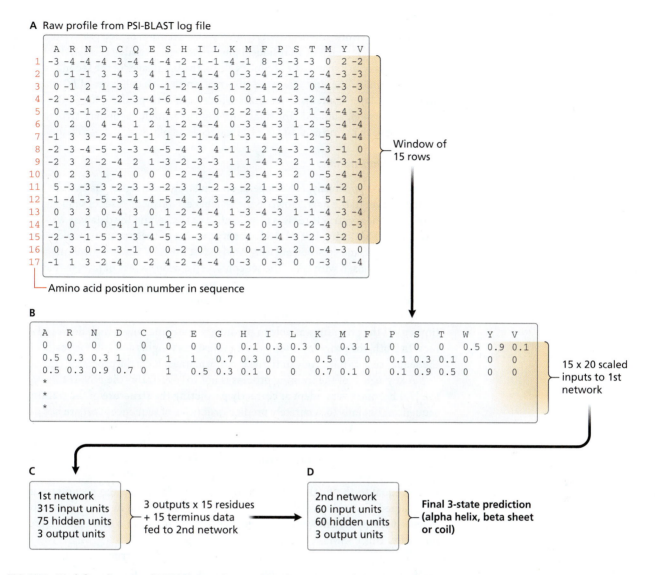

	A	R	N	D	C	Q	E	S	H	I	L	K	M	F	P	S	T	M	Y	V
1	-3	-4	-4	-4	-3	-4	-4	-4	-2	-1	-1	-4	-1	8	-5	-3	-3	0	2	-2
2	0	-1	-1	3	-4	3	4	1	-1	-4	-4	0	-3	-4	-2	-1	-2	-4	-3	-3
3	0	-1	2	1	-3	4	0	-1	-2	-4	-3	1	-2	-4	-2	2	0	-4	-3	-3
4	-2	-3	-4	-5	-2	-3	-4	-6	-4	0	6	0	0	-1	-4	-3	-2	-4	-2	0
5	0	-3	-1	-2	-3	0	-2	4	-3	-3	0	-2	-2	-4	-3	3	1	-4	-4	-3
6	0	2	0	4	-4	1	2	1	-2	-4	-4	0	-3	-4	-3	1	-2	-5	-4	-4
7	-1	3	3	-2	-4	-1	-1	1	-2	-1	-4	1	-3	-4	-3	1	-2	-5	-4	-4
8	-2	-3	-4	-5	-3	-3	-4	-5	-4	3	4	-1	1	2	-4	-3	-2	-3	-1	0
9	-2	3	2	-2	-4	2	1	-3	-2	-3	-3	1	1	-4	-3	2	1	-4	-3	-1
10	0	2	3	1	-4	0	0	0	-2	-4	-4	1	-3	-4	-3	2	0	-5	-4	-4
11	5	-3	-3	-3	-2	-3	-3	-2	-3	1	-2	-3	-2	1	-3	0	1	-4	-2	0
12	-1	-4	-3	-5	-3	-4	-4	-5	-4	3	3	-4	2	3	-5	-3	-2	5	-1	2
13	0	3	3	0	-4	3	0	1	-2	-4	-4	1	-3	-4	-3	1	-1	-4	-3	-4
14	-1	0	1	0	-4	1	-1	-1	-2	-4	-3	5	-2	0	-3	0	-2	-4	0	-3
15	-2	-3	-1	-5	-3	-3	-4	-5	-4	-3	4	0	4	2	-4	-3	-2	-3	-2	0
16	0	3	0	-2	-3	-1	0	0	-2	0	0	1	0	-1	-3	2	0	-4	-3	0
17	-1	1	3	-2	-4	0	-2	4	-2	-4	-4	0	-3	0	-3	0	0	-3	0	-4

Window of
15 rows

└── Amino acid position number in sequence

B

A	R	N	D	C	Q	E	G	H	I	L	K	M	F	P	S	T	W	Y	V
0	0	0	0	0	0	0	0	0.1	0.3	0.3	0	0.3	1	0	0	0	0.5	0.9	0.1
0.5	0.3	0.3	1	0	1	1	0.7	0.3	0	0	0.5	0	0	0.1	0.3	0.1	0	0	0
0.5	0.3	0.9	0.7	0	1	0.5	0.3	0.1	0	0	0.7	0.1	0	0.1	0.9	0.5	0	0	0
*																			
*																			
*																			

15 x 20 scaled
inputs to 1st
network

C

┌─────────────────┐
│ 1st network   │
│ 315 input units │
│ 75 hidden units │
│ 3 output units │
└─────────────────┘

3 outputs x 15 residues
+ 15 terminus data
fed to 2nd network

D

┌─────────────────┐
│ 2nd network   │
│ 60 input units │
│ 60 hidden units │
│ 3 output units │
└─────────────────┘

Final 3-state prediction
(alpha helix, beta sheet
or coil)

FIG. 7-18. Work flow diagram of PSIPRED neural network prediction program.
A. A query sequence is used as input for the program. The sequence is used by PSI-BLAST to generate a PSSM after three iterations of searches of a database containing several non-redundant structures. The PSSM ("Raw profile from PSI-BLAST Log File") is generated after the third iteration. A 15-amino acid window of the PSSM is scaled to a 0–1 value range. **B.** Four rows of scaled data from the PSSM are shown. **C.** The scaled data on 15 amino acid residues plus a unit indicating whether each amino acid is located at a terminus of the protein is entered into the input layer of the neural network. The hidden layer contains 75 hidden units. The hidden units relay information to three output units. Each unit outputs a value—one output for a helix, one output for a strand, and one output for a coil. For each amino acid in the 15-residue window, the three outputs from the first network plus data on whether the residue is a terminus of the protein is fed into a 60-unit input layer. **D.** The second network (filtering network) has a 60-unit input layer, and a 60-unit hidden layer. The 60 units in the hidden layer relay information to three output units. The three output units decide whether the residue is a helix, strand, or coil. The secondary structure with the highest value is reported to the user.

The 300 values are created from the amino acid sequence (which come from 15 amino acid positions in the window × 20 scores for each position). There are 15 additional values that register whether each of the original 15 positions in the window is a terminus (amino or carboxyl) of the protein chain. This means there are a total of 315 values. Each value is mathematically adjusted to vary between 0 and 1 and transmitted to 315 input units in the 1st network. Within the first network, values from the input layer are transmitted to a hidden layer of 75 units. The output values from the hidden units are linear functions that have been altered due to weights (more on this below). The values from the hidden layer are transmitted to the output layer, which consists of three units: one for helix, one for strand, and one for coil.

The second stage of PSIPRED consists of a second NN. This second NN, called the filtering network, filters outputs from the first network. Scaled predictions of one of three states (for each of 15 residues) plus whether the residue is located at the end of the sequence are sent to the second NN. So, the input layer of the second NN consists of 60 units. The units transmit information to a hidden layer of 60 units. The final output layer contains three units where each unit predicts the probability of α-helix, β-strand, and coil. The final output gives a single secondary structure prediction for each amino acid and the confidence level of the output scaled 0–9.

Neural network training

As mentioned previously, NNs depend on adjusting the weights that will be applied from one layer to the next. How are the weights established? A set of sequences with known secondary structures was placed into a database, and the database was used to create the PSSM using PSI-BLAST. Ninety percent of these sequences were used to "train" the NN, and 10% percent were used to test the NN. The sequences in the training set were chosen with care. The set contained a wide variety of protein folds and a minimum of duplicate structures, which ensures that the PSIPRED is not biased toward predicting a particular structure. During training, the weights were adjusted so that the error in the output was reduced. This training process was repeated several times until the program was able to predict the outcome with a high degree of accuracy. In truth, the program automatically adjusts its weights and automatically reruns the sequence data iteratively until the bioinformaticist halts the process.

A key aspect of the training process is not to "overtrain" the NN. If this occurs, the NN becomes very adept at correctly predicting the structure of the training set sequences, but fails to accurately predict structures of sequences that are not part of the training set. An overtrained NN becomes useless when it comes to predicting the outcome of new sequences. To determine whether the NN can correctly predict the secondary structures of other sequences, the test data is used (the 10% of the sequences that were not used for training). If the accuracy of the prediction of the test data is equal to the final accuracy of the training data, then the program was not overtrained. Once the weights are set, the PSIPRED neural network is not altered. When a user inputs a new protein sequence into the program, a new PSSM is created on the fly. This PSSM is used as input into the first NN.

The PSIPRED program is able to predict secondary structures with 73% Q_3 score and a 72% Sov score. Importantly, these scores were obtained for structures that were experimentally solved and were not used in the training or test data set for PSIPRED. A typical PSIPRED output is shown in Figure 7-19.

Prediction Method 4: Homology Modeling
Introduction to homology modeling

In a manner similar to secondary structure prediction, the approach of homology modeling is based on the principle that sequence determines structure. Two proteins

```
PSIPRED PREDICTION RESULTS
Key

Conf:  Confidence (0=low, 9=high)
Pred:  Predicted secondary structure (H=helix, E=strand, C=coil)
  AA:  Target sequence
Conf:  923788850068899998538983213555268822788714786424388875156215
Pred:  CCEEEEEEEHHHHHHHHHHHCCCCCCHHHHHHCCCCCEEEEECCCCCCHHHHHHHCCCCCC
  AA:  KDIQLLNVSYDPTRELYEQYNKAFSAHWKQETGDNVVIDQSHGSQGKQATSSVINGIEAD
            10        20        30        40        50        60
```

FIG. 7-19. PSIPRED prediction result from a 60-residue sequence.

with high similarity are expected to form the same structure. Early work with globin and cytochrome C homologs gives credence to this principle. Structures of protein orthologs indicate that they are strikingly similar with the exception of a few segments. Segments that do not have similar structures often have different residues in those segments. Given these observations, a protein sequence of unknown structure is likely to have a structure similar to a protein of known structure if their sequences are highly similar. Homology modeling is a structure prediction method that uses a protein template of known structure to build a structural model of a sequence with unknown structure.

Steps to homology modeling

There are eight steps to homology modeling.

STEP 1. **Search and select a template.** A template is a sequence with a structure that has been experimentally determined, usually by X-ray crystallography. The query sequence has unknown structure. One can build a structure model from the template if the template sequence has a high percentage of identity to the query sequence. Another possibility is to build a structure model from a template that is comprised of an average structure of several similar sequences that have been experimentally determined. A third possibility is to use fragments from different structures to create a unique structure template.

Models built on a template from a single structure will give accurate results if there is sufficiently high sequence identity (~50%) between it and the query sequence. The advantage over the average template derived from several sequences is that distortion due to averaging structures will not be incorporated into the model. Models built on a template derived from an average structure may be more realistic if the identity to any one template sequence is lower than 50%, and if there is more than one structure with reasonable identity. If a model is based on separate fragments from homologous structures, the software program must join the fragments together, which can create errors. Structure templates are derived from the PDB and can be found using a BLAST search of the PDB using the query sequence.

STEP 2. **Create a multiple sequence alignment that includes the query and template sequences.** Once a suitable template has been identified, it is critical to properly align the sequence with the query sequence. One can improve alignment from the initial BLAST search by adding more homologs (even if their structures are not known) to create a multiple sequence alignment. Alignment of several homologs helps to properly align the query and the template because conserved domains will easily be identified. Other information that can help is knowledge of the residues that constitute the active site. In a family of homologs, the active site residues are likely to be conserved. Furthermore, insertions and deletions within the aligned sequences should not fall within a run of α-helical or β-strand residues. These secondary structures are rarely broken in conserved sequences.

STEP 3. **Produce a model of the structurally conserved segments of the query sequence.** The structurally conserved regions are first modeled by transferring the X, Y, and Z coordinates of matched atoms from the template to the query sequence. Initially, the backbone atoms are joined together to form peptide bonds that display correct torsional angles. It is likely that not all of the side chain atoms of the query sequence will be modeled at this step because they will not always match the template side chain sequence. In the protein to be modeled, regions with insertions and deletions, visible in the multiple sequence alignment, are most problematic and are dealt with later (see step 5). This initial model is called the modeled core.

STEP 4. **Check the modeled core for errors.** If the modeled core has interruptions in conserved secondary structures, it is best to realign the query sequence and the template sequence (step 2) and proceed again.

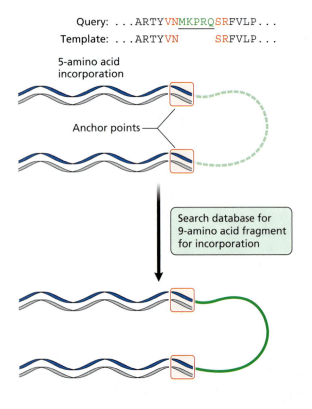

Query: ...ARTYVN<u>MKPRQ</u>SRFVLP...
Template: ...ARTYVN   SRFVLP...

5-amino acid incorporation

Anchor points

Search database for 9-amino acid fragment for incorporation

FIG. 7-20. Incorporation of a nine-residue fragment from PDB to insert a loop into a query structure. The sequence alignment shows that the query has an extra five-residue insertion that must be incorporated into the structure. The anchor points are two residues (red) that bracket the insertion. The schematic structure shows the query (blue) structure matched to the template (black) structure. The red bordered sections are the anchor points. The nine-residue query used to obtain the nine-residue fragment has the sequence VNMKPRQSR. The fragment obtained from the PDB is placed as loop onto the modeled core (green).

STEP 5. Add loops. Loops are regions that typically consist of insertions or deletions in the multiple sequence alignment. They are the most varied in sequence. However, in some cases loops may have important functions for the protein and must be carefully modeled, as is the case for p53. In the p53 DNA binding domain, the loops at the ends of the antiparallel β-strands interact with DNA targets (see Box 7-1). Let's start our discussion with insertion loops that must be inserted in the modeled core. These are regions of the query sequence that line up with gaps in the template sequence in the multiple sequence alignment. If there are other structure homologs that have sequences that match the query sequence in the loop area, those coordinates can be used to create a loop in the modeled core. If other structure homologs do not exist, then the loop is built from a database of structure fragments (created from the PDB).

To build a loop from a database of structure fragments, the homology modeling program does the following. Let's say that a loop of five residues is necessary for insertion into the modeled core. A fragment database containing nine residue-length structures is searched with nine residues of the query sequence (Figure 7-20). The nine-residue query contains the five-residue loop region. The structure of the sequence with the highest identity (or similarity) across the central five residues is used for loop building. Ends of the modeled core to which the loop must be incorporated are regarded as anchor points. The anchor points are typically two residues at each end. The RMSD of the Cα carbons of the anchor point residues and the ends of the fragments is calculated. The structure coordinates of the fragment that gives the lowest RMSD with the anchor points and has the least steric interference with the modeled core is used. All insertion loops are treated in this manner.

A deletion within the query sequence is a region where a gap is placed in the query sequence to optimally align it with the template sequences in the multiple sequence alignment. For a short deletion, a peptide bond is created between the amino acids that bridge the deletion in the modeled core, and the energy is minimized to reduce angle strain. If the deletion is too large, it is very difficult to create a peptide bond that connects the amino acids on either end.

STEP 6. Use rotamer libraries to model nonidentical residue side chains. Rotamers are particular side chain orientations. Studies have been conducted to analyze the rotamers in proteins that have high sequence identity. In matched residues, the rotamers are nearly always identical. Thus, when the aligned residues between the query and template sequences are identical, the atomic coordinates can be transferred from the template structure to the query structure. When the residues are not identical, homology modeling programs use a database of rotamer libraries. Rotamer libraries hold coordinates of the most common rotamers. These rotamers minimize steric clashes of side chain atoms with other atoms in the protein. One can distinguish different rotamers of the same side chain by the torsion angles.

Figure-7-21A shows torsion angles for glutamine bonded to two other residues. The φ angle is the angle of rotation about the N-Cα bond in the backbone. The ψ angle is the angle of rotation about the Cα-C$_{carbonyl}$ bond in the backbone. These angles dictate the secondary structure of this segment of the protein. The torsion angles χ_1, χ_2, and χ_3 describe the side chain orientation.

Most side chains in a protein are limited to a just a few of the many possible torsion angles. For example, for Leu there are two torsion angles, χ_1, χ_2 (Figure 7-21B) and there are two common Leu rotamers that account for

89% of the rotamers found in protein structures. One rotamer that accounts for 65% of the leucines has $\chi_1 = 292.0$, $\chi_2 = 168.5$. The other rotamer that accounts for 24% of the leucines has $\chi_1 = 183.0$, $\chi_2 = 66.6$. Often the rotamer favored depends on the backbone secondary structure. For homology modeling, the rotamer chosen by the program depends on backbone secondary structure and energy calculations to limit steric clashes.

STEP 7. Minimize energy to reduce errors. After the loop regions have been incorporated into the model, the loops should be energy minimized. Energy minimization is applied to the loops and junctions areas where the loops join the modeled core. At this point we have a complete homology model.

STEP 8. Check the homology model for errors. There are software programs available to analyze the homology model for errors. Some assess the quality by comparing the conformations of the residues to those derived from high-resolution structures. It is good practice to assess the quality of the template as well. "Errors" detected in the template (from faulty PDB structures) will likely have been transferred onto the homology model, thereby compromising the structure prediction.

One effective approach to test the accuracy of the homology model is to compare the frequency of finding two residues near to one another in the homology model to the frequency of finding the same two residues in known structures, even if those structures are unrelated to the model. The distribution of inter-residue distances in known protein structures for all 20×20 pairs of amino acids is calculated. For each pair within the query sequence, a probability distribution as a function of separation in space in the homology model is calculated. For example, consider the pair Glu/Leu. For each Glu/Leu pair in known structures, the distances between their $C\beta$ atoms and their relative positions in the sequences of the homology model is compared. If the homology model distance distribution is significantly different than the known structures' distance distribution, the model should be revisited and corrected.

Prediction Method 5: Threading

What happens when you wish to know the tertiary structure of a protein, but you are unable to do homology searching because your protein sequence does not meet the threshold of 50% identity with a protein of known structure? Interestingly, it is still likely that there are structures in the PDB that are similar to your protein. Surprisingly, there are proteins that share low identity but have very similar structures. Let's take a look at two such proteins. In Figure 7-22A is a global alignment of leghemoglobin from *Lupinus luteus* (European yellow lupin plant) and hemoglobin alpha chain from *Homo sapiens*. The shared sequence identity is only 16%, which is considered to be below the minimum identity to claim that the two proteins are homologs (see Chapter 3). However, as Figure 7-22B clearly shows, X-ray crystallographic structure analysis demonstrates that the two proteins have very similar structures, suggesting that they likely have a common ancestor. This means that, starting with a common ancestor gene, accumulated mutations in the two homologs resulted in very different sequences but the structures were maintained.[21] Here, we

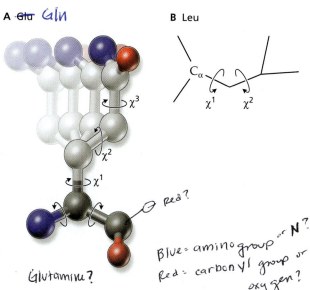

A ~~Glu~~ Gln   **B** Leu

Glutamine?

Blue = amino group or N?
Red = carbonyl group or oxygen?

FIG. 7-21. Torsion angles of ~~glutamate~~. glutamine **A.** Φ is the angle of rotation about the $N-C\alpha$ bond in the backbone. Ψ is the angle of rotation about the $C\alpha-C_{carbonyl}$ bond in the backbone. χ_1 is the angle of rotation about the $C\alpha-C\beta$ bond in the side chain. χ_2 is the angle of rotation about the $C\beta-C\gamma$ bond in the side chain. χ_3 is the angle of rotation about the $C\gamma-C\delta$ bond in the side chain. **B.** Torsion angles χ_1 and χ_2 in Leu.

---

[21] Another argument for this observation is that the two structures are the result of convergent evolution. At the molecular level, two unrelated genes may sustain mutations over millions of years such that they produce proteins of the same structure and function. This does not appear to be the case here.

A

```
leghemoglobin      1   GALTESQAALVKSSWEEFNANIPKH----THRFFILVLEIAPAAKDLFSF   46
                       .|:.:....||::|.:..|:...::    ..|.|:..    |..|..|..
human alpha chain  1   -VLSPADKTNVKAAWGKVGAHAGEYGAEALERMFLSF----PTTKTYFPH   45

leghemoglobin     47   LKGTSEVPQNNPELQAHAGKVFKLVYEAAIQLEVTGVVVTDATLKNLGSV   96
                       .   ::..:..::::.|..||..:.:.|    |..|......|..|..:
human alpha chain 46   F----DLSHGSAQVKGHGKKVADALTNA-----VAHVDDMPNALSALSDL   86

leghemaglobin     97   HVSK-GVADAHFPVVKEAILKTIKEVVGAKWSEELNSAWTIAYDELAIVI  145
                       |..|  .|...:|.::...:|.|:....:|:::..:::......::.|:
human alpha chain 87   HAHKLRVDPVNFKLLSHCLLVTLAAHLPAEFTPAVHASLDKFLASVSTVL  136

leghemoglobin    146   KKEMDDAA    153
                       ..:..
human alpha chain 137  TSKYR---    141

Identity:     26/158  (16.5%)
Similarity:   60/158  (38.0%)
Gaps:         22/158  (13.9%)
Score: 41.5
```

B

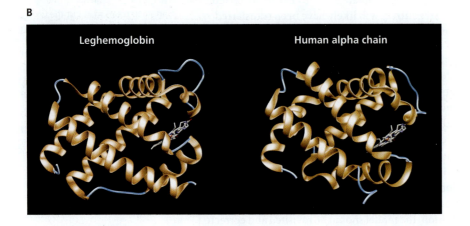

FIG. 7-22. Comparison of leghemoglobin from yellow lupin plant and alpha chain of hemoglobin from humans. **A.** Global alignment of leghemoglobin sequence from *Lupinus luteus* (PDB 2GDM) and alpha chain hemoglobin sequence from *Homo sapiens* (PDB 1GDX). **B.** Structures of leghemoglobin and alpha chain hemoglobin polypeptides solved by X-ray crystallography. Structures produced with Chimera version 1.9.

are comparing proteins from two species that diverged from a common ancestor more than one billion years ago. In fact, the function of the two proteins is maintained as well: both transport oxygen. The yellow lupin plant uses leghemoglobin to supply oxygen to bacteria, called rhizobia, in the root nodules. The bacteria use the oxygen to synthesize the energy molecule ATP so they can survive. Plants need the bacteria because they convert atmospheric nitrogen (N_2) into ammonium (NH_4^+) in a process known as nitrogen fixation. In humans, the alpha chain is part of a tetramer in the blood hemoglobin, which is necessary for transporting oxygen from the lungs to the tissues. Threading is a process that predicts a protein structure using folding patterns stored in protein structure-containing databases.[22] Threading does not strictly depend on pairwise sequence identities.

Threading requires a database that contains organized protein folds from the PDB. One such database is CATH (class, architecture, topology, homologous superfamily). In CATH, protein domains are organized into superfamilies, each of which contains domains with similar structures. There are more than 250,000 sequences organized into 2,000 superfamilies. This means that many different sequences can

---

[22] The term "threading" was coined by Jones, Taylor, and Thornton (1992).

fold into nearly the same structures. Threading can be conceptualized as pulling a string of query residues through a specific fold and examining the compatibility of each residue with that fold. The query sequence is aligned to each sequence in the library of folds from the CATH database. The compatibility of the sequence with that structure is calculated. This process is repeated for all folds in the library. Key to determining the compatibility is the degree to which the threaded amino acids are solvated in the local environment (solvation potential). The threading program returns to the user a ranked list of structural templates that are most compatible with the query sequence's predicted structure. Each structural template has a PDB file number, and the user can choose which template to use as a model for building a structure of your query sequence. Similar to homology modeling, at this point the query sequence is folded into a tertiary structure using the user-selected template as a scaffold.

SUMMARY

There is tremendous opportunity for bioinformaticians to contribute to bridging the gap in our knowledge of protein sequence and protein structure. Computational methods, such as quantum mechanical methods and molecular mechanics methods, are still too costly, in the computational sense, to predict secondary or tertiary structures on long sequences. The combination of computational methods, networked computers, and comparison with structures with similar sequences will likely help bioinformaticians predict tertiary structures with greater accuracy in the future. Protein structure prediction methods are becoming more accurate, in large part due to usage of the vast number of structures that have been experimentally solved. The most successful prediction programs are neural network programs designed to predict secondary structures. Their prediction accuracies often range from 70% to 80%. It may be likely that this is the upper limit to secondary structure prediction from a linear sequence because the tertiary structure might have some influence on secondary structure. This would mean that there is an upper limit in our ability to predict secondary structures strictly from sequence. The most successful tertiary structure prediction programs create homology models from structures, known as templates, with >50% sequence identity. If such obvious templates are not available, threading is an option. In threading, query sequences are compared to each amino acid in protein structures in a fold library. Query amino acids that can form structures compatible with a particular structure in the fold library and with the local solvent environment associated with that structure are scored. The structural templates (PDB files) producing the best scores are identified. The user selects one of the templates to use as a scaffold for building a structure from the query sequence.

EXERCISES

1. **Find the complete sequence of wild-type human p53 and perform a secondary structure prediction with PSIPRED, GOR, Chou-Fasman, or other secondary structure prediction algorithm (the instructor may wish to assign students different secondary structure predictors). Go to the Protein Data Bank and obtain the file for the p53 crystal structure (1TSR). There are three identical p53 chains in the file: A, B, and C. Choose the A chain for this exercise. In the remarks records of the file, you will find assignments of secondary structures for the residues. These will be named either "helix" or "sheet." For amino acids in the structure that are not classified as "helix" or "sheet," assume that they adopt a "coil" structure. Create a line graph that places the residue sequence in one row and the known secondary structure from the PDB file for that sequence in the next row (the choices are H for helix, E for β-strand, and C for coil). Create a third row on the line graph that shows the predicted structure for each**

residue. The 1TSR file only contains the DNA binding domain of p53 so you will be able to cover only about half of the protein. Calculate the Q_3 score by hand.

2. **Obtain structures of non-DNA binding domain segments of p53 from the Protein Data Bank (in different PDB files), and report the solved secondary structures of those segments. Calculate the Q_3 score of your prediction program for all solved segments of the p53 protein.**

3. **Open the PDB file 1TSR (p53) in a molecular viewer. Display chain B in surface display mode with electrostatic surface or coulombic surface coloring. You will see sections colored blue. Which amino acid side chains would you expect to find in these sections?**

4. **MDM2 is an inhibitor of p53. The crystal structure of the p53 binding domain of MDM2 was solved giving rise to the possibility that an inhibitor of**

MDM2 binding to p53 could restore p53 activity in some cancers. Approximately 7% of cancers have high levels of MDM2 and normal p53, making these cancers potential targets for "anti-MDM2" therapy. 1RV1 is a PDB file that contains three nearly identical MDM2 polypeptides bound to experimental drug inhibitors. Use a molecular viewer to remove all nonprotein molecules from the file. Create two files, one with chain A and one with chain C. Next, calculate the RMSD for superimposed A chain and C chain from these two files. Compare all nonhydrogen protein atoms in the chains (not just the backbone atoms).

5. Answer the following questions True or False. If false, explain why.

   a. Homology modeling requires alignment of a target to a template.
   b. Homology modeling can be applied to any protein sequence.

6. You have a protein sequence and you want to know its structure in a short time. You perform BLAST and PSI-BLAST searches of the PDB and you find a sequence with 15% amino acid identity to your protein. The *E*-value is 0.5. Which of the following options would you choose to achieve your goal?

   a. Use X-ray crystallography.
   b. Use NMR spectroscopy.
   c. Submit your sequence to a protein structure prediction server that performs homology modeling.
   d. Submit your sequence to a protein structure prediction server that performs molecular mechanics modeling.

7. pDomThreader is one of the many threading programs that predict structures that show very little sequence identity to proteins with known structures. One interesting human protein that appears to promote breast cancer cell proliferation is t-DARRP (accession number Q9UD71-2 in the UniProtKB/Swiss-Prot database). At the time of this writing, no experimental structures have been reported on this interesting protein. Can you use pDomThreader to predict the structure of this protein? What is the PDB file number of the template chosen by ProDomThreader for your query sequence (tDARRP)? What is the name of the protein template?

ANSWER TO THOUGHT QUESTION

7-1. Serine 94

Note that the orientation of the molecule in your answer may be different from that shown in the figure.

REFERENCES

Anderson, H. L. 1986. "Metropolis, Monte Carlo and the Maniac." *Los Alamos Science* (Fall): 96–107.

Anfinsen, C. B. 1973. "Principles That Govern the Folding of Protein Chains." *Science* 181: 223–230.

Baker, D., Z. Popović, and F. Players. 2010. "Predicting Protein Structures with a Multiplayer Online Game." *Nature* 466: 756–760.

Bryngelson, J. D., J. N. Onuchic, N. D. Socci, and P.G. Wolynes. 1995. "Funnels, Pathways, and the Energy Landscape of Protein Folding: A Synthesis." *Proteins* 21: 167–195.

Cho, Y., S. Gorina, P. D. Jeffrey, and N. P. Pavletich. 1994. "Crystal Structure of a p53 Tumor Suppressor-DNA Complex: Understanding Tumorigenic Mutations." *Science* 265: 346–355.

Chou, P. Y., and G. D. Fasman. 1978. "Prediction of Protein Conformation." *Annual Review of Biochemistry* 47: 251–276.

Cooper, S., F. Khatib, A. Treuille, J. Barbero, J. Lee, M. Beenen, A. Leaver-Fay, D. Baker, Z. Popović, and F. Players. 2010. "Predicting Protein Structures with a Multiplayer Online Game." *Nature* 466: 756–760.

Friend, S. 1994. "p53: A Glimpse at the Puppet Behind the Shadow Play." *Science* 265: 334–336.

Harder, T., W. Boomsma, M. Paluszewski, J. Frellsen, K. E. Johansson, and T. Hamelryck. 2010. "Beyond Rotamers: A Generative, Probabilistic Model of Side Chains in Proteins." *BMC Bioinformatics* 11: 306.

Hehre, W. 2003. *A Guide to Molecular Mechanics and Quantum Mechanical Calculations*. Irvine, CA: Wavefunction.

Jones, D. T., W. R. Taylor, and J. M. Thornton. 1992. "A New Approach to Protein Fold Recognition." *Nature* 358: 86–89.

Jones, D. T. 1999. "Protein Secondary Structure Prediction Based on Position-Specific Scoring Matrices." *Journal of Molecular Biology* 292: 195–202.

Kendrew, J. C., R. E. Dickerson, B. E. Strandberg, R. G. Hart, D. R. Davies, D. C. Phillips, and V. C. Shore. 1960. "Structure of Myoglobin: A Three-Dimensional Fourier Synthesis at 2 Å. Resolution." *Nature* 185: 422–427.

Kessel, A., and N. Ben-Tal. 2011. *Introduction to Proteins.* London: CRC Press, Taylor & Francis Group.

Kloczkowski, A., K. L. Ting, R. L. Jernigan, and J. Garnier. 2002. "Combining the GOR V Algorithm with Evolutionary Information for Protein Secondary Structure Prediction from Amino Acid Sequence." *Proteins* 49, 154–166.

Kussie, P. H., S. Gorina, V. Marechal, B. Elenbaas, J. Moreau, A. J. Levine, and N. P. Pavletich. 1996. "Structure of the MDM2 Oncoprotein Bound to the p53 Tumor Suppressor Transactivation Domain." *Science* 274: 948–953.

Kwan, A. H., M. Mobli, P. R. Gooley, G. F. King, and J. P. Mackay. 2011. "Macromolecular NMR for the Non-spectroscopist." *FEBS Journal* 278: 687–703.

Lesk, A. M. 2008. *Introduction to Bioinformatics*, 3rd ed. Oxford, UK: Oxford University Press.

Levinthal, C. 1969. "How to Fold Graciously." *Mössbaun Spectroscopy in Biological Systems Proceedings University of Illinois Bulletin* 67: 22–24.

Melero, R., S. Rajagopalan, M. Lázaro, A. C. Joerger, T. Brandt, D. B. Veprintsev, G. Lasso, D. Gil, S. H. Scheres, J. M. Carazo, A. R. Fersht, and M. Valle. 2011. "Electron Microscopy Studies on the Quaternary Structure of p53 Reveal Different Binding Modes for p53 Tetramers in Complex with DNA." *Proceedings of the National Academy of Sciences USA* 108: 557–562.

Moult, J. 1999. "Predicting Protein Three-Dimensional Structure." *Current Opinion in Biotechnology* 10: 583–588.

Persistence of Vision Pty. Ltd. 2004. Persistence of Vision Raytracer (Version 3.6) [Computer software]. Retrieved from http://www.povray.org/download/.

Pettersen, E. F., T. D. Goddard, C. C. Huang, G. S. Couch, D. M. Greenblatt, E. C. Meng, and T. E. Ferrin. 2004. "UCSF Chimera—A Visualization System for Exploratory Research and Analysis." *Journal of Computational Chemistry* 25: 1605–1612.

White, F. H., and C. B. Anfinsen. 1959. "Some Relationships of Structure to Function in Ribonuclease." *Annals of the New York Academy of Sciences* 81: 515–523.

Zemla, A., C. Venclovas, K. Fidelis, and B. Rost. 1999. "A Modified Definition of Sov, a Segment-Based Measure for Protein Secondary Structure Prediction Assessment." *Proteins* 34: 220–223.

Zvelebil, M., and J. O. Baum. 2008. *Understanding Bioinformatics*. New York: Garland Science, Taylor & Francis Group.

Zwanzig, R., A. Szabo, and B. Bagchi. 1992. "Levinthal's Paradox." *Proceedings of the National Academy of Sciences USA* 89: 20–22.

8

PHYLOGENETICS

8.1 INTRODUCTION

In Chapter 6 we discussed how local sequence alignment programs are useful in discovering sequences that are similar in evolutionarily distant species. Ortholog sequences from two distantly related species are usually conserved because the genes are necessary to carry out functions that both species require for survival. At the molecular level, sequence conservation is required for structure and function. We learned that some sequences may even have diverged to the point where their functions are different, yet they show sufficient similarity such that we can be fairly certain they were derived from a common ancestor gene. In this chapter, you will learn how to use sequence information to categorize how species are related to each other. Phylogenetics is the utilization of sequence information to create evolutionary histories of species. Sequence information from recently extinct human subspecies Neanderthal and Denisovan has become available. Using phylogenetic software programs, we can show at what point in time these human subspecies shared a common ancestor with us.

Bear in mind that mutation, natural selection, genetic drift, and gene flow cause new species to arise (we discussed mutation and natural selection in Chapter 3). Prior to our ability to identify DNA mutations directly, the field of paleontology was developed to study evolutionary histories (phylogeny). We will explore how paleontology, together with sequence information, contributes to the study of evolutionary histories. Before we delve into the analysis of sequence information, it is essential to garner a sense of appreciation for the development of this older science of paleontology.

Paleontology is the study of prehistoric life. It includes the study of fossils to create evolutionary histories. The origin of paleontology can be traced to George Cuvier (1769–1832), a French naturalist who first proposed that some animal species became extinct. In 1796 he read his first paper on the subject, "Mémoires sur les espèces d'éléphants vivants et fossils [Notes on the Species of Living and Fossil Elephants]," to the French National Institute.[1] Cuvier analyzed the fossil remains of elephants, a mammoth, and an unknown species known then as the Ohio animal. His paper established that Indian and African elephants were different species and that the mammoth did not belong to either the Indian or African elephant species and was, in fact, an extinct species. In a later paper he explained that the Ohio animal was another extinct species that he named mastodon. He proposed that catastrophic events caused mass extinction of mammoths and mastodons and, by extension, other animal species. He developed a theory he named catastrophism, which states that cataclysmic events cause elimination of many animal species.[2] His catastrophism theory was controversial in that it was difficult for scientists at the time to consider that entire species could be eliminated, but subsequent analysis of thousands of other fossils has now established that the earth was inhabited with an abundance of species that are now extinct. In fact, it has been estimated that since life first began on earth approximately 3.8 billion years ago, more than 99% of living species have become extinct—some through catastrophic events and others through lack of survival skills against superior competition.

For our discussion of phylogenetics, it is important to understand that fossil evidence suggests that life has existed for 3.8 billion years, but anatomically modern humans (i.e., modern *Homo sapiens*) appeared only within the last 200,000 years. This means that the period of our existence as a species is a small fraction of the period of life on earth—this fraction is 0.000052. To appreciate the relative briefness of human existence, consider the following comparison. Picture an adult with his arms stretched out from the sides of his body so that they are parallel to the floor (Figure 8-1). Let's say that the time of existence of life on earth (3.8 billion years) is represented by the distance from the end of one hand to the end of the other hand. The relative period of human existence is less than the length of the fingernail over-hanging one of the outstretched fingers (200,000 years). Such a brief existence makes it difficult (perhaps impossible) for us to witness the evolution of new species of complex organisms.[3] The fossil record presents us with a crude timeframe of the evolution of species. This timeframe lacks many speciation events (i.e., events in history when two species evolved from a common ancestor) because we simply do not have enough fossil data to give us a more detailed picture. Starting in the 1960s, molecular sequence data from living organisms, first from proteins and later from DNA, provided another means of measuring the timing of evolutionary events.

phylogeny The description of evolutionary relationships of organisms, or groups of organisms.

phylogenetics A type of phylogeny that uses character data (observable traits) to describe evolutionary relationships. One type of character data that can be used to describe evolutionary relationships is sequence information. When sequence is extensively used, it is called molecular phylogenetics.

phylogenetic tree (phylogram) A diagram that portrays evolutionary relationships in which the branch lengths are proportional to evolutionary time.

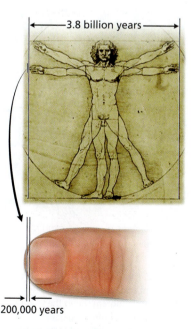

FIG. 8-1. Life has existed on earth for 3.8 billion years, and modern *Homo sapiens* has existed approximately 200,000 years.

---

[1] Cuvier's paper was published in 1800.

[2] Examples of catastrophic events include tsunamis, earthquakes, and meteor impacts.

[3] On the other hand, the evolution of viruses, due to their short life cycles, can be studied with modern sequencing methods. For further reading on this subject, see Gibbs, Calisher, and García-Arenal (1995).

This chapter explores bioinformatics methods applied to molecular sequence data that can be used to study the evolution of species.

8.2 PHYLOGENY AND PHYLOGENETICS

Phylogeny is the description of evolutionary relationships of organisms, or groups of organisms. Phylogenetics uses character data (observable traits) to analyze such relationships. Character data can be anatomy, morphology, sequence changes, and so on. Character data can be further delineated into character traits and character states. A character trait is one that is observed in all organisms that are to be compared. Take, for example, a specific bone in the body. A character state is a measurable property of the bone. This could be length, mass, density, and so on. It is desirable for the character state to be variable among the organisms being compared. Another character trait is a protein sequence. The corresponding character state is the percent differences in the sequences among organisms. When sequence data are extensively used to study phylogeny, we call this field molecular phylogenetics. A phylogenetic tree is a diagram, also known as a phylogram, that portrays evolutionary relationships such that branch lengths are proportional to evolutionary time. A cladogram also shows evolutionary relationships, but the branch lengths are not proportional to evolutionary time. Figure 8-2 shows a cladogram from the 19th century that attempts to describe how living species are related to each other.[4]

Charles Darwin, the primary architect of natural selection theory, also used trees to explain the relationships between species. In 1837 Darwin drew a tree in one of his scientific notebooks that resembles the trees that modern evolutionary biologists use to describe relationships between species (Figure 8-3). The lines in the tree are branches, and each branch tip, or leaf, represents a species. In Darwin's tree notation, the branch tip labeled "1" represents an extinct species from which other species arose. This is the ancestral species common

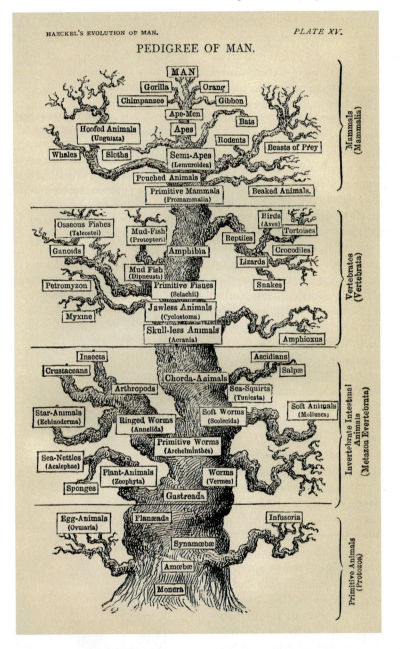

FIG. 8-2. The pedigree of man from the viewpoint of one 19th-century biologist. This early cladogram shows the relationships of species from the perspective of the 19th-century biologist Ernst Haeckel.

---

[4] This cladogram was drawn by the German biologist Ernst Haeckel (1834–1919). Prior to the wide availability of fossil data, most biologists classified species based on morphological data. Haeckel believed that during the period of development from fertilization to birth, organisms go through a series of apparent morphological stages, and that in the earlier stages of development they resemble their evolutionary ancestors. Haeckel coined the phrase "ontogeny recapitulates phylogeny." Ontogeny means the development of form. His theory, called recapitulation theory, was very popular at the beginning of the 20th century, but has waned quite a bit in recent years.

to all species in Darwin's tree. Aside from the branch tip labeled "1," there are other branch tips with no stems (one is labeled "n.s."). Branch tips with no stems are extinct species that do not give rise to other species. Branch tips with stems (one is labeled "s") are living species grouped into closely related species. In this figure, groups of living species are labeled A, B, C, and D. For example, in group B the four tips represent four highly related species. All species within a group evolved from a common ancestral species, which is located at a common point (node) on the tree (one node is labeled with a lowercase "b" in Figure 8-3). Ancestral species are extinct species that gave rise to other species.

The sum of the lengths of branches from two tips to a common node represents the evolutionary distance between two species (i.e., the amount of evolution that occurred since the two species shared a common ancestor); the longer the branches, the greater the evolutionary distance. A group containing species connected to a common node is a clade. Groups separated by numerous nodes in the tree, such as groups A and B in Darwin's tree, contain intergroup species that are evolutionarily distant because the sum of the branch lengths that connect them is relatively long. Scientists who study evolution continue to use trees to describe evolutionary relationships. We will explore phylogenetic tree creation later in this chapter, but first we will turn our attention to the concept of the molecular clock.

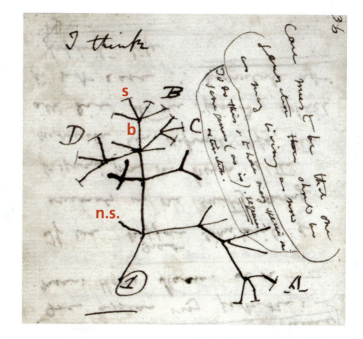

FIG. 8-3. A tree from Darwin's notebook drawn in 1837 (except "b," "s," and "n.s.," which was inserted by the authors of this text for instructional purposes). This image shows Darwin's first tree depicting species phylogeny. In fact, some contend that this is the first phylogenetic tree ever created. Stems at the ends of some of the branches indicate that these are existing species. The "stem" notation is no longer used.

Molecular Clocks

Now we will introduce an important development in the study of evolution: molecular clocks. In the early 1960s, when protein sequences became available for the first time, Linus Pauling and Emile Zuckerkandl (Caltech, Pasadena, California) compared hemoglobin sequences from different species and observed amino acid differences. From evolutionary theory they reasoned that the amino acid differences were due to mutations in the DNA that occurred as the different species evolved. They compared the amino acid changes to the estimated times of divergence of the species from common ancestor species based on the fossil record. Time of divergence (also known as divergence time) is the number of years between the present and the time when two species last shared a common ancestor in the past. For example, it is estimated that humans and chimpanzees last shared a common ancestor approximately five million years ago (MYA), thus the divergence time of humans and chimpanzees is 5 MYA. In their studies, Pauling and Zuckerkandl used divergence times estimated from the fossil record. Based on hemoglobin amino acid sequences obtained from different animals, they suggested that there existed a rough correlation between the number of amino acid differences in existing species and their estimated divergence times. In 1965 Pauling and Zuckerkandl fully developed their theory, which they named the molecular evolutionary clock (now known simply as the molecular clock). In essence, their theory states that protein sequences change at a roughly constant rate, so that the number of observed amino acid residue substitutions is linearly proportional to divergence times.

Figure 8-4 shows that the amino acid residue substitutions in the alpha subunit of hemoglobin are approximately linearly proportional to divergence times of human to each of five species: cow, platypus, chicken, carp, and shark. The divergence times

evolutionary distance The estimated number of substitutions that have occurred since two species shared a common ancestor genome.

clade A group of species connected to a common node.

molecular clock (molecular evolutionary clock) A theory that states that the rate of observable sequence change is linearly proportional to divergence times.

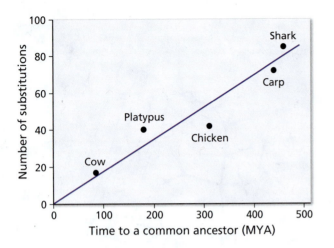

FIG. 8-4. Plot of observed amino acid substitutions as a function of divergence time in hemoglobins. Amino acid differences between human and listed species hemoglobin polypeptides were calculated. The differences were plotted as a function of estimated divergence time between human and listed species. Number of substitutions (to human) is the number of amino acid differences between the human hemoglobin polypeptides and the animal hemoglobin species specified in the graph. Time to common ancestor is in millions of years ago (MYA).

divergence time (time of divergence) Period of time in which two species last shared a common ancestor.

are listed in units of millions of years ago and indicate the times in which human and the listed species last shared a common ancestor.

The linear dependence of observed amino acid substitutions[5] on time means that protein sequence (and, by inference, DNA sequence) differences can be used to measure the timing of evolutionary events. In Figure 8-4 the *x*-axis represents the divergence times of human and animal species measured in millions of years. The *y*-axis represents the number of observed amino acid substitutions (i.e., differences) in the animal species compared to the human hemoglobin sequence. The slope of the plot is called the observed amino acid substitution rate. It should be emphasized that this rate is *not* equal to the mutation rate because it is highly probable that a large proportion of mutations do not survive to the next generation and, therefore, are never observed. Also, because protein sequences are considered, only nonsynonymous substitutions (nucleotide substitutions that result in a change in the amino acid residue) are observed. The rate is also not a true amino acid substitution rate, because a double substitution at a single site could have occurred (i.e., Ala → Gly → Ser) and the first substitution is not accounted for. For these reasons it is best to call this an observed substitution rate. Even given these caveats, it was clear that the observed substitution rate of hemoglobin sequence is fairly constant over short evolutionary time periods (less than 400 MYA).

When scientists sequenced hemoglobin from organisms that diverged from humans in the distant past (significantly greater than 400 MYA), they found that the observed frequency of amino acid substitution was lower than expected. In other words, the amino acid substitution rate did not appear to be constant when comparing sequences with divergence times of more than 400 MYA. This was primarily due to multiple amino acid substitutions at the same sites. A mathematical equation (described below) was used to correct for multiple amino acid substitutions.

Another difficulty with the molecular clock was its application to other proteins. To compare amino acid substitution rates in different proteins you need to consider that different proteins have different lengths. A longer protein would have a higher probability of having a substitution than a shorter protein because it has more possible sites for substitution. To compare the substitution rate of one protein to another, the amino acid residue substitution data were normalized to a common length of 100 residues. Richard Dickerson (Caltech) used an equation to account for multiple substitutions and to normalize the sequence lengths. To determine the corrected number of substitutions, *m*, he applied the following equation:

$$n/100 = 1 - e^{-m/100} \tag{8.1}$$

where *n* is the number of observed substitutions in a 100-amino-acid-residue stretch of polypeptide and *m* is the corrected number of substitutions that occurred in a 100-residue polypeptide.

This equation can be rearranged as follows:

$$m = -100 * \ln[1 - (n/100)] \tag{8.2}$$

---

[5] Recall from Chapter 3 that a substitution is the replacement of one type of amino acid by another type. It is also defined as the replacement of one type of nucleotide in a DNA sequence by another type due to a point mutation and subsequent fixation of the mutation in the population. Nucleotide substitutions are further divided into two categories. Synonymous substitution is a nucleotide substitution that does not lead to a change in the amino acid sequence (due to degeneracy of the genetic code). Nonsynonymous substitution does lead to change in the amino acid sequence.

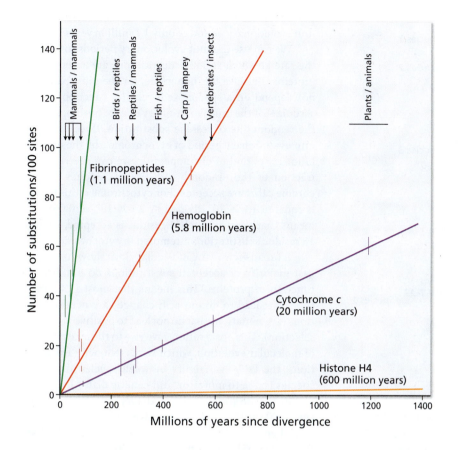

FIG. 8-5. Rates of corrected number of amino acid residue substitutions of four proteins. The corrected number of substitutions per 100 sites plotted as a function of divergence times between different species. The proteins are fibrinopeptides (green), hemoglobin (red), cytochrome *c* (blue) and histone H4 (brown). Times of divergence of major groups of organisms are shown at the top. Values in parentheses displayed under protein names are unit evolutionary periods.

Two important points can be understood from the plot in Figure 8-5. First, an analysis of the corrected amino acid substitution rates shows that each protein has a characteristic rate. Second, rates determined using ortholog sequences from organisms distantly related to each other (common ancestor species existed >400 MYA) are approximately equal to the rates determined using sequences from organisms that are evolutionarily near to one another. This indicates that the mathematical correction properly accounts for multiple substitutions. One way to capture the substitution rate in a single value for easy comparison is to calculate the unit evolutionary period. The unit evolutionary period is the time required for the protein sequence to change by 1%. In Figure 8-5 the longest unit evolutionary period is that of histone H4 (600 million years). It takes 600 million years for 1% of the amino acid residues to be substituted in histone H4. The shortest unit evolutionary period is that of the fibrino-peptide (1.1 million years).

At a deeper level of analysis the unit evolutionary period does not actually reflect the rate of substitution. Rather, it is the rate of substitution that is *accepted* into the protein. Acceptance depends on the extent to which sequence changes alter the protein's function. Substitutions are accepted if they do not greatly deteriorate a protein's function. Histone H4 has a central role in packaging DNA into chromosomes. Almost any change to the residues in H4 will affect its ability to perform this function, and therefore one might predict a low observed substitution rate for H4, and a correspondingly long unit evolutionary period. Fossil data suggests that cows and peas are two species that have a divergence time of ~1.2 billion years ago (i.e., this is when they last shared a common ancestor). Their H4s differ by only two residues (the length of H4 in both species is 102 residues).

Let's take a look at the other extreme. Fibrinopeptide is approximately 20 residues in length. It is removed from a larger polypeptide called fibrinogen during the blood clotting process. After removal, fibrinopeptide is degraded, so there is no strong

unit evolutionary period Time required for a protein sequence to change by 1%.

unit evolutionary period.

- time for a 1% change in sequence
^ protein

A

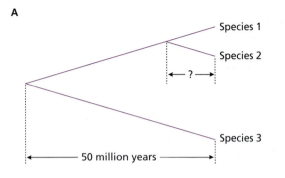

B

Sequence from modern species 1: CAATCGATCG
Sequence from modern species 3: CAATTTATTT

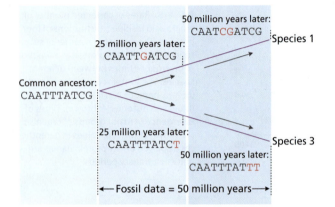

C

Sequence from modern species 1: CAATCGATCG
Sequence from modern species 2: CAATCGATCT

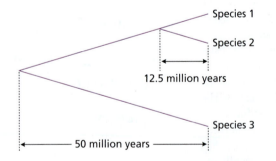

FIG. 8-6. Example of molecular clock used to determine the divergence time.
A. Phylogenetic tree depicting three species, two of which are known to have a divergence time of 50 MYA. **B.** DNA sequences from Species 1 and 3 differ by four nucleotides. **C.** DNA sequences from Species 1 and 2 differ by one nucleotide.

selective pressure to maintain its sequence, resulting in a unit evolutionary period of only 1.1 million years.

We can use the unit evolutionary periods to estimate the rate at which residue substitutions are accepted into proteins. Because fibrinopeptide, for the most part, does not depend on a specific sequence for function, we can assume that the unit evolutionary rate for fibrinopeptide is the random rate of residue substitution. If we divide the unit evolutionary period of cytochrome *c* by the unit evolutionary period of fibrinopeptide, we can determine the fraction of the randomly substituted residues in cytochrome *c* that are accepted. For cytochrome *c*, this fraction is equal to 1.1×10^6 divided by 20×10^6, or 0.055. This means that approximately 1 residue is accepted for every 18 residue substitutions attempted in cytochrome *c*.

A broad view of the molecular clock shows us that the vast majority of accepted substitutions do not alter the function of proteins. This means that most DNA mutations that lead to amino acid changes are neutral mutations (i.e., mutations that do not lead to positive or negative selection). This observation gave rise to the neutral theory of molecular evolution, which states that, with few exceptions, the DNA variability between species and within species is due to mutation and genetic drift.[6] Prominent scientists who developed this theory are Motoo Kimura (National Institute of Genetics, Mishima, Japan), Masatoshi Nei (Pennsylvania State University, University Park, Pennsylvania) and Tomoko Ohta (also at the National Institute of Genetics in Japan). See Chapter 3 for more information on neutral mutations.

Calibrating the molecular clock

After correcting for multiple substitutions, molecular clocks appear to have substitution rates that are constant. This property can be used to time speciation events when fossil data is lacking. To do this, we need to calibrate the molecular clock with a speciation event that has been established with fossil data (or some other independent means). Once calibrated, the molecular clock can estimate the timing of speciation events where there is no fossil data.

For example, let's imagine that there is a phylogenetic tree that contains three species (Figure 8-6A). The ortholog DNA sequences for a particular gene from all three species are available. From the fossil data, the divergence time at which Species 1 and Species 3 shared a common ancestor is estimated to be 50 million years ago. This divergence time can be used to calibrate the molecular clock. The calibrated molecular clock can then be used to calculate the divergence time of Species 1 and Species 2.

---

[6] Genetic drift is defined as random fluctuations in the numbers of gene variants in a population. Genetic drift takes place when the occurrence of variant forms of a gene, called alleles, increases and decreases by chance over time. These variations in the presence of alleles are measured as changes in allele frequencies. From http://www.nature.com/scitable/definition/random-genetic-drift-genetic-drift-201 (accessed July 25, 2014).

Imagine that a 10-nucleotide length of DNA found in Species 1 and in Species 3 differs by four nucleotides (Figure 8-6B). As mentioned above, independent fossil data indicates that these two species diverged from a common ancestor 50 million years ago.[7]

This molecular clock can be calibrated using the substitution rate equation

$$r = K/2t \qquad (8.3)$$

where r is the rate of substitution per site per year, K is the number of substitutions per site,[8] and t is the divergence time (in millions of years) of two sequences from a common ancestor sequence.

The molecular clock data that can be inserted into equation 8.3 is:

K = 4 substitutions/10 nucleotides, which equals 0.4 substitutions per site

t = 50 million years

Thus, r = 0.4/(2∗50) = 0.004 substitutions/[(site)∗(millions of years)]

This is the observed substitution rate for a particular sequence calibrated with the fossil data of the ancestor common to Species 1 and 3. The calibrated molecular clock can now be used to calculate the time of divergence of Species 1 and 2. Recall that there is no fossil data to time the speciation event that gave rise to Species 1 and Species 2. Equation 8.3 can be rearranged to solve for t:

$$t = K/2r \qquad (8.4)$$

One substitution in 10 nucleotides between Species 1 and 2 was observed. This gives K = 0.1. We know that r = 0.004 substitutions/[(site)∗(millions of years)]. Therefore,

$$t = 0.1/2(0.004) = 12.5 \text{ millions of years.}$$

This means that the speciation event that gave rise to Species 1 and 2 occurred approximately 12.5 million years ago (Figure 8-6C).

Considerations to bear in mind when using molecular clocks

Molecular clock theory can be a powerful tool to time speciation events. Conclusions drawn from the theory depend on a constant rate of substitution. However, as with any theory, there are limitations that one must consider:

1. The rate of substitution can vary depending on the organism. For example, virus genes undergo substitutions much more rapidly than do other organisms (Figure 8-7). The nucleotide substitution rate in RNA viruses is ~10^{-3} nucleotide substitutions per site per year (i.e., 10^{-3} substitutions/(site∗year)). The nucleotide substitution rate of nuclear DNA in mammals is ~10^{-9} nucleotide substitutions per site per year. The difference is a reflection of the inherent

---

[7] This means that there has been 2∗50 million years of evolution (i.e., 100 million years) between the two sequences. This is the evolutionary distance that separates the two sequences.

[8] In this simple example we are using the observed number of substitutions per site. The actual number of substitutions per site (accounting for multiple mutations at a single site) is what we really require to calculate an accurate value for the rate. This is a difficult number to obtain. Later in the chapter we will show different mathematical approaches for estimating the actual number of substitutions per site.

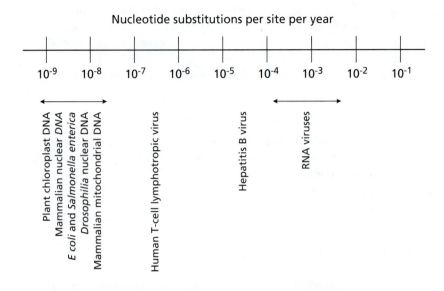

FIG. 8-7. Rates of observed nucleotide substitution in organisms. Plant chloroplast, mammalian nuclear, bacteria, and drosophila nuclear DNAs have the lowest substitution rates. Mammalian mitochondrial DNA substitution rate is approximately 2×10^{-8}. The human T cell lymphotropic virus (an RNA virus) and hepatitis B virus (a DNA virus) have intermediate substitution rates. Most RNA viruses have high substitution rates.

error rate of the polymerases copying the organism's genetic material and the degree of selection pressure required to maintain a sequence and maintain fitness of the organism.

2. The rate of substitution can vary depending on the location within a gene that is analyzed. Substitution rates for amino acid residues associated with the active sites of enzymes are lower than those for other residues in enzymes. For the alpha globin polypeptide in hemoglobin, the heme-binding amino acid substitution rate is 0.17×10^{-9} substitutions per site per year.[9] Heme is located where oxygen binds the protein. For the alpha globin polypeptide surface residues, the substitution rate is higher at 1.35×10^{-9} substitutions per site per year.

3. The molecular clock is applicable only when the gene retains its function over evolutionary time. When genes become nonfunctional, the substitution rate quickly increases because of the lack of selection pressure. Many species contain pseudogenes. Pseudogenes are genes that have mutated to the point where they no longer create protein products. Pseudogene sequences change much more quickly than their functional gene counterparts.

4. The molecular clock assumes that the nucleotide composition of each species is the same. This is not always true. For example, the % G + C content of a gene may change from species to species. This is especially true for microorganisms that live in extreme environments. Some species have % G + C as low as 15% and others have % G + C as high as 75%. Often the reason for the varied G + C content is not known. The molecular clock would not be constant when such a difference in composition occurs between species because the different % G + C's would cause an overall change in all sequences in the genomes of these organisms, including the types of codons used for amino acids in genes (i.e., codon usage). The rate of nucleotide substitution would show an abrupt change when comparing one species with another.

[9] Kimura and Ohta (1973).

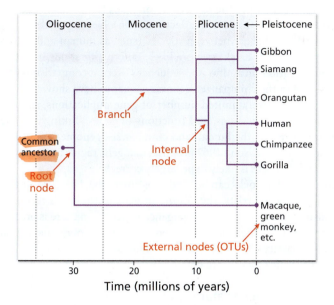

FIG. 8-8. Phylogenetic tree showing the divergence times of some hominoids estimated from Wilson's protein analysis data. Hominoids include the following species: gibbon, siamang, orangutan, human, chimpanzee, and gorilla. The ancestor species, common to all species listed in the table, is shown at the root node. Other terms are defined in the text. At the top of the tree are geological periods dating from the Oligocene to Pleistocene epochs. These periods are not formally part of the phylogenetic tree but are included to orient us to the timeframe of speciation events. The Pleistocene epoch ended 11,700 years ago, at which time the Holocene epoch started. We are now living in the Holocene epoch.

Given these caveats, the molecular clock theory must be handled with care. Scientists have developed several equations to mathematically correct for these and other concerns that influence the substitution rate. Aside from the Dickerson correction, other mathematical corrections will be discussed later in this chapter.

Application of molecular clock to dating the evolution of hominoid species

Now we are in a position to use the molecular clock to estimate the time of a speciation event. In 1967 Allan Wilson (University of California, Berkeley) used a method somewhat similar to amino acid residue sequencing (called immunological microcomplementation) to quantify residue differences in protein orthologs of serum albumin. The goal was to use the molecular clock to measure the times of divergence of humanoid species (humans, chimpanzees, and gorillas). Orthologs are used because they are coded by the same gene in different species. Most of the changes in ortholog sequences are due to neutral mutations, which do not affect the function of the protein. Residue differences in serum albumin from different primate species were used to calculate the substitution rate in Wilson's molecular clock. The molecular clock was calibrated, based on the fossil record, on data suggesting that the common ancestor of Old World monkeys and hominoids existed 30 MYA. From serum albumin substitution rate data, Wilson created a phylogenetic tree that showed that human, chimpanzee, and gorilla diverged from one another approximately 5 MYA (Figure 8-8). In 1967 this was an audacious claim because some paleontologists had estimated that the divergence time of human and chimpanzee was 15–30 million years ago. Modern DNA sequencing techniques have largely confirmed Wilson's phylogenetic tree.

A few years later, Wilson went on to demonstrate that a wide variety of human and chimpanzee protein orthologs differ, on average, by only 1%.[10] Given this small difference, he postulated that it is likely that the vast majority of chimpanzee proteins

---

[10] This is a number that still holds true today after comparing the protein coding sequences (exome) in humans to chimpanzees (see Varki and Altheide 2005).

and human proteins perform the same functions (i.e., the amino acid differences are due to neutral mutations that do not affect function). He proposed that morphology differences, whether they are between species (such as human and chimpanzee) or within a species (such as dog) are more likely due to alterations in gene regulation rather than to alterations in amino acid sequences. Recent comparisons of the human genome sequence to the chimpanzee genome sequence have shown that the two genomes differ in a relatively limited number of gene duplications, single nucleotide polymorphisms, gene inversions, and functions of genes. Taking all genetic differences into consideration, the human and chimpanzee genomes are 96% genetically identical. Remarkably, Wilson's 1967 proposal that gene regulation accounts for most morphological differences is likely to be largely correct.[11]

In sum, molecular clocks can be used to estimate the date of common ancestors for which no fossil data exists, thus filling in gaps in the fossil record. Molecular clocks can also be used to estimate divergence times when there is no obvious morphological change (this is particularly important for microorganisms that are difficult to visually classify). We will return to molecular clocks, but first let's analyze Wilson's phylogenetic tree in more detail.

Phylogenetic Tree Nomenclature

We can use Wilson's study to illustrate the nomenclature associated with phylogenetic trees.[12] In Figure 8-8 the horizontal lines, called branches, represent time in millions of years. The branch lengths are proportional to the time between speciation events. The points where lines converge are called nodes. Internal nodes represent speciation events. In other words, it is a point in time when two or more species arise from a single species. The vertical lines serve only to connect branches to nodes, so their length has no information. The external nodes at the branch tips are the species from which data is used to construct the tree. Usually these are existing species or groups of species, and they are called operational taxonomic units (OTUs). In Figure 8-8 there are six hominoid species and one group of Old World monkey species that represent a total of seven OTUs. A clade is a group of species that descend from an internal node of a tree. A clade is a subset of species that are more closely related to each other than to any other species in the tree that is not included in the subset. For example, human, chimpanzee, and gorilla plus their most recent common ancestral species form a clade.

If we turn our attention to the internal nodes, we find that usually two species arise from one. This is known as bifurcation from a common ancestor. When more than two species arise from a common ancestor it is called multifurcation. Multifurcation may indicate that more than one bifurcation occurred in close temporal proximity. Often, multifurcation is depicted in phylogenetic trees when scientists do not have enough data to distinguish the timing of bifurcations (i.e., the order of the bifurcation events are not known). The branch at the extreme left of the phylogenetic tree ends in a single node called the root node (labeled "Common Ancestor"). The root node is located at the bifurcation on the branch that connects this tree to the Tree of Life. The Tree of Life is a large tree that shows the evolutionary relationships of all organisms; we will discuss it later in this chapter.

By definition, trees must contain species that all descended from a single original species (as long as they are or were living organisms present on Earth, we assume that life had one origin). Trees may or may not specify the order of speciation events. If a tree specifies the order of all speciation events, it is called a rooted tree. Rooted trees contain a group of species called the outgroup, which is usually a group of species relatively unrelated to the species of interest. The presence of an outgroup in a

operational taxonomic unit (OTU) A species or group of species at the external node of a phylogenetic tree.

root node A bifurcation point within a phylogenetic tree. The single branch that bifurcates connects to the Tree of Life.

rooted tree A phylogenetic tree that specifies the order of all speciation events. Rooted trees contain an outgroup, which is required to order the speciation events.

unrooted tree A phylogenetic tree that has no root and therefore does not show an order of speciation.

outgroup A species or group of species approximately equally unrelated to other species in a rooted tree.

[Handwritten margin notes:]
- *S O... mutations in miRNA, promoter sequences... etc.?*
- *A limitation on the construction of phylograms using UPGMA is that if the methods used to determine molecular clock can those truly account for selection factors that involve Δ mutation rates?*
- *For all we know, A could be unchanged b/c of no environmental stressor, so it should be...*

---

[11] Varki and Altheide (2005).

[12] A phylogenetic tree is also known as a phylogram. A cladogram is similar to a phylogenetic tree, but the branch lengths are not meant to depict the evolutionary distance between related species.

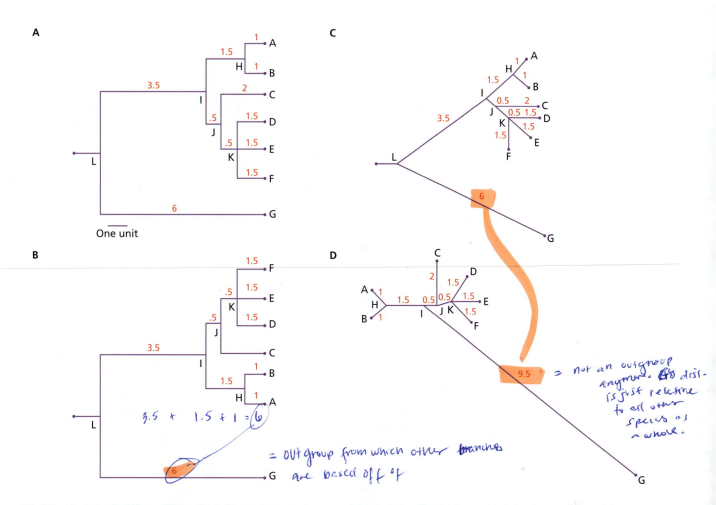

FIG. 8-9. Alternate depictions of Wilson's phylogenetic tree.
A. The OTUs are A, B, C, D, E, F, and G. The internal nodes are H, I, J, K, and the root node is L. Red numbers are in units of evolutionary time defined here as one unit equaling five million years. G is the outgroup species. **B.** Another possible rendition of the phylogenetic tree in panel A. Here the OTUs are rotated about the internal node I. This tree is equivalent to the tree in panel A. **C.** Another rendering of the phylogenetic tree where the tree is redrawn without the vertical lines. Here the vertical lines are removed and the tree is recreated with only branches and nodes. **D.** The unrooted tree created from the phylogenetic tree. Note the absence of the root node.

tree is needed to determine the relative order of speciation events. An outgroup is an OTU approximately equally distantly related to other OTUs in the tree.[13] Wilson used the Old World monkeys as the outgroup because fossil data showed that the common ancestor between Old World monkeys and other external node species existed in the distant past (30 MYA) and is equally distant to other living species in the tree. In an unrooted tree, no outgroup is designated, so the order of all the events in the tree is not provided by the arrangement of the branches.

Wilson used molecular data of serum albumin and fossil evidence to create his rooted phylogenetic tree. We can use this tree to illustrate other principles of trees. Figure 8-9A shows a simplified version of Wilson's phylogenetic tree. Figure 8-9B is an equivalent tree with the species A, B, C, D, E, and F rotated about the internal node I. A bioinformatics tree-drawing program could produce either tree, and both are equivalent. In other words, they have the same topology. Figure 8-9C shows another depiction of the tree. This time the vertical lines are removed, and only the nodes and branches remain. This tree is also equivalent to the first two trees discussed. Because all the information about relationships is contained within the nodes of the tree, the

---

[13] This external node that falls outside the other external nodes is called the "outgroup." The external nodes that do not belong to the outgroup belong to the "ingroup."

branches can be rotated 360 degrees (like a mobile) around any node without changing the relationships on the tree. The only way to change the tree topology (i.e., relationship information present in the structure of the tree) is by cutting off a branch and attaching it somewhere else in the tree.

A deeper discussion of the term "clade" is merited. Clade is a fairly flexible term. As long as the portion of the tree includes an ancestor and its descendants, it can be called a clade. In Figure 8-9, one clade is comprised of members A, B, and H, and another clade is comprised of members C, D, E, F, J, and K. It should be noted that we have labeled the internal nodes for instructional purposes. In most trees the internal nodes are not labeled but the species in the internal nodes are inferred by the clade. So the clade comprised of members A, B, and H is simply referred to as the clade with members A and B. Figure 8-9D shows an unrooted tree. Here the root node is removed and the distance between nodes I and G is the sum of the distance from I to L and L to G (distance equals 9.5 units). The unrooted tree does not specify how the tree attaches to the Tree of Life.

How to Tell if Sequences in Two Lineages Are Undergoing Sequence Substitution at Nearly Equal Rates

For the molecular clock to be useful for estimating divergence times, two species' lineages must exhibit nearly equal sequence substitution rates from a common ancestor. One test used to determine if two species have nearly equal rates is the Tajima relative rate test (Tajima test), named after its creator Fumio Tajima (National Institute of Genetics, Mishima, Japan). The Tajima test requires three sequences. Two sequences are from genes to be checked for equal substitution rates. The third sequence is from a potential outgroup. Recall that an outgroup is a species that is approximately equally distant from other species in the tree. It is important that the two species in the ingroup are more closely related to each other than either is to the outgroup species. If the two sequences in the ingroup are approximately equidistant to the outgroup, then the two sequences in the ingroup have nearly equal substitution rates.

To illustrate how the Tajima test works let's take a look at the phylogenetic tree in Figure 8-10. This phylogenetic tree shows the evolutionary relationship of mitochondrial DNA in hermit crabs. The tree was constructed assuming a constant rate of nucleotide substitution in the mitochondrial DNA of these species. The blue arrow points to the potential outgroup species *Pagurus acadianus* (the hairy hermit crab). The two potential mitochondrial DNA sequences we wish to test for equal substitution rates are from *Clibanarius vittatus* (thinstripe hermit crab) and *Coenobita spinosus* (terrestrial hermit crab). From the phylogenetic tree it appears that the latter two sequences are equally distant from the hairy hermit crab (that is, they have equal

Tajima relative rate test (Tajima test)
A calculation used to determine if two sequences are undergoing similar rates of substitution. In the calculation, substitution frequencies are computed by comparing sequences to an outgroup sequence.

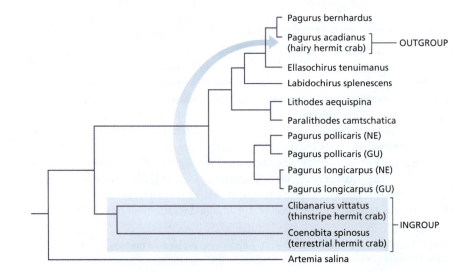

FIG. 8-10. Phylogenetic tree to be used to illustrate principles of the Tajima test. The two genes to be compared are from *Clibanarius vittatus* and *Coenobita spinosus* (sp.). The outgroup species is *Pagurus acadianus*.

substitution rates), but to be sure we need to perform the Tajima test. To do so, we compare the nucleotide sequences of three species (this is just an illustrative example; we would typically perform such an analysis on a much longer sequence):

Thinstripe: AATGAATGGTTGGACGAAAAACACACTGTTTC

Terrestrial: TATGAAAGGTCGAACGAGTGATAGACTGTCTC

Hairy (outgroup):   AATGAAAGGTTGGACAAAGTATCATCTGTTTC

m_{jij}:   1   1   1   1   1   = 5

m_{ijj}:   1   1   = 2

Yellow highlighted positions are identical in all three sequences, so there are no substitution rate differences at these positions. Similarly, there are no substitution rate differences at positions highlighted in pink because all three sequences are different (the Tajima test is concerned only with substitution rate *differences* between the ingroup when compared to the outgroup). The tallies for m_{jij} and m_{ijj} show alterations in substitution rates at each nonhighlighted position. We first need to sum the alterations as follows:

$$m_1 = \Sigma m_{jij} = n_{AGA} + n_{ACA} + n_{ATA} + n_{GAG} + n_{GCG} + n_{GTG} + n_{CAC} + n_{CGC} + n_{CTC} + n_{TAT} + n_{TGT} + n_{TCT} = 5$$

$$m_2 = \Sigma m_{ijj} = n_{AGG} + n_{ACC} + n_{ATT} + n_{GAA} + n_{GCC} + n_{GTT} + n_{CAA} + n_{CGG} + n_{CTT} + n_{TAA} + n_{TGG} + n_{TCC} = 2$$

where m_1 equals the number of sites in which nucleotides in terrestrial differ from both thinstripe and hairy, and

where m_2 equals the number of sites in which nucleotides in thinstripe differ from both terrestrial and hairy, and

where n_{YXY} and n_{XYY} are the values of specific nucleotide substitutions.

If $m_1 = m_2$, then the substitution rates are equal and we can apply the molecular clock theory. If m_1 and m_2 are not equal, then we need to check if the inequality falls outside the acceptable range. The Tajima test uses a statistical test called the chi-squared (χ^2) test to determine whether the substitution rates are nearly equal. A χ^2-test is a statistical test commonly used to compare observed data with data we would expect to obtain according to a specific hypothesis. In this case, the hypothesis is that the substitution rates are equal, and that any observed differences in substitution rates are only due to chance. This hypothesis is called a "null hypothesis." Once χ^2 is calculated, a table is used to find a corresponding P-value. If the P-value is greater than the significance level cutoff of 0.05, the null hypothesis can be accepted and any deviations can be accounted for by chance alone. In that case, the substitution rates may be considered nearly equal. (A more detailed discussion of P-values in the context of sequence alignments can be found in Chapter 12.)

The equation to calculate χ^2 in the Tajima test was adapted to this specific application of molecular clocks as shown below. To apply the data from our example in the χ^2-test, we do the following:

$$\chi^2 = (m_1 - m_2)^2 / (m_1 + m_2) = (5 - 2)^2 / (5 + 2) = 1.28$$

Now we need to look up this χ^2 value in a stat test table (Table 8-1) to determine if, given one degree of freedom, this P-value is less than 0.05. If $P < 0.05$, then we can reject the molecular clock hypothesis (at the 5% level). According to the table, the P-value is greater than 0.25. Because this P-value is > 0.05, we cannot reject the

TABLE 8-1. STAT TEST TABLE SHOWING *P*-VALUES FOR DATA WITH ONE DEGREE OF FREEDOM

	P-VALUES FOR DATA WITH ONE DEGREE OF FREEDOM											
	0.25	0.20	0.15	0.10	0.05	0.025	0.02	0.01	0.005	0.0025	0.001	0.0005
χ^2 value =	1.32	1.64	2.07	2.71	3.84	5.02	5.41	6.63	7.88	9.14	10.83	12.12

TABLE 8-2. HOMINOIDS AND OLD WORLD MONKEYS USED FOR PHYLOGENETIC TREE CONSTRUCTION

HOMINOIDS	OLD WORLD MONKEYS
Modern *H. sapiens*	Mitred leaf monkey
Chimpanzee	Hanuman langur
Pygmy chimp	
Gorilla	
Orangutan	
Neanderthal	
Denisovan	

molecular clock hypothesis. In other words, it is reasonable to apply the molecular clock hypothesis in this case. The Tajima test shows that for these sequences there is an approximately equal substitution rate, which suggests that thinstripe and terrestrial hermit crabs sequences are evolving at nearly the same rate.

Let's examine another example. Say that $m_1 = 50$ and $m_2 = 30$, then $\chi^2 = (50 - 30)^2/(50 + 30) = 5.00$. According to our stat test table, the *P*-value is between 0.05 and 0.025. In this case we reject the molecular evolutionary clock hypothesis at the 5% level, and we should not apply the molecular clock hypothesis. The substitution rates of the two sequences appear to be different. There are online tools available to precisely calculate *P*-values.[14]

DNA, RNA, and Protein-Based Trees

Wilson used micro-complementation data in serum albumins to create his phylogenetic tree. Although his results were quite accurate for the time, it is now common for scientists to use protein sequence or DNA sequence information to assess evolutionary relationships of species. When thinking about which sequences to use for tree construction, it is important to consider the span of years of evolution that will be covered. For example, if we chose histone H4 sequence to describe the evolutionary relationships between hominoids, we would have some difficulty. The unit evolutionary period for histone H4 protein is 600 million years. This clock is too slow for distinguishing hominoids because the hominoids evolved within the past five million years. We would detect no change in the histone H4 protein sequence over this short time span. However, this protein might be useful for determining the evolutionary relationships between plants, animals, and fungi, which diverged more than a billion years ago.

We will use the small subunit ribosomal RNA (12S) gene sequence from mitochondria as an example for describing evolutionary relationships between some hominoids. This gene can be thought of as a fast-ticking molecular clock. We will use these sequences to create a phylogenetic tree of some hominoids and two Old World monkey species (see Table 8-2). Prior to creating our phylogenetic tree, we will work

---

[14] For more information on the Tajima relative rate test, see Tajima (1993).

TABLE 8-3. COMMON PHYLOGENETIC TREE-RECONSTRUCTION METHODS AND SOFTWARE PROGRAMS THAT IMPLEMENT THESE METHODS

METHOD	CONVERSION OF ALIGNMENT TO DISTANCE MATRIX	RESULTING TREE TYPE	SINGLE TREE (ST) OR MULTIPLE TREES (MT)	SOFTWARE PROGRAM TO IMPLEMENT METHOD
UPGMA	Yes	Ultrametric, additive, rooted	ST	Neighbor,[1] MEGA3
Neighbor-joining	Yes	Additive, unrooted	ST	Neighbor,[1] MEGA3
Fitch-Margoliash	Yes	Additive, unrooted	ST	Fitch[1]
Minimum evolution	Yes	Additive, unrooted	MT	MEGA3
Quartet puzzling	No	Additive, unrooted	MT	Tree-Puzzle
Maximum parsimony	No	Additive, unrooted/ cladogram	MT	Dnapars,[1] Protpars,[1] MEGA3
Maximum likelihood	No	Additive, unrooted	MT	Dnaml,[1] Proml,[1] RaxML, Garli, PHYML
Bayesian	No	Additive, unrooted	MT	MrBayes

Adapted from M. Zvelebil and J. O. Baum, *Understanding Bioinformatics* (New York: Garland Science, Taylor & Francis Group, 2008).
[1] Part of the PHYLIP suite of software programs.

through the steps to develop a phylogenetic tree in a general format. Then we will return to the small subunit ribosomal RNA gene to run through the steps with this specific example of generating a phylogenetic tree for some primates.

8.3 TWO CLASSES OF TREE-GENERATION METHODS

Phylogenetic tree generation methods can be roughly divided into two classes. One class generates a single tree from a distance matrix. Another class generates multiple trees and attempts to choose the best tree given an evolutionary model. The field is still developing because of the difficulty in reconstructing evolutionary history from sequences derived from existing species (or recently extinct species). Each method has its advocates and critics. Table 8-3 lists some tree-generation methods and software packages that implement them. We will discuss in detail the UPGMA method and neighbor-joining method, both of which generate a single tree from a distance matrix. Those interested in other methods can look at the sources listed in the back of this chapter.

We are now ready to generate our first phylogenetic tree. We will use the UPGMA method. It requires three steps: (1) sequence alignment, (2) generation of distance matrix, and (3) generation of phylogenetic tree. The robustness of the tree topology will be assessed with the bootstrapping method.

Unweighted Pair Group Method with Arithmetic Mean (UPGMA)
Generating a sequence alignment

The UPGMA method is a distance-based method that requires sequence alignment to compute evolutionary distances.[15] The quality of sequence alignment is critical for distance-based methods. Sometimes, after sequence alignment, it is necessary to edit the sequences to remove regions that have gaps of varying lengths. The gaps can cause the sequences to slightly misalign and render the resulting distance matrix

bootstrapping A statistical method that uses rearranged character data to measure the robustness of the topology of a phylogenetic tree. Each clade of a tree may be assigned a bootstrap value, which is based on the percentage of bootstrapping pseudoreplicates that match the given clade topology in the phylogenetic tree.

[15] UPGMA first introduced by Sokal and Michener in 1958.

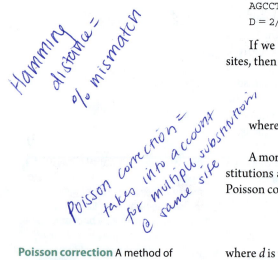

```
                    10          20        30
                    |     |     |     |    |
       Human  ATGGTGAGGAGCAGGCAAATGTGCAATACC
       Chimp  ATGGTGAGGAGCAGGCAAATGTGCAATACC
   Orangutan  ATGGTGAGGAGCAGGCAAATGTGCAATACC
  Rh. monkey  ------------------ATGTGCAATACC
      Rabbit  ---------------CAAATGTGCAATACC
    Bushbaby  ------------------ATGTGCAGTACC
```

After editing

```
                          10
                    |     |
       Human  ATGTGCAATACC
       Chimp  ATGTGCAATACC
   Orangutan  ATGTGCAATACC
  Rh. monkey  ATGTGCAATACC
      Rabbit  ATGTGCAATACC
    Bushbaby  ATGTGCAGTACC
```

FIG. 8-11. Removal of multiple gaps in aligned sequences. Six MDM2 gene sequences were aligned with Clustal Omega (version 1.1.0). Columns of aligned sequences were removed to create an edited alignment with the JalView (version 2.7) sequence editor.

UPGMA (unweighted pair group method with arithmetic mean) A clustering method that uses a distance matrix to create a phylogenetic tree. UPGMA assumes a constant molecular clock for all species in the tree. It creates an ultrametric tree.

Hamming distance = % mismatch

Poisson correction = takes into account for multiple substitutions @ same site

Poisson correction A method of generating a rate of sequence substitution that accounts for multiple substitutions.

unreliable. It is best to use a multiple sequence alignment software program that can export its output to a sequence editor because the editor can be used to remove columns in the alignment where there are variable numbers of gaps. An example of editing a sequence alignment to remove gaps is shown in Figure 8-11.

The distance matrix: Hamming distance, Poisson correction, and Jukes-Cantor model

The UPGMA method requires the creation of a distance matrix from aligned sequences. The distance matrix quantitatively describes the dissimilarity of sequences in a pairwise fashion. The simplest approach to creating a distance matrix is to align pairs of sequences and count the number of differences: the higher the number of differences, the greater the evolutionary distance. The degree of distance, D, can be described by the following equation:

$$D = n/N \tag{8.5}$$

where n is the number of sites at which there are differences, and N is the number of nucleotides (or amino acids) in the alignment.

This calculation for D is sometimes called the Hamming distance. The Hamming distance is defined as the number of positions with mismatching residues between two sequences of equal length divided by the length of the aligned sequences. For example:

```
AGGCTTTTCA
|| |||| ||
AGCCTTCTCA
D = 2/10 = 0.2
```

If we do not take into account other factors, such as multiple substitutions at sites, then the following equation holds:

$$d = D \tag{8.6}$$

where d is the evolutionary distance and D is the Hamming distance.

A more sophisticated measure of evolutionary distance accounts for multiple substitutions at sites. One measure that takes multiple substitutions into account is the Poisson correction. The Poisson correction is described by the following equation:

$$d = -\ln(1 - D) \tag{8.7}$$

where d is the evolutionary distance and D is the Hamming distance. Equation 8.7 is similar to equation 8.2, the equation Dickerson used for molecular clock analysis of different proteins. Dickerson used the number of changes occurring in a 100-amino-acid-residue segment of polypeptide. As in the Dickerson equation, dividing by the length of the aligned sequences allows comparisons between sequences of different lengths.

A third measure of evolutionary distance is the Jukes-Cantor model (JC model). To describe this model, it is best to think of a single site in an alignment of nucleotides. The JC model can be used for amino acid alignments, but it is simpler to derive this equation for use with nucleotides.[16] The four possibilities at a single site are A, C, G, or T. The model assumes that the four nucleotides are present in equal proportions and that the rate of substitution from one nucleotide to any of the other three nucleotides is equal. If the rate of substitution for one nucleotide is α, then the rate of change of one nucleotide to any other nucleotide is 3α. To calculate the average number of changes, the rate of change is multiplied by the length of time. An equation describing the average number of changes, d (or evolutionary distance), is:

$$d = 2t \times 3\alpha = 6\alpha t \tag{8.8}$$

where $2t$ is the total time of divergence of a common ancestor to two species (see equation 8.3) and 3α is the rate of change of one nucleotide to any other nucleotide.

The rate α is difficult to measure, so we must calculate the probability, D, that two species will have a different nucleotide at a particular site. D is the probability that two sequences will change at a particular site. The equation describing D is:

$$D = 3/4 - (3/4)e^{-8\alpha t} \tag{8.9}$$

Let's analyze this equation a bit further. If t is very small, the exponential can be expanded to first order in t, which means that $D \approx 6\alpha t$. This is expected, because D is equal to d (equation 8.8) when the evolutionary time is very short. Under these circumstances, the length of time (t) is too short for multiple substitutions to occur. On the other hand, if t is large, the limit of D is 3/4, which is more in line with multiple substitutions:

This term tends to 0 when t is large

$$D = 3/4 - (3/4)e^{-8\alpha t}$$

$$D = 3/4$$

In fact, if two random sequences containing equal frequencies of four bases are aligned, then on average, 3/4 of the sites will differ between the two sequences. In other words, D tends to 3/4 when the t is large, because evolutionary time is very large.[17]

Because equations 8.8 and 8.9 are functions of αt, this term can be eliminated by substitution.

Equation 8.9 can be rearranged as follows:

$$\ln(1 - (4/3)D) = -8\alpha t \tag{8.10}$$

Equation 8.7 can be rearranged as follows:

$$\alpha t = d/6 \tag{8.11}$$

Substitution gives:

$$d = -3/4 \ln(1 - (4/3)D) \tag{8.12}$$

where d is the evolutionary distance (sometimes called the "JC distance" when the JC model is used) between sequences. A more formal derivation of this equation can be found in Chapter 12.

Jukes-Cantor model A model of evolutionary distance that predicts the actual number of nucleotide (amino acid residue) replacements based on the observed fraction of sites that differ between two gene sequences.

---

[16] The general form of the JC model is $d = -b \ln(1 - (D)/(b))$. The constant $b = 3/4$ for nucleotides and $b = 19/20$ for amino acid residues.

[17] Alternatively, 1/4 of the nucleotides will align by chance. Recall our example of aligning two nucleotide sequences in a dot plot in Chapter 5. When we used a window of 1 nucleotide, the random chance of getting a match was 1/4.

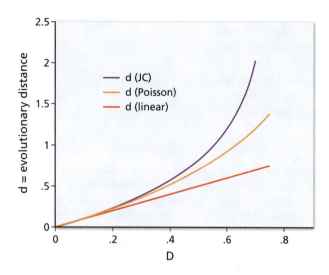

FIG. 8-12. Evolutionary distance, *d*, plotted as a function of *D*. For linear and Poisson correction models, *D* is the fraction of sites that differ. For the JC model, *D* is the probability that two sequences will have a different nucleotide at a particular site.

In this section we have discussed three equations that calculate evolutionary distance *d* dependent on a value of *D*. Figure 8-12 shows a plot of the evolutionary distances as a function of *D*. The linear rate (Hamming distance) does not account for multiple substitutions. The Poisson correction accounts for multiple substitutions at sites. The JC model also accounts for multiple substitutions at sites. There are other calculations that compute evolutionary distance, and these will be discussed later in the chapter.

Thought Question 8-1

Consider an alignment where 3 out of 60 aligned nucleotides differ. Calculate *d* using the Hamming distance equation, the Poisson correction equation, and the JC model equation.

Thought Question 8-2

Consider an alignment where 30 out of 60 aligned nucleotides differ. Calculate *d* using the Hamming distance equation, the Poisson correction equation, and the JC model equation.

Now that we have reviewed methods for calculating evolutionary distance, we will explore how to use this distance to create phylogenetic trees.

Clustering: The UPGMA distance-based method

A clustering algorithm can be used to create a phylogenetic tree. It uses evolutionary distances (discussed in the preceding section) of sequences to create the tree. Sequences with the shortest evolutionary distances are connected to form a cluster. The cluster gets larger as more sequences and other clusters are added. The algorithm stops once all sequences are connected into one large cluster. Clustering algorithms operate in the following general manner:

1. Join the closest two sequences to form a cluster.
2. Recalculate the evolutionary distances between the cluster and the remaining sequences.
3. Join the closest two sequences or join the closest cluster and sequence.
4. Recalculate the evolutionary distances between the clusters and the remaining sequences.
5. Repeat steps 3 and 4 until all sequences are connected in a single cluster.

As stated earlier, UPGMA is an acronym for unweighted pair group method with arithmetic mean. We will provide a simple evolutionary distance matrix (or distance matrix, for short) to demonstrate the UPGMA method. The distance matrix compares five sequences, A, B, C, D, and E, in pairwise fashion (Figure 8-13A). The values in the matrix are evolutionary distances calculated using any acceptable method.[18] The schematic diagram shows spots representing each sequence. The distances between spots are proportional to the calculated evolutionary distances between the sequences.

---

[18] Later in the chapter we will use the UPGMA method to generate a tree with 12S RNA gene sequences. In that example, we will use the JC model to generate the evolutionary distance values for our distance matrix.

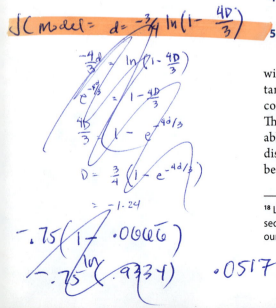

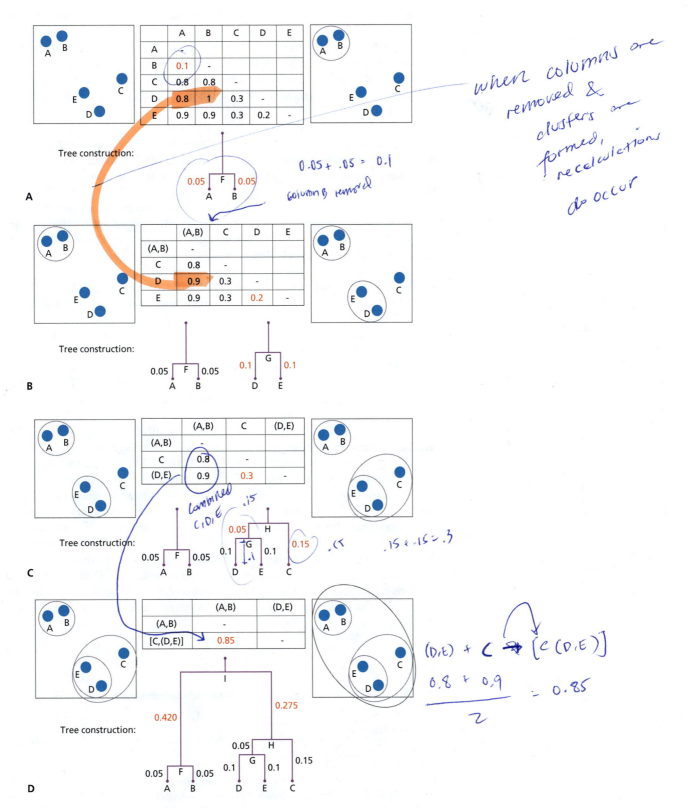

FIG. 8-13. An example of the UPGMA method of phylogenetic tree construction.

According to our distance matrix, the sequence pair A, B shows the least distance at 0.1. In other words, the sequences of A and B are the most similar of the pairs. We can draw a circle that groups the A, B spots to signify clustering. During the first part of tree construction a clade is created by the software program. The clade shows that A and B diverged from an ancestor sequence, F. The distance between F and each of its two descendant sequences is the value in the distance matrix divided by 2, which is 0.1/2 = 0.05.

In the next step, a new distance matrix is created that includes the cluster (A, B) (Figure 8-13B). In the new distance matrix the individual sequences of A and B are removed and replaced by the cluster (A, B). The distance between two clusters is the mean of the distances between the species in the two clusters. For example, in the first matrix the distance of A to D is 0.8 and the distance of B to D is 1.0. The mean distance is (0.8 + 1.0)/2 = 0.9. In the second matrix the distance between cluster (A, B) and D is 0.9. Once the values are calculated in the second matrix, the sequence pair with the lowest distance value is identified. This pair is D, E with a distance of 0.2. The D, E sequences are clustered and a clade is created by the tree creation program consisting of D, E and the ancestor sequence G. The distance between G and each of its two descendant sequences is 0.2/2 = 0.1. So far, clade D, E, G has been created with the length of the branches equal to twice the length of the branches in clade A, B, F.

In the next step, a new distance matrix is created and the shortest distance identified is between cluster (D, E) and sequence C (Figure 8-13C). Clade D, E, G and C have a common ancestor named H. The branch distance between H and C is 0.3/2 = 0.15. The sum of the branch distances between H and G and G and D (or E) is 0.15. Because the branch distance between G and D is 0.1, the branch distance between H and G must be 0.05.

In the final step, a cluster is created connecting all clades (Figure 8-13D). The last distance matrix contains one value connecting clusters (A, B) and [C, (D, E)], 0.85. The clade connects clades A, B, F and C, D, G, F, H to an ancestor sequence I. The distance from I to each OTU is 0.85/2 = 0.425. The UPGMA method produces an ultrametric tree. An ultrametric tree is a rooted additive tree where the external nodes are all equally distant from the root. The fact that equal distances are created implies a constant substitution rate for all genes in the tree. In other words, the molecular clock is constant throughout the entire phylogenetic tree. Deeper analysis of sequence data and fossil records suggests that the molecular clock is not constant and must take into account multiple substitutions and other caveats. Other tree creation methods take into account the fact that the molecular clock may not always be constant. We are now ready to apply the UPGMA method to analysis of mitochondrial DNA sequences.

Thought Question 8-3

Show how you would calculate the distance between cluster (A, B) and C.

Application of UPGMA to analysis of mitochondria DNA sequences in hominoids and Old World monkeys

We are now ready to apply the UPGMA method of tree construction to primate species. The hominoids are a category of primates that are divided into two groups, the lesser hominoids and the greater hominoids. The lesser hominoids are comprised of gibbon and siamang. The greater hominoids are comprised of modern *H. sapiens*, pygmy chimpanzee,[19] chimpanzee, orangutan, and gorilla. Extinct subspecies

---

[19] An alternate name for the pygmy chimpanzee is bonobo.

BOX 8-1 | **A CLOSER LOOK**

What Do We Know about Neanderthal and Denisovan?

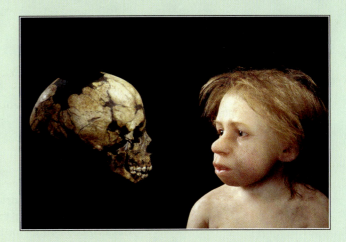

Homo erectus is a hominoid species that is estimated to have existed two million years ago. *Homo erectus* fossils have been found in different areas of Africa and Asia. The oldest fossils were found in Southeast Asia, but it is thought that this species originated in Africa. "Peking man" and "Java man" are examples of *Homo erectus* fossils. The *Homo erectus* skull reveals a receding forehead, pronounced supraorbital ridge, and projected face and jaws. These hominoids made advanced stone tools, known as Acheulian tools, that include hand axes, choppers, borers, and scrapers. *Homo erectus* wore clothing, built fires, lived in caves, and obtained food by hunting and scavenging. Interestingly, no weapons have been unearthed at sites where *Homo erectus* fossils or their artifacts have been unearthed. *Homo erectus* appears to have been replaced by Neanderthal, Denisovan, and modern *Homo sapiens*.

Evidence for Neanderthal (an alternate spelling is Neandertal) was discovered in the Neander Valley in Germany in 1856 by workers who uncovered a skull cap in the limestone. Other skeletal remains were also recovered in this area. The bones were given to Johann Karl Fuhlrott, who shared the find with Hermann Schaafhausen. In 1857 the two scientists jointly announced the bones as an example of "an early stage of human evolution linking man to an ape-like ancestor." More than 400 examples of Neanderthal individuals have since been uncovered. Fossil evidence shows that the face of Neanderthal projected slightly, the chin receded, and the skull had heavy brow ridges.

Neanderthal tools were made from bone and stone. Scrapers and spear points have been recovered and appear to be distinct from the tools of *Homo erectus*, both in quality and in mode of production. Interestingly, skeletons of elderly Neanderthal individuals and of Neanderthal individuals with healed fractures appear to demonstrate that they cared for the aged and the sick. They buried their dead, and it appears that food, weapons, and flowers may have been intentionally left in Neanderthal graves. This may mean that they had a notion of an afterlife. Evidence suggests that Neanderthals lived in Europe and the Middle East

and that they coexisted in the same geographical areas as modern *Homo sapiens* for approximately 10,000 years. Modern *Homo sapiens* individuals of European and Asian descent contain 1–4% of Neanderthal DNA in their genomes indicating that the two subspecies interbred. It is not clear why Neanderthal became extinct but it is known that Neanderthal coexisted with a second *Homo sapiens* subspecies—*Denisova hominin*.

We know much less about *Denisova hominin* (also known as *Homo sapiens Altai*), which we will refer to as Denisovan. A juvenile finger bone fragment was discovered in the Denisova cave in the Altai Mountains of Siberia in 2010. More recently, a molar (wisdom tooth) and toe bone were discovered. Incredibly, enough DNA was recovered from the finger bone fragment to obtain a significant amount of nuclear DNA sequence and a complete mitochondrial DNA sequence. The data suggests that Denisovan is a subspecies of modern *Homo sapiens* and is distinct from Neanderthal. Sequence comparisons indicate that Denisovan and Neanderthal are more closely related to each other than either is to modern *Homo sapiens*. Data presented in this chapter from mitochondrial gene sequence comparisons suggests that Neanderthal is more closely related to modern *Homo sapiens* than is Denisovan, but this is still an open question.

There is evidence that interbreeding occurred between Denisovan and a colony of modern *Homo sapiens* that now inhabit the island Papua New Guinea—the Melanesians. It is estimated that the colony of modern *Homo sapiens* did

continued

BOX 8-1 | **A CLOSER LOOK**

What Do We Know about Neanderthal and Denisovan?

continued

not arrive in Papua New Guinea until approximately 50,000 years ago. This suggests that Denisovan interbred with a colony of Homo sapiens that migrated to Papua New Guinea. It is also possible that Denisovan previously inhabited the island, interbred with an incoming *Homo sapiens*

colony, and became extinct. Much more research must be conducted to fill in the complex picture about the evolution of modern *Homo sapiens*. Sequence comparisons will undoubtedly continue to play a major role in advancing this fascinating area of phylogenetics.

within the greater hominoids are Neanderthal and Denisovan (Box 8-1).[20] The fact that we have mitochondrial and genomic DNA sequences from Neanderthal and Denisovan is due to the painstaking efforts of many scientists, including Svante Pääbo (Max Planck Institute for Evolutionary Anthropology, Leipzig, Germany; Box 8-2). These scientists developed methods to sequence samples from bone tissue preserved for tens of thousands of years and to distinguish this prehistoric DNA from contaminating DNA.

Neanderthal and Denisovan are subspecies of *H. sapiens* that became extinct 27,000 years ago. This is a very recent extinction in terms of evolutionary time periods, so we will count these subspecies as existing OTUs in our analysis. Monkeys are primates but not hominoids. There are two categories of monkeys: Old World monkeys and New World monkeys. We will use two Old World monkey species in our analysis: mitred leaf and hanuman langur (Table 8-2). The UPGMA method requires an outgroup, so one of the Old World monkey species will be designated as the outgroup by this method.

To begin our analysis we need to choose appropriate ortholog sequences. The species we are comparing are somewhat morphologically similar (they are all primates), and fossil evidence suggests that they diverged from a common ancestor approximately 30 million years ago.[21] We should choose a DNA or protein sequence that has a molecular clock that runs quickly enough to capture differences in species that evolved from a common ancestor fairly recently. Estimates from fossils, tool use, and other data suggest that modern *Homo sapiens* and Neanderthals diverged

---

[20] There is some debate as to whether Neanderthal should be considered a separate species from modern *H. sapiens*. In theory, two species should not be capable of producing viable progeny. Evidence from the Neanderthal genome sequence suggests that there was interbreeding with modern *H. sapiens*. Between 1% and 4% of the human genome from modern *H. sapiens* that populate Europe and Asia is derived from Neanderthal DNA. Thus, it is more appropriate to designate Neanderthal as a subspecies of *H. sapiens*. The proper name is *Homo sapiens neanderthalensis*. Denisovan is another group of *H. sapiens*, discovered in 2010 in the Altai mountains in Siberia. They are distinct from Neanderthal and appear to have interbred with modern *H. sapiens* as well. From 4% to 6% of modern *H. sapiens* genome in some individuals of Far East Asia is shared with Denisovan. We consider Denisovan as a subspecies of *H. sapiens*. The proper name is *Denisova hominin* or *Homo sapiens Altai*.

[21] There is, in fact, some morphological diversity within the primates, and hundreds of morphological traits have been used to propose phylogenetic relationships of primates leading to different phylogenetic trees. Gene sequence comparisons have helped to clarify some, but not all, of these relationships, especially when two nodes in the tree are very close in time. A case in point is the use of genome sequences to establish relationships between humans, pygmy chimpanzee (bonobos), and chimpanzees. Depending on the region of the genomes being compared, different phylogenetic trees are produced.

BOX 8-2 || **SCIENTIST SPOTLIGHT**

Svante Pääbo

By his own admission, Svante Pääbo has a romantic fascination with ancient Egypt. As an undergraduate studying at the University of Uppsala in Sweden, he worked two summers cataloging Egyptian pottery shards at the Mediterranean Museum in Stockholm. After his undergraduate degree he continued his studies at the University of Uppsala, where he pursued a Ph.D. in cell research in the laboratory of Per Pettersson. For his thesis work he studied adenovirus proteins, one of which inhibits the transport of transplantation antigens to the cell surface. This inhibition permits the virus-infected cells to evade the immune system. While studying virus proteins in graduate school, his fascination with ancient Egypt did not wane and he enrolled in Coptic language courses (the Coptic language is thought to be close to the language spoken by ancient Egyptians). As Pääbo became proficient at working with recombinant DNA, he began to formulate the idea of fusing his love of ancient Egypt with his knowledge of modern molecular biology. He wanted to know if it was possible to study DNA sequences from Egyptian mummies with the idea that those sequences may reveal how ancient Egyptians were related to one another and to people today. At the time, no one had attempted to isolate ancient DNA.

As a graduate student he decided to try some experiments that were outside the scope of his thesis work (at first without the knowledge of his thesis advisor!). He did these experiments late at night and on weekends. He first attempted to see if DNA could be isolated from mummified tissue. He purchased some calf liver from a supermarket and subjected it to heat and desiccation—techniques used by ancient Egyptians to mummify tissues. He found that he could extract DNA from his "mummified" calf liver, which he detected by gel electrophoresis and ethidium bromide staining. When he attempted the same technique on tissues from Egyptian mummies, he was unsuccessful. Undeterred, he tried another approach to detect mummy DNA. He rehydrated the mummy cartilage tissue and prepared slices for microscopic analysis. He stained the samples with a DNA-specific dye and found that some cell nuclei did indeed contain DNA. He was able to isolate the DNA from the cartilage, clone it into bacteria, and show that the sequence was human because it had a human specific DNA marker (an *Alu* sequence). Carbon dating of the mummy sample would prove that it was 2,400 years old. This was the first success that Svante Pääbo had in detecting and sequencing DNA from ancient tissue. Importantly, Pääbo developed several collaborations as he progressed in his research—collaborations that would prove critical as he attempted to sequence DNA from tissues considerably older than 2,400 years.

Buoyed by his success at sequencing ancient DNA, Pääbo assembled an international consortium to study prehistoric DNA samples. He eventually became director of the then newly created Department of Genetics at the Max Planck Institute in Leipzig, Germany. He built a special laboratory designed to reduce human contamination of prehistoric tissue. By 2012, the consortium had sequenced mitochondrial and genomic DNA from Neanderthal and Denisova bone tissue. Incredibly, the source of the Denisova bone tissue was a fingertip bone a few millimeters in length and a molar tooth from a girl (or two separate girls). The sequence of the Denisova tissue suggests that this human subspecies shared a common ancestor with modern humans

continued

continued

earlier than Neanderthal. Several techniques were developed to sequence prehistoric DNA. These included methods to distinguish contaminating modern human DNA from prehistoric DNA and methods to assemble genome sequences from extremely short reads. Importantly, Pääbo insisted that an independent laboratory confirm some of the key experimental results of the consortium prior to publication. The fact that we know about the existence of the Denisova subspecies without a fossil record marks the beginning of a new frontier in the study of prehistoric humans.

REFERENCE

Pääbo, S. 2014. *Neanderthal Man, in Search of Lost Genomes*. New York: Basic Books.

between 500,000 and 800,000 years ago. Estimates for the time of divergence of chimp and pygmy chimp range from 600,000 years to 1.8 million years ago.

The 12S ribosomal RNA gene (rDNA) from mitochondria may be suitable for this analysis. This rDNA gene is a DNA segment that codes for an RNA that is necessary to translate other RNA molecules into protein inside mitochondria.[22] Recall that observed substitution rates in DNA are higher than in proteins. In animals, mitochondrial DNA rarely undergoes large mutational events such as recombination, but it has a higher point mutation rate than nuclear DNA. This is likely due to the presence of oxygen radicals generated in close proximity to the DNA. In addition, mitochondrial DNA undergoes more rounds of replication (and is therefore more prone to substitutions) than nuclear DNA during the lifetime of a cell. Within the mitochondria, the product of rDNA, the 12S ribosomal RNA, must fold into a particular structure and associate with proteins to function properly. This means that there are many sites that are conserved over long evolutionary time periods. However, there are some variable sites that are substituted without affecting function (neutral mutations). These variable sites are necessary to create the distance matrix for phylogenetic tree construction.

To create the distance matrix we must obtain the sequences of rDNA required for the analysis. To do this, the modern *Homo sapiens* rDNA sequence is retrieved from GenBank. This sequence is used as a query to perform a nucleotide BLAST analysis of GenBank in order to capture rDNA sequences from other species. Usually, the rDNA sequences are embedded in the complete mitochondrial genomes of the organisms. By reading the GenBank flat file (see Chapter 2), the segment of the genome that codes for 12S rDNA genes can be easily identified. The rDNA sequences

---

[22] Recall from Chapter 1 that mitochondria are the powerhouses of the cell. They convert oxygen to water and create the energy currency, ATP, that allows the organism to live. Mitochondria are organelles that are likely to have been free-living bacteria billions of years ago. It is theorized that bacteria were engulfed by eukaryotic cells and became specialized to supply ATP to the eukaryotic cell. The engulfed bacteria evolved into mitochondria. Mitochondria have their own DNA, which they use to create RNA and proteins for a few specialized functions within the mitochondria. Mitochondria divide and can create many copies of themselves inside the cell. Mitochondria are always inherited from females. The egg supplies the mitochondria that constitute the precursor mitochondria to all cells for the progeny.

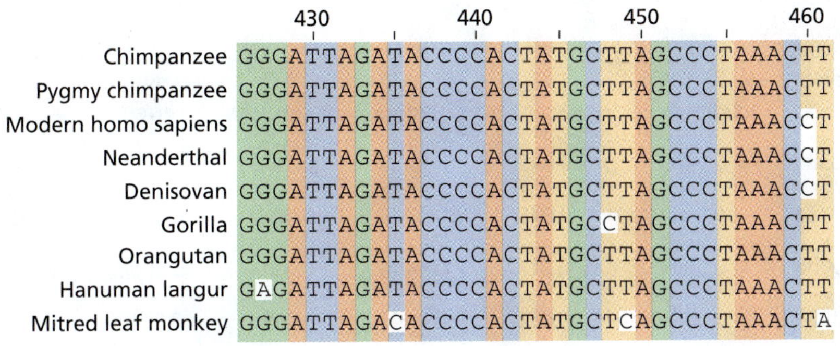

FIG. 8-14. Part of an alignment of the mitochondrial 12S ribosomal RNA genes from hominoids and monkeys. The length of the sequences was approximately 950 nucleotides.

of the species are captured in FASTA format and pasted into a single text file separated by appropriate headers. The next step is to align the rDNA genes with a multiple sequence alignment program (Figure 8-14).[23] The alignment is exported from the software program in PHYLIP format.[24] PHYLIP (phylogenetic inference programs), originally started by Joseph Felsenstein's group (University of Washington, Seattle), is a suite of software programs used for phylogenetic analysis.

From the aligned sequences, a distance matrix is then created using the JC model. Recall that the JC model accounts for multiple substitutions at single sites. The DNADIST program in the PHYLIP suite is used to create the distance matrix (Table 8-4). This is a symmetrical matrix—the values in the bottom left half of the matrix mirror the values in the top right half of the matrix. Note that the distance between a species and itself is always zero, which is evident along the long diagonal axis in the matrix. The two sequences that diverged the least are those from human and Neanderthal (shaded green). This is consistent with data from the fossil record. The two sequences that diverged the most are from orangutan and hanuman langur (shaded red).

The Neighbor software program in the PHYLIP suite is used to convert the distance matrix into a phylogenetic tree with the UPGMA method. This is done in two steps. In the first step, the distance matrix is converted to a Newick tree format:[25]

(((((chimp:0.00694,pyg_chimp:0.00694):0.01439,

((mod_h_sap:0.00422,neander:0.00422):0.00213,

denisova:0.00634):0.01498):0.00149,

gorilla:0.02281):0.02338,orangutan:0.04619):0.05081,

(han_langur:0.05081,mit_lf_mon:0.05081):0.04619);

---

[23] We used Clustal Omega for multiple sequence alignment.

[24] PHYLIP format is a common format that is available in multiple sequence alignment programs. PHYLIP format is necessary for use of PHYLIP software programs. If you choose to use other phylogenetic analysis software programs, such as PAUP or MEGA, be sure to save your sequence alignment in the format suited for those software programs.

[25] Newick tree format (or Newick notation or New Hampshire tree format) is a way of representing graph-theoretical trees with edge lengths using parentheses and commas. It was adopted by James Archie, William H. E. Day, Joseph Felsenstein, Wayne Maddison, Christopher Meacham, F. James Rohlf, and David Swofford, at two meetings in 1986, the second of which was at Newick's restaurant in Dover, New Hampshire, United States.

TABLE 8-4. EVOLUTIONARY DISTANCE MATRIX

	CHIMPANZEE	PYGMY CHIMP	MODERN *HOMO SAPIENS*	NEANDERTHAL	DENISOVAN	GORILLA	ORANGUTAN	HANUMAN LANGUR	MITRED LEAF MONKEY
Chimpanzee	0.0000	0.0139	0.0389	0.0400	0.0422	0.0435	0.0941	0.2000	0.1817
Pygmy chimp	0.0139	0.0000	0.0434	0.0457	0.0457	0.0445	0.0967	0.1972	0.1763
Homo sapiens	0.0389	0.0434	0.0000	0.0084	0.0127	0.0444	0.0914	0.2004	0.1876
Neanderthal	0.0400	0.0457	0.0084	0.0000	0.0127	0.0467	0.0890	0.2018	0.1903
Denisovan	0.0422	0.0457	0.0127	0.0127	0.0000	0.0489	0.0961	0.2018	0.1876
Gorilla	0.0435	0.0445	0.0444	0.0467	0.0489	0.0000	0.0870	0.1983	0.1748
Orangutan	0.0941	0.0967	0.0914	0.0890	0.0961	0.0870	0.0000	0.2158	0.2027
H. langur	0.2000	0.1972	0.2004	0.2018	0.2018	0.1983	0.2158	0.0000	0.1016
M. L. monkey	0.1817	0.1763	0.1876	0.1903	0.1876	0.1748	0.2027	0.1016	0.0000

Note: Hominoids and monkey mitochondrial 12S rDNA gene distances calculated with the JC model. The values are rounded to four digits after the decimal point.

Newick tree format (also known as Newick notation or New Hampshire tree format) A list of rules and distances used by phylogenetic tree drawing programs to create phylogenetic trees.

Newick tree format contains the instructions necessary for the phylogenetic tree drawing program to draw the tree. In the second step, a phylogenetic tree is created from these instructions. There is a phylogenetic tree drawing program in the PHYLIP suite, but there are other programs available online that create trees as well.[26]

From the Newick tree format we can envision the different parts of the drawn phylogenetic tree. The format shows that the distance from chimp and pyg_chimp to a common ancestor is 0.00694 substitutions/site. This is exactly half the distance separating chimp and pyg-chimp found in the distance matrix (0.0139) (Table 8-4). This is expected because UPGMA is an ultrametric tree-forming algorithm that assumes the same rate of substitution within each line descending from a common ancestor. Similarly, the Newick tree format shows that the distance from modern *Homo sapiens* and Neanderthal to a common ancestor is 0.00422. This is exactly half the distance found in the distance matrix (0.0084) for these two species. Other values show the distances from clades to common ancestors and are calculated during the clustering process.

Distances in the Newick tree format are used to create the phylogenetic tree shown in Figure 8-15. The tree shows that modern *Homo sapiens* are most closely related to Neanderthal and slightly more distantly related to Denisovan. The closest living species to human are the chimp and pygmy chimp. As expected, the Old World monkeys (hanuman langur and mitred leaf monkey) are the most distantly related to human. The divergence times can be calculated from the branch lengths displayed in the Newick-formatted data. As mentioned earlier, Allan Wilson's data and many other subsequent studies show that human and chimp diverged approximately five million years ago. From the values of the distances in the Newick tree format, the distance from chimps to the chimp-human common ancestor can be calculated as follows. The distance from the chimp to chimp–pygmy chimp common ancestor is 0.00694. The distance from the chimp–pygmy chimp common ancestor to the chimp-human common ancestor is 0.01439. So the distance from the chimp to the chimp-human common ancestor is

---

[26] The software program TreeDraw was used to create this phylogenetic tree from the Newick tree format.

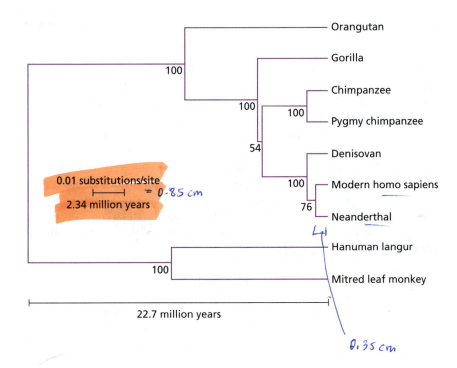

0.01 substitutions/site = 0.85 cm
2.34 million years

22.7 million years

0.35 cm

FIG. 8-15. Phylogenetic tree created with the UPGMA method. This is an ultrametric additive tree that assumes a strict molecular clock for all species compared in the tree. The scale near the bottom left within the tree corresponds to the average evolutionary distance (in substitutions per site) obtained from the distance matrix. Based on the estimated substitution rate of 4.266×10^{-9} substitutions/(site*year), the substitution per site scale can be converted to a time period. The entire distance from the OTUs to the root is 22.7 million years. Numbers at nodes are bootstrap values. The values indicate the number of times the group consisting of the species which are to the right of that bifurcation occurred among the trees, out of 100 trees.

$$\frac{2.34 \, my}{0.01 \, subs/site} \quad 0.35$$

$$\frac{0.01 \, subs/site}{0.85 \, cm} \circ 0.35 \, cm = 0.0041 \, subs$$

$$0.00694 + 0.01439 = 0.02133 \text{ substitutions/site.}$$

We can determine the substitution rate per year by dividing this value by five million years. This gives us 0.004266 substitutions/(site*million years). Now we have the data to calculate the approximate time in which modern *Homo sapiens* diverged from Neanderthal. From the values of distances in the Newick tree form we find that the distance from mod. H. sapiens to mod. H. sapiens-Neanderthal common ancestor is 0.00422. The following equation is used to calculate modern *Homo sapiens*-Neanderthal divergence time:

Mod. *Homo sapiens*-Neanderthal divergence time =

Mod. *Homo sapiens*–Neanderthal substitutions/site $\times$ (site*million years)/0.004266 substitutions

$$= 0.00422 \times (1/0.004266) = 0.989 \text{ million years}$$

This value falls somewhat outside the upper limit of the time range estimated by the fossil record and other data by ~20%, but this remains an active area of research. The apparent discrepancy indicates that there may be problems with the JC model of calculation, the UPGMA method in general, or both. The UPGMA method makes two assumptions involving evolutionary processes that can contribute to error:

1. The rate of the molecular clock is the same for all species analyzed.
2. The number of random substitutions that occur is uniform for all species analyzed.

Both of these assumptions have been shown to be not always true. The rate of the molecular clock is *approximately* constant for different species but not exactly. If the molecular clock is the same and there were no random substitutions, then all of the distances from the hominoid species to the monkey species would be identical, and this is what the phylogenetic tree in Figure 8-15 implies. But in fact the distances are not identical. The bordered section of the distance matrix in Table 8-4 shows that the calculated distances from the hominoids to the monkey range from 0.1748 to

0.2158. UPGMA takes the average of these distances (0.194) and applies that average to all hominoid OTUs.

Before we leave this subject, let's ask one more question. Can we use the calculated UPGMA tree to calculate the number of nucleotide substitutions per site per year? The answer is yes, and, furthermore, we can compare this calculated rate to the experimentally observed number of nucleotide substitutions per site per year, which is ~2×10^{-8} for mitochondrial DNA.[27] Let's work through our data of rDNA distances to see if we get a value close to 2×10^{-8} nucleotide substitutions per site per year.

From our previous calculation, the distance between chimp and its common ancestor to human is 0.02133. This corresponds to 0.02133 substitutions/site.

The time of separation of the two genes from a common ancestor is known. We will use five million years for the amount of time that separates a common ancestor to human and chimp, based on previous molecular clock data from the analysis of many genes. From the following calculation we get

$$0.02133/5,000,000 =$$

4.266×10^{-9} substitutions/(site*year). This value is within an order of magnitude of the experimentally measured mitochondrial gene nucleotide substitution rate (2×10^{-8} substitutions/site*year).

Thought Question 8-4

We calculated the distance from chimp to the chimp-human common ancestor to be 0.02133 substitutions per site. Using the Newick-formatted distances used for this phylogenetic tree, we could also have calculated the distance from human to the chimp-human common ancestor. Can you do this calculation?

Bootstrap Analysis

You might ask if there are methods to show if the tree we produced is correct. In the laboratory, a scientist would repeat an experiment many times to help answer that question. However, in this case, there is only one original set of data—the sequences used to generate the phylogenetic tree in the first place. Under these circumstances, a useful approach is to simulate new sets of data based on the original set of data. This statistical method is called bootstrap analysis, or bootstrapping. Once a large number of data sets are simulated through bootstrapping, one can ask whether an experimental result is due to chance.

Bootstrapping is a common method for measuring the robustness of the *topology* of phylogenetic tree. Topology is the branching order of the taxa. Bootstrapping does not measure the accuracy of the branch lengths. The idea of applying bootstrapping to phylogenetic data was first introduced by Joseph Felsenstein in 1985. The basic idea is to repeat the tree-building process with altered sequences from the multiple sequence alignment. The aligned sequences are altered in the following way. Columns of nucleotides from the original alignment are randomly selected and used to build a sequence alignment that is the same size (i.e., same number of columns and same number of rows) as the original alignment. Columns are selected at random and could be selected more than once from the original alignment. This is called sampling with replacement. Any column may appear multiple times, or not at all, in the new alignment. The new alignments are called pseudoreplicates (Figure 8-16).

---

[27] From Brown, George, and Wilson (1979). One must note, however, that there is quite a bit of variability in substitution rate depending on the organism analyzed and the particular region of DNA tested (see Nabholz, Glémin, and Galtier 2008). Our calculated substitution rate appears to be a bit on the low side, which may be due to the particular gene we have chosen to analyze.

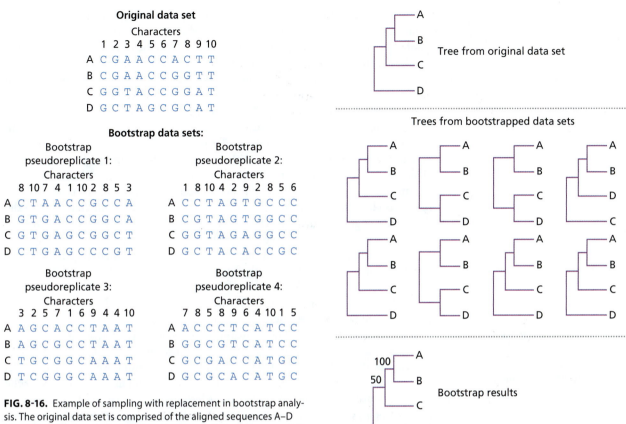

Original data set

Characters

	1	2	3	4	5	6	7	8	9	10
A	C	G	A	A	C	C	A	C	T	T
B	C	G	A	A	C	C	G	G	T	T
C	G	G	T	A	C	C	G	G	A	T
D	G	C	T	A	G	C	G	C	A	T

Bootstrap data sets:

Bootstrap pseudoreplicate 1:

Characters

	8	10	7	4	1	10	2	8	5	3
A	C	T	A	A	C	C	G	C	C	A
B	G	T	G	A	C	C	G	G	C	A
C	G	T	G	A	G	C	G	G	C	T
D	C	T	G	A	G	C	C	C	G	T

Bootstrap pseudoreplicate 2:

Characters

	1	8	10	4	2	9	2	8	5	6
A	C	C	T	A	G	T	G	C	C	C
B	C	G	T	A	G	T	G	G	C	C
C	G	G	T	A	G	A	G	G	C	C
D	G	C	T	A	C	A	C	C	G	C

Bootstrap pseudoreplicate 3:

Characters

	3	2	5	7	1	6	9	4	4	10
A	A	G	C	A	C	C	T	A	A	T
B	A	G	C	G	C	C	T	A	A	T
C	T	G	C	G	G	C	A	A	A	T
D	T	C	G	G	G	C	A	A	A	T

Bootstrap pseudoreplicate 4:

Characters

	7	8	5	8	9	6	4	10	1	5
A	A	C	C	C	T	C	A	T	C	C
B	G	G	C	G	T	C	A	T	C	C
C	G	C	G	A	C	C	A	T	G	C
D	G	C	G	C	A	C	A	T	G	C

FIG. 8-16. Example of sampling with replacement in bootstrap analysis. The original data set is comprised of the aligned sequences A–D (from four different species) used to create the original distance matrix. Bootstrap data sets contain four pseudoreplicates for illustration. Typically, 100 or 1000 pseudoreplicates are created.

FIG. 8-17. Demonstration of bootstrap analysis.

Tree construction is repeated for all pseudoreplicates until 100 or 1,000 trees are generated. The bootstrap-generated trees are compared to the original tree—specifically, each clade in the bootstrap-generated trees is compared to its corresponding clade in the original tree. If the topologies of the all of the bootstrap-generated clades match the clade topology of the original tree, then the bootstrap value is 100 (for 100%). A value of 70–75 is considered to be the lower range of acceptable bootstrap values. All clades in the tree are analyzed independently. This independent analysis of clades makes sense because there may be clades of the tree that exhibit high topology robustness and other parts that do not. A simple example of how a tree's topology is tested by bootstrap analysis is shown in Figure 8-17.

The top panel is a phylogenetic tree created from a distance matrix derived from the original data set. The middle panel shows eight trees created from bootstrap data sets. The bottom panel shows the percent of trees with a clade containing OTUs A and B (clade AB). To calculate the bootstrap value the number of trees with identical topology for the clade AB in the middle panel is divided by the total number of trees in this panel. To get the bootstrap value, this fraction is multiplied by 100. All eight trees have the same topology for clade AB. This means the topology for this clade is extremely robust. The bootstrap value of the clade containing the arrangement of OTUs A, B, and C (clade ABC) is 50, which falls into an unacceptable range and is considered unreliable. The bootstrap values are placed at the appropriate nodes on the tree. OTU D is considered the outgroup.[28]

---

[28] Note that in some software programs the raw bootstrap value is placed at the nodes instead of a percentage value.

Now we can return to our analysis of the primate phylogenetic tree generated in Figure 8-15. Bootstrap analysis was performed on the data to generate 100 trees.[29] Bootstrap values are shown to the left of each node in the tree. The value shows the number of trees in which the species to the right of the bifurcation appear (100 is the maximum value). There is a value of 76 at the node that bifurcates into modern *Homo sapiens* and Neanderthal. This means that in 24 of the trees either *Homo sapiens* or Neanderthal were replaced by Denisovan. These are the only possibilities because the clade representing *Homo sapiens*, Neanderthal, and Denisovan is found in all 100 trees. A concern about this particular phylogenetic tree stems from the clade containing *Homo sapiens*, Neanderthal, Denisovan, chimp, and pygmy chimp. This clade was found in only 54 of the 100 trees. This suggests that the clade containing chimp and pygmy chimp is found in another section of the tree in some cases (perhaps this clade diverged from an ancestor in common with orangutan prior to the divergence of gorilla). It is important to bear in mind the limitations of the data and evolutionary model; only one gene (rDNA) was analyzed, and the JC model was used to build the distance matrix. In the next section other distance matrix generation and tree-building methods will be briefly discussed.

Other Substitution Rate Models: Kimura Two-Parameter Model and Gamma Distance Model

Kimura two-parameter model
A nucleotide substitution model that explicitly considers that transversions are less frequent than transitions.

In this chapter we discussed four substitution rate models: Dickerson, linear, Poisson correction, and Jukes-Cantor. There are other substitution models to estimate distances, and we will highlight just a few. The first is the Kimura two-parameter model described by Motoo Kimura. The Kimura two-parameter model accounts for the fact that a purine-pyrimidine mutation is less frequent than a purine-purine or pyrimidine-pyrimidine mutation in DNA. Recall from Chapter 3 that a mutation in which a purine mutates to pyrimidine or vice versa is called transversion. A mutation in which a purine mutates to another purine or in which a pyrimidine mutates to another pyrimidine is called a transition. A transversion is more likely to cause a nonsynonymous substitution in a protein and is less frequent than a transition. For this reason, in the Kimura model a transversion results in a longer evolutionary distance than a transition. Even in the nonprotein coding regions of the genome, a transversion is disfavored because of the larger perturbation to the DNA structure after the mutation and because DNA repair systems are not as efficient in catching the error created during DNA synthesis.

gamma distance model A nucleotide substitution model that considers a variable rate of substitution due to relative frequencies of nucleotides in genomes.

Another substitution rate model, the gamma distance, accounts for rate variability between sites. As an example, bacteria can live in extreme environments (high or low temperatures, high or low pressures, high or low salinity) and their genomes have evolved so that their %G + C content allows the genomes to handle these environments. The high %G + C content makes it more difficult for DNA strands to come apart at high temperatures. The gamma distance can account for substitution rate variations in DNA regions that code for proteins. The third position of codons in the DNA can often be substituted without changing the amino acid. In bacteria with high %G + C content, DNA usually has G or a C in the third position to preserve genome stability, but not C or T. Genome stability thus *limits* the types of substitutions one observes in these organisms. This means that bacteria living at high temperatures have relatively low substitution rate variability. On the other hand, bacteria living at moderate temperature with a %G + C content of approximately 50% have a

---

[29] The bootstrap analysis was carried out using the PHYLIP suite of software programs (version 3.69). Briefly, SEQBOOT was used to generate 100 psuedoreplicates of the original aligned sequence data set. DNADIST was used to create 100 distance matrices using the JC model. Neighbor was used to generate 100 UPGMA trees, and CONSENSE was used evaluate the trees and create a consensus tree. CONSENSE generates a consensus tree with bootstrap values at each internal node. The generated consensus tree had the same topology as the phylogenetic tree created from the original data set. The bootstrap values were retrieved from the consensus tree and placed on the tree shown in Figure 8-15.

high substitution rate variability. Thomas Uzzell and Kendall Corbin (both at Yale University, New Haven, Connecticut) reported that the gamma distance d_Γ can account for substitution rate differences. Similar to the JC model, when the Hamming distance is low ($D < 0.2$), the d_Γ is approximately equal to D. When the Hamming distance is higher than 0.2, the d_Γ can increase significantly. The degree of the increase depends on the alpha factor, a. The equation for the gamma distance d_Γ is:

$$d_\Gamma = a[(1 - D)^{-1/a} - 1] \qquad \textbf{equation 8.1}$$

where a is the alpha parameter and D is the Hamming distance. For nucleotide sequence data that code for proteins, the estimated values of a are typically in the range of 0.2–3.5. To model a low variable substitution rate, a is set to greater than 1. To model a high variable substitution rate, a is set to less than 1.

Neighbor-Joining Method

In the neighbor-joining (NJ) method, clustering is used to create an additive unrooted tree. The researcher may choose to *display* the tree as a rooted tree by designating an outgroup. But in fact, it is still an unrooted tree. The display simply bends the unrooted tree into a form such that the designated outgroup species is placed on a branch separate from the rest of the sequences. The NJ method was developed by Naruya Saitou and Masatoshi Nei (University of Texas Health Science Center, San Antonio) in 1987. Unlike UPGMA, the NJ method does not assume that all sequences have the same rate of substitution, and thus it is thought to be a more realistic view of evolution. Saitou and Nei created their model with the idea that you can determine the *minimum* number of substitutions necessary to bring about evolution. In the NJ method all branch lengths in the tree should sum to the shortest possible overall length.

Figure 8-18A shows the NJ method-generated unrooted phylogenetic tree of the primate DNA sequences coding for the 12S ribosomal RNA. Let's look closely at this unrooted phylogenetic tree. Notice that there are long branches connecting the Old World monkeys to the hominoids. The longest set of branches connects the Old World monkey hanuman langur to the hominoid orangutan (red dotted line traces this set). This is expected given that the distance matrix shows that the evolutionary distance is longest for these two species (0.2158) (see Table 8-4). Of the branches that connect Old World monkeys to hominoids, the shortest set of branches connects mitred leaf monkey to gorilla (0.1748), as shown by the blue dotted line in Figure 8-18A. By measuring the branch lengths with a ruler, it is easy to see that the measured distances of the branches in Figure 8-18A are proportional to the values listed in Table 8-4. In another example, the shortest set of branches connects modern *H. sapiens* to Neanderthal, which reflects the short evolutionary distance between these species in the distance matrix (0.0084).

The NJ tree can be depicted in a format that is easier to read (Figure 8-18B). This form of the NJ tree is created by setting the mitred leaf monkey as the outgroup in the

neighbor-joining method
An evolutionary distance matrix-based clustering algorithm that creates a phylogenetic tree using the shortest possible evolutionary time. Pairs of OTUs are identified that minimize total branch length within the tree. At each stage of the clustering process the total branch length is minimized.

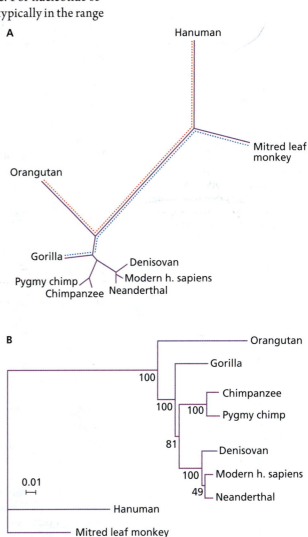

FIG. 8-18. Neighbor-joining phylogenetic tree created from JC model distance matrix.

A. The unrooted tree was created with TreeView. **B.** This phylogenetic tree was created by TreeView. The Newick format tree used for input into TreeView was created with Neighbor software program in the PHYLIP suite of software programs. Hanuman langur was denoted as the outgroup in the Neighbor software program. The TreeView output was modified to have a depiction similar to the UPGMA tree shown in Figure 8-15. The modification did not alter branch lengths or the types of species in the clades. Modification was created with CS6 Adobe Photoshop. Bootstrap analysis was performed with appropriate software programs from PHYLIP (version 3.69). Note that the bootstrap value to the left of the clade containing orangutan also applies to the clade containing mitred leaf monkey and han langur.

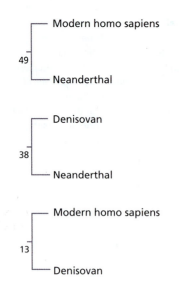

FIG. 8-19. Bootstrap values for alternate topologies of one clade of the NJ tree. Bootstrap analysis of the primate NJ tree was performed, and all topologies of the clade containing any two of the modern *Homo sapiens*, Neanderthal, and Denisovan are presented. The top clade is from the consensus tree. The clades in the middle and bottom were rejected by the boot-strap analysis software program but are shown here for comparison purposes.

recent African origin hypothesis
An explanation of the origin of modern *Homo sapiens* that supposes that a single common ancestor arose in sub-Saharan Africa.

display. In this format it is easy to compare the tree generated by the NJ to the tree generated by the UPGMA method (see Figure 8-15). Both methods generate trees with identical topologies, which gives us more confidence in the evolutionary relationships of these species. However, the bootstrap values are somewhat different. The NJ tree shows bootstrap values of 100 for most clades. The bootstrap value for the clade containing chimp, pygmy chimp, Denisovan, Neanderthal, and modern *Homo sapiens* is 81. This compares favorably with the same clade in the UPGMA tree, where the bootstrap value is 54, and therefore gives us more confidence in the arrangement of this clade. Of more concern is the clade containing Neanderthal and modern *Homo sapiens*. Its bootstrap value is only 49, whereas in the UPGMA tree it is 76. In this tree, Neanderthal and modern *Homo sapiens* are more closely related to each other than either is to Denisovan. Analysis of the output from the software program that calculated the bootstrap values indicates that two other tree topologies for this clade are obtained, each at lower bootstrap values (Figure 8-19). In one clade with a bootstrap value of 38, Neanderthal and Denisovan are more closely related. In another clade with a bootstrap value of 13, modern *Homo sapiens* and Denisovan are more closely related. Analysis of the bootstrap values from the UPGMA tree and the NJ tree indicates that there is some concern that Neanderthal is more closely related to Denisovan than to modern *Homo sapiens*. To resolve this issue it may be prudent to repeat the analysis using other genes from these species.

Another approach to verify conclusions on the evolutionary relationships of these species is to use other methods to create phylogenetic trees such as maximum likelihood or Bayesian methods. The mathematical basis of maximum likelihood and Bayesian methods are fleshed out in Chapters 11 and 12. Phylogenetic tree generation from these methods will not be discussed here, but interested readers may consult Lemey, Salemi, and Vandamme (2009).

8.4 APPLICATION OF PHYLOGENETICS TO STUDIES OF THE ORIGIN OF MODERN HUMANS

We have discussed the molecular clock hypothesis and explored how it can be used to order phylogenetic events within primates. By calibrating the clock to paleontological data, we can use the clock to estimate the time of the origin of modern *Homo sapiens* and how this species ended up inhabiting the varied regions of our planet. In 2000, Svante Pääbo and Ulf Gyllensten (University of Uppsala, Uppsala, Sweden) combined their research teams to sequence mitochondrial genomes of 53 individuals. The individuals were chosen from 14 different native language backgrounds in order to collect a high degree of genetic diversity in the genomes. The mitochondrial genomes ranged from 16,558 to 16,576 nucleotide base pairs in length. Gorilla and chimp mitochondrial genomes were used as outgroups. One segment of the genome, called the D-loop, was removed from analysis because its substitution rate did not proceed in a clocklike fashion. The researchers used the divergence time between humans and chimps (5 MYA) to calibrate their molecular clock. They found that the substitution rate was, on average, 1.70×10^{-8} substitutions per site per year.[30]

There are two major hypotheses as to how modern humans evolved. Both agree that one archaic human species, *Homo erectus*, migrated out of Africa approximately 2 MYA and, based on findings of fossil remains, populated Georgia (in Asia), India, Sri Lanka, China, and Java (in Indonesia). Eventually, *Homo erectus* was replaced by modern *Homo sapiens* (modern *H. sapiens*). One hypothesis, called the recent African

---

[30] This substitution rate is approximately four times higher than what we calculated for 12S ribosomal RNA gene of the mitochondria. This difference could be a reflection of the methods used to calculate the distance matrix, or it could be due to differences in substitution rates in a gene specific manner. Overall, the fact that the two values are within an order of magnitude of each other gives confidence that the values are close to the actual rate of substitution.

origin hypothesis, states that modern *H. sapiens* originated in sub-Saharan Africa approximately 200,000 years ago. Starting at approximately 100,000 years ago modern *H. sapiens* spread to the rest of the planet. It is believed that they first migrated to the east coast of Africa and continued to follow the coast northward. They migrated east through the Middle East, to India, and eventually settled in Australia. Later, the modern *H. sapiens* population spread to other parts of the Old World, including East Asia, Central Asia, the Middle East, and Europe. Migration to the New World (the Americas) occurred even later. As modern *H. sapiens* migrated, they replaced other archaic humans that evolved from *Homo erectus* (among the archaic human species were Neanderthal and Denisovan).

The second hypothesis, called the multiregional hypothesis, suggests that the emergence of modern *H. sapiens* occurred in different parts of the world independently. The replacement of archaic humans took place at approximately the same time in all areas, including in Africa. The recent African origin hypothesis hinges on the idea that individuals from a small colony of Africans, perhaps from a specific region within Africa, migrated out within the last 50,000 to 100,000 years. If relatively few individuals gave rise to non-African people (i.e., those who populated the remainder of the planet), then the descendants of that colony would have little DNA diversity relative to all of the people that remained in Africa. Descendants of that colony are the Asian, Australian, European, New World people, and some Africans (the part of that colony that remained in Africa). You would expect that the most recent common ancestor of these people existed 50,000 to 100,000 years ago.

The multiregional hypothesis suggests that modern *H. sapiens* evolved from the more primitive *Homo erectus* independently within several geographical locations on the planet. An approximately equal amount of DNA diversity in modern individuals would be consistent with the multiregional hypothesis. Furthermore, this hypothesis would predict that the most recent common ancestor of *all* people should have existed approximately 200,000 years ago, when modern *H. sapiens* first appeared.

The phylogenetic tree in Figure 8-20 shows that the data collected from 53 individuals is consistent with the recent African origin hypothesis. The neighbor-joining tree indicates that the most recent common ancestor of all Asians, all Australians, all Europeans, all New World people, and *some* Africans lived approximately 52,000 years ago. The branches leading to this common ancestor are colored blue, red, and green. The few Africans that shared this common ancestor (individuals 33–38) may be descendants of the part of the colony that remained in Africa. Because genetic divergence should occur after migration of the people out of Africa, this is the minimum bound for migration out of Africa. It is quite possible that the migration occurred a bit more than 52,000 years ago.

What about the other modern *H. sapiens* colonies that did not migrate out of Africa? One would expect long branches from descendants of those colonies to common ancestors, indicating high DNA diversity. Analysis of the data in Figure 8-20 shows that many branches (colored purple) from Africans (individuals 42, 48, 52, and 53) are longer than those of the clade that recently migrated out of Africa, confirming that some Africans are distantly related to each other and that their lineages existed for a longer time in Africa relative to descendants from other locations. The level of mitochondrial DNA sequence diversity shows that African mitochondrial DNA has had more time to undergo substitutions than non-African mitochondrial DNA. The amount of sequence diversity among all Africans is 3.7×10^{-3} nucleotide substitutions per site, whereas the amount of sequence diversity among non-Africans is only 1.7×10^{-3} nucleotide substitutions per site. This is also consistent with the recent African origin hypothesis.

In the study, an average mitochondrial DNA substitution rate in non-Africans was calculated from the distance matrix. Using this substitution rate and a generation time of 20 years per generation, it was estimated that a significant population expansion took place approximately 38,500 years ago. That expansion accounts for the non-African population living today.

multiregional hypothesis An explanation of the origin of modern *Homo sapiens* that supposes that more than one common ancestor was located in different geographical locations.

more sequence diversity in these longer branches

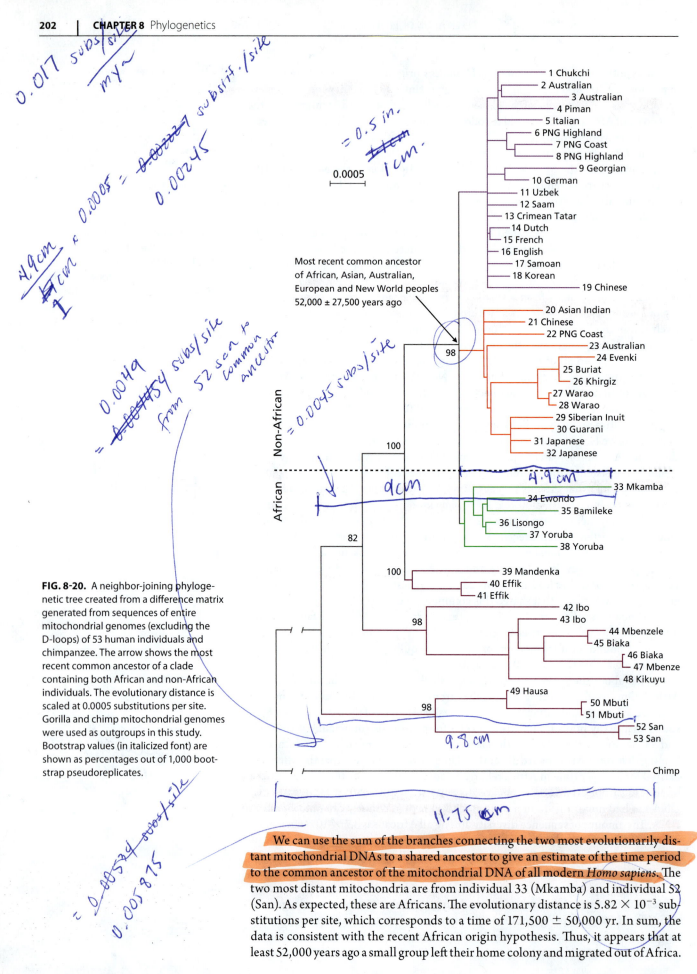

FIG. 8-20. A neighbor-joining phylogenetic tree created from a difference matrix generated from sequences of entire mitochondrial genomes (excluding the D-loops) of 53 human individuals and chimpanzee. The arrow shows the most recent common ancestor of a clade containing both African and non-African individuals. The evolutionary distance is scaled at 0.0005 substitutions per site. Gorilla and chimp mitochondrial genomes were used as outgroups in this study. Bootstrap values (in italicized font) are shown as percentages out of 1,000 bootstrap pseudoreplicates.

We can use the sum of the branches connecting the two most evolutionarily distant mitochondrial DNAs to a shared ancestor to give an estimate of the time period to the common ancestor of the mitochondrial DNA of all modern *Homo sapiens*. The two most distant mitochondria are from individual 33 (Mkamba) and individual 52 (San). As expected, these are Africans. The evolutionary distance is 5.82×10^{-3} substitutions per site, which corresponds to a time of $171,500 \pm 50,000$ yr. In sum, the data is consistent with the recent African origin hypothesis. Thus, it appears that at least 52,000 years ago a small group left their home colony and migrated out of Africa.

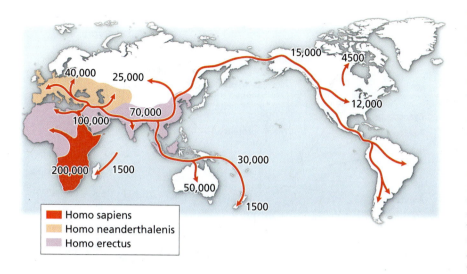

FIG. 8-21. Origin and dispersal of modern humans. Red areas represent the initial location of modern *Homo sapiens*. Red arrows represent migration patterns of these people. Yellow represents areas populated by Neanderthal. Pink represents areas inhabited by Homo erectus. The origin of modern humans occurred approximately 170,000 years ago in the southwest region of Africa. The first migration of people out of Africa occurred at least 52,000 years ago (here it is depicted as 100,000 years ago). In the first migration, people populated the Middle East, India, southeast Asia, and, eventually, Australia over a period of 50,000 years. Succeeding this first migration wave, people populated Europe, Northern Asia, and, finally, the Americas. As the migrations proceeded, earlier hominids and Neanderthal were replaced. The last Neanderthal existed approximately 27,000 years ago.

Their descendants replaced other archaic humans living in the Old World. The New World was populated later than the Old World. A diagram showing migration patterns consistent with this and other phylogenetic analyses is shown in Figure 8-21.[31]

8.5 PHYLOGENETIC TREE OF LIFE

We discussed the evolution of some primates with an emphasis on the evolution of modern *H. sapiens*, so now might be a good time to take a broader view of the relationships of all organisms. Until protein sequencing and DNA sequencing technology was developed, biologists largely used morphology and the fossil record to create phylogenetic trees. One such tree from the 19th century was shown earlier in this chapter (see Figure 8-2). Carl Woese (University of Illinois, Urbana-Champaign), was one of the first to use DNA sequence data to create phylogenetic trees. He specifically used DNA sequences of 16S ribosomal RNA genes (rDNAs) to build his phylogenetic trees in the early 1980s.[32] Before Woese, there were two known basal categories of life: eukaryotes and prokaryotes.[33] When Woese compared rDNAs from different microorganisms, he discovered a third category: archaea, which led him to propose three domains of life named bacteria, archaea, and eukaryota (also known as euckarya and eucarya). His discovery shifted the way we view life. Figure 8-22 shows a phylogenetic tree of life based on Carl Woese's analysis of substitutions in rDNA sequences. He named the tree the universal phylogenetic tree.

8.6 *TP53* GENE FAMILY MEMBERS IN DIFFERENT SPECIES

We have discussed in detail a few methods to study the phylogeny of primates and introduced the larger perspective of all of life in Woese's depiction of the Tree of Life. Let's now turn our attention to the relationship between species evolution and the *TP53* family orthologs. Here, we will take a moment to analyze a composite picture

---

[31] Some recent data from nuclear gene analysis suggests that there may have been more than one migration out of Africa.

[32] A more accurate term is small ribosomal RNA gene. In bacteria and archaea the size of the RNA produced from this gene is 16S and in eukaryotes the size is 18S.

[33] Before Woese, it was also believed that the evolutionary distance between plants and animals was as great as the evolutionary distance between animals and prokaryotes. It is now clear that plants and animals are phylogenetically close and both belong in the eukaryota domain (see Figure 8-22).

FIG. 8-22. Phylogenetic Tree of Life based on Carl Woese's 16S rRNA gene sequence comparisons.

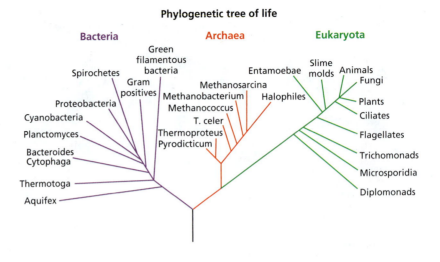

Phylogenetic tree of life

of a phylogeny that consists of a phylogenetic tree and a cladogram. The phylogeny OTUs will be analyzed for the presence of *TP53* family members in the genomes.

An analysis of the *TP53* gene family in different species gives us insight into the evolution of the family members' genes and functions. In the phylogeny shown in Figure 8-23, tree branch lengths are proportional to the genome-wide nucleotide substitution rates for jawed bony vertebrates (all OTUs above the dotted line). A cladogram is shown for other species (all OTUs below the dotted line) where branch lengths are not proportional to genome-wide nucleotide substitution rates. The genomes of the OTUs in the tree were also scanned for the presence of *TP53* gene family members. The column to the right of the tree shows the *TP53* paralogs present in each species. When only one *TP53* gene family member is detected it is referred to as "homolog," followed by the name of the gene in that particular species.

The *TP53* family of genes consists of three paralogs: *TP53*, *TP63*, and *TP73*. The three paralogs are maintained as distinct separate genes in most of the vertebrates (although some exceptions are apparent: alpaca and elephant to name a couple). In the elephant shark and lamprey (both are jawless vertebrates), there are three and two *TP53* family members, respectively. In other species (nonvertebrates), with the exception of yeast, there is usually only one family member. Sequence comparison studies with mammalian protein sequences shows that the putative protein sequences in nonvertebrates are more similar to *TP63* and *TP73*, rather than to *TP53*. The overall picture is that a single *TP63/TP73* gene in the common ancestor to vertebrates underwent duplications to give rise to three distinctive family members: *TP53*, *TP63*, and *TP73*. The three family members are somewhat specialized in their functions, but all share the ability to transcribe genes that initiate programmed cell death.

It is not uncommon for there to be a single gene homolog in nonvertebrates and more than one gene homolog in vertebrates. The common ancestor of vertebrates is thought to have existed approximately 400–600 million years ago. At this point in evolutionary history, there were one or two whole genome duplication events that drastically increased the size of many genomes.[34] Those duplication events likely resulted in three *TP53* family members in modern vertebrates.

Most nonvertebrates contain just one *TP53* family member, but there are a few nonvertebrate species such as mosquito, sea anemone, and choanoflagellates that contain novel paralogs of the invertebrate *TP63/TP73* gene. The novel paralogs likely arise from duplications that are independent of the genome-wide duplications

---

[34] For more information on whole genome duplications, see Dehal and Boore (2005); Kasahara (2007).

Phylogenetic tree (from whole genome substitution rates) and cladogram

Tp53 family members

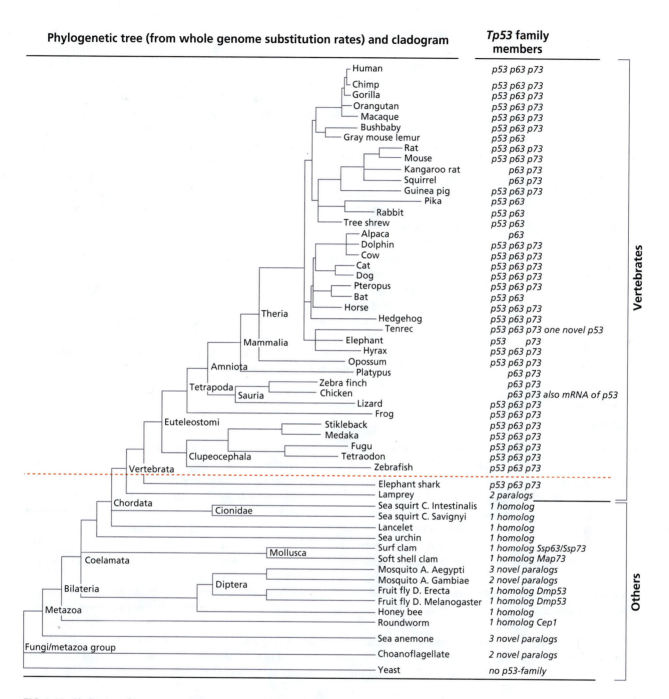

FIG. 8-23. Phylogeny of genomes and a listing of *TP53* family genes in selected species. Whole genome nucleotide substitution rates were used to create the phylogenetic tree above the dotted line. Below the red dotted line the species relationships are shown in a cladogram. *TP53* family members identified in the genome of each species are listed.

that gave rise to *TP53*, *TP63*, and *TP73* in the vertebrates. Researchers have not been able to detect a *TP53* family gene in yeast, sponge, or bacteria. The apparent absence of this gene in these species may be due to the fact that these species do not have a *TP53* homolog or it is quite possible that the *TP53* homolog is unrecognizable in these species due to extensive substitutions.

The common theme for *TP53* family members is programmed cell death because all three *TP53* family members instigate this process, but for different purposes. In the nonvertebrates, the p63/p73 protein appears to function as a protector against DNA damage in germ cells. This function is maintained in sea anemone, clams, insects, as well as the vertebrates. On the other hand, the p53 protein primarily causes cell death in response to cell stress, such as DNA damage, and its main purpose is to inhibit tumor formation. The p73 protein has some limited tumor suppressor activity, but its major function is to instigate cell death during brain development. Mice without sufficient amounts of p73 have neurological disorders—possibly due to the buildup of damaged brain cells that should have been removed during development. The p63 protein controls epithelial cell development. Mice without p63 die at birth, have no skin, and display limb abnormalities. Missense mutations in p63 have been found in some human conditions including ectrodactyly (loss of one or more central digits of the hand or foot), ectodermal dysplasia (abnormalities of ectodermal structures: teeth, hair, etc.), and cleft lip/palate syndrome. The relationship between the apoptosis function and the development functions of p63 is still under investigation.

A common theme retained among all three p53 protein family members is the protection of germ cells. In humans, p53 activates the transcription of a gene (called leukemia inhibitory factor, or LIF) that helps the fertilized oocyte to implant into the mother's uterine wall. Implantation is required for embryo development. In mice, specific p73 splice form deletions cause females to be infertile, and they show less ovulated oocytes. Again in mice, specific p63 splice form deletions result in female infertility and early menopause. The data suggests that protection of germ cells may be the universal function of the p53 protein family.

SUMMARY

Evolutionary relationships can be depicted as phylogenetic trees. Prior to molecular data, fossil evidence and morphology were the primary means by which phylogenetic trees were built. Beginning in the 1960s, protein and DNA sequence data were shown to be useful in describing phylogenies. The differences in sequence data roughly correlate with evolutionary distances between species. The molecular clock is an amino acid or nucleotide substitution rate that can be used to determine evolutionary events such as speciation. The molecular clock must be calibrated with fossil data to be useful in timing speciation events. The substitution rate depends on the specific proteins and DNA sequences used. Furthermore, there are mathematical corrections that should be used to account for multiple substitutions and other considerations. Computer software programs can be used to create sequence alignments of molecular sequence data, create evolutionary distance matrices, and create phylogenetic trees. Bootstrapping is one method to determine if the tree topology is robust. Phylogenetic trees show that Neanderthal and modern *Homo sapiens* diverged approximately 500,000–800,000 years ago. The common ancestor of Denisovan and modern *Homo sapiens* existed approximately 1–1.3 million years ago. Phylogenetic analysis of mitochondrial genomes of modern humans from 52 diverse individuals informs us about the migration patterns of modern humans out of Africa. DNA sequence analysis of different species was instrumental in the discovery of a third domain of life known as archaea that is distinct from eukaryota and bacteria.

EXERCISES

1. An example of an operational taxonomic unit (OTU) is:

   a. Multiple sequence alignment
   b. Protein sequence
   c. Clade
   d. Node

2. In the Kimura two-parameter model distance matrix calculation, transversions are weighted more heavily than transitions. In other words, the calculated evolutionary distance is longer when a transversion is detected at a particular site. Why?

3. According to the phylogenetic tree in Figure 8-20, which two individuals within the Africans have the *most* evolutionarily distant mitochondrial DNA? Calculate the approximate time (in years) when the two individuals shared a most recent common ancestor. The most recent common ancestor for humans and chimps existed five million years ago and the evolutionary

distance from humans to this ancestor is 0.085 substitutions/site. How does this reconcile with the claim that common ancestor of modern humans existed 171,000 ± 50,000 years ago?

4. Construct a neighbor-joining phylogenetic tree using the mitochondrial gene that codes for 16S ribosomal RNA. Use the JC model to create the distance matrix. Include the following organisms in your analysis: modern *H. sapiens*, chimpanzee, pygmy chimpanzee, Neanderthal, Denisovan, gorilla, orangutan, mitred leaf monkey, and hanuman langur. Construct your tree such that the mitred leaf monkey sequence is displayed as the outgroup.

The accession number of the gene sequence of modern *H. sapiens* 16S ribosomal RNA mitochondrial gene is NC_012920.

GCTAAACCTAGCCCCAAACCCACTCCACCTTACTA
CCAGACAACCTTAGCCAAACCATTTACCCAAATAA

AGTATAGGCGATAGAAATTGAAACCTGGCGCAATA
GATATAGTACCGCAAGGGAAAGATGAAAAATTATA

ACCAAGCATAATATAGCAAGGACTAACCCCTATACC
TTCTGCATAATGAATTAACTAGAAATAACTTTGC

AAGGAGAGCCAAAGCTAAGACCCCCGAAAC-
CAGACGAGCTACCTAAGAACAGCTAAAAGAGCA-
CACCCGT

CTATGTAGCAAAATAGTGGGAAGATTTATAGGTAGAGG-
CGACAAACCTACCGAGCCTGGTGATAGCTGGT

TGTCCAAGATAGAATCTTAGTTCAACTTTAAATTTG
CCCACAGAACCCTCTAAATCCCCTTGTAAATTTA

ACTGTTAGTCCAAAGAGGAACAGCTCTTTGGACAC
TAGGAAAAAACCTTGTAGAGAGAGTAAAAAATTTA

ACACCCATAGTAGGCCTAAAAGCAGCCACCAATTA-
AGAAAGCGTTCAAGCTCAACACCCACTACCTAAAA

AATCCCAAACATATAACTGAACTCCTCACACCCAAT
TGGACCAATCTATCACCCTATAGAAGAACTAATG

TTAGTATAAGTAACATGAAAACATTCTCCTCCGCAT
AAGCCTGCGTCAGATTAAAACACTGAACTGACAA

TTAACAGCCCAATATCTACAATCAACCAACAAGT-
CATTATTACCCTCACTGTCAACCCAACACAGGCATG

CTCATAAGGAAAGGTTAAAAAAAGTAAAAGGAAC
TCGGCAAATCTTACCCCGCCTGTTTACCAAAAACAT

CACCTCTAGCATCACCAGTATTAGAGGCACCGCCT-
GCCCAGTGACACATGTTTAACGGCCGCGGTACCCT

AACCGTGCAAAGGTAGCATAATCACTTGTTCCTTA-
AATAGGGACCTGTATGAATGGCTCCACGAGGGTTC

AGCTGTCTCTTACTTTTAACCAGTGAAATTGACCTG
CCCGTGAAGAGGCGGGCATAACACAGCAAGACGA

GAAGACCCTATGGAGCTTTAATTTATTAATGCAAA-
CAGTACCTAACAAACCCACAGGTCCTAAACTACCA

AACCTGCATTAAAAATTTCGGTTGGGGCGACCTCG
GAGCAGAACCCAACCTCCGAGCAGTACATGCTAAG

ACTTCACCAGTCAAAGCGAACTACTATACTCAATT-
GATCCAATAACTTGACCAACGGAACAAGTTACCCT

AGGGATAACAGCGCAATCCTATTCTAGAGTCCATAT
CAACAATAGGGTTTACGACCTCGATGTTGGATCA

GGACATCCCGATGGTGCAGCCGCTATTAAAGGTTC-
GTTTGTTCAACGATTAAAGTCCTACGTGATCTGAG

TTCAGACCGGAGTAATCCAGGTCGGTTTCTATCTAC
NTTCAAATTCCTCCCTGTACGAAAGGACAAGAGA

AATAAGGCCTACTTCACAAAGCGCCTTCCCCCGTA-
AATGATATCATCTCAACTTAGTATTATACCCACAC

CCACCCAAGAACAGGGTTT

a. What two sequences have the shortest evolutionary distance between them?
b. What two sequences have the longest evolutionary distance between them?
c. Give the Newick-formatted output of the NJ tree.
d. Confirm that the evolutionary distance between hanuman langur and mitred leaf monkey in the distance matrix equals the distance shown in the Newick-formatted output. (Note: use only five significant figures for the value from the distance matrix.)
e. If you assume that modern *H. sapiens* and chimpanzee diverged from a common ancestor 5 MYA, what is the amount of time in which *H. sapiens* and Neanderthal diverged from a common ancestor? Use your data to answer this question. Is this a realistic estimate for the number of years when modern *Homo sapiens* and Neanderthal evolved from a common ancestor?
f. Calculate the substitution rate per site per year in the 16S ribosomal gene assuming that modern *H. sapiens* and chimpanzee shared a common ancestor 5 MYA. Is this a realistic substitution rate?
g. Show the tree produced from Newick-formatted output.

5. Perform bootstrap analysis of the tree you created in your answer in #4. Remember that this is a neighbor-joining tree. Show the consensus tree with the bootstrap values.

6. Are any of these trees topologically equivalent?

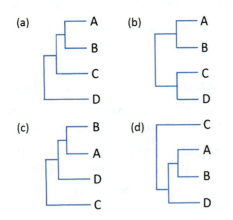

(a) A B C D (b) A B C D

(c) B A D C (d) C A B D

7. Construct a UPGMA tree for the following distance matrix. Show values of branch lengths on the tree.

	A	B	C	D
A	0			
B	9	0		
C	7	8	0	
D	5	10	8	0

8. Use Tajima's test on the three sequences shown below. Sequences 1 and 2 are to be tested to determine if they are evolving at the same rate from the outgroup sequence 3. Sequence 1 is from gorilla, sequence 2 is from orangutan, and sequence 3 is from mitred leaf monkey. All sequences are from 16S ribosomal RNA mitochondrial gene.

Seq. 1: AACCAAACCATTCACCCAAACAAAGTATAG GCGATAGAAATTACAATCCG

Seq. 2: AATCAAACCATTTACCCAAATAAAGTATAG GCGATAGAAATTGTAAATCG

Seq. 3: ATTAAATCATTCACAAATATAAAAGTATGGG TGATAGAAAGTCTACTCTG

9. In Figure 8-18B the branch length of the pygmy chimpanzee is longer than the chimpanzee branch length to their common ancestor. What is your interpretation of this difference?

ANSWERS TO THOUGHT QUESTIONS

8-1. The Hamming distance, D, is 3/60 = 0.05, which also equals the evolutionary distance, d; so $d = 0.05$.

> Poisson correction equation: $d = -\ln(1 - .05)$ = .051
> The JC model equation: $d = -3/4\ln[1 - (4 \times 0.05/3)] = 0.052$

Note that the two equations that account for multiple substitutions show only a small deviation away from the Hamming distance because the evolutionary distance of the two sequences is not large.

8-2. The Hamming distance, D, is 30/60 = 0.5, which also equals the evolutionary distance, d; so $d = 0.5$.

> The Poisson correction equation gives $d = 0.69$
> JC model equation gives $d = 0.82$

Note, Poisson and JC model equations result in larger evolutionary distances than the Hamming distance because they account for multiple substitutions.

8-3. The distance of A to C is 0.8, and the distance of B to C is 0.8. The mean distance is (0.8 + 0.8)/2 = 0.8. Thus, the distance of cluster (A, B) to C is 0.8.

8-4.

> Distance = distance of human to human-neand common ancestor
> + distance of human-neand common ancestor to human-neand.-denis common ancestor
> + distance of human-neand-denis common ancestor to chimp-pyg chimp common ancestor.
> = .00422 + .00213 + .01498 = .02133

REFERENCES

Barton, N. H., D. E. G. Briggs, J. A. Eisen, D. B. Goldstein, and N. H. Patel. 2007. "Phylogenetic Reconstruction." Chap. 27 in *Evolution*. Cold Spring Harbor, NY: Cold Spring Harbor Press. http://evolution-textbook.org/content/free/contents/ch27 .html. Accessed September 14, 2012.

Belyi V. A., P. Ak, E. Markert, H. Wang, W. Hu, A. Puzio-Kuter, and A. J. Levine. 2010. "The Origins and Evolution of the p53 Family of Genes." *Cold Spring Harbor Perspectives in Biology* 2: a001198.

Brown, W. M., M. George Jr., and A. C. Wilson. 1979. "Rapid Evolution of Animal Mitochondrial DNA." *Proceedings of the National Academy of Sciences USA* 76, 1967–1971.

Darwin, C. 1837–1838. Notebook B: [Transmutation of Species (1837–1838)]. Transcribed by Kees Rookmaaker. In *The Complete Works of Charles Darwin Online*. Edited by J. van Wyhe. http://darwin-online.org.uk/content/frameset? pageseq=1&itemID=CUL-DAR121.-&viewtype=text. Accessed September 2, 2012.

Dawkins, R. 2004. *The Ancestor's Tale: A Pilgrimage to the Dawn of Evolution*. New York: Houghton Mifflin.

Dehal, P., and J. L. Boore. 2005. "Two Rounds of Whole Genome Duplication in the Ancestral Vertebrate." *PLoS Biology* 3: e314.

Dickerson, R. E. 1971. "The Structures of Cytochrome C and the Rates of Evolution." *Journal of Molecular Evolution* 1: 26–45.

Gibbs, A. J., C. H. Calisher, and F. García-Arenal. 1995. *Molecular Basis of Virus Evolution*. Cambridge, UK: Cambridge University Press.

Higgs, P. G., and T. K. Attwood. 2005. *Bioinformatics and Molecular Evolution*. Malden, MA: Blackwell Publishing.

Hu, W., Z. Feng, A. K. Teresky, and A. J. Levine. 2007. "p53 Regulates Maternal Reproduction Through LIF." *Nature* 450: 721–724.

Ingman, M., H. Kaessmann, S. Pääbo, and U. Gyllensten. 2000. "Mitochondrial Genome Variation and the Origin of Modern Humans." *Nature* 408: 708–713.

Kasahara, M. 2007. "The 2R Hypothesis: An Update." *Current Opinion in Immunology* 19: 547–552.

Kimura, M. 1968. "Evolutionary Rate at the Molecular Level." *Nature* 217: 624–626.

Kimura, M., and T. Ohta. 1973. "Mutation and Evolution at the Molecular Level." *Genetics* (Supplement) 73: 19–35.

Lane, D. P., A. Madhumalar, A. P. Lee, B. H. Tay, C. Verma, S. Brenner, and B. Venkatesh. 2011. "Conservation of All Three p53 Family Members and Mdm2 and Mdm4 in the Cartilaginous Fish." *Cell Cycle* 10: 4272–4279.

Lemey, P., M. Salemi, and A.-M. Vandamme. 2009. *The Phylogenetic Handbook: A Practical Approach to Phylogenetic Analysis and Hypothesis Testing*, 2nd ed. Cambridge, UK: Cambridge University Press.

Llorens, C., R. Futami, M. Vicente-Ripolles, and A. Moya. 2008. "The Alignment Format Converter Server 1.0." *Biotechvana Bioinformatics Biotechvana*, Valencia. SCR: AFC.

Margoliash, E. 1963. "Primary Structure and Evolution of Cytochrome C." *Proceedings of the National Academy of Sciences USA* 50: 672–679.

Nabholz, B., S. Glémin, and N. Galtier. 2008. "Strong Variations of Mitochondrial Mutation Rate Across Mammals—the Longevity Hypothesis." *Molecular Biology and Evolution* 25: 120–130.

Nei, M. 2005. "Selectionism and Neutralism in Molecular Evolution." *Molecular Biology and Evolution* 22: 2318–2342.

Nei, M., Y. Suzuki, and M. Nozawa. 2010. "The Neutral Theory of Molecular Evolution in the Genomic Era." *Annual Review of Genomics and Human Genetics* 11, 265–289.

Ohno, S. 1970. *Evolution by Gene Duplication*. New York: Springer-Verlag.

Ohta, T. 1973. "Slightly Deleterious Mutant Substitutions in Evolution." *Nature* 246: 96–98.

Pääbo, S. 2014. *Neanderthal Man, in Search of Lost Genomes*. New York: Basic Books.

Pevsner, J. 2009. *Bioinformatics and Functional Genomics*. Hoboken, NJ: John Wiley and Sons.

Sarich, V. M., and A. C. Wilson. 1967. "Immunological Time Scale for Hominid Evolution." *Science* 158: 1200–1203.

Sokal, R. R., and C. D. Michener. 1958. "A Statistical Method for Evaluating Systematic Relationships." *The University of Kansas Scientific Bulletin* 38: 1409–1438.

Soltis, P. S., and D. E. Soltis. 2003. "Applying the Bootstrap in Phylogeny Reconstruction." *Statistical Science* 18: 256–267.

Tajima, F. 1993." Simple Methods for Testing the Molecular Evolutionary Clock Hypothesis." *Genetics* 135: 599–607.

Uzzell, T., and K. W. Corbin. 1971. "Fitting Discrete Probability Distributions to Evolutionary Events." *Science* 172: 1089–1096.

Varki, A., and T. K. Altheide. 2005. "Comparing the Human and Chimpanzee Genomes: Searching for Needles in a Haystack." *Genome Research* 15: 1746–1758.

Voet, D., and J. G. Voet. 2011. *Biochemistry*, 4th ed. Hoboken, NJ: John Wiley and Sons.

Woese, C. R., O. Kandler, and M. L. Wheelis. 1990. "Towards a Natural System of Organisms: Proposal for the Domains Archaea, Bacteria, and Eucarya." *Proceedings of the National Academy of Sciences USA* 87, 4576–4579.

Zuckerkandl, E., and L. Pauling. 1962. "Molecular Disease, Evolution, and Genetic Heterogeneity." In *Horizons in Biochemistry*, edited by M. Kasha and B. Pullman. New York: Academic Press.

Zuckerkandl, E., and L. Pauling. 1965. "Evolutionary Divergence and Convergence in Proteins." In *Evolving Genes and Proteins*, edited by B. Bryson and H. Vogel, 97–166. New York: Academic Press.

Zvelebil, M., and J. O. Baum. 2008. *Understanding Bioinformatics*. New York: Garland Science, Taylor & Francis Group.

GENOMICS

9.1 INTRODUCTION

The structure of the DNA double helix was published in 1953, but it would not be until the mid-1970s that the chemical and biochemical techniques would advance to the level where the order of nucleotides (i.e., the sequence) could be determined directly. In 1976, the first RNA virus genome, MS2, containing 3,569 nucleotides was completely sequenced by Walter Fiers's laboratory (University of Ghent, Belgium). A year later, Fred Sanger's group (MRC Laboratory, Cambridge, United Kingdom) sequenced the first DNA genome from a DNA virus named PhiX174, which infects bacteria. PhiX174 is a single stranded DNA virus (also known as bacteriophage, or simply, phage) that consists of 5,386 nucleotides. Once the sequence was elucidated, the locations on the virus genome could be precisely matched with those of the protein coding sequences of viral genes. This seminal achievement was made possible by pioneers who developed DNA sequencing technology, namely Sanger and the team of Walter Gilbert and Allan Maxam (Harvard University, Cambridge, Massachusetts).[1] By 1977, it was possible to identify the order of 150 to 200 nucleotides of DNA in a single experiment. This kind of experiment is known as a "read."

---

[1] In 1980, Sanger and Gilbert shared one-half of the Nobel Prize in Chemistry for their contributions to DNA sequencing technology. The other half of the Nobel Prize was awarded to Paul Berg (Stanford University) for his fundamental studies of the biochemistry of nucleic acids, with particular regard to recombinant DNA.

DNA sequencing technology improved in terms of longer reads, greater reliability, and, of great importance, significantly lower costs. By 2001, a draft of the general human genome sequence was completed, and today it is becoming increasingly affordable for individuals to have their own genomes sequenced (approximately 3×10^9 nucleotide base pairs per haploid). What will we do with this information? Can the genome sequence tell us if we are susceptible to certain diseases? Can it be used to reliably distinguish one individual from another? Can it reveal personality traits, intelligence, or athleticism? We will explore these issues in this chapter, but first let's lay out a specific research scenario to guide our exploration through this emerging field of genomics, the science of genome sequences.

Let's say you just sequenced the genome of a new microorganism that you collected from a container attached to a jet plane traveling at 30,000 feet above sea level. How do you categorize the organism? What method should you use to compare your newly sequenced genome to other genomes? What software programs do you use? How do you find the genes in your "high-altitude" genome? After you finish reading this chapter you should be able to answer these questions.

phage (also known as bacteriophage) A DNA or RNA virus that attacks bacteria.

read (also known as sequencing read) An experimentally determined sequence obtained from a single sequencing run on a segment of DNA. Reads can be 25 to 1,000 nucleotides in length depending on the sequencing method.

9.2 DNA SEQUENCING: DIDEOXY METHOD

Fred Sanger developed the dideoxy method (also known as the chain termination method or Sanger method) of sequencing, which is capable of sequencing up to ~1,000 nucleotides in one read. The dideoxy method was the dominant sequencing technique for the next 30 years of DNA analysis, and, although some aspects of the method have been replaced by newer sequencing technologies (called next-generation sequencing, or next-gen sequencing, or just NGS), many of its principles have carried over to these newer technologies. To gain insight into how genomes are sequenced it will be helpful to review the dideoxy method of DNA sequencing, followed by an introduction to next-generation sequencing.

Dideoxy Nucleotides

We begin our discussion of genomics with the technology of DNA sequencing. Dideoxy DNA sequencing is named for the set of nonnatural nucleotides called dideoxynucleotides (ddNTPs). They are chemically synthesized so that the hydroxyl group (–OH) on the 3' carbon of the sugar found in nature is replaced with a hydrogen (–H). Figure 9-1A shows a strand of natural DNA where two nucleotides are linked through the 3' carbon of the upper nucleotide and the 5' carbon of the lower nucleotide. In dideoxy DNA sequencing a special nucleotide called a dideoxynucleotide is used. The dideoxynucleotide lacks the oxygen bound to the 3' carbon and therefore cannot attach to another nucleotide at this position. Figure 9-1B shows a comparison between dideoxynucleotides (ddNTPs) and the natural deoxynucleotide dATP.

In dideoxy sequencing a DNA polymerase enzyme adds ddNTPs, along with natural nucleotides (dATP, dTTP, dGTP, dCTP), into a growing chain of DNA

dideoxy sequencing (also known as Sanger sequencing or chain termination sequencing) A method of DNA sequencing that uses a DNA polymerase to incorporate unnatural dideoxyribose-containing nucleotides (ddNTPs) to terminate DNA synthesis. The ddNTPs have fluorescent tags that allow them to be detected by a laser-fluorescence detector. This was the dominant method of DNA sequencing from the early 1980s to 2010.

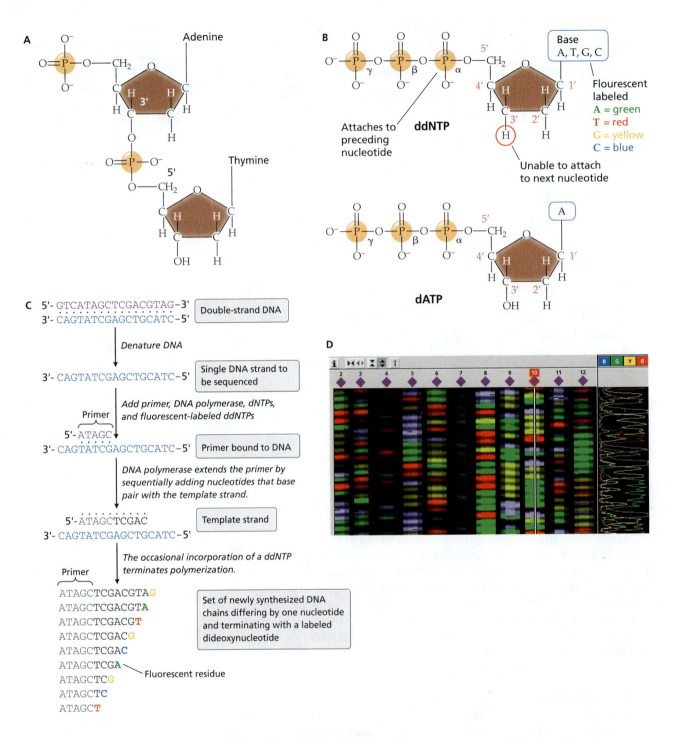

FIG. 9-1. Dideoxy sequencing.
A. Natural strand of DNA. **B.** Dideoxynucleotides and deoxyadenosine nucleotide (dATP). **C.** The dideoxy DNA sequencing procedure. **D.** The laser-produced color of DNA chains from several sequencing reactions each with a different primer. The colors of bands in lane 10 have been quantified and their densities are presented in the right.

through their α phosphoryl groups. If the polymerase incorporates a natural nucleotide into a chain the subsequent nucleotide attaches at the preceding nucleotide's 3'-OH group. Alternatively, if the polymerase incorporates ddNTP there is no 3'-OH group at the end of the DNA chain, so the chain cannot be extended with a subsequent nucleotide.

Step-by-Step Procedure of DNA Sequencing

Figure 9-1C shows a step-by-step procedure of DNA sequencing. The double strand DNA to be sequenced is first denatured to separate the two strands (either by increasing the temperature or adding an alkaline solution). A small oligonucleotide primer is added to the test tube along with the polymerase, the four natural deoxynucleotides, and four fluorescently labeled ddNTPs. The primer, chemically synthesized for just a few dollars at a commercial company, is designed so that its nucleotides are complementary to the part of the single strand of DNA that will be sequenced. The single strand DNA is called the template. The DNA polymerase begins to add natural dNTPs onto the primer as it travels along the template. The synthesized dNTPs are complementary to the template. When the polymerase adds a ddNTP instead of a dNTP, the synthesis terminates because the ddNTP has no 3'OH. The amount of ddNTP in the test tube is low enough so that ddNTPs are only occasionally incorporated into a population of growing DNA chains. This process creates DNA chains that differ in length by only one nucleotide, with the terminal nucleotide always a fluorescently labeled ddNTP.

Electrophoresis

Once DNA synthesis is stopped (by the addition of detergents that denature the polymerase enzyme that synthesizes the DNA), the DNA chains tagged with fluorescent dyes at their ends are separated by size in a gel through a process known as electrophoresis. In electrophoresis, a voltage gradient is created across the gel causing negatively charged DNA chains to travel to the positive end (anode) of the gel. The gel, which has the consistency of Jello, offers some physical resistance so that the shorter DNA chains migrate more quickly than larger chains. The sequencing gel has enough resolution to separate chains that differ by just one nucleotide. A laser light applied to the gel surface excites the fluorescent labels in the chains, and a fluorescence detector records the position, color, and intensity of the labels. The position, color, and relative intensity of each DNA chain is recorded, and a software program lists the sequence according to the position and color of the chain. Figure 9-1D shows the DNA chains separated on a gel and "read" with a laser-fluorescence detector. The "read" sequence is from the DNA that extended from the primer. The template (complementary strand) is the DNA from the organism.

Thought Question 9-1

Can you give the sequence of the DNA strand displayed in lane 10 (i.e., sequencing strand) in Figure 9-1D? What is the sequence of the natural (template) DNA strand? Do not forget to indicate the 5' and 3' ends in your answer.

9.3 POLYMERASE CHAIN REACTION (PCR)

Before we proceed to our discussion of next generation sequencing technologies, it is necessary to review a technique that exponentially increases (amplifies) the number of DNA molecules.[2] The technique, invented by Kerry Mullis (Cetus Corporation, Emeryville, California), is named polymerase chain reaction, or simply PCR.[3] The essential components of PCR are a target sequence (i.e., a sequence within the DNA

polymerase chain reaction (PCR)
A method to amplify the level of DNA in an exponential manner. The method is used extensively in molecular biology and in other fields.

---

[2] Amplification, or DNA amplification, is also sometimes used to describe a mutation event that results in the one or more DNA duplication events that occur in cells.

[3] Two scientists were awarded the Nobel Prize in Chemistry in 1993 for contributions to the developments of methods within DNA-based chemistry. One was Kerry Mullis, for his invention of the polymerase chain reaction (PCR) method. The other was Michael Smith, for his fundamental contributions to the establishment of oligonucleotide-based, site-directed mutagenesis and its development for protein studies.

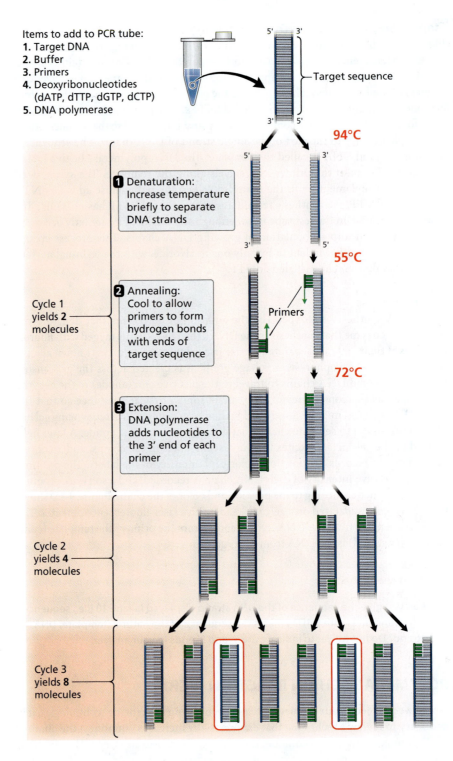

Items to add to PCR tube:
1. Target DNA
2. Buffer
3. Primers
4. Deoxyribonucleotides
   (dATP, dTTP, dGTP, dCTP)
5. DNA polymerase

Target sequence

94°C

1 Denaturation:
Increase temperature
briefly to separate
DNA strands

55°C

Cycle 1
yields **2**
molecules

2 Annealing:
Cool to allow
primers to form
hydrogen bonds
with ends of
target sequence

Primers

72°C

3 Extension:
DNA polymerase
adds nucleotides to
the 3′ end of each
primer

Cycle 2
yields **4**
molecules

Cycle 3
yields **8**
molecules

FIG. 9-2. Polymerase chain reaction steps.
At the end of the third cycle, the two DNA
molecules in white boxes are exact copies
of the target sequence. After several
cycles, the vast majority of the DNA
molecules are exact copies of the target
sequence.

molecule to be amplified), four deoxynucleotides (dATP, dTTP, dGTP, and dCTP),
oligonucleotide primers that complement the ends of the target sequence, and
heat-resistant DNA polymerase.[4] All of the contents are mixed in one tube and placed
in a thermocycler, a machine that rapidly changes the temperature of the tube.

Figure 9-2 shows what happens if you start with one DNA molecule containing
the target sequence in the tube. In our example, the temperature toggles from 94°C

---

[4] A buffered solution and Mg^{2+} ions are also necessary.

(denaturation step) to 55°C (annealing step) to 72°C (extension step), and then repeats this cycle two more times.[5] At 94°C the heat denatures the DNA containing the target sequence. As the temperature decreases to 55°C the oligonucleotide primers anneal to the ends of the target sequence in the DNA. At 72°C, the optimal temperature for DNA polymerase activity, the deoxynucleotides are synthesized into new strands. At the end of the first cycle a copy of each strand of the target sequence is produced and one target sequence has become two. To begin the second cycle, the temperature is again increased to 94°C to denature the newly formed duplex DNA strands. At the end of three cycles eight double-stranded DNAs are produced, each containing the target sequence. After 30 cycles, billions of nearly identical copies of DNA molecules containing the target sequence are produced. In order to be able to detect the fluorescent signals used in next-generation DNA sequencing, DNA amplification, typically via PCR, is usually necessary.

amplification (also known as DNA amplification) DNA copy number is increased several fold. This can occur in some parts of a chromosome in cancers. Amplification is also used to describe the process of polymerase chain reaction (PCR).

9.4 DNA SEQUENCING: NEXT-GENERATION (NEXT-GEN) SEQUENCING TECHNOLOGIES

At the time of this writing, there are few technologies that are evolving as rapidly as DNA sequencing. Much like the fast-paced development of the personal computer of the 1980s and 1990s, there are tremendous benefits to making DNA sequencing time- and resource-efficient and cost-effective. Personalized medicine will undoubtedly grow and, increasingly, will be dependent on DNA sequencing to properly diagnose diseases. Crime scene investigations will continue to require DNA sequencing to help recreate events and identify potential suspects. Identification of food sources (such as type of fish offered at restaurants) will increasingly depend on DNA sequencing. Progress in our study of ancient and prehistoric life is also linked to DNA sequencing. To meet such an intense and growing need for DNA sequencing, next-generation (next-gen) sequencing technologies are being developed.

next-gen sequencing Post-dideoxy sequencing methods that quickly sequence DNA samples. Many of these methods use the principle of sequencing by synthesizing DNA strands one nucleotide at a time in parallel.

We might ask, what are the categories of resources required to tackle next-gen sequencing? Broadly speaking, we need to think about the consumable reagents for sequencing reactions, the hardware for capturing the target DNA, the hardware for reading the signals associated with nucleotide sequencing reactions, and the software for aligning the sequence reads. Another category is human resources. There is a great need for skilled bioinformaticians to properly interpret sequencing data. Understanding the nuts and bolts of next-gen sequencing is essential for bioinformaticians to make significant contributions in this hot area. One thing more before we explore this topic—the prominent technologies described here may be replaced by new technologies in the near future. As much as possible, we will emphasize the principles of the technologies that will likely carry over to the "next-next-gen" sequencing technologies.

Common Themes in Next-Gen Sequencing Technologies

Most next-gen sequencing technologies have the following in common:

1. *Library formation.* The DNA to be sequenced is randomly fragmented and adapters (oligonucleotides of known sequence) are ligated to the ends. The adaptors will be used to hybridize to other oligonucleotides.

2. *Library amplification.* Fragmented DNA is captured on a solid surface such as a bead or glass surface such that single strand DNA molecules are spatially separated.

---

[5] These temperatures and the periods of time held at each temperature can vary depending on the nature of the experiment. These parameters vary depending on the percent G+C in the target sequence, the base composition of the primers, the length of the target sequence, and the type of heat-resistant DNA polymerase.

The captured DNA molecules are amplified (see Figure 9-2) to increase the signal level emitted during the sequencing reaction.

3. *Sequencing reaction.* There is a direct detection of each nucleotide base as it is incorporated into an extending DNA strand. The extending DNA strand is complementary to the amplified DNA.

4. *Massively parallel sequencing.* Hundreds of thousands to hundreds of millions of sequencing reactions are recorded during a sequence read.

5. *Relatively short read lengths.* Compared to dideoxy sequencing reactions followed by electrophoresis, the lengths of the sequence reads are usually short.

6. *Digital base calling.* The signals captured from the reads are quantitative, so that numbers of DNA molecules can be compared.

7. *Paired end reads.* Both ends of every amplified DNA fragment are sequenced. If the fragment is long and the middle part cannot be sequenced, the paired end reads allow the fragment to be aligned to a reference genome that has been previously sequenced.

One strong advantage of next-gen sequencing over traditional dideoxy sequencing is that it does not require knowledge of any part of the template sequence. Oligonucleotide adaptors attached to the ends of DNA provide the necessary complementary sequence for hybridization to the sequencing primers.

Ion Semiconductor Sequencing

Ion semiconductor sequencing is a method to detect incorporating nucleotides with an ion-sensing semiconductor. When a nucleotide is incorporated into a growing DNA strand it releases a pyrophosphate and a hydrogen ion (proton) as products (Figure 9-3). The hydrogen ion is released from the –OH group (bordered in red in this figure) at each instance of nucleotide incorporation. Hydrogen ion is detected in the solution by an ion sensor. The company Ion Torrent is currently the leader in ion semiconductor sequencing. Hydrogen ion detection is predicated on a similar older technology called pyrosequencing, which detects the pyrophosphate product produced by nucleotide insertion. The level of pyrophosphate is measured by enzyme-linked production of luciferin oxidation that emits light. Pyrosequencing was developed by Mostafa Ronaghi and Pål Nyrén (Royal Institute of Technology, Stockholm)[6] and was commercialized by 454 Life Sciences. Let's go through the steps of ion semiconductor sequencing technology (Figure 9-4).

STEP 1. DNA is fragmented to appropriate lengths for subsequent steps.

STEP 2. The ends of the fragments are enzymatically modified so that each end has a different adaptor ligated to it.

STEP 3. DNA molecules are denatured so that they are single stranded. Single stranded DNA molecules with adaptor "A" at one end and adaptor "B" at the other end are selected from other single stranded DNA molecules that failed to attach to adaptors.

STEP 4. Single stranded molecules are incubated with 28 micron diameter beads such that only one strand (a single molecule) is associated with each bead. Each bead attaches to one strand by hybridization to the adaptor. The beads are suspended in an emulsion such that each bead with its attached strand forms an aqueous microenvironment in a sea of oil. Each bead together with its aqueous microenvironment is called a microreactor. Each microreactor contains DNA

---

[6] Ronaghi, Karamohamed, Pettersson, Uhlén, and Nyrén (1996).

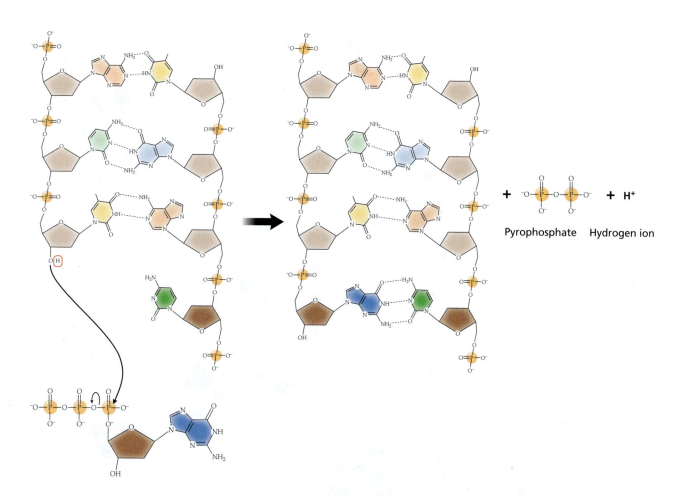

FIG. 9-3. Incorporation of a nucleotide into DNA strand during synthesis releases pyrophosphate and hydrogen ion.

polymerase, primers that bind to adaptors, and nucleotides necessary for DNA amplification through PCR.

STEP 5. Hundreds of thousands of microreactors in a single suspension are placed in a thermocycler, and single strand DNA molecules are amplified. This step is called emulsion PCR (emPCR).

STEP 6. The oil is separated from the microreactors.

STEP 7. The beads attached to amplified DNA are dispersed onto plates. The plates have small wells, each tailored to fit only one bead. The plates are centrifuged to settle the beads, and the primers are hybridized to the single stranded amplified DNA. Each well is connected to a semiconductor that can detect ion production in the form of a voltage change.

STEP 8. Nucleotides are added one at a time together with DNA polymerase. If the nucleotide incorporates into the DNA chain, the hydrogen ion is released and quantified by the semiconducting surface in the well. After each reaction the nucleotide/polymerase mixture is removed and a different nucleotide/polymerase is added for the next reaction. After four different nucleotide/polymerase rounds (i.e., a round for each A, T, C, and G), the process is repeated. Figure 9-4 (last step) shows that A is the first nucleotide in the

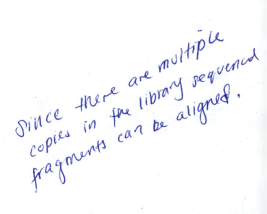

Nanol
Microfluidic
Droplet generation

Since there are multiple copies in the library sequenced fragments can be aligned.

ADAPTERS / PRIMERS

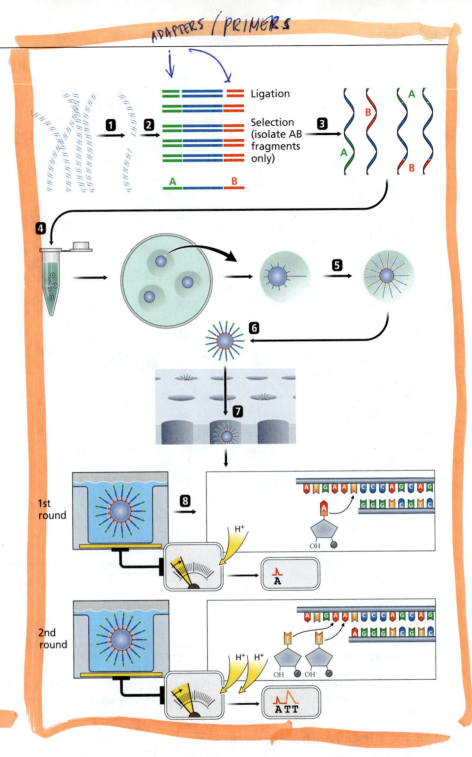

FIG. 9-4. Next-generation sequencing. Steps to ion semiconductor sequencing.

sequence (after the primer has bound) in one particular well. After flushing out A, T is added. T incorporation results in a twofold higher signal in the output because two Ts are incorporated. Of course, sequencing is done in parallel, so there are $10^5 - 10^7$ beads being sequenced simultaneously.

Every technology has inherent errors. The main source of error with ion semiconductor sequencing occurs when runs of a single base type are encountered (homopolymers). This gives the possibility of indel errors because it is difficult to accurately correlate large signals with the proper number of nucleotides. For example, if there is a run of seven A's, the sequencer may give an output of six A's or eight A's. In one study, the indel error rate was reported as 2.84%.[7] Other errors such as base substitutions appear to be quite low.

---

[7] Bragg, Stone, Butler, Hugenholtz, and Tyson (2013).

There are several applications for this DNA sequencing technology. One is to sequence genomes; another is to quantify RNA transcripts. RNA transcripts can be reverse transcribed into cDNA, fragmented, and sequenced as described above. We call this procedure RNA-seq (for RNA sequencing), and its application has significantly increased our knowledge of the transcriptome (see Chapter 10). We call ion semiconductor sequencing a "sequencing-by-synthesis" technology because at each synthesis step the identification and quantity of nucleotides incorporated are recorded.

Another next-gen sequencer that captures DNA sequences in a sequencing-by-synthesis fashion has been developed by Illumina. In the Illumina system, four nucleotides, each with a different fluorescent label, are added in a step-by-step fashion as the DNA polymerase synthesizes the DNA strand. A laser causes the label to fluoresce, and, depending on the wavelength of the emitted light, the incorporated nucleotide is identified. In Chapter 10 we give the details of how the Illumina system operates in RNA-seq, but it can be used to sequence DNA directly as well.

Nanopore-Based Sequencing

Nanopore-based sequencing was developed by Oxford Nanopore Technologies to sequence DNA without an amplification step. As mentioned earlier, amplification is typically required to increase the signal to the point where it can be detected. Nanopore-based sequencing takes advantage of two phenomena: (1) ion flow across single protein pores (channels) can be quantified, and (2) helicases can separate strands of DNA. At the time of this writing, nanopore-based sequencing is still in the developmental stage.

Figure 9-5A is a diagram showing a protein pore embedded in a membrane. The protein pore is created by a modified form of α-hemolysin-a membrane protein with a pore diameter of 1 nanometer. This diameter is approximately the width necessary to allow single strand DNA molecules, as well as smaller ions, to pass through. If smaller ions flow through the pore, the current stays constant. If the pore is impeded by a larger molecule, the current is decreased until the larger molecule goes through the pore. The molecule shape dictates the degree to which the current decreases. Nanopore-based sequencing can distinguish A, T, C, and G nucleotides in a DNA molecule by monitoring the current profile. A helicase enzyme is attached to the end of double strand DNA (Figure 9-5B). The helicase-DNA complex binds to the top of the protein pore and separates the DNA strands. One strand threads through the protein pore as ion current is monitored (Figure 9-5C). The ion current profile is used to sequence the single strand of DNA. A software program utilizes a hidden Markov model to translate the current profile into a sequence.[8]

We have reviewed just a few of the many next-gen sequencing technologies available. In general, there is an effort to link sequencing to technology developments in electrical engineering so that sequence information is easily converted into digital output. Now that we have discussed a bit about next-gen sequencing, we will turn our attention to genome analysis.

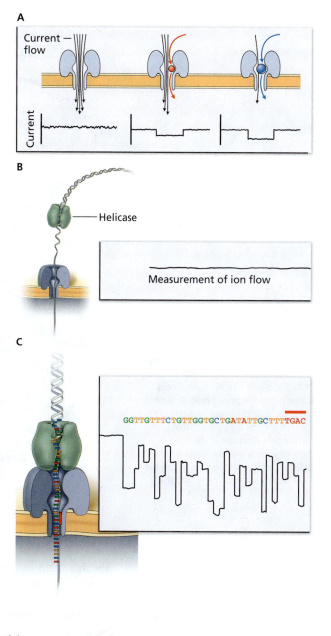

FIG. 9-5. Nanopore-based sequencing.

---

[8] See Chapter 12 for a detailed discussion of the structure of hidden Markov models.

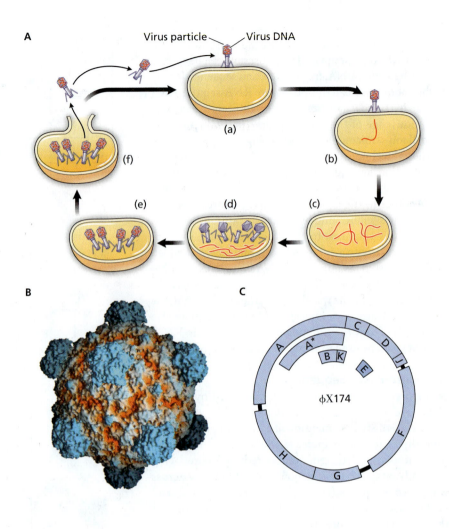

FIG. 9-6. The PhiX174 phage.
A. Phage life cycle. **B.** The PhiX174 phage coat is shown (from PDF file number 2BPA). The single stranded genome is located inside the phage coat. **C.** Genetic map of the PhiX174 phage genome. The segment lengths are proportional to the number of nucleotides in the genes. (The Greek letter φ is phi. Many early published papers on this subject use the Greek letter instead of the word "phi.")

origin of replication The location within the genome or plasmid where DNA replication starts.

9.5 THE PHIX174 BACTERIOPHAGE GENOME

The first DNA genome sequenced was that of a virus, a phage to be specific, which consists of a protein coat (named a virus capsid or phage coat) and a genome inside the coat.[9] Figure 9-6A shows, in broad outline, how a phage infects bacteria. The phage binds to the outside of the bacterium (a) and "injects" its genome into the cell, leaving its protein coat attached to the bacterium wall (b). Phage DNA utilizes bacterial host proteins to produce many copies of the phage DNA (c), and to express phage proteins (d). If the bacterium fails to destroy the phage DNA, thousands of progeny phages assemble inside the bacterium (e) and, eventually, create holes in the bacterium cell wall (f). The released phages then go on to infect other bacteria, leaving the dead host bacterium behind.

Let's explore the genome of the PhiX174 phage (accession number NC_001422) a bit further. It is a single stranded circular DNA that contains 11 genes. The genes are named after letters in the alphabet and have the following order in the genome: *A-A∗-B-K-C-D-E-J-F-G-H*. Most of the gene functions are well characterized. When the single stranded circular DNA of the phage is inside the cell, the bacterium unwittingly creates a complementary strand, which results in a double strand DNA (dsDNA). The origin of replication of dsDNA is located at approximately 10 o'clock on the phage genome map (Figure 9-6C) and is the starting point for the creation of many copies of the dsDNA. The dsDNA copies are then separated into single strands,

---

[9] There are RNA bacteriophages as well.

and phage assembly begins. One of the two separated DNA strands is the coding strand, and it is selected and packaged into the phage coat.

Each gene within the genome has a specific role. Genes *A* and *C* are required for separating dsDNA into single strand DNAs inside the bacterium. Genes *B*, *D*, and *J* are required for phage assembly. Genes *F*, *G*, and *H* code for proteins that constitute the phage coat. Gene *E* is necessary for lysing the host bacterium to allow phages to leave the bacterium. The protein expressed from gene *A*∗ is a truncated version of the protein expressed from gene *A*. The N-terminus of protein A∗ is coded from a start codon (ATG) that is internal to gene *A*. Protein A∗ is suspected to inhibit bacterial DNA replication, thereby allowing the virus to usurp the bacterium's replication machinery to copy the double strand phage DNA. The *K* gene appears to dictate the number of phages produced per bacterium.

When the PhiX174 genome was sequenced, the amino acid sequences of some phage proteins were already *known*. This was important because the amino acid sequences of these proteins (and the genetic code) helped confirm the DNA sequencing results. The genome sequence also revealed that the origin of replication is rich in A and T nucleotides, which makes sense because dsDNA dissociates more easily at AT-rich sequences than at GC-rich sequences. This is one reason why most origins of replications are AT-rich.

The phage genome, together with data from previous work in the field, was used to locate some of the gene promoters and almost all of the codons that mark the start and end of genes. Importantly, the genome sequence showed that there are overlapping genes—now known to be a common property of viruses because of selection pressures to maintain small genomes (overlapping genes are rarely seen in eukaryotic genomes). As we have seen above for protein A and protein A∗, two genes can overlap if one of the protein products is a truncated version of the other. Two genes can also overlap if they use different reading frames for protein production. In PhiX174, genes *B*, *K*, and *E* overlap other genes by utilizing different reading frames (see Figure 9-6C). In double strand genomes in other organisms, two genes may overlap when the coding sequence for one gene resides on one strand and the coding sequence of the overlapping gene resides on the complementary strand.

9.6 THE GENOME OF *HAEMOPHILUS INFLUENZAE* Rd AND THE WHOLE GENOME SHOTGUN SEQUENCING APPROACH

Viruses are a good place to start studying genomes. Their genomes are small, and, in the case of PhiX174, prior knowledge of some of the viral protein sequences could be used to confirm the DNA sequences through back translation.[10] As the cost of DNA sequencing dropped, scientists turned their attention to larger genomes: those from organisms that contain all the genes required to divide and live independently. Arguably, the simplest class of these "free-living" organisms is bacteria.[11] In 1995, Craig Venter and his colleagues at the Institute for Genomic Research (Gaithersburg, Maryland) reported the sequence of the *Haemophilus influenzae* Rd.[12] Like all bacteria, the *H. influenzae* genome is composed of double strand DNA. Although most *H. influenzae* bacteria are able to infect humans, the Rd serotype is a laboratory strain that is unable to infect humans. There are at least six other serotypes of *H. influenzae* (strains a–f). Of these, type b strains are the most harmful, sometimes causing

---

[10] Protein sequences were back-translated into potential codons. The DNA sequences could be checked with the potential codons to lower the possibility of errors.

[11] Some might contend that archaea are the simplest free-living organisms. Archaea, bacteria, and eukaryotes constitute the three major domains of life (see Chapter 8).

[12] This bacterial strain should not to be confused with influenza virus, which is an RNA virus that can cause the flu.

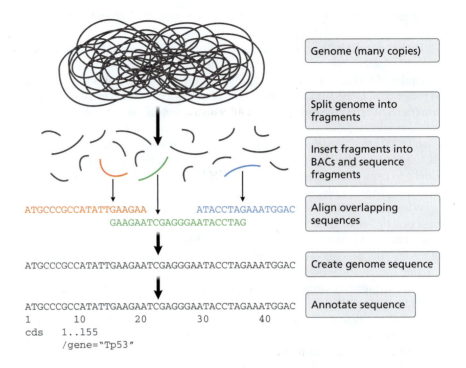

	Genome (many copies)
	Split genome into fragments
	Insert fragments into BACs and sequence fragments
	Align overlapping sequences
	Create genome sequence
	Annotate sequence

```
ATGCCCGCCATATTGAAGAA            ATACCTAGAAATGGAC
          GAAGAATCGAGGGAATACCTAG

ATGCCCGCCATATTGAAGAATCGAGGGAATACCTAGAAATGGAC

ATGCCCGCCATATTGAAGAATCGAGGGAATACCTAGAAATGGAC
1       10        20        30        40
cds     1..155
        /gene="Tp53"
```

FIG. 9-7. The whole genome shotgun sequencing strategy.

meningitis in children. Venter's group decided to work with the innocuous Rd strain to safely and quickly develop a general strategy for sequencing large genomes. The *H. influenzae* Rd genome is 1,830,137 base pairs, approximately 3,500 times larger than the PhiX174 genome. Let's begin with the strategy of how to sequence such a large genome.

Whole Genome Shotgun Approach

whole genome shotgun sequencing (WGSS) A sequencing method that includes random fragmentation of many copies of a genome. The fragments are sequenced and assembled to produce the entire genome sequence.

Venter used whole genome shotgun sequencing (WGSS), which is the same strategy that Sanger used to sequence the phage DNA but is much harder to execute on large genomes due to the complexity of assembling the sequenced DNA fragments. WGSS takes its name from the way a shotgun fires fine pellets (called "shot") in a spray that strikes a target in many random locations. Similarly, in WGSS the genome (actually, many copies of the genome) is randomly fragmented into shorter pieces of DNA (Figure 9-7). This random fragmentation process is often carried out by a machine, called a sonicator, which produces ultrasonic waves. The short DNA pieces are placed into bacterial artificial chromosomes (BACs). BACs are DNAs of known sequence capable of accepting DNA fragments of 150 to 350 kilobases. The BACs are stored in bacteria such that there is only one BAC clone per bacterium. As bacteria replicate they also make copies of the BAC, which is necessary for the sequencing to be successful because the high amount of DNA is required.

Once the clones are grown and the BAC DNA is isolated, the DNA is sequenced using the dideoxy DNA sequencing method (WGSS was developed before next-gen sequencing). By computer, the end of the sequence of one fragment is aligned with the beginning of the sequence of another fragment. This process continues until the entire genome sequence is assembled. The final genome sequence is then annotated with computer software programs to describe the features of different regions of the genome (see Chapter 2 for examples of sequence annotations). This description is a simplified version of the WGSS strategy Venter's group used to sequence and characterize the *H. influenzae* Rd genome.

One of the challenges with the WGSS approach is the "blind" sequencing step. Because there is no way to know which BAC contains which DNA fragment, it is possible that the same fragment will be sequenced repeatedly. Recall that many copies of the genome are used to generate DNA fragments, so to ensure that the entire genome is sequenced at least once, many DNA fragments must be sequenced. One can estimate the amount of sequencing required to ensure that almost the entire genome has been sequenced. Eric Lander (Massachusetts Institute of Technology, Cambridge) and Michael Waterman (University of Southern California, Los Angeles) showed that one can estimate the probability, P_0, that a nucleotide is *not* sequenced in a genome by assuming a Poisson probability distribution function:[13]

$$P_0 = e^{-c} \tag{9.1}$$

where c is the fold coverage of the genome, which is defined as

$$c = LN/G \tag{9.2}$$

where L is the length of DNA fragment sequenced,

where N is the total number of DNA fragments of length L,

and where G is the target sequence length (in our case, the length of the genome).

We can use equations 9.1 and 9.2 to calculate the number of nucleotides from the *H. influenzae* Rd genome that must be sequenced in order to cover 99.99% of the genome.

First we rearrange the terms of equation 9.1 to isolate the term c to one side of the equation,

$$c = -lnP_0$$

Next, if we want only a 0.01% chance of not sequencing a nucleotide in the genome, we want

$$P_0 = 1 - .9999 = 0.0001$$

The fold coverage $c = -ln\, 0.0001 = 9.2$

To obtain the number of nucleotides that must be sequenced to obtain 99.99% of the sequence we multiply the fold coverage by the number of nucleotides in the genome. For the *H. influenzae* Rd genome we need to sequence 9.2 × 1,830,137 nucleotides (or 16,837,260 nucleotides) to be sure that 99.99% of the *H. influenzae* Rd genome is sequenced at least once.

Although the sequencing step is long and laborious, the major challenge is assembling the fragments in the correct order. Assembly requires the use of software programs that correctly match the ends of the fragments. When the sequence has high complexity (a high level of variability) this is not difficult. However, when the sequence has low complexity (a low level of variability), assembly becomes more difficult because it is hard to know where an overlap begins and ends. Venter reported that the *H. influenzae* genome was sequenced with approximately sixfold coverage, meaning that 99.75% of the genome was sequenced. To confirm that the sequence was correct, the Venter group used sequence comparisons with some *H. influenzae* genes already deposited in GenBank. Perhaps not surprisingly, some sequence differences were noted. It is possible that the differences are due to sequencing errors by the Venter group, but it is also possible that the differences are due to sequencing errors by the original depositors of the *H. influenza*, or due to bacterial strain differences. Now it is time to take a closer look at what was found in the *H. influenzae* genome.

---

[13] Poisson distribution probability function is discussed further in Chapter 11.

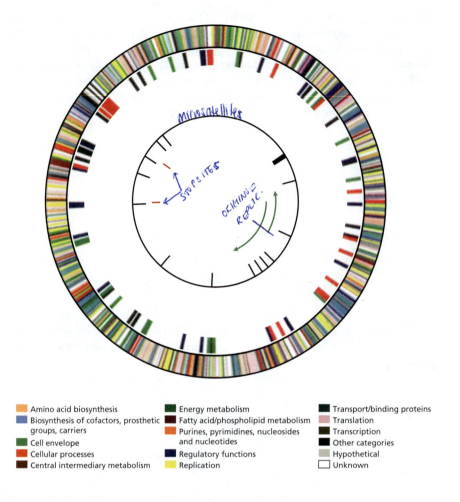

FIG. 9-8. The *H. influenzae* genome map. Outer ring length corresponds to 1,830,137 nucleotides, of which approximately 1.6×10^6 nucleotides (84%) code for protein or functional RNAs. Each colored segment in the outer ring corresponds to a gene. The average gene length is 900 nucleotides. The key of gene categories is shown below the genome map.

■	Amino acid biosynthesis	■	Energy metabolism	■	Transport/binding proteins
■	Biosynthesis of cofactors, prosthetic groups, carriers	■	Fatty acid/phospholipid metabolism	■	Translation
■	Cell envelope	■	Purines, pyrimidines, nucleosides and nucleotides	■	Transcription
■	Cellular processes	■	Regulatory functions	■	Other categories
■	Central intermediary metabolism	■	Replication	■	Hypothetical
				□	Unknown

Thought Question 9-2

How many nucleotides did Venter's group sequence in order to ensure that 99.75% of the *H. influenzae* genome was sequenced?

Haemophilus influenzae Rd Genome

Like the PhiX176 genome, the *H. influenzae* genome is circular. Figure 9-8 shows a map of the *H. influenzae* genome as four concentric rings with each ring depicting a different characteristic of the genome. The outer circle (the first and largest ring) shows the locations of the genes within the genome. The circumference of the outer ring corresponds to the genome length—1,830,137 nucleotides—and the segments within this ring correspond to genes segregated into 14 color-coded categories (see key at bottom of figure).[14] For example, genes involved in energy metabolism are dark

[14] The putative proteins are also more finely segregated into 102 biological role categories. These will not be discussed here. Searching for genes requires the use of software programs that search for, in the case of proteins, regions that contain open reading frames and upstream sequences that bind to ribosomes. Venter's group used GeneMark—a software program that uses a Markov chain model and Bayes-based algorithm (see Chapter 11 for the theoretical basis of Bayes' theorem and Chapter 12 for the theoretical basis of the Markov chain model). GeneMark was "trained" to detect *H. influenzae* sequences using the limited number of coding and noncoding sequences previously deposited into GenBank. Preliminary work showed that GeneMark was accurate in more than 91% of the coding sequences tested (again, from previously sequenced DNA). Predicted genes were used as queries in a search of a database, called the non-redundant bacterial protein database (NRBP). NRBP contains bacterial sequences from GenBank and Swiss-Prot (now renamed UniProt) that were culled to remove redundant sequences. To perform the search, the *H. influenzae* predicted genes were translated from all reading frames into protein sequences with the BLAZE program. The translated sequences

green segments. The total number of predicted genes is 1,743 with an average length of 900 nucleotides, which means that approximately 84% of the genome codes for proteins and functional RNAs.[15] Of the 1,743 predicted genes, 1,007 match genes with known functions and the vast majority of the predicted genes code for proteins.

Of the matched proteins, most are involved in translating mRNA into protein (14%, color-coded pink), but there are many other categories to which the proteins belong. The remaining proteins either match hypothetical proteins with no known function (347), or have no match with any protein in GenBank or UniProtKB (389).

The overall percentage of G+C in the *H. influenzae* genome is 38%. The Venter group chose this genome, in part, because the percent G+C is close to that of the human genome (41%) and the ultimate goal was to sequence the human genome. Areas particularly enriched in G+C and A+T were measured with a software program that captures a 5,000-nucleotide-length window and slides the window in one nucleotide increments around the genome (see Chapters 5 and 13 for discussions of sliding window programs). The sliding window program was used to create the next smallest ring (the second ring) in Figure 9-8 that shows areas that are relatively rich in G+C content (>42% red, >40% blue) and areas relatively rich in A+T content (>66% black and >64% green). One large G+C-rich area is located at approximately 10 o'clock (red area) on the genome. This region codes for several proteins that are similar to proteins coded by a phage called mu and suggests that, at one time, mu's genome (52% G+C) integrated into *H. influenza*'s genome. This is an example of lateral transfer of genes, which is common in microorganisms (see Chapter 3). The third ring in Figure 9-8 shows the locations of repeating sequences (known as repeats). The repeats in this genome are CTGGCT, GTCT, ATT, AATGGC, TTGA, TTGG, TTTA, TTATC, TGAC, TCGTC, AACC, TTGC, CAAT, and CCAA. All free-living organisms have repeats in their genomes.

To locate the putative origin of replication, a sequence comparison was made between the *H. influenzae* genome and the known origin of replication in *E. coli*. Sequence similarity analysis suggests that the *H. influenzae* origin is likely to be located at approximately 4 o'clock on the genome. This putative origin is shown in the smallest ring (fourth ring in Figure 9-8) between the bases of two curved green arrows. The green arrows represent the direction of movement of replication forks away from the origin. Replication termination is likely to occur at one of two possible regions denoted by the red bars—one located near 9 o'clock and another near 10 o'clock.

The *H. influenzae* Rd genome reveals clues as to why it is *not* infectious. In the pathogenic serotype b strain (the one associated with meningitis), eight genes are required for adhesion of bacteria to the human cells. These genes are absent from the *H. influenzae* Rd genome, suggesting that *H. influenzae* Rd is unable to gain entry into human cells and this may be the reason it is not infectious.

It is also interesting to analyze genes that are unique to *H. influenzae* Rd (i.e., they do not match to any other sequence in GenBank). These genes were identified because they have open reading frames and ribosomal binding sequences upstream of potential translation start sites. Open reading frames have a start codon and a stop codon and a run of codons that can, in principle, create a polypeptide. Interestingly, some of these unique putative genes are related to each other within the Rd genome and thereby comprise a gene family. One family codes for two putative proteins that share 75% identity, and another family codes for two putative proteins that share 30% identity. Venter also identified putative proteins that appear to be membrane proteins

---

matched to the NRBP with PRAZE, a software program that employs a modified Smith-Waterman algorithm.

[15] Functional RNAs are transfer RNAs and ribosomal RNAs. This list will likely be expanded soon because we are discovering that there are many other RNAs with important functions.

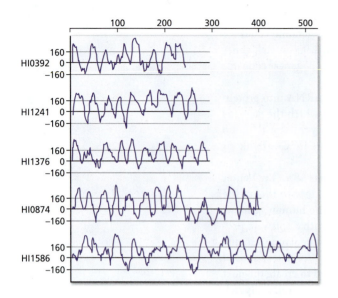

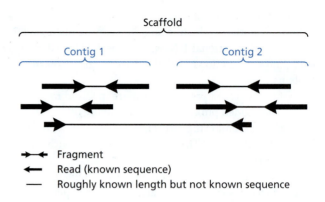

FIG. 9-10. Contigs and scaffolds. Contigs contain sequence information. Scaffolds contain the order of contigs and the size of the gap between the contigs. Image from http://genome.jgi-psf.org/help/scaffolds.html (accessed March 21, 2013).

FIG. 9-9. Hydropathy plots of five putative proteins of unknown function from the *H. influenzae* genome.

but have no sequence similarity with proteins in GenBank. Recall from Chapter 5 that the Kyte-Doolittle software program, also called the hydropathy program, is useful in predicting membrane-spanning regions of membrane proteins. Venter used the hydropathy program to predict the number of membrane-spanning regions in these proteins. Figure 9-9 shows a hydropathy analysis of five hypothetical membrane proteins from *H. influenzae* Rd.

Sequencing the *H. influenzae* Rd genome was a major accomplishment. It showed that WGSS is a valid approach to sequencing large genomes, and the lessons learned helped pave the way to use WGSS to sequence even larger genomes.

9.7 GENOME ASSEMBLY AND ANNOTATION

open reading frame A segment of DNA that has the potential to code for a protein because it begins with a start codon (codes for Met) and there are no stop codons for a relatively long stretch of DNA.

assembly (also known as genome assembly) Ordering of sequenced DNA segments.

Genome assembly is the process of taking short DNA sequences and ordering them so that they create the genome sequence. A genome assembly is composed of contigs and scaffolds. A contig (short for contiguous sequence) is a set of overlapping DNA sequences reads (recall that a read is an experimentally derived sequence of 1,000 nucleotides or less) collapsed into a consensus sequence (one sequence). Scaffolds are regions of genomes that contain correctly ordered and correctly oriented linked contigs (Figure 9-10). Scaffolds may contain gaps in sequence information, but are accurate in terms of the lengths of the DNA regions. A standard draft genome assembly consists of scaffolds that in aggregate is less than 90% complete in sequence and a high-quality draft genome assembly is at least 90% complete. Because it takes many sequencing runs to obtain a high-quality draft assembly, you must decide at what percent of genome completeness to begin to annotate the genome. The current measures for describing the quality of the genome assembly are contig N50 and scaffold N50, described below.

Contig N50 and Scaffold N50

Contig N50 is the number of contigs it takes to cover 50% of the genome. The 50% cutoff value is determined as follows: (1) order contigs of the genome by length starting from the longest, (2) sum the lengths of all contigs in the assembly, (3) take one-half of this sum and set as the cutoff value. The contig N50 is the number of contigs required to meet the cutoff value. The scaffold N50 is determined in the same fashion using scaffold lengths instead of contig lengths. If the scaffold N50 length is equal to or greater than the size of the median gene (including introns if this is a eukaryotic genome) then the genome may be annotated. Median gene lengths are roughly proportional to genome size. If we know the genome size then we can estimate the gene length.

BOX 9-1 | **SCIENTIST SPOTLIGHT**

J. Craig Venter

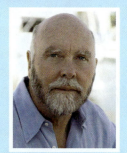

J. CRAIG VENTER was one of the first scientists to sequence large genomes including the human genome, but he came to the genomics field by a somewhat circuitous route. He spent his early childhood in Millbrae, just south of San Francisco, and attended Mills High School. By his own account he hated tests and rote memorization and did poorly in academics. After barely graduating from high school Venter took on a series of occupations: night clerk, truck driver, baggage handler. He moved to Southern California and spent many hours surfing. His lifestyle was interrupted when he received a military draft notice. He joined the navy medic corps and soon found himself near the front lines in Da Nang, Vietnam, during the Vietnam War. While serving, due to loneliness and the stress of treating injured and dying soldiers, he attempted to commit suicide by swimming out to sea. More than a mile offshore, he had a sudden change of heart and swam back, just barely getting back alive. He was exhausted but gained a determination to build a productive life. After he completed his military duty he attended San Mateo Community College for two years and continued his undergraduate studies at the University of California, San Diego.

While a student in San Diego, he conducted research on the effects of adrenaline and other catecholamines on cultured heart cells in the laboratory of Nathan O. Kaplan. Subsequent to earning his B.S. in biochemistry, he stayed on at UC San Diego and earned his Ph.D. in physiology and pharmacology—publishing several papers on the effects of small molecules on cultured cells. In 1975, Venter joined the faculty at the State University of New York at Buffalo. After nine years at Buffalo he joined the National Institutes of Health (NIH) in Bethesda, Maryland. At the NIH he adopted, with much effort, the relatively new technique of dideoxy DNA sequencing and published the sequence of the human adrenaline receptor. He then developed a procedure to rapidly partially sequence thousands of mRNAs (actually, he sequenced the DNA copies of the mRNAs called cDNAs). Up until then there had been only approximately 2,000 human genes sequenced. Venter's group reached the point where they could partially sequence 20 to 60 new human genes a day! These partial sequences are known as expressed sequence tags (ESTs) because the cDNAs are derived from genes that are expressed and the cDNAs contain a segment known as the 3′ untranslated region (3′ UTRs). These 3′ UTRs are located just downstream of the protein termination codon and constitute a unique tag for every gene.

Upon establishing his group as capable of efficiently sequencing genes, he proposed whole genome shotgun sequencing as a method to sequence large genomes but had trouble securing public funding for this massive project. He turned to the private sector for funding and crafted a business plan that created The Institute for Genome Research (TIGR)—a not-for-profit research institute that was funded by the for-profit company Human Genome Sciences. He successively headed other genome sequencing organizations—Celera and the J. Craig Venter Institute (JCVI). At Celera his team published the sequence of the human genome, and later, at JCVI, his team published the first diploid sequence of the human genome (his own). A diploid sequence is the sequence of both chromosomes one receives from parents—one from the father and one from the mother. His autobiography, *A Life Decoded* (2007) is unique in some respects. As one would expect, he describes his life and how

continued

continued

he became interested in science and genomics. Interestingly, in the book he also analyzes his own genes and offers suggestions as to how polymorphisms and mutations in his genes may influence his personality traits and his medical issues. In the future, such a genome analysis may become an accepted practice for all people.

REFERENCE

Venter, J. C. 2007. *A Life Decoded*. New York: Penguin Group.

synteny The similar arrangement of genes in the genomes of two species that share a common ancestor.

Bacterial Genome Annotation Systems

Once the contig N50 and scaffold N50 criteria have been met, we are ready to annotate our genome. We will use a bacterial genome assembly as an example. Because there are thousands of sequenced bacterial genomes, there is a large database of bacterial genome sequences to compare to newly sequenced genomes. This large database has aided the process of annotating new bacterial genomes, and there are several online software programs available for this purpose. We will present a general view of how these annotation systems operate, but it is important to bear in mind that specific programs will use variations of this general approach to genome sequence annotation.

First, we upload FASTA sequences from contigs to search for conserved genes. Conserved genes such as those that code for basic metabolic enzymes do not change much from species to species, so they can be analyzed for codon usage patterns. The patterns will help the program identify genes that are specific for the bacterial genome we are annotating. Second, the software program analyzes the contig DNA sequences for RNA polymerase binding sites used to begin transcription of functional RNA coding sequences. The software program also analyzes the contig DNA sequences for ribosome binding sites within protein coding genes. Repeats are annotated and predicted protein sequences generated. Third, predicted protein sequences are used to query the Uniprot database using the BLASTp software program. When hits are obtained, protein feature information is retrieved from the Uniprot database to annotate the bacterial genome. Fourth, enzymes are classified by comparing the predicted protein products of the bacterial genome to a database of enzymes with associated position-specific scoring matrices (PSSMs) tailored for each enzyme. Fifth, regions containing a series of genes in close proximity to each other are compared to other bacterial genomes to search for regions of synteny. Synteny is defined as the similar arrangement of genes in the genomes of two species that share a common ancestor. If regions of synteny are detected, then previously unassigned genes are assigned annotations based on their relative positions to genes with assigned function.

9.8 GENOME COMPARISONS

Synteny Dot Plot

Now that we have grounding in sequencing and an example of a sequenced and annotated bacterial genome, it is appropriate to compare two bacterial genomes. One way to compare genomes is to create dot plots—similar to the dot plots that compare individual proteins sequences (see Chapter 5). In Figure 9-11 we show a comparison of two substrain bacterial species of *E. coli* K-12, which inhabit the human intestine, in a synteny dot plot. Each green dot represents a region of DNA that exhibits synteny between the two genomes. To generate the synteny dot plot the following steps are followed:

1. All protein coding regions are extracted from each genome.

2. The sequences are BLASTed against each other to identify similar gene pairs.

3. Similar gene pairs are analyzed to determine if they share a collinear order in their respective genomes. In other words, let's say that a, b, and c are three genes in tandem in one genome, and A, B, and C are three similar genes in tandem in a second genome. Then a, b, and c and A, B, and C share a collinear order in the two genomes.

4. If similar gene pairs share a collinear order, they are colored green. If the order is not collinear, the similar gene pairs are colored gray.

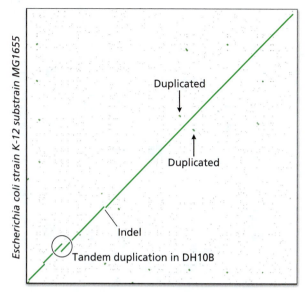

FIG. 9-11. Synteny dot plot comparing two substrains of bacteria *E. coli* strain K-12.

Comparison of E. coli Substrain DH10B to E. coli Substrain MG1655

In Figure 9-11, the *x*-axis of the dot plot is the nucleotide sequence of one genome (substrain DH10B) and the *y*-axis represents the nucleotide sequence of the second genome (substrain MG1655). The origins of replication in both genomes are located at the intersection of the two axes. For the most part, the two substrains share significant synteny as indicated by the green diagonal in the dot plot. A close inspection of the synteny dot plot shows that some segments of synteny are duplicated. If both genomes share a duplicated segment one will detect green dots equally distant from the diagonal. When the duplicated segments are distant from each other the green dots are farther away from the green diagonal. Areas in the genomes where there are large indels—insertions or deletions of nucleotides in one genome relative to the other—are depicted as breaks in the diagonal. If one genome has a set of repeated genes relative to the other genome, it is called tandem duplication. This will create a break at the location of the tandem duplication and a partial overlap of synteny regions. This type of genome analysis only compares genes and will not show areas of the genome that do not code for proteins. However, one should bear in mind that in bacteria, protein coding regions constitute at least 80% of the genome so much of the genome is represented in this synteny dot plot. The two *E. coli* genomes show extensive synteny suggesting that they recently evolved from a common ancestor. From molecular clock and other phylogenetic methods (see Chapter 8), it is estimated that the two substrains evolved from a common ancestor less than five million years ago.

9.9 THE HUMAN GENOME

We are now in a position to explore the human genome. A draft of the human genome was published in 2001 by two groups—one from the Human Genome Consortium, an international collaborative effort funded by the public largely through taxes and headed by Francis Collins (National Institutes of Health, Bethesda, Maryland); and the other from Celera Genomics Corporation, a company headed by Craig Venter (Rockville, Maryland). In 2003, both groups published completed human genomes (with more extensive coverage). Before we go into the details of the human genome, a little background about its general characteristics is warranted.

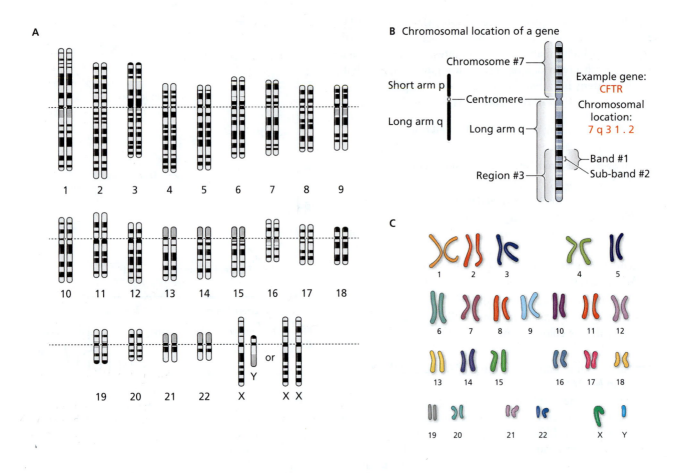

A

1 2 3 4 5 6 7 8 9

10 11 12 13 14 15 16 17 18

19 20 21 22 X or X X

B Chromosomal location of a gene

Chromosome #7

Short arm p

Centromere

Long arm q

Long arm q

Region #3

Example gene:
CFTR
Chromosomal
location:
7 q 3 1 . 2

Band #1
Sub-band #2

C

1 2 3 4 5
6 7 8 9 10 11 12
13 14 15 16 17 18
19 20 21 22 X Y

FIG. 9-12. Human chromosome pairs. A. Schematic figure of Giemsa-stained human chromosomes. The central horizontal line divides chromosomes at their centromeres. Areas above the centromeres are the p-arms, and areas below the centromeres are the q-arms. **B.** Chromosomal location of a gene according to its cytogenetic banding pattern. **C.** FISH-stained human chromosomes.

General Characteristics of the Human Genome

The number of base pairs in the human genome is ~3.3×10^9.[16] Unlike most bacteria, the human genome is linear (as is true for most eukaryotic organisms). The linear genome is distributed among 22 chromosome pairs of autosomal chromosomes and one pair of sex chromosomes. The 22 autosomal chromosome pairs are numbered 1–22 starting from the largest chromosome (chromosome 1) and ending with chromosome 21.[17] We use the term "chromosome pairs" because the human is a diploid organism. It receives one set of chromosomes from the mother and another set from the father. The genes in each set are almost identical, with the exception of polymorphisms. In a diploid human cell, then, there are ~6.6×10^9 base pairs of DNA.

In addition to the autosomal chromosomes, males have an X-Y pair of sex chromosomes and females have an X-X pair of sex chromosomes. Chromosomes can be treated with Giemsa stain, which darkens regions of the DNA that correspond to areas that are relatively AT-rich or they can be treated with oligonucleotides attached to fluorescent dyes that are specific for each chromosome (Figure 9-12). Giemsa-stained chromosomes have a banding pattern, visible with a simple microscope, that remains consistent within a species. The staining pattern is called a cytological banding map of the human genome. A centromere divides each chromosome into two unequal portions, or two arms. The short arm is the p-arm and the long arm is

---

[16] The actual number for one genome project is 3,320,602,130 base pairs. From Ensembl genome browser, assembly GRCh37.p10, February 2009.

[17] Chromosome measurements were originally made using the light microscope from which it appeared that chromosome 22 was the smallest autosomal chromosome. After the human genome was sequenced, chromosome 21 turned out to be smaller than chromosome 22. Chromosome 21 has 50 million base pairs and chromosome 22 has 56 million base pairs.

the q-arm.[18] Starting from the centromere, the bands are numbered until the ends of the chromosome (telomeres) are reached.

The bands direct the researcher to general areas of gene locations. For example, the *CFTR* gene cytological banding region, or cytogenetic location, is chromosome 7q31.2, which means chromosome 7, q-arm, region 3, band 1, sub-band 2 (Figure 9-12B). Sometimes a gene is located very close to the telomere. In such a case the gene is called pter or qter. For example, the *MIC6* gene cytogenetic location is 17qter, meaning it is located near the telomere on the q-arm of chromosome 17. You can more specifically pinpoint the location of a gene using genomic coordinates. Genomic coordinates specify the nucleotide number starting from the telomere of the p-arm and ending at the telomere of the q-arm. The genomic coordinates of *TP53* (which resides on chromosome 17) are 17:7,571,719–7,590,867.[19]

There are other staining techniques besides the Giemsa technique. A fluorescent dye staining technique, called fluorescence in situ hybridization (FISH), uses single stranded oligonucleotides that are attached to fluorescent dyes (Figure 9-12C). Chromosomes are attached to a microscope slide and denatured to separate the DNA strands. The separated strands are treated with single stranded tagged oligonucleotides that complement part of the chromosomes. When light of a particular wavelength is aimed at the chromosomes the dyes fluoresce. A set of oligonucleotides are designed to complement genes located at different regions of a particular chromosome. The set is tagged with one dye for a particular chromosome. In this way chromosomes can be "painted" with different colors.

Once the human genome was sequenced, the first somewhat striking discovery was the limited number of genes coded by the human genome—approximately 20,000,[20] which, if you only consider the exons, constitutes less than 2% of the genome sequence. To be specific, this 2% of the genome, called the exome, is comprised of the exons of protein-coding genes (CDS) and "functional" RNAs (tRNAs and ribosomal RNAs). The human exome has nearly the same number of genes coded by the house mouse *Mus musculus* and the roundworm *Caenorhabditis elegans* and fewer than the number of genes coded by the water flea, *Daphnia pulex* (31,000 genes), and rice, *Oryza sativa* (28,236 genes). Genomic comparisons have recently shown that the types of genes coded by humans and other mammals are nearly the same. Allan Wilson's pioneering studies in the late 1960s and early '70s showed that the morphological characteristics of organisms are dictated, for the most part, by the regulation of genes and not the number or types of genes (see Chapter 8). Our current knowledge of the genomes of different species adds credence to Wilson's prescient conclusions.

Comparison of the Human Genome to the Chimpanzee Genome

To demonstrate the relatedness of mammalian genomes, a comparison of the human genome and the chimpanzee genome is instructive. Whereas the human genome has 22 autosomal chromosome pairs, the genome of chimpanzee has 23. There is significant sequence similarity between the human and chimpanzee chromosomes. The reason why the chimpanzee has one additional chromosome is because human chromosome 2 has two almost equal portions, each of which is similar to a chimpanzee chromosome. One portion of human chromosome 2 is similar to chimpanzee chromosome 2A, and another portion is similar to chimpanzee chromosome 2B. Many of the similarities between human and chimpanzee genomes were previously discovered by FISH. Figure 9-13 is one way of showing the similarities of the two genomes. Solid thick lines show the similarities between the two species across each chromosome.

centromere A protein-DNA complex located on a chromosome. The centromere divides the chromosome into two unequal segments. The longer segment is the q-arm, and the short segment is the p-arm. During mitosis spindle fibers attach to the centromere, and the two chromatids of the chromosome are pulled away from each other.

telomere The end of a linear chromosome composed of repeated nucleotide sequences.

fluorescence in situ hybridization (FISH) A method of staining chromosomes with oligonucleotides attached to molecules that fluoresce when exposed to light of a particular wavelength.

exome The portion of the genome that codes for protein and "functional" RNAs such as tRNAs and rRNAs. In humans, the exome constitutes less than 2% of the entire genome.

---

[18] "p" in p-arm for "petit," and "q" in q-arm for "queue."

[19] These coordinates vary slightly depending on the version of the human genome assembly.

[20] According to one genome project the actual number is 20,848 protein coding genes (Ensembl genome browser, assembly GRCh37.p10, February 2009).

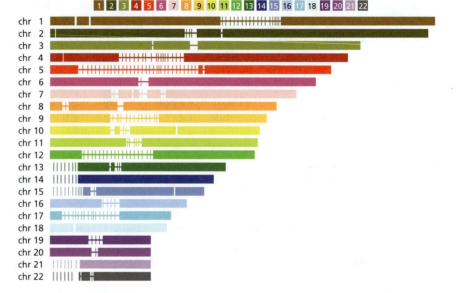

The color code identifies the chimpanzee chromosome numbers

FIG. 9-13. Genome sequences reveal the extent of similarity shared by human and chimpanzee genes in the autosomal chromosomes. Horizontal bars represent human chromosomes 1–22. The colors show the corresponding chromosomes from the chimpanzee. Each color represents a separate chimpanzee chromosome number (see color key on top). Note that chimpanzee chromosomes 2A and 2B are both colored the same and labeled as "chr 2." Thin horizontal lines within each human chromosome correspond to regions that are not similar to chimpanzee genome.

inversion (also known as DNA inversion) The reversal of a DNA segment in one genome relative to another genome.

The thin lines represent areas on the human chromosome that are not similar to the corresponding chimpanzee chromosome. Areas where there is no thick or thin horizontal line are those that had not been sequenced sufficiently well in either the human or chimpanzee genome. This study shows all possible areas of the genomes, including those that do not code for proteins.

We can also use a synteny dot plot to compare human and chimpanzee genomes (Figure 9-14). The human chromosome numbers and chimp chromosome numbers are listed on the x-axis and y-axis respectively. Human chromosomes are listed in order of size beginning from chromosome 1 to chromosome 21. Included in the list are the X and Y chromosomes. The green diagonal lines leading from the origin of the dot plot to the upper right corner suggest that there is extensive synteny between the two genomes. The black circles identify regions of synteny between human chromosome 2 and chimp chromosomes 2A and 2B. The red circle identifies one DNA segment that is inverted between the two species. DNA inversion is a reversal of a DNA segment in one genome relative to the other genome.

Thought Question 9-3

There are four other major inversions in humans relative to chimpanzees. Which human chromosomes contain these inversions?

The chimpanzee genome is approximately the same size as the human genome ($\sim 3.0 \times 10^9$ base pairs) and codes for 19,000 genes.[21] The percentage of G+C in both genomes is approximately 40% (actually, there is not much variation in percentage of G+C across all mammalian species). A comparison of the protein-coding regions of humans and chimpanzees shows that, on average, there are approximately 1–2 amino acid changes in orthologs shared by the two species (however, the p53 protein sequences from the two species are identical). The most widely varying families of proteins between the two species are those involved in reproduction and those involved in defense against pathogens.

[21] From Ensembl genome browser, assembly CHIMP2.1.4, February 2011, the number of genes is 18,759. This genome has 6x coverage.

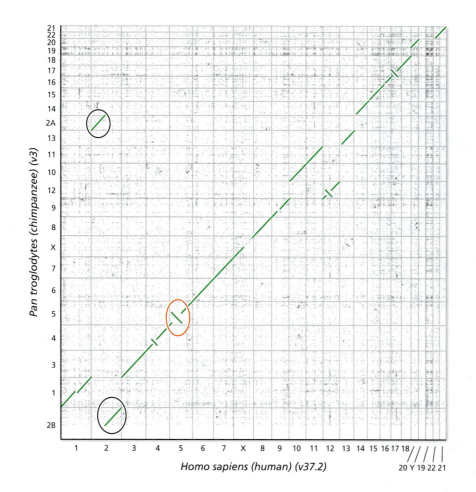

FIG. 9-14. Dot plot comparison of human and chimpanzee genes.

Detailed Analysis of the Human Genome Landscape

At present, the study of the human genome is an extremely active area of research. We will discuss our current understanding of the genome, but it is important to bear in mind that changes in our understanding will likely occur in the near future. Because more than 98% of the human genome does not code for proteins, tRNAs and rRNAs, how do we classify this large portion? We will start by describing some general features and then discuss specific regions previously classified by Lev Patrushev and Igor Minkevich (Russian Academy of Sciences, Moscow).

Areas rich in genes appear to be concentrated in random locations along the genome with long segments of non-coding DNA between gene-rich areas. The genes are found in areas that are relatively high in G+C and the non-coding DNA segments are found in areas that are relatively high in A+T. Sandwiched between the gene-rich areas and the long segments of non-coding DNA are repeats composed of Gs and Cs (up to 30,000 base pairs in length). Within the three billion base pairs of the human genome there are approximately 10–15 million locations where single base DNA differences occur.[22] These are called single nucleotide polymorphisms (SNPs). Overall, the DNA differences between individual human genomes amount to less than 0.083 percent, or 1 base difference for every 1,200 bases.[23] Genetically, humans are very homogeneous compared to other primates (with sequenced genomes).

single nucleotide polymorphism (SNP) A single base difference between one individual and another at the same position in the genome. In humans, this occurs in approximately 1 nucleotide in every 1,200 nucleotides.

---

[22] "International HapMap Project," http://hapmap.ncbi.nlm.nih.gov/whatishapmap.html.en (accessed December 5, 2015); Pevsner (2009).

[23] "International HapMap Project," http://hapmap.ncbi.nlm.nih.gov/whatishapmap.html.en (accessed December 5, 2015).

Composition of human genome

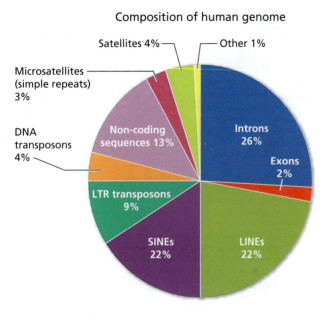

FIG. 9-15. Pie chart showing the composition of the human genome.

LINES =
Long dnterspersed
Elements

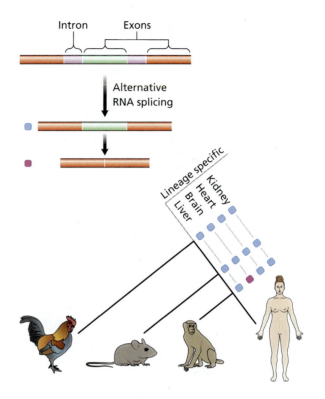

FIG. 9-16. Lineage specific alternative splicing.

Introns, exons, and alternative splicing of primary transcripts

Recall from Chapter 2 that eukaryotic genes are transcribed into RNAs called primary transcripts, which undergo splicing to remove intron sequences. The introns, DNA sequences between the exons of genes, comprise 26% of the human genome (Figure 9-15). Alternative splicing of the primary transcript frequently produces different protein forms that arise from the same gene. The human genome has approximately 7.8 introns per gene[24] and produces an average of 4.4 to 6.6 alternatively spliced transcripts per gene.[25] At least 70% of human genes undergo alternative splicing of primary transcripts, resulting in the production of 90,000 to 100,000 proteins with different sequences, although many families of proteins produced from single genes share a subset of their sequences. One thought is that the complexity of alternative splicing may distinguish one species from another.

The laboratories of Chris Burge (MIT, Cambridge, Massachusetts) and Benjamin Blencowe (University of Toronto, Canada) measured the number of alternatively spliced forms of genes from different vertebrate species and found a few hundred genes that undergo more complex splicing as a function of lineage type (Figure 9-16).[26] For example, a gene in birds may undergo the same splicing in all of its organs whether it is expressed in brain, heart, kidney, or liver. However, in primates the same gene expresses a new spliced form of the transcript in the brain, but not in other organs.

[24] Sakharkar, Chow, and Kangueane (2004).

[25] Pan et al. (2008) reported 88,000 to 133,000 alternative splicing events in human tissues. These values were divided by the number of genes in the human genome, 20,000, to arrive at 4.4 to 6.6 alternatively spliced mRNAs per gene.

[26] A lineage is a series of species, each formed from successive speciation events; birds, for example, constitute a lineage.

This is called lineage-specific splicing. The set of genes that undergo lineage-specific complex alternative splicing is enriched with proteins that bind to DNA and RNA. Altogether, the data suggests that more complex splicing occurs in primates in comparison to other vertebrate species.

Transposon-derived repeats

Approximately 50% of the human genome appears to be—or was, at one time—transposons (Figure 9-15). Transposons are capable of copying a segment of DNA and integrating that DNA into another part of the genome. Here, we will not distinguish between active and inactive transposons, but the vast majority of transposons in the human genome are thought to be inactive. Transposons can be divided into two major categories: class I and class II.

There are vast areas of the genome that are comprised of tandem repeats of transposons (also known as interspersed repeats) generated by class I and class II transposons. Let's begin with class I transposons. Recall from Chapter 3 that class I transposons are retrotransposons that are first transcribed from DNA into RNA by RNA polymerases. The RNA is then reverse transcribed into DNA, and the reverse transcribed DNA is integrated into the genome (a process called retrotransposition). Long interspersed elements (LINEs) are DNA repeats derived from class I transposon retrotransposition and code for an enzyme with reverse transcriptase activity and other potential proteins. The size range of LINEs is 1,000–7,000 base pairs, and LINEs constitute 22% of the human genome. A second type of class I transposon-derived repeats is short interspersed elements (SINEs), which include *Alu* repeats. *Alu* repeats are only found in primates and have a size range of 100–500 base pairs. SINEs do not code for their own reverse transcriptase and constitute 16% of the human genome. A third type of class I transposons is long terminal repeat (LTR) transposons. LTR transposons code for a retrovirus-like reverse transcriptase (as opposed to the reverse transcriptases in LINEs and SINEs) and contain sequences called long terminal repeats at their 5′ and 3′ ends. LTR transposons range in size from 3,000–12,000 base pairs and constitute 9% of the human genome. Class II transposons are DNA transposons that do not go through an RNA intermediate. Instead, DNA is directly copied into a DNA segment and the DNA segment integrates into the genome. DNA transposons are 500–17,500 base pairs in length. DNA transposons constitute 4% of the human genome and were first discovered in maize by Barbara McClintock (see Chapter 3).

Comparative genomic analysis has found that more than 10,000 class I retrotransposons (primarily LINEs) have incorporated into the human genome over the past six million years. These relatively new LINEs distinguish the human genome from the chimpanzee genome and may contribute to observed (phenotypic) species differences. Interestingly, some class I retrotransposons are currently active and their insertions have led to more than 60 cases of genetic disorders.[27]

Satellites and microsatellites

Repeats in transposons are considered moderately repetitive DNA. Two other categories of repeats are satellites and microsatellites, which are considered to be highly repetitive. Satellites are segments of 4–200 nucleotides that are repeated in tandem up to a total of hundreds of kilobases.[28] Satellites constitute 4% of the total DNA in the genome. Subclasses of satellite DNA are macrosatellites and minisatellites.

short interspersed elements (SINEs) DNA that was reverse transcribed from RNA. The RNAs, originally generated by RNA polymerase III, were 5S rRNA, tRNA, and others.

long interspersed elements (LINEs) DNA repeats derived from class I transposon retrotransposition. LINEs code, or did at one time code, for enzymes with reverse transcriptase activity and other proteins.

long terminal repeat (LTR) transposons DNA that is similar to retrovirus reverse transcriptase and retrovirus repeats.

satellites Segments of 4–200 nucleotides that are repeated in tandem up to a total of hundreds of kilobases. They are called satellites because when mammalian DNA is centrifuged through a CsCl density gradient, satellite DNA migrates to a position in the gradient that is different from the major DNA fraction.

---

[27] Chen, Férec, and Cooper (2006).

[28] Satellites are so named because during cesium chloride density purification of total DNA from cell nuclei some purified DNA forms a shoulder of the main fraction of DNA. The shoulder has a different buoyant density from the main fraction because it has repeated sequences. Because the shoulder is on the side of the main fraction it is considered satellite DNA.

microsatellites Segments of DNA, known as simple repeats, that have a range of 1–3 nucleotides that are repeated in tandem in blocks of up to 200 nucleotides in length. The lengths of these blocks often vary from individual to individual and are used for DNA profiling.

pseudogene A gene that does not produce a protein or functional RNA such as tRNA or rRNA. Some pseudogenes express RNAs that influence protein expression by binding to microRNAs.

Their distinctions are based on the composition and number of repeating units. As a rule, it appears that satellite DNA segments are not transcribed.

Microsatellites (treated here as being distinct from satellites), also known as simple repeats, have a range of 1–3 nucleotides and are arranged in blocks of up to 200 nucleotides in length. Unlike satellites, microsatellites are often transcribed and, in a few cases, are found in exons. The *huntingtin* gene contains a microsatellite composed of (CAG)n, which results in a polyglutamine stretch in the huntingtin protein (see Chapter 3). Microsatellites constitute approximately 3% of the human genome outside of the exome and are the basis for human DNA fingerprinting, also known as DNA profiling (see Box 9-2). The lengths of the microsatellites can vary from one individual to another, creating an opportunity to distinguish two individuals. DNA fingerprint evidence has been used to exonerate more than 300 individuals wrongly convicted of crimes.[29] Sadly, the average length of time such individuals are incarcerated prior to regaining freedom is 13.5 years.

Pseudogenes

Pseudogenes are genes deemed nonfunctional because they do not create a protein or a functional RNA (such as tRNA or rRNA). As we shall see in a moment, the term "pseudogene" may be a misnomer because some pseudogenes express RNAs that influence protein expression. Pseudogenes were discovered in 1977 by Jacq, Miller, and Brownlee (MRC Laboratory, Cambridge, United Kingdom) in frog tissue. Now approximately 18,000 have been identified in the human genome.[30] Pseudogenes are related to protein-coding genes in the human genome and are typically identified by sequence comparison to protein-coding genes. Pseudogenes have a premature stop codon or frameshift mutation that precludes them from producing protein products.

In some cases RNAs transcribed from pseudogenes can affect their protein-coding gene counterparts. For example, *PTEN* is a tumor suppressor gene that codes for PTEN protein. In a cell, if the PTEN protein level is too low there is a higher propensity for cell growth, which eventually can lead to cancer. A natural microRNA within the cell hybridizes to the PTEN transcript, causing the transcript to be unstable, resulting in a low level of PTEN protein. Interestingly, a pseudogene of *PTEN*, named *PTENP1*, expresses RNA that competes with PTEN mRNA for hybridization to the microRNA. The competition helps sustain PTEN mRNA levels to maintain normal PTEN protein levels. Some cancers show deletion of the *PTENP1* pseudogene, which effectively lowers PTEN protein levels. The cancer data gives credence to the hypothesis that *PTENP1* provides an important function in sustaining PTEN protein levels.

Three types of pseudogenes exist, defined by the mechanisms by which they are generated: processed, nonprocessed, and unitary. Processed pseudogenes are retrotransposed. An mRNA of a gene is reversed-transcribed by reverse transcriptase from a LINE into a DNA (without introns) and inserted back into the genome at a different location. Processed pseudogenes lack promoters and introns and often contain poly(dA) tails. They lack the 5′ end of their normal gene counterparts in their promoters due to the nature of the LINE reverse transcriptase system. Because they are missing functional promoters, processed pseudogenes are not commonly transcribed.

Nonprocessed pseudogenes arise from genomic duplications of their normal gene counterpart. These duplicated pseudogenes typically contain promoter sequences and introns, but have acquired mutations after duplication that result in their inability to produce proteins. Unitary pseudogenes, also known as disabled genes, are single genes that were once functional but have been mutated to such an extent that they no longer create proteins.

---

[29] Data from Innocence Project (2014).

[30] Data from human genome sequence build 76; see http://www.pseudogene.org/.

BOX 9-2 | A CLOSER LOOK

DNA Fingerprinting (DNA Profiling)

DNA fingerprinting, also known as DNA profiling, relies on differences in lengths of microsatellite repeats between individuals. It was first developed by Alec Jeffreys (University of Leicester, United Kingdom) in 1984. If there are two individuals with $(CAG)_n$ repeats, one individual may have 100 repeats (i.e., 300 nucleotides) in the microsatellite and another may have 50 repeats in the microsatellite. Such highly variable microsatellites across individuals are called short tandem repeats (STRs). In crime scene investigations, typical sources of DNA for DNA fingerprinting include white blood cells, hair follicles, saliva, and semen. The DNA is partially purified from the source tissue, and oligonucleotide primers are used to amplify the segment of DNA containing the microsatellites. The oligonucleotides are tagged with fluorescent labels, and their sequences are specific for DNA just upstream and downstream of the microsatellites. The oligonucleotide primers are in pairs. For each pair, one primer binds upstream of a microsatellite and the other binds downstream of a microsatellite. The microsatellites are amplified by polymerase chain reaction (PCR). The amplified DNA is separated on a medium of either agarose gel or capillary column. The distance traveled in the medium is inversely proportional to the length of the amplified DNA. Two individuals' DNA fingerprints can be distinguished by relative migration distances of the amplified DNA in the separating medium.

Let's take a hypothetical example where DNA samples have been collected from two individuals suspected of committing a crime. A separate DNA sample was collected at the crime scene. One STR, named D3S1358 (D3 for short), exists at chromosome location 3p21.31. The repeating unit is TCTA, which may be repeated a number of times depending on the individual. Usually, several STRs from different parts of the genome are investigated at the same time. Here, vWA and FGA repeats are also investigated to characterize the evidence DNA, suspect 1's DNA, and suspect 2's DNA.

DNA from the hypothetical crime scene (labeled Evidence) was used as a target for amplification by pairs of oligonucleotide primers specific for the DNA region just outside of the microsatellites. The first pair of primers specific

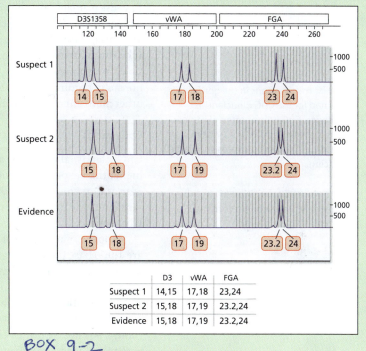

	D3	vWA	FGA
Suspect 1	14,15	17,18	23,24
Suspect 2	15,18	17,19	23.2,24
Evidence	15,18	17,19	23.2,24

for D3 amplified DNA with 15 and 18 repeats. Because humans are diploid organisms, it is possible that individuals are heterogeneous for D3. Each chromosome contributes an allele, so there are two alleles that can contribute to heterogeneity with an individual. From the Evidence DNA, the primers specific for D3 resulted in amplified DNA with 15 and 18 repeats. The primers specific for vWA amplified DNA with 17 and 19 repeats and the primers specific for FGA amplified DNA with 23.2 and 24 repeats. Suspect 1 and suspect 2 DNA samples were analyzed with the same sets of primers. From the capillary electrophoretogram above, it is clear that the DNA of suspect 2 matches the Evidence DNA. Let's summarize the capillary electrophoretogram data on the D3 repeats. The following is a listing of the number of D3 repeats obtained from each sample collected in the case:

SUSPECT 1:
First allele: $(TCTA)_{14}$
Second allele: $(TCTA)_{15}$

SUSPECT 2 AND EVIDENCE:
First allele: $(TCTA)_{15}$
Second allele: $(TCTA)_{18}$

One example of a unitary pseudogene found in humans and other primates is the *gulonolactone oxidase* pseudogene (*GULOP*). Interestingly, a functioning *gulonolactone oxidase* gene is present in nonprimate mammals, but in primates this genes appears to have been mutated and inactivated approximately 63 million years ago. The protein product in nonprimates is an enzyme essential for vitamin C synthesis. Because primates lost the *gulonolactone oxidase* gene, vitamin C is an essential nutrient for them. Perhaps because ancestors of modern primates received sufficient vitamin C through diet, it was not necessary to synthesize gulonolactone oxidase. It has been argued that it may have been evolutionarily advantageous to allow mutations to accumulate in the *gulonolactone oxidase* gene because resources and energy normally required for vitamin C synthesis in the body could be directed to make other more important molecules for primate survival.

CpG islands

CpG islands, also known as CG islands or CGI, are DNA regions that contain a higher than expected frequency of CpG sites. There are approximately 25,000 CpG islands that constitute 0.70% of the human genome. The "p" between the C and G represents the phosphodiester bond that connects the two nucleotides on the same DNA strand.

Approximately 70% of human promoters have CpG islands, which, on average, are 1,000 base pairs in length and reside in areas that are, in general, GC-rich. CpG islands undergo an enzymatic modification called methylation on the cytidine base. In methylation, a $-CH_3$ (methyl) group is attached to the cytidine base causing an alteration in the DNA structure. Methylation correlates with inhibition of transcription of genes, also known as gene silencing, near the CpG islands. Although it is not clear if DNA methylation is the initiation event in gene silencing, it is required for its maintenance. Removal of methyl groups from the CpG islands increases gene transcription. Software programs are available that can calculate whether regions have CpG islands that undergo methylation. The most successful programs use databases of experimentally determined CpG methylation sites and "learn" the sequence patterns that predict methylation. More recently, CpG islands have been found in areas that do not correspond to genes but, nevertheless, correlate with transcription initiation. They may in fact control transcription of non-coding RNA molecules. The "methylome" is the sum of the methylation sites of the entire genome. Sometimes, permanent methylation patterns can be maintained through cell division and can even be vertically transmitted from parents to offspring. Epigenetics is a change in gene expression without altering the DNA sequence. Because methylation alters gene expression without altering the nucleotide sequence, methylation is a form of epigenetics. In cancer cells many CpG islands undergo methylation pattern alterations and are correlated to abnormal gene expression.

Copy number variation (CNV)

Genomic structural variation is defined as an inversion or copy number variation. Copy number variation is the most common form of genomic structural variation. Copy number variation (CNV) is a DNA segment of at least 50 base pairs present at a variable copy number in comparison to a reference genome with the usual copy number of $N = 2$ (diploid). In 2004, CNV was shown to be widespread in human genomes and represents a significant source of genetic alterations in the population.[31] In diploid organisms, such as humans, we expect two copies of each gene ($N = 2$), but CNV results from a deletion, duplication, triplication, or higher order of copying of a DNA

CpG islands Regions in the genome enriched for the CpG dinucleotide sequence. CpG islands often exist near promoters. Methylation of cytosines within CpG islands correlates with gene silencing.

methylome The location of all of the methylated CpG sequences in the genome.

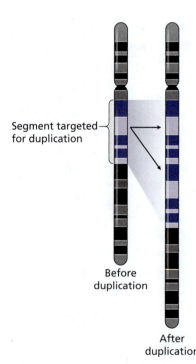

Segment targeted for duplication

Before duplication

After duplication

FIG. 9-17. An example of a copy number variation.

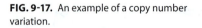

---

[31] A. J. Iafrate et al. (2004); J. Sebat et al. (2004).

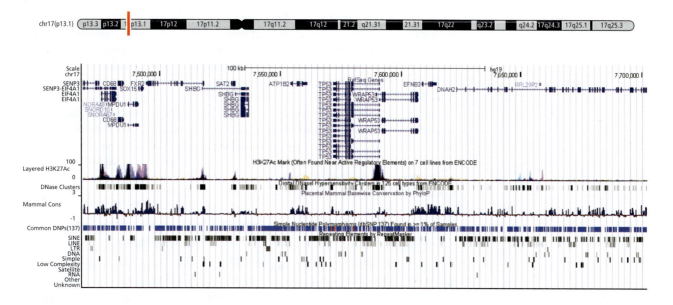

FIG. 9-18. A view of 233,929 nucleotides from chromosome 17p13.1.

segment (Figure 9-17). It has been estimated that 5–30% of the human genome is prone to CNV.[32]

CNVs are thought to arise by one of four mechanisms. One is nonallelic homologous recombination, which occurs during DNA replication during mitosis (somatic cell division) and meiosis (germ cell–specific cell division). A second mechanism is nonhomologous end-joining, a DNA repair process that responds to double strand breaks. A third mechanism is fork stalling and template switching, a process that can occur during DNA replication. A fourth mechanism is LINE 1 retrotransposition, where RNA transcripts are reverse transcribed into DNA and inserted into the genome.

If CNVs encompass genes, then there may be abnormal levels of protein expressed due to duplication or deletion of the gene. This is called a gene dosage effect, and it can play a role in some diseases. For example, CNVs resulting in high levels of epidermal growth factor receptor have been reported to be associated with non-small cell lung cancers.[33] CNVs resulting in low levels of CD16 cell surface immunoglobulin receptor (FCGR3B) may predispose patients with lupus erythematosus to immunologically mediated renal disease.[34]

copy number variation (CNV) A DNA segment that is present at a number that is different from the number found in a reference genome with a usual copy number. For humans and other diploid organisms the usual copy number of the reference genome is two.

9.10 REGION OF THE HUMAN GENOME THAT ENCOMPASSES THE *TP53* GENE

General Comments on the Region Encoding the *TP53* Gene

Now that we have discussed the landscape of the human genome, we are in a position to explore a specific region of the human genome on chromosome 17. This will give insight into the power of genomic analysis on just a few genes at this location. We will view 240 kilobases of the region of chromosome 17p13.1—a region that contains the *TP53* gene and two flanking genes (Figure 9-18).

[32] This range is wide because it is technically difficult to measure the lengths of large segments that are deleted or duplicated. See F. Zhang, W. Gu, M. E. Hurles, and J. R. Lupski (2009).

[33] F. Cappuzzo et al. (2005). In this study the increased level of EGFR allowed increased sensitivity to a chemotherapy drug that specifically targets this protein.

[34] T. J. Aitman et al. (2006).

Figure 9-18 shows the *TP53* gene and its flanking sequences as viewed by the University of California Santa Cruz Genome Browser (February 2009 version). This segment of chromosome 17p13.1 is lightly Giemsa-stained, meaning it is relatively GC-rich. Near the top of the figure is a schematic representation of chromosome 17, which has a narrow, pinched region signifying the centromere. The two telomeres of this chromosome are at the extreme left and extreme right of the chromosome (not shown). The red bordered region is expanded below the chromosome. The scale of this expanded region gives the length that corresponds to 100 kilobase pairs and within it the light vertical lines are spaced at 2.5 kilobase intervals. The base pair numbers listed below increase from left to right in the figure (starting with 7,500,000 and ending with 7,700,000). Under the base pair numbers, are horizontal "tracks" that display information about this segment of chromosome 17.

Tracks That Display Information about the *TP53* Region of the Genome
RefSeq Genes track

The first of these horizontal tracks is the RefSeq Genes track, which presents genes and their alternatively spliced forms. The central gene, *TP53*, shows 15 lines, each with arrows pointed to the left. The arrows point toward the direction of transcription. This means that the coding strand of DNA is the bottom strand in this region of the genome (by convention, the bottom strand is 5′ to 3′ going from right to left). Each line represents one of 15 alternatively spliced *TP53* RNAs. The blocks laid on top of the RNAs are the exon sequences retained in the p53 mRNAs. The small blocks are untranslated segments of the mRNAs, the 5′ and 3′ untranslated regions (5′ UTRs and 3′ UTRs). The large blocks are translated regions, and the thin lines are introns.

The genes upstream and downstream of TP53

The segment of the chromosome just to the left of *TP53* is *ATP1B2*, which codes for a polypeptide subunit of a sodium ion/potassium pump that requires ATP. The subunit is part of a complex of proteins required to maintain a voltage gradient (charge difference) across the cell membrane. The gene just to the right of *TP53* is *WRAP53*. *WRAP53* is WD-repeat antisense p53. A WD-repeat is a beta-strand structure in the shape of a propeller that terminates with a Trp-Asp dipeptide. Careful inspection of the figure shows that the second spliced variant WRAP53 mRNA can complement the TP53 mRNAs. When *TP53* and *WRAP53* are transcribed, this spliced variant WRAP53 mRNA binds to TP53 mRNAs resulting in lower levels of p53 protein. The protein product of *WRAP53* is part of the enzyme known as telomerase, which maintains the proper lengths of the telomere regions of chromosomes. *WRAP53* is an example of a gene with more than one function.

Histone 3 Lysine 27 Acetylation (H3K27Ac) track

Below the RefSeq Genes track is the Histone 3 Lysine 27 Acetylation (H3K27Ac) track. This track shows the locations along the genome where histones are acetylated. Recall that the DNA of the genome is wrapped around histones. Histones bind to and pack DNA so that DNA can fit into the cell nucleus. Local unpacking of DNA, also known as chromosome remodeling, is necessary for transcription because transcription factors must have access to this region to initiate transcription (Figure 9-19). Histones undergo enzymatic modifications to either pack or unpack DNA. One modification that unpacks DNA is placement of an acetyl group (CH_3–CO–) onto lysine number 27 of histone polypeptide 3. This modification correlates strongly to sites just upstream to genes that are transcribed. The H3K27Ac track shows a series of peaks with slightly different colors. Each color corresponds to one of the seven cell lines used for the experimental study.[35] The peak color and height correlates to the

---

[35] We need to be cautious deriving information from cell lines, because they are often cancerous and therefore their expression patterns may not entirely reflect normal cell behavior. Nonetheless, this track gives us clues as to locations of these markers for active transcription.

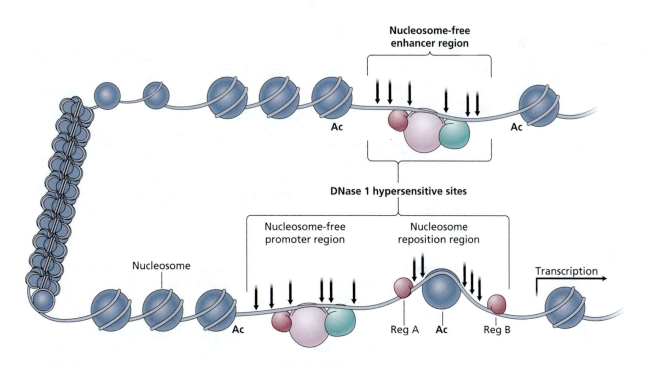

FIG. 9-19. Demonstration of histone acetylation effects on chromatin remodeling and DNAse I hypersensitivity footprinting.

finding that H3K27Ac was found in more cell lines (darker color) with higher levels of acetylation (peak heights are taller). Just upstream of the *TP53* gene in Figure 9-18 the peaks are dark and tall, suggesting that p53 transcription is very high in these cell lines. The p53 proteins expressed in the human cell lines used for data collection are likely to be mutated and, therefore, are nonfunctional in terms of suppressing tumors. This transcription factor binding region may also control WRAP53 expression because it is located upstream of the *WRAP53* gene as well (on the top strand of the DNA).

DNAse I hypersensitivity clusters track

The next track is called DNAse I hypersensitivity clusters track. Similar to the H3K27Ac track, this track shows the promoter regions of genes that are transcriptionally active. DNAse I hypersensitivity mapping is a method to locate regions that unpack from histones so that the DNA can access transcription factors (Figure 9-19). In an experimental setting, DNAse I cleaves double strand DNA unpacked from histones. Cleaved sites are called DNAse I hypersensitivity sites. In practice, a population of DNA molecules isolated from nuclei is treated with DNAse I so that each DNA molecule is cut once on average. Areas where there are many cuts are called hypersensitive sites. The UCSC genome browser shows the results of DNAse I hypersensitivity experiments performed separately on genomes from 125 human cell types. Some cells are from cell lines, and others are from tissues. DNAse I hypersensitivity clusters localize to the same areas as H3K27Ac sites. This is not unexpected because histone 3 acetylation unpacks DNA from histones, exposing the DNA to DNAse I for cutting. As expected, there is a DNAse I hypersensitivity cluster located in the *TP53* promoter region, as noted by the relatively dark region upstream of the *TP53* gene (see Figure 9-18).

Placental Mammals Basewise Conservation Track

The next track, named the Placental Mammals Basewise Conservation track, is a measure of the nucleotide substitution rates among 46 vertebrate species relative to

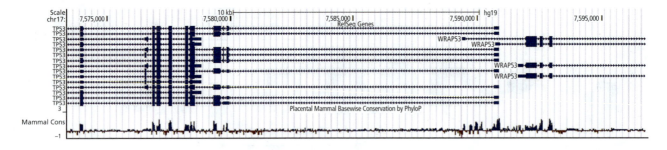

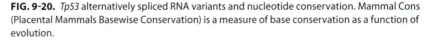

FIG. 9-20. *Tp53* alternatively spliced RNA variants and nucleotide conservation. Mammal Cons (Placental Mammals Basewise Conservation) is a measure of base conservation as a function of evolution.

the mammalian substitution rate. Peaks correspond to nucleotides that remain relatively more conserved than a neutral mutation drift model during evolution. Valleys correspond to nucleotides that are substituted at a higher than expected rate compared to a neutral mutation drift model. In Figure 9-20, a zoomed-in view of this track shows that peaks correlate with exon sequences and valleys correlate somewhat with intron sequences. One expects protein-coding regions to have fewer substitutions than nonprotein coding sequences because there is evolutionary pressure to maintain protein-coding regions. This pressure is relaxed for nonprotein coding regions because DNA in introns do not code for amino acid residues.

Simple nucleotide polymorphisms track

The next track maps locations of *simple* nucleotide polymorphisms. Not to be confused with single nucleotide polymorphisms (SNPs), these simple nucleotide polymorphisms include base changes and small indels. These are changes that occur in ≥1% of the human population. In Figure 9-18 the blue horizontal bars refer to single nucleotide polymorphisms and the red horizontal bars refer to small indels.

Repeats track

The last track maps locations of repeats in this DNA segment. Categories of repeats include SINEs, LINEs, LTR transposons, simple repeats (microsatellites), low complexity regions, satellites (macro- and mini-), and others. A magnified view of the *TP53* gene shows that repeated regions are not located in exons of the gene (not shown). These repeats were identified by the Repeat Masker program, which operates by comparing the DNA sequence to a database of repeats, named RepBase. Matches are classified as one of the repeat categories listed above.

Repeats must be handled with care to avoid erroneous annotations of genes. To identify genes in newly sequenced genomes we perform sequence alignments to sequences in GenBank. One of the difficulties in this approach is that repeated sequences within the sequenced genome will match with repeated sequences in GenBank. Even if repeat sequences are located in different genes (or outside of genes), such matches would be highly scored and could lead to incorrect gene matches. To avoid this problem we need to first run the query genome through the Masker program. The output of the Masker program shows the genome sequence with its repeated nucleotides listed as N or X, indicating that those nucleotides are masked. The masked sequences can then be used as a query for other databases, for example RefSeq Gene database (recall that RefSeq Gene database is a subset of GenBank). Alignment programs that align masked genome sequences with sequences in other databases ignore the masked sequences and, therefore, are not considered for sequence alignments.

9.11 THE HAPLOTYPE MAP

Now that we have discussed details of human genome annotation we will explore another perspective of the human genome, the haplotype. Recall that the human genome is a diploid organism. In other words, all human cells have two complete sets of genes (with the exception of germ cells). This is why we have 23 *pairs* of chromosomes in each cell. Haploid refers to a single set of chromosomes present in the germ cells (the sperm or egg cells), whereas diploid refers to the double set of chromosomes in somatic cells. Humans have 23 chromosomes in their germ cells. The genotype describes the set of genes (or more specifically, set of alleles) of an organism. The phenotype is the physical appearance of the organism—that is in large part dictated by the genotype.

What Is a Haplotype?

The haplotype is a way of specifying a specific region within a chromosome. Haplotypes are usually distinguished by closely linked markers on a chromosome. A marker is a unique feature of a segment of DNA. Markers can be closely linked genes. For example, human leukocyte antigens (HLA) are genes that often inherited together because they are located close to one another on chromosome 6. HLAs constitute the major histocompatibility complex (MHC) and their gene products make up part of the immune system. When tissues are donated from one individual to another, the HLAs of the donor and the recipient must match or nearly match so that there is no tissue rejection (formally known as graft vs. host disease). Three HLA genes are HLA-A, HLA-B, and HLA-DR. In the world population, there are at least 59 different HLA-As, 118 different HLA-Bs, and 1,635 different HLA-DRs. An individual may have two distinct HLA haplotypes for these specific genes—one for each chromosome 6. For example, one HLA haplotype is 10, 15, and 120 and the other is 3, 14, and 10, where the numbers represent HLA-A, HLA-B, and HLA-DR haplotypes, respectively. Because there are many possible combinations of HLAs, there are thousands of HLA haplotypes in the world.

Haplotypes Can Be Specified by Markers Derived from SNPs, Indels, and CNVs

We just discussed haplotypes in terms of linked genes. Another way of specifying a haplotype is with markers derived from SNPs, indels, and CNVs. There are worldwide efforts aimed at cataloging haplotypes from different populations. Associating a haplotype with a particular disease can help pinpoint the genes involved in that disease. The idea of associating haplotypes with diseases has led to large projects known as genome wide association studies (GWAS). In GWAS the genome is scanned for haplotypes that are associated with diseases. The genomes of organisms with a disease are compared to genomes of non-disease-bearing organisms. In the case of humans, more than 500,000 different linked SNPs have been cataloged in thousands of individuals involved in these studies.

As stated previously, the genomes within the human species differ by approximately 0.083%, which amounts to about one in every 1,200 bases. In the context of SNPs one individual might have an A at one location and another individual might have a G in the same location. Recall that in the p53 protein, there are two possible wild-type protein sequences expressed (see Chapter 3). One has a Pro at codon 72, and the other has an Arg at this position, the result of a single nucleotide polymorphism in the genome. Each of these genome sequences is called an allele, and the collection of alleles constitutes an individual's genotype. If a particular polymorphism is associated with a disease one can attempt to track it. In this example, there is a single nucleotide polymorphism, which creates two different alleles. The SNP that creates the amino acid change at codon 72 is part of a group of SNPs that together comprise a haplotype.

genotype The set of alleles contained within the genome.

phenotype The physical appearance of an organism that is largely dictated by the genotype.

haplotype A set of linked genetic markers located on the same chromosome.

haplotype = a set of SNPs that mark a certain phenotype

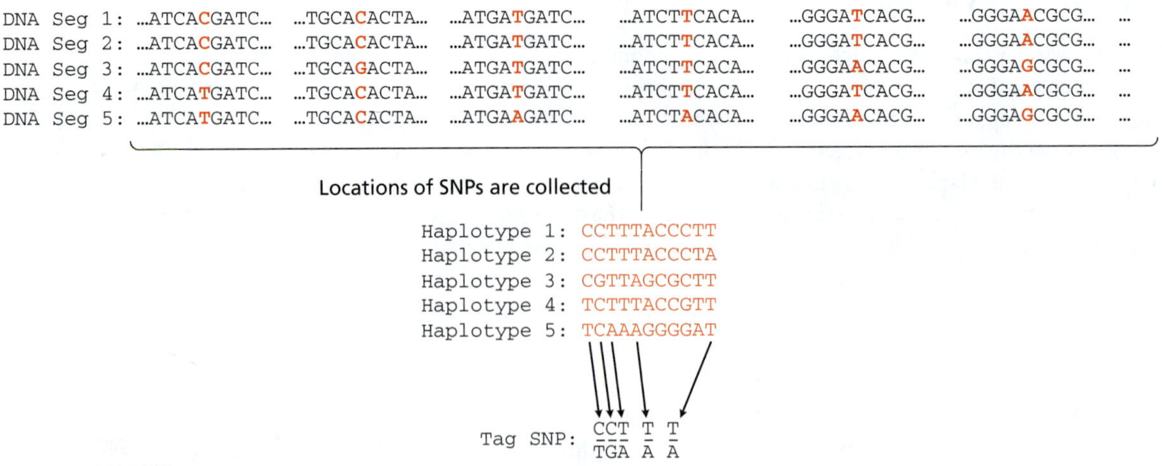

```
DNA Seg 1:  ...ATCACGATC...   ...TGCACACTA...   ...ATGATGATC...   ...ATCTTCACA...   ...GGGATCACG...   ...GGGAACGCG...   ...
DNA Seg 2:  ...ATCACGATC...   ...TGCACACTA...   ...ATGATGATC...   ...ATCTTCACA...   ...GGGATCACG...   ...GGGAACGCG...   ...
DNA Seg 3:  ...ATCACGATC...   ...TGCAGACTA...   ...ATGATGATC...   ...ATCTTCACA...   ...GGGAACACG...   ...GGGAGCGCG...   ...
DNA Seg 4:  ...ATCATGATC...   ...TGCACACTA...   ...ATGATGATC...   ...ATCTTCACA...   ...GGGATCACG...   ...GGGAACGCG...   ...
DNA Seg 5:  ...ATCATGATC...   ...TGCACACTA...   ...ATGAAGATC...   ...ATCTACACA...   ...GGGAACACG...   ...GGGAGCGCG...   ...
```

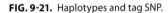

Locations of SNPs are collected

```
Haplotype 1:  CCTTTACCCTT
Haplotype 2:  CCTTTACCCTA
Haplotype 3:  CGTTAGCGCTT
Haplotype 4:  TCTTTACCGTT
Haplotype 5:  TCAAAGGGGAT
```

$$\text{Tag SNP:} \quad \frac{\text{CCT}}{\text{TGA}} \ \frac{\text{T}}{\text{A}} \ \frac{\text{T}}{\text{A}}$$

FIG. 9-21. Haplotypes and tag SNP.

Tag SNPs

It is estimated that there are 10–15 million SNPs in the human genome. Instead of testing each one for its association with a disease, it is easier to keep track of sets of SNPs. Sets of SNPs can be inherited. A haplotype is considered to be a set of linked genetic markers, so we can also consider a haplotype to be a set of linked SNPs. Figure 9-21 shows five sequences from nearly identical segments of DNA (for example, from chromosome 17p13.1) from five individuals. There are 11 SNPs (in red letters) within aligned DNA segments from these individuals. These and other polymorphisms are collected across chromosome 17. A subset of these polymorphisms can be identified that distinguishes one haplotype from another. This subset is collectively named tag SNP, defined as a minimum SNP set to identify (or tag) a haplotype. Recall from our discussion of HLA that an individual will have two haplotypes because of the diploid status. The tag SNP constitutes a haplotype. In Figure 9-21 we can see that the tag SNP from these individuals is C/T, C/G, T/A, T/A, and T/A. For each individual, analysis of this segment of the chromosome for these nucleotides, at these locations, is sufficient for distinguishing each of the five haplotypes. If individuals with haplotype 1 (with tag SNP set "CCTTT") have higher risk for a disease, then it is likely that the gene or genes associated with their linked SNPs can be investigated for contribution to the disease. In addition, genetic tests can be performed on the general population for prevalence of haplotype 1. If detected in an individual, haplotype 1 could be a predictor of risk for disease susceptibility. The HapMap database stores SNPs and tag SNPs collected from different populations in the world. Tag SNPs are more convenient to use for tracking diseases than individual SNPs because there are only 300,000 to 600,000, compared to more than 10 million individual SNPs.

Thought Question 9-4

For the haplotypes listed in Figure 9-21 the tag SNP is C/T, C/G, T/A, T/A, and T/A at positions 1, 2, 3, 5, and 11. Let's say you claim that the tag SNP is C/T, C/G, T/A, and T/A at positions 1, 2, 3, 5, and 10. Would you still be able to distinguish each individual with this tag SNP?

How Did Haplotypes Originate?

You might ask, where did these haplotypes originate? Why should you share a haplotype with anyone but your relatives? The answer is that all humans are fairly related

HapMap database A collection of SNPs and tag SNP information from different ethnic populations worldwide.

recombination (also known as crossing over) A process that occurs during meiosis where two parental chromosomes exchange DNA. This results in offspring with genotypes that differ from those of their parents.

Haplotype closely linked markers [handwritten margin note]

to one another. Humans, being diploid organisms, have pairs of chromosomes. One half of each pair is inherited from the father, and the other half is inherited from the mother. The germ cells of the mother and the father (each containing a haploid set of chromosomes) in the form of eggs and sperm combine to create diploid cells. Germ cells themselves are derived from diploid cells. During the derivation of germ cells from diploid cells a process called recombination (or crossing over) occurs. In recombination, pieces of each half of a chromosome pair exchange DNA segments resulting in hybrid (mixed) chromosome pairs. One half of the hybrid chromosome pairs are segregated into germ cells, creating a haploid state in a process called meiosis. Each germ cell contains a genome that has undergone a unique recombination event. The hybrid chromosomes are passed onto the next generation.

The recombination process is uneven: some DNA segments that recombine are long, and others are short. With each generation of offspring the recombination process leads to a greater degree of hybridization. Interestingly, the Y chromosome has no mate, so it does not undergo recombination (there are 68 genes coded by the Y chromosome). Generation after generation of Y chromosomes (through the male line) stay the same. Haplotypes are markers in the DNA segments that, because they are physically close to one another within the chromosome, frequently stay together (linked) during the recombination process. When the distance between two markers is long, there is a greater chance that they will be separated (unlinked) during recombination. To understand the variation of haplotypes in humans we need to consider the origin of our species.

Recall from Chapter 8 that all humans descended from modern *Homo sapiens* who lived in Africa approximately 200,000 years ago. The reason that humans differ only, in terms of SNPs, from one another in 0.083% of the genome is because, as a species, there has not been sufficient time for mutations to significantly vary human genomes. Instead, variation of current human genomes is primarily due the fact that, over the course of many generations, DNA segments of ancestral chromosomes (originating in Africa) are mixed through recombination events (Figure 9-22). Some DNA segments are shared by living humans because these DNA segments were not thoroughly mixed through successive recombination events. These DNA segments are the haplotypes that survive today. Furthermore, recall that a very small colony of humans migrated out of Africa 50,000–100,000 years ago; that colony carried with it a part, but not all, of the variation of the ancestral population.

What effect does this history have on haplotypes? The haplotypes of current modern humans outside Africa are a subset of haplotypes of current modern humans inside Africa. Because the successive generations of the small colony mated with each other, the DNA recombination did not result in much change in haplotypes (similar to inbreeding). At the time the colony migrated out of Africa the general African population was more varied in its markers so that when the people who remained in Africa mated with each other there were a higher number of haplotypes available. The result of this is that today, the DNA segments containing haplotypes in humans outside of Africa tend to be less varied than the DNA segments containing haplotypes in humans inside of Africa.

As humans migrated to different locations in the world the frequency of haplotype prevalence varied from one region to another. The variation of this frequency is

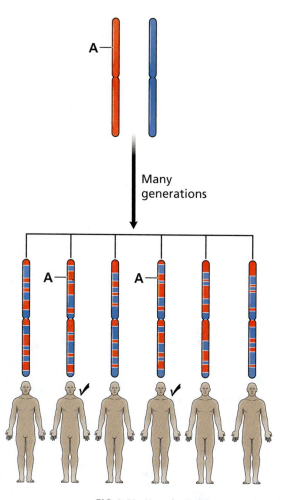

FIG. 9-22. Hypothetical demonstration of the origin of the majority of haplotypes. The ancestral chromosome pair existed in Africa approximately 200,000 years ago. "A" is a particular haplotype found on the chromosome on the left. Due to recombination and the passage of chromosomes from parent to offspring, there is mixing of the DNA segments. Two individuals in the current population retain the "A" haplotype. The "A" haplotype can be distinguished from other haplotypes by tag SNP analysis.

TABLE 9-1. FREQUENCIES OF THE C/C, C/G, AND G/G IN POPULATIONS AT SNP RS1042522 (WITHIN CODON 72 OF *TP53*)

POPULATION	GENOTYPE FREQUENCIES									TOTAL
	GENOTYPE	FREQ	COUNT	GENOTYPE	FREQ	COUNT	GENOTYPE	FREQ	COUNT	
CEU	C/C	0.083	5	C/G	0.300	18	G/G	0.617	37	60
CHB	C/C	0.200	9	C/G	0.578	26	G/G	0.222	10	45
JPT	C/C	0.227	10	C/G	0.364	16	G/G	0.409	18	44
YRI	C/C	0.458	27	C/G	0.424	25	G/G	0.119	7	59

Source: HapMap database.

due to natural selection and random chance.[36] These processes result in a given haplotype occurring at different frequencies in different populations. The greatest frequency differences are found between populations that are geographically separated with little likelihood of exchanging DNA through mating. In some instances, DNA mutations give rise to new haplotypes that have not had sufficient time to spread widely beyond the geographic region in which they originated.

The HapMap Database

The HapMap database collects SNPs and tag SNP information from different populations in the world. One SNP it has collected is associated with the *TP53* gene. From Chapter 3 we know that *TP53* codes for either a C or a G in codon 72. The frequency in which a C or a G (called SNP rs1042522) is in the p53 allele depends on the population (Table 9-1). There are three possible base combinations in diploid cells: C/C, C/G, or G/G. When there is a C present, the amino acid coded is Pro, and when there is a G present, the amino acid coded is Arg. When both are present, the diploid cells produce equal amounts of Pro and Arg p53 polypeptides. A few studies have shown that when Arg/Arg is present there is a slight increase in cancer risk, and when Pro/Pro is present there is a slight increase in lifespan. The increase in lifespan may be due to decreased cancer risk. Interestingly, the Pro/Pro alleles correlate with a more moderate waistline, blood pressure increase with aging, and more effective glucose clearance from blood. In the HapMap database four fairly isolated populations were investigated for polymorphisms at codon 72: CEU (Utah residents with ancestry from northern and western Europe), CHB (Han Chinese in Beijing, China), JPT (Japanese in Tokyo, Japan), and YRI (Yoruba in Ibadan, Nigeria). Although the number of samples is small, the results suggest that there are different frequencies of C/C, C/G, and G/G in these populations.

9.12 PRACTICAL APPLICATION OF TAG SNP, SNP, AND MUTATION ANALYSES

One potential benefit of efficient and cost-effective sequencing is the possibility of using tag SNP, SNP, and mutation to develop risk reports on health, ancestry, and other traits for individuals. One company named 23andMe, Inc. (hereafter named 23andMe) offers ancestry and health-risk assessment by performing DNA SNP and mutation analysis for any individual. The SNP and mutations analyzed can be assessed for the probability of health risks, inherited conditions, traits, and drug response. For relatively low cost, 23andMe sends a kit to your home. A plastic test tube with a specialized lid is in the kit. The lid has detergents and a buffer solution to break open the cells. You simply spit into the test tube and attach the lid. The spit dislodges enough epithelial cells from your mouth to supply a sufficient amount of DNA for

---

[36] A random chance event is, for example, a war or natural disaster that destroys a local population.

SHOW RESULTS FOR [Your name ▼]   SEE NEW AND RECENTLY UPDATED REPORTS »

Health Risks (122) ❓

↑ ELEVATED RISKS	YOUR RISK	AVERAGE RISK
Type 2 Diabetes	32.8%	25.7%
Ulcerative Colitis	1.7%	0.8%
Crohn's Disease	0.70%	0.53%
Celiac Disease	0.22%	0.12%

more »

↓ DECREASED RISKS	YOUR RISK	AVERAGE RISK
Alzheimer's Disease	4.3%	7.2%

more »

See all 122 risk reports…

Health Risks (122) ❓

Traits (60) ❓

REPORT	RESULT
Alcohol Flush Reaction	Does Not Flush
Bitter Taste Perception	Can Taste
Earwax Type	Wet
Eye Color	Likely Brown
Hair Curl ✕	Straighter Hair on Average

See all 60 traits…

Inherited Conditions (53) ❓

REPORT	RESULT
Hemochromatosis (HFE-related)	Variant Present
ARSACS	Variant Absent
Agenesis of the Corpus Callosum with Peripheral Neuropathy (ACCPN)	Variant Absent
Alpha-1 Antitrypsin Deficiency	Variant Absent
Autosomal Recessive Polycystic Kidney Disease	Variant Absent
BRCA Cancer Mutations (Selected)	Variant Absent
Beta Thalassemia	Variant Absent
Bloom's Syndrome	Variant Absent

See all 53 carrier status…

Drug Response (25) ❓

REPORT	RESULT
Clopidogrel (Plavix®) Efficacy	Reduced
Proton Pump Inhibitor (PPI) Metabolism	Intermediate
Response to Hepatitis C Treatment	Reduced
Abacavir Hypersensitivity	Typical
Alcohol Consumption, Smoking and Risk of Esophageal Cancer	Typical

See all 25 drug response…

several sequencing analyses. When the lid seals the top of the tube, the detergents and buffer mix with the spit. You shake the test tube a few times to allow proper cell lysis and DNA preservation, and you send the kit back to the company. Within a few weeks your inherited conditions, ancestry data, trait data, and health-risk data are delivered to you online. We will call this a personal genetic report. Figure 9-23 shows a snippet of information that was made available for one anonymous customer.[37] Of course these are estimated risks based on probabilities and proper professional follow-up is necessary to confirm these risks.

Let's look at just a few entries (circled in red in Figure 9-23) in the personal genetic report of our anonymous customer. The report suggests that s/he is at slightly increased risk for Type 2 diabetes (32.8%) compared to the average U.S. resident's risk (which is already quite high at 25.7%). The individual has more than twice the risk of having ulcerative colitis (average U.S. resident risk is 0.8%). Although this customer's risk for ulcerative colitis is not that great (1.7%), s/he may consider being on

FIG. 9-23. Health risk, inherited conditions, traits, and drug response reports from 23andMe. Circled information is discussed in the text. Note that more than 200 report entries are available for this customer.

[37] In November 2013 the Federal Drug Administration (FDA) requested that 23andMe *severely limit* supplying health risk information to its customers. The concern of the FDA was that health-risk information from 23andMe was given without a report of the false-positive or false-negative rates in the DNA sequence tests performed. If, for example, an individual receives news that the BRCA1 variant is present, it might prompt the individual to seek a double mastectomy based on this knowledge without verifying the result. If the BRCA1 report is a false positive, it would be disastrous for the individual. More recently, 23andMe has complied with the FDA request and, at the time of this writing, has begun to supply health risk information again.

the lookout for signs of this disease. In the category "Inherited Conditions," the customer should be relieved that common variants (i.e., mutations) of the *BRCA1* gene are absent from the genome. *BRCA1* is a common breast cancer gene susceptibility gene with a very high penetrance. Females with a common variant have a 50–70% chance of getting breast cancer by the age of 70. Males must also be concerned because they run the risk of passing the *BRCA1* variant to their daughters. These are just a few of the more than 200 genetic characteristics that are available through this service.

As we live through this genomic era, we have the capability of gaining access to a wealth of information about our genes. This brings challenges that will require thoughtful debate and development of balanced policies and laws. For instance, we will probably need to weigh the right to have access to this detailed information against the burden it places on the medical professional community. Of course, there is great concern that, should a personal genetic report fall into the hands of future employers and insurance companies, genetic discrimination could result. A personal genetic report may empower individuals to be more proactive in living a healthy lifestyle and avoiding choices that increase the risk of diseases to which they are already somewhat genetically predisposed. On the other hand, such information can be psychologically crippling or could cause individuals to seek several expensive follow-up medical tests to rule out having the disease. At the end of the day, members of our society might ask if we have the right to know our own genes.

9.13   WHAT IS THE SMALLEST GENOME?

We began this chapter by introducing two viral genomes, MS2 and PhiX126, the first genomes sequenced, which each have fewer than 6,000 nucleotides. It might be appropriate to close the chapter with an analysis of the smallest genomes sequenced. At the time of writing this chapter, the smallest viral genomes were found in a single stranded DNA virus family known as *Circoviridae* (1,800–3,800 nucleotides). The genomes of this virus family have only two open reading frames with two known protein products—a DNA replication protein and a capsid (coat) protein. By definition, viruses depend on a host organism's genes for some aspects of their life cycle, so they can be quite small.

As described in this chapter, the first prokaryotic genome sequenced was from the free-living organism *Haemophilus influenzae* Rd (1,830,137 base pairs). An interesting question is, what is the minimum number of genes free-living organisms require to survive and reproduce? Such knowledge may prove useful for creating custom-made organisms that can be genetically modified to express a specific product or to metabolize hazardous materials. In addition, there is inherent curiosity about the minimum gene requirement needed to sustain an organism. In 2006, Atsushi Nakabachi and Masahira Hattori (RIKEN Institute, Japan) in collaboration with Nancy Moran (University of Arizona, Tuscon) reported the genome sequence of the bacteria *Carsonella ruddii*. *C. ruddii* is an endosymbiont bacterium that resides in the cells of the abdomen of a psyllid (jumping plant louse)—an insect that uses plant sap as its source of nutrition.

An endosymbiont is an organism that lives in the body or cells of another organism. The *C. ruddii* genome has only 159,662 base pairs and contains 182 predicted open reading frames. The genome size is less than 10% the size of the *H. influenzae* Rd bacterial genome. Of the predicted genes, 35% are involved in translation and 17% are involved in amino acid metabolism. The gene density of the genome is very high, with 97% of the genome devoted to genes that code for protein and functional RNAs (less than 2% of the human genome is devoted to these tasks). Like the PhiX174 virus, there are many overlapping genes in the *C. ruddii* genome. Incredibly, there are no genes that appear to code for lipid and nucleotide metabolism. How, then, does the bacterium create its cell membrane or synthesize nucleotides for RNA and DNA?

It is likely that these molecules are imported from the host, the psyllid. This brings into question whether this organism is really free-living.

C. ruddii appears to be an organism that supplies psyllids with essential amino acids—hence the reason that many of its genes are involved in amino acid metabolism. Given the lack of genes necessary to synthesize lipids and nucleic acids, *C. ruddii* may be in an evolutionary transition phase. It may be transitioning from a free-living organism to an organelle, which is a specialized membrane bound subcellular compartment that carries out essential functions for the cell. It is likely that the mitochondrion and the chloroplast are two organelles that were once bacterial endosymbionts—living in the cells of hosts.

The lack of important classes of genes in the genome of *C. ruddii* illustrates that it might be difficult to pinpoint the minimum number of genes necessary to sustain a free-living organism. There is likely to be a spectrum of environmental niches that sustain organisms. As organisms adapt to their particular niche they shed the genes they do not need. Simply isolating the organism with the fewest genes does not give the minimum number of genes required to sustain a free-living organism because it does not take symbiosis into account. To arrive at the minimum number of genes for such an organism we need to develop a more precise definition of free-living organism that scientists can agree on.

SUMMARY

Genomes contain genes and instructions for regulation of gene expression. In genomes lie the keys to understanding how species differ. Many mammals (including humans) contain approximately 20,000 genes, but primates may have a more complex scheme for alternatively splicing the transcripts of genes amongst their different tissues. Searching the genome for clues about what makes us human is an area of intense research and largely lies within the instructions for regulating gene expression. The landscape of the human genome indicates that less than 2% of the DNA codes for proteins and functional RNAs (ribosomal RNAs and tRNAs). Interestingly, up to 70% of the genome is transcribed and it is possible these RNAs may contribute to gene expression regulation.[38] Such non-coding RNAs will likely shed light on new modes of gene regulation. In humans, introns and repetitive sequences comprise the majority of the genome. Interestingly, some of the repetitive DNA can be used as

a DNA fingerprint or DNA profile to distinguish one individual from another. Sequence information of the first human genomes is the result of dideoxy sequencing (or modified forms of this technique). Next-generation sequencing techniques continue to be refined to be more efficient and cost-effective. It is likely that soon, most individuals will be able to afford to have their exomes sequenced. Annotation of new genomes, especially prokaryotic genomes, is becoming more commonplace because of the significant number of genomes that have already been sequenced and annotated. Software programs that detect repeat sequences and open reading frames in prokaryotes are readily available.

Future research into the human genome will undoubtedly be focused on answering the question "What makes us human?" The HapMap database is a catalog of simple nucleotide polymorphisms and copy number variations in the human genome. The HapMap database includes tag SNPs that can be used to track haplotypes. The haplotypes can then be used in genome-wide association studies to correlate haplotypes with diseases and traits.

---

[38] This is a somewhat controversial issue. See ENCODE Project Consortium (2012) and opposing arguments by Graur et al. (2013).

EXERCISES

1. **Draw a synteny plot of two genomes that have the following gene arrangements:**

   a. Genome 1: origin-A-B-C-D-E
   Genome 2: origin-A-B-non-coding intervening sequence-C-D-D-E

   b. Genome 1: origin-A-B-C-D-E
   Genome 2: origin-A-B-D-C-E

   Note that in Genome 2 there is an inversion.

   c. Genome 1: origin-A-B-C-D-A
   Genome 2: origin-A-B-C-D-A

2. **What was the first genome sequenced?**

3. **Craig Venter's group reported sixfold coverage of the *H. influenzae* Rd genome. Given that the genome is 1,830,137 base pairs, calculate the number of base pairs that are likely to remain unsequenced.**

4. **There is a human MDM2 pseudogene located on chromosome 2. Using available Internet software tools, can you describe to which class this pseudogene belongs?**

5. **Look up the sequence of the FGA repeat. Give the sequences of the two FGA alleles for suspect 2 in Box 9-2. Why does the total number of nucleotides in the repeated region not match the** nucleotide number in Box 9-2 (~240 nucleotides according to the graph)?

6. **The instructor will assign you one of the following human diseases. Find a single gene that contributes to the disease. Give the name of the gene, describe its molecular function, and give its chromosome location (for example, *TP53* is located at chromosome 17p13.1). Give the name of the gene that is more telomeric (more toward the telomere) and the name of the gene that is more centromeric (more toward the centromere) than your assigned disease gene. Hint: start by exploring the database Online Mendelian Inheritance of Man (OMIM).**

Fragile X mental retardation syndrome	Lynch syndrome I	Cockayne syndrome, type A	Bloom syndrome	Huntington disease
Charcot-Marie-Tooth disease	Werner syndrome	Ataxia telangiectasia	Parkinson disease, autosomal recessive, juvenile	Familial adenomatous polyposis
Familial hypercholesterolemia, autosomal recessive	Retinoblastoma	Paroxysmal extreme pain disorder	Sickle cell anemia	Duchenne muscular dystrophy

7. **Annotate the bacterial genome sequence available on the publisher's website. You may use any online annotation tool. Do not forget to use RepeatMasker to mask repeat sequences. In addition answer the following:**

   a. What is the name of the species of the genome?
   b. What is the size of the genome?
   c. Approximately how many genes does this genome contain?
   d. Determine the approximate locations of large repeat sequences in this genome.

ANSWERS TO THOUGHT QUESTIONS

9-1. Steps to answering this question: (1) Record colors of bands starting from the bottom of the gel; (2) Use the color key to associate each color with a nucleotide base; (3) Write the complementary sequence. The complementary sequence is the natural (template) strand.

   Color key A = Green (G), T = Red (R),
   G = Yellow (Y), C = Blue (B)

```
           color:  RGYBGGGGYYBGBRYYYBGY
sequencing strand:  5'TAGCAAAAGGCACTGGGCAG3'
   template strand:  3'ATCGTTTTCCGTGACCCGTC5'
```

9-2. If there was sixfold coverage of the genome, then the number of nucleotides sequenced is equal to 6 × 1,830,137 nucleotides, which is 10,980,822 nucleotides.

9-3. Chromosomes 4, 12, 17, 18.

9-4. No. Given your tag SNP, here are the haplotypes for each individual at positions 1, 2, 3, 5, and 10:

   Individual #1: C, C, T, T, T
   Individual #2: C, C, T, T, T
   Individual #3: C, G, T, A, T
   Individual #4: T, C, T, T, T
   Individual #5: T, C, A, A, A

This is not a true tag SNP because individuals #1 and #2 have the same haplotype.

REFERENCES

Aitman, T. J., et al. 2006. "Copy Number Polymorphism in Fcgr3 Predisposes to Glomerulonephritis in Rats and Humans." *Nature* 439: 851–855.

Barbosa-Morais, N. L., M. Irimia, Q. Pan, H. Y. Xiong, S. Gueroussov, L. J. Lee, V. Slobodeniuc, C. Kutter, S. Watt, R. Colak, T. Kim, C. M. Misquitta-Ali, M. D. Wilson, P. M. Kim, D. T. Odom, B. J. Frey, and B. J. Blencowe. 2012. "The Evolutionary Landscape of Alternative Splicing in Vertebrate Species." *Science* 338: 1587–1593.

Bhasin, M., H. Zhang, E. L. Reinherz, and P. A. Reche. 2005. "Prediction of Methylated CpGs in DNA Sequences Using a Support Vector Machine." *FEBS Letters* 579: 4302–4308.

Bojesen, S. E., and B. G. Nordestgaard. 2008. "The Common Germline Arg72Pro Polymorphism of p53 and Increased Longevity in Humans." *Cell Cycle* 7: 158–163.

Borodovsky, M., and J. McIninch. 1993. "GeneMark: Parallel Gene Recognition for Both DNA Strands." *Computers & Chemistry* 17: 123–133.

Bragg, L. M., G. Stone, M. K. Butler, P. Hugenholtz, and G. W. Tyson. 2013. "Shining a Light on Dark Sequencing: Characterising Errors in Ion Torrent PGM Data." *PLoS Computational Biology* 9: e1003031.

Cappuzzo, F., et al. 2005. "Epidermal Growth Factor Receptor Gene and Protein and Gefitinib Sensitivity in Non-small-Cell Lung Cancer." *Journal of the National Cancer Institute* 97: 643–655.

Chen, J. M., C. Férec, and D. N. Cooper. 2006. "LINE-1 Endonuclease-Dependent Retrotranspositional Events Causing Human Genetic Disease: Mutation Detection Bias and Multiple Mechanisms of Target Gene Disruption." *Journal of Biomedicine and Biotechnology* 2006: 1–9.

Chimpanzee Sequencing and Analysis Consortium. 2005. "Initial Sequence of the Chimpanzee Genome and Comparison with the Human Genome." *Nature* 437: 69–87.

Cordaux, R., and M. A. Batzer. 2009. "The Impact of Retrotransposons on Human Genome Evolution." *Nature Reviews Genetics* 10: 691–703.

Creyghton, M. P., A. W. Cheng, G. G. Welstead, T. Kooistra, B. W. Carey, E. J. Steine, J. Hanna, M. A. Lodato, G. M. Frampton, P. A. Sharp, L. A. Boyer, R. A. Young, and R. Jaenisch. 2010. "Histone H3K27ac Separates Active from Poised Enhancers and Predicts Developmental State." *Proceedings of the National Academy of Sciences USA* 107, 21931–21936.

Deaton, A. M., and A. Bird. 2011. "CpG Islands and the Regulation of Transcription." *Genes and Development* 25: 1010–1022.

ENCODE Project Consortium. 2012. "An Integrated Encyclopedia of DNA Elements in the Human Genome." *Nature* 489: 57–74.

Fiers, W., R. Contreras, F. Duerinck, G. Haegeman, D. Iserentant, J. Merregaert, W. Min Jou, F. Molemans, A. Raeymaekers, A. Van den Berghe, G. Volckaert, and M. Ysebaert. 1976. "Complete Nucleotide Sequence of Bacteriophage MS2 RNA: Primary and Secondary Structure of the Replicase Gene." *Nature* 260: 500–507.

Fleishmann, R. D., et al. 1995. "Whole Genome Random Sequencing and Assembly of *Haemophilus influenzae* Rd." *Science* 269: 496–512.

Gibson, G., and S. V. Muse. 2009. *A Primer of Genome Science*, 3rd ed. Sunderland, MA: Sinauer and Associates.

Graur, D., Y. Zheng, N. Price, R. B. Azevedo, R. A. Zufall, and E. Elhaik. 2013. "On the Immortality of Television Sets: 'Function' in the Human Genome According to the Evolution-Free Gospel of ENCODE." *Genome Biology and Evolution* 5: 578–590.

Grigoriev, I. V., et al. 2012. "The Genome Portal of the Department of Energy Joint Genome Institute." *Nucleic Acids Research* 40 (Database issue): D26–32.

Han, J. Y, G. K. Lee, D. H. Jang, S. Y. Lee, and J. S. Lee. 2008. "Association of p53 Codon 72 Polymorphism and MDM2 SNP309 with Clinical Outcome of Advanced Nonsmall Cell Lung Cancer." *Cancer* 113: 799–807.

Hesselberth, J. R., X. Chen, Z. Zhang, P. J. Sabo, R. Sandstrom, A. P. Reynolds, R. E. Thurman, S. Neph, M. S. Kuehn, W. S. Noble, S. Fields, and J. A. Stamatoyannopoulos. 2009. "Global Mapping of Protein-DNA Interactions In Vivo by Digital Genomic Footprinting." *Nature Methods* 6: 283–289.

Iafrate, A. J., L. Feuk, M. N. Rivera, M. L. Listewnik, P. K. Donahoe, Y. Qi, S. W. Scherer, and C. Lee. 2004. "Detection of Large-Scale Variation in the Human Genome." *Nature Genetics* 36: 949–951.

Innocence Project. 2014. "DNA Exonerations Nationwide." http://www.innocenceproject.org/Content/DNA_Exonerations_Nationwide.php. Accessed August 3.

Jacq, C., J. R. Miller, and G. G. Brownlee. 1977. "A Pseudogene Structure in 5S DNA of Xenopus Laevis." *Cell* 12: 109–120.

Jones, P. A. 2012. "Functions of DNA Methylation: Islands, Start Sites, Gene Bodies and Beyond." *Nature Review Genetics* 13: 484–492.

Kent, W. J., C. W. Sugnet, T. S. Furey, K. M. Roskin, T. H. Pringle, A. M. Zahler, and D. Haussler. 2002. "The Human Genome Browser at UCSC." *Genome Research* 12: 996–1006.

Kim, Y. J., J. Lee, and K. Han. 2012. "Transposable Elements: No More 'Junk DNA.'" *Genomics and Informatics* 10: 226–233.

Koboldt, D. 2012. "Mass Genomics: Medical Genomics in the Post Genome Era." http://massgenomics.org. Accessed April 1, 2013.

Lyons, E., and M. Freeling. 2008. "How to Usefully Compare Homologous Plant Genes and Chromosomes as DNA Sequences." *The Plant Journal* 53: 661–673.

Lyons, E., B. Pedersen, J. Kane, and M. Freeling. 2008. "The Value of Nonmodel Genomes and an Example Using SynMap Within CoGe to Dissect the Hexaploidy That Predates Rosids." *Tropical Plant Biology* 1: 181–190.

Médigue, C., and I. Moszer. 2007. "Annotation, Comparison and Databases for Hundreds of Bacterial Genomes." *Research in Microbiology* 158: 724–736.

Merkin, J., C. Russell, P. Chen, and C. B. Burge. 2012. "Evolutionary Dynamics of Gene and Isoform Regulation in Mammalian Tissues." *Science* 338: 1593–1599.

Nakabachi, A., A. Yamashita, H. Toh, H. Ishikawa, H. E. Dunbar, N. A. Moran, and M. Hattori. 2006. "The 160-Kilobase Genome of the Bacterial Endosymbiont Carsonella." *Science* 314: 267.

Nishikimi M., R. Fukuyama, S. Minoshima, N. Shimizu, and K. Yagi. 1994. "Cloning and Chromosomal Mapping of the Human Nonfunctional Gene for L-gulono-gamma-lactone Oxidase, the Enzyme for L-ascorbic Acid Biosynthesis Missing in Man." *Journal of Biological Chemistry* 269: 13685–13688.

Ozinsky, S. 2010. "Can DNA Profiling Be the Answer to Reduced Crime?" http://www.ohwatch.co.za/sheryls-oh-watch-blog/2010/01/09/can-dna-profiling-be-the-answer-to-reduced-crime/. Accessed April 1, 2013.

Pan, Q., O. Shai, L. J. Lee, B. J. Frey, and B. J. Blencowe. 2008. "Deep Surveying of Alternative Splicing Complexity in the Human Transcriptome by High-Throughput Sequencing." *Nature Genetics* 40: 1413–1415.

Papasaikas, P., and J. Valcárcel. 2012. "Evolution. Splicing in 4D." *Science* 338: 1547–1548.

Patrushev, L. I., and I. G. Minkevich. 2008. "The Problem of the Eukaryotic Genome Size." *Biochemistry (Moscow)* 73: 1519–1552.

Patthy, L. 2009. *Protein Evolution*, 2nd ed. Malden, MA: Blackwell Publishing.

Pevsner, J. 2009. *Bioinformatics and Functional Genomics*, 2nd ed. Hoboken, NJ: John Wiley and Sons.

Pratt, C. W., and K. Cornely. 2011. *Essential Biochemistry*, 2nd ed. Hoboken, NJ: John Wiley and Sons.

Pseudogene.org. 2014. http://www.pseudogene.org/ Accessed August 3.

Reiling, E., V. Lyssenko, J. M. Boer, S. Imholz, W. M. Verschuren, B. Isomaa, T. Tuomi, L. Groop, and M. E. Dollé. 2012. "Codon 72 Polymorphism (rs1042522) of *TP53* Is Associated with Changes in Diastolic Blood Pressure over Time." *European Journal of Human Genetics* 20: 696–700.

Ronaghi, M., S. Karamohamed, B. Pettersson, M. Uhlén, and P. Nyrén. 1996. "Real-Time DNA Sequencing Using Detection of Pyrophosphate Release." *Analytical Biochemistry* 242: 84–89.

Sakharkar, M. K., V. T. Chow, and P. Kangueane. 2004. "Distributions of Exons and Introns in the Human Genome." *In Silico Biology* 4: 387–393.

Sanger, F., G. M. Air, B. G. Barrell, N. L. Brown, A. R. Coulson, C. A. Fiddes, C. A. Hutchison, P. M. Slocombe, and M. Smith. 1977. "Nucleotide Sequence of Bacteriophage PhiX174 DNA." *Nature* 265: 687–695.

Sebat, J., et al. 2004. "Large-Scale Copy Number Polymorphism in the Human Genome." *Science* 305: 525–528.

Vallenet, D., L. Labarre, Z. Rouy, V. Barbe, S. Bocs, S. Cruveiller, A. Lajus, G. Pascal, C. Scarpelli, and C. Médigue. 2006. "MaGe: A Microbial Genome Annotation System Supported by Synteny Results." *Nucleic Acids Research* 34: 53–65.

Venter, J. C. 2007. *A Life Decoded*. New York: Penguin Group.

Zhang, F., W. Gu, M. E. Hurles, and J. R. Lupski. 2009. "Copy Number Variation in Human Health, Disease, and Evolution." *Annual Review of Genomics and Human Genetics* 10: 451–481.

TRANSCRIPT AND PROTEIN EXPRESSION ANALYSIS

10.1 INTRODUCTION

It is likely that sustained advances in DNA sequencing technology will continue to drive the genomics field for the foreseeable future. Additional genome sequences from organisms of different species, as well as from unique organisms within species, will be available for bioinformaticists to study and compare. The central dogma predicts that once the DNA sequence is known, both the RNA sequence and protein sequence can be deduced, although alternative splicing of transcripts and post-translational processing of proteins complicate this deduction. Notwithstanding these complications, knowledge of the genome presents an opportunity to gain some understanding of the identities of all expressed RNAs and proteins. The transcriptome is defined as the identity and quantity of the entire population of RNA expressed from the genome in a cell. Similarly, the proteome is the identity and quantity of the proteins expressed from the entire genome in a cell. In large measure, the transcriptome and proteome dictate the cell's qualities and, ultimately, the organism's phenotype. RNA and protein expression from genes have been studied for decades, often at the pace of one gene at a time. But with the easy availability of DNA sequences, it is now relatively straightforward to study RNA and protein expression of an entire genome in one experiment.

transcriptome The identity, length, and relative level of all transcripts expressed in a cell.

There are some caveats that must be considered when studying the transcriptome and proteome. Unlike the genome, which for the most part is relatively stable from cell to cell and from organism to organism in the same species, the transcriptome and proteome of cells vary depending on many factors. For example, the transcriptome and proteome of a bacterial cell will vary depending on the temperature, the availability of nutrients, the density of the bacteria population, the presence of toxins, and so on. Even if all environmental factors are the same and the cells are genetically identical, single-cell studies have shown that there is gene expression heterogeneity in the population due to position of the cells in space and differences in cells' life cycle stage. Although typical transcriptome and proteome analyses are conducted on a population of cells, single-cell variations should be kept in mind. In this chapter, you will learn about methods for studying the transcriptome and proteome and review some of the software programs and databases associated with these fields. As we proceed, let's take into consideration a scenario. Recall that p53 is a transcription factor that controls the expression of several genes in the genome. Prior to the advent of advanced sequencing and hybridization techniques, many of these genes, often identified one at a time, were found to directly contribute to p53's tumor suppressor activities. Is it possible, then, to identify all the transcripts in the human genome regulated by p53? What are the mechanisms that p53 uses to regulate these transcripts? Let's bear these questions in mind as we discuss transcript and protein expression analysis. Prior to tackling these questions, it will be helpful to review the history of gene expression research.

10.2 BASIC PRINCIPLES OF GENE EXPRESSION REGULATION

To have an appreciation of the importance of gene expression measurement, a brief background into the history of its development is necessary. In 1941, while at the University of Paris, Jacques Monod studied the growth of bacteria in liquid broth. He could follow the growth by measuring the cloudiness of the liquid broth. The more dense the bacteria population the more cloudy the broth became. Bacterial division could be quantitatively measured using a colorimeter, an instrument much like a spectrophotometer, which measures the amount of light that passes through the broth. Because the amount of light that passes through the broth decreases as the cloudiness, or optical density, increases, Monod could follow bacterial growth by monitoring the optical density of the broth. Monod placed two nutritional sugars in the broth—one was glucose (a monosaccharide), and the other was a more complex sugar called lactose (a disaccharide). He found that bacterial growth occurred in two phases, a phenomenon he named diauxie (roughly translated as biphasic). The dividing bacteria first consumed all the glucose, and, subsequently, the bacteria stopped dividing for a short period of time. After approximately 15 minutes, the bacteria began

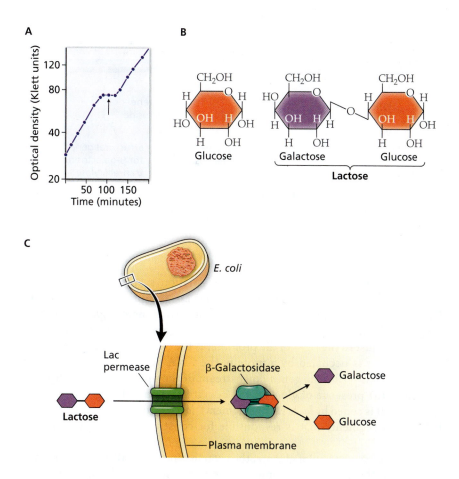

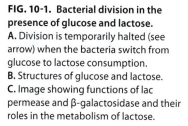

FIG. 10-1. Bacterial division in the presence of glucose and lactose.
A. Division is temporarily halted (see arrow) when the bacteria switch from glucose to lactose consumption.
B. Structures of glucose and lactose.
C. Image showing functions of lac permease and β-galactosidase and their roles in the metabolism of lactose.

to consume lactose and rapidly divided again. Figure 10-1A shows the growth curve of the bacterial culture in this experiment. Monod, joined by François Jacob at the Pasteur Institute in Paris, would spend the next 25 years exploring the molecular basis of diauxie.[1]

Through mutational and enzyme kinetic analyses, Monod and Jacob worked out the basic mechanisms of the gene regulatory system that accounts for the switch from metabolizing glucose to metabolizing lactose. The three genes responsible for import of lactose into the cell and its initial metabolism are *lacA*, *lacY*, and *lacZ*. The protein product of *lacA* is the enzyme acetyl transferase, which transfers an acetyl group from acetyl-CoA onto the hydroxyl group of the number six carbon of sugars. Its specific role in lactose metabolism remains unclear. The protein product of *lacY* is lactose permease, a membrane protein that imports lactose from the broth into the bacteria; and the protein product of *lacZ* is β-galactosidase, an enzyme that cleaves lactose into galactose and glucose—a necessary step in the metabolism of lactose (Figure 10-1B and C). The delay in bacterial division in the broth containing lactose (after the glucose has been consumed) is due to the time required to ramp up the expression of *lacY* and *lacZ* genes. *lacA*, *lacY*, and *lacZ* comprise part of a suite of genes that Monod and Jacob named the lac operon.

---

[1] Jacob and Monod used bacteria as their sole organism of study. Through mutation analysis, enzyme kinetic measurements, and deductive logic, they worked out the details of gene regulation. Their gene regulation model, the lac operon, is the bedrock on which more elaborate mechanisms of gene regulation have been elucidated. Incredibly, they worked out gene regulation in the absence of protein sequencing and DNA sequencing technologies. They, along with André Lwoff (in the field of virology), received the Nobel Prize in Medicine or Physiology in 1965 for their pioneering work.

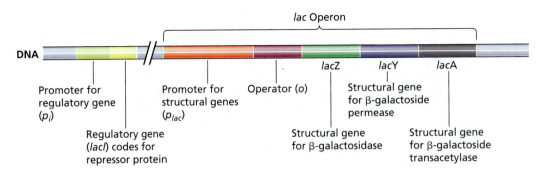

FIG. 10-2. The lac operon.

lac operon A bacterial gene regulatory system responsible for regulated import and hydrolysis of lactose.

The lac operon is a transcriptional regulation system that controls the import and hydrolysis of lactose. There are five DNA segments in the lac operon (Figure 10-2). In Figure 10-2, the red segment is the promoter p_{lac}, where RNA polymerase binds for transcription of the *lacZ*, *lacY*, and *lacA* genes downstream. The segment just to the right of the promoter is the operator, *o* (purple segment). If the operator is occupied by a protein, the RNA polymerase is blocked and is unable to transcribe the genes downstream of the operator. During bacterial division in the presence of glucose, the operator is bound to a protein called a repressor that is coded by *lacI* (yellow segment). *lacI* has its own promoter named p_i (light green segment) that is responsible for constitutively (constantly) transcribing *lacI*.[2]

During consumption of glucose, the protein repressor, called the LacI repressor, binds to the operator and prevents the transcription of downstream genes. Once glucose is depleted, a small amount of lactose is imported and converted into a closely related molecule known as allolactose.[3] Allolactose binds to the LacI repressor, triggering a conformational change in the repressor, which causes it to dissociate from the operator. The RNA polymerase then transcribes *lacZ*, *lacY*, and *lacA* of the lac operon, enabling the bacteria to metabolize lactose when glucose is absent. In sum, the lac operon is an inducible system and the *lacZ*, *lacY*, and *lacA* genes are inducible genes. This mode of gene induction is called derepression because a repressor is released from the operator. In contrast to an inducible system, a constitutive system is one in which a gene is constantly expressed from a promoter. There is no operator in a constitutive system. The inducible system is important because it allows the bacteria to conserve resources by not expressing the lac operon genes needlessly. Only when the need arises are these genes induced.

Aside from the derepression mode, there is another mode of induction named activation. In the lac operon, a protein named CAP (known either as catabolite

---

[2] *lacI* is located just upstream of the lac operon. However, experiments have shown that *lacI* could be located at other positions in the bacterial genome and remain effective in controlling the lac operon because its product, the repressor, is diffusible and can find its operator anywhere within the bacterial genome. The repressor protein is called a "trans" factor because the gene that codes for it is not dependent on it being in a specific location within the genome. On the other hand, the operator, *o*, is a segment of DNA and is considered a "cis" factor. A cis factor is required to be located at a particular position within the genome to be effective. Proteins are trans factors, and most, but not all, DNA segments are cis factors.

[3] It may seem confusing that lactose may be imported before the *lacY* is transcribed. The initial import process is not clear, but it is possible that a very low level of permease protein is present on the bacterial membrane surface to import small amounts of lactose.

activator protein or cAMP receptor protein) is capable of binding to a DNA segment between *lacI* and p_{lac} (not shown in Figure 10-2). CAP recruits RNA polymerase to the promoter and increases its ability to transcribe genes. CAP is most active when glucose levels drop. When glucose levels decrease, a small molecule named cyclic AMP (cAMP) is synthesized inside the cell and binds to CAP, altering its conformation such that it possesses high affinity for the CAP binding site upstream of p_{lac}. CAP binding to DNA causes RNA polymerase to bind more frequently to p_{lac}, thereby increasing the transcription of *lacZ*, *lacY*, and *lacA*. In summary, you must consider two modes of transcriptional regulation; the first mode controlled by a repressor is derepression, and the second mode controlled by an activator is called activation.

The principles of the lac operon system are applicable to other organisms as well. Promoters are DNA segments to which RNA polymerases bind. Repressor and activator binding sites are generally called response elements (or responsive elements); as you would expect, the DNA sequences of the response elements dictate which repressor or activator will bind. For example, E2F1 is an activator (sometimes called a transcriptional transactivator) expressed in mammalian cells that binds to the response element TTTCCCGC. There are several bioinformatics software programs available that can scan a segment of DNA for response elements (see exercise 1 at the end of this chapter). Scanning a DNA segment for potential response elements is usually the first step prior to experimentally identifying activators and repressors that control gene expression. Some proteins can act as both repressors and activators, depending on the types of protein modifications that they have. Given this potential dual role, repressors and activators are generally called transcription factors. For instance, the p53 protein is a transcription factor that can adopt an activator or repressor role depending on its post-translational modification pattern and the proteins bound to it.

In fact, research into other gene regulatory systems has shown that the arrangement of response elements relative to that of the promoter can differ from that of the lac operon, as they may be found far away from the promoter (Figure 10-3). In these cases, the three-dimensional structure of DNA brings the response elements into close proximity to the promoters, much like the active site of an enzyme may be composed of amino acid residues that come together from different regions of a

response element (also known as responsive element) Segment of DNA that binds to protein activators and repressors. Upon binding, the DNA-protein complex influences gene transcription rates, often by binding to RNA polymerase II.

enhancer A specialized response element located more than 200 nucleotides away from a gene promoter. The enhancer can be altered in its location and in its orientation relative to the promoter and still affect transcriptional activation.

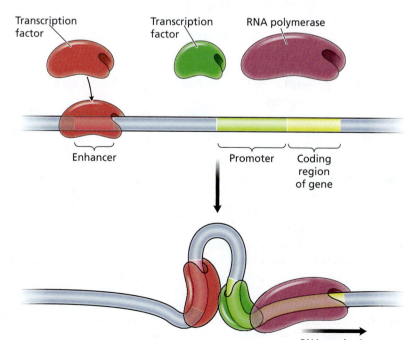

FIG. 10-3. The basics of regulation of transcription.

polypeptide to form a catalytic site. Some response elements are more than 200 base pairs away from the promoter. These response elements can be inverted within the DNA and even be placed in different parts of the genome up to thousands of base pairs away from the promoter and still affect transcription activity. Such orientation- and location-independent response elements are called enhancers. Now that we have covered some mechanisms for controlling inducible genes, we can discuss how to measure the rate and level of transcription. Then we can use such measurements to deepen our understanding of how the transcriptome responds to a changing cellular environment.

10.3 MEASUREMENT OF TRANSCRIPT LEVELS

Now that we have briefly reviewed the background of transcriptional regulation, let's explore RNA expression analysis with the methods of nuclear run-on: Northern, microarray, and RNA-seq. A theme shared by these methods is polynucleic acid hybridization, in which two single complementary strands can specifically bind each other and form a double strand hybrid. The hybrids can be composed of DNA:DNA, DNA:RNA, or RNA:RNA strands.

nuclear run-on An experimental method used to determine the rate of transcription. As DNA is transcribed, labeled nucleotides are incorporated into the transcripts. The transcripts are hybridized to oligonucleotides, and unhybridized molecules are washed away. From the amount of label incorporated into the transcripts within a specified time period the transcription rate can be calculated.

One parameter of gene expression is the rate of transcription. Early experiments, called nuclear run-on experiments, used radiolabeled nucleotides to measure the rate at which transcription takes place in nuclei isolated from cells. The rate at which a radiolabeled nucleotide (sometimes called a radionucleotide) is incorporated into a specific RNA is equal to the rate of transcription of that RNA. The identification of a specific RNA within the total population of nuclear RNA is accomplished by hybridization to a single strand of DNA of known complementary sequence. In nuclear run-on experiments, DNA with a known sequence is tightly bound to either chemically treated paper (called nitrocellulose) or to nylon membrane and denatured so that the DNA is single stranded. Nuclei from eukaryotic cells are treated with radionucleotides for a period of time, after which the transcription reaction is stopped by rapid protein denaturation. The radiolabeled transcript products are incubated with the nylon membrane-bound DNA. The temperature and salt concentration of the incubation medium is adjusted to produce only perfectly hybridized sequences. Transcript sequences that do not complement the DNA fail to bind to the DNA, and these nonhybridized transcripts are washed away. The amount of radioactivity in the surviving RNA:DNA hybrids can be assessed by measuring the amount of radioactivity by liquid scintillation counting or by densitometry measurement of X-ray film that has been exposed to the nylon membrane. This information, together with the length of time during which transcription occurred and the number of radionuclides expected to be incorporated into the molecule, are used to determine the rate of transcription.

Northern blotting An experimental method used to determine relative steady-state level and length of specific transcripts.

Nuclear run-on proved useful for determining the transcription rate; however, it was soon recognized that cells control the amount of transcript not only by increasing the rate of transcription, but also by controlling the rate of transcript degradation. It became important, therefore, to measure steady-state levels of transcripts. With Northern blotting, steady-state levels of transcripts of known sequence are measured by first isolating all RNAs from cells and separating them by size on an agarose or polyacrylamide gel by electrophoresis (Figure 10-4A).[4] In electrophoresis, a voltage gradient is created where one end of the gel is positively charged relative to the opposite end. RNAs travel toward the positive end and are separated by size. The separated RNAs are transferred from the gel to a nylon membrane. The nylon

---

[4] In practice, it is often the case that not all cellular RNA is used for Northern analyses. In eukaryotes, most RNAs that code for proteins undergo a modification called poly adenylation (polyA addition) on the 3' end. Scientists can use an affinity column purification procedure to isolate polyA mRNAs from other RNAs and then separate only the polyA mRNAs on agarose or polyacrylamide gels.

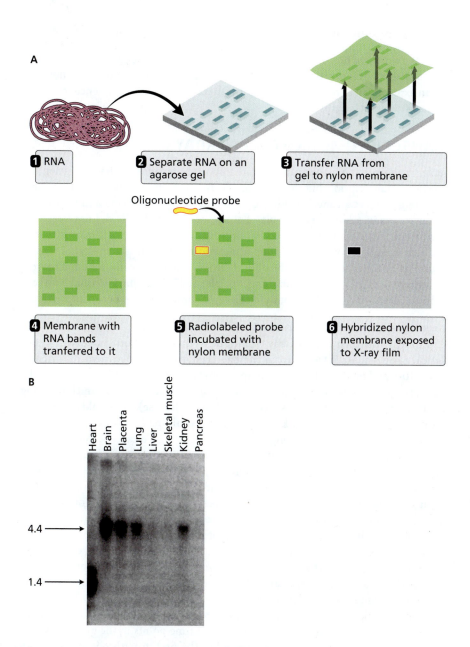

FIG. 10-4. Northern analysis.
A. Steps in Northern analysis method. In the gel, a different sample of RNA is run in each lane. Although the image shows that only three or four transcripts are separated in a lane, in practice there are thousands of transcripts applied to each lane, resulting is a smear of RNA. **B.** Northern analysis that identifies a specific transcript expressed in different tissues. The heart tissue expresses a transcript that is shorter than the transcripts expressed in other tissues. Numbers to the left of the blot represent the lengths of two standards (standard bands not shown). One standard is 4.4 kb and the other is 1.4 kb.

membrane is incubated with a radiolabeled oligonucleotide (or single strand radio-labeled cDNA) of known sequence that is complementary to the transcript of interest. The position of the hybridized nucleic acid is visualized with X-ray film or a phosphorimager.

The key to Northern blotting is that the sequence of the oligonucleotide is known and the position of the hybrid on the nylon membrane is known relative to RNA standards of known length. With this information, you can determine the identity and length of a particular RNA. Furthermore, the relative amount of a particular RNA expressed in different samples can be determined by measuring the densities of the hybrids (detected as bands) on the film (Figure 10-4B). With Northern blotting, many gene regulation systems were discovered, but the method proved to be somewhat cumbersome because it was difficult to process more than a few samples at the same time. With the availability of more genome sequences, researchers turned their attention to developing methods that would simultaneously measure transcripts from hundreds and even thousands of genes.

Thought Question 10-1

In the Northern blot shown in Figure 10-4B, the heart tissue expresses a transcript that is shorter than the transcript expressed in other tissues. Offer an explanation as to why two transcripts that share the same sequence could have different lengths.

10.4 TRANSCRIPTOME AND MICROARRAYS

Nuclear run-on and Northern analysis are important techniques that help scientists investigate transcription, but they both are fairly laborious and give insight into only a few transcripts at a time. The idea of simultaneously quantifying and identifying all transcripts in a biological sample of tissue or collected cells did not become a reality until the 1990s. In this period Pat Brown and his colleagues (Stanford University, Palo Alto, California) developed a method to measure steady-state levels of thousands of RNA transcripts (Box 10-1). This method, known as the microarray method, employs the same hybridization principle used in nuclear run-on and Northern analysis. Figure 10-5 shows the outline of a typical microarray experiment.

Stages of a Microarray Experiment

Let's apply Pat Brown's microarray technique to the lac operon system in a hypothetical experiment. If we perform the bacterial growth experiment outlined in Figure 10-1 and measure LacZ transcript, we would predict that its level would be high during the second growth phase (in the presence of lactose alone) and low during the first growth phase (in the presence of glucose and lactose). We would also predict that other transcripts, such as LacY and LacA, would be high during the second growth phase, because they belong to the same operon. We can use microarrays to determine the identity and levels of all gene transcripts that are differentially regulated under the two growth conditions. One way to create a microarray plate is by attaching single strand oligonucleotides to a plate. Each particular spot on the plate contains thousands of oligonucleotides with the same sequence. The oligonucleotides have sequences identical to short segments of the DNA coding strands, in other words, sequence that is the same as a short sequence of a specific RNA. Therefore, each spot specifies the transcript of a particular gene.[5] There are 4,290 genes that express proteins in *E. coli* (K-12 strain), so we will need a minimum of 4,290 different spots. The oligonucleotides are called the target DNAs and they can be synthesized directly on the plates made of glass, plastic, or chemically treated paper.

Once the microarray plate is spotted, we are ready to start stage 1 of the microarray experiment. We will collect bacteria at two time points in the growth experiment to find out which genes are expressed in the first growth phase and which genes are expressed in the second growth phase. To collect the bacteria under these conditions, two bacterial cultures are allowed to replicate in glucose/lactose broth. Bacteria are collected from one culture during the first growth phase, and other bacteria are collected during the second growth phase. In Figure 10-5, the bacteria culture on the left is collected from the first growth phase, and the bacterial culture on the right is collected from the second growth phase. In stage 2, equal numbers of bacteria from the two cultures are lysed by treatment with detergent or by sonication, and RNA is isolated. In stage 3, RNA is incubated with fluorescently tagged dNTPs and reverse transcriptase to generate cDNA. During reverse transcription, the number of fluorescently labeled cDNAs produced will correspond to the number of mRNA molecules present in the bacteria. The cDNAs representing the first growth phase culture are tagged with Cy3 (green)-tagged dNTPs, and the second growth phase

microarray technique A method to specify the level of all or a subset of transcripts in the organism. Usually, the method requires reverse transcription of RNA into cDNA, labeling the cDNA and hybridization of the cDNA to complementary single strand DNA molecules bound to a surface.

---

[5] The length of the DNA within a spot that specifies a gene may vary depending on the experiment. It is now typical that oligonucleotides with an approximate length of 30 nucleotides are used to specify a gene. Furthermore, in a microarray experiment more than one spot will specify a gene. In fact, one gene may be specified by four different oligonucleotides, each specific for a different region of the gene's transcript.

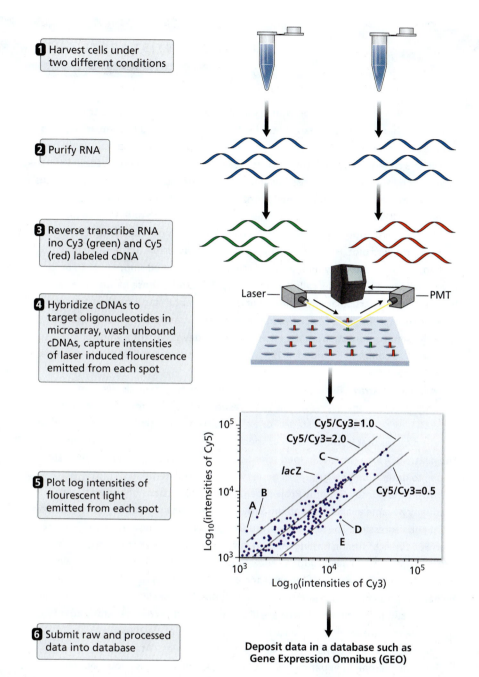

1 Harvest cells under two different conditions

2 Purify RNA

3 Reverse transcribe RNA ino Cy3 (green) and Cy5 (red) labeled cDNA

4 Hybridize cDNAs to target oligonucleotides in microarray, wash unbound cDNAs, capture intensities of laser induced flourescence emitted from each spot

Laser — PMT

5 Plot log intensities of flourescent light emitted from each spot

6 Submit raw and processed data into database

Deposit data in a database such as
Gene Expression Omnibus (GEO)

FIG. 10-5. Outline of stages for microarray analysis of RNA expression.

culture cDNAs are tagged with Cy5 (red)-tagged dNTPs. In stage 4, the tagged cDNAs are mixed, layered over the microarray plate, and allowed to hybridize to the target oligonucleotides attached to the plate. The attached target oligonucleotides are in excess of the added tagged cDNAs, so the number of labeled molecules that hybridize to any one spot represents the number of molecules of the mRNA with the complementary sequence. For each spot, a laser activates the fluorescently tagged cDNAs, and the intensities of the emitted light at the two wavelengths are captured by photomultiplier tubes (PMTs). The more RNAs transcribed from a particular gene, the more labeled cDNAs, the more target molecules in the spot are hybridized, and the higher the intensity. In stage 5, the $\log_{10}$ of emitted light intensities are plotted in a scatter plot, with each dot representing the two intensities of a single type of mRNA in the two cultures. In stage 6, the logs of the ratios of the emitted light intensities are submitted to a database. One database is the Gene Expression Omnibus sponsored by the NCBI, and another database is ArrayExpress sponsored by EMBL-EBI.

Patrick O. Brown

BOX 10-1 | **SCIENTIST SPOTLIGHT**

PAT BROWN, a professor of biochemistry at Stanford University, admits to grand schemes. When he set up his lab at Stanford in 1988, he wanted to compare a million people's genes to determine which ones make some people, for example, wallflowers and others cheerleaders. And this offbeat idea was only one of many sweeping studies of gene activity that were impossible at that time. Instead, scientists tended to focus on a favorite gene and deduce functions of that gene. But then Brown devised a way to monitor all of an organism's genes at one time to see which ones are active. He likens the old approach to looking at a couple of pixels on a screen and trying to deduce a movie's plot. His invention made all the pixels visible at the same time.

To study gene activity on this grand scale, Brown arrayed snippets of an organism's genes on a glass slide, called a microarray. To identify those genes that had been active in a particular environment or developmental stage, he then exposed the slide to fluorescently labeled cDNA, made by reverse-transcribing the RNA collected from the same organism, from the environment or developmental stage of interest. Affymetrix, a company founded in 1991, had a similar idea, but Brown was unaware of the competition.

With microarrays, Brown's group and collaborators systematically monitored changes in yeast gene expression as yeast responds to environmental changes or enters a new developmental stage. For example, he and others identified several hundred genes that become active at specific times between one cell division and the next. Such experiments helped scientists view genes in a new light. Scientists began to determine how groups of genes were co-regulated.

After the Human Genome Project began, large collections of human genes gradually became available, enabling Brown's group to study human tumor development and regression. The researchers studied patterns of gene expression in more than 500 types of tumor, finding marked differences in gene expression among tumors that are normally grouped together because they look the same. Brown's group analyzed thousands of human cancers to develop a system for classifying them by gene expression. Such a system has made it possible for clinicians to differentiate between, say, prostate tumors that are likely to respond to chemotherapy agents and those that are recalcitrant to such agents (LaPointe et al., 2004).

Sophisticated data analysis is needed for such studies. "At first, people were saying that microarray experiments would be ridiculous because you would get so much data that you couldn't make sense of it," Brown recalls. But the yeast experiments revealed that unknown genes often change their expression in sync with known genes, giving clues to their functions and the times at which they come into play. "So we had a general approach to organizing the data that allowed us to take advantage of its systematic nature," Brown says.

Brown's group devised many of the methods now used to systematically interpret and visualize microarray data. One of the tricks was to make movies that show gene expression patterns changing over time. "The fact that you can look at a whole movie enables you to see how coordinated and ordered the whole [developmental] program is, which you can't easily see by just looking at little bits of it," Brown said. By analyzing such programs, Brown is attempting to deduce the rules that govern the expression of each yeast gene and the processes that control the production, processing, transport, and breakdown of the corresponding proteins.

Pat Brown earned his Ph.D. in Biochemistry and M.D. from the University of Chicago, and he is a member of the National Academy of Sciences.

REFERENCE

Lapointe, J., C. Li, J. P. Higgins, M. van de Rijn, E. Bair, K. Montgomery, M. Ferrari, L. Egevad, W. Rayford, U. Bergerheim, P. Ekman, A. M. DeMarzo, R. Tibshirani, D. Botstein, P. O. Brown, J. D. Brooks, and J. R. Pollack. 2004. "Gene Expression Profiling Identifies Clinically Relevant Subtypes of Prostate Cancer." *Proceedings of the National Academy of Sciences USA* 101: 811–816.

Upon inspection of the scatter plot, you will find that most spot intensities fall near a diagonal line starting close to the origin of the plot, indicating that the relative abundance of most genes' transcripts are nearly identical in the two cultures.[6] Two lines parallel to the central diagonal line are also depicted in the scatter plot. Spots that fall within the two parallel lines correspond to mRNAs that are within a twofold difference in expression level between the two cultures. Spots that fall outside the two parallel lines correspond to mRNAs that show a greater than twofold difference, which is considered the minimum to register a significant alteration in RNA level. In this experiment, the LacZ mRNA level is greater than twofold higher in cultures from the second growth phase than in cultures from the first growth phase. You would expect that LacY and LacA mRNA levels (not shown) would show a similar difference in expression in the two cultures. Other mRNAs (labeled A, B, and C) are regulated in a manner similar to the LacZ mRNA and would be predicted to be involved in metabolizing lactose. There are also mRNAs that are at higher levels in the first growth phase than in the second growth phase (labeled D and E). You would expect that these mRNAs may be responsible for initial transport of glucose into the cells and are not needed when the switch is made to lactose.

In general, the relative level of each transcript derived from microarray experiments allows us to follow how these genes are regulated under specific cellular conditions. Similar expression patterns under different conditions tell us that the genes belong to the same functional pathways. Expression patterns are often measured by hierarchical cluster analysis. To do such an analysis, cells are placed in several different conditions and RNA levels are measured at each condition. If the same set of mRNAs show nearly identical relative levels under different conditions, it is likely that they are co-regulated, in other words, they lie within the same operon or are controlled by the same transcription factors. Hierarchical cluster analysis can be performed in a manner similar to the UPGMA methodology used to create evolutionary trees from an evolutionary distance matrix (see Chapter 8). For microarray data, Pearson correlation coefficients are calculated on the expression patterns of each possible pair of genes in the set of genes analyzed. The higher the coefficient value (the maximum is 1.0), the shorter the lengths of the branches connecting pairs of genes to a common node. Pairs of genes with closely matched correlation coefficients form a cluster that

---

[6] The diagonal is slightly off-center because the intensity of light emitted from Cy5 is slightly less than the intensity of light emitted from Cy3. It is possible to correct for this difference. It is also important to repeat the reverse transcription stage with the dyes switched to ensure that intensity differences are *not* due to cDNA fluorescence emission imbalance inherent with the dyes.

polycistronic RNA A transcript that codes for more than one protein. Polycistronic RNAs have more than one start codon and more than one stop codon.

heatmap A display that shows color gradations that correlate to level of gene expression.

can be displayed on a heatmap (see below). You would expect LacA, LacY, and LacZ mRNAs to be found in the same cluster with short branches connecting them to a common node.

In one microarray experiment conducted in Sydney Kutsu's laboratory at the University of California, Berkeley, the relative amounts of LacZ, LacA, and LacY mRNAs under lactose growth conditions were 57, 21, and 21 times greater, respectively, relative to the amount of these mRNAs under glucose growth conditions. The similar relative levels of mRNAs under the two growth conditions confirm that these three genes are regulated in nearly the same manner. In theory, the amounts of LacZ, LacA, and LacY mRNAs should be identical because *lacZ*, *lacA*, and *lacY* genes are transcribed into one transcript that contains coding sequences for all three genes. We call such a transcript polycistronic RNA. The higher relative level of the LacZ coding sequence mRNA over that of LacA and LacY observed by Kutsu's laboratory could be due to increased stability of the LacZ transcript region (other regions of the polycistronic RNA are degraded at a faster rate). If so, it would be interesting to know the mechanism that increases stability of LacZ transcript. Other growth experiments using different sugars would be needed to confirm that these three genes are consistently co-regulated.

Heatmaps

The microarray results in the scatter plot in Figure 10-5 can also be visualized in the form of a heatmap, which is a display that shows gradations of color that correlate with the expression level of genes. Heatmaps are created in different ways depending on the investigator's objectives. To compare gene expression across different genes, you can scale the colors of the heatmap. Typically these colors are red and green, and are scaled so that the color intensity matches the $\log_2$ of the relative intensity of the light emitted from Cy3 and Cy5 dyes collected by the photomultiplier tube.[7] The light intensities are often compared to those obtained from a control sample. In our case, the control sample can be the average expression of all genes in the microarray during the first and second growth phases. Most gene expressions are not expected to change during the course of the experiment. Using our example of glucose and lactose as sources of nutrients, a red color is displayed on the heatmap when the gene expression level is significantly higher than that from the control sample. When the expression level is significantly lower than that of the control sample, the heatmap will display a green color. When the expression level of a gene is near that that of the control sample, the heatmap will display a black color. Below is a depiction of a hypothetical heatmap comparing expressions of genes that are significantly different from the control sample in first or second growth phase. The brighter the heatmap color (green or red), the greater the difference in expression from the control sample.

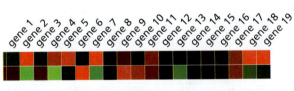

Thought Question 10-2

Which genes in the hypothetical heatmap could possibly be *lacZ*?

You might ask, what is considered to be a significant difference in expression from the control sample? In fact, there are no strict guidelines regarding the threshold for a significant change in expression. Some scientists have utilized as little as a twofold

---

[7] $\log_2$ is the log base 2.

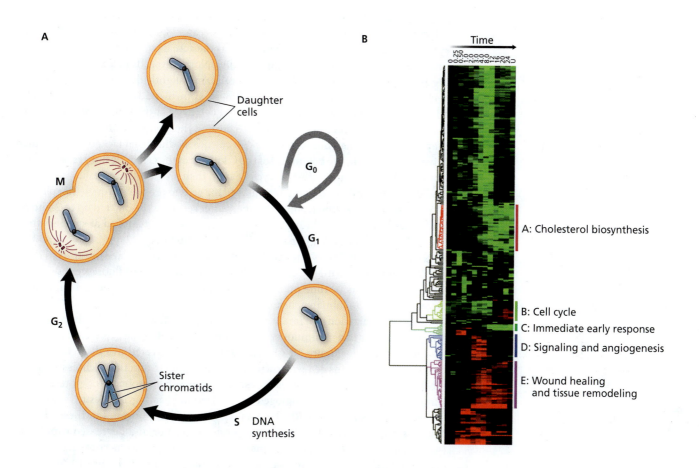

FIG. 10-6. RNA expression analysis as a function of cell cycle.
A. Cell cycle illustration **B.** Microarray analysis of human genes in human fibroblasts in response to serum addition. Time points are listed in units of hours. The column labeled "U" represents the relative RNA expression levels from asynchronous cells.

difference as the threshold for significance (see Figure 10-5), and some have utilized as large as an eightfold difference as the threshold (see Figure 10-6B).

Sometimes, if there are multiple sets of cellular conditions and only one control cellular condition, it is informative to scale the heatmap so that the lowest value for a particular mRNA (relative to the control) is displayed as green and the highest value for that same mRNA is displayed as red.[8] Let's say that one gene's expression level ranges threefold across different cellular conditions, but another gene's expression level ranges 10-fold across different cellular conditions. On the heatmap, color ranges for both genes would go from bright green to bright red. This scaling allows for a comparison of the levels of expression of particular genes across different cellular conditions but, unlike the heatmap example depicted above, does not allow for inter-gene expression comparison (i.e., comparison of the levels of multiple genes with each other). Microarray heatmap software programs usually have the flexibility of allowing the user to select the appropriate scale for the experimental objective—to compare gene expression across different samples or to compare intergene expression levels (see exercise 2 at the end of this chapter).

Cluster Analysis

We have used the microarray as a technique to follow the expression levels of three well-studied genes, *lacZ*, *lacY*, and *lacA*. We can use cluster analysis of microarray gene expression data to group many genes that are regulated in a similar manner. Patrick Brown and David Botstein (Stanford University) wanted to see if groups of genes show similar expression patterns in cells as those cells proceed through the

cell cycle A sequence of events that lead to cell division. The phases are G1 (gap 1), S (DNA synthesis), G2 (gap 2), and M (mitosis). Another phase is G0, a quiescent phase where cells exit the cell cycle for long periods of time.

---

[8] The heatmap colors do not necessarily match those of the fluorescent dyes obtained from the experiment.

cell cycle. The cell cycle is a sequence of events that lead to cell division (mitosis). The cell cycle is divided into phases named G1, S, G2, and M, where G1 is Gap 1, S is DNA synthesis, G2 is Gap 2, and M is mitosis (M is the actual partition of a single cell that produces two daughter cells) (Figure 10-6A). Normally, cells in a culture are dividing asynchronously: in other words, each cell is in a slightly different place in the cell cycle than all the others. For their experiment, Brown and Botstein needed to have all the cells going through each phase of the cell cycle at the same time. To do this, they exploited a cell culture technique to create synchronously dividing cells. First, they incubated human fibroblasts in a cell culture medium that lacked serum, which drives all the cells to the beginning of the G1 phase of the cell cycle but prevents them from going further through the cycle.[9] They then treated the arrested cells with medium containing serum, which stimulated the cells to proceed through the cell cycle phases, but now in a synchronous manner. After serum addition, and up to 24 hours later, as cells proceeded through these phases, cells were removed at specific times and prepared for microarray analysis. This preparation was accomplished by creating cDNAs from cell mRNAs and labeling them with Cy5 (red) fluorescent dye. These cDNAs were mixed with the cDNAs labeled with Cy3 (green) fluorescent dye from serum-starved cells (0 time point) and applied to a microarray spotted with 8,600 target cDNAs from human genes.

After hybridization and image analysis to measure fluorescence levels, only those genes that displayed a minimum of 3.0 $\log_2$ unit changes in gene expression were used for cluster analysis (Figure 10-6B). The scientists discovered five discrete clusters, each of which contains genes known to carry out similar functions: cholesterol biosynthesis, cell cycle, immediate early response (genes that respond quickly to external stimulation), signaling and angiogenesis, and wound healing and tissue remodeling (colored clusters in Figure 10-6B). You will notice that the genes in these clusters display similar patterns of increased expression (red) or/and decreased expression (green) in response to serum as a function of time. Other clusters contain genes that could not be classified into similar functions (black clusters in Figure 10-6B). From their cluster analysis, the researchers were able to associate specific groups of genes of known function with specific phases in the cell cycle. Genes of unknown function that fell into a black cluster with genes of known function suggests that those of unknown function may work in concert with the genes of known function.

Thought Question 10-3

> In the heatmap in Figure 10-6B, what is the relative difference in expression of dull green color in the experimental cell condition relative to the control cell condition (black color) if the dull green color corresponds to –3.0 $\log_2$ relative to the black color?

Practical Applications of Microarray Data

Microarrays offer us the ability to simultaneously monitor the expression patterns (or profiles) of thousands of genes in a single experiment. Several investigators used microarrays to generate gene expression patterns from cancer patient tissues undergoing chemotherapy and then monitored patient outcomes. The idea is that gene expression profiles of new cancer patients can predict patient response. If successful, clinicians can use gene expression data to determine a priori which patients are predicted to successfully respond to certain therapies. In one study by Sandrine Imbeaud (CNRS and Pierre and Marie Curie University, Villejuif, France), cancer tissue samples from colorectal cancer patients and matched normal tissue were processed for microarray expression analysis. Patients were treated with standard chemotherapy regimen (folinic acid, 5-fluorouracil, and irinotecan), and their therapeutic response

---

[9] Some researchers call this the G0 phase or quiescent phase. Cells will remain in the G0 phase unless conditions are favorable for entry into the G1 phase.

was measured in terms of drug sensitivity and drug resistance. Gene expression profiles of patients were compared to therapeutic responses, and a list of 679 gene expression profile differences was documented. This suggests that gene expression profiles can be used to predict response to chemotherapy. This study was conducted on only 13 patients, which means that larger studies are needed to confirm the results. Indeed, one review that reassessed several small microarray studies indicates that some reports of gene expression profiles that correlated with clinical outcomes could not be confirmed.[10] A combination of small sample size and data analysis errors contributed to erroneous conclusions. In addition to these errors, there are technical pitfalls that must be considered.

Considerations to Take in the Interpretation of Microarray Data

We must be cautious when drawing conclusions from microarray experiments because of the complicated nature in which gene expression is regulated. One consideration is alternative splicing. In eukaryotic cells, alternative splicing is common. Recall that p53 has at least 15 alternatively spliced forms of RNA, and not all RNAs create a full-length p53 protein of 393 residues. A typical microarray analysis would likely not detect these subtleties and may lead to a false impression of the amount (and form) of transcript expressed. Another consideration is the indirect step of measuring the RNA level. The RNA must first be reverse-transcribed into fluorescently tagged cDNA. It is known that reverse transcription may, in some cases, be incomplete and can vary from experiment to experiment—especially if the lengths of the mRNAs are long. Another consideration is the number of cells required to collect enough RNA to perform a microarray experiment. Typically, 1 to 5 μg of RNA is needed, which corresponds to ~10^7 eukaryotic cells—this amount is feasible when using cells cultured *in vitro* but is not a practical amount to obtain from tissues. To perform microarray analysis from small amounts of tissue, the cDNA must be amplified by PCR. This introduces another step in the analysis procedure and yet another potential source of error.

In accordance with the central dogma, you might expect that the relative level of mRNAs would correlate to the relative levels of corresponding proteins. However, several studies have shown that microarray and other forms of RNA level measurements (Northern analysis and SAGE, for example) often do not correlate to levels of corresponding proteins.[11] For example, one study using yeast cells found that the correlation coefficient between mRNA and protein was 0.74 (perfect correlation is 1.0), and another found the correlation coefficient to be 0.59. There appears to be greater correlation when the mRNA level and its corresponding protein level are abundant (i.e., those mRNAs that are highly expressed relative to other mRNAs).[12] When abundant mRNAs were included, two studies reported correlation coefficients 0.935 and 0.86. But when abundant mRNAs were excluded from the analysis, correlation coefficients from separate studies were 0.18, 0.21, 0.356, and 0.49. It appears that less abundant mRNAs show more complex protein expression patterns. Even in bacteria, an organism that undergoes very little alternative splicing, a study showed a correlation coefficient of only 0.77.[13] In fact, when single bacterial cells are examined, the mRNA level and its corresponding protein level are uncorrelated!

Protein Levels Can Be Controlled by Regulation of Degradation Rate

A researcher cannot attribute these low correlation coefficients solely to technology deficiencies, because there are biological reasons why RNA levels and protein levels

protein half-life The time it takes for 50% of a population of newly synthesized proteins to be degraded.

---

[10] Michiels, Koscielny, and Hill (2005).

[11] Waters, Pounds, and Thrall (2006).

[12] An example of an abundant mRNA found in fibroblasts is glyceraldehyde 3-phosphate dehydrogenase mRNA. Gygi, Rochon, Franza, and Aebersold (1999).

[13] Taniguchi, Choi, Li, Chen, Babu, Hearn, Emili, and Xie (2010).

may not correlate with each other. A primary cause is the complex regulation of protein degradation. Using p53 as an example, the level of p53 protein increases several-fold upon ionizing radiation treatment but p53 mRNAs fail to show any increase. It turns out that p53 protein level is regulated through manipulation of its protein degradation rate. The p53 protein half-life is approximately 20 minutes, which means that after 20 minutes, one-half of the original p53 population is degraded. As this degradation occurs, p53 is being replaced at a constant rate, resulting in a constant (steady-state) level of p53. As mentioned previously, MDM2 targets p53 for this rapid degradation. When ionizing radiation causes DNA damage, enzymes that recognize the damage modify p53 and MDM2, and then MDM2 can no longer promote the degradation of p53. Because p53 is being expressed at a constant rate, while at the same time the p53 degradation rate decreases, the steady state level of p53 rises.

This scenario is likely to be repeated for many other proteins with short half-lives. In sum, there are likely to be both methodological errors and biological explanations for the observation of the lack of correlation between RNA levels and protein levels across the transcriptome.

10.5 RNA-Seq (RNA SEQUENCING)

Advantages of RNA-Seq

RNA-seq is the abbreviated term for RNA sequencing, a more recent technology that can be used to quantify all RNAs in a sample. As an alternative to microarrays, it appears to be more quantitative and it is capable of revealing new RNAs. In fact, for organisms with small genomes, RNA-seq can be used to determine the transcriptome without prior knowledge of the genomes' sequences. Microarrays established the ability to measure thousands of transcript levels simultaneously and have gone a long way toward increasing our knowledge of the transcriptome and the sets of genes that are co-regulated. With RNA-seq, the entire RNA content of a sample can be sequenced quickly and efficiently.

RNA-seq has four advantages over microarrays. First, unlike in a typical microarray, RNA-seq can detect alternatively spliced forms of transcripts. Recall that when primary transcripts undergo splicing, two sequences that are separated in the genome end up being next to each other. In alternative splicing, an exon sequence in the primary transcript may be removed. If the target sequence in the microarray is complementary to the spliced out region of the primary transcript, the mRNA resulting from the primary transcript will not bind to the target DNA. Such an mRNA would not be detected by the microarray. In addition, RNA-seq can detect alternative transcription start and stop sites. These advantages are attributed to the fact that RNA-seq does not require prior knowledge of the sequence in the mRNA because it is not hybridization based; instead it sequences the RNA. Second, RNA-seq can be used to discover new genes. Because it does not presume a target sequence, RNA-seq can uncover transcripts from genes that were previously not known to exist. Third, the dynamic range of RNA-seq for transcript measurement is much greater than microarrays. Fourth, in microarrays, the likelihood of detection of transcripts is not equal for all transcripts. This bias occurs because the reverse transcription step in microarrays is inefficient when the mRNAs are long (>1 kb), which leads to an increased likelihood that the reverse transcriptase enzyme will fall off the mRNA template. This means that in microarray technology, if the target oligonucleotides are complementary to regions near the 3' end of the mRNA, there will be a bias toward detecting the mRNA. Conversely, there will be a negative bias if the target oligonucleotides are complementary to regions near the 5' end of the RNA.[14] RNA-seq alleviates these biases because the method reverse transcribes only ~50–100 nucleotide

---

[14] Some of this bias is removed by random priming during reverse transcription, as well as by including several target DNA oligonucleotides that detect different regions of each mRNA.

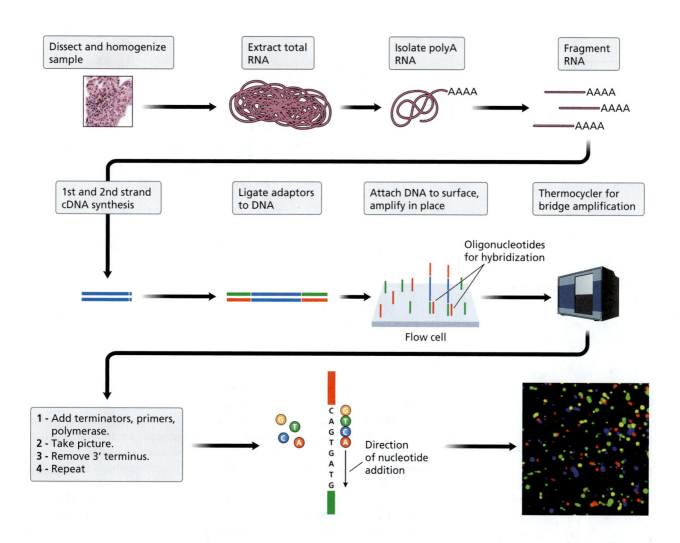

FIG. 10-7. Overview of RNA-seq method.

segments of RNA. There is very little chance that reverse transcriptase will fall off of the mRNA within 50–100 nucleotides of starting reverse transcription.

Overview of RNA-Seq Steps

How does RNA-seq detect and quantify RNA? In essence, RNA-seq requires RNA purification, reverse transcription of RNA into double strand DNA, PCR amplification of DNA, and DNA sequencing. In Chapter 9 we discussed next-gen DNA sequencing techniques in the context of genome sequencing. Here, in our discussion of RNA-seq, we introduce another popular next-gen DNA sequencing technique offered by the company Illumina. Of course, the Illumina next-gen DNA sequencing technique can be used to directly sequence DNA genomes. Conversely, the next-gen DNA sequencing techniques discussed in Chapter 9 (ion semiconductor or nanopore-based) can be used for RNA-seq as well.

Let's look at the RNA-seq flow chart in Figure 10-7. You need to start with 1 µg of RNA. To isolate mRNAs from eukaryotic cells, you can use a column with oligo-dT bound to the column matrix. The oligo-dT hybridizes to RNAs with polyA tails but does not hybridize to other RNAs such as ribosomal RNAs which lack polyA tails. The purified RNAs with polyA tails (the vast majority of mRNAs in eukaryotic cells) are fragmented into 50- to 100-nucleotide-long pieces. The RNAs are reverse-transcribed into double strand cDNAs. The ends of the cDNAs are ligated to synthetic double strand "adaptor" DNAs (with known sequences). The adaptors have segments that allow them to hybridize to oligonucleotides (of known sequence) that have been covalently bound to a surface inside a flow cell. The bound oligonucleotides have

bridge amplification A PCR amplification method that uses a DNA strand as a template. The DNA strand is hybridized to a primer bound to a surface. The second strand synthesis creates a free end that is complementary to a second primer bound to a surface. It also creates a bridge such that one end is bound to the surface and the second end is hybridized to a surface-bound primer. Successive rounds of PCR amplify the DNA.

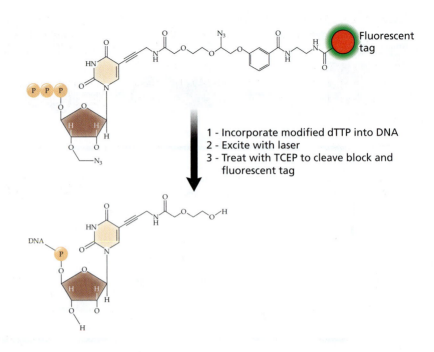

FIG. 10-8. Incorporation of modified thymine nucleotide into DNA for sequencing and cleavage at azido groups to remove 3′ OH block and fluorescent tag. TCEP is tris(2-carboxyethyl)phosphine.

1 - Incorporate modified dTTP into DNA
2 - Excite with laser
3 - Treat with TCEP to cleave block and fluorescent tag

two functions: they hybridize to the adaptors and are primers for bridge amplification (see below). Double strand cDNAs are denatured into single strands and diluted so that they are dispersed when they hybridize to the oligonucleotides bound to the flow cell. The flow cell is a chamber designed in a way so that small amounts of chemicals and enzymes can be added to the bound oligonucleotides and easily flushed out after a reaction is completed. The temperature of the flow cell can be altered to accommodate DNA amplification reactions.

The attached cDNAs are copied several times through bridge amplification to increase the signal-to-noise ratio when the cDNAs are sequenced at a later step. Hundreds of identical copies of single strand cDNAs are created from the cDNAs bound to the flow cell. Because the cDNAs were dispersed during the hybridization step, identical copies are grouped together in spots after amplification. An oligonucleotide sequencing primer is annealed to the adaptor at one end of the cDNAs, and a modified dideoxy sequencing method is performed.

In this sequencing method, four chemically modified dNTPs are added to the flow cell. Each dNTP contains two chemical moieties, called azido groups, one of which blocks attachment of the next nucleotide that the polymerase would incorporate (much like dideoxynucleotides). The other azido group creates a photolabile linkage to a fluorescent tag (Figure 10-8). Each of the four different dNTPs is linked to one of four different fluorescent tags, each of which emits a different color of light when excited by a laser. The sequencing is performed by repeated rounds of addition of only one nucleotide to the growing strand, each round including a detection of which nucleotide was added in each spot. Only one dNTP is incorporated into the growing complementary strand in each strand because of the azido block on the 3′ OH. A laser is used to excite the fluorescent tag on the added nucleotide, which leads to the emission of light. A picture of the glass plate is taken to record the pattern of colors in each spot after each round of sequencing. Next, the reducing agent[15] tris(2-carboxyethyl)phosphine is added to remove the block and the fluorescent tag. An example of the removal of the block and tag from modified dTTP is shown in Figure 10-8. The released block and tag moieties are flushed from the flow cell, and then fresh dNTPs and polymerase are added to the flow cell so that the sequencing strand is extended by one nucleotide.

---

[15] A reducing agent adds electrons to molecules.

The above process is repeated until 50–100 nucleotides are sequenced using a series of images of the glass plate. Each spot in the image represents a single mRNA fragment. A sequence obtained from a single spot is called a "read." If an RNA is abundant, many spots with overlapping reads will be recorded for that RNA. If an RNA is rare, few spots with overlapping reads will be recorded. The number of overlapping reads, normalized to the deduced length of the transcript, correlates to the level of particular RNAs in the sample.

Bridge Amplification

Prior to sequencing, the cDNA molecule must be amplified to create a signal of sufficient magnitude to be detected by a fluorescent microscope. An important technical advance that makes RNA-seq possible is bridge amplification. Bridge amplification is a DNA amplification technique (similar to polymerase chain reaction) that increases the number of copies of cDNAs in each spot on the flow cell. In RNA-seq, double strand cDNAs created from mRNAs are attached to adaptor oligonucleotides at both ends of the cDNA. The two strands of the modified cDNAs are separated by treatment with an alkaline solution. Single strand cDNAs are introduced into the flow cell previously populated with oligonucleotides (Figure 10-9, panel A). Two types of oligonucleotides are attached to the surface of the flow cell; one has

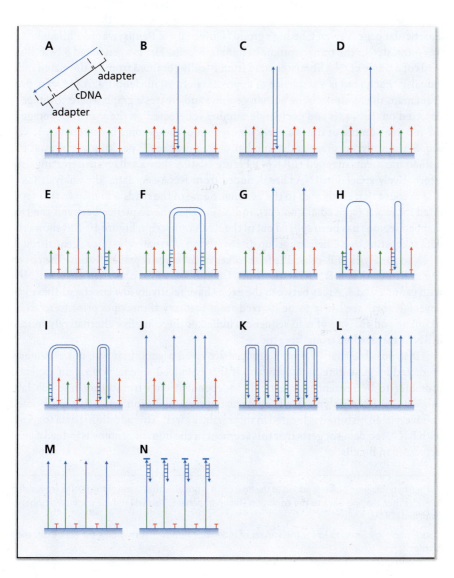

FIG. 10-9. Bridge amplification. Arrowheads represent the 3′ ends of the DNA strands.

coverage In RNA-seq it is the number of fragment reads per kilobase per million fragments mapped. Coverage gives the relative frequency a particular region of the genome is detected by RNA-seq.

a cleavable site depicted as a cross on the red oligonucleotide, and the other (green) is not cleavable. The cDNAs hybridize to the oligonucleotides through adaptor sequences on 3′ ends (panel B). The hybridized oligonucleotides are elongated with DNA polymerase and dNTPs (panel C). Formamide at 60°C is added to remove noncovalently attached cDNA strands (panel D). Hybridization buffer is introduced into the flow cell causing adaptor sequences at the ends of the cDNA stands to anneal to surface-bound oligonucleotides and form bridges (panel E). DNA polymerase and dNTPs are added to extend the oligonucleotides complementary to bridge DNAs (panel F). Hybrid DNA strands are separated with formamide (panel G) and allowed to hybridize with surface-bound oligonucleotides (panel H).[16] Hybridized oligonucleotides are extended, and DNA hybrids are separated for 35 cycles (panels I to L). Single strand DNAs attached to red oligonucleotides are cleaved and removed from the flow cells (panel M). The 3′ ends of the remaining single strand DNAs are treated with dideoxy dNTPs and an enzyme called terminal transferase. The treatment places single dideoxy dNTPs on the ends of the DNAs that prevent addition of sequencing dNTPs to these ends (panel N). DNA sequencing primers complementary to adaptor sequences are hybridized to the cDNAs, and the cDNAs are sequenced as described in the preceding section.

Analysis of an Experiment Using RNA-Seq

Now that we have discussed the technical details of RNA-seq, let's walk through one experiment where RNA-seq analysis is used to create a picture of the transcripts from a particular gene. Vivian Cheung's group (University of Pennsylvania, Philadelphia) was interested in the transcriptome of human B-cells. Her lab generated 879 million reads of a single cDNA library created from B-cells obtained from 20 unrelated individuals.[17] Each read is a sequence of approximately 50 nucleotides. A software program maps the reads onto the human genome and creates a graph where coverage is depicted on the y-axis and nucleotide number is depicted on the x-axis. Coverage is the relative number of times a particular region of the genome is detected by RNA-seq. Specifically, coverage is the number of fragment reads per kilobase of exon per million fragments mapped (FRKM). Figure 10-10A shows a diagram depicting how alternatively spliced mRNAs are deduced from RNA-seq data. The analysis takes into account that reads map to the genome. Some of the reads will map to two separated locations (spliced alignment), indicating that the sequence between the two mapped regions has been spliced out of the RNA transcript. Figure 10-10B shows the data suggesting that Cheung's group discovered an alternatively spliced gene through RNA-seq analysis of B-cells. The four spikes of coverage represent exons. The coverage indicates that mRNAs with exons 1 and 3 are more prevalent than mRNAs with exons 2 and 4. Areas between the exons have relatively low coverage. These low coverage areas are likely to be derived from primary transcripts prior to splicing. Mapping and analysis of read sequences indicates there are five alternatively spliced mRNAs transcribed from this gene.

It is possible that Cheung's group discovered a gene that was not previously predicted by gene detection software. If this is indeed a gene, you might expect a promoter to be present near exon 1. Independent data from other research labs showed that RNA polymerase II binds just upstream of exon 1 and that methylated and acetylated histone 3 is located in this region as well. This additional data together with RNA-seq data suggests that this segment of the human genome is indeed a gene expressed in B-cells.

proteome The identity and level of proteins expressed from the genome in a cell.

---

[16] When hybridization buffer is added to the flow cell it is possible that some of the cDNAs rehybridize to each other, leading back to the condition shown in panel F. This will reduce the final number of amplified cDNAs.

[17] Such pooling is used to lower the impact of individual trends that may bias the RNA expression profile.

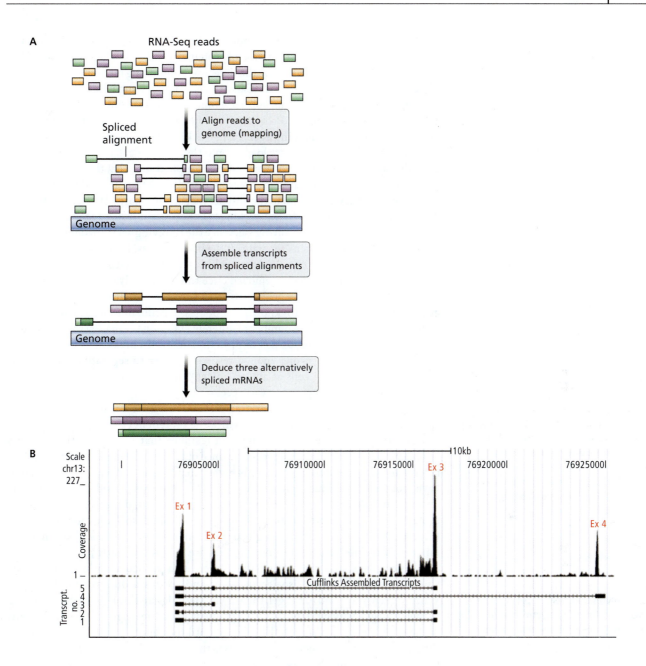

10.6 THE PROTEOME

The proteome is the identity and level of the proteins expressed from the entire genome in a cell. Like the genome, the proteome changes with cell type and cell environment. Analysis of the proteome becomes especially valuable when a cell type is compared to itself under different conditions. The identity of proteins expressed in one condition but not in another could give clues as to how such cells respond to these conditions. We will explain the technical aspects of a common method of proteome analysis and review its advantages and challenges.

Separation of Proteins and Quantification of Their Steady-State Levels: Two-Dimensional (2D) Gel Electrophoresis

Analysis of the proteome requires protein separation, relative protein level measurement, and identification. One method for accomplishing the first two steps of this process is two-dimensional (2D) gel electrophoresis. In 2D gel electrophoresis,

FIG. 10-10. RNA-seq mapping and analysis.

A. Diagram showing how RNA-seq reads are mapped onto the genome—a process that allows one to deduce the identities of alternatively spliced mRNAs. **B.** Quantification of RNA expression by RNA-seq method from human B-cells.

proteins are separated by isoelectric point in the first dimension and by mass in the second dimension. Efficient separation of proteins by mass in a gel was first achieved in 1971 by Ulrich Laemmli, who, at the time, was conducting his studies at the MRC Molecular Biology Laboratory at Cambridge, United Kingdom.[18] In Laemmli's system, a vertical discontinuous polyacrylamide slab gel is composed of a stacking gel on top of a resolving gel. Proteins are prepared for separation by boiling in a mixture of detergent, reducing agent, and tracking dye. The denatured proteins, without disulfide cross-links, are loaded onto the stacking gel, and the proteins migrate through the gel when a voltage gradient is applied such that the anode is on the bottom of the gel and the cathode is on top. Proteins migrate toward the bottom of the gel because they are bound to the anionic (negatively charged) detergent sodium dodecyl sulfate (SDS). Within the stacking gel, the proteins squeeze together (stack) as they migrate toward the resolving gel. Once the resolving gel is encountered by the stacked proteins, the proteins separate on the basis of mass. Proteins with high mass migrate slowly through the resolving gel, and proteins with low mass migrate quickly. The voltage gradient is removed once the tracking dye (very low mass) has migrated to the bottom of the gel. Stains can be used to visualize the proteins in the gel. Laemmli used a blue stain known as Coomassie Brilliant Blue, which detects as little as 0.2 μg, but there are a number of different stains on the market that can detect as little as 1 nanogram (0.001 μg) of protein. Laemmli's breakthrough in protein separation did not give strong enough resolving power to separate the thousands of proteins expressed in cells at any one given time. To increase resolving power, a second dimension of separation was necessary.

The second dimension of separating proteins in a 2D gel is by isoelectric point (pI). The isoelectric point is the pH at which the protein net charge is zero. Because proteins typically have many amino acid residues that form ions, they possess unique pIs. Proteins that have a preponderance of basic residues, such as Lys and Arg, relative to acidic residues will have pIs higher than 7; and conversely, proteins that have more acidic residues, such as Asp and Glu, than basic residues will have pIs lower than 7. When 2D gel electrophoresis is performed, denatured and reduced proteins are first separated by their pIs in a process called isoelectric focusing (IEF) (Figure 10-11). Isoelectric focusing takes place in a strip of gel that also contains a mixture of mobile ampholytes. Ampholytes are compounds that have both acidic and basic groups, and can act as buffers to create a particular localized pH. Each ampholyte has a different isoelectric point and creates a different localized pH. When voltage is applied to the gel, the ampholytes migrate through the gel according to their charges, which establishes a pH gradient along the length of the gel. Proteins in the sample migrate through the gel strip until they encounter a pH that matches their pI. The strip with separated proteins is then attached to a Laemmli gel and the proteins are separated by mass.

The 2D gel electrophoresis method was first developed by A. J. MacGillivray and D. Rickwood (Beatson Cancer Research Institute, Glasgow, United Kingdom) in 1974 and, a year later, was significantly advanced by Patrick O'Farrell (University of Colorado, Boulder) when he demonstrated that 1,100 proteins from *E. coli* could be resolved on a single gel (Figure 10-12). Once the gel is stained, the density (darkness) of the spots is a measure of the relative abundance of the proteins. One technical difficulty with the 2D gel system is the consistency of the IEF step. In the early stages of IEF development, ampholytes were of variable quality, leading to gels that were difficult to compare. The 2D gel system of resolving proteins advanced with the introduction of immobilized pH gradient strips (IPG strips) that create a consistent pH gradient. Another improvement was the use of sensitive fluorescent dyes, such as SYPRO Ruby Red stain that allow for the detection of very small quantities of proteins.

2D gel electrophoresis A method used to separate proteins by size and isoelectric point. The method is used in proteome studies because many proteins can be visualized at one time. In addition, relative levels of proteins can be easily measured and the proteins can be identified by subsequent techniques.

isoelectric focusing (IEF) A method used to separate proteins by isoelectric point. A strip of gel contains ampholytes that create a pH gradient. Proteins migrate by electrophoresis until they settle at a position where their net charges are neutral.

---

[18] Ulrich Laemmli later went on to become professor of biochemistry at the University of Geneva in Stockholm, Switzerland.

Two dimensional gel electrophoresis

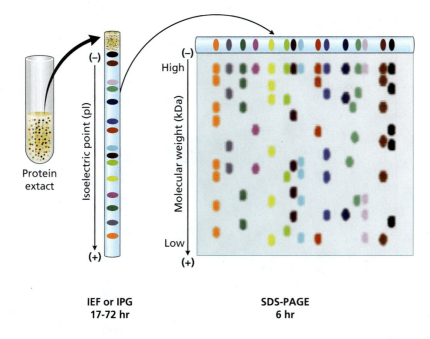

IEF or IPG
17-72 hr

SDS-PAGE
6 hr

FIG. 10-11. Two-dimensional gel electrophoresis method. Note that proteins are colored for the purpose of demonstrating their positions in the first and second dimensions. Proteins are not visible until after they are stained after the second dimension separation is completed.

Each spot in the 2D gel represents a different protein or a particular modified form of a protein. Proteins undergo modifications that can alter their mass and isoelectric point; for example, there may be two spots that have identical sequences but differ in phosphorylation or glycosylation. Proteome researchers are often interested in identifying changes in levels of proteins when cells are placed in different conditions or when cells are genetically altered. To detect such changes, much like microarray analysis, proteins must be collected from cells under two different sets of conditions. Proteins from each set of conditions are run separately on 2D gels, and the spots are compared. Software programs scan and measure spot intensities, and comparisons of spot intensities between the gels reveal whether proteins are up- or downregulated.

Identification of Proteins: Liquid Chromatography-Mass Spectrometry (LC-MS)

Now that we have described how to separate proteins and measure their relative levels, how do we identify them? To identify the separated proteins, liquid chromatography-mass spectrometry (LC-MS) is often employed. Proteins of interest are excised from the gel, destained to remove the bound stain, and digested with trypsin. In Figure 10-13, the protein circled in red is an example of a protein that can be excised from the 2D gel. This process can be scaled up to excise any number of proteins in parallel. Trypsin is known to clip proteins at the peptide bond after Lys or Arg. Peptides produced from trypsin digestion are separated by reverse phase column chromatography, also known as simply liquid chromatography (LC), and are introduced to a mass spectrometer (Figure 10-13). One type of mass spectrometer is the electron spray ionization time-of-flight (ESI-TOF) mass spectrometer. In ESI-TOF, solvent droplets containing peptides that elute from the chromatography column are charged with an electric current. As the solvent evaporates, the repulsive forces on the charges cause the droplets to break apart until the peptides are devoid of solvent. The positively charged peptides move to a flight tube, which

liquid chromatography mass spectrometry (LC-MS) A two-step method that first separates molecules by column chromatography. This step usually uses an HPLC. Eluant from the column is ionized and sent through a mass spectrometer. The charge-to-mass ratio of each molecule is determined with high precision.

peptide mass fingerprint Masses of peptides created from proteins by protease digestion are measured by mass spectrometry. Knowledge of the masses and the protease is used to identify the proteins.

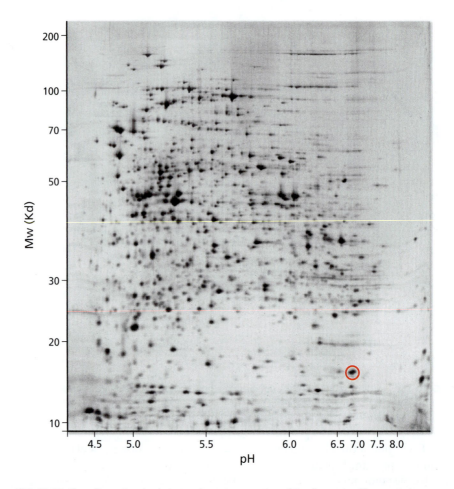

FIG. 10-12. Two-dimensional gel electrophoresis separation of *E. coli* proteins. The *x*-axis shows the pH scale, and the *y*-axis shows the mass scale (in kilodaltons). The protein circled in red has a pH of 6.8 and a mass of 16 kDal.

measures mass-to-charge ratios (*m/z*'s).[19] In time-of-flight mass spectrometers, *m/z*'s are determined by precise measurement of the time required for peptide ions to migrate from one end of a vacuum chamber to another in a flight tube. Positively charged peptide ions migrate through the flight tube because the opposite end of the chamber is negatively charged. The *m/z*'s are recorded in a mass spectrometry (MS) spectrum. The *m/z*'s of the peptide ions resulting from a trypsin digestion of a protein is called the peptide mass fingerprint of the protein.[20]

Most mass spectrometers have the capability of getting more information from a peptide ion by fragmenting the peptide ion (or parent ion) with a neutral gas such as He or Ar. Gas particles bombard the parent ion and sever the peptide bonds to

MS/MS spectrum A spectrum that plots relative abundance of molecules versus mass/charge (*m/z*) ratio. To generate the MS/MS spectrum, peptides are fragmented by a neutral gas. The masses of the fragments and the original peptide can be used to deduce the peptide sequence.

[19] Peptides may be negatively charged depending on the mode of the charging by the mass spectrophotometer. The electrical charge either adds protons (positive mode) or removes protons (negative mode).

[20] A single peptide may have more than one m/z depending on how many charges are created on the peptide. The general formula to follow is $(M + M_H*n)/n$, where M is the mass of the uncharged peptide, M_H is the mass of a proton, and n is the number of protons bound to the peptide. An uncharged peptide with $M = 800.00$ would, if singly protonated, have an *m/z* of 801.00. If the same peptide has two protons bound, the *m/z* would be 401.00.

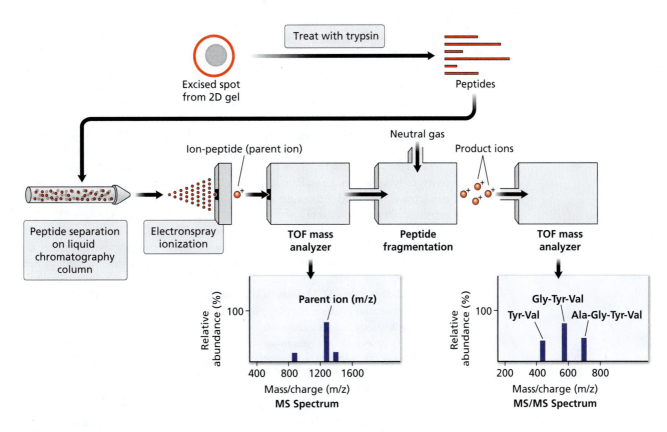

FIG. 10-13. Steps to proteome analysis by LC-MS.

produce product ions.[21] The *m/z*'s of product ions can be used to partially sequence the parent ion and confirm its identity. The spectrum that results from a fragmentation is called an MS/MS spectrum. The *m/z*'s of the parent ions together with their accompanying product ions are entered into a software program (such as Protein Prospector or MASCOT). The software program searches databases that contain translated genomes that have been "in silico" digested with trypsin (i.e., the computer has created sequences of peptides of each protein ending with Lys or Arg). If the gene coding for the protein is in the database, the software program records matches to the experimentally derived masses. In fact, a ranking of protein hits is listed as output by the software program.

Another type of mass spectrometer useful for peptide mass fingerprinting is matrix-assisted laser desorption ionization time-of-flight (MALDI-TOF). The trypsin digest is mixed with an acidic matrix and placed on a plate. A laser is used to vaporize and protonate peptides, and the peptide ions accelerate toward a negatively charged plate. The times of flight of the peptide ions are measured, and the *m/z*'s are recorded in a MS spectrum. MALDI-TOF typically produces singly protonated peptide ions, and ESI-TOF typically produces multiply charged peptide ions.

Advantages and Challenges of Current Proteome Analysis Techniques

Advances in mass spectrometry have made it possible to measure peptide masses with a precision of 0.1 dalton. If the mass spectrometer is properly calibrated, such precision should result in the accurate measurement of peptide masses. Once a protein has been identified by LC-MS, it is not difficult to quickly confirm its identity

---

[21] Peptide bonds are preferentially cleaved by gas bombardment. Less frequently, other covalent bonds in the parent ion are cleaved, such as those in the side chains.

by performing a Western blot with an antibody that specifically binds the protein (of course this requires that an antibody is available to detect the identified protein). Another attraction of this proteome method is that there are no indirect steps in the identification process, unlike microarray and RNA-seq methods that use reverse transcription (and usually DNA amplification). However, there are limitations to proteome analysis. One is that the composition of some proteins makes them unsuitable for separation on 2D gels. For example, membrane proteins are hydrophobic, causing them to form aggregates after extraction from cells. The aggregates do not resolve in the IEF stage of the procedure, making it difficult to separate membrane proteins into individual spots in the second dimension. Another limitation is sensitivity, especially in mammalian cells where the total number of proteins greatly exceeds a few thousand. Proteome analysis tends to readily identify highly expressed proteins, but not proteins of low abundance. Several reports have identified the same abundant proteins showing up as being differentially regulated in mammalian cells. Part of this conundrum is due to the limitations of sensitivity of the technique, and part is due to the multitude of spots on the 2D gel that actually have the same protein sequence (as previously mentioned, multiple spots are due to modifications of the same abundant protein). Low levels of protein may be detected by sensitive stains, but the sequential procedure of trypsin digestion, peptide recovery, and mass spectrometry loses significant amounts of sample at each of these steps. Given these limitations, proteome analysis in fact captures only a fraction of the proteome.

You may ask, what is the number of parent ion masses needed to positively identify a protein? Typically, m/z's of 8–15 peptides greater than eight amino acid residues in length are sufficient to identify average-sized proteins (with masses 5 to 100 kDa) in the protein databases that are available to the public. If you can narrow the search parameters in the software program to species type, approximate protein mass, and isoelectric point, fewer peptide ions may be necessary.

10.7 REGULATION OF p53-CONTROLLED GENES

We are now in a position to appreciate the complexity of gene transcription and protein level regulation. We can use p53-mediated transcription of its effector genes as an illustrative example of this complexity, but we should bear in mind that there are several other mechanisms by which gene expression can be controlled. The p53 protein was discovered to be involved in cancers in 1979, but its biochemical mechanism of action did not become clear until 1991 when Bert Vogelstein's group (Johns Hopkins University, Baltimore, Maryland) showed that p53 could bind to a DNA response element. Northern analyses were instrumental in showing which genes are activated when p53 levels increase. These genes include, among others, *PUMA*, *GADD45*, and *MDM2* (Figure 10-14, top panel). *PUMA* is required for instigating apoptosis (programmed cell death), *GADD45* is responsible for mediating cell cycle arrest, and *MDM2* is responsible for targeting p53 for degradation (as part of a negative feedback loop) so that p53 can be maintained at a low level in the absence of genotoxic events. These are just a few of the p53-target genes necessary for p53 tumor suppressor function and p53 regulation. Early microarray studies showed that p53 activates 104 genes out of the 6,000 genes available at the time. In addition to the biochemical activities listed above, the p53-activated genes are involved in repair of DNA, inhibition of blood vessel growth (anti-angiogenesis), embryo implantation in the uterine wall, and several other functions.

After this initial microarray analysis, a method that locates all response elements for a particular transcription factor in the genome (called ChIP-seq) uncovered 743 potential p53 response elements. Though these elements have not been rigorously tested, they are at locations on the human genome where p53 binds and likely regulates genes. Curiously, the early microarray analysis also showed that p53 caused, directly or indirectly, the repression of 54 genes. Exactly how p53 represses gene

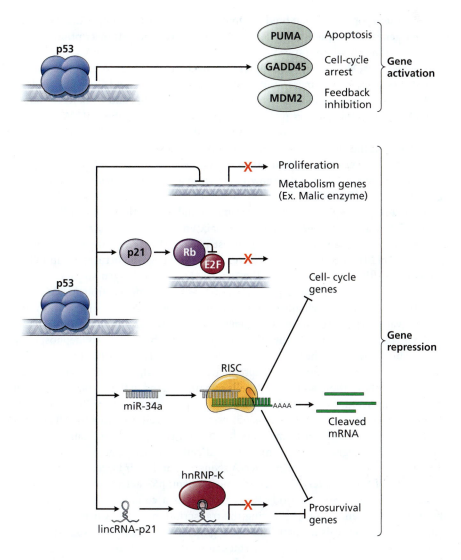

FIG. 10-14. p53-mediated regulation of gene expression.

expression remained unclear until scientists learned about additional ways gene expression could be inhibited, described below.

In parallel to studies that uncovered p53 target genes, Andrew Fire and Craig Mello, working with the nematode *C. elegans*, discovered a new class of RNAs, called microRNAs (miRNAs) or small interfering RNAs, approximately 20 nucleotides long, that silence gene expression by hybridizing to specific mRNAs.[22] The hybridized mRNAs are substrates for a multiprotein complex called RNA-induced silencing complex (RISC) within the cell cytoplasm. RISC recognizes and degrades the mRNAs and releases the miRNAs to bind more mRNAs. miRNAs also appear to act by inhibiting translation and sometimes even transcription as well. In addition, it was discovered that other RNAs, called large intergenic non-coding RNAs (lincRNAs), also regulate DNA expression. One particular lincRNA named lincRNA-p21 binds to a protein named heterogeneous nuclear ribonucleoprotein K (HNRPK) to form

microRNA (miRNA) RNA molecules approximately 20 nucleotides in length that silence gene expression by hybridizing to mRNAs. The targeted mRNAs are degraded by the RNA-induced silencing complex.

---

[22] Andrew Fire at Carnegie Institution of Washington and Craig Mello at the University of Massachusetts, Worcester, received the 2006 Nobel Prize in Physiology or Medicine for their discovery of microRNAs and their role in modulating transcript levels. Andrew Fire has since moved to Stanford University.

RNA-protein complexes that bind to DNA responsive elements near the promoters of some genes. Attachment of lincRNA-p21 appears to repress transcription of genes normally transcribed from such promoters.

When these RNA-mediated repression mechanisms came to light, researchers tested p53 for a role in increasing expression of miRNAs and lincRNA. Indeed, p53 increases the expression of an miRNA named miR-34a and a lincRNA named lincRNA-p21 by binding to p53-response elements within these RNA-encoding genes. In two cell types, it was discovered that p53 represses 422 and 960 genes, depending on the cell type. The number of p53 repressed genes is likely to depend on both the cell type and its environmental conditions.

Now there are at least four known distinct mechanisms by which p53 represses gene expression (Figure 10-14, bottom panel). The first is direct repression. Here, p53 binds to a response element near the promoter and lowers the ability of RNA polymerase II to transcribe genes. The gene for malic enzyme-2 (ME2) is repressed in this manner, which contributes to cell cycle arrest, partially by increasing p53 activity. The lower level of ME2 leads to decreased levels of NADPH, which, in turn, results in high levels of reactive oxygen species. The increased reactive oxygen species activates AMP-dependent kinase, which phosphorylates p53, releasing p53 from MDM2.

The other three mechanisms of p53-mediated repression are indirect. In one, p53 activates the transcription of the gene that expresses p21$^{cip-1/waf-1}$ enzyme, which prevents phosphorylation of the Rb protein. When Rb is relatively dephosphorylated, it binds tightly to the E2F1 transcription factor and inhibits the ability of E2F1 to bind to its response element and upregulate the transcription of genes responsible for cell growth. Without active E2F1, cells remain arrested in the G1 phase of the cell cycle. In another mechanism, p53 activates the transcription of miR-34a that is complementary to sequences on some genes essential for cell growth and cell survival (anti-apoptosis genes). Once miR-34a levels rise, the growth and survival genes are silenced by degrading their mRNAs, and cell cycle arrest or apoptosis ensues. There are databases that contain miRNA sequences and their targets (see exercise 5 at the end of this chapter). In the final mechanism, p53 activates transcription of lincRNA-p21, a 2,956-nucleotide RNA that forms a complex with HNRPK and binds to many genes, thus preventing the transcription of genes necessary for cell survival (anti-apoptosis factors). This small glimpse of p53's activities illustrates the complexity of expression regulation. You can imagine that this level of complexity, when expanded to include the thousands of transcription factors in the cell, creates a transcriptome that is extremely dynamic.

SUMMARY

In this chapter, we briefly discussed the history of gene expression research starting with the seminal breakthroughs made by Jacques Monod and François Jacob and their development of the lac operon model. Advances in technology have led to the simultaneous analysis of all transcripts expressed in a population of cells. The methods of analysis of the transcriptome are typically microarray and RNA-seq. Using p53 regulation of its target genes as one example, we have given insight to the complexity of expression analysis. The current methods of proteome analysis rely on 2D gel electrophoresis separation of proteins followed by mass spectrometry. More robust methods for proteome analysis will likely be an area of expansion in the future.

EXERCISES

1. **Transcription factor binding sites. There are a number of software programs that can scan a nucleotide sequence for responsive elements. Use one of these online software programs to scan the** *TP53* **promoter region for binding sites. Find the −1,000 to −1 nucleotide sequence upstream of the** *TP53* **primary transcript sequence. List 10 putative transcription factors that bind to this region. How**

many responsive elements that may bind to the transcription factor YY1 exist in this region?

2. **Microarrays. Download Heatmap Builder program (available for download from the publisher's website). Download heatbuildertest-mini as a text file. Make sure that your computer has .NET Framework Redistributable 1.1 (or later) installed. Use the program to generate a small microarray image with the heatbuildertest-mini text file. In the output, samples labeled "1" are from left ventricular heart tissues of 11 patients at the time in which a heart assist device was implanted in that area. Samples labeled "2" are from the left ventricular heart tissues of the same 11 patients after removal from patients during a heart transplantation procedure.**

   a. What is the range of RNA expression for "angiotensin receptor-like 1" mRNA under the conditions shown in the file?

   b. What is the range of RNA expression for "H3 histone, family 3A" mRNA?

   c. At the lowest value for angiotensin receptor-like 1 mRNA, what is the color?

   d. At the lowest value for H3 histone, family 3A, what is the color?

   e. What is the maximum fold-difference in expression for angiotensin receptor-like mRNA across the 22 samples? What is the maximum fold-difference in expression for H3 histone, family 3A mRNA across the 22 samples?

   f. Using the default form of the Heatmap Builder program, is the heatmap output more useful for comparing expression levels of different genes or more useful for comparing a particular gene expression across different samples?

   g. What can you say about the expression of angiotensin receptor-like 1 gene and H3 histone, family 3A gene before and after implantation of the heart assist devise?

      Now set the Heatmap Builder program to Dataset Normalizing Sorting mode and create the heatmap again.

   h. What can you say about the relative expression of angiotensin receptor-like 1 gene to H3 histone, family 3A gene?

   Reference for HeatMap Builder:

   King, J. Y., R. Ferrara, R. Tabibiazar, J. M. Spin, M. M. Chen, A. Kuchinsky, A. Vailaya, R. Kincaid, A. Tsalenko, D. X. Deng, A. Connolly, P. Zhang, E. Yang, C. Watt, Z. Yakhini, A. Ben-Dor, A. Adler, L. Bruhn, P. Tsao, T. Quertermous, and E. A. Ashley. 2005. "Pathway Analysis of Coronary Atherosclerosis." *Physiological Genomics* 23: 103–118.

Reference for data file used for HeatMap Builder:

Chen et al. 2003. "Novel Role for the Potent Endogenous Inotrope Apelin in Human Cardiac Dysfunction." *Circulation* 108: 1432–1439.

3. **Bridge amplification.**

   a. Come up with a mathematical expression for calculating the number of single strand DNA molecules in a sample using bridge amplification, prior to cleavage with a nuclease.

   b. If you started with 1 single stranded DNA molecule, what is the number of single stranded DNA molecules you would get after 35 cycles using bridge amplification?

4. **DNA sequencing. The image of the sequencing reactions in Figure 10-7 shows multiple spots, each with a different color.**

   a. What accounts for the different colors?

   b. Give plausible explanations for the different color intensities and different shapes.

5. **miRNAs. There are several databases that contain miRNA sequence information that are useful for gaining deeper insight into how miRNAs control cell characteristics. The p53 tumor suppressor can increase expression of miRNA-34a precursor. The miRNA precursor forms a stem-loop structure that is processed into mature miRNA-34a-5p and mature miRNA-34a-3p single strand RNAs. The 5p and 3p strands are nearly complementary. A frequent target of miRNA is the 3'UTR region of mRNAs. One transcript that is targeted is CDK4 mRNA. Use a miRNA database to provide the following information.**

   a. Show a putative stem-loop structure of the pre-miRNA-34a.

   b. Give the sequence of mature miRNA-34a-5p.

   c. Give the 3'UTR sequence of *CDK4* gene, and indicate the regions that are predicted to be targeted by miRNA-34a-5p.

   d. Show the sequence alignments of the miRNA-34a-5p and the target RNA regions.

   e. Is the targeting of CDK4 transcript by miRNA-34a-5p in line with the tumor suppressor function of p53? Explain.

6. **Below is an MS spectrum (generated from a matrix-assisted laser desorption ionization time-of-flight instrument).**

   a. From the masses of the peaks of the peptide ions, use an online software program to identify the protein and the species of the organism the protein is from. Hint: use the eight most abundant

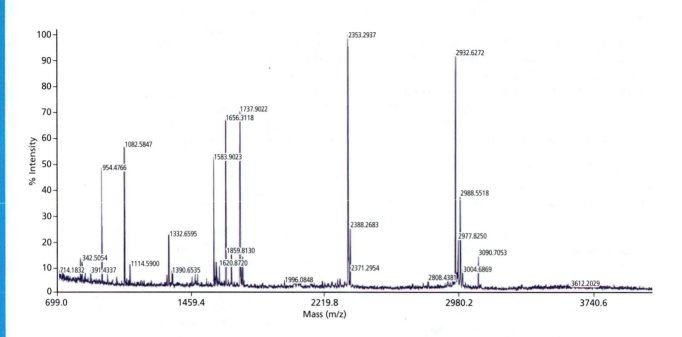

ions and assume the masses are monotopic ions (i.e., with only one proton).

b. What is the lowest precision required to successfully identify this protein?

c. What is the sequence of the peptide ion **2962.82**? What is the percent error in the experimental value assuming that the precision of the mass spectrometer instrument is 0.1 Da?

ANSWERS TO THOUGHT QUESTIONS

10-1. One explanation is that two transcripts can share a subset of sequences if they are transcribed from the same gene but undergo alternative splicing. In this particular analysis, the splicing that takes place in the heart tissue is different from the splicing that occurs in the other tissues. The longer transcript may contain extra RNA sequences not contained in the shorter transcript. The oligonucleotide probe is composed of a sequence that is common to both the long and short transcripts. If the oligonucleotide probe was specific to the extra RNA specific to the long transcript we would fail to detect the transcript in the heart tissue.

10-2. Gene 6, 9, or 10.

10-3. The cells express 2^3 (or eightfold) *less* transcript in the experimental condition compared to the control condition (in this case, the control is the 0 time point condition).

REFERENCES

Barsotti, A. M., and C. Prives. 2010. "Noncoding RNAs: The Missing 'Linc' in p53-Mediated Repression." *Cell* 142: 358–360.

Botcheva, K., S. R. McCorkle, W. R. McCombie, J. J. Dunn, and C. W. Anderson. 2011. "Distinct p53 Genomic Binding Patterns in Normal and Cancer-Derived Human Cells." *Cell Cycle* 10: 4237–4249.

Campbell, M., and L. H. Heyer. 2003. *Genomics, Proteomics & Bioinformatics*. San Francisco: Pearson Education.

Eisen, M. B., P. T. Spellman, P. O. Brown, and D. Botstein. 1998. "Cluster Analysis and Display of Genome-Wide Expression Patterns." *Proceedings of the National Academy of Sciences USA* 95: 14863–14868.

Futcher, B., G. I. Latter, P. Monardo, C. S. McLaughlin, and J. I. Garrels. 1999. "A Sampling of the Yeast Proteome." *Molecular Cell Biology* 19: 7357–7368.

Grabherr, M. G., et al. 2011. "Full-Length Transcriptome Assembly from RNA-Seq Data Without a Reference Genome." *Nature Biotechnology* 29: 644–652.

Graudens, E., V. Boulanger, C. Mollard, R. Mariage-Samson, X. Barlet, G. Grémy, C. Couillault, M. Lajémi, D. Piatier-Tonneau, P. Zaborski, E. Eveno, C. Auffray, and S. Imbeaud. 2006. "Deciphering Cellular States of Innate Tumor Drug Responses." *Genome Biology* 7: R19.

Gygi, S. P., Y. Rochon, B. R. Franza, and R. Aebersold. 1999. "Correlation Between Protein and mRNA Abundance in Yeast." *Molecular and Cellular Biology* 19: 1720–1730.

Hass, B. J., and M. C. Zody. 2010. "Advancing RNA-Seq Analysis." *Nature Biotechnology* 28, 421–423.

Jacob, F., and J. Monod. 1961. "Genetic Regulatory Mechanisms in the Synthesis of Proteins." *Journal of Molecular Biology* 3: 318–356.

Jiang, P., W. Du, A. Mancuso, K. E. Wellen, and X. Yang. 2013. "Reciprocal Regulation of p53 and Malic Enzymes Modulates Metabolism and Senescence." *Nature* 493: 689–693.

Kern, S. E., K. W. Kinzler, A. Bruskin, D. Jarosz, P. Friedman, C. Prives, and B. Vogelstein. 1991. "Identification of p53 as a Sequence-Specific DNA-Binding Protein." *Science* 252: 1708–1711.

Laemmli, U. K. 1970. "Cleavage of Structural Proteins During the Assembly of the Head of Bacteriophage T4." *Nature* 227: 680–685.

Lapointe, J., C. Li, J. P. Higgins, M. van de Rijn, E. Bair, K. Montgomery, M. Ferrari, L. Egevad, W. Rayford, U. Bergerheim, P. Ekman, A. M. DeMarzo, R. Tibshirani, D. Botstein, P. O. Brown, J. D. Brooks, and J. R. Pollack. 2004. "Gene Expression Profiling Identifies Clinically Relevant Subtypes of Prostate Cancer." *Proceedings of the National Academy of Sciences USA* 101: 811–816.

Lesk, A. M. 2008. *Introduction to Bioinformatics*, 3rd ed. Oxford, UK: Oxford University Press.

Loomis, W. F., Jr., and B. Magasanik. 1967. "Glucose-Lactose Diauxie in Escherichia coli." *Journal of Bacteriolology* 93: 1397–1401.

MacGillivray, A. J., and D. Rickwood. 1974. "The Heterogeneity of Mouse-Chromatin Nonhistone Proteins as Evidenced by Two-Dimensional Polyacrylamide-Gel Electrophoresis and Ion-Exchange Chromatography." *European Journal of Biochemistry* 41: 81–90.

Michiels, S., S. Koscielny, and C. Hill. 2005. "Prediction of Cancer Outcome with Microarrays: A Multiple Random Validation Strategy." *Lancet* 365: 488–492.

Monod, J. 1941. *Recherches sur la croissance des cultures bactériennes*. Paris: Hermann.

O'Farrell, P. H. 1975. "High Resolution Two-Dimensional Electrophoresis of Proteins." *Journal of Biological Chemistry* 250: 4007–4021.

Petrak, J., R. Ivanek, O. Toman, R. Cmejla, J. Cmejlova, D. Vyoral, J. Zivny, and C. D. Vulpe. 2008. "Déjà Vu in Proteomics. A Hit Parade of Repeatedly Identified Differentially Expressed Proteins." *Proteomics* 9: 1744–1749.

Pevsner, J. 2009. *Bioinformatics and Functional Genomics*, 2nd ed. Hoboken, NJ: John Wiley and Sons.

Purves, W. K., D. Sadava, G. H. Orians, and H. C. Heller. 2003. *Life: The Science of Biology*, 7th ed. Sunderland, MA: Sinauer Associates.

Relógio, A., C. Schwager, A. Richter, W. Ansorge, and J. Valcárcel. 2002. "Optimization of Oligonucleotide-Based DNA Microarrays." *Nucleic Acids Research* 30: e51.

Soupene, E., W. C. van Heeswijk, J. Plumbridge, V. Stewart, D. Bertenthal, H. Lee, G. Prasad, O. Paliy, P. Charernnoppakul, and S. Kustu. 2003. "Physiological Studies of Escherichia coli Strain MG1655: Growth Defects and Apparent Cross-Regulation of Gene Expression." *Journal of Bacteriology* 185: 5611–5626.

Taniguchi, Y., P. J. Choi, G. W. Li, H. Chen, M. Babu, J. Hearn, A. Emili, and X. S. Xie. 2010. "Quantifying E. coli Proteome and Transcriptome with Single-Molecule Sensitivity in Single Cells." *Science* 329: 533–538.

Toung, J. M., M. Morley, M. Li, and V. G. Cheung. 2011. "RNA-Sequence Analysis of Human B-Cells." *Genome Research* 21: 991–998.

Waters, K. M., J. G. Pounds, and B. C. Thrall. 2006. "Data Merging for Integrated Microarray and Proteomic Analysis." *Briefings in Functional Genomics and Proteomics* 5: 261–272.

11

AFTER STUDYING THIS CHAPTER,
YOU WILL:

- **Know basic operations on sets,
  such as union, intersection, and
  complement.**

- **Know the basic rules of probability
  and be able to apply them to
  compute probabilities of events
  from known probabilities of
  related events or by using
  appropriate counting rules.**

- **Know the meanings of
  dependence and independence.**

- **Know how to compute conditional
  probabilities from the definition or
  by using Bayes' law.**

- **Know how to use Bayes' law to
  compute posterior probabilities
  from prior probabilities and
  observed data.**

- **Know how the probability mass
  and the probability density
  functions describe the distribution
  of a random variable.**

BASIC PROBABILITY

11.1 INTRODUCTION

Many applications of bioinformatics, such as the interpretation of the *E*-value in the output of a BLAST search, hidden Markov models in algorithms for multiple sequence alignment, and the derivation of the Jukes-Cantor model of evolutionary distance, rely on tools and concepts from probability theory. In this chapter we will develop the basic tools of probability needed to understand these applications. If you have already taken an introductory probability or statistics course, then this chapter can serve as a refresher and reference. For students without such a background, this chapter will prepare you for Chapter 12, where we will develop more advanced techniques for the specific bioinformatics applications mentioned above.

11.2 THE BASICS OF PROBABILITY

Definitions and Basic Rules

Life science data is now recorded at a breakneck pace. Scientists who analyze the data must use statistics to verify whether their conclusions are valid, for which, in turn, they must understand the concept of probability. Probability theory as an area of mathematics originated from the desire to analyze games of chance. We will use the example of rolling dice to introduce definitions and notation. Rolling a die is not as unrelated to bioinformatics as it may sound because you can imagine rolling a four-sided die to create a random nucleotide sequence, or a 20-sided die to create a random protein sequence! Figure 11-1 shows an example of a four-sided DNA die. The letter that shows on the visible sides at the top of the die is the result of the roll.

When you roll the DNA die, you know that the possible *outcomes* can be the letters A, C, G, and T, but you do not know which of these outcomes you will see on any particular roll of such a die. This is a typical situation and is often referred to as the *experiment* in analogy to a biological experiment where we also do not know the outcome beforehand. We call the set of possible outcomes of the experiment the *sample space S*. In the case of a single DNA die roll, we have $S = \{A, C, G, T\}$. Sets are usually listed in braces, and if the list is short, all potential outcomes are listed.

A subset[1] of the sample space, that is, a collection of some or all of the outcomes, is called an *event*.[2] We say that *an event occurs* when any of the outcomes of the event occurs. We usually use capital letters to denote sets, and either use generic letters $A, B, C, \ldots$, or letters that remind us of the event the letter stands for.

Example 11.1. Let's simulate a random DNA sequence of length two by rolling a DNA die twice. In this experiment, the sample space S is given by

$$S = \{AA, AC, AG, AT, CA, CC, CG, CT, GA, GC, GG, GT, TA, TC, TG, TT\},$$

where the first letter in each pair indicates the result of the first die roll, and the second letter is the result of the second die roll. There are altogether 16 outcomes of the experiment (two die rolls) because there are 4 possibilities for the first die roll and 4 possibilities for the second one. Now let's say we are interested in whether at least one of the nucleotides is an A. This event can be written as the set of the following outcomes

$$A = \{AA, AC, AG, AT, CA, GA, TA\}.$$

If any of these seven outcomes occurs, then we would say the DNA sequence contains an A, or that the event A has occurred.

Because we use sets to describe events (collections of outcomes), we will review basic set theory. We say that a set B is a *subset* of a set A, denoted by $B \subseteq A$, if every outcome in set B is also an outcome in set A. The *intersection of A and B* is denoted by $A \cap B$ and consists of the outcomes that are in **both A and** B. The *union of A and B* is denoted by $A \cup B$ and consists of the outcomes that are in **either** A **or** B (or both). Finally, the *complement of A*, typically denoted by either A', $\bar{A}$, or A^c depending on the taste of the author, consists of the outcomes in the sample space S that are **not** in the set A. A set with no outcomes is called the *empty set*, and we denote it by $\emptyset$. The empty set is considered a subset of every set. Two sets A and B are *mutually exclusive* or *disjoint* if $A \cap B = \emptyset$.

Subsets, the union and intersection of sets, the complement of a set, and disjoint sets can be nicely visualized with a *Venn diagram*. Figure 11-2 shows these notions for generic sets A and B, where the relevant set is shaded. Note that the sets in question may be empty.

Example 11.2. A DNA die is rolled twice. Let A be the event that at least one A is rolled (see example 11.1), and B be the event that the two nucleotides are the same. Then

$$A = \{AA, AC, AG, AT, CA, GA, TA\}$$

$$B = \{AA, CC, GG, TT\}.$$

The set $A \cap B$ describes the event that there is at least one A **and** that both nucleotides are the same, so $A \cap B = \{AA\}$. It is a subset of both the set A and the set B as

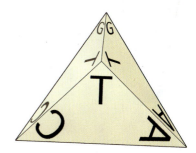

FIG. 11-1. A four-sided DNA die. The particular die roll shown resulted in the nucleotide T.

sample space The set (collection) of all possible outcomes or results of an experiment in probability theory.

event A set of outcomes of an experiment to which a probability can be assigned.

intersection of events The set of outcomes that are in every one of the events referred to.

union of events The set of outcomes that are in at least one of the events referred to.

complement of an event The set of outcomes that are in the sample space but not in the event.

Venn diagram A graphical tool for visualizing the union of events, the intersection of events, and the complement of an event.

---

[1] Unless specified to be a *proper* subset, a subset can be equal to the set of which it is a subset.

[2] The notion of sets shows up in many areas of mathematics. Depending on the area, sets are given different names. In probability theory, sets that list outcomes are commonly referred to as events, and we will use that convention.

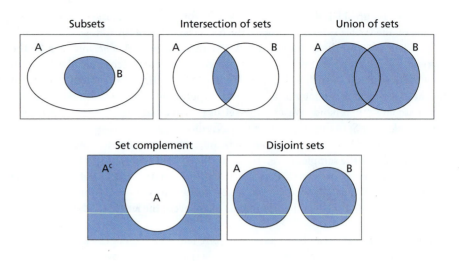

FIG. 11-2. Venn diagrams showing the subset relation $B \subseteq A$, the intersection $A \cap B$ of two sets, the union $A \cup B$ of two sets, the complement A^c of set A, and two disjoint sets, $A \cap B = \varnothing$.

both contain the outcome AA. The event $A \cup B$ describes the event that there is at least one A **or** that both nucleotides are the same, so

$A \cup B = \{$AA, AC, AG, AT, CA, CC, GA, GG, TA, TT$\}$.

Note that we do not list duplicate entries in the set $A \cup B$. Finally, B^c describes the event that the two nucleotides are not the same, so

$B^c = \{$AC, AG, AT, CA, CG, CT, GA, GC, GT, TA, TC, TG$\}$.

probability of an event A measure of the likelihood that the event will occur, that is, that any one of the outcomes in the event occurs.

Now that we have reviewed basic set theory, we are ready to define the probability of an event. Probabilities are often referred to as percentages, such as "there is a 20% chance of rain." If we express this percentage as a probability, we convert to decimal notation by recognizing that percent stands for "out of hundred." That is, we divide the percentage number by 100 and write it either as a fraction or in decimal notation. For example, 20% = 20/100 = 0.2. If some event cannot happen, which means there is no outcome that satisfies the description of the event, then we would say there is a 0% chance, and so the probability of the empty set is 0. The largest percentage chance would be 100%, which translates into 100/100 = 1, which is the probability of the sample space (because one outcome of the experiment has to happen!). These give us the first two of the basic rules of probability listed below.

Basic Rules of Probability

Let S be the sample space and let A be any subset of S. Then

RULE 1: $P(\varnothing) = 0$ (the impossible event) and $P(S) = 1$ (the sure event).

RULE 2: $0 \le P(A) \le 1$ (probabilities are numbers between 0 and 1, with 0 and 1 included).

RULE 3: If $S = \{s_1, s_2, \ldots, s_N\}$ is a finite set with N elements and all outcomes are *equally likely*, then $P(A) = |A|/N$, where $|A|$ denotes the number of elements in A.[3] Specifically, $P(\{s_i\}) = 1/N$ for all i.

---

[3] Just as the absolute value for numbers $|x|$ measures the magnitude or size of the number x, $|A|$ measures the magnitude or size of the set A as the number of outcomes that are in the set.

Rule 3 allows us to compute the probability of an event when we are in the special circumstance that all outcomes are equally likely, for example when we roll **fair** dice, or if we select items **at random**. Let's see how this works by computing the probability that a DNA sequence of length two has the same nucleotides in both positions. The relevant event is described by the set $B = \{AA, CC, GG, TT\}$ in example 11.2, and we also saw in example 11.1 that the sample space has $N = 16$ outcomes. Because the DNA die is fair, all outcomes are equally likely. There are four outcomes in B, so we have $|B| = 4$ and $P(B) = P(\text{the two nucleotides are the same}) = |B|/|S| = 4/16 = 0.25$.

However, if we are not in this special circumstance, we need to be able to compute probabilities in other ways. One way is to compute probabilities of events from probabilities of related events that are given or have already been computed. The rules given below are quite general, and we will encounter them frequently. Note that these rules apply to all events, whether the underlying outcomes are equally likely or not.

Additional Rules of Probability

For events $A \subseteq S$ and $B \subseteq S$ we have that

RULE 4: $P(A^c) = 1 - P(A)$.

RULE 5: $P(A \cup B) = P(A) + P(B) - P(A \cap B)$.

RULE 6: If A and B are disjoint, that is, $A \cap B = \varnothing$, then $P(A \cup B) = P(A) + P(B)$.

To remember these three rules we can use the Venn diagrams given in Figure 11-2. Because A and A^c together make up the sample space S and the two sets have no overlap, their probabilities have to add to the probability of the sample space, namely 1. This reasoning gives us rule 4. To remember rule 5, imagine adding up the probabilities of the sets A and B to get the combined probability of A **or** B. If the two sets overlap, then the intersection probability $P(A \cap B)$ has been counted twice, and therefore must be subtracted from the sum of the individual probabilities. Lastly, rule 6 is a special case of rule 5 in which A and B are disjoint (have no overlap, see Figure 11-2), so $P(A \cap B) = 0$.

Because many probabilities in bioinformatics applications arise from events comprised of equally likely outcomes, we will now look at ways to count the number of outcomes in an event or the sample space, so that we can compute probabilities according to rule 3.

Counting Methods When Order Matters

We now present several methods for counting outcomes that can be described as *ordered selections* or as *selections with order*. Examples of such outcomes are codons (three-letter sequences of nucleotides) or proteins (sequences of amino acids). The easiest counting method is the multiplication principle, a method that is very intuitive and that we have already used in example 11.1 to compute that there are 16 possible outcomes when rolling a DNA die twice.

Multiplication Principle

Let $E_1, E_2, \ldots, E_k$ be sets with $n_1, n_2, \ldots, n_k$ elements, respectively. Then there are a total of $n_1 \cdot n_2 \cdots n_k$ ways in which we can first choose an element of E_1, then an element of $E_2, \ldots,$ and finally an element of E_k.

We illustrate this principle with a few examples that relate to biology.

Example 11.3. DNA sequences consist of strings of the nucleotides A, C, G, and T. Codons are DNA segments that consist of three letters and code for amino acids (see Figure 1-11).

multiplication principle The principle that if there are a ways of doing something and b ways of doing another thing, then there are $a \cdot b$ ways of doing both things.

(a) How many different codons are there?

There are four choices for selecting the first nucleotide, four choices for selecting the second nucleotide, and four choices for selecting the third nucleotide. In the language of the multiplication principle this corresponds to three sets (one for each selection) consisting of the four nucleotides, that is, $E_1 = E_2 = E_3 = \{A, C, G, T\}$, with $n_1 = n_2 = n_3 = 4$ choices per selection. We can visualize this process of selecting the letters of the DNA sequence by marking each selection with a horizontal bar, then writing below it the set of possible choices, and above it the number of these possible choices. To obtain the final result, we multiply the number of possible choices for each of the three positions of the codon.

$$\underset{A,C,G,T}{4} \cdot \underset{A,C,G,T}{4} \cdot \underset{A,C,G,T}{4} = 4^3 = 64.$$

(b) How many codons start with **T** and end with **C**?

$$1 \cdot \underset{A,C,G,T}{4} \cdot 1 = 4.$$
$$\underset{T}{} \qquad \underset{C}{}$$

(c) How many codons consist of only A, C, and T nucleotides and have no repeated nucleotides?

$$\underset{A,C,T}{3} \cdot \underset{\text{all but first choice}}{2} \cdot \underset{\text{remaining choice}}{1} = 6.$$

The answer in part (c) of example 11.3 is a product of successive integers, which occurs frequently in counting problems. We will use the short-hand notation $n!$ (pronounced n factorial) for the product $n \cdot (n-1) \cdot (n-2) \cdots 2 \cdot 1$ of consecutive integers from 1 to n and define $0! = 1$.

The multiplication principle and the resulting outcomes (selections), in this case nucleotide sequences, can also be visualized by using a *tree diagram*. Figure 11-3 shows the diagram associated with part (c) of example 11.3. From an arbitrary starting point, draw as many branches as there are choices for the first element of the selection, in our case, the first nucleotide. Because there are three choices, namely A, C, and T, we draw three branches and label their end points with the choices. For the next element, the second nucleotide, there are two choices no matter what the first nucleotide happens to be, so we draw two branches from each of the labels of the first nucleotide, and then label the end points of the branches with the respective choices. Finally, there is just one choice for the last nucleotide in each case, so we draw a single branch from each of the labels for the second nucleotide and label its end point with the remaining nucleotide. Finally, we read off the possible codons by following the various paths along the branches. Obviously, tree diagrams are useful only when the number of choices is small, because they become unwieldy very quickly as the number of choices increases.

Example 11.3 showcases two types of counting that occur so often that there is a formula for the general case of each. In example 11.3 part (a), we selected $k = 3$ letters from the same set of $n = 4$ letters; that is, repetition of letters was allowed. In general, we have

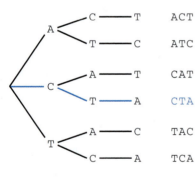

FIG. 11-3. Tree diagram showing the possible codons that do not contain a G nucleotide and have no repeated nucleotides. The outcomes can be read off by following the paths along the tree branches as shown in blue for the outcome CTA.

$$\text{Number of ways to select } k \text{ elements with repetition}$$
$$\text{and with order} = \underbrace{n \cdot n \cdots n}_{k \text{ factors}} = n^k. \qquad (11.1)$$

On the other hand, in part (c) of example 11.3 we selected $k = 3$ letters from a set of $n = 3$ possible letters, with the restriction that repetition of letters was not allowed. In general, the number of ways to select $k \le n$ elements without repetition and with order from a set of n elements is given by

$$P_{n,k} = n \cdot (n-1) \cdot (n-2) \cdots (n-k+1), \qquad (11.2)$$

because there are n choices for the first element, $n - 1$ (all but the element already selected) choices for the second element, and so on. Multiplying both the numerator and denominator in the expression for $P_{n,k}$ by $(n - k)!$, we obtain a more compact formula:

$$
\begin{aligned}
P_{n,k} &= n \cdot (n-1) \cdot (n-2) \cdots (n-k+1) \\
&= \frac{n \cdot (n-1) \cdot (n-2) \cdots (n-k+1) \cdot (n-k) \cdot (n-k-1) \cdots 2 \cdot 1}{(n-k) \cdot (n-k-1) \cdots 2 \cdot 1} \\
&= \frac{n!}{(n-k)!}
\end{aligned}
\tag{11.3}
$$

The letter P is used because ordered selections with distinct elements are called *permutations*. Specifically, $P_{n,k}$ denotes the number of ways one can select k items from n possible choices, when the order of selection matters. Both equations 11.2 and 11.3 are used interchangeably, depending on which one is easier to compute. For small values of n and k, equation 11.3 is quicker because most calculators have a button to compute the factorial of a number. However, as factorials become very large very quickly, for larger values of n and k the formula in equation 11.2 might be the only one your calculator can handle. Let's look at a few additional examples.

Example 11.4. How many four-letter DNA sequences have *at least* one G nucleotide? This is an instance where direct counting is tedious. Counting sequences that contain at least one G nucleotide means counting sequences containing exactly one G, exactly two Gs, exactly three Gs, or exactly four Gs. These are non-overlapping cases or events, so we could try to count each case separately and then add up their respective counts. However, it is much easier to look at the complement of the event. If A denotes the event "at least one G," then A^c is the event of not containing any G nucleotide (recall the Venn diagram describing A and A^c in Figure 11-2). The complement consists of only one case, and we can count the number of four-letter DNA sequences without a G very easily as follows:

$$
\underset{\text{A,C,T}}{\underline{3}} \cdot \underset{\text{A,C,T}}{\underline{3}} \cdot \underset{\text{A,C,T}}{\underline{3}} \cdot \underset{\text{A,C,T}}{\underline{3}} = 3^4
$$

Alternatively, we can use equation 11.1 for ordered selections with $k = 4$ elements from a set of $n = 3$ elements (A, C, and T) with repetition, for a total of $n^k = 3^4 = 81$ such sequences. This is the count for the event A^c, so to answer the question, we need to compute the number of sequences in the sample space and then subtract the number of sequences in A^c. The sample space consists of all four-letter DNA sequences, so $|S| = 4^4 = 256$ ($n = 4$ letters, $k = 4$ elements). Therefore, there are $|A| = |S| - |A^c| = 256 - 81 = 175$ four-letter DNA sequences that have at least one G nucleotide.

This example shows that when a counting question involves the terms *at least, at most, more than,* or *less than* in its description, it is worthwhile to consider counting the outcomes in the complementary event. If we had taken the longer (and incidentally, at this point harder) route of computing the number of DNA sequences that contain exactly one, two, three, or four G nucleotides, then we would have used a second general principle that is useful for counting equally likely outcomes, the addition principle.

Addition Principle

If the outcomes being counted can be divided into different (non-overlapping) cases, then the total number of outcomes is the sum of the numbers of outcomes in each group.

We will use both the multiplication and addition principles in an example concerning p53.

addition principle The principle that when outcomes can be divided into non-overlapping cases, the total number of outcomes is the sum of the number of outcomes in each case.

Example 11.5. In Chapter 3 you learned that human p53 binds to a DNA consensus sequence consisting of two copies of the 10-base motif PuPuPuC(A/T)(T/A)-GPyPyPy, separated by 0 to 13 nucleotides, where each Pu individually is either A or G, each Py individually is either C or T, and (A/T)(T/A) is either the pair AT or the pair TA.[4] Let's compute how many of the possible sequences have 5 nucleotides between the two motifs.

We start by counting the number of sequences that make up each of the two motifs. Using the multiplication principle, we can compute the number of sequences as follows:

$$\underset{\text{A,G}}{2} \cdot \underset{\text{A,G}}{2} \cdot \underset{\text{A,G}}{2} \cdot \underset{\text{G}}{1} \cdot \underset{\text{AT,TA}}{2} \cdot \underset{\text{G}}{1} \cdot \underset{\text{C,T}}{2} \cdot \underset{\text{C,T}}{2} \cdot \underset{\text{C,T}}{2} = 2^7.$$

Note that because we must have either an AT or a TA pair in positions 5 and 6 of the sequence, we have combined these two positions in our count. To count the five-letter DNA sequences in the center of the sequence, we use equation 11.1 (repetition is allowed) with $n = 4$ and $k = 5$. Putting the three parts together, we obtain that the total number of such sequences is

$$\underset{\text{left motif}}{2^7} \cdot \underset{\text{center}}{4^5} \cdot \underset{\text{right motif}}{2^7} = 16{,}777{,}216.$$

To count all the theoretically possible sequences to which human p53 could bind, we follow the steps above for sequences that have center parts of lengths $0, 1, \ldots,$ 13, and then use the addition principle to obtain the total count. The total count (see exercise 9) indicates that there are 1,466,015,498,240 sequences that human p53 could bind to. These are the theoretically possible sequences, but not all of them may exist in nature.

Counting Methods When Order Does Not Matter

For some selections **order does not matter**. It matters only whether an element of a set has been selected or not. A typical example is the selection of a committee. If all committee members have the same job, then all that matters is whether an individual is selected or not, not the order in which the selection has happened. We refer to unordered selections as *combinations*. You might wonder what this has to do with bioinformatics, but we will see that we can use combinations in a clever and unexpected way to count ordered sequences, such as DNA sequences, and we will need this trick when we derive the probabilities for the binomial distribution, which we will encounter shortly.

It can be shown that the number of combinations (unordered selections) of k objects selected from a set of n objects is given by

$$C_{n,k} = \frac{n!}{(n-k)! \cdot k!} \tag{11.4}$$

where $k \leq n$. This quantity is referred to as the *binomial coefficient* and is also denoted by $\binom{n}{k}$, pronounced "n choose k."

Example 11.6. Four graduate students, eight seniors, and ten juniors have applied to work in the lab of a famous chemistry professor. If the professor randomly selects six students for a team to work on the new project, how likely is it that the team consists of two graduate students, two seniors, and two juniors?

Because the professor selects the team at random, all teams are equally likely, so we can compute the desired probability using basic probability rule 3. The sample space S consists of all possible teams that the professor might select. The total number

---

[4] There are many sequences in the human genome that may conform to this consensus sequence. Some of these sequences will bind to and be regulated by p53.

of possible teams is given by $C_{22,6}$, because there are a total of 22 students from which six are chosen, so[5]

$$|S| = \frac{22!}{16! \cdot 6!} = \frac{22 \cdot 21 \cdot 20 \cdot 19 \cdot 18 \cdot 17 \cdot 16 \cdot 15 \cdots 2 \cdot 1}{(16 \cdot 15 \cdots 2 \cdot 1) \cdot (6 \cdot 5 \cdot 4 \cdot 3 \cdot 2 \cdot 1)}$$

$$= \frac{22 \cdot 21 \cdot 20 \cdot 19 \cdot 18 \cdot 17 \cdot \cancel{16 \cdot 15 \cdots 2 \cdot 1}}{\cancel{(16 \cdot 15 \cdots 2 \cdot 1)} \cdot (6 \cdot 5 \cdot 4 \cdot 3 \cdot 2 \cdot 1)} = \frac{22 \cdot 21 \cdot 20 \cdot 19 \cdot 18 \cdot 17}{6 \cdot 5 \cdot 4 \cdot 3 \cdot 2 \cdot 1} = 74{,}613.$$

Next we need to count the outcomes in event A, that is, the number of teams that have exactly two students each from the graduate students, seniors, and juniors. We can use the multiplication principle to count these as follows: Let E_1 be the set of all possible pairs of graduate students, of which there are $C_{4,2}$, so $n_1 = C_{4,2} = 6$. Similarly, let E_2 and E_3 be the sets of all possible pairs of seniors and juniors, respectively. We then have $n_2 = C_{8,2} = 28$ and $n_3 = C_{10,2} = 45$, so

$$|A| = \underbrace{C_{4,2}}_{\text{Graduates}} \cdot \underbrace{C_{8,2}}_{\text{Seniors}} \cdot \underbrace{C_{10,2}}_{\text{Juniors}} = 6 \cdot 28 \cdot 45 = 7{,}560.$$

Altogether, the probability that the team has an equal number of students from each group is $|A|/|S| = 7{,}560/74{,}613 = 0.101 \approx 10\%$.

Independence

After having learned some useful counting techniques that we can apply in the case of equally likely outcomes, we now return to the general case where outcomes may not be equally likely, and where we need to be given the probability of certain related events to be able to compute probabilities of the events we care about. For example, if we know that we are in a CpG island (see Chapter 9) of the genome, then we can no longer assume that the nucleotides are all equally likely.

We have already encountered one formula that can be used in the general case, namely the formula for the probability of the union of two events A and B. Knowing the probabilities of the two events A and B and the probability of their intersection, we can then compute $P(A \cup B)$. This formula becomes very simple when the two events are disjoint (no overlap), in which case the probability of the union is just the sum of the individual probabilities (see rule 6).

We will now look at a formula for the probability of the intersection of two events. For intersection probabilities, there is also a property that makes life simple, and that is *independence*. Independence of two events captures the idea that knowing whether or not one of the events has occurred does not influence the probability that the second event will occur. We will illustrate the notion of independence in an example before giving the formal definition.

In example 11.3 we computed the total number of codons to be $|S| = 4 \cdot 4 \cdot 4 = 64$, and the number of codons starting with T and ending with C to be $1 \cdot 4 \cdot 1 = 4$. If we let A denote the event that the codon starts with T and ends with C, and we assume that the occurrence of each letter in the DNA sequence is equally likely (is selected at random), then

$$P(A) = \frac{|A|}{|S|} = \frac{1 \cdot 4 \cdot 1}{4 \cdot 4 \cdot 4} = \frac{1}{4} \cdot 1 \cdot \frac{1}{4}.$$

We can think of the right-hand side of this equation as the product of three probabilities. Specifically, if we let A_1, A_2, and A_3 denote the three successive nucleotides, then

$$P(A_1 = T) = \frac{1}{4}, \qquad P(A_2 = \text{any letter}) = 1, \qquad P(A_3 = C) = \frac{1}{4},$$

independence and dependence
Two events are independent if the occurrence of one event does not change the probability that the other one occurs. If the events are not independent, then they are called dependent.

---

[5] Note that, if needed, we can always divide out the bigger factorial to compute the result using a calculator.

and so

$$P(A) = P(A_1 = T) \cdot P(A_2 = \text{any letter}) \cdot P(A_3 = C).$$

On the other hand, we know that the event A is the intersection of three events, namely

$$A = (A_1 = T) \cap (A_2 = \text{any letter}) \cap (A_3 = C),$$

so by combining these two facts we have

$$P(A) = P((A_1 = T) \cap (A_2 = \text{any letter}) \cap (A_3 = C))$$
$$= P(A_1 = T) \cdot P(A_2 = \text{any letter}) \cdot P(A_3 = C).$$

This result hints at a simple rule: The probability of an intersection of events is the product of the probabilities of the individual events, as long as the occurrence of one event has no influence on the probability that the other events occur and vice versa. Turning this idea around leads to the following definition of independence: Two events A and B are *independent* if and only if $P(A \cap B) = P(A)P(B)$. Here the phrase "if and only if" means that the sentence can be read in two ways: (1) if A and B are independent then $P(A \cap B) = P(A)P(B)$, and (2) if $P(A \cap B) = P(A)P(B)$, then A and B are independent.

Typical examples of independent events include successive tosses of a coin, successive rolls of dice, and the formation of zygotes, where alleles are received independently from the father and mother.

Example 11.7. Consider a dominant gene where plants with genotype BB or Bb have phenotype tall, and those with genotype bb have phenotype short. What are the genotypes of the offspring of a cross between a BB plant (Plant 1) and a Bb plant (Plant 2)? How likely is each genotype?

We can visualize the inheritance of alleles in a *Punnett square* as shown in Figure 11-4. We list the two alleles that determine the genotypes of the two plants at the top and on the left, then fill in the possible pairings of the alleles.

Two of the four offspring have genotype BB, the other two have genotype Bb. Because all offspring are equally likely to have formed, we have that $P(\text{BB}) = P(\text{Bb}) = 2/4 = 1/2$. Due to random mating, the genes from Plant 1 and Plant 2 are independent of each other, so we could have computed the probability of the genotypes also by using independence:

$$P(\text{BB}) = P((B \text{ from Plant } 1) \cap (B \text{ from Plant } 2))$$
$$= P(B \text{ from Plant } 1) \cdot P(B \text{ from Plant } 2) = 1 \cdot (0.5) = 0.5,$$

giving the same result. When dealing with more than one gene, this method is easier to use, as the Punnett square for multiple genes becomes very large.

FIG. 11-4. Punnett square to determine the genotypes of offspring of a cross between BB and Bb plants.

Dependence

How do we deal with situations where events are not independent? Consider color-blindness, a sex-linked inherited trait which is more common among men than women. One special type of color blindness, protanopia (inability to see the color red), has a prevalence of 1.01% among white men and 0.02% among white women. These percentages are examples of *conditional probabilities*, because they quantify the probability of one event, in this case that a randomly chosen white adult is color-blind, based on the known status of another event (the condition), here that this person is male or female.

Example 11.8. Assume that we have a population of 100,000 white people, consisting of 50,000 men and 50,000 women. Of the men, 505 are color-blind, whereas only

10 of the women are color-blind. What is the probability that a randomly chosen **person** is color-blind? How does this compare to the probability that a randomly chosen **woman** is color-blind?

Let's first define the relevant events. We denote the event that a person is color-blind by Cb, and that the person is a man or a woman by M or W, respectively. Then the probability that a randomly chosen person is color-blind is computed as

$$P(Cb) = \frac{505 + 10}{100,000} = \frac{515}{100,000} = 0.0052 = 0.52\%. \tag{11.5}$$

On the other hand, the conditional probability that a person is color-blind if we know (given) that the person is a woman is computed as the number of color-blind women divided by the total female population:

$$P(Cb\,|\,W) = \frac{10}{50,000} = 0.0002 = 0.02\%, \tag{11.6}$$

where the | is read as "given that." Notice that here we have reduced (shrunk) the sample space to consist of just the women, not the whole population, in order to take into account the condition.

Figure 11-5 shows the Venn diagrams that are associated with the computation of these two probabilities. On the left is a Venn diagram showing the full sample space consisting of 50,000 men, including 505 who are color-blind, and 50,000 women, including 10 who are color-blind. On the right is a Venn diagram showing the reduced sample space consisting of just the women. In general, if the sample space has equally likely outcomes and we want to compute a conditional probability, then we use the reduced sample space that consists of the event that we condition on.

How would we define a conditional probability not based on counts, but instead based on probabilities of related events? Intuitively, we would like to replace counts by appropriate probabilities. Dividing both numerator and denominator of the fraction in equation 11.6 by 100,000 (the size of the full sample space) gives us a clue:

$$P(Cb|W) = \frac{10}{50,000} = \frac{10/100,000}{50,000/100,000} = \frac{P(Cb \cap W)}{P(W)}.$$

We have succeeded in expressing the conditional probability as a fraction of related probabilities, namely the probability of the intersection of the two events and the probability of the condition. The reasoning used in this example leads to the following formal definition of conditional probability: For two events A and B with $P(B) \neq 0$, the conditional probability of A given B is defined as

$$P(A|B) = \frac{P(A \cap B)}{P(B)}. \tag{11.7}$$

conditional probability The probability of an event given that another event has occurred.

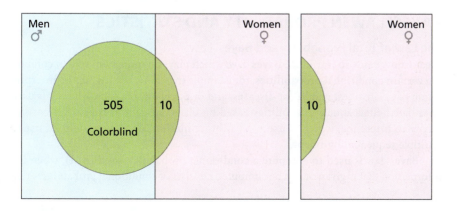

FIG. 11-5. Venn diagrams illustrating the concept of reducing the sample space for conditional probabilities. On the left, the entire rectangle represents the full sample space consisting of 100,000 people. The green circle represents the 515 color-blind people, of which 10 are color-blind women. The diagram on the right represents the reduced sample space consisting of only the 50,000 women.

The take-home message here is that we can always compute conditional probabilities by using equation 11.7, whether we start with counts of equally likely outcomes or with probabilities. In the case of equally likely outcomes we also have the option to compute the conditional probability directly from the counts using the reduced sample space.

Note that we can rearrange equation 11.7 to obtain a new formula for computing intersection probabilities, namely

$$P(A \cap B) = P(B)P(A|B). \tag{11.8}$$

This formula makes sense if we think about an intersection of events as a two-step process, where one of the events occurs first, followed by the second. The probability $P(A \cap B)$ that both events A and B occur, say that a randomly chosen person is both color-blind and female, equals the probability $P(B)$ that of one of the events (say being female) occurs, multiplied by the probability $P(A|B)$ that the second event (being color-blind) will occur given that the first one has occurred. Note that when we compute an intersection probability in this way, we can use either of the two events as the condition; which one we choose depends on the context and the probabilities that are provided or can be computed.

Typical instances in bioinformatics where conditional probabilities arise are due to the variation in observed nucleotide frequencies depending on the region of the genome under study. In a GC-rich region (see Chapter 9), we are more likely to observe a C or G nucleotide than in a normal region. We can express the probability that a C or G nucleotide is observed as a conditional probability, with the region as the condition.

Example 11.9. Researchers are interested in GC-rich regions of DNA sequences such as CpG islands because they often correlate with promoter regions for genes. It is known that the average percentage GC content (that is, occurrence of C or G) in human CpG islands is 63.8%, and that the overall percentage of CpG islands in the human genome is 1.9%. What is the probability that a randomly chosen nucleotide is a C nucleotide in a CpG island?

Let CpG stand for the event that the randomly chosen nucleotide is in a CpG island, C stand for the event that it is a C, and for simplicity assume that there are equal amounts of C and G nucleotides in CpG islands. What we are asked to find is the intersection probability $P(CpG \cap C)$. What we know are the probabilities $P(CpG) = 0.019$ and $P(C|CpG) = 0.319$ (half of 0.638). (The second probability is a conditional probability because it indicates the probability of observing a C or a G nucleotide on the condition that it is in a CpG island.) Using equation 11.8, we obtain

$$P(CpG \cap C) = P(CpG) \cdot P(C|CpG) = 0.019 \cdot 0.319 = 0.0061 \approx 0.6\%,$$

that is, the chance of observing a C that is in a CpG island is about 0.6%.

11.3 BAYES' LAW IN PROBABILITY AND STATISTICS

The Law of Total Probability and Bayes' Law

Bayes' law A rule to compute a conditional probability that is at the heart of Bayesian inference.

We are now ready to introduce Bayes' law, which plays an important role in computing certain conditional probabilities, for example in genetic testing, in the interpretation of positive test results for diseases, and in Bayesian statistics, where it is used to update beliefs about probabilities based on observed data. It is also the basis of Bayesian inference, which is used to calibrate hidden Markov models for use in multiple sequence alignment.

Bayes' law is used to compute a conditional probability, say $P(B|A)$, when the information that is given or can be computed consists of conditional probabilities that include the conditional probability $P(A|B)$ where the roles of A and B are reversed. Let's look at the general setup for situations where Bayes' law is useful. Assume that the sample space S (represented by the rectangle in Figure 11-6) is completely split up into non-empty non-overlapping subsets $B_1, B_2, \ldots, B_k$ like a jigsaw puzzle. In applications, each of the subsets will represent a different case. Now let's look at an event A in this sample space, represented by the shaded oval in Figure 11-6. Because the sets B_i are *mutually disjoint* (that is, $B_i \cap B_j = \varnothing$ for all i and j) we can see that

$$A = (A \cap B_1) \cup (A \cap B_2) \cup \ldots \cup (A \cap B_k).$$

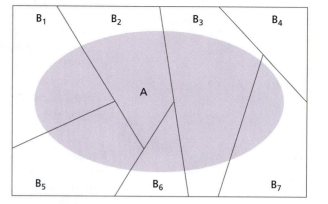

If sets B_i are mutually disjoint then sets $A \cap B_i$ must also be mutually disjoint, so

FIG. 11-6. A partition of the sample space S into sets $B_1, \ldots, B_7$ and the associated partition of an event A (shaded oval).

$$P(A) = P(A \cap B_1) + P(A \cap B_2) + \ldots + P(A \cap B_k) = \sum_{j=1}^{k} P(A \cap B_j), \qquad (11.9)$$

by probability rule 6. Using equation 11.8, the rule for computing intersection probabilities, then gives

$$P(A) = P(A|B_1)P(B_1) + P(A|B_2)P(B_2) + \ldots + P(A|B_k)P(B_k) \qquad (11.10)$$

$$= \sum_{j=1}^{k} P(A|B_j)P(B_j).$$

Equations 11.9 and 11.10 are referred to as the **law of total probability**; in both cases, sets B_i must form a *partition*, that is, be non-overlapping and together encompass the entire sample space. The two versions of the law of total probability give two different ways of computing the probability of an event A. Which of the two formulas is used depends on the form in which information is provided: equation 11.9 is used with intersection probabilities, whereas equation 11.10 is used with conditional probabilities. Note that $\sum_{j=1}^{k} P(A \cap B_j)$ and $\sum_{j=1}^{k} P(A|B_j)P(B_j)$ are mathematical shorthand notations for the respective sums. For example, $\sum_{j=1}^{k} P(A \cap B_j)$ can be read as sum of terms $P(A \cap B_j)$ where j takes integer values from 1 to k.

law of total probability The rule that the total probability of an event can be computed as a weighted average of conditional probabilities.

Example 11.10. You are working in a lab and your task is to stain a large number of cells. It is known that new cells will stain with probability 0.95, one-day-old cells will stain with probability 0.90, and two-day-old cells will stain with probability 0.80. Suppose that 60% of the cells are new cells, 25% are one day old, and 15% are two days old. Find the probability that a randomly selected cell will stain properly.

Because we know how likely a cell will stain depending on its age, we split up the sample space of all cells according to their age. If we let D_0, D_1, and D_2 be the events that the cell is new, one day old, or two days old, respectively, and C the event that the cell stains properly, then using the law of total probability (specifically equation 11.10) we have

$$P(C) = P(C|D_0)P(D_0) + P(C|D_1)P(D_1) + P(C|D_2)P(D_2)$$

$$= (0.95)(0.60) + (0.90)(0.25) + (0.80)(0.15) = 0.915 \approx 92\%.$$

Therefore, the probability that a randomly chosen cell will stain properly is about 92%.

Thought Question 11-1

> For the example above, can you (without any computations) give a range within which the probability that a randomly selected cell will stain properly must lie?

The law of total probability is also used in computations of certain conditional probabilities.

Example 11.11. In answering a question on a multiple choice test with five possible answers, a student either knows the answer or guesses randomly among the five answers. Assume that the student knows the answer to the question with probability 0.5. What is the probability that the student actually knew the answer, given that s/he answered correctly?

Let C denote the event that the answer is correct, and K denote the event that the student knows the answer. Then we want to compute $P(K|C)$. We are given that $P(K) = 0.5$, and can infer that $P(C|K) = 1$, as the student will answer correctly on the test with a probability of 1 if s/he knows the answer. We start by using the definition of conditional probability to obtain

$$P(K|C) = \frac{P(K \cap C)}{P(C)}.$$

We do not have either of the two probabilities on the right side of the equation, but we can use equation 11.8 to transform the intersection probability in the numerator, and use equation 11.10, the law of total probability, to compute the probability in the denominator. Here the two cases that partition the sample space are "the student knows the answer" and "the student does not know the answer," and so we obtain

$$P(K|C) = \frac{P(K \cap C)}{P(C)} = \frac{P(C|K)P(K)}{P(C|K)\,P(K) + P(C|K^c)\,P(K^c)}$$

$$= \frac{1 \cdot (0.5)}{1 \cdot (0.5) + (0.2)\,(0.5)} = 0.83.$$

This tells us that the probability of the student actually knowing the answer when s/he selected the correct answer on the test is 83%.

Note the structure of the question posed in example 11.11. We are to find a conditional probability $P(B|A)$ (here $P(K|C)$) when the information that is provided or can be computed consists of conditional probabilities that include the conditional probability where the roles of A and B are reversed, namely $P(A|B)$ (here $P(C|K)$). This type of question shows up so frequently that the formula used to compute it has been given a name. It is called Bayes' theorem or Bayes' law, named after Thomas Bayes, an English mathematician and Presbyterian minister who lived from 1701 to 1761. The name should sound familiar—in Table 8-3 you encountered a software package to compute phylogenetic trees named MrBayes.

Bayes' Law

For any event A and partition $\{B_1, B_2, \ldots, B_k\}$ of S,

$$P(B_i|A) = \frac{P(A \cap B_i)}{P(A)} = \frac{P(A|B_i)P(B_i)}{\sum_{j=1}^{k} P(A|B_j)P(B_j)}. \qquad (11.11)$$

Figure 11-7 illustrates equation 11.11 for the case when there are five sets B_i. We draw a square with side length 1 and divide the width of the square into intervals of length $P(B_1)$, $P(B_2), \ldots, P(B_5)$. These intervals become the bases of rectangles whose heights are $P(A|B_1)$, $P(A|B_2)$, ..., $P(A|B_5)$. The area of each rectangle is then $P(A \cap B_i)$ for $i = 1, \ldots, 5$. Adding the areas of the shaded rectangles gives $P(A)$ by the law of total probability (see equation 11.9), and $P(B_3|A)$ is given by the ratio of the area of the rectangle associated with B_3 (dark shaded) to the total shaded area.

When the partition consists of just two sets or cases (as in example 11.11), equation 11.11 simplifies to:

$$P(B|A) = \frac{P(A|B)\,P(B)}{P(A|B)\,P(B) + P(A|B^c)\,P(B^c)}. \quad (11.12)$$

Example 11.12. An entomologist spots an individual of what might be a rare subspecies of beetle that has a very distinctive pattern on its back. In the rare subspecies, 97% of the beetles have this distinctive pattern, whereas only 3% of the beetles in the common subspecies have this pattern. The rare subspecies accounts for 0.2% of the total population of beetles. How likely is it that the entomologist did see a beetle from the rare subspecies?

Using Bayes' law with two cases (rare and common subspecies), we get

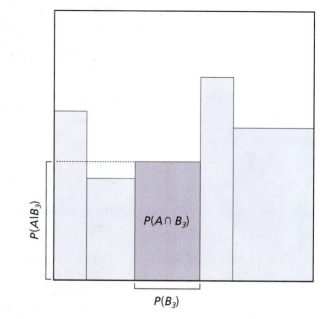

FIG. 11-7. Diagram illustrating Bayes' law for $P(B_3|A)$. The area of the large square equals 1, and the areas of the shaded rectangles equal $P(A \cap B_j)$ for $j = 1, 2, \ldots, 5$, respectively. The probability $P(B_3|A)$ is the ratio of the area of the dark shaded rectangle, which is equal to $P(A|B_3)P(B_3)$, to the total shaded area, which equals $P(A) = \sum_{j=1}^{k}[P(A|B_j)P(B_j)]$.

$$
\begin{aligned}
P(\text{rare}|\text{pattern}) &= \frac{P(\text{pattern}|\text{rare})P(\text{rare})}{P(\text{pattern}|\text{rare})P(\text{rare}) + P(\text{pattern}|\text{common})P(\text{common})} \\[2mm]
&= \frac{(0.97)(0.002)}{(0.97)(0.002)+(0.03)(.998)} \approx 6.1\%,
\end{aligned}
$$

that is, the probability that the entomologist saw a beetle from the rare subspecies is about 6%.

Another typical application of Bayes' law is the evaluation of clinical test results. Most tests for diseases such as breast cancer are not perfect. They usually detect the disease with high probability, but not 100%, leading to false negatives (a negative test result when the disease is present). The test may also produce false positives (a positive test result when the patient is disease free). Thus, the crucial question for a patient and her doctor when a test comes back positive is the likelihood that the patient actually has the disease. We will look at an example.

Example 11.13. Suppose that a certain rare disease occurs in half a percent of the general population. A blood test for the disease exists and is 98% effective (correctly detects presence of the disease). However, the test also produces false positives in 5% of disease-free patients. If you are a patient and the test result comes back positive, what is the probability that you have the disease?

Let $+$ stand for a positive test result and D for the event that you have the disease. Using Bayes' law we obtain

$$
\begin{aligned}
P(D|+) &= \frac{P(+|D)P(D)}{P(+|D)P(D) + P(+|D^c)P(D^c)} \\[2mm]
&= \frac{(0.98)(0.005)}{(0.98)(0.005) + (0.05)(0.995)} \approx 9\%,
\end{aligned}
$$

that is, you have a 9% probability of having the disease. This result may be somewhat surprising—after all, the test result was positive! So what is happening? Let's visualize these percentages in numbers, to see whether it all makes sense. Say we start with 100,000 people. Then we expect $0.005 \times 100,000 = 500$ people to have the disease. If everybody took the test, then there would be $0.98 \times 500 = 490$ people with a correctly detected positive test result, and $0.05 \times 99,500 = 4,975$ incorrect positive test results. Now we use the definition of conditional probability (equation 11.7) to compute

$$P(D|+) = \frac{P(+\cap D)}{P(+)} = \frac{490/100,000}{(490+4,975)/100,000} \approx 9\%.$$

In this case, false positives are so much more numerous than true positives that the probability of actually having the disease remains small even though the test is very effective. This result cautions against testing for very rare diseases, as the benefits are often negated by negative side effects of the test or of unnecessary treatment for false positive cases. However, when the disease is not as rare then the situation is quite different, as we will see below.

Assume now that the disease affects 10% of the population and all the probabilities for test results remain the same. What is now the probability of having the disease when the test result is positive? Using Bayes' law, we have that

$$P(D|+) = \frac{P(+|D)P(D)}{P(+|D)P(D) + P(+|D^c)P(D^c)}$$
$$= \frac{(0.98)(0.10)}{(0.98)(0.10) + (0.05)(0.90)} \approx 69\%.$$

Note that in both examples the fact of a positive test result indicated a greater likelihood of actually having the disease relative to that of the general population. The difference in the two cases is that for rare diseases, even a positive test result may not increase the likelihood of having the disease to levels that in and of themselves would justify strong action. Typically, in the case of rare diseases positive test results are followed by additional tests (which might be more expensive and/or have side effects) to confirm or refute the positive test result.

We now turn to the applications of Bayes' law in statistics.

Bayesian Inference

In Table 8-3, a program called MrBayes was mentioned that can compute phylogenetic trees using simulations of posterior probabilities. In this section, we will discuss Bayesian inference and the use of posterior probabilities computed from observed data to decide between several hypotheses such as those that describe the branching behavior in the phylogenetic trees. We will denote the n hypotheses by $H_1, H_2, \ldots, H_n$, and the observed data by D. The observation can consist of a single outcome or of a sequence of outcomes. If we consider particular hypotheses, we may give them names that make them easy to remember. To make this general setup more concrete, we will look at an example.

Bayesian inference A method to update the likelihood of a hypothesis based on observed data. For example, the likelihood of a particular genetic makeup of parents can be computed based on the genetic makeup of the children.

Example 11.14. You are asked to play the following game: Your host presents you with two apparently identical coins, one is a fair coin, the other one is biased and shows heads with probability 2/3 and tails with probability 1/3. Only your host can tell which one is which. You are to pick a coin and toss it 10 times, and then guess whether it is the fair coin or the biased coin. If your guess is correct you will receive a prize. How could you use probability theory to help you win the prize?

Luckily, you have just learned about Bayes' law. In this case, there are two hypotheses: $H_1 = F =$ "the fair coin was selected" and $H_2 = B =$ "the biased coin was selected." The data in this case would be the sequence of the 10 coin tosses, for example

D = HTHTHHHHTHT. In other words, the first coin came up heads, the second came up tails, and so on. What you need to compute is $P(F|D)$ and $P(B|D)$, the adjusted probabilities of having selected the fair or biased coin, respectively, in light of the observed data. If $P(F|D) > P(B|D)$, then you would guess that the fair coin had been selected, and otherwise your guess would be that the biased coin had been selected. In what follows, we will discuss in detail how to compute these probabilities.

Because the coins are equally likely to be picked, at the start we have that $P(F) = P(B) = 1/2$. These two probabilities together are referred to as the *initial* or *prior distribution*, as they reflect the probabilities initially, or prior to, observing any data. Processing the data one toss at a time, let's now look at the probabilities of interest after the first toss, which came up heads:

$$P(F|H) = \frac{P(H|F)P(F)}{P(H|F)P(F)+P(H|B)P(B)} \tag{11.13}$$

$$P(B|H) = \frac{P(H|B)P(B)}{P(H|F)P(F)+P(H|B)P(B)}$$

To compute these two probabilities, which together give the *posterior distribution*, we need to compute $P(H|F)$ and $P(H|B)$, probabilities that give us the *likelihood of observing the data*, in this case heads on the first toss, under each of the two hypotheses. We obtain that $P(H|F) = 1/2$ and $P(H|B) = 2/3$. We can now compute the posterior probabilities as follows:

$$P(F|H) = \frac{P(H|F)P(F)}{P(H|F)P(F)+P(H|B)P(B)} = \frac{(1/2)(1/2)}{(1/2)(1/2)+(2/3)(1/2)} = \frac{3}{7},$$

$$P(B|H) = \frac{P(H|B)P(B)}{P(H|F)P(F)+P(H|B)P(B)} = \frac{(2/3)(1/2)}{(1/2)(1/2)+(2/3)(1/2)} = \frac{4}{7}.$$

Does this result make intuitive sense? Because heads is more likely to occur with the biased coin, observing an H on the first toss (or any toss) increases the posterior probability of hypothesis B (biased coin), and correspondingly decreases the posterior probability of hypothesis F (fair coin). Note also that the two posterior probabilities add to one as we have only two hypotheses. We can therefore compute the second probability much more easily as we will see below.

Now the coin is tossed a second time and we perform a computation of the posterior or adjusted probabilities, based on the second observation. What should be used as the initial distribution? The original one, $P(F) = P(B) = 1/2$, or $P(F|H) = 3/7, P(B|H) = 4/7$, which takes into account the outcome of the first coin toss? The smart choice is to use the latter one as it already incorporates information about the coin that we have observed. We now can perform a similar computation for the second coin toss by replacing H by T (outcome of the second toss) in equation 11.13 and by using the posterior distribution after the first toss as the prior distribution of the second toss, $P(F) = 3/7, P(B) = 4/7$. Using $P(T|F) = 1/2$ and $P(T|B) = 1/3$, we have

$$P(F|T) = \frac{P(T|F)P(F)}{P(T|F)P(F)+P(T|B)Pv(B)} = \frac{(1/2)(3/7)}{(1/2)(3/7)+(1/3)(4/7)} = \frac{9}{17},$$

$$P(B|T) = 1 - P(F|T) = \frac{8}{17}.$$

We can continue these computations based on the observed outcomes of the individual tosses. Table 11-1 shows the prior probabilities at each step, the observed outcome, and the computed posterior probabilities, which then become the initial probabilities for the next step. The final posterior probabilities, shown in blue, take all 10 coin toss outcomes into account.

TABLE 11-1. STEPS IN THE COMPUTATIONS OF THE BAYESIAN INFERENCE. THE POSTERIOR PROBABILITIES IN THE TWO RIGHTMOST COLUMNS BECOME THE PRIOR PROBABILITIES FOR THE NEXT STEP, LISTED IN THE TWO LEFTMOST COLUMNS

PRIOR PROBABILITIES			POSTERIOR PROBABILITIES	
$P(F)$	$P(B)$	OUTCOME	$P(F|\text{OUTCOME})$	$P(B|\text{OUTCOME})$
0.500	0.500	H	0.429	0.571
0.429	0.571	T	0.529	0.471
0.529	0.471	H	0.458	0.542
0.458	0.542	T	0.559	0.441
0.559	0.441	H	0.487	0.513
0.487	0.513	H	0.416	0.584
0.416	0.584	H	0.348	0.652
0.348	0.652	T	0.445	0.555
0.445	0.555	H	0.375	0.625
0.375	0.625	T	0.474	0.526

Because $P(F|\text{HTHTHHHTHT}) = 0.474$ and $P(B|\text{HTHTHHHTHT}) = 0.526$, we conclude that it is more likely that the biased coin was chosen. In fact, it can be proven that the same posterior probabilities result when taking the whole sequence of coin tosses as the data all at once and performing a single application of Bayes' law, instead of computing posterior probabilities for each single observation one at a time. In exercise 16 you will work through an example of this important fact about Bayesian inference.

We now summarize the general setup for Bayesian inference.

Bayesian Inference

For a given set of n hypotheses $H_1, H_2, \ldots, H_n$, we start out with the *initial* or *prior distribution*

$$P(H_1), P(H_2), P(H_3), \ldots, P(H_n),$$

which reflects our initial belief about how likely it is for each hypothesis to be true prior to observing the data. When no prior knowledge exists with regard to the hypotheses, then all hypotheses are assumed to be equally likely, that is, $P(H_i) = 1/n$ for $i = 1, 2, \ldots, n$.

In addition, we also need to know (or be able to compute) the list of likelihoods of observing the data D if in fact hypothesis H_i is true for i = 1, 2, ..., n, that is,

$$P(D|H_1), P(D|H_2), \ldots, P(D|H_n).$$

In example 11.14, $H_1 = F$ and $H_2 = B$, the prior distribution is given by $P(F) = P(B) = 1/2$, the likelihoods of observing H are $P(H|F) = 1/2, P(H|B) = 2/3$, and the likelihoods of observing T are given by $P(T|F) = 1/2, P(T|B) = 1/3$.

We want to compute the *posterior distribution*

$$P(H_1|D), P(H_2|D), P(H_3|D), \ldots, P(H_n|D),$$

prior and posterior probabilities
In Bayesian inference, the probabilities before the observation of new data and after the observed data have been used to reassess the probabilities, respectively.

Simulated outcome	P(F|D)
Biased	
H	0.429
H	0.360
H	0.297
H	0.240
T	0.322
H	0.263
H	0.211
T	0.286
H	0.231
H	0.184

Simulated outcome	P(F|D)
Biased	
H	0.429
H	0.360
H	0.297
T	0.388
T	0.487
H	0.416
H	0.348
T	0.445
H	0.375
H	0.311

Simulated outcome	P(F|D)
Fair	
H	0.429
T	0.529
H	0.458
T	0.559
H	0.487
H	0.416
H	0.348
T	0.445
H	0.375
H	0.311

FIG. 11-8. Output of three Excel simulations of 10 coin tosses each and computation of $P(F|D)$. In the first two simulations a biased coin was selected, and in the third simulation a fair coin was selected.

which reflects our adjusted belief about the probabilities of each of the hypotheses after having observed the data D. The posterior probability of a specific hypothesis H_i for $i = 1, 2, \ldots, n$ is computed using Bayes' law:

$$P(H_i|D) = \frac{P(D|H_i)P(H_i)}{\sum_{j=1}^{n} P(D|H_j)P(H_j)} \text{ for } i = 1, 2, \ldots, n.$$

If a subsequent observation is made, then the posterior probabilities from the previous observation, "the new reality," will serve as the prior probabilities for the computations based on the next observation. In example 11.14, the observed data for the first step is H and the posterior distribution is $P(F|H) = 3/7$ and $P(B|H) = 4/7$. Note that the probabilities of the prior distribution, the likelihoods, and the posterior distribution, respectively, each have to add up to 1. This is why we were able to compute $P(B|H)$ simply as $P(B|H) = 1 - P(F|H)$ in example 11.14.[6]

We now return to the setup of example 11.14 and explore Bayesian inference using simulations of 10 coin tosses.[7] Figure 11-8 shows the output of three such simulations of 10 coin flips, together with the information as to which coin was selected in the beginning, fair or biased. The entries in each row show the observed outcome D and the computed posterior probability that the simulated coin is fair, $P(F|D)$.

In the simulation depicted in the left table of Figure 11-8, in which the biased coin was selected, we correctly obtain a low posterior probability 0.184 that the fair coin was chosen. However, in the other two simulations, the posterior probability $P(F|D)$ displayed in the last row is the same, namely 0.311, even though in one simulation the biased coin was chosen, and in the other one the fair coin was selected.

Thought Question 11-2

Can you guess why the final posterior probabilities (values in the last row) of the second and third simulations shown in Figure 11-8 are the same even though in one case the biased coin was selected and in the other case the fair coin was? Hint: In what way are the two simulations alike, but different from the first simulation?

---

[6] Likewise, when there are n different hypotheses, the posterior probability of the last hypothesis, H_n, can be computed as $P(H_n|D) = 1 - \sum_{i=1}^{n-1} P(H_i|D)$.

[7] The Excel spreadsheet for this simulation is posted on the website for this book. A similar program can be written in Matlab or any other programming language.

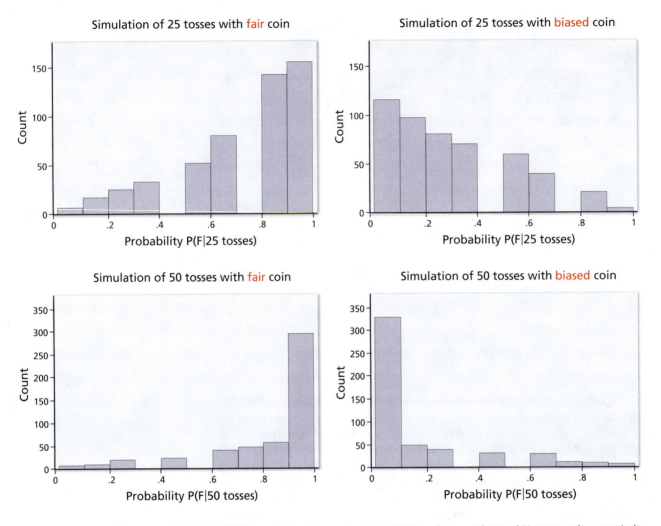

FIG. 11-9. Histograms of posterior probabilities $P(F|D)$ that the fair coin was selected in 1,000 simulations with 25 and 50 tosses each, respectively. On the horizontal axis, values for the posterior probability are grouped in intervals of width 0.1, and the vertical axis shows the number of simulations with values of the posterior probabilities in these intervals.

Computing posterior probabilities becomes more accurate (that is, the proper coin is more likely to be detected) if more coin tosses are used to compute the posterior probability $P(F|D)$. The histograms of Figure 11-9 display the results of 1,000 simulations, where each simulation consisted of either 25 (top two histograms) or 50 (bottom two histograms) coin tosses. For each simulation, the posterior probability $P(F|D)$ was computed. The histograms on the left show these posterior probabilities for simulations in which the fair coin was selected, whereas the histograms on the right show the posterior probabilities for the simulations in which the biased coin was selected. The vertical axes in each histogram show the number of times the computed probability $P(F|D)$ falls into the respective interval of width 0.1. As you would expect, when the fair coin was chosen, the majority of the posterior probabilities were closer to 1, whereas in the simulations in which the biased coin was chosen, the majority of the posterior probabilities were closer to 0. With more coin tosses this pattern becomes more pronounced, showing that more coin tosses (that is, more data) gives better results and a higher probability of detecting the correct coin.

We have looked at Bayesian inference in the familiar setting of coin tosses, and have seen how the observed data can change and improve our beliefs about the likelihood of a hypothesis. We will encounter a similar situation in the next chapter when discussing multiple sequence alignment and hidden Markov models.

We now will turn our attention to random variables, which will get us closer to one of our goals—namely, defining and interpreting the *E*-value generated by sequence alignment programs.

11.4 RANDOM VARIABLES

So far we have associated probabilities with individual outcomes or events. In many applications, however, we are not interested in the particular outcomes or events, but in a numerical value associated with them. For example, we may not care in which order the tails in a sequence of coin tosses appeared but just want to know the total number of tails. In this instance, we have associated with each outcome (consisting of a sequence of heads and tails) a numerical value, namely the number of tails. We may then ask, for example, "How likely is it to obtain at least two tails in a sequence of 10 coin tosses?" In order to deal with this type of situation, we introduce the notion of a *random variable*, which associates a numerical value with each random outcome. These values together with their associated probabilities are called the *distribution of the random variable*. We distinguish random variables according to the type of their values as either *discrete* or *continuous*. A discrete random variable has a countable number of distinct values, whereas a continuous random variable has possible values from an interval and often consists of a measurement. For example, the number of tails in a sequence of heads and tails is a discrete random variable, whereas the height, weight, and the body mass index computed from the height and weight of a randomly selected person are continuous random variables.[8] We start by discussing discrete random variables.

random variable A variable that can take on a set of possible different values each with an associated probability.

continuous random variable A random variable with values that lie in a continuous interval or collection of intervals.

discrete random variable A random variable with values that form a finite or countable set.

Discrete Random Variables

Examples of discrete random variables are quantities that arise from counting, such as the number of heads or tails in coin tosses, or the number of times a G nucleotide shows up in a DNA sequence of a given length.

Example 11.15. We define X to be the number of tails in three coin tosses of a fair coin, that is, X "counts" the number of tails. The random variable X can take on the four values $x = 0$, $x = 1$, $x = 2$, and $x = 3$. Note that when we refer to the random variable we use capital letters, and when we refer to actual values that the random variable can take on, we use lowercase letters. Once we have determined the values, we can compute the associated probabilities that the random variable X will have the value x, $P(X = x)$. To do so, we collect all the outcomes that result in the same value of the random variable, in this example, the same number of tails, and then determine the probability of the set consisting of all these outcomes. For example, the event $A =$ one tail was tossed $= \{HHT, HTH, THH\}$ has probability $P(A) = P(X = 1) = 3/8$ as we have eight equally likely outcomes, three of which belong to event A. We can summarize the results in a table. Note that we use $p(x)$ as a shorthand notation for $P(X = x)$.

	HHH	HHT, HTH, THH	TTH, THT, HTT	TTT
x	0	1	2	3
$p(x)$	1/8	3/8	3/8	1/8

---

[8] Values are said to be countable if they can be written as a list that contains every value. This list can contain a finite number of values or be infinitely long. In contrast, values from an interval, like points from a contiguous segment of the number line, cannot be written as a list.

Extracting the relevant information and converting the fractions to decimal notation reduces to

x	0	1	2	3
p(x)	0.125	0.375	0.375	0.125

Note that it does not matter whether we write the values of $p(x)$ as fractions or in decimal notation. Fractions can have the advantage of being more precise, but decimals might be easier to use in computations.

distribution The description of a random variable, consisting of the possible values and their probabilities. In the case of a discrete random variable, this description takes the form of a probability mass function; in the case of a continuous random variable, a probability density function.

We have derived what is called the *probability distribution*, or for short, the *distribution*, of the discrete random variable X. For a discrete random variable, the probability distribution consists of a listing of the values of the random variable, together with the probabilities that each of the values will occur. If there are only a few values, the probability distribution may be given as a table. When there are many values, the probabilities of the different values are usually given as a formula. Note that all values of the *probability mass function*, which is the list of the probabilities, have to satisfy certain properties that follow from our basic rules for probabilities: Specifically, all individual values of $p(x)$ must be between 0 and 1, and if they are all added up, the sum must be equal to one.

Thought Question 11-3

Explain why the probabilities of the values of a discrete random variable have to add up to 1.

probability mass function A function that gives the probabilities that a discrete random variable is exactly equal to each of its possible values.

Bernoulli random variable The simplest discrete random variable, with just two possible values, namely 0 and 1.

Bernoulli trial An experiment that has only two possible outcomes.

We can visualize a probability mass function by drawing vertical bars whose heights equal the probability of the corresponding values on the horizontal axis. For the random variable of example 11.15, the probability mass function is shown in Figure 11-10.

The simplest discrete random variable is the *Bernoulli* random variable, which has exactly two values, 0 and 1. It is named after Jacob Bernoulli (1654–1705), one of the many prominent mathematicians of the Swiss Bernoulli family. The Bernoulli random variable is used to model any experiment that has exactly two possible outcomes (such as yes and no, true and false, heads and tails), and such an experiment is called a *Bernoulli trial*. The positive outcome is called a *success*, and we use the parameter p to refer to the probability of success, $P(X = 1)$. The **distribution of the Bernoulli random variable** is then given by

$$P(X = 1) = p \text{ and } P(X = 0) = 1 - p.$$

Note that the two outcomes $X = 0$ and $X = 1$ do not have to be equally likely, as can be seen in example 11.16, where we use a sequence of Bernoulli trials to solve a problem that arises in bioinformatics.

Example 11.16. In a DNA sequence with 20 nucleotides, what is the probability that the sequence contains exactly three G nucleotides?

For the purpose of solving this problem, we can represent any DNA sequence as consisting of just two letters, G and N (for "not G"). For example, consider a sequence with Gs in the 3rd, 15th and 16th positions, which we can represent as NNGNNNNNNNNNNNGGNNNN.

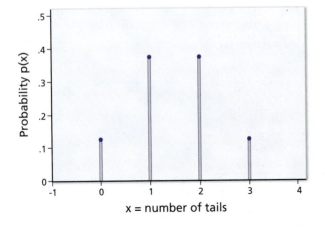

FIG. 11-10. Probability mass function for the number of tails in three tosses of a fair coin. The height of the vertical line at a value x, here number of tails, represents the probability that this value occurs.

We assume that the four nucleotides A, C, G, and T are equally likely at each position, independently of the nucleotides at the other positions. In this case, which nucleotide appears at each position can be treated as the outcome of a Bernoulli trial, with $p = 1/4$ for outcome G, and $1 - p = 3/4$ for outcome N. The entire sequence describes an intersection event based on the 20 Bernoulli trials—namely, that G does not appear at the first and second positions, G appears at the third position, and so on. The probability of this event is the product of the individual probabilities (due to the independence assumption):

$$P(\text{NNGNNNNNNNNNNNNNNGGNNNN}) = \frac{3}{4}\frac{3}{4}\frac{1}{4}\frac{3}{4}\cdots\frac{3}{4}\frac{1}{4}\frac{1}{4}\frac{3}{4}\frac{3}{4}\frac{3}{4}\frac{3}{4} = \left(\frac{1}{4}\right)^3\left(\frac{3}{4}\right)^{17}.$$

Note that for computing the probability of this sequence with Gs in specific positions, all that mattered was how many Ns and Gs there were. Therefore, each sequence with exactly three Gs and 17 Ns will occur with the same probability. Using equation 11.4, we know that there are $C_{20,3} = \binom{20}{3} = \frac{20!}{17!3!}$ different possibilities as to where among the 20 positions the three Gs can occur,[9] and so, using the addition principle, we obtain that

$$P(\text{exactly 3 Gs}) = \frac{20!}{17!3!}\left(\frac{1}{4}\right)^3\left(\frac{3}{4}\right)^{17} = 0.133896 \approx 13.4\%.$$

Therefore, approximately 13.4% of all possible DNA sequences with 20 nucleotides will contain exactly three Gs.

Example 11.16 leads us to the *binomial random variable*, which counts the number of successes in a fixed number n of independent Bernoulli trials, each with the same success probability p. If there are n trials, then the binomial random variable can take integer values from 0 (no success) to n (all successes). The probability that there are exactly k successes is the product of $C_{n,k}$, the number of ways that k of the n trials can be "selected" for success, and $p^k(1 - p)^{n-k}$, which is the probability that any such sequence of exactly k successes in n trials occurs. Taken together, the **distribution of the binomial random variable** X is given by

$$P(X = k) = \frac{n!}{(n-k)!k!}p^k\left(1 - p\right)^{n-k} \quad \text{for } k = 0, \dots, n. \tag{11.14}$$

binomial random variable A discrete random variable that counts the number of successes in a fixed number of Bernoulli trials that all have the same success probability.

Note that once we know n and p, we know the complete probability distribution. We therefore say that the binomial random variable and distribution have *parameters* n and p. We will look at an example that illustrates the use of the binomial distribution.

Example 11.17. You are looking at an RNA sequence consisting of 100 codons that code for amino acids. What is the probability that such a sequence has exactly 10 codons that code for threonine (Thr)?

What you are looking for is a count of the instances of codons for threonine. If we assume that all codons are equally likely to occur at any position within the RNA sequence,[10] then each of the 100 codons either codes for threonine or not. Therefore, the number of codons that code for threonine is a binomial random variable where the number of trials is $n = 100$. The success probability p is the probability that a codon codes for threonine. Because four of the 64 codons code for threonine,

---

[9] Imagine we have a hat with slips of paper numbered 1 through 20 and we randomly select 3 of these for the locations of the Gs. Each such selection will describe one of the possible ways in which the three Gs can occur.

[10] This is not an entirely realistic assumption as each species has its own codon bias, but is assumed here for the sake of illustration.

$p = 4/64 = 1/16 = 0.0625, 1 - p = 0.9375$, and therefore the probability that there are exactly $k = 10$ codons that code for threonine in an RNA sequence with $n = 100$ codons is given by

$$P\left(\text{exactly 10 codons that code for Thr}\right) = \frac{100!}{90!10!}(0.0625)^{10}(0.9375)^{90} = 0.04726.$$

Because the factorials that show up in this probability are too big for a calculator to compute, we have used Excel's function BINOM.DIST to compute the desired probability. The relevant cell entry is =BINOM.DIST(10, 100, 0.0625, FALSE), which returns as answer 0.04726. Thus, there is about a 4.7% probability of seeing 10 codons that code for Thr in an RNA sequence of 100 codons.

We often want to characterize the distribution of a random variable with a parameter, a single numeric value that captures a certain aspect of the distribution. The two most common such parameters are the mean and the variance. The *mean* of a random variable is the average of the values weighted according to their probability, whereas the *variance* (or its square root, the *standard deviation*) characterizes the spread of the data. Let's consider a mean that is very familiar to students, namely the grade point average, or GPA.

mean of a distribution The mean of all the possible values of the random variable described by the distribution, weighted by how likely they are.

Example 11.18. Assume that you have received three As, one B, and two Cs in the last academic term, and that all courses carry the same number of units. To compute your term GPA, you first convert the letter grades into their numerical scores, $A = 4.0, B = 3.0$, and $C = 2.0$. Then you add the numerical scores of the six courses and divide by six, so

$$GPA = \frac{4.0 + 4.0 + 4.0 + 3.0 + 2.0 + 2.0}{6} = \frac{19.0}{6} = 3.16667 \approx 3.17.$$

variance of a distribution The variance of all the possible values of the random variable described by the distribution, weighted by how likely they are.

standard deviation The square root of a variance, used to express how spread out the corresponding values are in the units of those values.

We can rewrite this fraction by combining like grades, to obtain

$$GPA = \frac{3\cdot(4.0) + 1\cdot(3.0) + 2\cdot(2.0)}{6} = \frac{3}{6}\cdot(4.0) + \frac{1}{6}\cdot(3.0) + \frac{2}{6}\cdot(2.0).$$

Writing the GPA in this form we see that each grade is "weighted" according to its proportion within the set of grades, as is appropriate. For example, the grade 4.0 is weighted by its proportion 3/6.

So how does this relate to discrete random variables? We can think of the course grade as a random variable X that has three values, namely 2.0, 3.0, and 4.0, with respective probabilities 2/6, 1/6, and 3/6. To compute the average, we multiply each value with its associated probability and then add up the resulting values. This example leads to the following definition:

The mean μ (pronounced mew) of the distribution is computed as the **weighted sum** of the values according to their probability to occur:

$$\mu = \sum_{i=1}^{N} x_i \cdot p(x_i).$$

expected value An alternative name for the mean of a distribution.

Note that we also refer to the mean as the *average* or *expected value* of the random variable and also denote it by $E[X]$.

Let's use this definition to compute the average number of tails in three coin tosses.

Example 11.19. To determine the average number of tails in three coin tosses, we look at the probability distribution (derived in example 11.15)

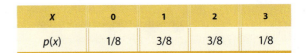

X	0	1	2	3
p(x)	1/8	3/8	3/8	1/8

and compute

$$\mu = 0 \cdot \frac{1}{8} + 1 \cdot \frac{3}{8} + 2 \cdot \frac{3}{8} + 3 \cdot \frac{1}{8} = \frac{12}{8} = 1.5.$$

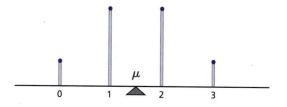

FIG. 11-11. Illustration of the mean (average value) as the balancing point of a weightless rod with weights $p(x)$ at each value x of the random variable.

Obviously, three coin tosses cannot result in 1.5 tails, so what does it mean that there are an average of 1.5 tails in three coin tosses? It means that if the experiment were performed many times, then there would be roughly 1.5 tails per three coin tosses, or half a tail per toss.

We can visualize the mean as follows: Imagine the real line as a weightless rod. At each value x of the discrete random variable, attach a weight of size $p(x)$ to the rod. Then the rod is balanced if a fulcrum is put at the value of x that corresponds to the mean. Figure 11-11 illustrates this idea for the random variable of example 11.19.

This visualization also can explain why the average is not necessarily one of the values of the random variable—the fulcrum may not balance at any of the values the random variable can take. However, the fulcrum must always be at a value that lies between the smallest and largest values of the random variable.

Just as we have computed the mean of the number of tails in three coin tosses, we can compute the mean of more general random variables such as the Bernoulli and binomial random variables. The Bernoulli random variable has just two values, 0 and 1, so its mean is computed as

$$\mu = 0 \cdot (1-p) + 1 \cdot p = p,$$

whereas the mean of the binomial random variable is

$$\mu = \sum_{k=0}^{n} k \binom{n}{k} p^k (1-p)^{n-k} = np. \qquad (11.15)$$

The first equality in equation 11.15 is the definition of the mean, whereas the second equality results from algebraic simplification.[11] The result should feel quite intuitive, especially if we write equation 11.15 as $p = \mu/n$, which states that the probability of success in a single trial equals the average number of successes in an experiment divided by the number of trials in that experiment. We can use equation 11.15 to compute the average number of codons that code for Thr in an RNA sequence consisting of 100 codons.

Example 11.20. In example 11.17, we identified the number of codons that code for Thr to be a binomial random variable with $n = 100$ and $p = 0.0625$, so the average number of such codons is $n \cdot p = 6.25$, which explains the low probability of observing 10 such codons in an RNA sequence consisting of 100 codons.

Now let's look at a second characteristic of a distribution, namely its variance. This quantity is important because the mean does not tell the whole story. We can have radically different random variables that have nevertheless the same mean, as in the following example.

Example 11.21. Let X and Y be two random variables, where X has values $-1, 0$, and 1 with probabilities $p(-1) = p(1) = 1/4$, and $p(0) = 1/2$, and Y has values -10 and 10, each with probability $1/2$. The probability mass functions of these two random variables are shown in Figure 11-12.

---

[11] Equation 11.15 can be derived using algebraic manipulation, specifically use of the binomial formula $(a+b)^n = \sum_{k=0}^{n} \binom{n}{k} a^k b^{n-k}$. To see a step-by-step derivation of the mean of the binomial random variable, see, for example, Ghahramani (2004, 194–195).

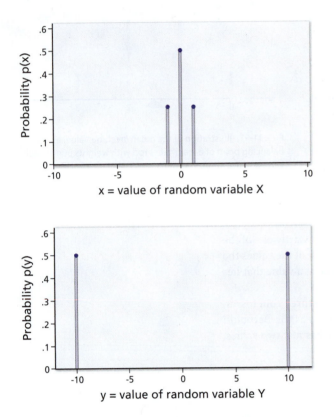

FIG. 11-12. Probability mass functions of two random variables with the same mean, but different variation.

If we let μ_X and μ_Y denote the means of the two random variables, then

$$\mu_X = (-1)\cdot\frac{1}{4} + 0\cdot\frac{1}{2} + 1\cdot\frac{1}{4} = 0,$$

and

$$\mu_Y = (-10)\cdot\frac{1}{2} + 10\cdot\frac{1}{2} = 0.$$

The means are the same, but the distributions are very different. Specifically, the distribution of Y is considerably more "spread out." That is, a typical value of Y is more likely to be farther away from the mean μ_Y than a typical value of X is from its mean μ_X. We say that Y has greater variation than X.

So how could we capture this variation in a single numerical value? Because deviations above and below the mean contribute equally to variation, a logical choice would be to use the weighted average of the absolute deviations (or distances) of values x_i from mean μ to measure the variation. However, advanced theoretical considerations lead to averaging the squared deviation, namely $(x_i - \mu)^2$ instead.

The **variance** σ^2 is then defined as the **average squared deviation from the mean**, or

$$\sigma^2 = \sum_{i=1}^{N}(x_i - \mu)^2 p(x_i),$$

where x_i denotes the i-th value of the random variable X that has a total of N values.[12] Since many random variables carry physically meaningful units, we often use the square root of the variance, the *standard deviation* σ, which quantifies the variation in the units of the random variable.

Let's compute the variances of the two random variables of example 11.21:

$$\sigma_X^2 = (-1-0)^2\cdot\frac{1}{4} + (0-0)^2\cdot\frac{1}{2} + (1-0)^2\cdot\frac{1}{4} = \frac{1}{4} + \frac{1}{4} = 0.5,$$

and

$$\sigma_Y^2 = (-10-0)^2\cdot\frac{1}{2} + (10-0)^2\cdot\frac{1}{2} = \frac{1}{2}10^2 + \frac{1}{2}10^2 = 100.$$

The corresponding standard deviations are $\sigma_X = \sqrt{0.5} \approx 0.71$ and $\sigma_Y = \sqrt{100} = 10$, values that are consistent with the fact that the values of the variable X are much closer to their mean than those of the random variable Y.

Thought Question 11-4

What does it mean when the variance of a random variable equals 0?

[12] Note that if all values of the random variable are equally likely, then $\sigma^2 = \frac{1}{N}\sum_{i=1}^{N}(x_i - \mu)^2$. This formula is very similar to another formula you may have encountered, namely for the *sample variance*, $s^2 = \frac{1}{N-1}\sum_{i=1}^{N}(x_i - \bar{x})^2$. Here the x_i are data values, $\bar{x}$ is the mean of these values, and s^2 is the computed estimate of σ^2, the true variance. Note that the denominator in this latter equation is $N - 1$ rather than N.

We can also compute the variances of the general Bernoulli and binomial random variables. The variance of the Bernoulli random variable is

$$\sigma^2 = (0-p)^2(1-p) + (1-p)^2 p = p^2(1-p) + (1-2p+p^2)p = p - p^2 = p(1-p),$$

so the standard deviation is given by $\sigma = \sqrt{p(1-p)}$. For the binomial distribution, the variance is

$$\sigma^2 = \sum_{k=0}^{n}(k-np)^2 \binom{n}{k} p^k (1-p)^{n-k} = np(1-p).$$

The standard deviation is $\sigma = \sqrt{np(1-p)}$.

There are additional named discrete random variables, for example the geometric distribution, which counts the number of Bernoulli trials until the first success occurs, and the *Poisson distribution*.[13] The latter distribution arises as a limiting case of the binomial distribution when the number of trials n becomes large while the expected number of successes $n \cdot p$ remains constant (that is, the success probability p becomes correspondingly small). Specifically, the Poisson distribution is a good approximation of the binomial distribution when $n \cdot p < 10$ and $p < 0.1$. The relevant parameter of the Poisson distribution in this case is $\beta = np$. Here is the definition of the Poisson distribution:

$$P(X = k) = \frac{\beta^k}{k!} \cdot e^{-\beta} \text{ for } k = 0, 1, 2, \ldots \tag{11.16}$$

where $e \approx 2.718$ is Euler's constant and $\beta > 0$ is the parameter of the distribution. Note that for the Poisson distribution, $\mu = \beta$ and $\sigma^2 = \beta$, that is, the mean and the variance of this distribution are the same.

The setup in example 11.17 is a good candidate for using the Poisson distribution to approximate the binomial distribution because $p = 0.0625 < 0.1$ and $n \cdot p = 100 \cdot 0.0625 = 6.25 < 10$. Let's redo example 11.17 using the Poisson distribution.

Example 11.22. We want to count the number of codons that code for Thr, which is modeled by a binomial random variable X with parameters $n = 100$ and $p = 0.0625$. Because $p < 0.1$ and $n \cdot p < 10$, we can approximate X by a Poisson random variable Y with parameter $\beta = n \cdot p = 6.25$. Then the probability that there are exactly $k = 10$ codons coding for Thr in the 100 codon RNA sequence is approximated by

$$P(Y = 10) = \frac{6.25^{10}}{10!} \cdot e^{-6.25} = 0.04838.$$

The percentage error between the Poisson approximation and the exact result from the binomial distribution obtained in example 11.17 is $[(0.04838 - 0.04726)/0.04726] \cdot 100\% \approx 2.37\%$, indicating a good approximation. Note how much easier it is to compute this probability using the Poisson distribution, which can even be done on a calculator.

Even though the Poisson distribution arose as a limiting case of the binomial distribution, it is a useful distribution in its own right. In general, a discrete random variable with a Poisson distribution counts occurrences of events (= successes) over specified time periods when these events occur independently, and all have the same average rate of occurrence. In the context of bioinformatics, the Poisson distribution is used for example to determine the amount of sequencing needed in the whole

Poisson random variable A discrete random variable that that can be used to count the number of events occurring in a fixed interval of time if these events occur with a fixed average rate and independently of each other. A Poisson random variable is used in the Jukes-Cantor model to count nucleotide changes occurring over time.

---

[13] The Poisson (pronounced PWAH-sahn) distribution is named after the French mathematician Simon Denis Poisson (1781–1840), even though the distribution had been published in 1711 by his countryman Abraham de Moivre.

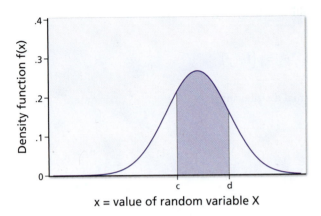

FIG. 11-13. Probability $P(c \leq X \leq d)$ computed as area under the density function $f(x)$ over the interval [c, d].

probability density function

A function that is used to determine the probability of a continuous random variable to take on a given range of values.

genome shotgun sequencing approach (see Chapter 9), to model the number of high scoring segment pairs when a query sequence is used to detect related sequences in a large database of sequences, and in the derivation of the Jukes-Cantor model.

We now will discuss continuous random variables, specifically the normal and the extreme value distributions. The latter distribution is the basis for computing the E-value that tells us about the significance of a sequence alignment.

Continuous Random Variables

Random variables that take values in an interval [a, b] (where a and b are real numbers, but we also allow $a = -\infty$ and/or $b = \infty$) are called *continuous random variables*. Because there are infinitely many distinct values in an interval, we can no longer list the probabilities of individual values of the random variable. Instead, the probability mass function is replaced by the *probability density function* $f(x)$, and the probability $P(c \leq X \leq d)$ is defined as the area under $f(x)$ over the interval from c to d, that is, the area bounded by the the probability density function $f(x)$, the horizontal axis, and the vertical lines $x = c$ and $x = d$ as shown in Figure 11-13. At first glance, this may seem to be a somewhat strange definition, so example 11.23 and the discussion that follows illustrate how mathematicians arrived at this definition. Note that the area of a region is the same whether the bounding curve is included or not, because a line has no width and therefore no area. As a result, the following probabilities are all the same for a continuous random variable:

$$P\big(c \leq X \leq d\big) = P\big(c < X \leq d\big) = P\big(c \leq X < d\big) = P\big(c < X < d\big).$$

Example 11.23. Suppose that you have taken 2,000 measurements of a quantity that can be modeled by a continuous random variable that ranges from −3.3 to 3.3, and that you have grouped this data in intervals of width 0.1, with the value of the lower bound included and the value of the upper bound excluded (to be consistent as to where the interval boundary values are counted). A part of the data is shown in the table below, where for example the interval 0.2–0.3 contains values x that satisfy $0.2 \leq x < 0.3$:

INTERVAL	0.2–0.3	0.3–0.4	0.4–0.5	0.5–0.6	0.6–0.7
COUNT	89	68	84	55	59

Now let's assume that you want to determine the probability that a new measurement lies in the interval [0.2, 0.7).[14] It is natural to estimate the probability that the unknown random variable takes a value in the interval [0.2, 0.7) as the proportion of measurements that lie in this interval, that is,

$$P\big(0.2 \leq X < 0.7\big) \approx \big(89 + 68 + 84 + 55 + 59\big) / 2000 = 355/2000 = 0.1775 \approx 0.18.$$

so there is about an 18% chance that the random variable takes a value in [0.2, 0.7).

Let's visualize these counts in a histogram, where for each interval we draw a bar of height equal to the count for that interval, as shown in the left panel of Figure 11-14. Now suppose that we scale the bar heights in such a way that the **area** of each bar

---

[14] For an interval of values, a bracket indicates that the associated value is included in the interval, whereas a parenthesis indicates that the value is not included. For example, [0.2, 0.7) translates into $0.2 \leq x < 0.7$.

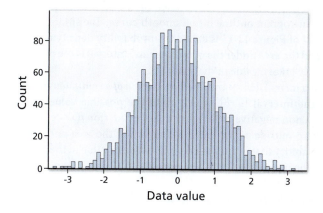

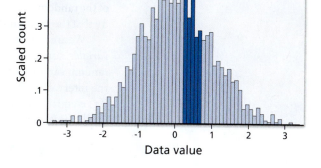

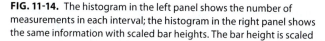

FIG. 11-14. The histogram in the left panel shows the number of measurements in each interval; the histogram in the right panel shows the same information with scaled bar heights. The bar height is scaled such that the areas of the bars correspond to the respective interval probabilities. The area shaded in dark blue gives the probability that the random variable takes a value in the interval [0.2, 0.7].

gives the *interval probability*, that is, the probability that an observed value of the random variable lies in that interval. What is the correct scaled height? We know that

Interval probability = Area of bar = Scaled bar height · Interval width,

where the interval probability can be approximated by the number of data values within the interval divided by total number of data values. Solving for the scaled bar height we obtain

Scaled bar height = Interval probability/Interval width.

Let's apply this result to the data from example 11.23. The probability that a value is in the interval [0.2, 0.3) is approximated by the interval count divided by the total number of data values, or 89/2000 = 0.0445. To make the area of the bar equal to the interval probability, we scale the bar height to 0.0445/0.1 = 0.445, where 0.1 = 0.3 – 0.2 is the interval width.

The scaled histogram for example 11.23 is shown in the right panel of Figure 11-14. In this histogram, the bars associated with the interval [0.2, 0.7] are shown in dark blue. Together their area equals the interval probability 0.1775 that the random variable takes a value in the interval [0.2, 0.7]. In addition, the total area of all histogram bars equals one, as adding up the probabilities of the individual intervals gives us the probability that the random variable has a value in the interval [–3.3, 3.3], which is its full range.

If we repeat this process with larger and larger data sets, which ultimately allow for narrower intervals, the outline of the histogram will become progressively smoother, while preserving the property that the area of the bars over any interval equals the probability that the random variable takes a value in that interval. In the top panel of Figure 11-15 is a histogram with 200,000 computer-generated data values from the same continuous random variable as in Figure 11-14, grouped into intervals of width 0.1. It is clear that taking

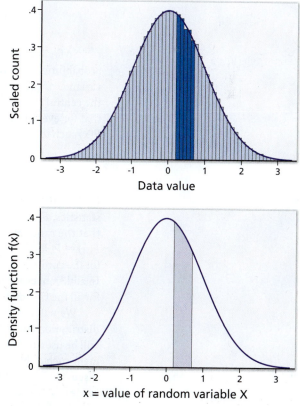

FIG. 11.15 Scaled histogram of 200,000 measurements together with the associated probability density function. The area shaded in dark blue in the bottom panel gives the probability that the random variable has values in the interval [0.2, 0.7].

even more data values and grouping them into smaller and smaller intervals would eventually transform the histogram outline into a smooth curve. The smooth curve shown in the bottom panel of Figure 11-15 is called the probability density function of the random variable, and the area under the curve over any interval is the probability that the random variable takes a value in that interval.

We are now ready for a formal definition. The *distribution of a continuous random variable* consists of (1) the interval $[a, b]$, which gives the possible values of the random variable, and (2) a non-negative probability density function $f(x)$ defined on the interval $[a, b]$ (and zero outside of this interval) such that the area underneath $f(x)$ is equal to 1. The probability that a random variable X takes a value in the interval $[c, d]$ that is contained in the interval $[a, b]$ is then given by:

$$P(c \leq X \leq d) = \text{area under } f(x) \text{ over the interval } [c, d] = \int_c^d f(x)\, dx. \quad (11.17)$$

The last expression is called the definite integral of $f(x)$ with respect to x over the interval $[c, d]$, and it represents the desired area under the function $f(x)$. The graph in the bottom panel of Figure 11-15 shows the area corresponding to the probability $P(0.2 \leq X \leq 0.7)$. Note that we always have that $P(a \leq X \leq b) = 1$, because $P(a \leq X \leq b)$ gives the probability of *all* values of the random variable X.

Just as in the case of discrete random variables, certain continuous distributions and their density functions play an important role in a variety of applications. Before we look at any specific distributions, we will give formulas for the mean μ and the variance σ^2 of continuous random variables. Recall that for discrete random variables we computed these values as weighted averages using sums. For continuous random variables, the sums are replaced by integrals, and the probability mass function is replaced by the probability density function. Specifically,

$$\mu = \int_a^b x \cdot f(x)\, dx \quad \text{and} \quad \sigma^2 = \int_a^b (x - \mu)^2 \cdot f(x)\, dx.$$

So far, we have not said anything about how to compute the integrals that give us probabilities that arise from a continuous random variable, and the mean and the variance of such a random variable. Finding methods to compute integrals is one of the central themes of calculus. Notably, the fundamental theorem of calculus states that for every function $f(x)$ defined on an interval $[a, b]$, there is a related function $F(x)$ such that

$$\int_a^b f(x)\, dx = F(b) - F(a). \quad (11.18)$$

$F(x)$ is called the antiderivative of $f(x)$, and in the context of probability theory and statistics, $F(x)$ is called the *cumulative distribution function*. It gives the probability that the random variable X takes a value less than or equal to x, in other words, $F(x) = P(X \leq x)$. Fortunately, for the distributions of interest to us, either a formula for the cumulative distribution function $F(x)$ or a table of its values has been derived. In addition, every statistics software package (such as R) "knows" the function $F(x)$ for all the common distributions.

We will now look at two important continuous distributions, the *extreme value distribution* and the *normal* or *Gaussian distribution*. The extreme value distribution will be used to find the E-value that is used to determine the significance of an alignment such as between the human p53 sequence and a potential yeast p53 sequence discussed in the introduction. The normal distribution plays an important role in a range of statistical applications, because random measurement errors in physical sciences and many traits of living organisms tend to have normal distributions. Examples of the latter include the height and weight of human and animal populations, as observed by Sir Francis Galton (1822–1911), a cousin of Charles Darwin. We start by discussing the normal distribution.

The Normal Distribution

The *normal* or *Gaussian distribution with parameters μ and* σ^2 is defined on the interval $(-\infty, \infty)$ and has probability density function

$$f(x) = \frac{1}{\sigma\sqrt{2\pi}} \exp\left(\frac{-(x-\mu)^2}{2\sigma^2}\right),$$

where μ is the mean and σ^2 is the variance of the distribution. Note that in this formula, the exponential function e^x is written as $\exp(x)$. This notation is used in mathematical formulas when there are relatively complicated expressions in the exponent, as in this case.

A special case of the normal distribution is the *standard normal distribution*, which has mean $\mu = 0$ and standard deviation $\sigma = 1$. By convention, a standard normal random variable is always denoted by the letter Z with probability density function $\phi(z)$ and cumulative distribution function $\phi(z)$. (Recall that we denoted random variables by uppercase letters, and their values by lowercase letters.) The probability density function of any normal random variable is centered at its mean μ and is symmetric[15] with respect to μ. The standard deviation σ describes the spread of the density function—the greater σ, the more spread out and flatter the graph of the probability density function is. Figure 11-16 shows the graphs of the probability density functions of the standard normal random variable and a normal random variable with mean μ = 2 and standard deviation $\sigma = 1.5$.

An unfortunate fact about all normal distributions (standard and otherwise) is that there is no explicit formula for the cumulative distribution function. That is, if $f(x)$ is the probability density function of a normal distribution, then we can get $F(x)$ only from mathematical tables created using numerical approximation methods. This leads to the question: Do we need different tables for all combinations of μ and σ values? Luckily, any normal random variable can be transformed into a corresponding standard normal random variable, and therefore, we need only **one** table.

So how does the transformation work? If x is a value of **any** (not necessarily normal) random variable X with mean μ and standard deviation σ, then the corresponding *z-value* or *z-score* of the standardized random variable Z is given by

$$z = \frac{\text{value} - \text{mean}}{\text{standard deviation}} = \frac{x - \mu}{\sigma} \qquad (11.19)$$

We illustrate this transformation in the case of a normal random variable in Figure 11-17. The top graph depicts the probability density function of a normal random variable X with mean $\mu = 2$ and standard deviation $\sigma = 1.5$. The area of the shaded region gives the probability that the random variable X takes a value between 2 (the mean) and 3.5 (one standard deviation above the mean). The bottom graph represents the probability density function of the standard normal random variable Z, with a corresponding shaded region. The lower bound of this region is at the z-value corresponding to $x = 2$, namely

$$z = \frac{x - \mu}{\sigma} = \frac{2 - 2}{1.5} = 0$$

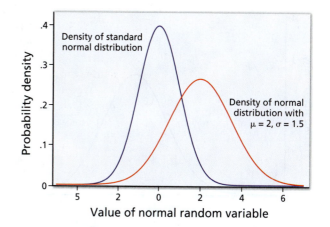

FIG. 11-16. The graphs of the probability density functions of a standard normal random variable (in blue) and a normal random variable with mean $\mu = 2$ and standard deviation $\sigma = 1.5$ (in red).

normal (Gaussian) distribution
A very important probability distribution in statistics. Physical quantities (such as measurement errors) that result from the actions of many independent processes are typically normally distributed.

z-score Expresses how many standard deviations the value of a random variable lies above or below the mean of that variable.

---

[15] That is, the portions of the graph of $f(x)$ to the left and right of the vertical line passing through $x = \mu$ are mirror images of each other. Mathematically, $f(\mu + x) = f(\mu - x)$ for all x.

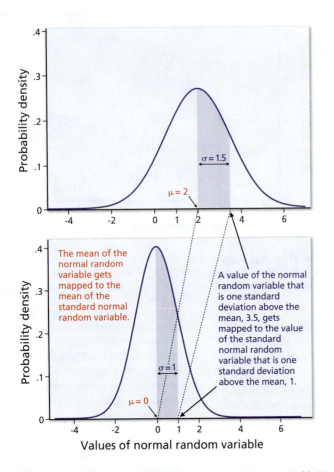

The mean of the normal random variable gets mapped to the mean of the standard normal random variable.

A value of the normal random variable that is one standard deviation above the mean, 3.5, gets mapped to the value of the standard normal random variable that is one standard deviation above the mean, 1.

FIG. 11-17. Illustration of how the values of a normal random variable X with mean $\mu = 2$ and standard deviation $\sigma = 1.5$ are mapped to values of a standard normal random variable Z with mean $\mu = 0$ and standard deviation $\sigma = 1$.

The answer is not a coincidence; an x-value corresponding to the mean of the random variable X always maps to a z-value of 0, the mean of the standardized random variable Z. The z-value of the upper bound of this region is the z-value corresponding to $x = 3.5$, namely

$$z = \frac{3.5 - 2}{1.5} = \frac{1.5}{1.5} = 1$$

Here, the x-value corresponding to one standard deviation above the mean of X maps to a value of $z = 1$, which is one standard deviation above the mean of Z. So we see that the z-value tells us how many standard deviations σ the x-value is above or below its mean μ. Intuitively, we can say that the z-value measures the distance between an x-value and the mean μ of the random variable X using a ruler marked in terms of the standard deviation σ of X. Let's now return to the subject of probability. The areas of the two shaded regions in Figure 11-17 equal the probabilities that random variables X and Z take values in the respective intervals, that is, $2 \leq X \leq 3.5$ and $0 \leq Z \leq 1$. Each of these intervals contains values between the mean and one standard deviation above the mean of the respective random variable. It is a central fact of normal distributions that these two regions have identical area, and hence the corresponding probabilities are also the same. More generally, for a random variable X with mean μ and standard deviation σ, the area under the probability density function $f(x)$ over the interval $[c, d]$ equals the area under the probability density function $\phi(z)$ over the interval $[z_c, z_d]$, where $z_c = (c - \mu)/\sigma$ and $z_d = (d - \mu)/\sigma$. We then have

$$P(c \leq X \leq d) = \int_c^d f(x)\,dx$$

$$= \int_{z_c}^{z_d} \phi(z)\,dz = P(z_c \leq Z \leq z_d)$$

$$= \phi(z_d) - \phi(z_c), \tag{11.20}$$

where the first and third equalities follow from equation 11.17 and the last equality follows from equation 11.18.[16] The values of $\phi(z)$ (cumulative distribution function of the standard normal distribution) can be found in statistics books or on the Internet. They are incorporated into all statistics software packages and can also be computed in Excel. Let's see how this process works in an application.

Example 11.24. The heights of American women aged 18 to 24 can be approximated by a normal distribution with mean 65.5 inches and standard deviation 2.5 inches. What is the probability that a randomly chosen American woman in this age bracket has a height of less than 62 inches?

Let X be the height in inches. Then we want to compute $P(X < 62)$. Therefore, the x-value is 62, so its corresponding z-score is $z = (x - \mu)/\sigma = (62 - 65.5)/(2.5) = (-3.5)/(2.5) = -1.4$. Therefore, we have $P(X < 62) = P(Z < -1.4) = \phi(-1.4)$. Note that the transformation tells us that a height of 62 inches corresponds to a value that is 1.4 standard deviations below the mean.

---

[16] The second equality results from an integration technique called "u-substitution" where $(x - \mu)/\sigma$ is replaced by z.

Now that we have translated the problem into the "Z universe," that is, into a problem for a standard normal random variable, we can look up $\phi(-1.4)$ in a table of values of $\phi(z)$. Alternatively, we can use Excel's function NORM.DIST to compute the desired probability. The relevant cell entry is $=$ NORM.DIST($-1.4,0,1$,TRUE), which returns the value 0.0808, indicating an 8.1% probability that a randomly selected American woman aged 18 to 24 years is less than 62 inches tall.[17] We can express this probability in another way, namely that 8.1% of 18- to 24-year-old American women are less than 62 inches tall. This "statistical" view is justified as follows:

$$P(X < 62) = \frac{\text{Number of women} < 62'' \text{ tall}}{\text{Total number of women}} \cdot 100\%$$
$$= \text{Percentage of women} < 62'' \text{ tall.}$$

More generally, we can compute the percentage of values taken by a normal random variable that are within one, two, or three standard deviations of the mean by computing the probability that a value of the normal random variable falls into the respective intervals $(\mu - k \cdot \sigma, \mu + k \cdot \sigma)$ for $k = 1, 2, 3$. For **any** normal random variable X with mean μ and standard deviation σ we have that

- $P(\mu - \sigma \leq X \leq \mu + \sigma) = P(-1 \leq Z \leq 1) = 0.683$. That is, about 68% of values are within one standard deviation of the mean;

- $P(\mu - 2\sigma \leq X \leq \mu + 2\sigma) = P(-2 \leq Z \leq 2) = 0.954$. That is, about 95% of values are within two standard deviations of the mean;

- $P(\mu - 3\sigma \leq X \leq \mu + 3\sigma) = P(-3 \leq Z \leq 3) = 0.997$. That is, about 99.7% of values are within three standard deviations of the mean.

The last result tells us that almost all values taken by a normal random variable lie within three standard deviations of the mean, with only about 0.15% of values below and 0.15% of values above this range. Figure 11-18 presents this information graphically. These facts about the percentages of values that lie within one, two, and three standard deviations from the mean are referred to in statistics as the *empirical rule*, or *68-95-99.7 rule*.

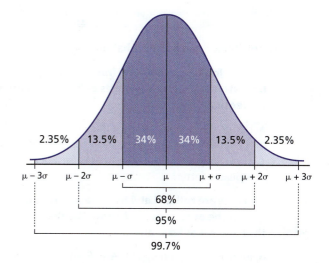

FIG. 11-18. Probabilities that a value of a normal random variable lies within one, two, and three standard deviations of the mean. This forms the basis for the 68-95-99.7 rule in statistics.

SUMMARY

In this chapter we started with a definition of the probability of an outcome, and then learned how to use counting and other probability rules to compute the probability of events. We discussed independence of events, which we exploit in DNA and RNA modeling when we assume that the probability that a nucleotide or amino acid will be observed is the same across the sequence. However, we know that in certain areas of the genome, such as in a GC-rich region, the probabilities are not the same, which leads us to use conditional probability in such cases. Conditional probabilities are also used in Bayesian inference in the computation of posterior probabilities.

We also discussed discrete and continuous random variables and their distributions, specifically the binomial, Poisson, and normal distributions. We are now ready to study more advanced probability tools that are used in specific bioinformatics applications, such as the the extreme value distribution, hidden Markov models, and the Jukes-Cantor model.

---

[17] If we use Excel, then we do not have to do the transformation to the standard normal random variable. In fact, if we use the cell entry $=$ NORM.DIST($62,65.5,2.5$,TRUE), then Excel will perform the transformation automatically.

EXERCISES

Sample Space and Basic Probability

1. **You are looking at a sliding frame of size 5 on the DNA strand, that is, sequences of five consecutive nucleotides.**

   a. Define the sample space for this experiment.
   b. How many elements are in the sample space?
   c. How many five letter sequences consist of a single repeated letter?

2. **Two dice are rolled. Let *A* be the event that the sum of the outcomes is odd and *B* be the event that there is at least one 1.**

   a. Interpret the events $A \cap B$, $A^c \cap B$, and $A^c \cap B^c$ both by describing the events in words, as well as by writing out the corresponding set of outcomes.
   b. Compute the probabilities of each of the five events A, B, $A \cap B$, $A^c \cap B$, and $A^c \cap B^c$.

Counting Methods

3. **Codons are three-letter nucleotides, consisting of the four nucleotides A, C, G, and T.**

   a. How many codons start with either A or T and end with either C or G, and have different nucleotides in the first and second positions?
   b. Draw the tree diagram for these codons.
   c. Use the tree diagram from part (b) to count how many codons start with either A or T and end with either C or G and have no repeated letters. (Note that these are not so easy to count using the methods we have learned so far.)

4. **In Chapter 6, in the description of the BLAST algorithm, it was claimed that the table of three-letter amino acid words contains 8,000 entries. Explain how this number is computed.**

5. **Given 10 different amino acids, how many pentapeptides (peptides of length 5) can you make if**

   a. each amino acid can be used only once?
   b. amino acids can be repeated?

6. **Let *S* be the set of binary sequences (i.e., sequences consisting of zeros and ones) of length 12.**

   a. How many elements are in *S*?
   b. How many elements in *S* have exactly eight zeros (and four ones)?

7. **Let *S* be the set of DNA strings of length 8 (i.e., sequences consisting of the four nucleotides A, T, C, and G).**

   a. How many elements are in *S*?
   b. What is the probability that a DNA string of length 8 has exactly two A nucleotides?

8. **Gametes carry 23 single chromosomes, each of which is equally likely to come from the father or the mother. Assuming there is no crossover, how likely is it that a certain gamete has 12 paternal and 11 maternal chromosomes?**

9. **In example 11.5 we investigated DNA sequences to which human p53 binds. They consist of two copies of the 10-base motif PuPuPuC(A/T)(T/A)GPyPyPy, separated by 0 to 13 nucleotides, where each Pu individually is either A or G, each P individually is either C or T, and (A/T)(T/A) is either the pair AT or the pair TA. We previously computed the number of DNA sequences that have 5 nucleotides in the center string. Compute the total number of DNA sequences of this type (allowing for 0 to 13 center nucleotides) by counting the number of DNA sequences for each of the different cases (number of center nucleotides) and then using the addition principle.**

Independence, Dependence, and Bayes' Law

10. **When synthesizing DNA in a test tube, each nucleotide has a very small probability of being replicated incorrectly. Suppose that a DNA sequence consisting of 64 letters is replicated and that each letter has a probability of 0.001 of being copied incorrectly, independently of the other letters. What is the probability that the sequence is replicated without errors?**

11. **Using the information of example 11.8, compute the probability that a randomly selected white adult man is color-blind.**

12. **Assume that protanopia (inability to see the color red) has a prevalence of 1.01% among white men and 0.02% among white women. Compute the probability that a randomly selected person is color-blind if the population consists of 100,000 males and 80,000 women. Compare your result to example 11.8. Can you explain the difference?**

13. **Genes occur in different versions, called alleles. In human blood, there are three alleles, A, B, and O, where A and B are dominant over O, and A and B are co-dominant. Therefore, there are four different blood types (phenotypes): A (either genotype AA or AO), B (either genotype BB or BO), AB, and O (OO). During fertilization, both parents randomly contribute one of their alleles. The frequencies of the four different blood types vary across populations. Based on the information by the Red Cross about the prevalence of the four phenotypes**

for the U.S. Asian population,[18] the allele frequencies of A, B, and O can be computed as 18.8%, 17.6%, and 63.6%, respectively. If a mother and son from this population have blood type AB and the father has blood type B, what is the probability that the father's genotype is BB?

14. For *Drosophila melanogaster* (a kind of fruit fly), B, the gray body, is dominant over b, the black body, and V, the wild-type wing is dominant over v, the vestigal (or very small) wing. A geneticist, T.H. Morgan, observed that 42% of the offspring were BbVv, 41% of the offspring were bbvv, 9% of the offspring were Bbvv, and 8% of the offspring were bbVv when mating *Drosophila* of genotype BbVv with *Drosophila* of genotype bbvv. Based on these results, should we expect that the body color and wing-type genes of *Drosophila* are independent? Why or why not?

15. A laboratory blood test is 99% effective in detecting a certain disease when it is, in fact, present. However, the test also yields a false positive result for 1% of the healthy persons tested. If 0.5% of the population actually has the disease, what is the probability that a person has the disease given that the test result is positive?

16. There are two indistinguishable coins on a table. One is a fair coin with $P(H) = P(T) = 1/2$, and the other is a biased coin with $P(H) = 2/3$ and $P(T) = 1/3$. One of the coins is chosen at random and tossed five times, resulting in the sequence HHTHT. You are to decide based on the data whether the coin that was tossed is the fair coin or the biased coin. Let *F* stand for the event that the fair coin was chosen, and *B* for the event that the biased coin was chosen.

   a. In example 11.14, we computed the posterior probabilities based on seeing a head on the first toss. Use these posterior probabilities $P(F|H)$ and $P(B|H)$ as the prior probabilities for the second toss, reflecting that your view about the likelihood of the coin being fair or biased has changed as a result of seeing the first head. Compute the new posterior probabilities after seeing the second head, which reflect an updated assessment of the likelihood that the coin is fair or biased after seeing the results of two coin tosses.

   b. Now compute the posterior probabilities after two coin tosses in a different way. Rather than updating the likelihoods after the first toss, use the results of the first two coin tosses together as

the data, that is, compute $P(F|HH)$ and $P(B|HH)$. Recall that coin tosses are independent events, thus $P(HH) = P(H)P(H)$. Your answers should be the same as in part (a).

   c. The fact that the results in (a) and (b) are identical suggests that one can use the data from the five coin tosses at once to compute the posterior probabilities in a single step, rather than updating posterior probabilities one coin toss at a time and using the newly computed posterior probabilities as prior probabilities for the next step. Use this fact to compute the posterior probabilities $P(F|HHTHT)$ and $P(B|HHTHT)$ after seeing this sequence of five coin tosses.

Random Variables

17. Compute the mean μ and the variance σ^2 of a discrete random variable with values $\{-2,0,1,3,4\}$, all occurring with equal probability.

18. **a.** Compute the variance of the random variable *X* of example 11.19.

   b. There is an alternative formula for computing the variance of a random variable, namely

$$\sigma^2 = \left(\sum_x x^2 \cdot p(x) \right) - \mu^2.$$

   Compute the variance in part (a) by using this alternative formula.

19. Determine how long (= number of codons) an RNA sequence has to be to have an average of twelve codons that code for Thr.

20. Compute the expected number of G nucleotides that occur in a DNA sequence of length 20 if all four nucleotides are equally likely to occur.

21. For an RNA sequence consisting of 100 codons, use the Poisson approximation of a binomial random variable to compute the probability of observing

   a. no codon that codes for Thr;
   b. at least one codon that codes for Thr;
   c. three codons that code for Thr.

22. **a.** Use Excel or any statistics program to compute the actual binomial probabilities of the events in exercise 21.

   b. Compute the percentage errors of the approximate probabilities computed in exercise 21. Recall that the percentage error of an approximation is computed as

$$\frac{\text{approximate value} - \text{actual value}}{\text{actual value}} \cdot 100\%$$

---

[18] American Red Cross (2012).

23. **William Sealy Gosset,[19] a chemist at the Arthur Guinness and Son brewery in Dublin, rediscovered the Poisson distribution when working on yeast cell counts measured with a haemacytometer. Gosset published under a pseudonym, "Student," and is remembered through Student's t-distribution used in statistical applications. Here is his yeast cell count data for 400 squares of the haemacytometer:[20]**

Number of cells	0	1	2	3	4	5
Frequency	213	128	37	18	3	1

Let X denote the number of yeast cells per square. To test how well the data fits a Poisson distribution for X, perform the following tasks.

a. Use the relative frequency (= frequency/total) of the zero count to estimate the parameter β of the Poisson distribution. (Recall that $P(X = 0) = e^{-\beta}$.)

b. With the parameter β found in (a), compute $P(X = k)$ for $k = 0,1,\ldots,5$. How close are these values to the relative frequencies? Use the

percentage error (see exercise 22) to assess the difference, where the relative frequencies are to be used as the actual value, and the predicted values as the approximations.

24. **For a normal random variable X with mean $\mu = 2$ and standard deviation $\sigma = 1.5$, compute the z-scores that correspond to the x-values 4, 0.5, and –2.**

25. **Let X be a random variable that has normal distribution with the stated mean and standard deviation. Use Excel or any other statistical software to compute the probabilities below without first transforming the respective probability into a probability of the standard normal random variable. Round your answers to four decimal places.**

   a. $\mu = 2$, $\sigma = 6$, $P(-10 \leq X \leq 14) =$ —
   b. $\mu = -5$, $\sigma = 1$, $P(-8 \leq X \leq -2) =$ —
   c. $\mu = 35$, $\sigma = 2$, $P(33 \leq X \leq 37) =$ —

   Are you surprised by your answers? Compare your results to Figure 11-18 and the bulleted list related to it and explain why you obtained your answers.

26. **When measuring the mRNA expression level of a gene, we find a mean expression level of 1,000 units with a standard deviation of 50 units. Assuming that the measurements have normal distribution, what is the probability that a measurement is within 25 units of its mean?**

---

[19] International Encyclopedia of the Social Sciences. "Gosset, William Sealy." http://www.encyclopedia.com/topic/William_Sealy_Gosset.aspx (accessed May 3, 2013).
[20] http://support.sas.com/documentation/cdl/en/statug/63962/HTML/default/viewer.htm#statug_fmm_a0000000321.htm (accessed May 3, 2013).

ANSWERS TO THOUGHT QUESTIONS

11-1. Because the lowest probability that a cell will stain properly is 80% (two-day-old cells) and the highest probability is 95% (new cells), the probability that a randomly selected cell will stain properly must lie between those two extremes, in other words, it must be between 80% and 95%. Moreover, because the majority of cells (60%) are new, and the probability that a randomly selected cell will stain properly is a weighted average, we should not be surprised that it is closer to 95% than to 80%.

11-2. The second and third simulations had the same number of T outcomes, whereas the first simulation had one fewer T. Intuitively, we know that each tail T makes a case for the fair coin, whereas each head H makes a case

for the biased coin. In fact it can be proven that the final posterior probability depends only on the total numbers of heads and tails and not their order.

11-3. Each outcome has been mapped to one and only one value of the random variable. The sum of the probabilities of the individual outcomes is 1 because their union is the sample space. So the probabilities of the values of the random variable must also add up to 1. If they do not, then either some value of the random variable is missing or there is an algebra mistake.

11-4. A variance of 0 indicates that there is no variation from the mean, that is, all values of the random variable are identical. So the variable is not random at all, as a single value occurs with probability 1.

REFERENCES

American Red Cross. 2012. "Blood Types." Accessed December 14. http:// www.redcrossblood.org/learn-about-blood/ blood-types.

DeGroot, M. H., and M. J. Schervish 2011. *Probability and Statistics*, 4th ed. Boston, MA: Addison Wesley.

Gardiner-Garden, M., and M. Frommer 1987. "CpG Islands in Vertebrate Genomes." *Journal Molecular Biology* 196, no. 2: 261–82, doi:10.1016/0022-2836(87)90689-9. PubMed (PMID 3656447).

Ghahramani, S. 2004. *Fundamentals of Probability, with Stochastic Processes*, 3rd ed. Upper Saddle River, NJ: Prentice Hall.

International Encyclopedia of the Social Sciences. "Gosset, William Sealy." Accessed May 3, 2013. http://www.encyclopedia.com/topic/William_Sealy_Gosset.aspx.

Runkel, P. 2012, March 12. "Beer, Statistics, and Quality." *The Minitab Blog*. http:// blog.minitab.com/blog/statistics-and-quality-data-analysis/beer-statistics-and-quality.

Tredoux, G. 2015. "Sir Francis Galton." Accessed December 8. http://galton.org/.

Wilkinson, D. J. 2007. "Bayesian Methods in Bioinformatics and Computational Systems Biology". *Briefings in Bioinformatics* 8, no. 2: 109–116.

- **Know how to compute and interpret *P*- and *E*-values to evaluate sequence alignments.**

- **Know the main characteristic of a Markov process (probability of current state dependent only on previous state), and be able to translate information about a Markov process into a state diagram and the associated transition matrix.**

- **Know how to compute the probability that a particular sequence of states resulting from a Markov process occurs.**

- **Know the structure of hidden Markov models (HMMs) and understand how HMMs are used in sequence alignment.**

- **Understand the derivation of the Jukes-Cantor model of evolutionary distance.**

ADVANCED PROBABILITY FOR BIOINFORMATICS APPLICATIONS

12.1 INTRODUCTION

In Chapter 11, we covered the basic tools of probability needed to understand the specific bioinformatics applications we will cover in this chapter. For example, in Chapter 6 you learned that BLAST can be used to find sequences related to a gene of interest. A scientist who performed a BLAST search of the yeast genome with the human p53 sequence as a query might find a "hit" for a yeast gene that had not been described previously. However, before announcing at a bioinformatics conference that the "hit" is a p53 homolog, this claim must be substantiated via an *E*-value, which indicates how likely the similarity score of the alignment arose from a commonality of function or evolutionary history between the two genes, rather than from chance. To interpret the *E*-value and assess the significance of alignment we will introduce the extreme value distribution, which describes the scores of high scoring segment pairs (HSPs).

Another instance where a bioinformaticist relies on probability theory is when using algorithms for multiple sequence alignment. We will introduce hidden Markov models that learn from known sequence alignments and encode that knowledge in their underlying structures. Once "trained," such a model can then be used to perform alignments on new sequences.

Finally, we will use tools from probability theory to derive an equation based on the Jukes-Cantor model to determine evolutionary distance (which cannot be observed directly) from the fraction of sites that differ between two sequences (which can be observed).

Even though you may not develop new models or algorithms, it is important for you to understand what goes on "behind the scenes." Also, to work with a mathematician or programmer you need to know the basic mathematical vocabulary. One example of a successful collaboration of life scientists with a mathematician is the story of Michael Waterman (see Box 12-1), one of the two developers of the Smith-Waterman algorithm. His background is in applied mathematics, but his most often cited papers are those in which he used mathematics to answer questions related to bioinformatics. In fact, when he wrote some of his most famous papers, bioinformatics as a field did not yet exist, and he was instrumental in laying the groundwork for its development.

| BOX 12-1 | **SCIENTIST SPOTLIGHT** | Michael Waterman |

Michael Waterman was born in Oregon in 1942, and earned a B.S. degree in mathematics from Oregon State University and a doctorate in statistics and probability from Michigan State University. He is one of the founders and current leaders in the area of computational biology, applying mathematics, statistics, and computer science techniques to various problems in molecular biology. His work has contributed to some of the most widely used tools in the field. Waterman's path from (in his own words) "innocent mathematician" to bioinformaticist was one of chance, of being at the right place at the right time with the right group of people.

In 1974, having run out of funding for his theoretical mathematics work, Waterman accepted an invitation from Bill Beyer to come to Los Alamos National Laboratories and join an NSF-funded summer project to study molecular biology and evolution. Los Alamos was a fertile ground for interdisciplinary collaboration. Stan Ulam, a well-known mathematician, thought that there was mathematics to be found in the new biological sciences. He was instrumental in bringing together Beyer, Temple Smith (originally a physicist), and Waterman. The latter two settled in for the summer in a little windowless office behind the security fence, and at the end of the summer they had written two papers, one on sequence alignment and one on molecular evolution. At the time, there was no field called computational biology, and so it took a long time for the papers to be published (1976 and 1978, respectively). In fact, their 1976 paper (with Beyer) on sequence metrics was so far out of the mainstream of

continued

continued

mathematics that the referees did not know what to make of it. Luckily, the editor of the highly selective and respected journal *Advances in Mathematics* recognized the importance of the paper and published it. The collaboration between Waterman and Smith continued and resulted in the Smith-Waterman algorithm, developed in 1981, which is the basis for many sequence comparison programs. In 1988, Waterman and Eric Lander (originally an economist) published a landmark paper describing a mathematical model for fingerprint mapping. This work formed one of the theoretical cornerstones for many of the later DNA mapping and sequencing projects, especially the Human Genome Project. Waterman also authored one of the earliest textbooks in the field, *Introduction to Computational Biology*, and is coauthor (with R. C. Deonier and S. Tavaré) of *Computational Genome Analysis: An Introduction*.

Michael Waterman has been honored many times for his role in founding the area of bioinformatics, including fellowships in four societies, election to four national and international academies of science and engineering, a Guggenheim Fellowship, and the first Celera Genomics Fellowship. He is a founding editor of the *Journal of Computational Biology* and was one of the founders of the international Research in Computational Biology (RECOMB) conference. He currently holds an endowed associates chair in biological sciences, mathematics, and computer science at the University of Southern California in Los Angeles. Since 2008, he also is a chair professor at Tsinghua University in Beijing.

http://www.cmb.usc.edu/people/msw/Waterman.html

12.2 EXTREME VALUE DISTRIBUTION

We start our exploration of tools from probability theory that apply to specific bioinformatics applications with the extreme value distribution. It models the distribution of the maximum value of a *random sample*, where a random sample of size n consists of a set of n random variables that have a common distribution and are independent of each other. Let's make this abstract concept concrete by using an example.

Example 12.1. Suppose we select 25 college women at random and measure their heights. What is the probability that the tallest woman's height is more than 70 inches?

As discussed in example 11.24, the heights of college women follow a normal distribution with a mean of 65.5 inches and a standard deviation of 2.5 inches. The height of the tallest woman in a randomly selected group of 25 college women is also a random quantity that has its own distribution. What is that distribution? In mathematical terminology, the heights of the 25 women are random variables $X_1, X_2, \ldots, X_{25}$, and the tallest woman's height is a random variable $Y = \max\{X_1, X_2, \ldots, X_{25}\}$. We can derive the distribution of the random variable Y based on the distribution common to all of the individual random variables X_i. If we let X stand for the height in inches of a randomly selected woman with the distribution from example 11.24, then the cumulative distribution function of the random variable Y for $y = 70$ is given by

$$F(70) = P(Y \le 70) = P\left(\max\{X_1, X_2, \ldots, X_{25}\} \le 70\right)$$

$$= P\left((X_1 \le 70) \cap (X_2 \le 70) \cap \cdots \cap (X_{25} \le 70)\right)$$

$$= P(X_1 \le 70) \cdot P(X_2 \le 70) \cdots P(X_{25} \le 70) = \left[P(X \le 70)\right]^{25}$$

$$= \left[P\left(Z \le \frac{70 - 65.5}{2.5}\right)\right]^{25} = \left[P(Z \le 1.8)\right]^{25} = 0.964^{25} = 0.40.$$

In the second line of this derivation, we use the fact that if the tallest woman in the selected group is at most 70 inches tall, then all women in that group also must have heights that are at most 70 inches tall. In the third line, we use the fact that because they were randomly selected we can assume their heights to be independent, so the intersection probability equals the product of the individual probabilities. Finally, we translate the problem expressed in terms of a specific normal distribution into one expressed in terms of the standard normal distribution and obtain the value of the cumulative probability from a table. The probability we seek, $P(Y > 70)$, can then be computed as $P(Y > 70) = 1 - P(Y \le 70) = 1 - 0.40 = 0.60$. Altogether, the probability that the tallest woman is more than 70 inches tall is about 60%.

Now let's look at how the two distributions compare, one for random variable X, the height in inches of a single college woman, and the other for random variable Y, the height in inches of the tallest of 25 college women. Figure 12-1 shows the probability density functions of these two distributions.[1] As we would expect, the density function of Y is shifted to the right because greater values are more likely. Also, at first glance it appears that Y, like X, has a normal distribution, albeit taller and narrower. Upon closer inspection, however, we see that the probability density function of Y is not quite symmetrical. The right side tail is thicker than the left. This shape is characteristic of the extreme value distribution, which Y approximates.

We are now ready for a formal definition. The *extreme value distribution with parameters* $\hat{K} > 0$ *and* $\lambda > 0$ is defined on the interval $(-\infty, \infty)$ and has probability density function[2]

$$f(x) = \hat{K}\lambda \exp\left(-\lambda x - \hat{K}\exp(-\lambda x)\right), \tag{12.1}$$

with mean μ and standard deviation σ given by

$$\mu = \frac{\ln(\hat{K}) + \gamma}{\lambda} \quad \text{and} \quad \sigma = \frac{\pi}{\sqrt{6}}\frac{1}{\lambda}, \tag{12.2}$$

where $\gamma \approx 0.5772$ is the Euler-Mascheroni constant. The cumulative distribution function is given by

$$F(x) = \exp\left(\hat{K}e^{-\lambda x}\right). \tag{12.3}$$

It is a very cool fact that the extreme value distribution can be used to approximate the distribution of *any* random variable that is the maximum value of a **large**

extreme value distribution
A distribution of a random variable that is itself the maximum of a set of random variables, used in bioinformatics to assess whether an alignment score indicates biological significance of the alignment.

---

[1] We obtain the probability density function of the height of the tallest woman for a generic value of x by computing $F(x)$ as above for $x = 70$ and then taking the derivative with respect to x.

[2] Mathematicians use a different form of the density function, namely $f(x) = \frac{1}{b}\exp\left(\frac{a-x}{b} - \exp\left(\frac{a-x}{b}\right)\right)$ with parameters $a > 0$ and $b > 0$. It can be shown that the two forms of the probability density function are the same, with $\lambda = 1/b$ and $\hat{K} = \exp(a/b)$.

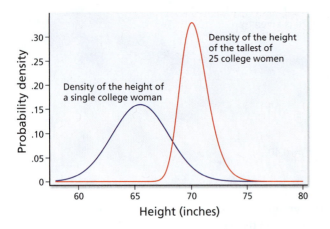

FIG. 12-1. Probability density functions of the height of a single college woman (blue curve) and the height of the tallest of 25 college women (red curve).

sample mean The mean of observed values of a random variable, typically used as an estimate of the mean of the distribution of that random variable.

random sample regardless of what the common distribution of the individual random variables is. This means that we do not actually need to know what that common underlying distribution is.

However, we do need to determine the parameters $\hat{K}$ and λ of the specific extreme value distribution that best models the random variable of interest. Suppose, for example, we have a large set of values x distributed according to an extreme value distribution. One way to find the the parameters $\hat{K}$ and λ is to exploit the fact that the *sample mean* $\bar{x}$ and the *sample standard deviation s*, computed from the values of a sample of size N as

$$\bar{x} = \frac{1}{N}\sum_{i=1}^{N} x_i \quad \text{and} \quad s = \sqrt{\frac{\sum_{i=1}^{N}(x_i - \bar{x})^2}{N-1}}, \tag{12.4}$$

are the best estimators of the theoretical (population) mean μ and standard deviation σ, respectively.

To make use of this fact for an extreme value distribution, we start by solving equation 12.2 for the parameters $\hat{K}$ and λ in terms of μ and σ (see exercise 1 at the end of this chapter) and obtain:

$$\lambda = \frac{\pi}{\sqrt{6}} \cdot \frac{1}{\sigma} = \frac{1.2825}{\sigma} \quad \text{and} \quad \hat{K} = \exp\left(1.2825 \cdot \frac{\mu}{\sigma} - \gamma\right). \tag{12.5}$$

Substituting $\bar{x}$ and s for μ and σ, respectively, in equation 12.5 results in expressions that allow us to estimate λ and $\hat{K}$ from sample data:

$$\lambda \approx \frac{1.2825}{s} \quad \text{and} \quad \hat{K} \approx \exp\left(1.2825 \cdot \frac{\bar{x}}{s} - 0.5772\right). \tag{12.6}$$

Example 12.2. We return to the example of the college women, whose heights can be modeled using a normal random variable with a mean of 65.5 inches and a standard deviation of 2.5 inches. Let's assume that you have randomly selected 30 college campuses from across the nation. At each campus, you randomly selected 25 college women whose height you measured in inches, and then determined the height of the tallest woman. Thus you have a sample of size 30 for the random variable modeling the height of the tallest woman of a group of 25 randomly selected women. Assume that the data you have obtained is as follows:

71.84	69.73	69.47	69.71	71.75	70.91	69.99	71.13	66.80	69.15
68.58	70.78	70.09	70.67	69.07	69.69	71.36	70.19	68.85	70.64
69.73	68.77	68.07	71.57	71.27	72.74	70.32	69.82	70.50	69.81

Here $x_1 = 71.84$, $x_2 = 69.73$, and so on. Using equation 12.4 we obtain that

$$\bar{x} = \frac{1}{30}\left(71.84 + 69.73 + \cdots + 69.81\right) = \frac{2103}{30} = 70.10$$

and

$$s^2 = \frac{(71.84 - 70.10)^2 + \cdots + (69.81 - 70.10)^2}{29} = \frac{44.6312}{29} = 1.5390.$$

Taking square roots in the last equality gives a sample standard deviation of $s = 1.2406$. Now we can compute the parameters for the relevant extreme value distribution using equation 12.6, namely

$$\lambda = \frac{1.2825}{1.2406} = 1.03$$

and

$$\hat{K} = \exp\left(1.2825 \cdot \frac{70.10}{1.2406} - 0.5772\right) = \exp(71.8904) = 1.666 \times 10^{31}.$$

Note that we have kept additional decimal places in intermediate computations and rounded only at the end of the computation for the values of λ and $\hat{K}$. Due to the size of $\hat{K}$, we carry an additional significant digit.

Having estimated the parameter values of λ and $\hat{K}$ from our data, we now have two different probability density functions of the random variable that gives the height of the tallest woman of a group of 25 college women, namely

1. the approximate extreme value distribution with parameters $\hat{K} = 1.666 \times 10^{31}$ and $\lambda = 1.03$ derived in example 12.2 and

2. the exact distribution based on the known distribution of the heights of the women (see example 12.1 and discussion afterward).

The graph in Figure 12-2A shows the approximate and the exact probability density functions together. As you can see, the two functions have a similar shape, but are not identical. The reasons that the approximate distribution is not as close to the exact distribution as one would hope are the small sample size of 30 and the fact that the size of the group of which the tallest height is determined is also small ($m = 25$). If we use larger values for the size of the groups and the sample size then we get a better agreement between the exact distribution and the extreme value distribution that approximates it. The graph in Figure 12-2B shows the distributions when the tallest woman is chosen from a group of 1,000 women and the sample size is 10,000 (= number of campuses). The corresponding parameter estimates are $\lambda = 1.48$ and $\hat{K} = 8.35 \times 10^{46}$.

An alternative method for estimating the parameters directly from the data without computing intermediate values such as the sample mean and the sample standard deviation is the *maximum likelihood method*.[3] This method selects the parameter values of the assumed distribution, here λ and $\hat{K}$ of the extreme value distribution, that maximize the probability of observing the given data. For large sample sizes both methods are known to converge to the true parameter values and therefore will produce (theoretically) the same values for $\hat{K}$ and λ. To illustrate this

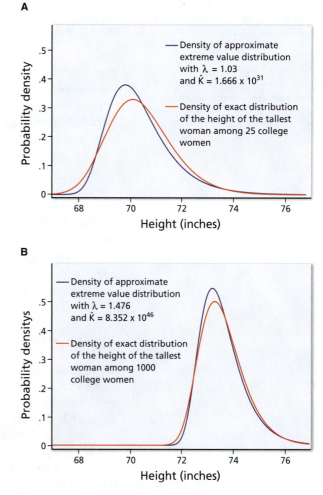

FIG. 12-2. A. Plots of the extreme value probability density function with $\lambda = 1.03$ and $\hat{\kappa} = 1.666 \times 10^{31}$ (in blue) and the exact probability density function of the height of the tallest woman in a group of 25 college women (in red). **B.** Plots of the extreme value probability density function with $\lambda = 1.476$ and $\hat{\kappa} = 8.352 \times 10^{46}$ (in blue) and the exact probability density function of the height of the tallest woman in a group of 1,000 college women (in red). Parameters λ and $\hat{\kappa}$ were estimated from 30 data points (30 groups of 25 women each) in the first case and from 10,000 data points (10,000 groups of 1,000 women each) in the second case.

---

[3] This method is widely used in bioinformatics, but is beyond the scope of this book.

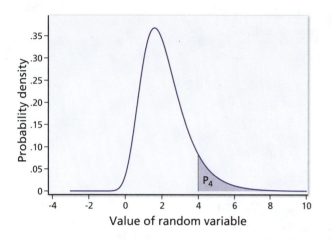

FIG. 12-3. The right-tail P-value $P_4 = P(X \geq 4)$ of the extreme value distribution with $\hat{\kappa} = 5$ and $\lambda = 1$ equals the area of the shaded region.

right-tail P-value The probability of seeing a value of a random variable that is as large as or larger than a given value.

fact, the authors created 10,000 pseudo-random values of a random variable that follows an extreme value distribution with $\hat{K} = 1.769 \times 10^7$ and $\lambda = 0.267$ and then estimated the parameters $\hat{K}$ and λ using sample mean and sample standard deviation as well as the maximum likelihood method. Repeating this process 200 times and averaging the respective values resulted in estimates of $\hat{K} = 1.760 \times 10^7$ and $\lambda = 0.267$ using the sample mean and sample standard deviation, whereas the maximum likelihood method resulted in estimates of $\hat{K} = 1.763 \times 10^7$ and $\lambda = 0.267$. The maximum likelihood method will make another appearance when we discuss the extreme value distribution as it relates to BLAST and also when we describe the hidden Markov model for multiple sequence alignments.

Before we look at the role these distributions play in bioinformatics applications, especially BLAST and the interpretation of its results, we need one additional concept, the right-tail P-value.

Right-Tail P-value

For any random variable X, the *right-tail P-value* associated with a value x is denoted by P_x and defined as $P_x = P(X \geq x)$. That is, P_x gives the probability that the random variable X takes a value that is at least as large as x. From the definition of the cumulative distribution function $F(x)$, we have that

$$P_x = P(X \geq x) = 1 - P(X < x) = 1 - F(x).$$

Example 12.3. If the random variable X has standard normal distribution, then $P_x = 1 - \Phi(x)$. On the other hand, if the random variable X has an extreme value distribution with parameters $\hat{K}$ and λ, then $P_x = P(X \geq x) = 1 - \exp\left(-\hat{K} \cdot e^{-\lambda x}\right)$. Figure 12-3 illustrates the right-tail P-value P_4 associated with $x = 4$ of the extreme value distribution with $\hat{K} = 5$ and $\lambda = 1$.

We are now ready to evaluate the significance of a BLAST sequence alignment.

12.3   SIGNIFICANCE OF ALIGNMENTS

We opened this chapter with the question about how to determine whether a BLAST "hit" between a query sequence and a subject sequence in a database is significant (biologically meaningful). What a bioinformaticist wants to do is to quantify how likely a certain similarity score is to arise out of pure chance. Is a top score of 100 unusual (i.e., has a low probability)? If so, then the score of 100 would support the hypothesis that the alignment is not between the query sequence and a random sequence, but rather between the query sequence and a related sequence.

Historically, bioinformaticists thought that local alignment scores followed a normal distribution, and so they used the z-score or z-value defined in equation 11.19 to assess biological significance. Recall that a positive z-score indicates how many standard deviations a value of a given random variable is above the mean of the distribution of that random variable. A z-score of 0 indicates that the value equals the average of the distribution, and might have well arisen by chance. On the other hand, higher positive z-scores correspond to values that are less likely to occur by chance, and the notion of how likely a score is can be made precise by computing its associated right-tail P-value. Under the assumption of the normal distribution, a BLAST similarity score with a z-score of 3 would have a P-value of $1 - \Phi(3) \approx 0.15\% = 1.5 \times 10^{-3}$

(see Figure 11-18), indicating biological significance according to Lesk (2008). In contrast, a well-established rule of thumb based on the similarity scores of sequences known to be related asserted that a z-score of 5 is needed to infer biological significance. This discrepancy led to the realization that BLAST similarity scores are not normally distributed.[4]

In fact, it has been shown mathematically that when a scoring matrix is used to evaluate sequence alignments, the highest score reached from a search of random data follows an extreme value distribution. That is, if the random variable X models the BLAST **top hit score** from alignment between the query sequence and random ("un-related") sequences, then the probability density function of X is given by equation (12.1). This should not be a surprise because X is the **highest** similarity score from among all the high scoring segment pairs (HSPs) in a sequence alignment, there are **many** of these scores, and they are **independent** and **have the same distribution**.

Parameters $\hat{K}$ and λ of equation 12.1 depend on the particular scoring matrix used to assign similarity scores and on the respective sizes of the query and subject sequence (or subject database). More specifically, $\hat{K}$ is given by $\hat{K} = K \cdot m \cdot n$, where K depends on the scoring matrix, m is the effective length of the query sequence, and n is the effective length of the database.[5] We can evaluate a particular BLAST top hit with *similarity score s* (not to be confused with the symbol s for the sample standard deviation) by using the right-tail probability P_s derived in example 12.3

$$P_s = P(X \geq s) = 1 - \exp\left(-K \cdot n \cdot m \cdot e^{-\lambda s}\right) \tag{12.7}$$

To compute this probability, we need to know the values of the parameters K and λ for the particular scoring matrix. FASTA estimates these parameters from the scores generated by actual database searches. BLAST estimates them from pseudo-random data. That is, using a standard protein amino acid composition, a large number of random amino acid sequence pairs are generated and the maximum local alignment score of each pair is computed using the Smith-Waterman algorithm.[6] Estimates of K and λ are then computed from these scores using the maximum likelihood method.

Let's look at an example of a partial BLAST output. Figure 12-4 shows the HSP with the highest similarity score above the threshold score A. BLAST reports the *bit score s'*, the similarity score s, as well as the Expect (or E-) value. For example, in Figure 12-4, the HSP has a bit score of 34.3, a similarity score s of 77, and an E-value of 0.020. In addition, the full output of the BLAST search will also report the effective length of the query ($= m$), the effective length of the database ($= n$), as well as $\hat{K}$ and λ.

> **bit score** A measure of the similarity between sequences that does not depend on the size of the database that was searched.

Notably missing from the output is a P-value. In example 12.4 we will compute the P-value of this top hit. We will then discuss how to interpret the P-value as a measure of biological significance. Finally we will define and discuss the E-value and the bit score, which are related to the P-value and have largely replaced it for assessing sequence alignment results.

Example 12.4. Figure 12-4 shows a partial BLAST output containing the alignment with the highest similarity score. The full report states that the effective length of the query is $m = 305$, the effective length of the database is $n = 1,414,501$, so the size of the effective search space is $n \cdot m = 431,422,805$. The BLOSUM62 matrix with $\hat{K} = 0.0410$ and $\lambda = 0.267$ for gapped alignments was used to compute similarity scores.

---

[4] To avoid this common pitfall, numerous tests are available to determine whether a given data set follows a normal distribution.

[5] Effective lengths are less than total lengths and take into account the fact that, for example, the 50th nucleotide in a query sequence cannot be aligned with any of the first 49 nucleotides of the subject sequence.

[6] This technique, whereby pseudo-random data is created by repeating a calculation with randomized inputs, is called the Monte Carlo method, after the famous casino in Monaco.

RecName: Full=Transcription factor cep-1; AltName: Full=C.elegans p53-like protein 1
Sequence ID: splQ20646.2|CEP1_CAEEL **Length:** 644 **Number of Matches:** 1

Range 1: 337 to 427 GenPept Graphics

Score	Expect	Method	Identities	Positives	Gaps
34.3 bits (77)	0.020	Compositional matrix adjust.	28/91 (31%)	43/91 (47%)	14/91 (15%)

```
Query   224   EVGSDCTTIHY----------NYMCNSSCMGGMNRRPILTI-ITLEDSSGN-LLGRNSFE   271
              E  GS T I Y          +MC   C+   +RR + + + L+D +GN +L      +
Subject 337   EKGSTFTLIMYPGAVQANFDIIFMCQEKCLDLDDRRKTMCLAVFLDDENGNEILHAYIKQ   396

Query   272   VRVCACPGRDRRT--EEENLRKKGEPHHELP   300
              VR+ A P RD +    E  E+ ++K    ELP
Subject 397   VRIVAYPRRDWKNFCEREDAKQKDFRFPELP   427
```

FIG. 12-4. Partial BLAST output of human p53 alignment to the UniProtKP/Swiss-Prot database (database restricted to *C. elegans*). BLAST reports a bit score of 34.3, a similarity score of 77, and an *E*-value of 0.020.

The specific alignment shown in Figure 12-4 has a similarity score s of 77, so equation 12.7 gives

$$P_{77} = P(X \geq 77) = 1 - \exp\left(-0.0410 \cdot 431,422,805 \cdot e^{-0.267 \cdot 77}\right) = 0.0206.$$

This probability is reasonably small, making it likely that the two sequences are related. We can also compare the similarity score to the average similarity score for this distribution, namely $\mu = (\ln(0.0410 \cdot 431,422,805) + 0.5772)/0.267 = 64.67$ (using equation 12.2). The actual score of 77 is quite a bit higher than the average score of 65, which explains the small P-value. You will compute just how much higher s is than μ in terms of the standard deviation σ in exercise 2.

Now that we know how to compute P-values, we need to understand how they are interpreted by bioinformaticists. From the definition of the P-value it is clear that the smaller the P-value, the less likely an outcome has occurred just by chance and therefore, the higher the likelihood that the sequences are related. But how small is small enough? Table 12-1 gives a rough guide to interpreting P-values (from Lesk 2008).

In example 12.4, we obtained that $P_{77} = 2.06 \times 10^{-2}$, thus the two sequences are likely to be distant relatives.

Typically, a local alignment program such as BLAST will instead report the ***E-value*** associated with score s, or E_s, defined as the *expected number of HSPs with a score of at least s*. Here too, small values of E_s are more desirable than big values.

How is this E-value computed? It is by definition the expected value of the random variable Y that counts the *number* of HSPs from the alignment of random sequences that have a score S that exceeds s. Be careful not to confuse the distribution of the HSP similarity score S (a random variable with unknown distribution) with the discrete random variable Y that counts the number of HSP similarity scores that have a certain characteristic, namely being at least s. Each HSP either has a score

E-value For a given similarity score s, the expected number of HSPs with at least score s, assuming random sequences. Used as a measure to assess whether similarity score s could have been obtained by random chance.

TABLE 12-1. GUIDE TO INTERPRETING *P*-VALUES

P-VALUE	BIOLOGICAL SIGNIFICANCE
$\leq 10^{-100}$	Exact match
$10^{-100} - 10^{-50}$	Sequences nearly identical, for example, alleles or SNPs
$10^{-50} - 10^{-10}$	Closely related sequences, homology certain
$10^{-5} - 10^{-1}$	Usually distant relatives
$>10^{-1}$	Match probably insignificant

that exceeds s or not, independently of the others, and therefore Y is a binomial random variable, but one with a twist: the success probability $p = P(S \geq s)$ is computed from the distribution of the similarity score S. Note that although we have a formula for $P(X \geq s)$, the probability that the *top score* in an alignment with random data will equal or exceed s, we do not have one for $P(S \geq s)$, the probability that an arbitrary HSP score S would equal or exceed that value. Fortunately, we will not need to know $P(S \geq s)$ to find the E-value.

Because we have large values of m and n, and therefore a large number of potential hits that lead to HSP scores, we can approximate the binomial random variable Y by a Poisson random variable that has the same expected value as the binomial random variable. Recall that for the Poisson distribution with parameter β (see equation 11.16) the expected number of occurrences is also β. Because E_s is by definition the expected number of HSPs with a similarity score of at least s, it therefore equals the value of β for the Poisson distribution. From equation 11.16 we have that

$$P\big(\text{exactly } k \text{ HSP scores} \geq s\big) = P(Y = k) = \frac{(E_s)^k}{k!} \cdot \exp(-E_s),$$

and the probability of seeing at least one such HSP is given by

$$P\big(\text{at least one HSP score} \geq s\big) = 1 - P\big(\text{no HSP score} \geq s\big) = 1 - P(Y = 0)$$
$$= 1 - \frac{(E_s)^0}{0!} \cdot \exp(-E_s) = 1 - \exp(-E_s). \tag{12.8}$$

On the other hand, we have from equation 12.7 that the probability of observing an HSP similarity score of s or higher, that is, seeing at least one such HSP, is equal to

$$P\big(\text{at least one HSP score} \geq s\big) = P\big(\text{maximum HSP score} \geq s\big) = P(X \geq s)$$
$$= 1 - \exp\big(-\hat{K}e^{-\lambda s}\big). \tag{12.9}$$

The first equality in equation 12.9 follows because when there is an HSP score that is at least s, then the maximum score also has to be at least s, and vice versa. This equality allows us to use the known distribution of the maximal alignment score X to determine the E-value of any HSP similarity score, whether it is the maximal one or not. Comparing the right-hand sides of equations 12.8 and 12.9 we see that the blue expressions have to be the same, so

$$E_s = \hat{K}e^{-\lambda s} = K \cdot m \cdot n \cdot e^{-\lambda s}. \tag{12.10}$$

For the BLAST search shown in Figure 12-4, the E-value is

$$E_{77} = 0.0410 \cdot 431,422,805 \cdot e^{-0.267 \cdot 77} = 0.0208.$$

The discrepancy to the E-value of 0.020 shown in Figure 12-4 comes from the fact that the values of K and λ used here are rounded. Note that the E-value is identical to the P-value calculated in example 12.4 to three decimal places.

Let's have a closer look at the formula for E_s. If either m (= effective length of query) or n (= effective length of database) is doubled, that should double the number of HSP similarity scores that exceed a given score s, as there are now twice as many opportunities for such scores. On the other hand, because HSP similarity scores arise from the addition of the similarity values of adjacent amino acid pairs, attaining a similarity score of $2s$ is roughly equivalent to having two adjacent HSPs each with similarity score s (that combine into a single HSP with similarity score $2s$), which should be considerably more difficult to obtain than an HSP score of s. These facts are appropriately modeled because E_s increases linearly with m and n and decreases exponentially with s.

So what does the E-value tell us in terms of biological significance? It is important to realize that BLAST does not test for homology, but rather that the E-value points to potential relatedness, which then must be substantiated by other means. As a consequence, although BLAST can be used to infer that two sequences are unrelated, it is not sufficient in and of itself to declare that they are related. As for interpreting E-values, note that the P-value and the E-value can be computed from each other using equation 12.8. Thus, knowing the P-value gives the corresponding E-value and vice versa. The reason that BLAST (and other programs) report the E-value instead of the P-value is twofold: For small values when homology is a distinct possibility, the E- and P-values are nearly identical, so it does not matter which one is used. However, when they are large and differ, it is easier to work with E-values such as $E = 5$ or $E = 10$ instead of their corresponding P-values, namely 0.993 and 0.99995, respectively.

Besides the similarity score s, BLAST also reports a bit score s'. Like the E-value, the bit score expresses how rare—and hence indicative of biological significance—the similarity score is. The *bit score s'* is defined as the base 2 logarithm of the size of a search space $m \cdot n$ of random data that would have to be scanned in order to see on average **one** HSP with similarity score s, that is, for this size search space we would have $E_s = 1$. The bit score s' has an important advantage over the similarity score s in that it is independent of the size of the actual database used, and thus bit scores can be compared across different database searches.

Let's see how the bit score s' and the similarity score s relate to each other, given that HSPs follow an extreme value distribution with parameters $\hat{K}$ and λ. To derive a formula for s', we start with equation 12.10 and set $E_s = 1$ (by definition of the bit score):

$$1 = E_s = K \cdot m \cdot n \cdot e^{-\lambda s}.$$

Solving for the size of the search space, we obtain

$$m \cdot n = \frac{1}{K \cdot e^{-\lambda s}} = K^{-1} e^{\lambda s},$$

and hence[7]

$$s' = \log_2 (m \cdot n) = \log_2 \left(K^{-1} e^{\lambda s} \right) = \frac{\ln\left(K^{-1} e^{\lambda s}\right)}{\ln 2} = \frac{\ln\left(K^{-1}\right) + \ln\left(e^{\lambda s}\right)}{\ln 2} = \frac{\lambda s - \ln K}{\ln 2}$$

Overall, we have that

$$s' = \frac{\lambda s - \ln K}{\ln 2} \tag{12.11}$$

This is the formula promised. Using the parameter values in example 12.4, we can compute the bit score s' corresponding to the similarity score $s = 77$:

$$s' = \frac{\lambda s - \ln K}{\ln 2} = \frac{0.267 \cdot 77 - \ln 0.0410}{0.693147} = \frac{0.267 \cdot 77 - (-3.19418)}{0.693147} = 34.27,$$

which agrees with the bit score (rounded to one decimal place) displayed for the HSP in Figure 12-4.

In addition, we can compute the E-value from the bit score:

$$E_{s'} = m \cdot n \cdot 2^{-s'}. \tag{12.12}$$

---

[7] We use these facts about logarithms in this derivation:
$\ln_a b = (\log_c b)/(\log_c a)$ $\ln a = \log_e a$ $\log_a(b \cdot c) = \log_a b + \log_a c$ $\log_a(b^c) = c \cdot \log_a b$ $\ln e^a = a$
which hold for all acceptable values of a, b, and c.

In our example we obtain that

$$E_{s'} = m \cdot n \cdot 2^{-s'} = 431{,}422{,}805 \cdot 2^{-34.27} = 0.0208$$

as before. The E-value is the same, whether it is computed from the similarity score s using equation 12.10 or from the associated bit score s' using equation 12.12.

Let's investigate the bit score a little bit more. You might wonder why the bit score is defined as a base 2 logarithm. First, database sizes can be quite large and the logarithm allows large numbers to be expressed compactly. Second, computer memory is measured in binary terms, so the base 2 logarithm is the natural choice. Let's see what the bit score in example 12.4 tells us about the size of the search space needed to see on average one HSP with similarity score 77. The bit score of $s' = 34.27$ corresponds to a search space size of $m \cdot n = 2^{s'} = 2^{34.27} \approx 20{,}700{,}000{,}000$. The actual search space for the BLAST search in example 12.4 is 431,422,805 with $E_{77} = 0.0208$, that is, the actual search space is quite a bit smaller than the one needed to have an E-value of 1.

12.4 STOCHASTIC PROCESSES

When a random value is observed repeatedly over time, we refer to the sequence of observations as a *stochastic process*, where "process" indicates that we have repeated observations, and "stochastic" indicates that the observed value is random. Note that to be considered random, an observation need not be entirely unpredictable, but must not be completely determined by some formula or pattern.

Stochastic processes are categorized as discrete or continuous, both with respect to the value observed, and with respect to how it is observed. For example, a physical property such as body mass or air temperature is said to be continuous because it can take any value over a range of values, like numbers on a segment of the number line. So the corresponding stochastic process is said to be *continuous-valued*. In contrast, the particular allele observed at a gene site can take only specific values, making the corresponding stochastic process *discrete-valued*. Another kind of discrete value is a count, such as the number of mutations that occur at a gene site over a specified period of time, because this number may only be an integer. In summary, continuous-valued and discrete-valued refer to the type of random variable that underlies the stochastic process.

Now we consider how the stochastic process is observed. Suppose, for example, that we observe the size of a bacterial population in a culture once every hour, or we record the percentage of individuals in each generation of a plant population that display a particular phenotype. These are examples of what is called a *discrete-time* stochastic process. In such a process, we denote the nth observation as X_n. Note that it is quite common to refer to X_n as the *state of the process at time n*, even when, as in the second example, the index does not actually refer to time. On the other hand, suppose that the value of interest is observed continuously over time, the way an electrocardiogram (EKG) records heart electrical activity, then the corresponding stochastic process is said to be *continuous-time*. Continuous-time observations are denoted by $X(t)$ to make explicit reference to the "time" variable.

An example of a continuous-time stochastic process in bioinformatics is the occurrence of sequence alterations of a genome via neutral mutations (see Chapter 8) at a specific site. These alterations, including mutations that are later reversed as well as multiple mutations at the same site, accumulate over time to create genetic distance between individuals who are descendants of a common ancestor. We will use a discrete-valued, continuous-time stochastic process to model the number of mutations at a specific site in the derivation of the Jukes-Cantor model of evolutionary distance.

Markov Chains

Let us now turn our attention to a special class of discrete-time stochastic processes, namely Markov chains, which lie at the heart of *hidden Markov models,* or HMMs. HMMs are widely used in pattern-seeking applications, such as speech recognition and numerous biological applications. These include constructing genetic linkage maps, distinguishing coding from non-coding regions of DNA, domain profiling, and scoring sequence alignments. HMMs can be "trained" to detect a wide variety of patterns, giving them a broader range of capabilities than, for example, the PAM250 scoring matrix, which is built on a fixed set of closely related sequences.

A *Markov chain* is a discrete-valued, discrete-time stochastic process used to model a phenomenon, called the *system,* that is known to have a finite set of possible outcomes at each observation, called the *states* of the system. Furthermore, the likelihood that any particular outcome (state) is observed depends only on the prior observed state. That is, the system does not "remember" its history of earlier states. For example, a DNA sequence can be considered a system in which the observations are not separated by time, but rather by position, consisting of the observed nucleotides at consecutive positions along the DNA sequence. Each consecutive nucleotide in the sequence is an observation, and the states of the system consist of the four nucleotides A, C, G, and T. The likelihood that the system transits from the current state to any particular subsequent state (e.g., that after an A the next nucleotide will be a C) depends only on the current state (A in this case).

Here is the formal description of the ingredients of a Markov chain with N states and *state space* $S = \{s_1, s_2, s_3, \ldots, s_N\}$. The random variables $X_0, X_1, X_2, \ldots$ describe the *state* of the chain at the ith observation (which is also referred to as "time" i) for $i = 0,1,2,\ldots$, where the 0th observation refers to the initial state. We denote the *transition probabilities* $P(X_{k+1} = s_j | X_k = s_i)$ of movement from state s_i to state s_j by $p_{i,j}$ and compactly organize these probabilities in the *transition matrix* $P = (p_{i,j})$, where the indices i and j each range from 1 to N. We can visualize any Markov chain with a *state diagram* (see Figure 12-5). The nodes of the diagram correspond to the possible states (labeled 1,2,3, and 4). Arrows between the nodes indicate allowed transitions from one state to the other, accompanied by the respective transition probabilities. Let's look at an example of a Markov chain to become familiar with these definitions and the notation.

Example 12.5. Suppose a molecule moves between four cells that are arranged in a ring. From one observation to the next, the molecule remains in the current cell with probability 0.9. If it leaves, the molecule is equally likely to move to either of the two adjacent cells (see state diagram in Figure 12-5). This situation can be described by a Markov chain. The states of the system are the cells in which the molecule may be located, denoted by the numbers 1 through 4, so $S = \{1, 2, 3, 4\}$. The molecule remains in the current cell (that is, in the same state) with probability 0.9, so $p_{i,i} = 0.9$ for $i = 1,\ldots, 4$. If the molecule leaves the cell, it is equally likely to go to either of the two adjacent cells, that is, with probability $(1 - 0.9)/2 = 0.05$. So $p_{1,4} = p_{1,2} = p_{2,1} = p_{2,3} = p_{3,2} = p_{3,4} = p_{4,3} = p_{4,1} = 0.05$. All other transitions, such as the transition between cell 1 and cell 3, do not occur, so they have zero probability and therefore no arrow. If we let the row indices of the matrix refer to the current state, and the column indices to the next state, then the transition matrix P is given by

$$P = \begin{pmatrix} 0.9 & 0.05 & 0 & 0.05 \\ 0.05 & 0.9 & 0.05 & 0 \\ 0 & 0.05 & 0.9 & 0.05 \\ 0.05 & 0 & 0.05 & 0.9 \end{pmatrix}$$

Note that the sum of the entries in each row of the matrix P equals 1, because the row sum equals the total probability of going from a given state s_i to all of the states, including state s_i. A matrix with this property is called a *stochastic* matrix. The matrix

---

Markov chain A discrete-time model that represents transitions of an object from one condition (state) to another without memory. That is, the probability of a state depends only on the prior state and not on the sequence of states that preceded it.

transition probabilities The probabilities that describe how likely it is for each state in a Markov chain to transition to each other possible subsequent state in one time step.

state diagram A graph that shows the transitions between states of a Markov chain and the probabilities of each transition.

transition matrix A matrix containing all the transition probabilities of a Markov chain.

entry colored in blue is $p_{1,1}$, and the one in red is $p_{4,3}$. The arrows corresponding to the colored matrix entries are shown in the same colors in the state diagram of Figure 12-5. Each node is a state, and each arrow indicates an allowed transition with its associated transition probability.

Thought Question 12-1

> If the four cells where arranged such that the molecule can move from any cell to any other cell with the same probabilities for remaining in the cell, how would the transition matrix change? Indicate the entries that change and their new values.

Now we can consider the following two questions:

- What is the probability of a specific *path*, that is, the probability that a particular sequence of states is visited in order? This question is at the heart of the hidden Markov models, as training of the model and the resulting multiple sequence alignment are based on maximizing path probabilities.

- What is the probability that a system in state s_i will move to state s_j in k steps? The ideas used to answer this question also show up in the derivation of the HMM.

Let's look at specific instances of these general questions in the context of the molecule movement from one cell to another discussed in example 12.5. An example of the first question is finding the probability that if a molecule starts out in cell 4, that it will move to cell 3, then back to cell 4, then remain in cell 4, move to cell 1, remain in cell 1, and finally move to cell 2. We represent a path as a list of the states interspersed with arrows to represent the movement from one state to the next. For example, the path specified above is written as: $4 \rightarrow 3 \rightarrow 4 \rightarrow 4 \rightarrow 1 \rightarrow 1 \rightarrow 2$. If we let X_i denote the cell or state after the ith movement, then $X_0 = 4$ (the initial state). What we want to compute is the probability that, starting at cell 4, the molecule will move to cell 3, AND then starting at cell 3 will move to cell 4, AND then starting at cell 4 will remain at cell 4, and so on, ending with a move to cell 2. All those "AND"s remind us that we are looking at an intersection probability, for which the relevant formula is given by equation 11.8, namely $P(A \cap B) = P(A|B)P(B)$. Consider just the first move. The probability that it occurs is given by

$$P(4 \rightarrow 3) = P(X_0 = 4 \cap X_1 = 3) = P(X_0 = 4)P(X_1 = 3|X_0 = 4) = 1 \cdot p_{4,3} = p_{4,3},$$

where we obtain $p_{4,3}$ from the transition matrix of this system. The probability that the first and second moves both occur can be written as

$$P(4 \rightarrow 3 \rightarrow 4) = P(4 \rightarrow 3 \cap X_2 = 4) = P(4 \rightarrow 3)P(X_2 = 4|4 \rightarrow 3)$$

$$= p_{4,3}P(X_2 = 4|(X_0 = 4 \cap X_1 = 3)) = p_{4,3}P(X_2 = 4|X_1 = 3) = p_{4,3} \cdot p_{3,4}$$

where we have used the *Markov property* that the transition probability depends only on the most recent state, not the whole history, in the second-to-last equality. Continuing on in this fashion and using the Markov property at each step, we have that the complete path probability is computed as

$$P(4 \rightarrow 3 \rightarrow 4 \rightarrow 4 \rightarrow 1 \rightarrow 1 \rightarrow 2) = p_{4,3} \cdot p_{3,4} \cdot p_{4,4} \cdot p_{4,1} \cdot p_{1,1} \cdot p_{1,2}$$

$$= (0.05) \cdot (0.05) \cdot (0.90) \cdot (0.05) \cdot (0.90) \cdot (0.05)$$

$$= 5.06 \times 10^{-6}.$$

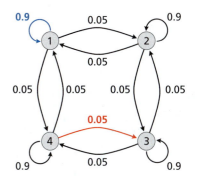

FIG. 12-5. State diagram of the Markov chain of example 12.5. Each node represents a state and each arrow an allowed transition. The number next to each arrow indicates the corresponding transition probability. For example, the blue arrow represents the transition from state 1 to 1— that is, the molecule remains in cell 1, which occurs with probability 0.9.

In general, **to compute the probability that a Markov chain follows a particular path we just multiply the corresponding transition probabilities.**

To answer the second type of question, we will use matrix multiplication. A matrix with n rows and m columns is said to be an $n \times m$ matrix. If C is an $n \times k$ matrix and D is a $k \times m$ matrix, then the resulting matrix has dimensions $n \times m$, and the element in row i and column j of the product matrix is the sum of products computed by multiplying the elements in row i of matrix C by the corresponding elements in column j of matrix D. That is, the (i, j) element in the product of an $n \times k$ matrix C and a $k \times m$ matrix D is given by

$$(C{\cdot}D)_{i,j} = \sum_{\ell=1}^{k} c_{i,\ell} \cdot d_{\ell,j}$$

where $c_{i,\ell}$ is the element in row i and column ℓ of matrix C, and $d_{\ell,j}$ is the element in row ℓ and column j of matrix D. Notice that in order to multiply the two matrices, the number of columns in matrix C has to match the number of rows in matrix D. Let's look at an example of a matrix multiplication.

Example 12.6. Let

$$C = \begin{pmatrix} -3 & -1 & 5 \\ 4 & 0 & 3 \end{pmatrix} \quad \text{and} \quad D = \begin{pmatrix} 1 & 6 & -1 & 3 \\ 4 & 6 & 5 & 6 \\ 0 & 1 & 3 & 6 \end{pmatrix}.$$

Because C is a 2×3 matrix and D is a 3×4 matrix, we can compute the product $C{\cdot}D$ (which results in a 2×4 matrix), but not the product $D{\cdot}C$ because the latter would have mismatched dimensions. We show how to compute the element in row 1, column 4 of $C{\cdot}D$ from the elements in the first row of C and the elements of the fourth column of D, which are highlighted in blue:

$$C \cdot D = \begin{pmatrix} -3 & -1 & 5 \\ 4 & 0 & 3 \end{pmatrix} \cdot \begin{pmatrix} 1 & 6 & -1 & 3 \\ 4 & 6 & 5 & 6 \\ 0 & 1 & 3 & 6 \end{pmatrix} = \begin{pmatrix} -7 & -19 & 13 & 15 \\ 4 & 27 & 5 & 30 \end{pmatrix}.$$

The relevant formula is $(C \cdot D)_{1,4} = (-3) \cdot 3 + (-1) \cdot 6 + 5 \cdot 6 = -9 - 6 + 30 = 15$.

Now we return to the second question, which asks for the probability that a path starting in state s_i will end in state s_j after k steps. We denote this k-step (transition) probability by $p_{i,j}^{(k)}$. If $k = 1$, then $p_{i,j}^{(1)} = p_{i,j}$. Let's see how we can compute $p_{1,4}^{(2)} = P(X_2 = 4 \mid X_0 = 1)$ for the Markov chain from example 12.5. The molecule starts in cell 1, moves to a cell j (which includes remaining in the current cell), and then moves from there to cell 4. Because there are four states, we have four potential paths, some of which may have zero probability. The possible paths and their corresponding probabilities are listed below:

$$
\begin{array}{ccccccc}
1 & \to & 1 & \to & 4 & \quad & p_{1,1} \cdot p_{1,4} \\
1 & \to & 2 & \to & 4 & \quad & p_{1,2} \cdot p_{2,4} \\
1 & \to & 3 & \to & 4 & \quad & p_{1,3} \cdot p_{3,4} \\
1 & \to & 4 & \to & 4 & \quad & p_{1,4} \cdot p_{4,4}
\end{array}
$$

These paths are mutually exclusive (disjoint union), so we can add the corresponding probabilities using the addition principle to obtain that

$$
\begin{aligned}
p_{1,4}^{(2)} &= p_{1,1} \cdot p_{1,4} + p_{1,2} \cdot p_{2,4} + p_{1,3} \cdot p_{3,4} + p_{1,4} \cdot p_{4,4} \\
&= (0.9) \cdot (0.05) + (0.05) \cdot 0 + 0 \cdot (0.05) + (0.05) \cdot (0.9) = 0.09.
\end{aligned}
$$

We can write this sum more compactly as $p_{1,4}^{(2)} = \sum_{l=1}^{4} p_{1,l} \cdot p_{l,4}$, which is exactly the value of the $(1, 4)$ position in the matrix that results from multiplying the transition matrix P by itself:

$$
\begin{pmatrix}
p_{1,1} & p_{1,2} & p_{1,3} & p_{1,4} \\
p_{2,1} & p_{2,2} & p_{2,3} & p_{2,4} \\
p_{3,1} & p_{3,2} & p_{3,3} & p_{3,4} \\
p_{4,1} & p_{4,2} & p_{4,3} & p_{4,4}
\end{pmatrix}
\begin{pmatrix}
p_{1,1} & p_{1,2} & p_{1,3} & p_{1,4} \\
p_{2,1} & p_{2,2} & p_{2,3} & p_{2,4} \\
p_{3,1} & p_{3,2} & p_{3,3} & p_{3,4} \\
p_{4,1} & p_{4,2} & p_{4,3} & p_{4,4}
\end{pmatrix}
$$

Therefore, $P^2 = P \cdot P$ contains the two-step transition probabilities. That is, $P_{(i,j)}^2 = p_{i,j}^{(2)}$ gives the probability that if the system starts in state i, then two steps later it will be in state j. For the transition matrix of example 12.5, we get

$$
P^2 = P \cdot P =
\begin{pmatrix}
0.815 & 0.09 & 0.005 & 0.09 \\
0.09 & 0.815 & 0.09 & 0.005 \\
0.005 & 0.09 & 0.815 & 0.09 \\
0.09 & 0.005 & 0.09 & 0.815
\end{pmatrix}
$$

and can read off the value of $p_{1,4}^{(2)}$ (displayed in blue) that we computed above by considering all the possible paths one by one. By a similar argument it can be shown that in general, **the matrix P^k contains the k-step transition probabilities** $p_{i,j}^{(k)} = (P^k)_{i,j}$. Typically one uses computer software to compute these higher powers of the transition matrix.[8] Once the desired power of the matrix P is computed, it is easy to read off any multistep transition probability. For example, here is the third power of the transition matrix P for the Markov chain in example 12.5:

$$
P^3 = P^2 \cdot P =
\begin{pmatrix}
0.7425 & 0.122 & 0.0135 & 0.122 \\
0.122 & 0.7425 & 0.122 & 0.0135 \\
0.0135 & 0.122 & 0.7425 & 0.122 \\
0.122 & 0.0135 & 0.122 & 0.7425
\end{pmatrix}.
$$

We can read off the probability that a three-step path starting in cell 3 will end in cell 2, $p_{3,2}^{(3)} = 0.122$ (displayed in blue) from the matrix P^3. Note that from P to P^2 to P^3, the probabilities of returning to the same cell (diagonal elements) decrease, whereas the probabilities of moving to a different cell increase in every case. As path length becomes very large, we would expect that the elements of P would all converge to 0.25. That is, regardless of where a molecule starts, in the long run it will occupy all the cells with equal probability, the cells being indistinguishable from each other.

To summarize, in this section we have laid the foundation for our discussion of hidden Markov models and their use in multiple sequence alignments. The path probabilities we have computed to answer the two questions that were posed in this section will show up prominently in the discussion of the profile hidden Markov model of multiple sequence alignment.

Hidden Markov Models

We now have the necessary tools to discuss hidden Markov models (HMMs). HMMs are a variation of the concept of a Markov chain. Instead of a state corresponding

hidden Markov model (HMM)
A probabilistic model that assigns likelihoods to all possible combinations of gaps, matches, and mismatches to determine the most likely multiple sequence alignment (MSA) or set of possible MSAs.

---

[8] For example, in *Mathematica*, the function MatrixPower[P, n] will compute the nth power of the matrix P, whereas in Matlab, P^n will give the same result.

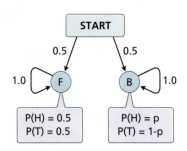

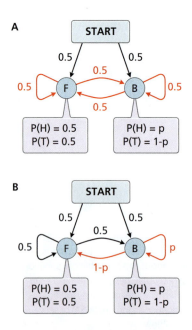

FIG. 12-6. State diagram of the HMM when one coin is initially selected at random and then tossed repeatedly. The F state indicates that the fair coin is flipped, whereas the B state indicates that the biased coin is flipped. The respective outcome probabilities are listed in the boxes that are attached to each state.

FIG. 12-7. A. State diagram of the HMM when the coin to be flipped is randomly selected for each toss. B. State diagram of the HMM when a second toss of the outcome coin determines whether to change coins or not.

to a unique outcome, each state of the *underlying Markov chain* corresponds to a distribution function over multiple outcomes. That is, each state can produce several outcomes, with probabilities specific to that state. HMMs allow us to model, for example, GC-rich regions of the genome where there is a higher probability of occurrence of a C or G nucleotide, compared to a normal region. They are also used for multiple sequence alignment.

Before we delve into the use of HMMs for multiple sequence alignment, let's use some simple thought experiments to familiarize ourselves with the structure of an HMM. We will again use the flipping of coins to model two random aspects: a *decision coin* for the movements from state to state and an *outcome coin* to decide the outcome for the state. We will also have a *start state* from which the transition to the first state is made.

Example 12.7. Let's say you are in a room with a barrier. Behind the barrier, a person is performing a coin-tossing experiment with two coins, a fair (F) one and a biased (B) one. All you are told are the outcomes of the individual tosses, say HHTHTTHH..., but not which coin was tossed. You are also told that the person initially flips the fair coin to decide which of two coins to use for the subsequent coin tosses that produce the sequence of heads and tails. (This is the experiment we discussed in example 11.14.) We can model this experiment as an HMM that has as its states the two coins, and associated with each state a probability distribution that gives the probabilities of heads and tails (in general, of each of the possible outcomes that the state produces). For the fair coin, the probability distribution is given by $P(H) = 0.5$ and $P(T) = 0.5$, whereas the probability distribution for the biased coin is given by $P(H) = p$ and $P(T) = 1 - p$ (with $p = 2/3$ in example 11.14). The decision coin for the initial state is the fair coin, which is why the probabilities from the start state to F and B equal 0.5. Since the chosen coin is used from then on, the probability to remain in the state is one. The HMM state diagram of this experiment is shown in Figure 12-6.

We now add a little more randomness to the experiment, allowing the coin tosser to switch between coins.

Example 12.8. In the first variation, the coin tosser initially flips a fair decision coin to select the coin for the first toss as before. However, instead of continuing to flip the initially selected coin, s/he randomly selects either the fair or the biased coin to be flipped for the next coin toss. We model this scenario by using a fair coin as the decision coin in each step. It randomly selects either the fair or the biased coin for the initial state and for each subsequent toss. The selected coin then becomes the outcome coin. Note that the Markov chain has the same states and the same probability distributions for each state as in example 12.7 because there is still a flip of either the fair or the biased coin to determine the outcome. What has changed are the transition probabilities from state to state. Instead of staying with probability one in the same state, there is now a 50-50 chance to transition to the other state. Figure 12-7A shows the HMM state diagram of this variation, with the changes from the HMM of example 12.7 shown in red.

In a second variation, after each outcome toss, the outcome coin is flipped again (that is, becomes the decision coin) to decide which coin is to be used next. If heads shows, the current coin will be used; if tails shows, then the person will switch to the other coin. The HMM state diagram of this version is shown in Figure 12-7B, with the changes from the first variation shown in red. Again, the states and their respective probability distributions have remained the same, but the transition probabilities between the states have changed.

Note that the HMM as described above has no designated end state, that is, the process can go on for any length of time. In addition, the probability distributions are the same in each step, which is a limitation if we want to model a process where there is a variation of the probabilities of the outcomes depending on time or position. For example, in a nucleotide or amino acid sequence, the likelihood of a particular outcome may change depending on where in the sequence we are. To model this more general process, we create distinct states for each outcome choice, such as the choice of nucleotide along a section of DNA. Each state may have individual outcome and transition probabilities. We will illustrate this generalization with a bioinformatics example, namely modeling a segment of a genome that could be either from a GC-rich region or from a normal region.

Example 12.9. Let's assume that in a GC-rich region of the genome, both C and G occur with probability 0.35, whereas A and T each occur with probability 0.15. To model this situation, we create two different types of states, one for GC-rich regions and one for normal regions. The GC-rich states have associated probability distribution $P(A) = P(T) = 0.15$, and $P(C) = P(G) = 0.35$, whereas the normal states have probability distribution $P(A) = P(C) = P(G) = P(T) = 0.25$. We assume that the regions are patchy—that is, it is more likely than not that successive nucleotides belong to the same type of region. Let's assume that the next nucleotide belongs to the same type of region with probability 0.75 and is from a different type of region with probability 0.25. Figure 12-8 shows the associated HMM, where we have denoted GC-rich states by R, and normal states by N. Unlike the models in examples 12.7 and 12.8, this model has individual N and R states corresponding to each position in the DNA sequence. We also explicitly added an end state to model a finite length DNA sequence.

Note that although the probability distributions are the same for both types of states in example 12.9, regardless of their position in the DNA sequence, the "stretched-out" form of the HMM allows us to model processes that have position-dependent probability distributions for each state, which is useful in multiple sequence alignment.

Now that we have this nice model, what can we do with it? Example 12.10 shows how we can use it to decide whether a particular observed sequence more likely arose from a normal or from a GC-rich region. To illustrate the approach, we will use the example of a short observed sequence to keep the computations to a manageable size.

Example 12.10. Let's suppose that we have observed the sequence TCGA and want to determine whether this DNA sequence belongs to a GC-rich region of the genome. That is, we want to compute the conditional probability

$$P\left(\text{GC-rich}|\text{TCGA}\right) = \frac{P\left(\text{TCGA} \cap \text{GC-rich}\right)}{P\left(\text{TCGA}\right)} \qquad (12.13)$$

We know that the observed nucleotide sequence results from a particular sequence of states of the underlying Markov chain, with each emitting (= producing) a specific

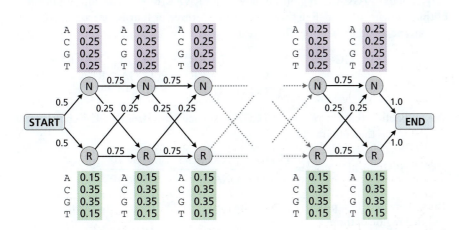

FIG. 12-8. State diagram of the HMM for a sequence that can switch between normal and GC-rich regions. The probability distributions associated with each state list the probabilities for A, C, G, and T nucleotides, respectively, from top to bottom.

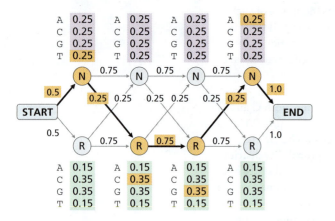

FIG. 12-9. State diagram of the HMM of a sequence that can switch between normal and GC-rich regions. The nodes and the transition probabilities corresponding to the path NRRN and the emission probabilities corresponding to the sequence TCGA are highlighted in yellow.

nucleotide. Therefore we need to take into account all the possible paths through the state diagram from the start state to the end state, the respective transition probabilities, and the respective *emission or outcome probabilities* of creating the observed nucleotides. From these probabilities we will be able to compute the probability that the observed sequence arises from a GC-rich region.

For each of the four nucleotide sites there are two possible states, so there are a total of $2^4 = 16$ possible paths (see equation 11.1) from the start state to the end state through the state diagram, namely NNNN, NNNR, NNRN, NRNN, RNNN, ..., RRRN, RRRR, where R denotes a GC-rich state and N a normal state. Let's consider the NRRN path. The probability that the NRRN path is followed and that the sequence TCGA is emitted is given by

$$P(\text{NRRN} \cap \text{TCGA}) = p_{\text{start,N}} \cdot P(\text{T}|\text{N}) \cdot p_{\text{N,R}} \cdot P(\text{C}|\text{R}) \cdot p_{\text{R,R}} \cdot P(\text{G}|\text{R}) \cdot p_{\text{R,N}} \cdot P(\text{A}|\text{N}) \cdot p_{\text{N,end}}$$
$$= (0.5) \cdot (0.25) \cdot (0.25) \cdot (0.35) \cdot (0.75) \cdot (0.35) \cdot (0.25) \cdot (0.25) \cdot 1$$
$$= 1.79 \times 10^{-4}$$

That is, the probability that both a particular path is taken and a specific sequence is created is the product of the state transition probabilities for the consecutive states in the path and the emission probabilities of the observed nucleotides, based on the respective states. We call such a probability a *sequence-path probability*. Figure 12-9 shows a state diagram in which the nodes and the associated transition probabilities of the path NRRN and the emission probabilities of the sequence TCGA are highlighted in yellow.

To compute the probabilities in equation 12.13, we must first perform calculations like the one for the path NRRN for each of the 16 possible paths. Figure 12-10 summarizes information for four of the 16 possible paths, and exercise 10 asks you to fill in the information on the 12 missing paths of Figure 12-10.

With computed values for all 16 sequence-path probabilities, we can obtain $P(\text{TCGA})$, the probability that the sequence TCGA would be observed at all, regardless of the region. This is simply the sum of the entries in a fully completed Figure 12-10 consisting of 16 sequence-path probabilities:

$$P(\text{TCGA}) = P(\text{TCGA} \cap \text{NNNN}) + P(\text{TCGA} \cap \text{NNNR}) + \cdots + P(\text{TCGA} \cap \text{RRRR})$$
$$= 3.45 \times 10^{-3}.$$

Next we need to compute $P(\text{TCGA} \cap \text{GC-rich})$. We first have to decide what it means for the sequence to come from a GC-rich region. Clearly the path RRRR corresponds to a GC-rich region, but how about RNRR, or RNNR? For the purposes of this problem, let us define that any path with three or four R's is from a GC-rich region, and all others are not. There are five such paths, giving us:

$$P(\text{TCGA} \cap \text{GC-rich}) = P(\text{TCGA} \cap \text{RRRR}) + P(\text{TCGA} \cap \text{NRRR}) + P(\text{TCGA} \cap \text{RNRR}) + P(\text{TCGA} \cap \text{RRNR}) + P(\text{TCGA} \cap \text{RRRN}) = 1.32 \times 10^{-3}.$$

The probability that the observed TCGA sequence arises from a GC-rich region of the genome is then obtained as

$$P(\text{GC-rich}|\text{TCGA}) = \frac{P(\text{TCGA} \cap \text{GC-rich})}{P(\text{TCGA})} = \frac{1.32 \times 10^{-3}}{3.45 \times 10^{-3}} = 0.38.$$

Thus, there is a 38% probability that the observed TCGA sequence comes from a GC-rich region of the genome.

We now discuss how HMMs are used in multiple sequence alignment of protein and DNA sequences (see Chapter 6). Multiple sequence alignment consists of identifying common underlying structures in proteins derived from DNA sequences that come from different species and perform the same function. Based on these underlying structures new sequences are identified that are likely to also perform this function. Krogh et. al. (1994) adapted an HMM used in speech recognition for this purpose. Their model, which is called the *profile HMM*, uses three types of states—namely, insert states, delete states, and match states—to represent the evolution of DNA sequences via the insertion, deletion, and substitution of nucleotides, respectively. Note that we can use HMMs for protein or DNA multiple sequence alignment, but for simplicity, we will use DNA sequence alignment in our example.

Consider, for example, an arbitrary four-nucleotide ancestral DNA sequence, say CTGA. A simple profile HMM for this sequence, shown in Figure 12-11, consists of four match states m_1, m_2, m_3, m_4, five insert states i_0, i_1, i_2, i_3, i_4, and four delete states d_1, d_2, d_3, d_4. In a profile HMM there is always one more insert state than there are match and delete states so that we can model a sequence for which the first nucleotides are not matched. Each insert and match state has a set of emission probabilities (not shown in Figure 12-11) that give the relative likelihoods that each of the four nucleic acids A, C, G, and T will appear at the site corresponding to that state. For example, we might have $P(A) = P(G) = 0.1$, $P(C) = 0.3$, and $P(T) = 0.5$. The delete state has a single outcome (with probability one), namely the deletion of a nucleotide site, which introduces a gap in the alignment.

The most straightforward way that this model would produce the CTGA sequence would be via the path $S \rightarrow m_1 \rightarrow m_2 \rightarrow m_3 \rightarrow m_4 \rightarrow E$, with emissions C, T, G, and A at the four successive match states. Over evolutionary time, we would expect that insertions, deletions, and changes in the nucleotides emitted at each site would result in changes to the sequence, modeled as different paths through the states of the HMM and different emissions at match and insert states. For example, deletion of the second site would be modeled by a path that passes through $m_1, d_2, m_3,$ and m_4. Subsequent addition of a nucleotide at the end of the sequence would be modeled by a path through $m_1, d_2, m_3, m_4,$ and i_4. In each case, which nucleotide is emitted at the match and insert states would be determined according to the emission probability distributions at the respective states.

The profile HMM can be used to perform multiple sequence alignments in the following fashion. First, a set of DNA sequences is identified that are known to

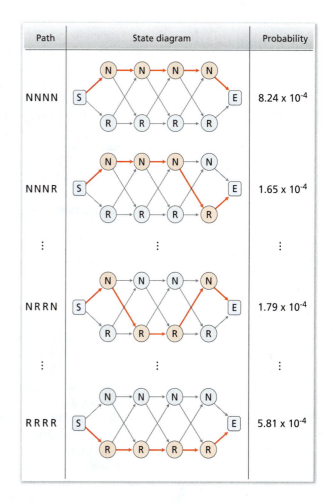

Path	State diagram	Probability
NNNN		8.24×10^{-4}
NNNR		1.65×10^{-4}
⋮	⋮	⋮
NRRN		1.79×10^{-4}
⋮	⋮	⋮
RRRR		5.81×10^{-4}

FIG. 12-10. Four possible paths through the state diagram with associated sequence-path probabilities. S denotes the start state and E denotes the end state. The respective path is marked in red.

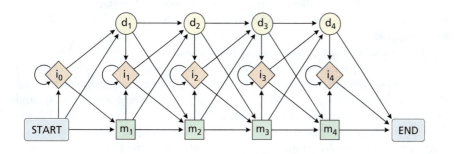

FIG. 12-11. Profile HMM model for multiple sequence alignments with four delete states, four match states, and five insert states. Adapted from Krogh et al. (1994).

perform a common function or produce a protein with a specific structure, and which are evolutionarily related. This is called the *training set*, comprised of *training sequences*. Second, the model size (= number of states) is chosen, plus preliminary values of the transition and emission probabilities. Third, these probabilities are optimized so as to maximize the chance that a random output from the model will be one of the training sequences. This process is called "training" the model. Finally, the trained model is used to align new sequences (sometimes called test sequences) that are not in the training set, with the goal of discovering whether the structure that underlies the training sequences is present in the new sequences. When aligning new sequences, calculations are made of how well each new sequence fits the trained model, that is, how likely it is that a random output of the trained model would be the test sequence. A good fit suggests that the sequence in question is evolutionarily related to the training sequences and performs the same function.

Example 12.11. Let's say we want to find *TP53* promoter regions in species for which they are not yet known. As a training set we use five sequences of known *TP53* promoters, namely human (532 bp), rat (527 bp), golden hamster (374 bp), pig (104 bp), and sheep (70 bp). It is customary to choose the model size N as the average sequence length of the training set, that is, $N = (532 + 527 + 374 + 104 + 70)/5 = 321.4 \approx 322$. So our model will have 322 match and delete states and 323 insert states.

The next step is to assign preliminary transition and emission probabilities. One way is to start with a blank slate, using equally likely transition and emission probabilities (similar to determining a prior probability in Bayesian inference). Because each state (except for the last match, delete, and insert states) has three outgoing arrows, each such arrow would be assigned a probability of 1/3. The last match, delete, and insert states each have two outgoing arrows, and each such arrow would have probability 1/2. All emission probabilities would equal 1/4.

Alternatively, instead of starting with a blank slate we may assign preliminary transmission and emission probabilities that exploit the known structure underlying the training sequences, to get a head start toward optimal values. For the five training sequences in example 12.11, alignment with the ClustalW algorithm results in a sequence alignment of which a segment is shown in Figure 12-12.

Looking at the individual columns of the multiple sequence alignment in Figure 12-12, we distinguish two types of columns. In the columns that are shaded, at least half of the sequences have a nucleotide at the particular site, whereas in the other columns, fewer than half of the species have a nucleotide. We associate the shaded columns with match or delete states, and the other three columns with a repeatedly visited insert state (each of the insert states in Figure 12-11 has a loop). Note that a hyphen in any of the shaded columns refers to a delete state (= gap), whereas a hyphen in the un-shaded columns is just a spacing tool. Also keep in mind that a match state does not require identical nucleotides for all the species.

Why does this assignment of states make sense? If we assume that all the sequences originate from a common ancestor, then for columns in the alignment where some species have nucleotides in the alignment and others do not, there are two explanations: either all the species without nucleotides had deletions, or all the species with nucleotides had insertions at the same position. If the majority of the species have nucleotides in the alignment, the most likely explanation is that the species without nucleotides in the alignment had deletions. Likewise, in a column where only a few species have a nucleotide, the most probable explanation is that those species had insertions.

Now we can assign preliminary transition and emission probabilities. In the alignment shown partially in Figure 12-12 there are 10 match states upstream of the segment shown. So the columns in the displayed segment correspond to match (and delete) states m_{11} through m_{18} and d_{11} through d_{18}, respectively, insert state i_{18}, and match (and delete) states m_{19} through m_{24} and d_{19} through d_{24}, respectively. We use the relative frequencies of nucleotides in the corresponding columns of the sequence alignment to determine the emission probabilities for the respective states.

$$m_{11} - m_{18} \qquad m_{19} - m_{24}$$
$$d_{11} - d_{18} \qquad i_{18} \quad d_{19} - d_{24}$$

Rat	CGCAG-TG---GCCCAC
Sheep	CGCAG-TG---GCC---
Pig	CGCAGGTA---GCT---
Human	TCCAGCTGAGAGCAAAC
Hamster	--CGGTTTCC-ACCAAT
	* * * *

FIG. 12-12. Segment of a multiple sequence alignment from ClustalW algorithm for *TP53* promoters in five species. The positions in the shaded areas correspond to match and delete states, and positions outside the shaded areas correspond to insertions. Note that a match state does not require that all species have the same nucleotide at that position. An asterisk (*)highlights columns in which the nucleotides are the same for all five species.

For transition probabilities between nucleotides at consecutive positions we count how many sequences have a specific transition, say from m_{23} to d_{24}, and then divide by the total number of transitions that leave state m_{23}.

Example 12.12. We will compute preliminary emission probabilities of match states m_{11}, m_{13}, and m_{24}, and insert state i_{18}, as well as transition probabilities from match state m_{21}, delete state d_{22}, and insert state i_{18}. We start with the emission probabilities of the match states.

- m_{11} (first column) has three Cs and one T, therefore $P(C|m_{11}) = 3/4$, $P(T|m_{11}) = 1/4$, and $P(A|m_{11}) = P(G|m_{11}) = 0$.
- m_{13} (third column) has only Cs, therefore $P(C|m_{13}) = 1$ and $P(A|m_{13}) = P(G|m_{13}) = P(T|m_{13}) = 0$.
- m_{24} (last column) has two Cs and one T, therefore $P(C|m_{24}) = 2/3$, $P(T|m_{24}) = 1/3$, and $P(A|m_{24}) = P(G|m_{24}) = 0$.

Now let's look at the probabilities of the insert state i_{18}. Repeated visits to the state produce a total of two As, two Cs, and one G. They all come from the same emission probability distribution, so we need to take all the entries in these three columns into account and obtain $P(A|i_{18}) = P(C|i_{18}) = 2/5$, $P(G|i_{18}) = 1/5$, and $P(T|i_{18}) = 0$.

Next we assign preliminary transition probabilities. Let's look at match state m_{21} (third column in second box). For three species (rat, human, and golden hamster) the next position is a match state, whereas for the other two species, a deletion takes place. No insertions take place. So we let $p_{m21,m22} = 3/5$, $p_{m21,d22} = 2/5$, and $p_{m21,i21} = 0$. Similarly, for d_{22}, we let $p_{d22,d23} = 1$ (because both delete states are followed by delete states), and $p_{d22,i22} = p_{d22,m23} = 0$. Finally, let's look at the transition probabilities from the insert state i_{18} to states i_{18}, m_{19}, and d_{19}, respectively. Of the five nucleotides in the insertion columns, three are followed by another insertion, (namely A followed by G, G followed by A, and C followed by C), so we let $p_{i18,i18} = 3/5$. The other two nucleotides are followed by a match state, as the gaps in the insertion columns are not deletions, but just a spacing tool to allow for the insertions, so $p_{i18,m19} = 2/5$, and $p_{i18,d19} = 0$.

Note that in example 12.12, some of the preliminary emission and transition probabilities are zero, which is not desirable. In the iterative techniques used to train the HMM, a zero probability does not allow for change to any other value. Knowing that our training sequences do not account for all the variations that could occur in related sequences, zero probabilities are to be avoided in the initial assignment of probabilities. So zero probabilities are replaced with small positive values and the nonzero probabilities are reduced accordingly. Because we are making only an initial assignment of probabilities at this point of the model derivation, we do not have to get the probabilities exactly right; we just need to have a reasonable assignment. The hope is that information resulting from the ClustalW alignment of the training sequences will either make the training phase faster or will result in a better model. This completes the initialization phase of the profile HMM.

The next phase is the training phase. During the training phase, the probabilities of the model are modified in such a way that the probabilities of obtaining training sequences as random output from the modified model are maximized. The first step is to compute sequence-path probabilities under the initial model. Again, we illustrate with the simple HMM shown in Figure 12-11.

Example 12.13. Suppose that we have initialized an HMM with $N = 4$, equally likely emission probabilities, and the following transition probabilities (where j ranges over all appropriate indices):

- $P(S \rightarrow m_1) = P(S \rightarrow i_0) = P(S \rightarrow d_1) = 1/3$.
- $P(m_j \rightarrow m_{j+1}) = 1/2$, $P(m_j \rightarrow d_{j+1}) = P(m_j \rightarrow i_j) = 1/4$.
- $P(i_j \rightarrow i_j) = 1/2$, $P(i_j \rightarrow d_{j+1}) = P(i_j \rightarrow m_{j+1}) = 1/4$.

- $P(d_j \to d_{j+1}) = P(d_j \to i_j) = 1/4, P(d_j \to m_{j+1}) = 1/2.$
- $P(i_4 \to E) = P(d_4 \to E) = P(m_4 \to E) = 1/2.$
- $P(d_4 \to i_4) = P(m_4 \to i_4) = 1/2.$

For ease of computations, let's look at a small segment, TGAGA, of the human training sequence. Table 12-2 shows three possible paths through the HMM that could have created the segment, together with the respective sequence-path probabilities. Among these, the first path is the most likely.

Notice that the sequence-path probabilities we computed in example 12.13 are conditional probabilities that a particular sequence is generated by the specified path given the current model (that is, current assignment of emission and transition probabilities and size of the HMM). To get the *sequence probability* that the sequence TGAGA will be generated by the model, denoted by $P(\text{TGAGA}|\text{model})$, we sum the TGAGA sequence-path probabilities over all possible paths (and there are many).

In general, if we let T_j denote the jth training sequence, then the sequence probability that the training sequence T_j is generated by the model is given by

$$P\big(T_j \mid \text{model}\big) = \sum_{\text{paths}} P\big((T_j \cap \text{path}) \mid \text{model}\big).$$

The goal of training is to maximize the overall probability that the set of training sequences $\{T_1, T_2, ..., T_k\}$ would be created by the model, which can be achieved by maximizing the product of the sequence probabilities over all training sequences. That is, we want to maximize

$$P\big(\{T_1, T_2, ..., T_k\} \mid \text{model}\big) = \prod_{j=1}^{k} P\big(T_j \mid \text{model}\big).$$

This overall probability is called the *likelihood of the model*.

Let's pause for a moment to assess where we are in the modeling process: we have a current model and a way to assess how well the current model produces the training sequences, namely $P(\{T_1, T_2, ..., T_k\}|\text{model})$. The next step is to improve the model, ideally to find the model that best fits the training sequences.

Due to the complexity of the computations and the large number of paths, there is no efficient way to directly calculate the best model. Instead, the Baum-Welch algorithm is used to find a "locally" best model from an initial model. That is, it starts from the initial assignments of transition and emission probabilities and then modifies them in gradual steps (iterations), each step chosen to improve the likelihood of the model. The closer the initial assignments are to ideal, the more likely an ideal result will be achieved. So it is generally a good idea to start with initial probabilities

TABLE 12-2. EXAMPLES OF POSSIBLE PATHS THROUGH THE HMM, THE OUTPUT THAT WOULD BE DISPLAYED IN A MULTIPLE SEQUENCE ALIGNMENT, AND THE ASSOCIATED SEQUENCE-PATH PROBABILITY

PATH	OUTPUT	SEQUENCE-PATH PROBABILITY
$i_0 i_0 m_1 d_2 m_3 m_4$	TGA–GA	$\left(\dfrac{1}{3} \cdot \dfrac{1}{2} \cdot \dfrac{1}{4} \cdot \dfrac{1}{4} \cdot \dfrac{1}{2} \cdot \dfrac{1}{2} \cdot \dfrac{1}{2}\right)\left(\dfrac{1}{4}\right)^5 = 1.27 \times 10^{-6}$
$m_1 i_1 m_2 i_2 d_3 m_4$	TGAG–A	$\left(\dfrac{1}{3} \cdot \dfrac{1}{4} \cdot \dfrac{1}{4} \cdot \dfrac{1}{4} \cdot \dfrac{1}{4} \cdot \dfrac{1}{2} \cdot \dfrac{1}{2}\right)\left(\dfrac{1}{4}\right)^5 = 3.18 \times 10^{-7}$
$m_1 m_2 d_3 d_4 i_4 i_4 i_4$	TG– –AGA	$\left(\dfrac{1}{3} \cdot \dfrac{1}{2} \cdot \dfrac{1}{4} \cdot \dfrac{1}{4} \cdot \dfrac{1}{4} \cdot \dfrac{1}{2} \cdot \dfrac{1}{2} \cdot \dfrac{1}{2}\right)\left(\dfrac{1}{4}\right)^5 = 3.18 \times 10^{-7}$

that reflect the known structure of the training sequences, rather than a blank slate. Here are the Baum-Welch algorithm steps:

1. Assign probabilities to the initial model.

2. Compute new transition and emission probabilities as follows:

   2.1. For all training sequences add up the sequence-path probabilities of all paths that include a given transition, say from i_3 to m_4. Let this be the preliminary new transition probability $p_{i3,m4}$. Once all the preliminary new transition probabilities have been computed, rescale the new transition probabilities leaving each state so that the sum of the probabilities leaving the state is one.[9] In this way, paths with a high probability of producing a training sequence are given greater weight, and consequently, transitions taken by those paths are given greater weight.

   2.2. Similarly add up all sequence-path probabilities that emit each particular nucleotide, say A, at a given state, say m_3, and let this be the new emission probability of $P(A|m_3)$. After all emission probabilities (here for A, C, G, and T) for that state have been assigned, rescale them to sum to one.

3. Replace the old emission and transition probabilities with the newly computed ones to obtain a new model.

4. Repeat steps 2 and 3 until the probabilities do not change more than some predetermined small threshold amount.

The bottleneck of the computation is step 2, but there are computational algorithms that can cut down the computational complexity. Typically, the iterative process takes fewer than 10 iterations, even for large models and large sets of training sequences.

Now that we have an optimized HMM, we can use it to perform a multiple sequence alignment of new sequences that are not in the training set. The first step in this alignment is to use the Viterbi algorithm to identify the most likely path through the model for each new sequence. This path is the one with the highest sequence-path probability based on the trained HMM. We will call this path and its associated sequence-path probability the *Viterbi path* and *Viterbi probability*, respectively. Let's look at an example of how the Viterbi paths are used to align sequences.

Example 12.14. Let's assume that we seek to align three new sequences GGCT, ACCGAT, and CT, to a model with $N = 4$ and that the Viterbi algorithm has been used to find the following Viterbi paths: for ACCGAT it is $i_0 \rightarrow m_1 \rightarrow d_2 \rightarrow m_3 \rightarrow i_3 \rightarrow i_3 \rightarrow m_4$; for GGCT it is $m_1 \rightarrow m_2 \rightarrow m_3 \rightarrow m_4$; and for CT it is $m_1 \rightarrow d_2 \rightarrow d_3 \rightarrow m_4$, where we have omitted the start and end states in each path. Then the multiple sequence alignment would be as follows, where gaps corresponding to insert states are labeled with dots rather than dashes to distinguish them from deletions:

i_0	m_1	m_2/d_2	m_3/d_3	i_3	i_3	m_4
A	C	–	C	G	A	T
.	G	G	C	.	.	T
.	C	–	–	.	.	T

Besides aligning the new sequences, we would like to quantify how well a new sequence fits the trained model, and hence how likely it is to share function and evolutionary history with the training sequences. To quantify goodness of fit (or more accurately, badness of fit) we define the <u>D</u>istance of the <u>S</u>equence to the <u>M</u>odel as follows:

$$\text{DSM (sequence)} = -\log(\text{Viterbi probability})$$
$$= -\log(P(\text{sequence} \cap \text{Viterbi path}|\text{model})).$$

---

[9] To make a sum of probabilities equal to 1, divide each of the probabilities by their sum.

This definition makes DSM a positive number that becomes smaller as the fit of the sequence to the model becomes better, in other words, the Viterbi probability becomes larger.

Example 12.15. For the sake of illustration, let's assume that Table 12-2 gives the complete list of sequence-path probabilities for the sequence TGAGA. The Viterbi algorithm would tell us that the Viterbi path for TGAGA is $i_0 \rightarrow i_0 \rightarrow m_1 \rightarrow d_2 \rightarrow m_3 \rightarrow m_4$, corresponding to the alignment TGA-GA. For this sequence and model we then have

$$\text{DSM}(\text{TGAGA}) = -\log(1.27 \times 10^{-6}) = 5.89$$

where log is the logarithm with respect to base 10.

Because the DSM is the negative logarithm of a sequence-path probability, which is itself a product of transition and emission probabilities, the DSM is also the sum of the negative logarithms of those probabilities because $\log(u \cdot v) = \log(u) + \log(v)$. So the DSM can be interpreted as a sum of penalties associated with each transmission and emission multiplier in the Viterbi probability. For example, we can rewrite the DSM for the sequence TGAGA as follows:

$\text{DSM}(\text{TGAGA})$

$= -\log(P(\text{TGAGA} \cap i_0 i_0 m_1 d_2 m_3 m_4 | \text{model}))$

$= -\log\left[\left(\dfrac{1}{3} \cdot \dfrac{1}{2} \cdot \dfrac{1}{4} \cdot \dfrac{1}{4} \cdot \dfrac{1}{2} \cdot \dfrac{1}{2} \cdot \dfrac{1}{2}\right) \cdot \left(\dfrac{1}{4}\right)^5\right]$

$= -\log\left(\dfrac{1}{3}\right) - \log\left(\dfrac{1}{2}\right) - \log\left(\dfrac{1}{4}\right) - \log\left(\dfrac{1}{4}\right) - \log\left(\dfrac{1}{2}\right) - \log\left(\dfrac{1}{2}\right) - \log\left(\dfrac{1}{2}\right) - \log\left(\dfrac{1}{4}\right)^5$

$= -\log\left(p_{\text{start},i_0}\right) - \log\left(p_{i_0,i_0}\right) - \log\left(\dfrac{1}{4}\right) - \log\left(p_{m_1,d_2}\right) - 3\log\left(\dfrac{1}{2}\right) - \log\left(\dfrac{1}{4}\right)^5$

In this example, $-\log\left(p_{\text{start},i_0}\right)$ corresponds to an insertion-initiation penalty, $-\log\left(p_{i_0,i_0}\right)$ corresponds to an insertion-extension penalty, and $-\log(p_{m_1,d_2})$ corresponds to a gap-initiation penalty, each at the respective position in the alignment. This shows that the HMM allows for these penalties to be position-dependent, unlike the fixed gap penalties in other alignment algorithms. In addition, penalty values are "learned" from the underlying structure of the training sequences rather than simply assigned. These features make the profile HMM a flexible tool that can be used not only for DNA sequence alignment, but also for protein alignment and for finding motifs.

We will now shift gears and look at an application of a continuous-time stochastic process, namely the derivation of the Jukes-Cantor model, which is used to estimate evolutionary distance from genomic data.

Poisson Process and Jukes-Cantor Model

Jukes-Cantor model A model of evolutionary distance that predicts the actual number of nucleotide (amino acid residue) replacements based on the observed fraction of sites that differ between two gene sequences.

In Chapter 11, we discussed the discrete Poisson random variable, which was derived as an approximation of the binomial random variable when the number of trials n is large and the success probability p is small. The corresponding Poisson distribution has parameter $\beta = n \cdot p$, and the expected value of the Poisson random variable is equal to β. Our goal is to model substitutions in the gene sequence, which can occur at any time, not just at discrete time steps. Therefore we will use a continuous-time stochastic version of the Poisson random variable, the *Poisson process*, to model this random process.

Suppose that starting from a point $t = 0$, we count the number of events that occur by time t, which we denote by $N(t)$. For each value of t, $N(t)$ is a random quantity with possible values $\{0,1,2,\ldots\}$. If this process satisfies three natural assumptions then $N(t)$ can be shown to have a Poisson distribution for every value of t. Here are the assumptions:

1. For any two time intervals Δ_1 and Δ_2 of equal lengths, the probability that k events occur in time interval Δ_1 equals the probability that k events occur in time interval Δ_2, that is, time intervals of equal lengths have identical event probabilities.

2. For any two time intervals that do not overlap, event probabilities are independent.

3. The probability that more than one event occurs in a very small time interval is essentially zero, that is, in a very small time interval either one event or no event occurs. In plain English, this means that no two events occur at exactly the same time.

If these assumptions are satisfied then the continuous-time process $N(t)$ is a Poisson process with parameter βt, where β equals the rate of occurrences per unit time interval or equivalently, the average number of occurrences in a unit time interval. That is, the equivalent of equation 11.16 for the Poisson process is given by

$$P\big(N(t) = k\big) = \frac{(\beta t)^k}{k!} e^{-\beta t}. \tag{12.14}$$

Note that what we observe is a discrete random variable (a count) but we observe it continuously. Figure 12-13 shows a particular example of a Poisson process $N(t)$ with $\beta = 1/5$, that is, on average 1/5 events occur in one time unit, which is equivalent to one occurrence (on average) every five time units.

We will apply this process to derive the Jukes-Cantor model for evolutionary distance (see Chapter 8). Before we do so, we will look at an example of a Poisson process.

Example 12.16. Suppose that on a summer evening, shooting stars are observed at a Poisson rate of one every 12 minutes. What is the probability that exactly three shooting stars are observed in a particular 30-minute interval? We need to determine the value of β. We make the obvious choice for time unit, namely one minute.

Then, because there is one shooting star every 12 minutes, there are 1/12 shooting stars per time unit, so $\beta = 1/12$, and the number of shooting stars in a time interval of length t is a Poisson random variable with parameter $(1/12)t$. Using equation 12.14 with $\beta t = (1/12) \cdot 30 = 2.5$, we have

$$P\big(N(30) = 3 \big) = \frac{2.5^3}{3!} e^{-2.5} = 0.21 \,,$$

that is, the probability of seeing exactly three shooting stars in a 30-minute interval is about 21%.

We now derive the Jukes-Cantor (JC) model. When we compare two sequences that come from a common ancestor, we are interested in the *evolutionary distance d*, which is the estimated number of mutations or substitutions that have taken place per site of the ancestor genome. The simplest model of sequence evolution, the Jukes-Cantor model, assumes that substitutions at each site are random and that each of these substitutions (A to C, A to G, A to T, C to A, and so on) happens at the same rate, denoted by α, which is constant over time. Substitutions are rare events, so we use a Poisson process for modeling the number of substitutions, as the three assumptions (equal probabilities for number of occurrences in equal time intervals, independence of occurrences in non-overlapping time intervals, and non-simultaneous occurrences) for a Poisson process are believed to be satisfied. Because the rate of each substitution is α, and

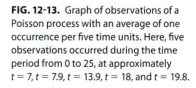

FIG. 12-13. Graph of observations of a Poisson process with an average of one occurrence per five time units. Here, five observations occurred during the time period from 0 to 25, at approximately $t = 7$, $t = 7.9$, $t = 13.9$, $t = 18$, and $t = 19.8$.

each nucleotide can change into any of the three other nucleotides, the total rate of change is 3α. For organisms that have split off from their common ancestor an amount of time t in the past, the overall time for changes is $2t$, as each branch independently changes over time period t. Thus, the JC distance (= expected number of substitutions per site) in such a sequence is

$$d = 3\alpha \cdot 2t = 6\alpha t \qquad (12.15)$$

This is a very nice and simple formula, but the problem is that we can observe neither α nor t. What we can observe is the fraction D of sites that differ between the two descendant sequences. If nucleotide changes were never reversed (for example A changing to G and then back to A), then we would have $d = D$. Because changes do get reversed, we know that the JC distance d is generally greater than D. So is there a formula for the JC distance d in terms of D? Indeed there is. We will derive an equation that connects the two quantities and that does not depend on α or t.

To do so, we model the nucleotide value at any particular site with a *Markov process*, the continuous-time version of a Markov chain, having four states unsurprisingly named A, C, G, and T. We will keep track of two processes: the continuous-time Markov process $X(t)$ that denotes the nucleotide value at a given site at time t, and the Poisson process $N(t)$ that counts the number of changes at this site that have occurred by time t. Figure 12-14 shows the state diagram of the Markov process.

The arrows between states are labeled $p_{AA}(t), p_{AC}(t), p_{AG}(t), \ldots, p_{TT}(t)$, where $p_{ij}(t) = P(X(t) = j | X(0) = i)$ is the probability that if the site has an i nucleotide at time 0, then it will have a j nucleotide at time t. We have labeled only one set of arrows, namely the ones that are leaving state A, which account for the case when the nucleotide at time 0 was an A. The arrows for the other states represent the corresponding probabilities. Because the substitution rates from one nucleotide to another are the same, we only have two distinct transition probabilities, namely those where the same nucleotide is present at time 0 and at time t, and those where there has been a change between time 0 and time t:

$$p_{AA}(t) = p_{CC}(t) = p_{GG}(t) = p_{TT}(t),$$
$$p_{AC}(t) = p_{AG}(t) = p_{AT}(t) = p_{CA}(t) = p_{CG}(t) = \ldots = p_{TC}(t) = p_{TG}(t). \qquad (12.16)$$

In addition, because the probabilities on arrows leaving any node, including the arrow that loops back, must sum to 1, we know that for node A,

$$p_{AA}(t) + p_{AC}(t) + p_{AG}(t) + p_{AT}(t) = 1, \qquad (12.17)$$

and likewise for the other nodes. Combining equations 12.16 and 12.17, we find that if we know $p_{AA}(t)$, then we can compute all of the transition probabilities—they are either equal to $p_{AA}(t)$ or to $(1 - p_{AA}(t))/3$. To find a formula for $p_{AA}(t)$, we set up an equation that connects the state of the Markov process at time t with the state of the Markov process a short time later, at time $t + \Delta t$.[10]

Recall how we computed two-step transition probabilities for the molecule moving among a circle of cells in example 12.5 by accounting for all the intermediate places the molecule could have been at time 1. We will use the same idea here. To determine the transition probability $p_{AA}(t + \Delta t)$ at time $t + \Delta t$, we check the nucleotide at our specified site at time t. That is, we write the transition probability $p_{AA}(t + \Delta t)$ as a two-step transition probability, accounting for all the possible paths, starting in state A at time 0 and ending in state A at time $t + \Delta t$, with an intermediate state at time t:

$$p_{AA}(t + \Delta t) = p_{AA}(t) \cdot p_{AA}(\Delta t) + p_{AC}(t) \cdot p_{CA}(\Delta t) + p_{AG}(t) \cdot p_{GA}(\Delta t)$$
$$+ p_{AT}(t) \cdot p_{TA}(\Delta t). \qquad (12.18)$$

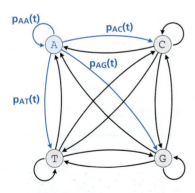

FIG. 12-14. State diagram of the Markov process $X(t)$ for the Jukes-Cantor model, with arrows of the form $p_{Aj}(t)$ for j = A, C, G, T indicated and labeled in blue.

---

[10] Students familiar with advanced mathematics will recognize that in what follows we develop and solve a differential equation for $p_{AA}(t)$.

Now we look at the terms in this equation. Because at most one substitution occurs in a small time period of length Δt (the number of substitutions follows a Poisson process), $p_{CA}(\Delta t)$ is the probability that there is exactly one substitution from C to A during time interval Δt. Because there are three equally probable substitutions (from C to A, C to G, and C to T), all included in the count by Poisson process $N(t)$, we have that

$$p_{CA}(\Delta t) = \frac{1}{3}P\big(N(\Delta t) = 1\big).$$

We can compute this probability using equation 12.14, where $N(t)$ has parameter $3\alpha\,\Delta t$ in light of the fact that substitutions occur at a total rate of 3α:

$$p_{CA}(\Delta t) = \frac{1}{3}P\big(N(\Delta t) = 1\big) = \frac{1}{3}\cdot\frac{(3\alpha\Delta t)}{1!}\exp(-3\alpha\Delta t) = \frac{1}{3}(3\alpha\Delta t)\exp(-3\alpha\Delta t).$$

Using Taylor expansion,[11] it can be shown that we can approximate the function $x\exp(-x)$ by x when the value of x is small, so we can estimate the quantity $\frac{1}{3}(3\alpha\Delta t)$ $\exp(-3\alpha\Delta t)$ by $\frac{1}{3}(3\alpha\Delta t) = \alpha\Delta t$. Using equations 12.16 and 12.17, we have that

$$p_{CA}(\Delta t) = p_{AC}(\Delta t) = p_{GA}(\Delta t) = p_{AG}(\Delta t) = p_{TA}(\Delta t) = p_{AT}(\Delta t) = \alpha\Delta t$$

and

$$p_{AA}(\Delta t) = 1 - \big(p_{AC}(\Delta t) + p_{AG}(\Delta t) + p_{AT}(\Delta t)\big) = 1 - 3\alpha\Delta t.$$

Substituting these two equations into equation 12.18 gives

$$\begin{aligned}
p_{AA}(t+\Delta t) &= p_{AA}(t)\cdot\big((1 - 3\alpha\Delta) + p_{AC}(t) + p_{AG}(t) + p_{AT}(t)\big)\cdot(\alpha\Delta t) \\
&= p_{AA}(t)\cdot(1 - 3\alpha\Delta t) + \big(1 - p_{AA}(t)\big)\cdot(\alpha\Delta) \\
&= p_{AA}(t)\cdot(1 - 4\alpha\Delta) + \alpha\Delta t \\
&= p_{AA}(t) + \Delta t\big(\alpha - 4\alpha\cdot p_{AA}(t)\big).
\end{aligned}$$

Subtracting $p_{AA}(t)$ from both sides and then dividing the equation by Δt gives that the average rate of change over the interval Δt is

$$\frac{p_{AA}(t+\Delta t) - p_{AA}(t)}{\Delta t} = \alpha - 4\alpha\cdot p_{AA}(t).$$

Because the average rate of change does not depend on the time interval Δt, the instantaneous rate of change $p'_{AA}(t)$ is the same as the average rate of change, so we have

$$p'_{AA}(t) = \alpha - 4\alpha\cdot p_{AA}(t).$$

This equation is a linear differential equation [12] for $p_{AA}(t)$ and any book on the subject lists a solution for this type of differential equation. Alternatively, a computer algebra system such as *Maple* or *Mathematica* can be used to obtain a solution. Using the initial condition $p_{AA}(0) = 1$ gives

$$p_{AA}(t) = \frac{3}{4}\exp(-4\alpha\cdot t) + \frac{1}{4}$$

---

[11] For any function $f(x)$, we have that $f(x) \approx f(0) + f'(0)\cdot x$ when x is close to zero. In this case, the function is $f(x) = xe^{-x}$ which has derivative $f'(x) = e^{-x}(1 - x)$.

[12] A differential equation contains both an unknown function and its derivative(s).

and therefore,

$$D = P(\text{nucleotide at the given site has changed in time period } 2t)$$

$$= 1 - p_{AA}(2t) = \frac{3}{4} - \frac{3}{4}\exp(-4\alpha \cdot 2t) = \frac{3}{4} - \frac{3}{4}\exp(-8\alpha \cdot t) \qquad (12.19)$$

We now have managed to express both the evolutionary distance d and the fraction of substitutions D in terms of $\alpha \cdot t$, so we solve for $\alpha \cdot t$ in equation 12.15 ($\alpha \cdot t = d/6$) and substitute the result into equation 12.19 to obtain

$$D = \frac{3}{4} - \frac{3}{4}\exp\left(-8 \cdot \frac{d}{6}\right) = \frac{3}{4} - \frac{3}{4}\exp\left(-\frac{4}{3}d\right)$$

Solving this equation for d using the natural logarithm function gives the desired result, an equation that computes the evolutionary distance d from the observed substitution frequency D:

$$d = -\frac{3}{4}\ln\left(1 - \frac{4}{3}D\right).$$

Using a continuous-time stochastic process and tools from basic probability has allowed us to deduce a quantity that cannot be observed from one that can be observed.

SUMMARY

We started the chapter with the definition of the extreme value distribution, which is used to compute the E-value that is reported in a BLAST search and is used by the bioinformaticist to decide whether an alignment is significant or not. Alternative measures are the z-score (now rarely used) and the P-value, which gives the probability of obtaining a score as high or higher than the BLAST score by chance.

Next we studied stochastic processes, that is, successive observations of random quantities. An important example of a discrete-time stochastic process is the Markov chain, which plays a prominent role in the hidden Markov model used for multiple sequence alignment algorithms. We investigated how the profile HMM is set up and how known sequences are used to train the model for use with new sequences.

Finally, we used a continuous-time Markov process and the stochastic version of the Poisson random variable to derive the equation for determining the evolutionary distance between two sequences from the fraction of sites that differ between them, based on the Jukes-Cantor model.

EXERCISES

Extreme Value Distribution and Significance of Alignment

1. **Show the details of the derivation of equation 12.5 from equation 12.2.**

2. **Compute the standard deviation σ of the extreme value distribution of example 12.4. Use your result to express how many standard deviations the similarity score $s = 77$ is above the mean $\mu = 65$.**

3. **The full report of the BLAST sequence alignment of human p53 to the UniProtKP/Swiss-Prot database reports the following information for the second highest scoring alignment: $s = 61$, $s' = 28.1$, and $E = 1.5$.**

   a. Use the parameter values given in example 12.4 to verify that the similarity score $s = 61$ has a bit score of 28.1 (rounded to one decimal place).

   b. Use the parameter values given in example 12.4 to verify the E-value for the second highest scoring alignment in two different ways: by using the similarity score $s = 61$ with equation 12.10 and by using the bit score $s' = 28.1$ with equation 12.12.

   c. How big a search space would you have to scan to see on average one HSP with similarity score 61?

   d. Now assume that you did not know the values of K and λ used for this search. Use equation 12.11 and the pairs of similarity and bit scores of the two highest scoring alignments to solve for K and λ. (Use the information from Figure 12-4 for the highest scoring alignment.) How do the values you computed compare to the values $K = 0.0410$ and $\lambda = 0.267$ reported by BLAST?

4. Equation 12.10 is also used to translate the (maximal) cut-off E-score, set by the system or entered by the user, into the associated (minimal) alignment cutoff score A. The derived alignment cutoff score A is then used to display only HSPs with similarity scores greater than A (see Figure 6-1).

  a. Solve equation (12.10) for s in terms of E_s.

  b. Using the parameter values of m, n, and K of example 12.4, compute the alignment cutoff scores A for these E-values: $E = 1$, $E = 0.1$, and $E = 0.01$.

5. Explain whether a small or a large bit score indicates significance of alignment.

Markov Chains and Hidden Markov Models

6. Assume that five cells are arranged in a line. A molecule will remain in the cell it currently occupies with probability 0.9, and will move to one of the two adjacent cells with equal probability. If it is at the end of the chain of cells, it moves with probability 0.1 to the single adjacent cell. State the transition matrix and draw the associated state diagram.

7. An $n \times n$ matrix with non-negative entries is called *stochastic* if the entries in each row sum to 1. The matrix is called *doubly stochastic* if each row and each column sums to 1.

  a. For each of the following matrices, determine whether it is stochastic or doubly stochastic.

$$P_1 = \begin{pmatrix} 0 & 1 & 0 \\ 1 & 0 & 0 \\ 0 & 0 & 1 \end{pmatrix} \qquad P_2 = \begin{pmatrix} 1/3 & 1/3 & 1/3 \\ 1/3 & 2/3 & 0 \\ 1/3 & 0 & 2/3 \end{pmatrix}$$

$$P_3 = \begin{pmatrix} 1/3 & 1/3 & 0 & 1/3 \\ 1/4 & 1/4 & 1/4 & 1/4 \\ 0 & 0 & 1 & 0 \\ 1/2 & 0 & 0 & 1/2 \end{pmatrix} \qquad P_4 = \begin{pmatrix} 1/2 & 0 & 0 & 0 & 1/2 \\ 1/3 & 0 & 1/3 & 1/3 & 0 \\ 0 & 1/3 & 0 & 1/3 & 1/3 \\ 0 & 0 & 0 & 1 & 0 \\ 1 & 0 & 0 & 0 & 0 \end{pmatrix}$$

  b. Draw the state diagrams of the Markov chains corresponding to these transition matrices.

  c. A matrix is called *regular* if the associated Markov chain is such that every state can be reached from every other state (not necessarily in one step), which means that there is at least one power of the matrix P that has only non-zero entries. Determine which of the matrices in part (a) is regular.

8. For the transition matrix of the Markov chain in example 12.5, we computed P^3 and obtained that $p_{1,4}^{(3)} = 0.122$. Verify this result by listing the seven paths of length three from state 1 to state 4 and computing their respective probabilities. The sum of these probabilities should equal 0.122.

9. Suppose you have a DNA sequence with heterogeneous base composition, modeled as a Markov chain with two states. State 1 generates an AT-rich sequence, whereas state 2 generates a GC-rich sequence. The DNA sequence stays in the same state with probability 0.99, and changes state with probability 0.01.

  a. Draw the state diagram.

  b. State the transition matrix.

  c. What is the probability of seeing the sequence $1 \rightarrow 1 \rightarrow 1 \rightarrow 1 \rightarrow 1 \rightarrow 2 \rightarrow 2 \rightarrow 2 \rightarrow 2 \rightarrow 1 \rightarrow 1$ if the chain starts in state 1?

  d. If the sequence is initially in state 1 with probability 0.8, that is, $P(X_0 = 1) = 0.8$, what is the probability that $X_7 = 1$? (Hint: Use the law of total probability to account for the initial state.)

10. In example 12.10, we computed some sequence-path probabilities of the HMM for observing the sequence TCGA, which may have arisen from the normal region, the GC-rich region, or a mixture of the two regions. There are 16 possible paths for the HMM, four of which were displayed in Figure 12-10. For the remaining 12 paths, list the path, the state diagram, and the associated sequence-path probability.

11. Use the segment given in Figure 12-12 to compute

  a. the emission probabilities for match states m_{12}, m_{16}, m_{18}, and m_{23};

  b. the transition probabilities from match states m_{12} (to states i_{12}, m_{13} and d_{13}) and m_{15} (to states i_{15}, m_{16} and d_{16}), respectively.

12. For the profile HMM and the set of p53 training sequences:

  a. Find three possible paths through the HMM for the sequence TGAGA that are different from the ones listed in Table 12-2. Try to find at least one path that has a higher sequence-path probability than the paths listed in Table 12-2.

  b. For each of the paths you found in part (a), compute the respective sequence-path probability and show the output of the multiple sequence alignment based on these paths.

  c. Compute the sum of the penalties (negative logs of the emission and transition probabilities) for the path with the highest sequence-path probability among the paths you found in part (a).

ANSWERS TO THOUGHT QUESTIONS

12-1. The entries on the diagonal, p_{11} through p_{44}, remain the same. All the other entries in the matrix equal

$0.1/3 = 1/30$, as the molecule can now move to three neighbors with equal probability.

REFERENCES

Altschul, S. F., W. Gish, W. Miller, W. E. Myers, and D. J. Lipman. 1990. "Basic Local Alignment Search Tool." *Journal of Molecular Biology* 215: 403–410.

Baldi, P., and S. Brunak. 2001. *Bioinformatics: The Machine Learning Approach*, 2nd ed. Cambridge, MA : MIT Press.

Clote, P., and R. Backofen. 2000. *Computational Molecular Biology—An Introduction*. New York: John Wiley & Sons Ltd.

DeGroot, M., and M. J. Schervish. 2011. *Probability and Statistics*, 4th ed. Boston, MA: Addison Wesley.

Deonier, R. C., S. Tavaré, and M. S. Waterman. 2004. *Computational Genome Analysis*. New York: Springer.

Eddy, S. R. 1996. "Hidden Markov Models." *Current Opinions in Structural Biology* 6: 361–365.

Eddy, S. R. 1997, November 14. "Maximum Likelihood Fitting of Extreme Value Distributions." http://selab.janelia.org/publications/Eddy97b/Eddy97b-techreport.pdf.

Ghahramani, S. 2004. *Fundamentals of Probability, with Stochastic Processes*, 3rd ed. Upper Saddle River, NJ: Prentice Hall.

Gibson, G., and S. V. Muse. 2009. *A Primer of Genome Science*, 3rd ed. Sunderland, MA: Sinauer Associates, Inc.

Gumbel, E. J. 1958. *Statistics of Extremes*. New York: Columbia University Press.

Karlin, S., and S. F. Altschul. 1990. "Methods for Assessing the Statistical Significance of Molecular Sequence Features by Using General Scoring Schemes." *Proccedings of the National Academy of Sciences USA* 87: 2264–2268.

Karlin, S., A. Dembo, and T. Kawabata. 1990. "Statistical Composition of High-Scoring Segments from Molecular Sequences." *Annals of Statistics* 18: 571–581.

Krogh, A., M. Brown, I. S. Mian, K. Sjölander, and D. Haussler. 1994. "Hidden Markov Models in Computational Biology—Applications to Protein Modeling." *Journal Molecular Biology* 235: 1501–1531.

Lesk, A. M. 2008. *Introduction to Bioinformatics*, 3rd ed. New York: Oxford University Press.

Mount, D. W. 2001. *Bioinformatics—Sequence and Genome Analysis*. Cold Spring Harbor, NY: Cold Spring Harbor Laboratory Press.

Nagle, R. K., E. B. Saff, and A. D. Snider. 2012. *Fundamentals of Differential Equations*, 8th ed. Boston, MA: Pearson.

National Center for Biotechnology Information. 2012. "The Statistics of Sequence Similarity Scores." Accessed November 23. http://www.ncbi.nlm.nih.gov/blast/tutorial/Altschul-1.html.

Prescott, P., and A. T. Walden. 1980. "Maximum Likelihood Estimation of General Extreme Value Distribution." *Biometrica* 67, 3: 723–724.

Rabiner, L. R., and B. H. Juang, 1986. "An Introduction to Hidden Markov Models." *IEEE ASSP Magazine* 3, 1: 4-16.

Smith, T. F., and M. S. Waterman. 1981. "Identification of Common Molecular Subsequences." *Journal Molecular Biology* 147: 195–197.

Thode, H. C., Jr. 2002. *Testing for Normality*. Boca Raton, FL: CRC Press.

Tramontano, A. 2007. *Introduction to Bioinformatics*. Boca Raton, FL: Chapman and Hall/CRC Press.

Waterman, M. S., T. F. Smith, and W. A. Beyer. 1976. "Some Biological Sequence Metrics." *Advances in Mathematics* 20: 367–387.

Waterman, M. S. and T. F. Smith. 1978. "On the Similarity of Dendrograms." *Journal Theoretical.Biology* 73:789–800.

Weisstein, E. W. 2012. "Extreme Value Distribution." *MathWorld—A Wolfram Web Resource*. Accessed November 12. http://mathworld.wolfram.com/ExtremeValueDistribution.html.

Wilkinson, D. J. 2007. "Bayesian Methods in Bioinformatics and Computational Systems Biology." *Briefings in Bioinformatics* 8, no. 2: 109–116.

PROGRAMMING BASICS AND APPLICATIONS TO BIOINFORMATICS

13.1   INTRODUCTION

In any field of study, it is important to develop a solid understanding of the basic fundamental concepts. Because bioinformatics is a cross-disciplinary field encompassing biology, mathematics, and computer science, it is necessary to understand the fundamental concepts in each of these fields. Chapter 1 provided a review of fundamentals of biology, and Chapters 11 and 12 provided an introduction to probability and its applications, both focusing on topics most relevant to the field of bioinformatics. In this chapter, we cover the fundamentals of programming using the Python programming language, a popular language for bioinformatics, and in Chapter 14, we cover the basics of designing a relatively simple bioinformatics tool: pairwise sequence alignment. With a basic understanding of biology, probability, and computer science, biologists become better users of bioinformatics tools and computer scientists become better developers of bioinformatics tools. Furthermore, with a shared understanding of these basics, biologists and computer scientists can effectively work together to develop and enhance bioinformatics tools to keep pace with and advance scientific discoveries in molecular biology and genomics.

If you have never programmed before, you might question the value of learning how to program if there are already existing bioinformatics tools available. An apt analogy is, why bother to learn to cook if there are restaurants available in the area? But as you know, restaurants may not serve the food you want or may be too expensive or take too much time. Likewise, if you have a specific problem that you need to solve, existing bioinformatics tools may not provide the desired functionality or perhaps they are too costly or take too long to perform the desired task. The need for new bioinformatics tools may be obvious to many, especially as the field continues to evolve. Less obvious though, is the need to learn the programming fundamentals given the many existing code modules readily available online. Again let's consider the cooking analogy. Although it may be faster to prepare a meal from pre-prepared foods such as frozen or boxed dinners, a true home-cooked meal often tastes better, is cheaper, and is more nutritious. Also, the number and variety of recipes that can be made using basic ingredients is essentially endless, whereas this is not the case when cooking with pre-prepared foods. The same is true for programming: once you have learned the basics, you can develop programs tailored to solve any number and variety of problems.

When learning to cook, it is a good idea to start with some existing recipes in order to learn the fundamentals of cooking. Likewise, when learning to program it is valuable to study existing programs (or algorithms, as we will explain shortly). However, if all you can do is follow a recipe, are you a chef? Chefs can alter existing recipes to create or enhance certain flavors and even create their own unique recipes. Likewise, expert programmers need to be able to alter existing programs to enhance the functionality or performance of an existing tool and to develop their own programs in order to solve new problems. How do programmers learn to do that? Just as beginning chefs need to start experimenting with their own recipes (even for simple dishes), beginning programmers need to start by writing their own programs. So as a beginning programmer, when trying to solve a problem, avoid the temptation to look for an existing solution online. Simply modifying existing solutions will not help you learn how to design programs; you need to develop your own solution to the problem. As you become an expert programmer, on the other hand, even though you can write all of the code on your own, readily available modules (found in software libraries) can help you more quickly create complex tools.

Up until this point we have been talking about programming using a specific programming language. However, as discussed in more detail later in this chapter, program design starts with having a good understanding of the problem you are trying to solve, followed by developing an *algorithm* to solve the problem. When writing a paper you start with a thorough understanding of the topic and then you develop an outline that helps you structure your thoughts without worrying about the proper syntax and grammar of sentences and paragraphs. Likewise, when developing a program you start with a thorough understanding of the problem you are trying to solve followed by developing an algorithm that is programming language independent. Algorithms can be used to describe how to solve the problem without the need to worry about the syntax and grammar of a specific programming language. Chapter 14 presents the various types of algorithms in more detail.

Although it may seem more logical to present algorithms before delving into the Python programming language, in this textbook we have decided to present programming basics using the Python language in this chapter and focus on algorithms

algorithm A set of steps to accomplish a task.

program A set of instructions that tell a computer what to do in order to perform a task.

syntax A set of rules that describe how to organize symbols to create correctly structured code as defined by a specific programming language. The code must have proper syntax in order for it to be correctly interpreted or compiled.

and how they can be used to develop a pairwise sequence alignment tool in Chapter 14. This ordering allows those who already know how to program in Python or any other programming language to skip this chapter if they so desire. On the other hand, this chapter is useful for those who have little or no programming experience or for those who already know how to program but want to learn Python. Note that it is not possible to teach the entire Python programming language within one chapter. Rather, the goal of this chapter is to provide you with enough programming knowledge and experience for you to be able to use Python to develop the pairwise sequence alignment bioinformatics tool in Chapter 14.

13.2 DEVELOPERS AND USERS WORK TOGETHER TO MAKE NEW DISCOVERIES

An early example of a biological problem that was solved using computer programs is the structure of membrane proteins. Structures of non-membrane proteins are relatively easy to elucidate in a wet laboratory through protein crystallization and X-ray diffraction. The difficulty in determining the structure of membrane proteins is that the protein segments associated with the membrane often fail to crystallize. As discussed in Chapter 5, Kyte and Doolittle created a simple software program that uses amino acid sequence information to predict the segments that associate with the membrane (see Box 13-1). The bacteriorhodopsin protein was used as an example in Kyte and Doolittle's classic paper that described their software program. A few years after publication of their program, the structure of bacteriorhodopsin was determined experimentally. Incredibly, the predicted and experimental structures matched. This spurred the development of more refined membrane protein structure prediction programs. Bioinformaticists routinely use such programs to make accurate

sequence (in computer science)
An ordered collection of objects or elements. There are three types of sequences in Python: lists, strings, and tuples. Sequences can be indexed to access specific elements and sliced to create subsequences.

| BOX 13-1 | **SCIENTIST SPOTLIGHT** | Russell F. Doolittle |

RUSSELL F. DOOLITTLE (born 1931) is a professor emeritus in the molecular biology and chemistry & biochemistry departments at the University of California, San Diego, focusing on structure and evolution of proteins. He earned a B.A. in biology from Wesleyan University in 1952, and an M.A. in education from Trinity College in 1957. In 1962, he received his Ph.D. in biochemistry at Harvard University, where he investigated blood clotting factors. He did postdoctoral research in Sweden.

In Sweden, Doolittle started a project to sequence the blood clotting factor fibrinogen from several large animal species. He was one of the first to create a phylogenetic tree from amino acid substitution data, which he derived from his fibrinogen protein sequences. Starting with Dayhoff's *Atlas of Protein Sequences*, Doolittle created a new database he named NEWAT (for "New Atlas"). His database was curated to remove redundant sequences, and he added many more sequences from scientists who were sequencing proteins throughout the world. Actually, his son, 11 years old at the time, typed in many of the sequences. Doolittle stored NEWAT on a DEC PDP11 computer and created simple programs for searching the database for identical sequences. For scientists who agreed to submit their sequence to NEWAT, Doolittle offered to search NEWAT for similar sequences. This was perhaps the first sequence searching service made available to scientists. The service fostered new collaborations between scientists who, as a result of Doolittle's searches, found out they worked on similar proteins.

continues

Russell F. Doolittle

BOX 13-1 | **SCIENTIST SPOTLIGHT**

continued

Notably, he co-developed the hydropathy index and the sliding window program to predict protein structure trends. He predicted, with astounding accuracy, the transmembrane segments of bacteriorhodopsin. At the time of this writing, the seminal paper of Kyte and Doolittle that describes the index and the sliding window program has been cited more than 19,000 times in the literature. This is one of the most highly cited bioinformatics articles ever written. Doolittle was elected to the National Academy of Sciences in 1984.

REFERENCES

Kyte, J., and R. Doolittle. 1982. "A Simple Method for Displaying the Hydropathic Character of a Protein." *Journal of Molecular Biology* 157: 105–132.

Jones, N. C., and P. A. Pevzner. 2004. *An Introduction to Bioinformatics Algorithms*. Cambridge, MA: MIT Press.

predictions on membrane protein structures. In this chapter, Python concepts will be introduced in the context of developing Kyte-Doolittle's sliding window tool.

As discussed in Chapter 5, the sliding window tool slides a window down the protein sequence, computing the average hydropathy of the amino acids within each window in order to identify hydrophilic and hydrophobic regions of a protein. The sliding window tool will take as inputs a protein sequence, a window size, and the hydropathy scale. The hydropathy scale (shown in Table 13-1) indicates the degrees to which amino acids are hydrophobic (water fearing) or hydrophilic (water loving). A score of 4.5 indicates the most hydrophobic, and a score of −4.5 indicates the most hydrophilic. The program will output the average hydropathy value for each window. This data can then be plotted, using a spreadsheet program, in order

TABLE 13-1. KYTE AND DOOLITTLE'S (1982) HYDROPATHY VALUES USED FOR CONSTRUCTING THE HYDROPATHY PLOTS

AMINO ACID	HYDROPATHY VALUE	AMINO ACID	HYDROPATHY VALUE
A	1.8	M	1.9
C	2.5	N	−3.5
D	−3.5	P	−1.6
E	−3.5	Q	−3.5
F	2.8	R	−4.5
G	−0.4	S	−0.8
H	−3.2	T	−0.7
I	4.5	V	4.2
K	−3.9	W	−0.9
L	3.8	Y	−1.3

to visually identify the hydrophobic and hydrophilic regions of a protein sequence to help predict the structure of the protein. Kyte and Doolittle's hydropathy plot for bacteriorhodopsin is shown in Figure 13-1.

While their initial study used a window size of 7 for identifying transmembrane regions in bacteriorhodopsin, Kyte and Doolittle subsequently experimented with window size to determine the best values for detecting transmembrane and surface region proteins. A transmembrane region of an integral membrane protein will consist of 18–20 hydrophobic amino acids, and thus, Kyte and Doolittle found that a window size of 19 was best for identifying transmembrane regions. If a window has an average hydropathy value greater than 1.6 it is predicted to be a transmembrane region. For globular proteins, the hydrophobic regions will be located on the inside of the protein and the hydrophilic regions will be located on the outside of the protein, interacting with the water in the cytosol. Kyte and Doolittle determined that a window size of 9 would best predict the hydrophilic surface regions on a globular protein. These surface regions can be identified by the valleys relative to the midline of the plot.

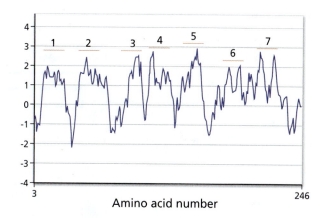

FIG. 13-1. Hydropathy plot. Kyte and Doolittle used a sliding window program (with a window length of 7) to predict the membrane spanning regions of bacteriorhodopsin from its amino acid sequence. The seven known membrane-spanning regions are numbered 1–7 in red on the plot. Note that this particular software program averaged the hydropathy values in the window (http://www.vivo.colostate.edu/molkit/hydropathy/index.html). The original program by Kyte and Doolittle summed the hydropathy values. J. Kyte and R. F. Doolittle, "A Simple Method for Displaying the Hydropathic Character of a Protein," *Journal of Molecular Biology* 157 (1982): 105–132.

13.3   WHY PYTHON?

DNA and protein sequences are represented as strings of nucleotides and amino acids, respectively, and thus, it is natural to select a language such as Python that has built-in support for strings. Moreover, Python is both easy to learn and a powerful language used by many developers for a wide array of application domains including web development, database access, graphical user interfaces (GUIs), and scientific and numeric computing. Python has an extensive standard library with solutions to many common programming problems and third-party modules for many tasks. Python integrates well with other languages such as C, C++, and Java.

Python is an object-oriented scripting language. Object-oriented means that programs are organized by their data (objects), and functions (methods) are written to manipulate that data. Object-oriented programming allows you to develop complex, robust programs. Because we are focusing on programming basics, we will not introduce the object-oriented programming paradigm but we will use Python's built-in objects and methods. Although Python can be compiled, it is often run as a script, which is a program that can be written and executed on the fly. Because of this convenience, Python programs can be written quickly and tested easily. Another benefit of Python is that it is portable: the same source code (program) can run on different platforms (operating systems) and the implementations remain unchanged.

Python is open source, which means that it is freely available. Information about Python, including documentation, tutorials, and downloads, can be found at the official Python programming language website (www.python.org). You can download Python for Windows, Mac OS X, Linux/Unix, and other platforms.[1] Python is a dynamic language that is continually evolving. For the basic introduction that we are providing in this book, we will use Python 2.7.2 on a Windows platform.[2] At this time, Python 2 is the predominant version and the one most widely supported by third-party packages and utilities including BioPython. BioPython is a set of

string A sequence of characters enclosed by quotes. A built-in data type in Python.

data (in computer science) Information that is stored in computer memory and manipulated by computer operations.

function A named encapsulated sequence of instructions that performs a task. Functions can be called by name, passed input data, and return results after performing the task.

object-oriented programming (OOP) language A programming language that provides a set of rules for creating and managing objects. Examples of OOP languages are Python, C++, and Java.

script A program or sequence of instructions that is interpreted by another program (a run-time environment) that will carry out the execution of the program. In contrast, *compiled* programs are executed directly by the computer hardware.

---

[1] Python runs on many platforms and virtual machines including .NET and the Java virtual machine.

[2] The examples shown in Chapters 13 and 14 are executed using Python 2.7.2 on a Windows 7 Professional platform.

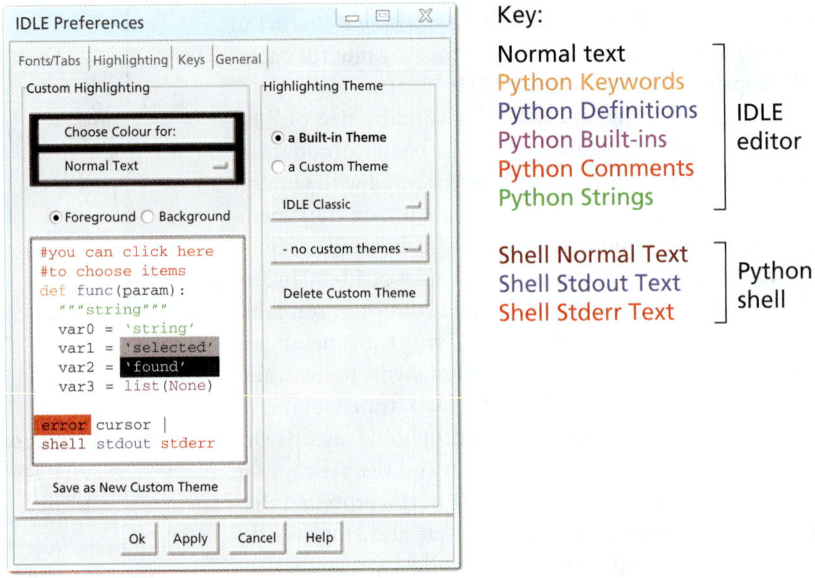

FIG. 13-2. The Classic Python colorization scheme in IDLE editor and Python Shell.

freely available tools for biological computing written by an international team of developers. You may be more familiar with other programming languages or another version of Python. If so, you may want to use them to create bioinformatics software programs. In either case, this chapter will be useful because it describes the features common in many programming languages that are necessary for creating bioinformatics software programs. Furthermore, the approach to programming presented here is useful for solving problems regardless of the language used.

13.4 GETTING STARTED WITH PYTHON

module A file containing Python definitions and statements. The file name is the module name with the suffix .py appended.

Python is distributed with IDLE (Integrated DeveLopment Environment), a cross-platform application that works on Windows, Mac OS X, and Linux/Unix.[3] IDLE has a multi-window text editor that can be used for developing Python programs. The IDLE editor colorizes Python code, which makes it easier to read and develop code. Figure 13-2 shows the IDLE Classic color scheme we will be using in our examples throughout this chapter. IDLE has a Python Shell for executing Python commands and modules. The Python Shell also supports Python colorizing. Note that a color can have different meanings in the IDLE editor and the Python Shell. For instance, the color blue is used for Python definitions in the IDLE editor and is used for standard output (stdout) text in the Python Shell. Likewise, the color red is used for Python comments in the IDLE editor and is used for standard error (stderr) text in the Python Shell.

Let's try the classic "Hello World!" program to see how to develop a simple Python module in IDLE. After starting IDLE, follow these instructions:

- From the File dropdown menu, select New Window (Ctrl+N)
- Type: print "Hello World!"
- From the File dropdown menu, select Save (Ctrl+S) and type the File name: hello.py
- From the Run dropdown menu, select Run Module (F5)

[3] Note that there are many more advanced IDEs available for Python; however, IDLE was chosen for the examples in this chapter because it is a cross-platform tool that is distributed with Python and thus should persist where other third-party IDEs may not. You are encouraged to investigate and use more advanced IDEs.

A

```
hello.py - C:\Users\Documents\Bioinformatics Text\Python Examples\hello.py

File  Edit  Format  Run  Options  Windows  Help

print "Hello World!"

                                                          Ln:1 Col:20
```

B

```
Python Shell

File  Edit  Shell  Debug  Options  Windows  Help

Python 2.7.2 (default, Jun 12 2011, 14:24:46) [MSC v.1500 64 bit (AMD64)] on win32
Type "copyright", "credits" or "license()" for more information.
>>> =============================== RESTART ===============================
>>>
Hello World!
>>>
                                                          Ln:6 Col:4
```

Figure 13-3A shows the Python module hello.py, and Figure 13-3B shows the output (called standard output, or stdout) shown in blue in the Python Shell. The ".py" filename suffix is used to indicate that this is a Python module. Files with a .py suffix will be Python colorized by the IDLE editor. When running the code outside of IDLE, the .py suffix also tells the operating system to invoke the Python interpreter. You can run the code either from the operating system prompt or by double-clicking on the file. In Windows, an interactive console window will pop up once you run the code. To prevent the console window from terminating when the script has finished executing, the *raw_input()* built-in function with a prompt requesting the user to press enter is included at the end of the program. Without this statement, the console window will terminate before the user has gotten a chance to see the output. Figure 13-4 shows the modified hello.py module and the Windows console window. Console windows (also known as dummy terminals) provide a simple user interface. There are a large number of graphical user interface (GUI) frameworks available for Python, but for beginners, a simple dummy terminal interface is a good place to start.

FIG. 13-3. Simple Python module run in the Python Shell.

A

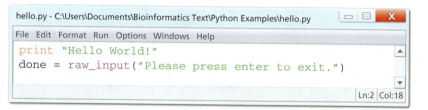

```
hello.py - C:\Users\Documents\Bioinformatics Text\Python Examples\hello.py

File  Edit  Format  Run  Options  Windows  Help

print "Hello World!"
done = raw_input("Please press enter to exit.")

                                                          Ln:2 Col:18
```

B

```
C:\Python27\python.exe

Hello World!
Please press enter to exit.
```

FIG. 13-4. Python programs can also be run by double-clicking on the file. **A.** Modified hello.py program. **B.** Windows console window.

FIG. 13-5. Python statements can be executed directly from the command prompt (">>>") of the Python Shell.

One of the reasons why Python is an easy language to learn is because you can test out Python statements directly from the command line prompt (">>>") of the Python Shell. Instead of creating a module to print "Hello World!," we could have typed the statement directly into the Python Shell as shown in Figure 13-5. Although the Python Shell is very useful for exploring new programming concepts, you should use the IDLE editor to develop Python modules for your bioinformatics applications (also referred to as tools). Why? Once you exit the Python Shell, your commands will be lost, whereas once you save your Python module, you can rerun it at any time.

The Python Shell is a powerful tool for developers. It features command recall, which allows you to copy commands by simply placing the cursor over the command and pressing the enter key. Once the command is copied you can press the enter key again to run it or you can modify the command and then press enter to run the modified command. The Python Shell also allows you to view all the variables that have been defined in the environment (either by running a module or through command line statements). We will start learning the basics of the Python programming language using the Python Shell.

13.5 DATA FLOW: REPRESENTING AND MANIPULATING DATA

The first thing to understand about programming is how data flows through a program through variables. As it flows through the program, data is manipulated by operators and functions.

Let's start with a simple example to illustrate how variables are used in a program. The example in Figure 13-6A prompts the user to enter two numbers and then prints (or outputs) the sum of the two numbers. Let's look at how this program will execute. On the first line of the program the user is prompted to input (enter) the first number, which is then stored in the variable called *Num1*. On the next line, the user is prompted to enter the second number, which is stored in the variable called *Num2*. On the third line of the program, the numbers stored in *Num1* and *Num2* are added, and the result is stored in a variable called *Sum*. On the fourth line, the string "The sum is:" is printed to the display (the comma prevents a new line from being inserted so that the next thing to be printed will be on the same line). The fifth line of the program prints the value stored in variable *Sum* to the display.

In this example, the program has been executed (run) twice in the console window as shown in Figures 13-6B and C. The first time the values entered by the user are 5 and 23, and the second time the values entered by the user are 17 and 538. Notice that the use of variables allows the same program to be run multiple times with different input data. Also notice how the data flows through the program using variables. For the first execution of the program, the value 5 is stored in variable *Num1* and the value 23 is stored in variable *Num2*. The values stored in these variables are then summed and the result value 28 is stored in variable *Sum*. The value in *Sum* is then printed to the display in the console window. The second time the program is run, the values 17 and 538 are stored in variables *Num1* and *Num2* and their sum 555, stored in the variable *Sum*, is printed to the display.

FIG. 13-6. Variables in Python are used to hold data, and operators such as add (+) can be used to operate on the variables. **A.** Simple Python program to add two numbers. **B.** Results of running the program in the console window where the two numbers entered by the user are 5 and 23. **C.** Results of running the program a second time in the console window with two different inputs, 17 and 538.

Variable Names

You should use meaningful variable names (identifiers) to make it easier to follow the flow of data. For example, consider the statement *x* = *9*. What does *x* refer to? On the other hand, in the statement *win_size* = *9*, it is clear we are referring to the window size. Note that in Python (as in most languages), the following restrictions apply to variable names: (1) they must consist of only letters, numbers, and underscores; (2) they must start with a letter or underscore; (3) they are case sensitive; and (4) they must not be a reserved word. Reserved words are words in Python that have special meaning, such built-in function names. If using a colorized editor, reserved words will be a color other than the regular black text. You can use underscore or capitalization for multi-word names, and abbreviations can be used to shorten long names. Also, Python has specific naming conventions to identify internal variables used by the interpreter, such as leading and tailing underscores, so programmers should avoid using these. Table 13-2 shows some variable naming conventions.

Data Types and Operators

One of the reasons Python is both easy to learn and powerful to use is that it has a rich set of built-in data types. Unlike variables in most languages, Python variables are not declared. That is, instead of declaring a variable to be an integer, floating point, string, and so on, the data type of a variable is determined by what type of value it is assigned (as shown in Table 13-3). When a variable is assigned, a new variable is

variable A symbolic name associated with a data storage location within a program. The value of a variable can be changed as the program executes.

identifier (in computer science) Variable name. It is important when coding to create meaningful identifiers that make it easy to follow the data flow.

TABLE 13-2. NAMING CONVENTIONS

NAMING CONVENTION	EXAMPLES
Variable names should be meaningful. In general, avoid single-letter names.	hydropathy_scale HydropathyScale hydro_scale
Use all uppercase letters for constants.	HYDROPHOBIC_THRESHOLD HYDROPHOBIC_THRESH
Avoid starting or ending names with underscores.	Avoid the following naming formats (where *X* can be any name): _X __X__ __X_

TABLE 13-3. PYTHON DATA TYPES USED IN THIS TEXTBOOK

DATA TYPE	EXAMPLE	MUTABLE OR IMMUTABLE
Integer	win_size = 9	*Immutable*
Floating point	hydro = 1.8	*Immutable*
String	win_seq = "PLSQETFSD"	*Immutable*
List	hydro_vals = [−1.6, 3.8, −0.8, −3.5, −3.5, −0.7, 2.8, −0.8, −3.5]	*Mutable*
Tuple	aa_properties = ("Alanine", 'A', 1,0,0, 1,0,0)	*Immutable*
Dictionary	hydro_scale = {'a': 1.8, 'c': 2.5, 'd': −3.5, 'e':−3.5, 'f':2.8, 'g':−0.4, 'h':−3.2, 'i':4.5, 'k':−3.9, 'l':3.8, 'm':1.9, 'n':−3.5, 'p':−1.6, 'q':−3.5, 'r':−4.5, 's':−0.8, 't':−0.7, 'v':4.2, 'w':−0.9, 'y':−1.3}	*Mutable*

Note: A complete list of Python data types can be found at www.python.org.

data type Defines the kind of a data value and determines how the value can be used in a programming language. Some data types are mutable, and some are immutable. Examples of data types are integer, floating point, and string.

integer A type of data in Python for representing whole numbers that can be positive, negative, or zero.

mutable (in computer science) Used to describe a type of data in Python that can be altered. Mutable data types include lists and dictionaries.

immutable (in computer science) Used to describe a type of data in Python that cannot be altered without creating a new location (variable). Immutable data types include integers, floats, strings, and tuples. For example, assume an integer variable x has been assigned and holds the value 5; the statement $x = x + 1$ actually creates a new variable called x in a new location and assigns it the value 6.

floating point number A format used to represent real numbers in computer systems.

created—that is, a new space in memory is allocated for it. Variable names can be reused, making it appear that the same variable is being reassigned a new value. In reality, what happens is that a new variable is created (allocated) and the variable name now refers to the new variable's location.

Table 13-3 lists the standard Python data types we will be using in this textbook and indicates whether the data type is mutable or immutable. A variable is considered mutable if a variable can be modified "in place"—that is, without reallocating it. A variable is considered immutable if a variable cannot be modified "in place." For example, because a list is mutable, you can add values to a list and modify existing values in a list. A string, on the other hand, is immutable, so you cannot add, delete, or change the characters within a string. We will discuss these examples in more detail as we introduce list and string data types.

The data type of a variable can be defined by assigning a constant value to the variable as shown in Table 13-3 or assigning an expression that evaluates to that data type. For example,

```
>>> win_avg = sum(hydro_vals)/len(hydro_vals)
>>> win_avg
−0.8666666666666667
```

In this example, sum and len are built-in functions where sum(hydro_vals) will sum the elements of the list hydro_vals and len(hydro_vals) will compute the length of the list. The data type of win_avg will be floating point because the sum of the hydro_vals is a floating point number divided by the length of the hydro_vals list which is an integer.

Numeric data types

Python has three numeric data types: integers, floating point numbers, and complex numbers (we will not be using complex numbers). In Python, the size of an integer is unlimited. In Python 2, if a numeric literal is too large to be represented as a plain integer it will be represented as a long integer. Long integers end with the character L (for example, *999999999999999999999L*). Floating point numbers represent real numbers and can be of the form *3.14159* or *314.159e-2* (scientific notation). For example,

```
>>> pi = 314.159e-2
>>> pi
3.14159
```

Numeric data types are immutable. The only way to change the value of a variable is to reassign it. For example, $x = x + 1$ will create a new variable x and assign it the value of the old variable x plus 1. Table 13-4 shows the basic operations that can be performed on integers and floating point numbers (as well as on other data types). Most operations, such as add (+), subtract (−), and multiply (∗), work on integers and floating point numbers as you would expect. Using the divide operator on integers will always produce an integer result in Python 2. The result of an integer divide, however, is a truncated quotient. This can be undesirable in some cases, such as computing the average of integers. There are several ways to produce a real number as a quotient. One way is to multiply the dividend (numerator) or divisor (denominator) by 1.0 before the division. For example, if x is 5 and y is 3,

```
>>> x/y
1
>>>1.0*x/y
1.6666666666666667
```

TABLE 13-4. BASIC PYTHON OPERATORS USED IN THIS TEXTBOOK

	DATA TYPE				

OPERATORS	**INTEGER**
	>>> x = 5
	>>> y = 3
		FLOATING POINT
		>>> f1 = 0.5
		>>> f2 = 3.2
			LIST
			>>> L1 = [1,2]
			>>> L2 = ['a', 'b']
				TUPLE
				>>> T1 = (1,2)
				>>> T2 = (3,4)
					STRING
					>>> str1 = "a"
					>>> str2 = "b"

+ **add** *(concatenate)*	>>> x + y 8	>>> f1 + f2 3.7	>>> L1 + L2 *[1,2,'a','b']*	>>> T1 + T2 *(1,2,3,4)*	>>> str1 + str2 *'ab'*
− **subtract**	>>> y − x −2	>>> f1 − f2 −2.7			
* **multiply** *(replicate)*	>>> y ∗ x 15	>>> f1 ∗ f2 1.6	>>> L1 ∗ 4 *[1,2,1,2,1,2,1,2]*	>>> T2 ∗ 2 *(3,4,3,4)*	>>> str1 ∗ 3 *'aaa'*
/ **divide**	>>> x/y 1	>>> f1/f2 0.15625			
% **modulus/** **remainder** *(format string)*	>>> x%y 2	>>> f2%f1 0.2			>>>"%d"%5 *'5'*
** **power**	>>> x ∗∗ y 125	>>>f2 ∗∗ f1 1.789			

For integers, the modulus operator (%) returns the remainder of the division of two numbers. For example, *5/3* is *1* with a remainder of *2*, or, *5%3* is *2*. The modulus operator can also be used for floating point numbers. For example, if *f1 = 0.5* and *f2 = 3.2*, then *f2%f1* will yield *0.2* because the integer part of *f2/f1* equals *6* with a remainder of *0.2*. The % operator has a special meaning for strings that will be explained later. The power operator (∗∗) is a useful built-in operator for integer and floating point numbers, where x∗∗y is the same as x^y.

Some operators in Python are overloaded, which means that they can have different meanings depending on the data type. For example, for integers and floating point numbers, the plus symbol (+) refers to the add operation. On the other hand, for tuples, strings, and lists, + refers to concatenation, ∗ refers to replication, and % refers to format string. These alternative operator meanings are listed in Table 13-4 in italics. As we discuss each data type, we will explore the alternative operator meanings in more detail.

Strings

The string data type is particularly useful for bioinformatics programs because DNA and proteins can be represented as character sequences of nucleotides and amino acids, respectively. Strings can contain any characters. Backslashes (\) are used as escape characters to invoke either the original interpretation of a character (e.g., \' can be used to put a single quote in a string delimited by single quotes) or an alternative interpretation of a character (e.g., \n refers to a newline character, and \t refers to a horizontal tab character). If you want to include a backslash in a string, it must be preceded with a backslash as well (e.g., \\). Strings can be delimited using either single quotes ('String') or double quotes ("String"), which makes it easy to include quotes within a string without escape characters. Use double quotes to delimit the string if the string includes a single quote and vice versa. For example,

```
>>> str1 = "Here's an example of a single quote in a string."
>>> str1
"Here's an example of a single quote in a string."
>>> str2 = 'He said "I got it!" once he understood.'
>>> str2
'He said "I got it!" once he understood.'
```

If you want to use both single and double quotes in a string or create multiline strings, you can use a triple quote. For example,

```
>>> str3 = """Triple quotes are really useful.
I can embed single quotes (') and double quotes (") within a string and
can have multiline strings."""
>>> str3
'Triple quotes are really useful.\nI can embed single quotes (\') and
double quotes (") within a string and\ncan have multiline strings.'
```

Note that str3 contains the escape characters \n and \', which mean newline and single quote, respectively.

There are numerous functions and methods for working with strings. Table 13-5 presents just a few of the common string operations. The concatenate operator (+) combines two or more strings into a new string, which is useful for creating strings on the fly. For example, when doing sequence alignment, if the program determined that a gap ('−') should be inserted into a partially aligned sequenced called *align_seq1*, it can insert the gap using the following statement: *align_seq1 = align_seq1+'−'*. Testing that statement in the Python Shell (assuming align_seq1 = "A-REMA" initially):

```
>>> align_seq1 = align_seq1+'−'
>>> align_seq1
'A-REMA-'
```

You can also insert numbers into the string using the str function and the concatenate operator. Continuing the example above:

```
>>> print "There are " + str(len(align_seq1) − align_seq1.count('−')) +
" amino acids."
There are 5 amino acids.
```

In this example, print is a Python keyword and refers to the print statement. The functions str and len convert a number to a string and compute the length of the string (the number of characters in the string), respectively. The term *align_seq1.count('−')* is a method. Methods are functions that are associated with a specific object, in this case, the string object. In this example, this method counts the number of occurrences of the character '−' in the string *align_seq1*.

You can also use formatted strings to insert values into strings. For example,

```
>>> print "There are %d amino acids." % ((len(align_seq1) −
align_seq1.count('−'))
There are 5 amino acids.
```

In this example, %d is a placeholder for a decimal number. The value of the expression after the % will be inserted into the string in place of the %d. Some other useful placeholders are %s for strings and %f for floating point numbers.

Indexing can be used to access specific characters within a string. An index is an integer that points to a particular character of a string. For example, referring to Table 13-5, *aa_seq[0]* refers to the character at index 0, which is the first character

TABLE 13-5. COMMON STRING OPERATIONS

aa_seq = "PLSQETFSD"		
EXPRESSION	**VALUE**	**PURPOSE**
len(aa_seq)	9	Compute the number of characters in aa_seq.
"hydropathy" + "scale"	'hydropathy scale'	Concatenate strings.
15*'*'	'***************'	Replicate string.
"%d amino acids"% len(aa_seq)	'9 amino acids'	Format strings.
"ACCTG" == "ACTG" "ACCTG" == "ACCTG"	False True	Test for equality.
'A' < 'T' 'C' < 'A'	True False	Test for alphabetical order.
aa_seq[0] aa_seq[1] aa_seq[-1]	'P' 'L' 'D'	Access character at index. • Indices start at 0. • Negative indices are relative to end of string.
aa_seq[0:3] aa_seq[3:] aa_seq[:5] aa_seq[-4:]	'PLS' 'QETFSD' 'PLSQE' 'TFSD'	Slice (string[first:last]) to create a substring that • Starts at the first index (default is beginning of string). • Ends at the last index minus 1 (default is end of string).
aa_seq.lower()	'plsqetfsd'	Convert to lowercase.
str1 = "Bioinformatics" str1.upper()	"BIOINFORMATICS"	Convert to uppercase.
aa_seq.replace('S','*')	'PL*QETF*D'	Within the string, replace occurrences of a substring with another substring.
aa_seq[len(aa_seq)]	Traceback (most recent call last): File "<pyshell#161>l", line 1, in <module> aa_seq[len(aa_seq)+1] IndexError: string index out of range	Generates a traceback message indicating that there is an index error because the string index cannot be greater than the string length minus 1.

of string *aa_seq* (i.e., 'P'). Negative indices are used to point to characters starting at the end of the string. Thus, an index of −1 refers to the last character of string. For example, *aa_seq[−1]* is 'D'.

A powerful string operation is slicing (shown in Table 13-5). In a slice, you can optionally specify the first and last indices. The slice starts at the first index and ends at the last index minus 1. If you do not specify the first index, it defaults to the beginning of the string (e.g., starting at index 0), and if you do not specify the last index, it defaults to the end of the string (e.g., ending at the last character of the string). Slicing is a very useful operation for creating a window of amino acids in sliding window programs. For example, assuming *aa_seq* = "*PLSQETFSD*":

```
>>> aa_seq
'PLSQETFSD'
>>> win_size = 3
>>> win_start = 0
>>> aa_seq[win_start:win_start+win_size]
'PLS'
```

The first window of size 3 in this sequence is 'PLS'. We can easily slide the window down the sequence by incrementing win_start. For example,

```
>>> win_start = win_start+1
>>> aa_seq[win_start:win_start+win_size]
'LSQ'
>>> win_start = win_start+1
>>> aa_seq[win_start:win_start+win_size]
'SQE'
```

Notice in the examples above that variables were given meaningful names. In this way, the slicing operation expression itself is clearly connected to the purpose of the program, which is to create a window of a specified length.

Strings are immutable, which means that they cannot be changed in place. For example,

```
>>> str1 = "Bioinformatics"
>>> str1.upper()
'BIOINFORMATICS'
>>> str1
'Bioinformatics'
```

The method returns the string converted to uppercase but cannot directly modify *str1*. To modify a string, you need to reassign it with the returned value as shown below.

```
>>> str1 = str1.upper()
>>> str1
'BIOINFORMATICS'
```

Note that you can use a different name for the modified string. For example, continuing this example (*str1* is now 'BIOINFORMATICS'),

```
>>> str2 = str1.lower()
>>> str2
'bioinformatics'
```

Lists

A list is a sequence of elements delimited by square brackets ([List]).[4] Elements in a list are separated by a comma ([element0, element1]). Like strings, lists can be indexed starting at index zero. Whereas every element in a string is a character, in a list each element can be a different data type (although typically they all have the same

---

[4] Lists are similar to arrays in other programming languages such as C and Java.

TABLE 13-6. COMMON LIST OPERATIONS

jumble = [−1.6, "A", 5, [1, 2, 3], '*']		
EXPRESSION	**VALUE**	**PURPOSE**
jumble[0] jumble[3]	−1.6 [1, 2, 3]	Access element at index (a list element can be any type, even another list).
jumble[−1] jumble[−3]	'*' 5	Access elements relative to the end of the list using a negative index (index −1 refers to the last element of the list, −2 to the second to last, and so on).
jumble[1:]	['A', 5, [1, 2, 3], '*']	Slice (list[first:last]) to create a sub-list that • Starts at the first index (default is beginning of list). • Ends at the last index minus 1 (default is end of list).
jumble[3][1]	2	Access values in a nested list using two indices.
"A" in jumble 6 not in jumble	True True	Test for membership in a list.
jumble.append(75)	[−1.6, "A", 5, [1, 2, 3], '*', 75]	Add to end of list (lists are mutable).
jumble[0] = 'q' jumble[3].append(4)	['q', 'A', 5, [1, 2, 3], '*', 75] ['q', 'A', 5, [1, 2, 3, 4], '*', 75]	Modify elements within a list.

data type). In Table 13-3, the list *hydro_vals* has elements that are all the same type (floating point), whereas the list *jumble* in Table 13-6 shows that lists can contain elements of different types.

You can use a list of lists to create a two-dimensional list, or a list of list of lists to create a three-dimensional list, and so on! In Chapter 14 we will use two-dimensional lists for scoring matrices during pairwise sequence alignment.

Lists are mutable, which means they can be modified in place. In other words, you can modify elements and add/delete elements to/from a list without having to reassign it. Table 13-6 shows some common list operations. Notice that, like strings, lists can be indexed and sliced.

list A comma-separated sequence of elements delimited by square brackets. Elements within a list can be of any data type, though usually they are the same data type.

Tuples

Tuples are similar to lists in that they are a sequence of elements that can be of any data type. The elements are also numbered, starting with zero. Unlike lists, though, tuples are immutable, so the tuple and elements within the tuple cannot be modified. Tuples are useful for passing information around the program and for creating data that cannot be modified during the execution of a program. Tuples are delimited by parentheses, in other words, (Tuple). Elements of a tuple are separated by commas, in other words, (element0, element1).

Table 13-7 lists the general characteristics of alpha amino acids. This information can be stored within the program as a tuple of tuples. Using nested tuples helps organize the information and allows for faster queries. For example, the first two rows of the table can be created by the following assignment, where *aa_properties* is a three dimensional tuple.

tuple An immutable sequence of elements of any data type. Tuples are delimited by parentheses, and elements are separated by commas. Useful for returning multiple values from a function, for assigning multiple values simultaneously, and for defining a constant set of values.

TABLE 13-7. GENERAL CHARACTERISTICS OF ALPHA AMINO ACIDS

RESIDUE	LETTER CODE	HYDROPHOBIC	AROMATIC	ALIPHATIC	SMALL	POLAR	CHARGED
Alanine	A	✓			✓		
Arginine	R					✓	✓
Asparagine	N				✓	✓	
Aspartate	D				✓	✓	✓
Cysteine	C	✓			✓	✓	
Glutamate	E					✓	✓
Glutamine	Q					✓	
Glycine	G	✓			✓	✓	
Histidine	H	✓	✓			✓	✓
Isoleucine	I	✓		✓			
Leucine	L	✓		✓			
Lysine	K	✓				✓	✓
Methionine	M	✓					
Phenylalanine	F	✓	✓				
Proline	P				✓		
Serine	S				✓	✓	
Threonine	T				✓	✓	
Tryptophan	W	✓	✓			✓	
Tyrosine	Y	✓	✓			✓	
Valine	V	✓		✓			

```
>>> aa_properties = (("Alanine", 'A', ("Hydrophobic", "Small")),
("Arginine", 'R', ("Polar", "Charged")))
```

This example highlights how Python supports nesting tuples and the convention for indexing into nested tuples (nested lists are similarly indexed). Actually, no special convention is needed for indexing into multidimensional tuples. If an element of a tuple happens to be another tuple, then it can also be indexed. For example, aa_properties is a tuple with two nested tuples, one for *Alanine* and one for *Arginine*. Each of these tuples has three elements, where the third element is another tuple. Thus, there are three levels of nesting.

```
>>> aa_properties
(('Alanine', 'A', ('Hydrophobic', 'Small')), ('Arginine', 'R', ('Polar', 'Charged')))
>>> aa_properties[1]
('Arginine', 'R', ('Polar', 'Charged'))
>>> aa_properties[1][2]
('Polar', 'Charged')
>>> aa_properties[1][2][0]
'Polar'
```

In this example, *'Polar'* is the **zeroth** element of a nested tuple (e.g., *('Polar', 'Charged')*) which itself is the **second** element within another nested tuple (e.g., *('Arginine', 'R', ('Polar', 'Charged')))* which is the **first** element of the tuple *aa_properties*.

Information can be retrieved from the tuple *aa_properties* by indexing, slicing, and querying. Here are some examples.

```
>>> aa_properties[1][0]
'Arginine'
>>> aa_properties[1][1]
'R'
>>> "Polar" in aa_properties[1][2]
True
>>> "Hydrophobic" in aa_properties[1][2]
False
>>> aa_properties[0][0:2]
('Alanine', 'A')
```

This last example illustrates slicing. Here, *aa_properties[0]* is the tuple *('Alanine', 'A', ('Hydrophobic', 'Small'))*. We can slice off the last element of the tuple using the expression *aa_properties[0][0:2]*. Recall that slices start at the first index (in this case, index 0) and end at one less than the last index (in this case, ending at index 2-1, which is 1), and thus, the result is *('Alanine', 'A')*.

Dictionaries

Dictionaries are a data type where information is organized by keys. Each key is mapped to a value. Dictionaries are similar to lists, but, whereas lists are ordered and indexed, dictionaries are not ordered and lookup into a dictionary is through keys. Keys can be any type. Like lists, the values stored in dictionaries can also be any type. Dictionaries are delimited by curly brackets, and keys and values are separated by a colon, in other words, {key:value}. Table 13-8 shows a dictionary containing the first five values in Kyte and Doolittle's hydropathy scale and common operations that can be performed on dictionaries.

dictionaries (keys and values)
A Python structure for organizing data *values* that can be looked up (accessed) using *keys*.

TABLE 13-8. COMMON DICTIONARY OPERATIONS

hydro_scale = {'a': 1.8, 'c': 2.5, 'd': −3.5, 'e':−3.5, 'f':2.8}		
EXPRESSION	**VALUE**	**PURPOSE**
hydro_scale.keys()	['a', 'c', 'e', 'd', 'f']	Generate list of keys in the dictionary.
hydro_scale.values()	[1.8, 2.5, −3.5, −3.5, 2.8]	Generate list of values in the dictionary.
hydro_scale['a'] hydro_scale['g']	1.8 Traceback (most recent call last): File "\<pyshell#143\>", line 1, in \<module\> hydro_scale['g'] KeyError: 'g'	Look up value using a valid key. Look up value using an invalid key.
hydro_scale.has_key('d') hydro_scale.has_key('y')	True False	Check for keys.
hydro_scale.update({'g':−0.4}) hydro_scale	{'a': 1.8, 'c': 2.5, 'e': −3.5, 'd': −3.5, 'g': −0.4, 'f': 2.8}	Add pair (key and value) to dictionary.
hydro_scale['a'] = "junk" hydro_scale	{'a': 'junk', 'c': 2.5, 'e': −3.5, 'd': −3.5, 'f': 2.8}	Modify values stored in the dictionary.

Dictionaries can be built from a list of keys and a list of values using the *zip()* function to create a list of tuple pairs of keys and values. The *dict()* function is used to convert the list into a dictionary. For example,

```
>>> keys = ['a', 'b', 'c']
>>> vals = [1, −8, "**"]
>>> zip(keys, vals)
[('a', 1), ('b', −8), ('c', '**')]
>>> dictionary = dict(zip(keys, vals))
>>> dictionary
{'a': 1, 'c': '**', 'b': −8}
```

Notice that dictionaries are not ordered (e.g., the order of pairs in the list of tuples is not the same as the order of pairs in the dictionary). Because dictionaries are not ordered, they cannot be indexed or sliced.

13.6 PUTTING IT TOGETHER: A SIMPLE PROGRAM TO LOOK UP THE HYDROPHOBICITY OF AN AMINO ACID

Let's combine everything that we have learned so far into a simple program to look up the hydrophobicity of an amino acid. As mentioned earlier, the Python Shell is a great tool for learning the language, but programs should be written in a Python module as shown in Figure 13-7A. Do not forget to save your program with a ".py" suffix (e.g., *hydro_lookup.py*) so that it can be run outside of the Python Shell, and so that the program is colorized, which makes it easier to read.

Whenever you write a program, you should include a header comment with a brief description of the program, your name as the author of the program, and the date the program was created. This information will be useful to anyone wanting to use or modify your code at a later date. Python comments start with a pound sign (#). In addition, you should include comments throughout your program to make it easier to understand.

Let's look at how the data flows through the program shown in Figure 13-7A. First, a dictionary, *hydro_scale*, is created with the hydropathy values for each amino acid. Next, the program prompts the user to input an amino acid. The *raw_input()* built-in function will print the prompt and return a string. In this program, the string will be assigned to the variable *aa*. Because the user could enter a lowercase or upper-case letter for the amino acid, *aa* is converted to lowercase because the dictionary keys are lowercase. Because strings are immutable, we reassign *aa*. Next, *aa* is used to look up the hydrophobicity value in the *hydro_scale* dictionary. The returned value is assigned to the variable *hydro*. The program then prints the amino acid (*aa*) and its hydrophobicity value (*hydro*) using the built-in function *print()* and a formatted string. Notice that the formatted string uses a tuple to group the *aa* and *hydro* values. %s is a placeholder in the formatted string for the *aa* string value, and %.2f is a place-holder in the formatted string for the *hydro* floating point value. The ".2" indicates that the floating point number should be printed with two digits after the decimal point.

Figure 13-7B and C show the results of running the program in the Python Shell and as a stand-alone program, respectively. Recall that to run a script in Windows, you need to add a request for input from the user at the end of the script (e.g., *dummy = raw_input("Press enter to exit.")*). This will keep the console window open until the user is ready to exit the program. The variable name *dummy* refers to the fact that the variable is actually never used. Running the program inside the Python Shell allows you to view the values of the variables. Notice that whenever the program is rerun in the Python Shell, it restarts the shell, erasing old values stored in the shell. Once a program is fully developed, it is usually run as a stand-alone tool.

A

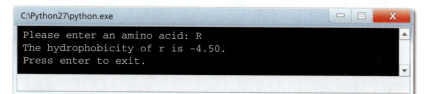

```
hydro_lookup.py - C:/Users/Documents/Bioinformatics Text/Python Examples/hydro_lookup.py

File  Edit  Format  Run  Options  Windows  Help

# Program to compute the hydrophobicity of an amino acid.
# Written by: Nancy Warter-Perez
# Date created: April 14, 2015

# Kyte-Doolittle Hydropathy scale
hydro_scale = {'a': 1.8, 'c':2.5, 'd': -3.5, 'e':-3.5, 'f':2.8, 'g':-0.4,
               'h':-3.2, 'i':4.5, 'k':-3.9, 'l':3.8, 'm':1.9, 'n':-3.5,
               'p':-1.6, 'q':-3.5, 'r':-4.5, 's':-0.8, 't':-0.7, 'v':4.2,
               'w':-0.9, 'y':-1.3}

aa = raw_input("Please enter an amino acid: ")
aa = aa.lower() # convert to lower case
hydro = hydro_scale[aa]
print("the hydrophobicity of %s is %.2f." %(aa, hydro))

dummy = raw_input("Press enter to exit.")

                                                            Ln: 9 | Col: 33
```

B

```
>>> ============================= RESTART =============================
>>>
Please enter an amino acid: A
The hydrophobicity of a is 1.80.
Press enter to exit.
>>> ============================= RESTART =============================
>>>
Please enter an amino acid: Y
The hydrophobicity of y is -1.30.
Press enter to exit.
>>> hydro_scale
{'a': 1.8, 'c':2.5, 'e': -3.5, 'd':-3.5, 'g':-0.4, 'f':2.8,'i':4.5, 'h':-3.2,
'k':-3.9, 'm':1.9, 'l':3.8, 'n':-3.5, 'q':-3.5, 'p':-1.6, 's':-0.8, 'r':-4.5,
't':-0.7, 'w':-0.9, 'v':4.2, 'y':-1.3}
>>> aa
'y'
>>> hydro
-1.3
>>>
```

C

```
C:\Python27\python.exe

Please enter an amino acid: R
The hydrophobicity of r is -4.50.
Press enter to exit.
```

FIG. 13-7. A. Simple program to look up the hydrophobicity of an amino acid. **B.** The Python Shell interactive window showing the inputs and outputs after running the program twice. **C.** The console window created when running the program outside of the Python Shell by clicking on the filename, *hydro_lookup.py.*

13.7 DECISION MAKING

Although we often think about computers being able to perform complex computations fast, it is the decision-making capabilities that make computers appear intelligent. In the simple hydrophobicity lookup example, what if the user typed an invalid amino acid? The program would display a traceback error and crash. The program can be made more robust (i.e., less prone to crashing) by first checking to see if the user entered a valid amino acid. Instead of continually rerunning the program to look up the hydrophobicity values of different amino acids, a more "intelligent" version of the program would ask if the user wanted to continue and, if so, loop back and execute the code again. Figure 13-8 shows the modified program with an if-else test and a while loop. We will refer to this example as we introduce these concepts in this section.

while loop Used for repeating a block of code until a defined condition is met.

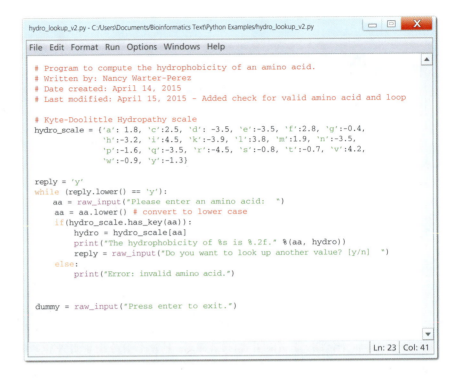

FIG. 13-8. User-friendly version of a modified hydrophobicity lookup program.

TABLE 13-9. RELATIONAL OPERATORS

RELATIONAL OPERATORS		x = 5, y = 3, z = 2	
SYMBOL	**MEANING**	**EXPRESSION**	**VALUE**
==	Equal	x == y	False
!=	Not equal	x != y	True
<	Less than	x < y	False
		−1 < x < 10	True
<=	Less than or equal	x − y <= z	True
>	Greater than	x − y > 0	True
		x > y > z	True
		'abc' > 'ab'	True
>=	Greater than or equal	z >= x	False
		'z' >= 'x'	True

Operations for Decision Making

Decisions are made based on a true-or-false value referred to as a Boolean value. Relational operators (Table 13-9) can be used for comparing values. Built-in functions and methods can also be used for decision making. For example, the built-in function *in* is used for testing membership in lists, strings, and tuples; and *has_key()* is used for testing if a dictionary has a given key. Logical operators (Table 13-10) can be used to combine decision-making operators. If you need to test if two or more conditions are true, use the *and* logical operator. If you need to test if at least one condition is true, use the *or* logical operator. If you want to test the opposite of the condition, use the *not* logical operator.

TABLE 13-10. LOGICAL OPERATORS

LOGICAL OPERATORS	x = 5, y = 3, z = 2	
SYMBOL	**EXPRESSION**	**VALUE**
and	x == 5 and y != 0 x == 5 and y == 2	True False
or	x == 5 or y == 2	True
not	not x == 5	False

Let's take a closer look at the relational operators in Table 13-9. Notice that as in most languages, in Python two equal signs are used to test for equality because one equal sign is used for assignment. Some symbols are not found on a keyboard, and so they are made of two symbols. These include tests for not equal (!=), greater than or equal (>=), and less than or equal (<=). Notice that the ordering of the two symbols is significant, in other words, >= is not the same as =>. Python allows you to use multiple relational operators in the same expression, which is useful for checking if a variable is within a range defined by a lower bound and an upper bound (e.g., −1 < x <= 10).

If-Tests

If-tests are used for making decisions. For example, in Figure 13-8 the program uses an if-test to determine if the user entered a valid amino acid. If the amino acid is valid, the program will display the hydrophobicity value and prompt to see if the user wants to continue. If the amino acid is not valid, then the program will display an error message and continue in the while loop (to be explained in the next section) to prompt the user for another value.

Table 13-11 shows the Python syntax, the grammatical rules and structural patterns, for different if-test formats. If-tests are conditional statements where the execution of the then and else actions depends on the conditional statement. Python *if*, *elif*, and *else* statements end with a colon to indicate that the statement that follows is dependent. All dependent statements are also indented. Thus, in Python, **white spaces are significant!** Python editors will automatically indent the code. If you type an *if statement* and press enter, but it does not automatically indent, then either you forgot the colon or you forgot to save the program with a *.py* suffix (in which case the editor will not apply the Python syntax and semantics rules). After typing the action statements, you will have to backspace to exit out of the indentation. Although some people are not accustomed to indenting their programs, indenting actually makes Python programs much easier to read and understand. When typing code in the interactive Python Shell, the shell editor will indent the next line of code after a colon. To exit the conditional statement indentation, press the enter key twice.

conditional statements (if-tests)
Statements in Python that perform different operations depending on whether the condition is true.

TABLE 13-11. SYNTAX OF COMMON IF-TEST STATEMENTS

IF-THEN	IF-THEN-ELSE	NESTED IFS

if *conditional expression:*
    then-action
	if *conditional expression:*
	    then-action
	else:
	    else-action
		if *conditional expression1:*
		    then-action1
		elif *conditional expression2:*
		    then-action2
		else:
		    else-action

A

```
hydro_query_v1.py - C:\Users\Documents\Bioinformatics Text\Python Example\hydro_query_v1.py      _ □ X

File  Edit  Format  Run  Options  Windows  Help

# Program to determine if an amino acid is hydrophobic.
# Written by: Nancy Warter-Perez
# Date created: April 15, 2015

# Kyte-Doolittle Hydropathy scale
hydro_scale = {'a': 1.8, 'c':2.5, 'd': -3.5, 'e':-3.5, 'f':2.8, 'g':-0.4,
               'h':-3.2, 'i':4.5, 'k':-3.9, 'l':3.8, 'm':1.9, 'n':-3.5,
               'p':-1.6, 'q':-3.5, 'r':-4.5, 's':-0.8, 't':-0.7, 'v':4.2,
               'w':-0.9, 'y':-1.3}

print "***Program to determine if an amino acid is hydrophobic.***\n"
aa = raw_input("\tPlease enter an amino acid:  ")
hydro = hydro_scale[aa.lower()]
if hydro > 0:
    print("\n\tAmino acid %s is hydrophobic.\n" % aa)

dummy = raw_input("Press enter to exit.")

                                                            Ln: 19  Col: 0
```

B

```
***Program to determine if an amino acid is hydrophobic.***

        Please enter an amino acid: L

        Amino acid L is hydrophobic.
Press enter to exit.
>>> ================================= RESTART =================================
>>>
***Program to determine if an amino acid is hydrophobic.***

        Please enter an amino acid: N
Press enter to exit.
```

FIG. 13-9. A. Program with if-then statement that checks if an amino acid is hydrophobic. **B.** Sample input/output from running the program twice in the Python Shell.

Figure 13-9 shows a simple program that uses an if-then statement to determine if an amino acid is hydrophobic. Let's consider the pseudo-code equivalent, where pseudo-code is code that is written without worrying about the syntax of a specific programming language such as Python. Pseudo-code for the if-then statement in the example in Figure 13-9 is:

> *If* the hydropathy value of an amino acid is greater than zero
>   *(then)* print the amino acid and indicate that it is hydrophobic.

Note that in Python the *then* is implied and not actually part of an if statement. This example program also shows how you can format your output to make it easier to read using newline (\n) and tab (\t) escape characters.

Figure 13-10 shows a modified version of this program that includes an if-then-else statement to determine if an amino acid is either hydrophobic or hydrophilic. Pseudo-code for the if-then-else statement in this example is:

> *If* the hydropathy value of an amino acid is greater than zero
>   *(then)* print the amino acid and indicate that it is hydrophobic
> *else* (its hydropathy value is less than or equal to zero)
>   print the amino acid and indicate that it is hydrophilic.

Note that the else does not actually need an else condition, as shown in parenthesis in the above pseudo-code, because when the if condition is false, the else condition is by default true.

Finally, Figure 13-11 shows a third version of the program that uses nested if statements (using the *elif* conditional statement) to determine the degree to which an

A

```
hydro_query_v2.py - C:\Users\Documents\Bioinformatics Text\Python Examples\hydro_query_v2.py

File  Edit  Format  Run  Options  Windows  Help

# Program to determine if an amino acid is hydrophobic OR hydrophilic
# Written by: Nancy Warter-Perez
# Date created: April 15, 2015
# Last modified: April 15, 2015 -- Modified program to output if hydrophilic

# Kyte-Doolittle Hydropathy scale
hydro_scale = {'a': 1.8, 'c':2.5, 'd': -3.5, 'e':-3.5, 'f':2.8, 'g':-0.4,
               'h':-3.2, 'i':4.5, 'k':-3.9, 'l':3.8, 'm':1.9, 'n':-3.5,
               'p':-1.6, 'q':-3.5, 'r':-4.5, 's':-0.8, 't':-0.7, 'v':4.2,
               'w':-0.9, 'y':-1.3}


print "***Program to determine if an amino acid is hydrophobic or hydrophilic. ***"
aa = raw_input("\tPlease enter an amino acid:  ")
hydro = hydro_scale[aa.lower()]
if hydro > 0:
    print("\n\tAmino acid %s is hydrophobic." % aa)
else:
    print("\n\tAmino acid %s is hydrophilic." % aa)


dummy = raw_input("Press enter to exit.")
```

Ln: 5 Col: 0

B

```
***Program to determine if an amino acid is hydrophobic or hydrophilic.***

        Please enter an amino acid: L

        Amino acid L is hydrophobic.
Press enter to exit.
>>> ================================ RESTART ================================
>>>
***Program to determine if an amino acid is hydrophobic or hydrophilic.***

        Please enter an amino acid: N

        Amino acid N is hydrophilic.
Press enter to exit.
```

FIG. 13-10. A. Program with if-then-else statement that checks if an amino acid is hydrophobic or hydrophilic. **B.** Sample input/output from running the program twice in the Python Shell.

amino acid is hydrophobic or hydrophilic using relational operators to specify ranges of hydrophobicity. Pseudo-code for the nested if statements in this example is:

If the hydropathy value of an amino acid is greater than 3
   (then) print the amino acid and indicate that it is strongly hydrophobic

else
   if, the hydropathy value of the amino acid is greater than 0
      (then) print the amino acid and indicate that it is hydrophobic

   else
      if the hydropathy value of the amino acid is greater than or equal to −3
         (then) print the amino acid and indicate that it is hydrophilic

      else (its hydropathy value must be less than −3)
         print the amino acid and indicate that it is strongly hydrophilic.

Note that Python uses the *elif statement* instead of "else if" to avoid needing deeper levels of indentation for nested if-else statements.

Nested if statements are also useful for implementing menu-driven interfaces. A menu-driven interface allows the user to select an option from a menu. Figure 13-12 shows a user-defined function with a nested if for setting the window size for a hydropathy sliding window program. A user-defined function starts with the keyword

A

```
hydro_query_v3.py - C:\Users\Documents\Bioinformatics Text\Python Examples\hydro_query_v3.py    ─ ☐ ✕

File  Edit  Format  Run  Options  Windows  Help

# Program to determine if an amino acid is hydrophobic OR hydrophilic
# Written by: Nancy Warter-Perez
# Date created: April 15, 2015
# Last modified: April 15, 2015 -- Modified program to determine degree of hydrophobicity

# Kyte-Doolittle Hydropathy scale
hydro_scale = {'a': 1.8, 'c':2.5, 'd': -3.5, 'e':-3.5, 'f':2.8, 'g':-0.4,
               'h':-3.2, 'i':4.5, 'k':-3.9, 'l':3.8, 'm':1.9, 'n':-3.5,
               'p':-1.6, 'q':-3.5, 'r':-4.5, 's':-0.8, 't':-0.7, 'v':4.2,
               'w':-0.9, 'y':-1.3}

print "***Program to determine the degree an amino acid is hydrophobic or hydrophilic.***\n"
aa = raw_input("\tPlease enter an amino acid:  ")
hydro = hydro_scale[aa.lower()]
if hydro > 3:
    print("\n\tAmino acid %s is strongly hydrophobic (hydropathy > 3)." % aa)
elif hydro > 0:
    print("\n\tAmino acid %s is weakly hydrophobic (0 < hydropathy <= 3)." % aa)
if hydro > 3:
    print("\n\tAmino acid %s is weakly hydrophilic (-3 <= hydropathy <= 0)." % aa)
else hydro > 3:
    print("\n\tAmino acid %s is strongly hydrophilic (hydropathy < -3)." % aa)


dummy = raw_input("Press enter to exit.")
                                                                    Ln: 26  Col: 0
```

B

```
***Program to determine the degree an amino acid is hydrophobic or hydrophilic.***

        Please enter an amino acid: V

        Amino acid V is strongly hydrophobic (hydropathy > 3).
Press enter to exit.
>>> ================================ RESTART ================================
>>>
***Program to determine the degree an amino acid is hydrophobic or hydrophilic.***

        Please enter an amino acid: M

        Amino acid M is weakly hydrophobic (0 hydropathy <= 3).
Press enter to exit.
>>> ================================ RESTART ================================
>>>
***Program to determine the degree an amino acid is hydrophobic or hydrophilic.***

        Please enter an amino acid: G

        Amino acid G is weakly hydrophilic (-3 hydropathy <= 0).
Press enter to exit.
>>> ================================ RESTART ================================
>>>
***Program to determine the degree an amino acid is hydrophobic or hydrophilic.***

        Please enter an amino acid: K

        Amino acid K is strongly hydrophilic (hydropathy < -3).
Press enter to exit.
```

FIG. 13-11. A. Program with nested if statements that checks the degree to which an amino acid is hydrophobic or hydrophilic. **B.** Sample input/output from running the program four times in the Python Shell.

A

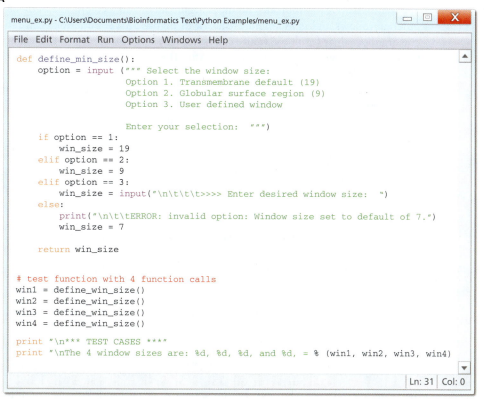

```python
def define_min_size():
    option = input (""" Select the window size:
                    Option 1. Transmembrane default (19)
                    Option 2. Globular surface region (9)
                    Option 3. User defined window

                    Enter your selection:  """)
    if option == 1:
        win_size = 19
    elif option == 2:
        win_size = 9
    elif option == 3:
        win_size = input("\n\t\t\t>>>> Enter desired window size:   ")
    else:
        print("\n\t\tERROR: invalid option: Window size set to default of 7.")
        win_size = 7

    return win_size


# test function with 4 function calls
win1 = define_win_size()
win2 = define_win_size()
win3 = define_win_size()
win4 = define_win_size()

print "\n*** TEST CASES ***"
print "\nThe 4 window sizes are: %d, %d, %d, and %d, = % (win1, win2, win3, win4)
```

Ln: 31 | Col: 0

B

```
Select the window size:
                Option 1.  Transmembrane default (19)
                Option 2.  Globular surface region (9)
                Option 3.  User defined window

                Enter your selection: 1
Select the window size:
                Option 1.  Transmembrane default (19)
                Option 2.  Globular surface region (9)
                Option 3.  User defined window

                Enter your selection: 2
Select the window size:
                Option 1.  Transmembrane default (19)
                Option 2.  Globular surface region (9)
                Option 3.  User defined window

                Enter your selection: 3
                    >>>> Enter desired window size: 15
Select the window size:
                Option 1.  Transmembrane default (19)
                Option 2.  Globular surface region (9)
                Option 3.  User defined window

                Enter your selection: 4

        ERROR: invalid option! Window size set to default of 7.

*** TEST CASES ***

The 4 window sizes are: 19, 9, 15, and 7.
```

FIG. 13-12. Nested if statements are useful for creating menu-driven interfaces.

def followed by the function name (e.g., define_win_size), a list of input arguments in parenthesis (e.g., () indicates no input arguments), and a colon. The body of the function follows and is indented as shown in Figure 13-12. Functions can return one or more values (multiple values can be returned using a tuple). In this example, the value of the window size is returned. After the function, there are four function calls to the *define_win_size()* function in order to test the different menu options. Figure 13-12B shows the inputs and outputs for a test case. Notice that triple-quoted strings are useful for generating the menu option display. Later in this chapter we will explore user-defined functions in more detail.

Conditional Expressions

Conditional expressions are similar to conditional statements except that they return a value. Thus, they are a useful shorthand. Conditional expressions have the following syntax:

$$x = \text{true\_value if condition else false\_value}$$

Consider the following conditional statement for computing the absolute value of a number:

```
if a > 0:
    abs_val = a
else:
    abs_val = −a
```

This conditional statement can be implemented using the following conditional expression:

```
abs_val = a if a > 0 else −a
```
where the condition is $a > 0$, the true_value is *a*, and the false_value is –*a*.

Loops

In most programs, the same task is performed over and over again. For example, in the sliding window program, the average hydropathy value is repetitively computed as the window slides across the amino acid sequence. Python provides two loop constructs for repetitively (or iteratively) executing a task (i.e., the loop body), a *while* loop and a *for* loop.

for loop A Python statement used to iterate over an ordered sequence such as a list, string, or tuple. As it iterates over the sequence, each value in the sequence can be operated upon.

While loops

The *while* loop is useful for executing a loop while the loop condition is true. For example, when reading data from an input file, a while loop can be used (e.g., while not at the end of the file, keep reading input data). In the user-friendly version of the hydropathy lookup program (Figure 13-8) a while loop is used to continually loop until the user enters a character other than 'y' or 'Y'.

The syntax for a while loop is:

```
while loop condition:
    loop body
```

The keyword *while* is followed by the loop condition and a colon (e.g., *while(reply. lower() == 'y'):*). The body of the loop is then indented beneath this statement. To end the loop body, stop indenting. During program execution, for every iteration of the loop, the loop condition (e.g., *reply.lower() == 'y'*) is tested, and if it is true, the loop body is executed. If the loop condition is false (e.g., reply is not 'y' or 'Y'), the loop will exit. When using while loops, if the loop condition is initially false, the loop body will never execute.

When writing a *while* loop it is important to (1) initialize the loop condition before the loop (e.g., *reply* = '*y*') and (2) update the loop condition in the body of the loop (e.g., *reply* = *raw_input*("*Do you want to look up another value? [y/n]*")). If the loop condition is not updated, then the loop may be an infinite loop that never exits. Notice in the user-friendly hydrophobicity program (Figure 13-8), reply is updated only in the *then* case because the user entered a correct amino acid and its hydrophobicity value has been displayed. The *else* case will be true if the user entered an invalid amino acid. In this case, the loop will automatically execute again (because reply is not updated), which means that the user will be prompted again to enter an amino acid.

Thought Question 13-1

Suppose you want to find the location (index) of the first occurrence of a *G* or *C* nucleotide in a DNA sequence that is stored in a string variable called *dna*. After finding the index, you want to display (print) it to the screen.

(a) First brainstorm how to solve the problem, and then describe your solution in words or pseudo-code. Pseudo-code is code that uses words like *if* and *while* but does not have to use the syntax of any specific programming language.
(b) Write the Python code segment to solve this problem using a while loop.

For loops

In Python a *for* loop is used to iterate over the items of a sequence such as a list, string, or tuple.

The syntax of a *for* loop is:

```
for x in sequence:
    loop body
```

where *x* is an iterating variable.

Consider this Python Shell example iterating over list:

```
>>> hydro_vals = [-1.6, 3.8, -0.8, -3.5, -3.5, -0.7, 2.8, -0.8, -3.5]
>>> for h in hydro_vals:
        print h
-1.6
3.8
-0.8
-3.5
-3.5
-0.7
2.8
-0.8
-3.5
```

This *for* loop iterates over the list *hydro_vals* and prints each element in the list on a separate line. In each iteration of the loop, the variable *h* is assigned to the next element in the list. To print the values on the same line, in Python 2 you can use a comma at the end of the print statement to suppress the newline. For example,

```
>>> for h in hydro_vals:
        print h,
-1.6 3.8 -0.8 -3.5 -3.5 -0.7 2.8 -0.8 -3.5
```

If we wanted to display the absolute value of the hydropathy values, we can use the following loop:

```
>>> for h in hydro_vals:
        print abs(h),
1.6 3.8 0.8 3.5 3.5 0.7 2.8 0.8 3.5
```

abs(h) is a built-in function that returns the absolute value of *h*.

Thought Question 13-2

Suppose you want to determine the percentage of the amino acids within a protein that are strongly hydrophobic (hydropathy value > 3) and display (print) the percentage to the screen.

(a) First brainstorm how to solve the problem and then describe your solution in words or pseudo-code. Again, pseudo-code uses words like *if* and *for* but does not have to use the syntax of any specific programming language.
(b) Write the Python code segment to solve this problem using a for loop.

Your solutions should assume that a string variable called *Protein* has already been created that contains the protein sequence and a dictionary variable *hydro_scale* has already been created that contains the Kyte-Doolittle hydropathy values of each amino acid.

List comprehensions

A convenient way to make lists on the fly is through list comprehensions. List comprehensions use *for* expressions to create lists. The syntax of a list comprehension is:

list_variable = [*expression* for *iteration_variable* in *sequence*]

For example, to create an empty list *L* of 10 values,

```
>>> L = [0 for i in range(10)]
>>> L
[0, 0, 0, 0, 0, 0, 0, 0, 0, 0]
```

where *range(10)* is a built-in function that produces a list of numbers starting from 0 up to 1 less than 10 (e.g., *range(10)* yields [0,1,2,3,4,5,6,7,8,9]).

Range is a useful function for creating lists. You can optionally specify a starting value other than 0 and a step other than 1. Here are a couple more examples of the range function.

```
>>> range(5)
[0, 1, 2, 3, 4]
>>> range(2,5)
[2, 3, 4]
>>> range(0,10,2)
[0, 2, 4, 6, 8]
```

In the last example, the value 2 in *range(0,10,2)* specifies the step.

Getting back to list comprehensions, we can automatically generate the *hydro_vals* list from the *win_seq* string and the *hydro_scale* dictionary as shown below.

```
>>> win_seq = "PLSQETFSD"
>>> hydro_scale = {'a': 1.8, 'c': 2.5, 'd': -3.5, 'e':-3.5, 'f':2.8,
'g':-0.4, 'h':-3.2, 'i':4.5, 'k':-3.9, 'l':3.8, 'm':1.9, 'n':-3.5, 'p':-1.6,
'q':-3.5, 'r':-4.5, 's':-0.8, 't':-0.7, 'v':4.2, 'w':-0.9, 'y':-1.3}
```

```
>>> hydro_vals = [hydro_scale[aa] for aa in win_seq.lower()]
>>> hydro_vals
[-1.6, 3.8, -0.8, -3.5, -3.5, -0.7, 2.8, -0.8, -3.5]
```

Here is another example of a list comprehension used to create a two-dimensional list (or a list of lists). Lists of lists are used to create matrices in Python. Scoring matrices used for sequence alignment are just one of the many uses of two-dimensional lists in bioinformatics.

```
>>> list_2D = [[3*row+col for col in range(3)] for row in range(4)]
>>> list_2D
[[0, 1, 2], [3, 4, 5], [6, 7, 8], [9, 10, 11]]
```

The expression used in this example (e.g., *3*row+col*) assigns each element of the two-dimensional list a unique integer that corresponds to its location in the list. When creating a two-dimensional list, the elements are usually assigned to some initial value which is dependent on how the list is used (often the initial value is 0).

To more easily view a two-dimensional list, a *for* loop can be used to display each list element (row) in the two-dimensional list.

```
>>> for R in list_2D: print R
[0, 1, 2]
[3, 4, 5]
[6, 7, 8]
[9, 10, 11]
```

Thought Question 13-3

The pseudo-code statement:

$$\text{for } i = 0 \text{ to } n-1$$
$$L_i = 0$$

can be used to create a list, L, of n elements where each element is initialized to zero.

(a) In Python programming, which of the following constructs should be used to create L?
   i. While loop
   ii. For loop
   iii. List comprehension
(b) Write Python code to create L using the construct you selected in part (a).

13.8 INPUT AND OUTPUT

By this point, you should have a basic understanding of the different data types, if-tests, loops, list comprehensions, and some built-in functions. With this rudimentary knowledge of Python we are almost ready to develop bioinformatics tools. One remaining area that is important to understand includes Python input and output (I/O) functions. There are two types of I/O functions: standard I/O and file I/O. Standard input refers to input from the keyboard and standard output refers to output displayed on the screen. We have already been using the common standard I/O functions *input*, *raw_input*, and *print* (Table 13-12) in the examples in this chapter.

The *raw_input* function always returns a string, whereas the *input* function will process the input to determine its data type and return a value of that data type.

TABLE 13-12. COMMON STANDARD INPUT/OUTPUT (I/O) FUNCTIONS

FUNCTION	INTERPRETATION
in = input(*promt_string*)	Use the optional prompt string to prompt the user for an input and return the input value which can be any data type.
in = raw_input(*promt_string*)	Use the optional prompt string to prompt the user for an input; raw_input always returns a string.
print x	Print the contents of variable x followed by a newline, where x can be any data type.
print x,	Print the contents of variable x but suppress the newline, where x can be any data type.
print *formatted string*	Print a formatted string.

Consider the following examples. In example A, *raw_input* returns the string '5', whereas in example B, *input* returns the integer 5. If you use *raw_input* you can use the *int( )* function to convert a string to an integer (example C).

>>> x = raw_input("Enter a number: ")   **(example A)**
Enter a number: 5
>>> x
'5'

>>> x = input("Enter a number: ")   **(example B)**
Enter a number: 5
>>> x
5

>>> x = raw_input("Enter a number: ")   **(example C)**
Enter a number: 5
>>> x
'5'
>>> x = int(x)
>>> x
5

Standard I/O is useful for building user interfaces, but many data-intensive programs rely on file I/O where input data can be read directly from a file and output data can be written directly to a file. The data in these files is often produced or used by other programs, and thus you often do not need to worry about nicely formatting data written to output files.

To open a file, the *open* function has the following syntax:

$$myfile = open('pathname', <mode>)$$

where *pathname* indicates the location of the file. If the file is in the same directory (folder) as the program, then the *pathname* is simply the name of the file. The *mode* indicates whether the file is to be read (i.e., it is an input file) or written (i.e., it is an output file). For reading, the *mode* is 'r'; for writing, the mode is 'w'. The variable *myfile* (also known as a file object) will contain information about the file, such as the location of the file and the last location read/written within the file, as it is read or written by the program.

TABLE 13-13. COMMON INPUT FILE FUNCTIONS

FUNCTION	INTERPRETATION
input = open ('file', 'r')	Open input file
S = input.read()	Read entire file into string S
S = input.read(N)	Read N bytes (N>= 1)
S = input.readline()	Read next line
L = input.readlines()	Read entire file into list of line strings

TABLE 13-14. COMMON OUTPUT FILE FUNCTIONS

FUNCTION	INTERPRETATION
output = open('file', 'w')	Create output file
output.write(S)	Write string S into file
output.writelines(L)	Write all line strings in list L into file
output.close()	Manual close (good habit)

For example, on a Windows platform the statement

infile = open("D:\\Docs\\test.txt", 'r')

will open up a file called *test.txt*, which resides in the *Docs* folder on the *D* drive. Note that the pathname uses two backslashes "\\" instead of one because backslash is an escape character in strings. Because the mode is 'r', the file will be an input file.

Here is an example of opening up an output file (mode is 'w') where the file *out.txt* will be created in the same directory as the Python program.

outfile = open("out.txt", 'w')

Tables 13-13 and 13-14 show common file input and output functions, respectively. It is good practice to close a file once the program is finished reading from or writing to the file. When reading a file, you have the option to either read the entire file all at once (using the *read()* or *readlines()* methods) or to read it one byte or one line at a time. If the input file is not too large, it is faster to read the entire file all at once. However, if the input file is large, then this will require a large amount of memory to store the data while the program is executing. In that case, it is probably best to use the *readline()* method to read it in one line at a time. Likewise, when writing to a file you can choose to write the data one line at a time (*writeline()*) or write a list of lines all at once (*writelines()*).

13.9 PROGRAM DESIGN: DEVELOPING KYTE-DOOLITTLE'S HYDROPATHY SLIDING WINDOW TOOL

Now that we have covered the Python basics, we will discuss the program design process (Figure 13-13) and apply the process to develop a Kyte-Doolittle hydropathy sliding window tool. The inputs to the tool are the window size and the filename of a file that contains a protein sequence. The program will slide the window down the

Program design process

1. Understand problem—develop sample input(s) and output(s)
    a. Identify input(s) and output(s)
    b. Work the problem by hand and create test case(s) with sample input(s) and output(s)

2. Develop algorithm—step-by-step, flow chart, or pseudo code
3. Hand test algorithm with sample input(s) and output(s)
4. Refine algorithm (repeating steps 3 and 4) until the algorithm can be coded in target language
5. Code in target language (e.g., Python)
6. Test program with sample input(s) and output(s)
7. Debug—find and fix mistakes in algorithm or coding, repeat steps 6 and 7 until program is working

FIG. 13-13. Seven steps in the program design process.

protein sequence one amino acid at a time. For each window, the program will compute the average hydropathy value. The output of the tool will be stored to a file named "result.txt". The output file will contain the results in a two-column format where the first column will be the index of the first amino acid in the window and the second column will have the corresponding average hydropathy value for that window. The data in each row will be separated by a tab. This data can be used by other programs for further analysis. For example, the hydropathy data can be plotted using a spreadsheet program to visualize the transmembrane regions of a protein.

Step 1: Understanding the Problem

debug Process of identifying and fixing errors (bugs) in a computer program.

The first step in the program design process is to understand the problem the program is supposed to solve. This is the most important step in the process. If you cannot solve the problem by hand, then how can you write a program that instructs the computer how to solve it? If you do not really understand the problem, you cannot test it properly and identify mistakes (this process is called debugging a program).

Let's first identify sample inputs and outputs. To test our program we will use a window size (*win_size*) of 7 and the following protein sequence: PLSQETFSDLWK-LLPENNVLS, which will be stored in an input file called *input1.txt*. Let's call this sequence *aa_seq* in our program. We will also use the dictionary *hydro_scale* that was first introduced in Table 13-3. The dictionary contains the Kyte-Doolittle's hydropathy scale (Table 13-1).

Figure 13-14 shows how we can solve the problem by hand. As we solve it by hand we will also try to understand how to implement it using Python. Because sequences are ordered, let's start by labeling the *aa_seq* with its indices (drawn above each letter in the sequence). So, *aa_seq[0]* = 'P', *aa_seq[1]* = 'L', and so on. Next, let's look at the zeroth window (window 0), which corresponds to the slice *aa_seq[0:6]*. The first window (window 1) is *aa_seq[1:7]*, the second window (window 2) is *aa_seq[2:8]*, and so on. The last window in the sequence (window 15) is *aa_seq[15:21]*. Notice that the window number (*win_num*) corresponds to the start index of each slice and the end of each slice is *win_num* + *window_size* (recall that slices go from the start index to the end index minus 1). To slide the window, all we need to do is increment the window number! To avoid going past the end of the sequence, the program should stop sliding the window when the *win_num* is equal to the length of the protein sequence minus the window size (*win_num* == *len(aa_seq)* − *win_size*).

For each window, we need to compute the average hydropathy value. First, look up the hydropathy value for each amino acid in the window and sum the hydropathy

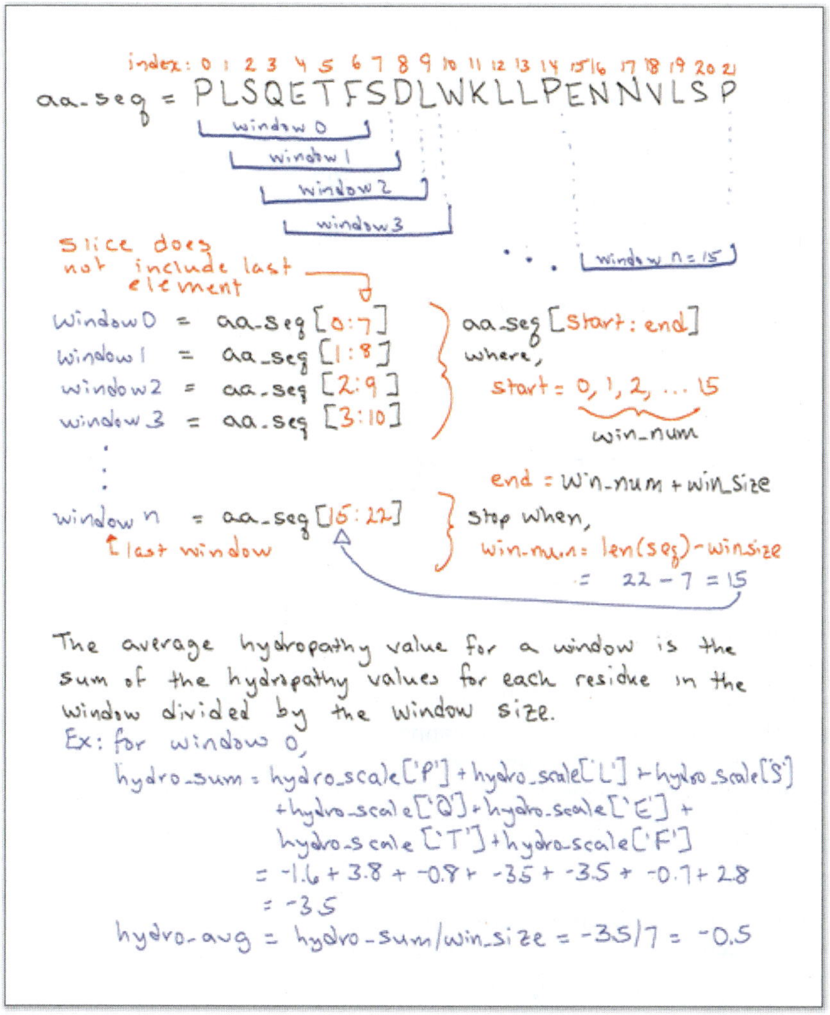

FIG. 13-14. Hand solution to the Kyte-Doolittle hydropathy sliding window problem (step 1 in the program design process). (Recall that Python slices do not include the element at the end of the slice. So, for the *string aa_seq*, the slice *aa_seq[0:7]* consists of characters *aa_seq[0]* to *aa_seq[6]* ('PLSQETF'), which corresponds to window 0.)

values (*hydro_sum*). Next, compute the average hydropathy (*hydro_avg*) by dividing *hydro_sum* by *win_size*. Table 13-15 shows the values for of *hydro_sum* and *hydro_avg* for *win_num* values: 0, 1, 2, 3, and 15. This information can be used during testing and debugging to verify that the program is generating the correct values.

Steps 2 Through 4: Develop and Refine Algorithm

The next steps in the program design process (steps 2 through 4) are to develop an algorithm to solve the problem. It may be tempting to skip the algorithm and directly write your program, but you will find that it will generally take you much longer to complete the program design process. Just like you need to write an outline for a paper to organize your thoughts and order the flow of ideas, you need to develop an algorithm that helps you organize the data and how to process it.

We start with a high-level algorithm (step 2), which typically corresponds to the steps that a person could easily follow to solve the problem. We then test that the algorithm is correct by using the sample inputs and outputs (step 3). Unlike a person, a computer is not intelligent, and so you must tell it exactly what to do. Thus, we continually refine and test the algorithm until we reach a point that a computer can solve the problem (step 4)—that is, until we reach a point where we can easily write the program in the target language, which in our case is Python. As you develop a better understanding of the Python language, step 4 will become easier. Figure 13-15 shows a high-level algorithm and a refined algorithm for the Kyte-Doolittle Hydropathy problem.

TABLE 13-15. SAMPLE INPUTS AND OUTPUTS GENERATED BY HAND (aa_seq = 'PLSQETFSDLWKLLPENNVLSP' AND win_num = 7)

win_num	aa_seq[win_num:win_num+win_size]	hydro_sum	hydro_avg
0	'PLSQETF'	−1.6 + 3.8 + −0.8 + −3.5 + −3.5 + −0.7 + 2.8 = −3.5	−0.5
1	'LSQETFS'	3.8 + −0.8 + −3.5 + −3.5 + −0.7 + 2.8 + −0.8 = −2.7	−0.39
2	'SQETFSD'	−0.8 + −3.5 + −3.5 + −0.7 + 2.8 + −0.8 + −3.5 = −10	−1.43
3	'QETFSDL'	−3.5 + −3.5 + −0.7 + 2.8 + −0.8 + −3.5 + 3.8 = −5.4	−0.77
n = 15	'ENNVLSP'	−3.5 + −3.5 + −3.5 + 4.2 + 3.8 + −0.8 + −1.6 = −4.9	−0.7

High-level algorithm (Step 2)	Refined algorithm (Step 4)
input the window size and protein sequence filename	display hydropathy sliding window banner prompt the user for the window size win_size = user defined value prompt the user for the protein sequence filename seq_filename = user defined value
read the input sequence from the protein sequence file	prot_file = open the protein sequence file aa_seq = read sequence from prot_file
open the output file (result.txt)	outfile = open the output file result.txt

compute and output the average hydropathy for each window	win_num = 0
	hydro_sum = 0
	while not past the end of the sequence (*win_num <= len(aa_seq)– win_size*):
	    for each amino acid in window (*aa_seq[win_num:win+win_size]*):
	        add hydropathy value of the amino acid to hydro_sum
	    hydro_avg = hydro_sum/win_size
	    output the win_num and hydro_avg to outfile
	increment win_num

close the output file	close outfile

FIG. 13-15. High-level and refined algorithms for the Kyte-Doolittle hydropathy sliding window tool (steps 2 and 4 in the program design process).

Just like there are different ways to solve a problem, you can develop different algorithms to solve the same problem. They can differ in simple ways, such as the names chosen for variables, to more significant differences in the way data is represented or processed. Some algorithms may be more efficient in terms of how much computer memory space they require. Other algorithms may run faster. As a beginning programmer, your primary concern is to develop an algorithm that produces the correct results. Just remember that there is not one right way to do it and that your algorithm can be different from someone else's. We will discuss analyzing algorithm efficiency in Chapter 14.

Step 5: Code in Target Language (Python)

After you have developed your refined algorithm, you are ready to code the algorithm using Python (step 5). Figure 13-16 shows the first version (i.e., first draft) of the Kyte-Doolittle Hydropathy program (Kyte_Doolittle_v1.py). Make sure to include a header comment that describes the purpose of the program, the developer's name, and the date that it was written.

Steps 6 and 7: Program Verification (Testing and Debugging)

Just as we needed to verify the algorithm, we also need to verify the program on sample inputs (step 6). It is rare that the first version of a program will work properly.

Kyte-Doolittle_v1.py - C:\Users\Documents\Bioinformatics Text\Python Examples\Kyte-Doolittle_v1.py — Module location

File Edit Format Run Options Windows Help

```
# Kyte-Doolittle Hydropathy sliding window program       ← Header comment
# Written by: Nancy Warter-Perez
# Date created: April 15, 2015

# Kyte-Doolittle Hydropathy scale
hydro_scale = {'a': 1.8, 'c':2.5, 'd': -3.5, 'e':-3.5, 'f':2.8, 'g':-0.4,   ← Dictionary of
               'h':-3.2, 'i':4.5, 'k':-3.9, 'l':3.8, 'm':1.9, 'n':-3.5,        hydropathy values
               'p':-1.6, 'q':-3.5, 'r':-4.5, 's':-0.8, 't':-0.7, 'v':4.2,
               'w':-0.9, 'y':-1.3}

print """ ********* Kyte-Doolittle Hydropathy Sliding Window Tool *********

Description:     This program implements a sliding window program where the     ← Program banner
                 average hydropathy value for each window is computed. The         lets the user know
                 window number and average hydropathy value for each window         how to use the
                 will be output to a tab-delimited text file (result.txt).          tool

Instructions:    1. Create an input text file with the desired protein sequence
                    List only amino acids in the sequence (using upper or
                    lower case symbols). Do not include any spaces or newline
                    characters.
                 2. Enter the filename of the protein sequence.
                 3. Enter the desired window size."""

prot_filename = raw_input("Protein sequence filename:  ")       ← Prompt user for
prot_file = open(prot_filename, 'r')                               inputs and
aa_seq = prot_file.read()                                          retrieve inputs
prot_file.close()
aa_seq = aa_seq.lower()
win_size = input("Window size:  ")
outfile = open("result.txt", 'w')                              ← Open output file

win_num = 0                                                    ← Compute average
hydro_sum = 0                                                     hydropathy value
while win_num <= len(aa_seq) - win_size:                          for each window
    for aa in aa_seq[win_num:win_num + win_size]:                 and output to
        hydro_sum = hydro_sum + hydro_scale[aa]                   result file
    hydro_avg = hydro_sum/win_size
    outfile.write("%d\t%f\n"%(win_num,hydro_avg))
    win_num += 1
outfile.close()                                               ← Close output file
print "The results have been stored to the result.txt file.\n"   and end program
dummy = raw_input("Press enter to exit.")
```

FIG. 13-16. Step 5—Code the algorithm in Python. This is the first version of the program (Kyte_Doolittle_v1.py).

```
protein.txt - C:/Users/Documents/Bioinformatics Text/Python Examples/protein.txt

File  Edit  Format  Run  Options  Windows  Help

PLSQETFSDLWKLLPENNVLSP
```

FIG. 13-17. Sample protein sequence input file (*protein.txt*).

There are different kinds of mistakes that program developers make. Mistakes in programs are typically called bugs. The simplest kind of bug to discover and fix is a syntax error. These types of errors arise because you did not follow the syntax (language rules). For example, you may have forgotten a colon at the end of a *while* statement.

Another type of bug arises from improperly coding the algorithm in the target language. As your knowledge of Python semantics and functionality improves, you will have fewer of these types of bugs. Figure 13-17 shows the sample protein sequence input file (*protein.txt*) used to test the Kyte-Doolittle Hydropathy program. Figure 13-18 shows the results of running the first version of the Kyte-Doolittle Hydropathy program in the interactive Python Shell. The interpreter detected an error as the program ran and it generated the traceback error. Although the format of traceback errors is not very user-friendly, the information that they provide is very useful in debugging programs. In this case, the traceback error indicates that on line 39 of the module, corresponding to the statement *hydro_sum = hydro_sum + hydro_scale[aa]*, it encountered a *KeyError*. The key being referred to is the amino acid variable *(aa)*, which is being used to look up the hydropathy value in the *hydro_scale* dictionary. The traceback error message indicates that the invalid key is '\n'. To see where this invalid key came from, we can query the contents of the *aa_seq* variable in the interactive window. The last character in the *aa_seq* string is the newline character ('\n'). Referring back to Figure 13-17, it is clear that there is a newline at the end of the string (put in automatically by the IDLE editor used to create the protein.txt file).

Figure 13-19 shows the second version of the program with this bug fixed. The original statement *aa_seq = aa_seq.lower()*, which converts the sequence to lowercase letters, has been modified to *aa_seq = aa_seq[:−1].lower()*. Before converting to lowercase, *aa_seq[:−1]*, will create a slice of *aa_seq* starting at index 0 and ending at the character before the last character in the string, effectively slicing off the newline character.

The next type of bug arises from a mistake in the original algorithm. Even though you hand-tested your algorithm, it is still possible for it to be incorrect because humans, being intelligent, will make assumptions that the computer cannot. Testing version 2 of the Kyte-Doolittle Hydropathy program (Figure 13-19), we see that the program runs without any traceback errors (Figure 13-20). However, if

```
********* Kyte-Doolittle Hydropathy Sliding Window Tool *********

Description:    This program implements a sliding window program where the
                average hydropathy value for each window is computed. The
                window number and average hydropathy value for each window
                will be output to a tab-delimited text file (result.txt).

Instructions:   1. Create an input text file with the desired protein sequence
                   List only amino acids in the sequence (using upper or
                   lower case symbols). Do not include any spaces or newline
                   characters.
                2. Enter the filename of the protein sequence.
                3. Enter the desired window size.
Protein sequence filename: protein.txt
Window size: 7

Traceback (most recent call last):
  File "C:/Users/nwarter/Documents/Bioinformatics Text/Python Examples/Kyte_Doolittle.py", line 39, in <module>
    hydro_sum = hydro_sum + hydro_scale[aa]
KeyError: '\n'
>>> aa_seq
'plsqetfsdlwkllpennvlsp\n'
```

FIG. 13-18. Step 6—Test first version of the program (Kyte_Doolittle_v1.py) with sample inputs. The traceback error message shows that there is a bug in the program.

```
# Kyte-Doolittle Hydropathy sliding window program
# Written by: Nancy Warter-Perez
# Date created: April 15, 2015

# Kyte-Doolittle Hydropathy scale
hydro_scale = {'a': 1.8, 'c':2.5, 'd': -3.5, 'e':-3.5, 'f':2.8, 'g':-0.4,
               'h':-3.2, 'i':4.5, 'k':-3.9, 'l':3.8, 'm':1.9, 'n':-3.5,
               'p':-1.6, 'q':-3.5, 'r':-4.5, 's':-0.8, 't':-0.7, 'v':4.2,
               'w':-0.9, 'y':-1.3}


print """ ********* Kyte-Doolittle Hydropathy Sliding Window Tool *********

Description:      This program implements a sliding window program where the
                  average hydropathy value for each window is computed. The
                  window number and average hydropathy value for each window
                  will be output to a tab-delimited text file (result.txt).

Instructions:     1. Create an input text file with the desired protein sequence
                     List only amino acids in the sequence (using upper or
                     lower case symbols). Do not include any spaces or newline
                     characters.
                  2. Enter the filename of the protein sequence.
                  3. Enter the desired window size."""

prot_filename = raw_input("Protein sequence filename:  ")
prot_file = open(prot_filename, 'r')
aa_seq = prot_file.read()
prot_file.close()
aa_seq = aa_seq[:-1].lower()  # BUG FIX -- strip off newline character at end of string
win_size = input("Window size:  ")
outfile = open("result.txt", 'w')

win_num = 0
hydro_sum = 0
while win_num <= len(aa_seq) - win_size:
    for aa in aa_seq[win_num:win_num + win_size]:
        hydro_sum = hydro_sum + hydro_scale[aa]
    hydro_avg = hydro_sum/win_size
    outfile.write("%d\t%f\n"%(win_num,hydro_avg))
    win_num += 1
outfile.close()
print "The results have been stored to the results.exe file.\n"
dummy = raw_input("Press enter to exit.")
```

FIG. 13-19. Second version of the program (Kyte_Doolittle_v2.py) (step 7). A bug in the program was fixed by removing (slicing off) the newline character from the end of *aa_seq string*.

```
>>> =============================== RESTART ===============================
>>>
Please enter an amino acid: A
 ********* Kyte-Doolittle Hydropathy Sliding Window Tool *********

Description:      This program implements a sliding window program where the
                  average hydropathy value for each window is computed. The
                  window number and average hydropathy value for each window
                  will be output to a tab-delimited text file (result.txt).

Instructions:     1. Create an input text file with the desired protein sequence
                     List only amino acids in the sequence (using upper or
                     lower case symbols). Do not include any spaces or newline characters.
                  2. Enter the filename of the protein sequence.
                  3. Enter the desired window size.
Protein sequence filename: protein.txt
Window size: 7
The results have been stored to the result.txt file.

Press enter to exit.
```

FIG. 13-20. Step 6—Continue testing. The program ran successfully without any traceback errors.

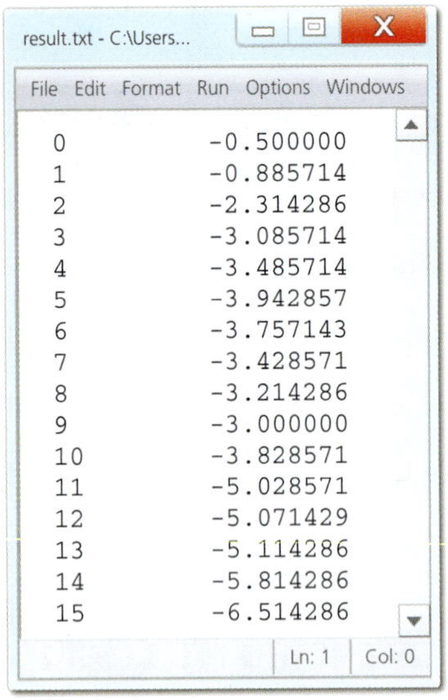

```
result.txt - C:\Users...
File  Edit  Format  Run  Options  Windows
   0           -0.500000
   1           -0.885714
   2           -2.314286
   3           -3.085714
   4           -3.485714
   5           -3.942857
   6           -3.757143
   7           -3.428571
   8           -3.214286
   9           -3.000000
  10           -3.828571
  11           -5.028571
  12           -5.071429
  13           -5.114286
  14           -5.814286
  15           -6.514286
                              Ln: 1   Col: 0
```

FIG. 13-21. Step 6—Verify sample output comparing with hand-computed output from Table 13-15. Sample output does not agree with hand-computed output.

we compare the results in the output file (Figure 13-21), we see that they do not agree with the hand-generated results (Table 13-15).

In some cases it may be easy to determine the mistake just by comparing the results to the expected results, and in other cases you will need to debug your program. In the debugging process, you need to trace the execution of the program to verify that the variables are being updated as expected. A simple debugging technique is to insert print statements at different points in the program. A better approach is to use the debugger provided with the development environment.

IDLE provides a relatively simple but effective debugger, shown in Figure 13-22. To invoke the debugger, in the IDLE Python Shell click on the *Debug* tab and select *Debugger*. The *Debug Control* window will pop up. In the IDLE editor window start debugging the module by clicking on the *Run* tab and selecting *Run Module* (or by clicking the F5 shortcut key). The program will start to run in debug mode, stopping at the first instruction of the module.

The *Debug Control* window (on left in Figure 13-22) has checkboxes that allow you to specify what information you want to view: *Stack*, *Locals*, *Globals*, and *Source*. The *Stack* area at the top shows the current instruction being executed (*hydro_sum = hydro_sum + hydro_scale[aa]*). If you are executing a program with user-defined functions, the stack window will also show the call stack of functions. We will explore this aspect of the debugger in more detail later in this chapter. The *Locals* area shows the current values of the local variables (*none*), which are variables defined and used within a function. The *Globals* area shows the current values of the global variables (*aa*, *aa_seq*, etc.), which are variables that can be used and modified by all functions within a module. If you click the *Source* check box it will pop up an IDLE editor window (on right in Figure 13-22) showing the instructions around the current instruction executing.

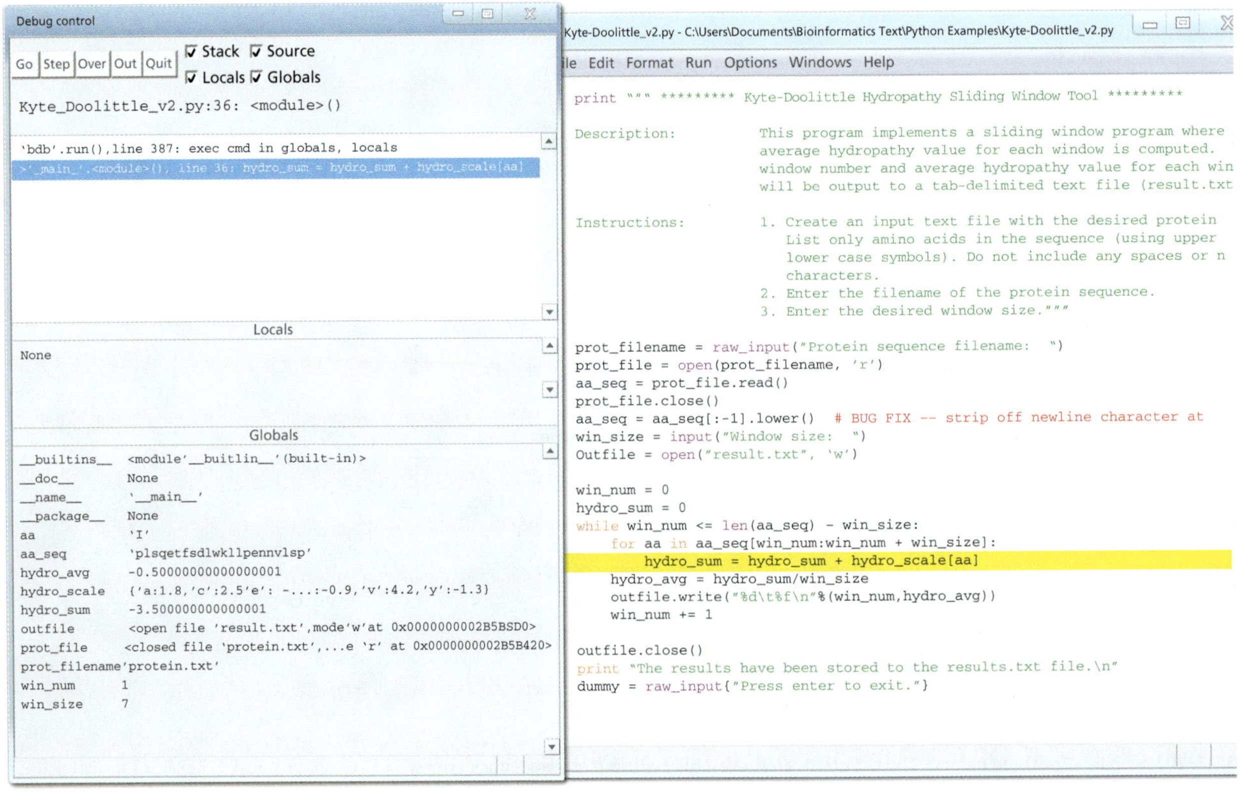

FIG. 13-22. Step 7—Debug the Kyte_Doolittle_v2.py program using IDLE debugger. The program is stopped at the beginning of computing *hydro_sum* for the second window (*win_num = 1* and *aa = 'l'*). Note *hydro_sum = −3.5*, but it should be zero.

There are five control buttons in the upper left-hand corner of the *Debug Control* window that allow you to select one of the following actions: *Go, Step, Over, Out,* and *Quit. Go* will execute the program until it reaches a breakpoint (a predetermined stopping point). If no breakpoint has been set, it will execute the entire module (assuming no traceback errors occur). If the program prompts the user for input, the debugger will wait until you have input the information in the IDLE Python Shell window. Setting a breakpoint is a good way to test small sections of the program to see if the execution is correct up to the breakpoint.

To set a breakpoint, put the cursor in the IDLE editor window over the point in the program where you want to stop, right- click the mouse, and select *Set Breakpoint.* After setting a breakpoint you can clear it if you no longer want to stop at that location in the program. This is accomplished by again putting the cursor at that point, right clicking the mouse, and selecting *Clear Breakpoint.* You can set multiple breakpoints in a program. After one or more breakpoints are set, you can click the *Go* button and the program will execute until it reaches the first breakpoint. Clicking *Go* again, the program will continue executing from that point until it reaches the next breakpoint. If the breakpoint is set in the body of a loop, the program may stop at the same point again, but it will actually be the next iteration of the loop. Breakpoints allow you to quickly jump to the location in the program where you think the bug may be occurring.

The next *Debug Control* button is *Step. Step* allows you to step instruction by instruction through the program. This is useful for pinpointing the exact location of the

local variable A variable with a local scope is visible only within the block of code in which it is defined. For example, variables defined within a function by default can be used only within the function.

global variable A variable defined outside of any function and can be used by any function.

```
# Kyte-Doolittle hydropathy sliding window program
# Written by: Nancy Warter-Perez
# Date created: April 16, 2015

# Kyte-Doolittle Hydropathy scale
hydro_scale = {'a': 1.8, 'c':2.5, 'd': -3.5, 'e':-3.5, 'f':2.8, 'g':-0.4,
               'h':-3.2, 'i':4.5, 'k':-3.9, 'l':3.8, 'm':1.9, 'n':-3.5,
               'p':-1.6, 'q':-3.5, 'r':-4.5, 's':-0.8, 't':-0.7, 'v':4.2,
               'w':-0.9, 'y':-1.3}

print """ ********* Kyte-Doolittle Hydropathy Sliding Window Tool *********

Description:    This program implements a sliding window program where the
                average hydropathy value for each window is computed. The
                window number and average hydropathy value for each window
                will be output to a tab-delimited text file (result.txt).

Instructions:   1. Create an input text file with the desired protein sequence
                   List only amino acids in the sequence (using upper or
                   lower case symbols). Do not include any spaces or newline
                   characters.
                2. Enter the filename of the protein sequence.
                3. Enter the desired window size."""

prot_filename = raw_input("Protein sequence filename:  ")
prot_file = open(prot_filename, 'r')
aa_seq = prot_file.read()
prot_file.close()
aa_seq = aa_seq[:-1].lower()  # BUG FIX -- strip off newline character at end of string
win_size = input("Window size:  ")
Outfile = open("result.txt", 'w')

win_num = 0
hydro_sum = 0  # BUG FIX -- need to initialize hydro_sum for every window
while win_num <= len(aa_seq) - win_size:
    for aa in aa_seq[win_num:win_num + win_size]:
        hydro_sum = hydro_sum + hydro_scale[aa]
    hydro_avg = hydro_sum/win_size
    outfile.write("%d\t%f\n"%(win_num,hydro_avg))
    win_num += 1
outfile.close()
print "The results have been stored to the result.txt file.\n"
dummy = raw_input("Press enter to exit.")
```

FIG. 13-23. The third version of the program (Kyte_Doolittle_v3.py) (step 7) fixed another bug by moving initialization of *hydro_sum* inside the *while* loop.

```
result.txt - C:\Users...                    ▢ ▢  X

File  Edit  Format  Run  Options  Windows

 0              -0.500000              ▲
 1              -0.385714
 2              -1.428571
 3              -0.771429
 4              -0.400000
 5              -0.457143
 6               0.185714
 7               0.328571
 8               0.214286
 9               0.214286
10              -0.828571
11              -1.200000
12              -0.042857
13              -0.042857
14              -0.700000
15              -0.700000              ▼

                          Ln: 1   Col: 0
```

FIG. 13-24. Step 6—Rerun the program (Kyte_Doolittle_v3.py) and compare results in output file *results.txt* with hand-computed results in Table 13-15. The results agree, and so the program is working properly.

bug. If you encounter a function while stepping through the program, you can step over it by clicking the *Over* button. Similarly if you are inside a function and want to get out easily, you can click the *Out* button. This is a very useful button because it is easy to accidentally step into a function that you do not want to debug (perhaps you have already fully tested that function and know that it works). Once you have discovered your bug, you can exit the debugger by clicking the *Quit* button.

To debug the second version of the Kyte-Doolittle Hydropathy program (Figure 13-22), we know the *hydro_avg* values are incorrect and we know that *hydro_avg* depends on *hydro_sum*, and so, we set a breakpoint on line 36 (*hydro_sum = hydro_sum + hydro_scale[aa]*). The computation for the first window appears to be fine, but when the computation for the second window starts (*win_num = 1*), we notice that *hydro_sum* is not zero. We have discovered the mistake in our algorithm; we initialize *hydro_sum* to zero only once instead of at the beginning of each window. Figure 13-23 shows the third version of the Kyte-Doolittle Hydropathy program with this bug fixed. The results from running this version are shown in Figure 13-24. Because the results agree with our hand-computed values in Table 13-15, we conclude that the program is working properly.

Thought Question 13-4

Whenever you develop a program, it will almost certainly not work correctly the first time, so it is important to learn how to debug your programs.

(a) Why it is important to have a really good understanding of the problem you are trying to solve when trying to debug your code?

(b) In which of the following steps of the program design process could mistakes be introduced, and how can those mistakes be found?

   i. Work the problem by hand, and create test case(s) with sample input(s) and output(s).

   ii. Develop an algorithm—step-by-step, flow chart, or pseudo-code.

   iii. Hand-test algorithm with sample input(s) and output(s).

   iv. Code in target language (e.g., Python).

   v. Test program with sample input(s) and output(s).

   vi. Debug—find and fix mistakes in algorithm or coding.

13.10 HIERARCHICAL DESIGN: FUNCTIONS AND MODULES

When solving difficult problems, it is usually a good idea to break them into smaller subproblems that are easier to understand, have fewer variables to consider, and thus are easier to solve. Then, the solutions to the subproblems can be combined into an overall solution. This is also how software engineers develop large software projects. The process is formally referred to as hierarchical design or modular development. In a hierarchy there is a top-level task that controls the overall flow, and then smaller subtasks that perform more specific tasks. Note that subtasks can also be broken into successively smaller subtasks as well. In addition to being beneficial for the individual developer, hierarchical design also allows large software projects to be divided into smaller tasks that can be developed in parallel by different scientists/developers. Hierarchical design requires well-defined specifications for each subtask including its inputs, outputs, and functionality. Fortunately, the Python programming language, like many other programming languages, has functions and modules to support modular development and hierarchical design.

Python Functions

A *function* is a piece of code that performs a specified task. Functions are *defined* in a program and then are used when the function is *called*. When a function is called, control of the program jumps to the function where the instructions in the function are executed. After the instructions in the function are executed, control *returns* back to the statement that called the function. A function can also call other functions to perform subtasks. Functions can be developed once and reused many times. A function can be called multiple times from within the same program, either from the same location within a loop or from different locations in the program. If it is a very common function that would be useful for many programs, it can be placed in a library or included as a built-in function within the language.

In a program, functions must be defined before they are called.[5] The syntax for a Python function definition is:

> def *function_name*(*function_parameter1, function_ parameter2, . . .*):
>   *function_body*
>   return *return_value*

Function names follow the same naming conventions as variables: they can include only letters, numbers, and underscores; they cannot start with a number; and they cannot be keywords. Inputs to a function are referred to as parameters, and they also follow the variable naming convention. When functions are called, parameters are assigned the values that are being passed to the function, which are called arguments. For example, in Figure 13-25, the function name is *adder* and the function's parameters are *a*, *b*, and *c*. In the example, the function is called four different times. The first time the function is called, *adder(1, 2, 3)*, the argument assigned to parameter *a* is *1*, to parameter *b* is *2*, and to parameter *c* is *3*. The function body contains Python statements that perform the desired task. In this example, the body of the function performs addition on the parameters *a*, *b*, and *c* and assigns the resulting sum to *x*.

A function can return zero or more values. Some functions perform a task that does not require a return value, such as printing a value to the computer's display. In the example in Figure 13-25, the value of the variable *x* is returned to the calling statement. In the first function call, the value of 6 (1 + 2 + 3 = 6) is returned. In this case, the returned value is directly passed as an argument to the *print* function, a built-in function,[6] to display the value on the screen.

Python functions are polymorphic, which means that the same function can work for arguments with different data types. Figure 13-25 shows how the adder function can be used for integers, floating-point numbers, strings, and lists. When integers or floating-point numbers are passed to the *adder* function, it adds the numbers and returns the sum. When strings and lists are passed to the *adder* function, it concatenates the values assigned to *a*, *b*, and *c* and returns the new string or list.

The scope of global and local variables

Recall that in Python, a variable is defined when it is assigned a value. For example, in Figure 13-25, the variable *x* is defined by the statement *x = a + b + c*. The *scope*

A

```
def adder (a,b,c):
    x = a + b + c
    return x

print "Functions in Python are polymorphic"
print "1. Intergers:",
print adder(1,2,3)
print "2. Floats:",
print adder(1.5, 3.8, -6.4)
print "3. Strings:",
print adder('ac', '-', 'tg')
print "4. Lists:",
print adder([2, 4, 6], [4, 5], [7, 11])
```

B

```
>>>
Functions in Python are polymorphic
1. Intergers: 6
2. Floats: -1.1
3. Strings: ac-tg
4. Lists: [2, 4, 6, 4, 5, 7, 11]
```

FIG. 13-25. Python functions are polymorphic. Supported data types depend on the operators in the function. **A.** Python function adder: to illustrate polymorphic nature of python functions, adder is called with integer, floating point, string, and list arguments. **B.** Sample output: the '+' operator can mean add or concatenate. (Note that the built-in print function is arbitrarily colored orange or purple because in Python 2.7 it is treated as both a keyword and a built-in function. So, do not be concerned if you notice the different colors used for the print function when viewing your own programs in IDLE. This is an artifact of the evolving nature of Python and this colorization problem is fixed in Python 3.)

---

[5] As we will see later in the chapter, recursive functions contain calls to the same function, but a recursive call cannot be the first call to the function.

[6] Note that *print* is an abnormal function in Python 2 because it does not require the use of parentheses around the arguments.

of a variable, the region of the program where it can be used, depends on where the variable is defined. Variables that are defined within a function are considered *local variables*. The scope of a local variable is the function body. Thus, local variables cannot be used outside of the function where they are defined. Parameters (inputs to functions) are also considered local variables.

Variables that are defined outside of a function (at the top level) are known as *global variables*. Global variables can be used anywhere in a Python program. Variables defined within a function can also be global variables if they are formally defined as global variables using the *global* operator. For example, to make the variable x global in the *adder* function, the statement "global x" could have been inserted before the $x = a + b + c$ statement. If x were a global variable, then its value would not need to be returned. In general, it is considered better programming practice to use local variables, passing values into functions and returning them from functions, than to use only global variables.

To gain a better understanding of the hierarchical structure that functions provide and the scope of local and global variables, we have modified the Kyte-Doolittle Hydropathy sliding window program as shown in Figure 13-26. The top level of the

```
*Kyte-Doolittle_v3.(with functions).py - C:/Users/Documents/Bioinformatics Text/Python Examples/Kyte-Doolittle_v3.(with functi...
File  Edit  Format  Run  Options  Windows  Help
# Kyte-Doolittle Hydropathy sliding window program (with functions)
# Written by: Nancy Warter-Perez
# Date created: April 16, 2015

# Kyte-Doolittle Hydropathy scale
hydro_scale = {'a': 1.8, 'c':2.5, 'd': -3.5, 'e':-3.5, 'f':2.8, 'g':-0.4,
               'h':-3.2, 'i':4.5, 'k':-3.9, 'l':3.8, 'm':1.9, 'n':-3.5,
               'p':-1.6, 'q':-3.5, 'r':-4.5, 's':-0.8, 't':-0.7, 'v':4.2,
               'w':-0.9, 'y':-1.3}

def user_input ():
    print """ ********* Kyte-Doolittle Hydropathy Sliding Window Tool *********

    Description:   This program implements a sliding window program where the
                   average hydropathy value for each window is computed. The
                   window number and average hydropathy value for each window
                   will be output to a tab-delimited text file (result.txt)."""

    prot_filename = raw_input("Protein sequence filename:  ")
    prot_file = open(prot_filename, 'r')
    aa_seq = prot_file.read()
    aa_seq = aa_seq[:-1].lower()  #strip off newline character and convert to lower case
    prot_file.close()
    win_size = input("Window size:  ")
    return(aa_seq, win_size)

def compute hydro(aa_seq, win_size):
    outfile = open("result.txt", 'w')
    win_num = 0
    while win_num <= len(aa_seq) - win_size:
        hydro_sum = 0
        for aa in aa_seq[win_num:win_num + win_size]:
            hydro_sum = hydro_sum + hydro_scale[aa]
        hydro_avg = hydro_sum/win_size
        outfile.write("%d\t%f\n"%(win_num,hydro_avg))
        win_num += 1
    outfile.close()
    print "The results have been stored to the result.txt file.\n"

(seq, win_size) = user_input()
compute_hydro(seq, win_size)
dummy = raw_input("Press enter to exit.")
                                                              Ln: 41  Col: 28
```

FIG. 13-26. Kyte-Doolittle Hydropathy sliding window program modified to include user-defined functions user_input and compute_hydro.

program now consists of the global variable definition of the dictionary *hydro_scale*, two function definitions (subtasks), *user_input* and *compute_hydro*, and three instructions. The first instruction calls the *user_input* function to get the amino acid sequence and window size. The second instruction calls the *compute_hydro* function, which, for the given sequence and window size, computes the hydrophobicity values and writes them to the output file. The third instruction is the call to the *raw_input* function, which prevents the pop-up terminal window from closing until the user inputs a value.

Let's take a look at how the local and global variables are used and how data is passed into and returned from the functions. The function *user_input* has no input parameters. Rather, it prompts the user to provide the inputs including the filename of the input sequence and the window size. The *user_input* function defines and uses four local variables, *prot_filename, prot_file, aa_seq,* and *win_size*. The *user_input* function has two return values, the input amino acid sequence and the window size. It passes the data in the *aa_seq* and *win_size* variables back to the calling function or top-level program using parentheses to group them in a tuple. In the top-level program, the values are assigned to the global variables *seq* and *win_size*, respectively.

The *compute_hydro* function has two input parameters, the amino acid sequence (*aa_seq*) and the window size (*win_size*). The data stored in the global variables *seq* and *win_size* are passed as arguments to the *aa_seq* and *win_size* parameters when the *compute_hydro* function is called. The *compute_hydro* function defines and uses five local variables, *outfile, win_num, hydro_sum, aa,* and *hydro_avg*. It also uses the global variable *hydro_scale*. The *compute_hydro* function has no return values. It outputs the average hydropathy values for each window to an output file. The function automatically returns when it reaches the end of the function body. This example illustrates how data can be passed around a program through variables.

Note that sometimes function arguments are constants (as shown in Figure 13-25) and sometimes the function arguments are variables (as shown in Figure 13-26). The benefit of using variables is that they are defined as the program executes. For example, *win_size* depends on the input value specified by the user. Note that parameter names do not have to be the same as the argument names, and, in fact, often are not. In this example, *win_size* is used to refer to both the parameter and argument variable, but *aa_seq* is the parameter name and *seq* is the name of the argument variable.

To gain a better understanding of function calls, local variables, and global variables, two breakpoints have been set in the Kyte-Doolittle Hydropathy sliding window program as shown in Figure 13-27 by the highlighted lines. As the program runs in the debugger it will stop at each breakpoint. Figure 13-28 shows the debug control window at the first breakpoint. Notice that the call stack, or stack for short, shows which functions are executing at that point in the program and the order in which functions have been called with the most recently called function, highlighted in blue, at the top of the stack. The call stack in Figure 13-28 shows the call stack growing down in the window so that the top of the stack is the last function shown. Looking at the call stack in Figure 13-28 we see that the main top-level script has called the *user_input* function, which is on the top of the call stack. The local variables defined in *user_input* are shown in the Locals window, and the global variables defined up to this point in the main script are shown in the Globals window. Note that the Globals window shows everything that is defined globally, including variables and functions.

Figure 13-29 shows the debug control window at the second breakpoint. Notice the call stack now shows that the main script has called the *compute_hydro* function. Now local variables defined in *compute_hydro* are shown in the Locals window and the global variables defined up to this point in the main script are shown in the Globals window.

Recursive functions

Functions that call themselves are called *recursive functions* and are often used to solve dynamic programming problems, such as pairwise sequence alignment, where

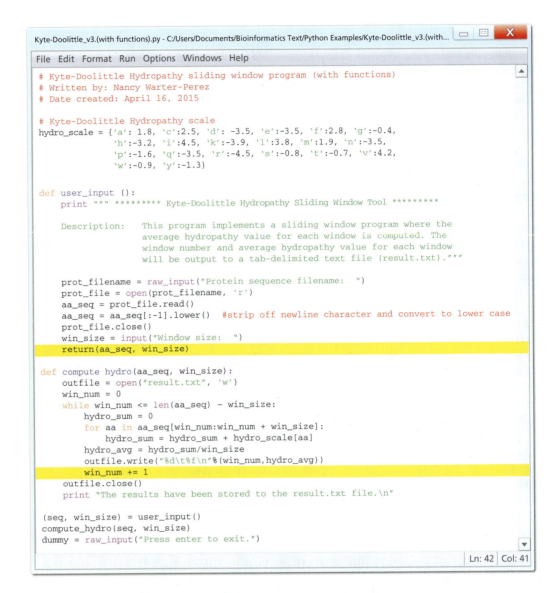

FIG. 13-27. Kyte-Doolittle Hydropathy sliding window program with two breakpoints set (highlighted lines). As the program runs in the debugger it will stop at each breakpoint.

the solution can be derived from solutions to smaller subproblems. As the function is called recursively, the parameters of the function define smaller and smaller subproblems to solve. When the stop condition is reached, an initial solution is determined based on the smallest subproblem. As the recursive function returns, the solution is incrementally formed from the solutions of the subproblems until the recursion ends and the entire solution is formed.

How does a recursive function execute? Think about putting plates on top of one another in a stack until someone says stop. After the person says stop, remove the plates from the stack one at a time until the stack is empty. Using the plate-stacking analogy, a plate corresponds to an instance of the function. Note that the statements in a function body do not change as a function executes; instead, just the values of the local variables change as different arguments are passed to the function. Thus, the function instance consists of the function's arguments and local variables. Each recursive function call creates another function instance (plate) that is put on the top of the stack. The function instance that is on the top of the stack contains the arguments and local variables for the currently executing function. A recursive

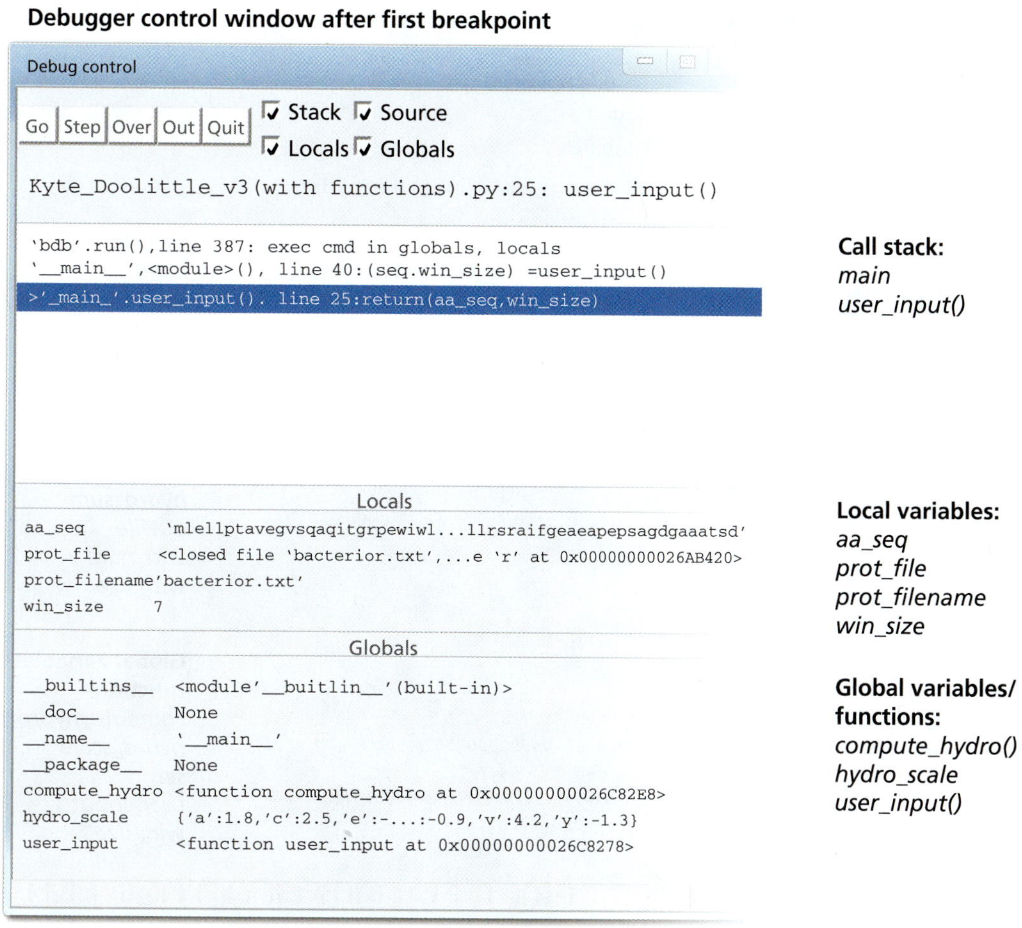

Debugger control window after first breakpoint

```
Debug control                                              ▭ ▢

 Go | Step | Over | Out | Quit    ☑ Stack ☑ Source
                                  ☑ Locals ☑ Globals

 Kyte_Doolittle_v3(with functions).py:25: user_input()

 'bdb'.run(),line 387: exec cmd in globals, locals
 '__main__',<module>(), line 40:(seq.win_size) =user_input()
>'_main_'.user_input(). line 25:return(aa_seq,win_size)




                          Locals
 aa_seq         'mlellptavegvsqaqitgrpewiwl...llrsraifgeaeapepsagdgaaatsd'
 prot_file      <closed file 'bacterior.txt',...e 'r' at 0x00000000026AB420>
 prot_filename'bacterior.txt'
 win_size      7
                          Globals

 __builtins__   <module'__buitlin__'(built-in)>
 __doc__        None
 __name__       '__main__'
 __package__    None
 compute_hydro <function compute_hydro at 0x00000000026C82E8>
 hydro_scale    {'a':1.8,'c':2.5,'e':-...:-0.9,'v':4.2,'y':-1.3}
 user_input     <function user_input at 0x00000000026C8278>

```

Call stack:
main
user_input()

Local variables:
aa_seq
prot_file
prot_filename
win_size

**Global variables/
functions:**
compute_hydro()
hydro_scale
user_input()

FIG. 13-28. Debug control window at the first breakpoint of the Kyte-Doolittle Hydropathy sliding window program.

function will continue to call itself until it reaches a defined stop condition, after which it returns out of the recursion. A function instance (plate) is removed from the stack when the function reaches a *return* statement or when it reaches the end of the function body.

As the program executes we can follow what is happening in the debugger. As the examples in Figures 13-28 and 13-29 illustrated, the call stack in the debugger shows which function calls are on the stack, and the Locals window shows only the values of the local variables of the function that is currently executing—that is, the function that is on the top of the stack.

Figure 13-30A shows a recursive function, *fn_rec*, to reverse a list. Figure 13-30B shows the values of *l* that are printed each time the function encounters the print statement. This example illustrates how partial solutions are formed, and how they can be used to form a complete solution. In Figure 13-30A we can see that a breakpoint is set at the return statement in *fn_rec*. Figure 13-30C shows the debug control window after the fifth recursive call. In the call stack we can see that the main script has called the *fn_rec* function and then the *fn_rec* function has called itself four times.

Let's take a closer look at what is happening as *fn_rec* executes. First, the function parameter, *x*, is the list that needs to be reversed. The first statement in the function checks to see if the list is empty, and, if so, the function returns an empty list. This is the stop condition. If the stop condition is false, then the next statement

Debugger control window after second breakpoint

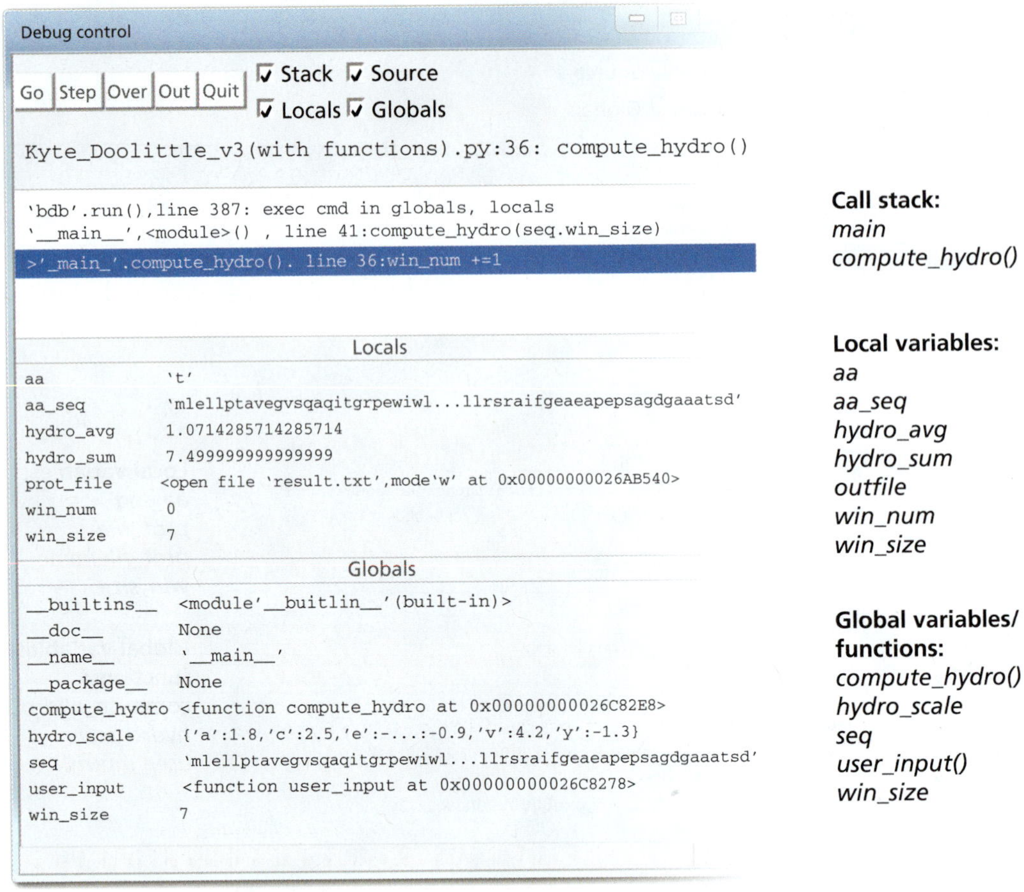

FIG. 13-29. Debug control window at the second breakpoint of the Kyte-Doolittle Hydropathy sliding window program.

recursively calls the same function. Note that the argument for the function call is *x*[1:], which is a slice of list *x* with the first element of the list removed. Thus, this is a subproblem of the initial problem. For example, when the function is initially called, the function parameter *x* is assigned the list [1,2,3,4]. The second time the function is called, the function parameter *x* is assigned the list [2,3,4]. The function continues to be recursively called with smaller and smaller lists until an empty list is passed. When the empty list is detected by the stop condition, it returns the initial solution, which in this example is an empty list.

Where does the program execution go when the function returns? Functions always return to the statement that calls the function. In this case, the function returns to the last statement that called the function and assigns the empty list to the local variable *l*. After assigning the local variable *l*, the next statement in the function body is to print the value of *l*. Thus, the first print statement encountered will print the empty list. The next statement in the function body is to return from the function. The return value is the list *l* appended with first element of list *x*. Now, this part is a little tricky because you have to know what the current value of *x* is. Recall that *x* changes each time the function is called. The value of *x* for the current function call that is on top of the stack is [4].

Using the plate analogy, Figure 13-31 shows the local variables on the stack as the function *fn_rec* executes, where the values of the currently executing function on the top of the stack are shown in blue (these are the values shown in the Locals window in the debugger). Figure 13-31A shows the values of *x* as the function *fn_rec*

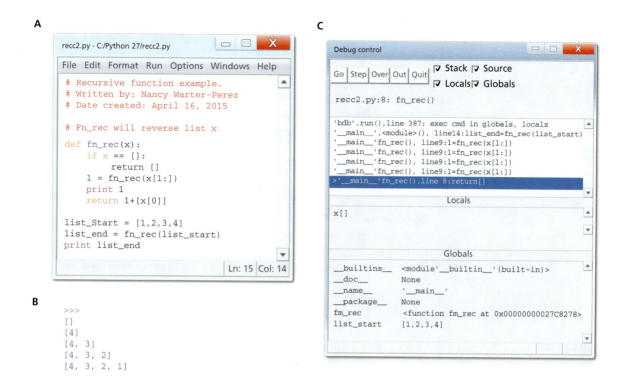

FIG. 13-30. A. A recursive function with breakpoints highlighted. **B.** The sample output after the program executes. **C.** The debug control window after the fifth recursive call to fn_rec.

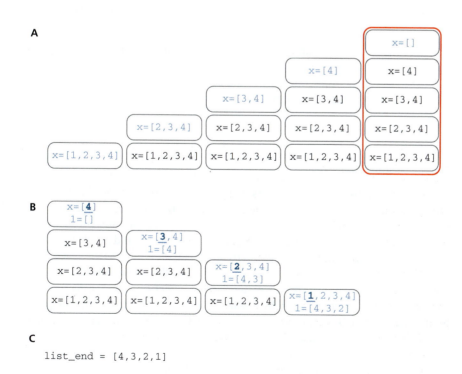

FIG. 13-31. Going into and out of a recursion can be viewed as creating a stack of plates (going into the recursion) and removing the plates from the stack (coming out of the recursion). In this analogy, a plate holds the values of the local variables for a specific instance of the recursive function. Each recursive function call adds a plate to the stack, and each return from a recursive function removes a plate from the stack. **A.** The local variables on the stack as function *fn_rec* is recursively called. **B.** The local variables on the stack as function *fn_rec* returns out of the recursion after the stop condition is reached. **C.** The final return value from function *fn_rec* is assigned to variable *list_end* in the main script.

is recursively called, and Figure 13-31B shows the values of *x* and *l* as the function returns. As a recursive program executes, statements before the recursive function call are executed before the call (going into the recursion) and statements after the recursive function call are executed after the call returns (coming out of the recursion). Thus, *l* is not shown in Figure 13-31A because it is created coming out of the recursion. You can see how the solution in *l* is formed from previous solutions by appending the first element of *x* (in bold and underlined in Figure 13-31B) to *l*. To help you correlate the stack of plates view with the debugger window view shown in Figure 13-30C, the red box in Figure 13-31A corresponds to the view of the stack after the fifth recursive call.

After *fn_rec* finishes executing, in the main script the return value is assigned to the variable *list_end* as shown in Figure 13-31C. Although recursion is a natural method for implementing dynamic programming, it is important to note that anything that can be solved recursively can also be solved iteratively using a *for* or *while* loop.

Thought Question 13-5

(a) Given the following recursive function, trace through the program by hand to determine what will be displayed to the screen after the function call *slice_str('abcd')* is executed. Recall that a function will return when it encounters a return statement or when it reaches the end of the code.

```
def slice_str(in_string):
    if in_string == '':          #if in_string is an empty string
        return
    print in_string
    slice_str(in_string[1:])
```

(b) Given the revised recursive function, trace through the program by hand to determine what will be displayed to the screen after function call *slice_str('abcd')* is executed.

```
def slice_str(in_string):
    if in_string == '':          #if in_string is an empty string
        return
    slice_str(in_string[1:])
    print in_string
```

(c) Why are the outputs different?

Built-in functions and methods

As mentioned earlier, Python supports both procedure-oriented and object-oriented programming paradigms. In procedure-oriented programming, data is passed to and from functions that operate upon the data, whereas in object-oriented programming, the data and functions are combined together using *classes*, and an *object* is an instance of a class. Although learning how to define classes is beyond the scope of this text, we have been using some built-in classes such as the string class. For example, the statement *sequence = "actgacc"* creates a string object named *sequence* with data *"actgacc"* that can be operated on by a wide variety of functions that are supported by the string class. In object-oriented programming, a function that is associated with a class is called a *method*. The syntax to call a method is:

$$String\_object\_name.method\_name()$$

Continuing our example, to determine the length of *sequence*, *sequence.len()* calls the *len()* method. The *len()* method returns the length of the string, which in this example is 7, because *sequence = "actgacc"* is a string with seven characters. In addition to methods for built-in data types, some useful functions are also built into the Python interpreter. Table 13-16 contains a list of built-in functions and methods that will be useful for developing the pairwise sequence alignment tool in Chapter 14.

method (in computer science)
A function that belongs to a class. Example *upper()* is a method for the string class. Given the object *str*, an instance of the string class that has been assigned a string value, the expression *str.upper()* will invoke the method *upper()* to convert every lowercase character in the string to an uppercase character.

TABLE 13-16. USEFUL BUILT-IN FUNCTIONS AND METHODS FOR BUILDING PAIRWISE SEQUENCE ALIGNMENT TOOL IN CHAPTER 14

FUNCTION OR METHOD	DESCRIPTION	EXAMPLE
File I/O Functions		
open(filename, [mode])	Open a file as either readable ('r') or writeable ('w'). filename is a string corresponding to the name of the file. Assumes the file is in the same directory as the script unless the complete path is provided. Returns a file object that is used to keep track of access into the file.	infile= open ('in.txt', 'r') outfile = open('D:\\Docs\\out.txt', 'w')
readlines()	Read entire file into list of line strings.	L = infile.readlines()
write(S)	Write string S into a file.	outfile.write(S)
close()	Manually close file.	infile.close() outfile.close()
String Methods		

lower()	Convert string to lowercase.	>>> n1 = "ACtG"
		>>> n1.lower()
		'actg'

upper()	Convert string to uppercase.	>>> n2 = "actG"
		>>> n2.upper()
		'ACTG'

split([sep])	Split a string into list of strings. Splits when a separator is encountered. Default separator is whitespace.	>>> str1 = "A R NS D"
		>>> str1.split()
		['A', 'R', 'NS', 'D']

startswith(prefix)	Returns **True** if string starts with the given prefix. Otherwise returns **False**.	>>> str2 = '*xyz'
		>>> str2.startswith('*')
		True
		>>> str2.startswith('x')
		False

isalpha()	Returns **True** if string has only alphabet characters. Otherwise returns **False.**	>>> str3 = 'a b 1 *'
		>>> str3.isalpha()
		False
		>>> str3[0].isalpha()
		True

Other Useful Functions		
int()	Convert to integer.	>>>int('−4') **−4** >>>int(3.56) **3**
float()	Convert to float.	>>>float('1.8') **1.8** >>>float(3) **3.0**

zip()	Zip together two or more lists.	>>> L1 = ['a','b','c']
		>>> L2 = [5,6,11]
		>>> zip(L1, L2)
		[('a',5), ('b',6), ('c',11)]

dict()	Convert a list of tuples (key,data) into a dictionary.	>>> keys = ['a', 'b', 'd']
		>>> vals = [1.8, 2.5, −3.5]
		>>> hydro = dict(zip(keys,vals))
		>>> hydro
		{'a': 1.8, 'b': 2.5, 'd': −3.5}

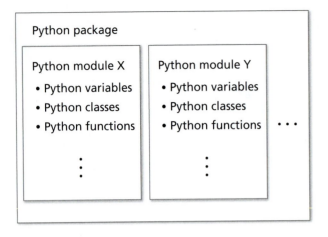

FIG. 13-32. Python packages contain modules (files) for organizing the code in an application.

Python Modules and Packages

When developing large applications, *modules* and *packages* provide a way for developers to logically organize their code. Python modules and packages provide a hierarchical framework, as shown in Figure 13-32, where a module is a file containing Python code and a package is a directory (also known as a folder) containing multiple modules. It is possible for a package to have subpackages just like a directory can have subdirectories. For a module, the filename is the module name with the suffix *.py* appended. Likewise, for a package, the name of the directory (or folder) is the package name. Code that is developed in one module or package can be *imported* into another module and used.

Large applications often have multiple developers working on different aspects of the design. By utilizing modules and packages, each developer can create his or her own code and easily use code created by others. Of course, co-developers have to collaborate to decide on the types of functions, classes, and variables that they will develop for others to use. One of the main advantages of Python modules for projects with multiple developers is that each module has its own namespace. This means that co-developers do not have to worry about global variable names clashes. In other words, the same name can be used for different global variables defined in different modules and not cause a problem. In this chapter, we will provide a brief introduction to modules including their use and naming conventions. More details about how to use Python modules and packages can be found at www.python.org.

We will use the Kyte-Doolittle sliding window program to illustrate how to use modules, as shown in Figure 13-33. Normally such a simple program would be contained in one module, but through this example we can see how variables and functions can be defined in different modules (files) and shared among different modules. Module KD_hydroscale (file *KD_hydroscale.py*) contains the *hydro_scale* dictionary variable (Figure 13-33A), module KD_input contains the *user_input* function (Figure 13-33B), module KD_compute contains the *compute_hydro* function (Figure 13-33C), and the top-level main module KD_main uses the *user_input* function from KD_input and the *compute_hydro* function from KD_compute (Figure 13-33D). To execute the program, the top-level KD_main module can either be run in the interpreter or run as a stand-alone script.

As mentioned, in order to use functions, classes, and global variables that are defined in another module, the module must first be imported. A module can import part or all of another module. Figure 13-33C shows how the KD_compute module imports the KD_hydroscale module in order to use the *hydro_scale* dictionary in the *compute_hydro* function. The import statement is

import KD_hydroscale

This will import the entire module (which in this case has only the *hydro_scale* dictionary). To refer to a function, variable, or class from another module, the module name must be included in the reference. The dot convention is used where the module name and variable name are concatenated using a period, which is referred to as a dot. For example, to refer to (or use) the *hydro_scale* dictionary defined in module KD_hydroscale, the following reference is used:

KD_hydroscale.hydro_scale[aa]

If you do not want to have to use the module name when referring to a function, variable, or class that is imported, you can use the *from* keyword to import part or all

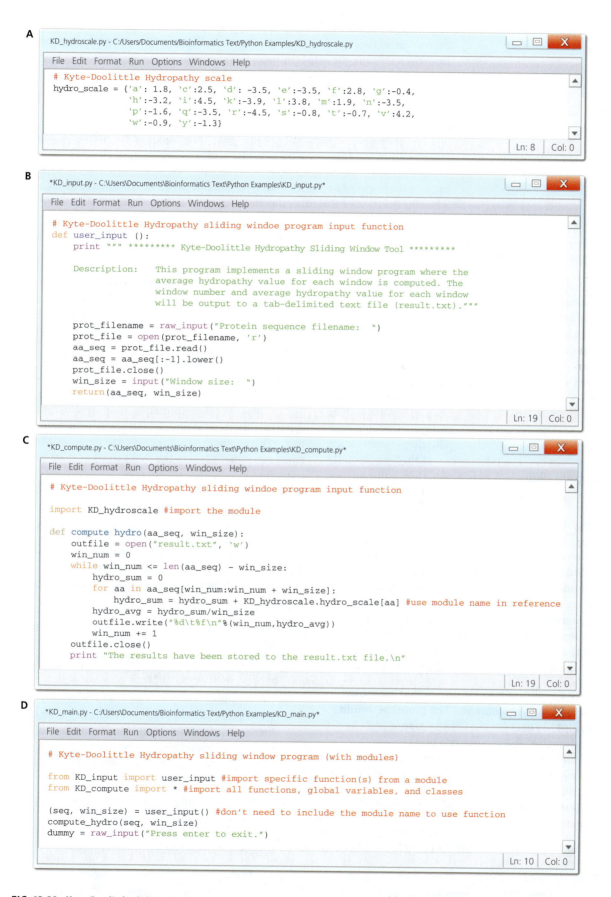

A

```
KD_hydroscale.py - C:/Users/Documents/Bioinformatics Text/Python Examples/KD_hydroscale.py
```
File Edit Format Run Options Windows Help
```
# Kyte-Doolittle Hydropathy scale
hydro_scale = {'a': 1.8, 'c':2.5, 'd': -3.5, 'e':-3.5, 'f':2.8, 'g':-0.4,
               'h':-3.2, 'i':4.5, 'k':-3.9, 'l':3.8, 'm':1.9, 'n':-3.5,
               'p':-1.6, 'q':-3.5, 'r':-4.5, 's':-0.8, 't':-0.7, 'v':4.2,
               'w':-0.9, 'y':-1.3}
```
Ln: 8 | Col: 0

B

```
*KD_input.py - C:\Users\Documents\Bioinformatics Text\Python Examples\KD_input.py*
```
File Edit Format Run Options Windows Help
```
# Kyte-Doolittle Hydropathy sliding windoe program input function
def user_input ():
    print """ ********* Kyte-Doolittle Hydropathy Sliding Window Tool *********

    Description:   This program implements a sliding window program where the
                   average hydropathy value for each window is computed. The
                   window number and average hydropathy value for each window
                   will be output to a tab-delimited text file (result.txt)."""

    prot_filename = raw_input("Protein sequence filename:  ")
    prot_file = open(prot_filename, 'r')
    aa_seq = prot_file.read()
    aa_seq = aa_seq[:-1].lower()
    prot_file.close()
    win_size = input("Window size:  ")
    return(aa_seq, win_size)
```
Ln: 19 | Col: 0

C

```
*KD_compute.py - C:\Users\Documents\Bioinformatics Text\Python Examples\KD_compute.py*
```
File Edit Format Run Options Windows Help
```
# Kyte-Doolittle Hydropathy sliding windoe program input function

import KD_hydroscale #import the module

def compute_hydro(aa_seq, win_size):
    outfile = open("result.txt", 'w')
    win_num = 0
    while win_num <= len(aa_seq) - win_size:
        hydro_sum = 0
        for aa in aa_seq[win_num:win_num + win_size]:
            hydro_sum = hydro_sum + KD_hydroscale.hydro_scale[aa] #use module name in reference
        hydro_avg = hydro_sum/win_size
        outfile.write("%d\t%f\n"%(win_num,hydro_avg))
        win_num += 1
    outfile.close()
    print "The results have been stored to the result.txt file.\n"
```
Ln: 19 | Col: 0

D

```
*KD_main.py - C:/Users/Documents/Bioinformatics Text/Python Examples/KD_main.py*
```
File Edit Format Run Options Windows Help
```
# Kyte-Doolittle Hydropathy sliding window program (with modules)

from KD_input import user_input #import specific function(s) from a module
from KD_compute import * #import all functions, global variables, and classes

(seq, win_size) = user_input() #don't need to include the module name to use function
compute_hydro(seq, win_size)
dummy = raw_input("Press enter to exit.")
```
Ln: 10 | Col: 0

FIG. 13-33. Kyte-Doolittle sliding window program split into modules. A. Module KD_hydroscale contains the *hydro_scale* dictionary. **B.** Module KD_input contains the *user_input* function. **C.** Module KD_compute imports the KD_hydroscale module because the hydro_scale dictionary is used by the *compute_hydro* function. **D.** The KD_main module imports functions from KD_input and KD_compute.

of a module. Figure 13-33D shows the top-level KD_main module, which imports the *user_input* function from the KD_input module and the *compute_hydro* function from the KD_compute using the following statements:

from KD_input import user_input
from KD_compute import *

The first statement imports just the *user_input* function, and the second statement imports all of the functions, classes, and global variables from the KD_compute module (in this case, though, there is only the *compute_hydro* function). Notice that because the keyword *from* is used here, the module names do not need to be included when the functions are used, as shown in Figure 13-33D.

SUMMARY

Programming may seem like a daunting task at first, but if you are good at problem solving, then you can program. It is just a matter of learning the language of the computer (Python in this case) to be able to tell the computer how to solve the problem. In this chapter we covered the basics of the Python programming language, from creating variables with meaningful names, operating on variables using operators and built-in functions, controlling the flow of the program through control statements such as if-tests and loops, to interfacing with the user through input and output functions. We also applied the program design process to develop a bioinformatics tool, the Kyte-Doolittle Hydropathy sliding widow program. During this process we saw the importance of fully understanding the problem we want our program to solve and we learned how to develop and refine an algorithm, how to code the algorithm in Python, and finally how to test and debug the program. Finally, we learned the benefits of hierarchical design using functions, modules, and packages.

In bioinformatics, whether you are a tool developer or a tool user, it is important to understand the program design process. Bioinformatics tool development is driven primarily by scientists who identify the need for a tool. With an understanding of how to program, the scientist can either develop the tool herself or work with a programmer to co-develop the tool.

EXERCISES

1. Given a DNA sequence stored in a string variable called *dna_seq*, write a list comprehension with a conditional expression that generates a list where a 1 is stored if there is a 'G' or 'C' in the sequence and a 0 is stored otherwise. You can assume that *dna_seq* uses only uppercase letters for the nucleotides A, C, T, and G. Assign the list you create to a variable called *GorC*.

2. Write a sliding window program that computes the %GC of DNA sequence. The program should prompt the user for the DNA sequence and the window size. You can assume that the window increment is 1—that is, as the program slides the window, it will move the window over one nucleotide.

   Test your program on the following DNA sequence:

   GAACTCATACGAATTCACGTCAGCCCATCGTGCCACGT

3. Type up and test the Kyte-Doolittle Hydropathy program provided in Figure 13-23. Get the amino acid sequence for bacteriorhodopsin. Experiment with different window sizes to try to re-create the graph shown in Figure 13-1. You will need to plot your data using a tool such as Windows Excel. Which window size gives you the most accurate result?

4. Make one or more of the suggested modifications to the hydropathy tool in Figure 13-23. Test your modified tool on the input sequence 'PLSQETFSDLWKLLPENNVLSP'.

   a. Often protein sequences are in a format (such as FASTA) that include extra characters (see Chapter 2 for details of FASTA format). Modify the tool to ignore non-amino acid characters in the input sequence. To test, add non-amino acid characters to the input test sequence.

   b. Modify the tool to give the user two options for inputting the protein sequence:
      i. By specifying the filename (already supported).
      ii. By directly providing a protein sequence.

   c. Modify the tool to give the user the option to dump the data to the screen as well as write it to the result.txt file. The screen dump format should have the following format:

WINDOW	WINDOW SEQUENCE	AVERAGE HYDROPHOBICITY
0	'PLSQETF'	−0.5
1	'LSQETFS'	−0.39
2	'SQETFSD'	−1.43

d. In the output file, result.txt, change the first column of data from index of the first amino acid in the window to the index of the mid amino acid in the window. Thus, for a window size of 7, the first window will correspond to the first through seventh amino acids and the midpoint will be the fourth amino acid so a 4 should be displayed. This is actually the way that Kyte and Doolittle output their data.

e. Add a menu-driven interface for inputting the window size by including the user-defined function *def_win_size()* from Figure 13-12 in your hydropathy sliding window tool. You should place the function at the top of the module and then insert a call to the function to assign *win_size*.

f. In order to compare results from executing the tool with different window sizes, modify the tool to include the window size as the first line in the output file. Because data files are often read by other programs, just put the value of the window size without any additional text. To make it easier for the user to keep track of the files for different window sizes, also add the window size to the result file name (e.g., *result_winsize7.txt*).

5. **Write a program to detect transmembrane regions in a protein. Your program should input the protein file and the results.txt file generated by the Kyte_Doolittle_v3.py program. Kyte and Doolittle found that to detect transmembrane regions in a protein, a window size of 19 is needed. Transmembrane regions are identified by average hydropathy scores greater than 1.6. Your program should output the following information in tabular form:**

***************TRANSMEMBRANE REGIONS***************

WINDOW	WINDOW SEQUENCE	AVERAGE HYDROPATHY (> 1.6)

Steps:

i. Write your program.
ii. Run the Kyte_Doolittle_v3.py program (see exercise 3) on the bacteriorhodopsin protein with a window size of 19.
iii. Test your program using the bacteriorhodopsin protein file with the result.txt file generated from step ii.

6. **Work with matrices and functions. In Python, matrices are 2-D lists (list of lists).**

a. Write a Python script to create a matrix with m columns and n rows initialized to 0, where m and n are variables. Test with m = 7 and n = 5. Hint: use nested list comprehensions.

b. Write a Python function *CreateMatrix(m,n)* to create a matrix with m columns and n rows initialized to 0. The function should return the matrix. To test, include the following code after the function definition:

```
M1 = CreateMatrix(7,5)
print M1
M2 = CreateMatrix(3,8)
print M2
```

c. Write a Python function *PrintMatrix(matrix)* to print the matrix row by row. For example, instead of using a Python *print* command where the output looks like this:

[1, 2, 3], [4, 5, 6], [7, 8, 9]

the output of your function should look like this:

[1, 2, 3]
[4, 5, 6]
[7, 8, 9]

d. Write a Python function *FindMin(matrix)* to find the minimum value in a matrix. The function should return a tuple with the minimum value, and the row index (min_i) and column index (min_j) should correspond to the location of the minimum in the matrix. To test, include the following code after the function definition:

```
testM = [[1, 2, 3], [−4, 7, −8], [20, −16, 100],
[4, 17, −3]]
(min_val, min_i, min_j) = FindMin(testM)
print min_val, min_i, min_j
```

7. **To gain a better understanding of how recursive functions work, write a recursive function *print_stars(n)* that will take the number of stars to print per line (*n*) and print the following pattern:**

Initial value of n = 4	Initial value of n = 3
****	***
***	**
**	*
*	*
*	**
**	***
***	
****	

Hints: The stop condition for the recurrence is reached when the *n* is equal to zero (*n* == 0). Notice how the numbers of stars to print decreases and then increases in a mirrored fashion. This can be achieved by using a print statement before and after the recursive call.

To gain practice using Python modules, place the recursive function in its own module (file). Name the file *stars.py*. In the same folder create a main Python script in a file named *main.py*. In the main script, import the stars module and call the *print_stars* function three times with different values of *n*.

ANSWERS TO THOUGHT QUESTIONS

13-1. The question asks you to write a code segment to display the index of the first G or C nucleotide in a DNA sequence stored in a variable called *dna*.

(a) Here is a solution (algorithm) written in pseudo-code.

> *index = 0*
> *while dna[index] is not 'G' **and** dna[index] is not 'C'*
> *   increment index*
> *print index*

Note: An "and" is needed for the while condition, because it will continue to loop while both conditions are true—that is, when a 'G' is not encountered and a 'C' is not encountered. When either a 'G' or a 'C' is encountered the loop will exit. Also note that the 'G' and 'C' are in quotes because they refer to constant values, the characters G and C, and not to variables named G and C.

(b) Here is a solution (Python code) using a while loop:

```
index = 0
while dna[index] != 'G' and dna[index] != 'C':
    index = index + 1
print("The index of the first G or C nucleotide is: %d" % index)
```

13-2. The question asks you to write a code segment to display the percentage of a protein that is strongly hydrophobic.

(a) Here is a solution (algorithm) written in pseudo-code.

> *count = 0*
> *for each amino acid in Protein*
> *   if the amino acid is strongly hydrophobic*
> *   count = count + 1*
> *percentage = 100 * count/(number of amino acids in the protein)*
> *display the percentage to the screen*

(b) Here is a solution (Python code) using a for loop:

```
count = 0        #count of strongly hydropho-
                 bic amino acids
for aa in Protein:
    if hydro_scale[aa] > 3:
        count = count + 1
percentage = (100.0 *count)/len(Protein)
```

```
print ("The percent strongly hydrophobic is: %.2f%%" % percentage)
```

Note: We are multiplying by 100.0 (a real number) to ensure that floating-point division is performed. If integer division is performed it will truncate any fractional part of the answer (so instead of printing 66.67% it would print 66%). The %.2f will print the number with two digits of precision after the decimal point. Two percent symbols (%%) are needed to display a % sign because the %symbol is a special character in formatted strings.

13-3.

(a) The answer is that any of the three constructs can be used, but a *list comprehension* is the easiest.

(b) Here is how to use a list comprehension to create a list, L, of n elements where each element is initialized to 0.

> L = [0 for i in range(n)]

Note that range(n) will create the list [0, 1, 2, . . . n−1], which is a list of n elements that corresponds to the indices of the list we are trying to create. For each index, the value of 0 is being assigned to the corresponding element at that index.

If your answer to part (a) was to use a *while* loop here is a possible solution:

```
L = []               #create an empty list
i = 0
while i < n:

    L.append(0)    #for each value of i from 0 to
                   n−1, append a 0 to L
    i = i + 1
```

If your answer to part (a) was to use a *for* loop here is a possible solution:

```
L = []            #create an empty list
for i in range(n):
    L.append(0)   #for each value of i in the
                  range 0 to n−1, append a 0 to L
```

13-4.

(a) In order to debug your code, you need to have a very good understanding of the problem you are trying to solve. As you debug your program, whether stepping instruction by instruction, function by function, or by setting breakpoints, for each code segment you are debugging you should know what the current values of the variables being used should be and what the values being produced by that segment of the code should be. For each code segment you are testing, there are three possible scenarios:

1. The current values are correct, and the values being produced are correct. Great! That code segment appears to be working (at least for this test case), so you can move on.

2. The current values are correct, but the values being produced are wrong. Aha! You have found a bug—that is, you know there is a mistake in that part of the code. Now you have to figure out why it is not giving you the expected value.

3. The current values are wrong. You need to determine which value is wrong, find out where that value was produced, and debug that part of the code. If more than one value is wrong, you may need to debug multiple parts of the code.

Mistakes can be found in all of them! Well sort of, let's look at each case individually.

i. When you work the problem by hand and create test case(s) with sample input(s) and output(s):
   - You may make some assumptions that you do not explicitly state. When developing the algorithm and code, these assumptions may not be true or you may need to explicitly write code to make them true (such as initializing variables).
   - Your test cases may not fully capture all the possible inputs. In this case, your program may work for some inputs but not all.
   - You may make a mistake in your test case. As you debug your program you may find that your outputs are not what you expected for the given inputs. You might think that you have found a bug in your code, but in fact your code could be correct and it is your hand solution that is wrong. You would likely find this mistake when you hand-test your algorithm.
   - If you do not understand the problem, it is also possible to solve the problem incorrectly and generate an incorrect output. In this case your code may generate results that match with your test case but the output is incorrect. In this case, the mistake is in your understanding of the problem.

ii. When you develop an algorithm—step-by-step, flow chart, or pseudo-code:
   - You may make a mistake in your algorithm.
   - You could create only a partial solution because you ignore some of the possible input cases.
   - You might ignore the assumptions made during the hand solution.

iii. When you hand-test an algorithm with sample input(s) and output(s):
   - You cannot introduce a mistake here, but you can propagate a mistake—that is, it is possible to not catch the mistake. If you make assumptions when testing your algorithm, you may not catch mistakes. It is important to act as a "dumb computer" when testing your algorithm and only do exactly what the algorithm states.

iv. When you code in target language (e.g., Python):
   - As you are learning a language, it is very easy to improperly code an algorithm. The problem stems from not understanding exactly how each Python code construct works. So, to avoid this problem, use the Python interactive shell to create simple test cases to test your understanding of the various Python constructs you plan to use.
   - You can also test out parts of your code before inserting it into your program (or during debugging). In the interactive shell you first need to create the variables to be used and assign them the values you would expect them to have at this point in the program. Then you can type and run the code segment in the interactive shell to see if it produces the correct value. For complex code segments, you can also do this by creating test Python scripts and running the scripts in the interactive shell.

v. When you test the program with sample input(s) and output(s):
   - Again, these tests may not be correct, so if the sample output from your hand solutions does not match the program output, first make sure that your sample solution is correct.
   - As mentioned before, you can make a mistake in testing if you have not considered all possible input cases. This does not mean that you have to exhaustively test your program: rather, you need to make sure that you have representative test cases for all possible scenarios. For example, if testing whether a value is greater than a threshold, you should test with numbers below the threshold, at the threshold, and above the threshold, and, if appropriate, test with both positive and negative values.

vi. When you debug to find and fix mistakes in algorithm or coding:

- Again, you cannot really introduce a bug by simply debugging your code. But you can introduce one when trying to fix a bug.
- Also, when you fix a bug, it may fix multiple problems that appear to be wrong with your program. So fix bugs one at a time (this is especially true for syntax errors).
- On the other hand, when you fix a bug, it may reveal another mistake in your program. Debugging is an iterative process!
- When the expected output does not match your program's output you have discovered a mistake/bug. To find the bug faster, think critically about how the outputs disagree and where in your code the incorrect output values are produced. Then go back and check the expected input values for that code segment. Use the ideas in part (a) to work backward through your program to track down your bug.

Note: Many students think it will be easier to write a program by modifying code that someone else has written. In fact, this often takes much longer because if you have not gone through the program design process, it is very difficult to test and debug your code.

13-5.

(a) The following would be displayed to the screen:

abcd

bcd

cd

d

(b) The following would be displayed to the screen:

d

cd

bcd

abcd

(c) The outputs are different because of where the print statement is placed. In part (a) the print statement is before the recursive function call, so the values are printed going into the recursion. In part (b) the print statement is after the recursive function call, so the values are printed coming out of the recursion. Thus in part (a) you see the values of in_string for the first, second, third, and fourth function calls, respectively. In part (b) you see the values of in_string for the fourth, third, second, and first function calls, respectively.

REFERENCES

Hetland, M. L. 2008. *Beginning Python from Novice to Professional*, 2nd ed. Berkeley: Apress.

Johnson, S., R. McCord, and L. Robinson. 2002, February 27. "A One Stop Site for Kyte Doolittle Hydropathy Plots." Accessed May 2, 2012. http://gcat.davidson.edu/rakarnik/KD.html.

Kinser, J. 2008. *Python for Bioinformatics*. Sudbury, MA: Jones and Bartlett.

Kyte, J., and R. Doolittle. 1982. "A Simple Method for Displaying the Hydropathic Character of a Protein." *Journal of Molecular Biology* 157: 105–132.

Model, Mitchell. 2009. *Bioinformatics Programming Using Python*. Sebastopol, CA: O'Reilly Media.

O'Brien, P. 2002. "ONLamp.com: Beginning Python for Bioinformatics." *ONLamp.com*. O'Reilley, 17 Oct. 2002. Web. 31 Mar. 2012. http://onlamp.com/lpt/a/2727.

DEVELOPING A BIOINFORMATICS TOOL

AFTER STUDYING THIS CHAPTER, YOU WILL:

- Understand how to design a bioinformatics tool

- Gain a deeper understanding of pairwise sequence alignment

- Gain an understanding of algorithms and algorithm complexity

- Understand the longest common subsequence (LCS) algorithm

- Be able to extend LCS to implement global and local pairwise sequence alignment

- Be able to develop a basic pairwise alignment tool with a simple, menu-driven interface

14.1 INTRODUCTION

Developing a bioinformatics tool will benefit biologists and computer scientists in different ways. If you are a biologist, developing a bioinformatics tool will give you a deeper understanding of how tools work, which ultimately will help you become a more effective tool user. In addition, because biologists are usually the ones who identify biological problems, a deeper understanding of algorithms and program design will enable you to work with other developers to create tools to solve these problems or even to develop your own tools. If you are a computer scientist or mathematician, developing a pairwise alignment tool will help you understand how to apply your knowledge of algorithms and program design to solve important biological problems. In Chapter 13 we presented an introduction to the Python programming language and provided an overview of the program design process. These concepts were used to develop a simple Kyte-Doolittle Hydropathy sliding window tool. In this chapter we discuss how to develop a more sophisticated bioinformatics tool for performing pairwise sequence alignment.

Recall from Chapter 5 that global alignment, originally proposed by Needleman and Wunsch, is used to align two protein (or nucleotide) sequences from end to end. This alignment is performed to determine if there is significant similarity, suggesting that the sequences have a common ancestor. Ends-free global alignment (also referred to as semi-global alignment) is a variation of global alignment that does not penalize gaps at the beginning or end. Ends-free global alignment is used for sequences of different lengths and for sequences where the N- and C-termini may differ but the majority of the sequences are similar. Sometimes sequences are globally very dissimilar, but they have regions of similarity called conserved regions that have either no substitutions or conservative substitutions. These conserved regions often indicate that the region has structural or functional importance. Chapter 5 presented Smith and Waterman's local alignment algorithm for identifying these regions of similarity.

In this chapter we will guide you through the development of a pairwise sequence alignment tool that implements global, ends-free global, and local alignment. Tools can be developed using any programming language, and this chapter will focus on the algorithms needed to implement the tool. However, because Python is a commonly used programming language for bioinformatics, we encourage you to apply the Python concepts covered in Chapter 13 to implement your pairwise sequence alignment tool.[1]

In this chapter we will start by analyzing the output report of an existing local sequence alignment tool, EMBOSS Water, to familiarize ourselves with its inputs, outputs, and functionality. Next, we will look at an overview of SPA, the simple pairwise alignment tool we will be developing. Before discussing the algorithms used in SPA, we will more formally introduce the concept of algorithms—taking a look at different ways to express algorithms. One well-known algorithm is the longest common subsequence (LCS) algorithm that can find the longest common sequence of characters between two strings. We will discuss the LCS algorithm and how it can be extended to implement local and global pairwise alignment. Next, we will see how to evaluate the complexity of algorithms—that is, how much memory and time they require. We will end the chapter with a discussion of possible extensions to the basic SPA tool.

14.2 ANALYSIS OF AN EXISTING TOOL: EMBOSS WATER LOCAL ALIGNMENT TOOL

As mentioned in Chapter 13, the most important step in the program design process is to understand the problem you are trying to solve and identify the inputs and outputs. Before developing our own tool, let's investigate an existing tool, the EMBOSS Water local alignment tool, to understand: (1) why such a tool is useful to biologists, (2) what are the various inputs needed to perform the alignment, (3) what are the different types of output data an alignment tool can provide to the user, and (4) how the alignment can be formatted to help the user interpret the resulting alignment.

As discussed in Chapter 3, sequence alignment is a useful tool for gaining a better understanding of the evolution of species and the function and structure of proteins. Figure 14-1 shows the output report of a local alignment performed by the

---

[1] The Python concepts presented in Chapter 13 are sufficient for implementing the pairwise sequence alignment tool. Although there are a variety of higher-level functions available in various packages, building a tool from the basics provides strong foundational knowledge and skills that can be used to develop many different types of tools.

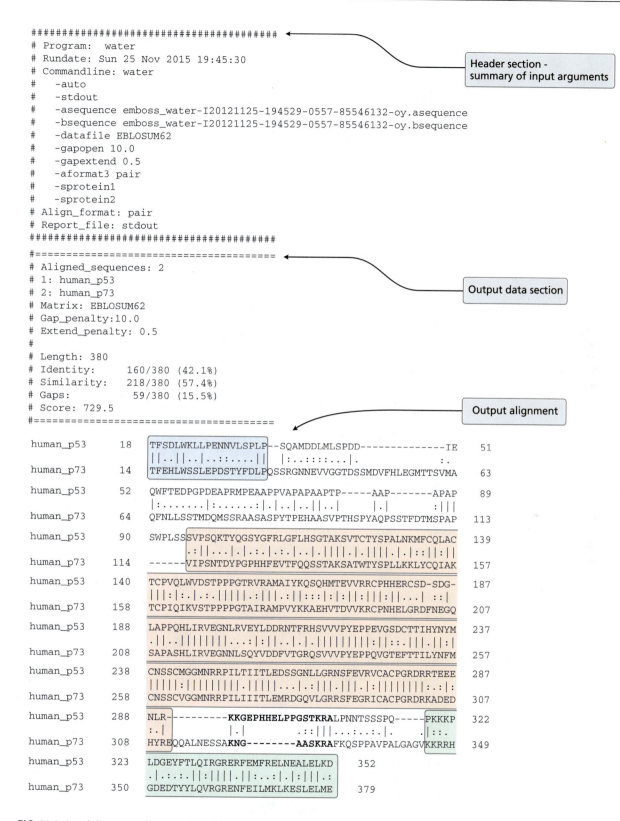

```
##########################################
# Program:  water
# Rundate: Sun 25 Nov 2015 19:45:30
# Commandline: water
#    -auto
#    -stdout
#    -asequence emboss_water-I20121125-194529-0557-85546132-oy.asequence
#    -bsequence emboss_water-I20121125-194529-0557-85546132-oy.bsequence
#    -datafile EBLOSUM62
#    -gapopen 10.0
#    -gapextend 0.5
#    -aformat3 pair
#    -sprotein1
#    -sprotein2
# Align_format: pair
# Report_file: stdout
##########################################

#=======================================
# Aligned_sequences: 2
# 1: human_p53
# 2: human_p73
# Matrix: EBLOSUM62
# Gap_penalty:10.0
# Extend_penalty: 0.5
#
# Length: 380
# Identity:     160/380 (42.1%)
# Similarity:   218/380 (57.4%)
# Gaps:          59/380 (15.5%)
# Score: 729.5
#=======================================
```

Header section - summary of input arguments

Output data section

Output alignment

```
human_p53       18  TFSDLWKLLPENNVLSPLP--SQAMDDLMLSPDD--------------IE   51
                    ||..||..|...|..:::...||  |:..::::...|.        :.
human_p73       14  TFEHLWSSLEPDSTYFDLPQSSRGNNEVVGGTDSSMDVFHLEGMTTSVMA   63

human_p53       52  QWFTEDPGPDEAPRMPEAAPPVAPAPAAPTP-----AAP-------APAP   89
                    |:.......|:......:|.|..|..||.|     |.|      :|||
human_p73       64  QFNLLSSTMDQMSSRAASASPYTPEHAASVPTHSPYAQPSSTFDTMSPAP   113

human_p53       90  SWPLSSSVPSQKTYQGSYGFRLGFLHSGTAKSVTCTYSPALNKMFCQLAC   139
                          .:||...|.|.:.|..:.|..|.||||.|||||.|.|.::||:||
human_p73      114  ------VIPSNTDYPGPHHFEVTFQQSSTAKSATWTYSPLLKKLYCQIAK   157

human_p53      140  TCPVQLWVDSTPPPGTRVRAMAIYKQSQHMTEVVRRCPHHERCSD-SDG-   187
                    |||:|:..|..:.|||||.:||||.:||::|:|:|:|||:||...  ::|
human_p73      158  TCPIQIKVSTPPPGTAIRAMPVYKKAEHVTDVVKRCPNHELGRDFNEGQ   207

human_p53      188  LAPPQHLIRVEGNLRVEYLDDRNTFRHSVVVPYEPPEVGSDCTTIHYNYM   237
                    .||..||||||||..::|:|.|.:||||:|||||||||:|:.:|||.||:|
human_p73      208  SAPASHLIRVEGNNLSQYVDDFVTGRQSVVVPYEPPQVGTEFTTILYNFM   257

human_p53      238  CNSSCMGGMNRRPILTIITLEDSSGNLLGRNSFEVRVCACPGRDRRTEEE   287
                    |||||:||||||||.|.|||||.:.|.|||.|:||||.|:||||||||::.|:
human_p73      258  CNSSCVGGMNRRPILIIITLEMRDGQVLGRRSFEGRICACPGRDRKADED   307

human_p53      288  NLR---------KKGEPHHELPPGSTKRALPNNTSSSPQ-----PKKP   322
                    :.|         |.|        .::|||...:...:.|.    .|:.:.
human_p73      308  HYREQQALNESSAKNG--------AASKRAFKQSPPAVPALGAGVKKRRH   349

human_p53      323  LDGEYFTLQIRGRERFEMFRELNEALELKD   352
                    .|.:.:.||:|||||.||:...:|.|:||||.:
human_p73      350  GDEDTYYLQVRGRENFEILMKLKESLELME   379
```

FIG. 14-1. Local alignment of human p53 and human p73 proteins using the EMBOSS Water program. Three conserved regions of structural and functional importance are highlighted: blue = transactivation domain, red = DNA binding domain, and green = oligomerization domain.

matrix (in computer science) A two-dimensional array or list. Can be formed in Python using list of lists—that is, a list of elements in which each element is another list.

EMBOSS Water tool on human p53, a tumor suppressor protein, and human p73, a protein necessary for embryo growth and development. Looking at the locally aligned region in Figure 14-1, there are three highly conserved regions that can be easily detected (which we have highlighted). The first conserved region in Figure 14-1, highlighted in blue, corresponds to the transactivation domain. The transactivation domain is important for interacting with other proteins that transcribe mRNAs that ultimately produce the proteins that will cause cell death. The second conserved region, highlighted in red, corresponds to the DNA binding domain. The third conserved region, highlighted in green, corresponds to the oligomerization domain, which helps the protein bind with itself because four polypeptides are needed to create the tetramer protein that binds to DNA. In this particular example, the functions of these three regions are well known to biologists, but if that were not the case, further wet lab experiments would need to be conducted to confirm their function.

It is common practice for computer scientists to include the input arguments at the beginning (header) of an output report so that experiments conducted using the tool are well defined and easily reproducible. From Figure 14-1 we can see that the EMBOSS Water tool takes as input the alignment sequences (sequence 1 is human p53, and sequence 2 is human p73), the substitution matrix (BLOSUM62), the gap opening and extension penalties, the alignment format (pair, referring to pairwise sequence alignment), and the report file (stdout, which refers to standard output which is typically the computer monitor).

Let's review why these inputs are needed to produce a meaningful alignment. First, we need to decide which type of alignment to perform. In this case we want to perform local alignment, so we are using the EMBOSS Water alignment tool, which performs only local alignment. In other tools that can perform multiple types of alignments, the type of the alignment can also be an input to the tool. Second, because we are performing pairwise sequence alignment, we need to input the two sequences to align. Third, in order to create an alignment that identifies the conserved regions among the two sequences, we need additional information that allows us to accurately account for substitutions, insertions, and deletions that have occurred over the evolution of these species. Recall from Chapter 3 that substitutions will result in mismatches in the alignment and insertions and deletions (indel mutation) will result in gaps in the aligned sequences.

As we learned in Chapter 4, some amino acid substitutions are more likely to occur than others. Substitution matrices, such as PAM and BLOSUM, provide a substitution score for each possible amino acid pair that indicates the likelihood of this substitution occurring. This score is used by the alignment algorithm to determine whether a mismatch is reasonable to allow in the alignment in order to create the best possible alignment.[2] In the example in Figure 14-1, we can see that the BLOSUM62 substitution matrix was used which is the default matrix for the EMBOSS Water tool.

Insertions and deletions, referred to as indel mutations, correspond to gaps in an alignment. Different models have been proposed to penalize gaps during alignment. Two common gap penalty models are linear gap penalty and affine gap penalty. In the linear gap penalty model, each gap is equally weighted with a fixed gap penalty, w. Thus, a gap of length L will have a gap penalty of $W(L) = w*L$. Although linear gaps are easy to implement, they overly penalize indel mutations. Recall the equation for the affine gap penalty, $W(L) = g_{open} + g_{ext}*L$. As discussed in Chapter 5, in general, the gap opening penalty is much higher than the gap extension penalty because the occurrence of an insertion or deletion mutation (indel mutation) is more significant

---

[2] Although the goal is to create the best possible alignment, the alignment algorithm will produce the highest scoring alignment for the given inputs, which, depending on a variety of factors, including the substitution matrix and gap penalty model used, may or may not produce the best alignment.

than the length of the mutation. In the alignment shown in Figure 14-1, the gap opening penalty is -10.0 and the gap extension penalty is -0.5.[3]

Referring to Figure 14-1, we can see that after the header, which provides a summary of the user inputs, the output report provides data about the output alignment followed by the alignment. The output data includes the length of the local alignment, the identity of the alignments (number of identical residues), the similarity of the alignments (number of similar residues; residues are similar when they have a substitution score greater than or equal to 1), the number of gaps, and the alignment score.

Notice that the alignment is longer than the width of a page, so it has been broken up into slices of 50 residues, including gaps. For each protein, the name of the aligned protein is shown, followed by the location of the beginning residue of that slice, followed by the alignment of the protein including gaps ('—'), and at the end is the sequence number corresponding to the last residue of that slice in the protein. In the EMBOSS pair alignment format, there is a markup line placed between the two sequences that uses a space to indicate gaps, a '.' to indicate residues with little or no similarity, a ':' to indicate similar residues (for a substitution score greater than or equal to 1.0), and a '|' to indicate an identity.

Thought Question 14-1

Consider this small snippet from the alignment shown in bold in Figure 14-1.

```
KKGEPHHELPPGSTKRA
| · |         . : : | | |
KNG--------AASKRA
```

(a) What is the starting residue number for each sequence in the snippet?
(b) Using the BLOSUM62 substitution matrix (Figure 4-14), what is the alignment score?
(c) What is the length of the alignment?
(d) What is the percent identity?
(e) What is the percent similarity?
(f) What is the percent gap?

14.3 OVERVIEW OF SPA: A SIMPLE PAIRWISE ALIGNMENT TOOL

Figure 14-2 shows an overview of a basic pairwise sequence alignment tool that we will call SPA, for simple pairwise alignment tool. SPA will have a simple user interface, a suite of alignment algorithms, and the ability to read in input sequences and substitution matrices and to display the alignment. Although SPA could be designed to provide many different options, we have selected a few that represent options in existing alignment tools. Our objective is to show a beginning programmer how to develop a useful tool. Those who have a strong foundation in computer science are encouraged to extend the simple tool by enhancing the input/output options, adding a graphical user interface, interfacing to an online database, and applying advanced algorithms for increased functionality or efficiency.

SPA's user interface will be a simple, text-based, menu-driven interface that prompts the user for the different options. The user will specify the type of alignment, the input sequence format, the scoring method to be used, the gap penalty method, and the output format. Options for alignment will be global, ends-free global, and

---

[3] Note that an affine gap penalty model can be used to model linear gap penalties by setting the g_{open} penalty to zero and g_{ext} to the fixed gap penalty value w.

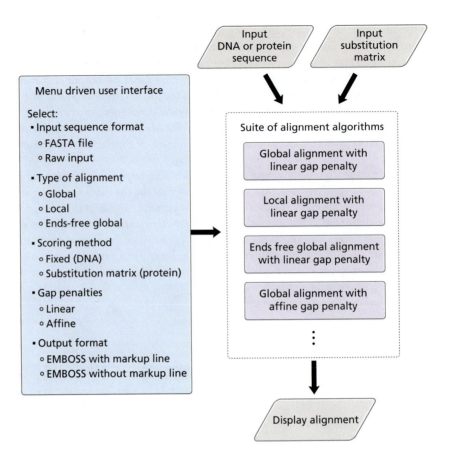

FIG. 14-2. Overview of SPA, a simple pairwise sequence alignment tool.

local. For the input sequence, the user will either enter raw sequences (without header information) or provide the name of a FASTA file. The scoring method will be either fixed for DNA sequences, or a substitution matrix for protein sequences. For fixed scoring, the user will be prompted to enter the match and mismatch score. For substitution matrix scoring, such as PAM or BLOSUM matrices, the user will be prompted to enter the substitution matrix filename.[4] Next, the user will specify the gap penalty model, either linear or affine. Lastly, the user will specify the output format, either EMBOSS with a markup line or EMBOSS without a markup line.

At the heart of SPA is the suite of alignment algorithms. The tool will support a modified version of Needleman-Wunsch's global alignment algorithm, an ends-free global alignment algorithm, and a Smith-Waterman local alignment algorithm. These algorithms will support linear gap penalties and can be implemented by extending the LCS algorithm. The global and local alignment algorithms will be discussed in detail in this chapter, whereas the algorithm for ends-free global alignment is an end-of-chapter exercise and the algorithm for integrating affine gap penalties is a challenge exercise for more advanced programmers.

14.4 ALGORITHMS

Before we delve into developing the SPA sequence alignment tool, it is important that we have a basic understanding of algorithms. There are two ways to define an algorithm. One definition is that an algorithm is a procedure for solving a computational

---

[4] Note that substitution matrix scoring can also be used for DNA sequences if the identity matrix is used.

problem in a finite number of steps. Computational algorithms can usually be proven correct, and their complexity, how much time and memory they require, can be analyzed (algorithm complexity will be discussed at the end of the chapter). Well-known computational algorithms have been developed to solve many generalized problems in computer science, mathematics, and computational sciences. It is a good idea to study these well-known algorithms because algorithms developed in one field can be adapted to solve a similar problem in another field. For example, in the field of computer science, LCS was developed to find the longest common subsequence between two strings. As mentioned previously, the LCS algorithm can be modified to develop algorithms for globally and locally aligning nucleotide and protein sequences.

More broadly, an algorithm can be defined as a procedure for solving a particular problem. As mentioned in Chapter 13, algorithms are an important step in the program design process. They should be used for organizing the overall flow of a program or function including steps for user input and output. Algorithms are programming language independent. A program can be developed from an algorithm using a wide variety of programming languages. Algorithms are typically written using one of the following formats:

1. Step-by-step
2. Flow chart
3. Pseudo-code

Step-by-step algorithms are similar to a cooking recipe or instructions to assemble a bookshelf. They can include decisions in which steps are skipped or repeated based on some condition. Flow charts use standard symbols, such as those shown in Figure 14-3, to capture the procedural or control flow of the algorithm. Flow charts are particularly useful when there are a lot of decisions, and hence more complex control flow, where skipping and repeating steps can be confusing. Flow charts use decision blocks to control the flow of the program where a decision can be true or false and, depending on the outcome of the decision, different operations are performed. Pseudo-code uses a syntax that is similar to a programming language but can be understood without knowledge of a specific language. The high-level and refined algorithms for the Kyte-Doolittle hydropathy sliding window tool shown in Chapter 13 were written in pseudo-code. When developing the pairwise alignment tool, we will use all three of these algorithm formats to illustrate how each format is used. Typically though, depending on personal preference, a programmer will choose to use only one of these formats to express their algorithms.

14.5 ALGORITHMS FOR SPA

A top-level algorithm for the simple pairwise sequence alignment tool SPA is given in Figure 14-4. The flow chart shows that the program will first gather the input arguments from the user, and then, based on the alignment type, it will perform the appropriate alignment, and finally, it will print the alignment in the desired format. The program then will ask if the user wants to perform another alignment, and if so, will loop back to the beginning to repeat the process. If the user does not want to continue, the program will stop. In the algorithm shown in Figure 14-4, bold font is used to identify the subtasks (or functions) that need to be performed. Algorithms will be shown for each of these subtasks. The decision of what to make into a subtask is up to the program developer. Typically, if there are multiple steps involved, it is often useful to create a subtask. For example, "Input sequence format" is a

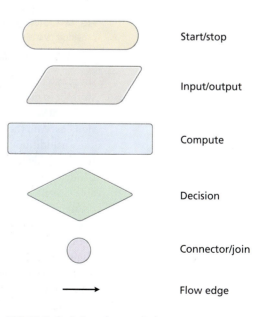

Start/stop

Input/output

Compute

Decision

Connector/join

Flow edge

FIG. 14-3. Basic flow chart symbols.

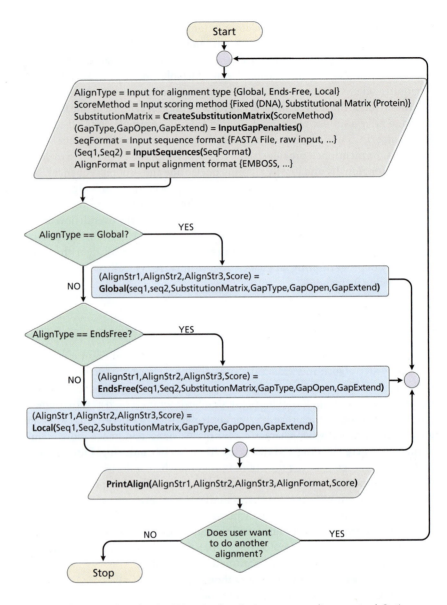

FIG. 14-4. Flow chart algorithm for SPA, a simple pairwise sequence alignment tool. Options provided by the menu-driven interface are shown within curly braces {}.

simple step that prompts the user for the sequence format and reads in the value, and stores it in global variable *SeqFormat*. On the other hand, a subtask has been created for **InputSequences** (Figure 14-5) because it involves several steps including a decision based on the sequence format.

In addition to showing the overall flow of tasks for the program (called the control flow), algorithms also show the flow of the data through variables. For example, the data about the scoring method is taken as input from the user, stored in the variable *ScoreMethod* and then passed to the function **CreateSubstitutionMatrix** to create the substitution matrix. After the substitution matrix is created, the data (the matrix) is passed back from the **CreateSubstitutionMatrix** function and assigned to the *SimMatrix* variable. Later, the data in the *SimMatrix* variable is passed to the alignment functions.

It is important to note that there are many ways to solve this problem and the algorithm provided in Figure 14-4 is one possible solution. Computer scientists use their creativity and knowledge to design programs, deciding the overall flow, how to divide a task into smaller subtasks, and what variable names to use. This text is

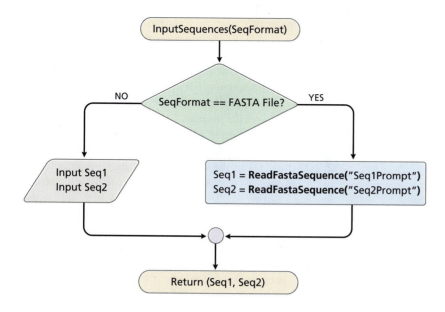

FIG. 14-5. Algorithm for InputSequences function.

written so that students who are new to programming can put together the basics presented to create their own tool. Beginning programmers may choose to closely follow the given algorithms (though you are encouraged to choose variable names that are meaningful to you), whereas more experienced programmers will undoubtedly want to apply their own knowledge and creativity to design their own tool.

Input Sequences

A simple algorithm for inputting two sequences to align is shown in Figure 14-5. The function will take as input the sequence format stored in variable *SeqFormat*, which is the format specified by the user for both sequences. If the user wants to use sequences stored in a file in FASTA format, then another function is called. If not, then the user is prompted to input the raw sequences—that is, sequences without any identifying information. Notice the hierarchical design of the algorithm as the problem is being divided into smaller and smaller subtasks.

A sequence in FASTA format begins with a single-line description starting with the greater-than symbol, with subsequent lines containing the sequence (see Chapter 2 for introduction to FASTA format). This single-line description is called the header or the definition line. The sequence below the definition line can be expressed using uppercase or lowercase letters.[5] It is recommended that all text lines be shorter than 80 characters in length. Blank lines are not allowed in the middle of a FASTA input. Here is an example of p53 sequence in FASTA format:

>gi|10720197|sp|Q9WUR6|P53_CAVPO CELLULAR TUMOR ANTIGEN P53
MEEPHSDLSIEPPLSQETFSDLWKLLPENNVLSDSLSPPMDHLLLSPEEVASWLGENP
DGDGHVSAAPVSEAPTSAGPALVAPAPATSWPLSSSVPSHKPYRGSYGFEVHFLKS
GTAKSVTCTYSPGLNKLFCQLAKTCPVQVWVESPPPPGTRVRALAIYKKSQHMTEV
VRRCPHHERCSDSDGLAPPQHLIRVEGNLHAEYVDDRTTFRHSVVVPYEPPEVGS
DCTTIHYNYMCNSSCMGGMNRRPILTIITLEDSSGKLLGRDSFEVRVCACPGRDR
RTEEENFRKKGGLCPEPTPGNIKRALPTSTSSSPQPKKKPLDAEYFTLKIRGRKNFEIL
REINEALEFKDAQTEKEPGESRPHSSYPKSKKGQSTSCHKKLMFKREGLDSD

---

[5] After reading the sequences, to simplify your comparisons, your program should convert them all to uppercase (the canonical representation).

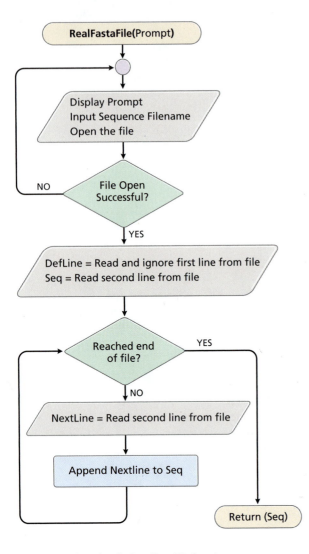

FIG. 14-6. Algorithm for ReadFastaFile function.

The algorithm for the **ReadFastaFile** function is shown in Figure 14-6. The function takes as input a prompt that will be displayed to the user. In this way, the same function can be used to input both sequences and a unique prompt can be given for each sequence. For example, the first time the function is called, the prompt could be "Enter filename of the first sequence: " and the prompt for second time the function is called could be: "Enter filename of the second sequence: ". After inputting the filename, the algorithm tries to open the file. If the file is in the same folder as the program, then the program should be able to open the file. If the user gave an invalid filename or if the file is in another folder and the user gave an invalid path to open the file, then an error will occur.[6] If an error occurs, the algorithm loops back and tries again. After the file is successfully opened, the FASTA definition line is read into the DefLine variable and, in this case, is ignored.[7] Then the subsequent lines are read in and appended to the sequence. When the end of the file is reached, the function returns the sequence.

Create Substitution Matrix

In addition to the two input sequences, the pairwise alignment tool needs to know how to score matches and mismatches due to substitutions. Typically, DNA sequence alignment uses a fixed score for matches and mismatches. For protein sequence alignment, PAM and BLOSUM substitution matrices are often used. Figure 14-7 shows a simple algorithm for creating the substitution matrices from either the PAM or BLOSUM file, or, if fixed scoring is used, from the match and mismatch scores provided by the user. For reference, the PAM150 substitution matrix file is shown in Figure 14-8. These files contain a header with information about how and when the matrix was created. The lines in the header are preceded by a pound symbol ('#'). A Python language-specific algorithm for reading in a PAM or BLOSUM matrix from a file is shown in Figure 14-9. It is Python specific because it uses Python data types including dictionaries, strings, and lists and Python functions such as slice and zip. To illustrate the different types of algorithm formats, the **InputSubstitutionMatrix** algorithm uses a step-by-step format rather than a flow chart format.

The algorithm describes how to create a dictionary of dictionaries to build the substitution matrix. The basic idea is to convert each row of the matrix into a dictionary, where the keys are the amino acid symbols and the values are the substitution scores. As each row is read in and converted into a dictionary, it is appended to a list of dictionaries. After all of the rows have been read in, the substitution matrix (*SubstitutionMatrix*), a dictionary of dictionaries, is created using the amino acid values as keys and the list of dictionaries as values.

Let's go back and look at each step of the algorithm shown in Figure 14-9. After creating the empty list to hold the row dictionaries (step 1), the algorithm prompts the user for the substitution scoring matrix file name (step 2), opens the file and reads in and ignores the header comment lines which start with a '#' character (step 3), and

---

[6] An error will also occur if the file is currently being used by another program.

[7] A useful extension would be to extract the information from the FASTA definition line, such as the sequence name, to use when displaying the alignment.

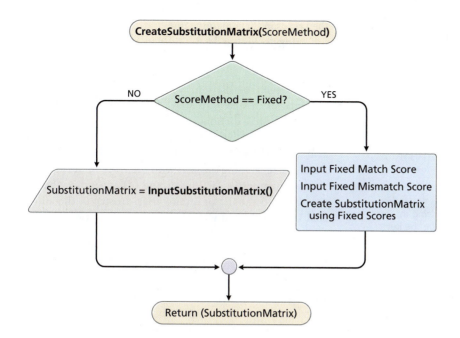

FIG. 14-7. Algorithm for CreateSubstitutionMatrix function.

```
#
# This matrix was produced by "pam" Version 1.0.6 [28-Jul-93]
#
# PAM 150 substitution matrix, scale = ln(2)/2 = 0.346574
#
# Expected score = -1.25, Entropy = 0.754 bits
#
# Lowest score = -7, Highest score = 12
#

   A  R  N  D  C  Q  E  G  H  I  L  K  M  F  P  S  T  W  Y  V  B  Z  X  *
A  3 -2  0  0 -2 -1  0  1 -2 -1 -2 -2 -1 -4  1  1  1 -6 -3  0  0  0 -1 -7
R -2  6 -1 -2 -4  1 -2 -3  1 -2 -3  3 -1 -4 -1 -1 -3  1 -4 -3 -2  0 -1 -7
N  0 -1  3  2 -4  0  1  0  2 -2 -3  1 -2 -4 -1  1  0 -4 -2 -2  3  1 -1 -7
D  0 -2  2  4 -6  1  3  0  0 -3 -5 -1 -3 -6 -2  0 -1 -7 -4 -3  3  2 -1 -7
C -2 -4 -4 -6  9 -6 -6 -4 -3 -2 -6 -6 -5 -5 -3  0 -3 -7  0 -2 -5 -6 -3 -7
Q -1  1  0  1 -6  5  2 -2  3 -3 -2  0 -1 -5  0 -1 -1 -5 -4 -2  1  4 -1 -7
E  0 -2  1  3 -6  2  4 -1  0 -2 -4 -1 -2 -6 -1 -1 -1 -7 -4 -2  2  4 -1 -7
G  1 -3  0  0 -4 -2 -1  4 -3 -3 -4 -2 -3 -5 -1  1 -1 -7 -5 -2  0 -1 -1 -7
H -2  1  2  0 -3  3  0 -3  6 -3 -2 -1 -3 -2 -1 -1 -2 -3  0 -3  1  1 -1 -7
I -1 -2 -2 -3 -2 -3 -2 -3 -3  5  1 -2  2  0 -3 -2  0 -5 -2  3 -2 -2 -1 -7
L -2 -3 -3 -5 -6 -2 -4 -4 -2  1  5 -3  3  1 -3 -3 -2 -2 -2  1 -4 -3 -2 -7
K -2 -3 -3 -5 -6  0 -1 -2 -1 -2 -3  4  0 -6 -2 -1  0 -4 -4 -3  0  0 -1 -7
M -1 -1 -2 -3 -5 -1 -2 -3 -3  2  3  0  7 -1 -3 -2 -1 -5 -3  1 -3 -2 -1 -7
F -4 -4 -4 -6 -5 -5 -6 -5 -2  0  1 -6 -1  7 -5 -3 -3 -1  5 -2 -5 -5 -3 -7
P  1 -1 -1 -2 -3  0 -1 -1 -1 -3 -3 -2 -3 -5  6  1  0 -6 -5 -2 -2 -1 -1 -7
S  1 -1  1  0  0 -1 -1  1  1 -2 -3 -1 -2 -3  1  2  1 -2 -3 -1  0 -1  0 -7
T  1 -2  0 -1 -3 -1 -1 -1 -2  0 -2  0 -1 -3  0  1  4 -5 -3  0  0 -1 -1 -7
W -6 -1 -4 -7 -7 -5 -7 -7 -3 -5 -2 -4 -5 -1 -6 -2 -5 12 -1 -6 -5 -6 -4 -7
Y -3 -4 -2 -4  0 -4 -4 -5  0 -2 -2 -4 -3  5 -5 -3 -3 -1  8 -3 -3 -4 -3 -7
V  0 -3 -2 -3 -2 -2 -2 -2 -3  3  1 -3  1 -2 -2 -1  0 -6 -3  4 -2 -2 -1 -7
B  0 -2  3  3 -5  1  2  0  1 -2 -4  0 -2 -5 -2  0  0 -5 -3 -2  3  2 -1 -7
Z  0  0  1  2 -6  4  4 -1  1 -2 -3  0 -2 -5 -1 -1 -1 -6 -4 -2  2  4 -1 -7
X -1 -1 -1 -1 -3 -1 -1 -1 -1 -1 -2 -1 -1 -3 -1  0 -1 -4 -3 -1 -1 -1 -1 -7
* -7 -7 -7 -7 -7 -7 -7 -7 -7 -7 -7 -7 -7 -7 -7 -7 -7 -7 -7 -7 -7 -7 -7  1
```

FIG. 14-8. PAM150 substitution matrix.

InputSubsitutionMatrix()

Step 1. Create an empty list that will be used for a list of dictionaries.

Step 2. Prompt the user for the substitution matrix filename.

Step 3. Open the file and ignore the header lines by reading in and ignoring each line until you reach a line that does not start with "#".

Step 4. Starting with the first line that does not start with a "#", read in the rest of the lines in the file and append them to a list of strings.

Step 5. The first string that does not start with a "#" is the string of amino acid symbols. Split this string into a list of amino acid characters. These will be the keys for the dictionary.

Step 6. For the remaining strings in the list of strings:

Step 6a. Slice off the first character (amino acid).

Step 6b. For the rest of the string split into individual numbers and convert to a list of intergers. This is your data for your dictionary.

Step 6c. Zip the keys and data together and convert into a dictionary.

Step 6d. Append the dictionary to the list of dictionaries.

Step 7. After you have read all lines, create the SubstitutionMatrix, which is a dictionary of dictionaries, by zipping the keys and the list of dictionaries and converting into a dictionary.

Step 8. Return SubstitutionMatrix.

FIG. 14-9. Step-by-step algorithm for InputSubstitutionMatrix function.

then reads in the remaining lines into a list of strings (step 4). Referring to Figure 14-8, you can see that the first non-header line contains the amino acid symbols that will be converted into a list of keys (step 5). This list of keys is used to create each row dictionary in step 6 and the dictionary of dictionaries in step 7. In step 6, each row of the matrix is converted into a dictionary by: (a) stripping off the leading amino acid (which is already recorded in the list of amino acid keys), (b) splitting the string of numbers into a list of numbers, (c) zipping the keys with the score values and converting them into a dictionary, and (d) appending the dictionary onto the list of dictionaries. In step 7, the amino acid keys are zipped with this list of dictionaries and then converted in the *SubstitutionMatrix*, which is a dictionary of dictionaries. The *SubstitutionMatrix* is returned to the calling function (**CreateSubstitutionMatrix**) in step 8.

Let's look at the result of executing the **InputSubstitutionMatrix** algorithm on the PAM150 matrix. After the PAM150 substitution matrix has been read into the dictionary variable *SubstitutionMatrix*, the substitution score of amino acids *A* and *K* can be looked up using the following reference:

$$SubstitutionMatrix['A']['K'],$$

which will yield the score -2. This corresponds to one cell in the matrix. To see how a row in the matrix is represented as a dictionary, let's look at *SubstitutionMatrix['A']* which yields:

{'*': −7, 'A': 3, 'C': −2, 'B': 0, 'E': 0, 'D': 0, 'G': 1, 'F': −4, 'I': −1, 'H': −2, 'K': −2, 'M': −1, 'L': −2, 'N': 0, 'Q': −1, 'P': 1, 'S': 1, 'R': −2, 'T': 1, 'W': −6, 'V': 0, 'Y': −3, 'X': −1, 'Z': 0}

Referring to Figure 14-8, this dictionary corresponds to the row in the matrix for amino acid 'A'. Recall that dictionaries use keys to look up data, and thus they are not ordered. This is why even though the values are correct, they are not in the same order as they are in the matrix in Figure 14-8. Let's check a couple of the values to verify that the dictionary is correct. As mentioned, the data with amino acid key 'K' is −2, which is correct because the score of 'A' and 'K' is −2 in the PAM150 matrix. Likewise, the data with amino acid key 'F' is −4, which is also correct because the score of 'A' and 'F' is −4.

To see what the matrix looks like as a dictionary of dictionaries, let's look at a portion of the matrix stored in the *SubstitutionMatrix* variable:

{'*': {'*': 1, 'A': −7, 'C': −7, 'B': −7, 'E': −7, 'D': −7, 'G': −7, 'F': −7, 'I': −7, 'H': −7, 'K': −7, 'M': −7, 'L': −7, 'N': −7, 'Q': −7, 'P': −7, 'S': −7, 'R': −7, 'T': −7, 'W': −7, 'V': −7, 'Y': −7, 'X': −7, 'Z': −7}, 'A': {'*': −7, 'A': 3, 'C': −2, 'B': 0, 'E': 0, 'D': 0, 'G': 1, 'F': −4, 'I': −1, 'H': −2, 'K': −2, 'M': −1, 'L': −2, 'N': 0, 'Q': −1, 'P': 1, 'S': 1, 'R': −2, 'T': 1, 'W': −6, 'V': 0, 'Y': −3, 'X': −1, 'Z': 0}, 'C': {'*': −7, 'A': −2, 'C': 9, 'B': −5, 'E': −6, 'D': −6, 'G': −4, 'F': −5, 'I': −2, 'H': −3, 'K': −6, 'M': −5, 'L': −6, 'N': −4, 'Q': −6, 'P': −3, 'S': 0, 'R': −4, 'T': −3, 'W': −7, 'V': −2, 'Y': 0, 'X': −3, 'Z': −6}, ...

This corresponds to three rows of the PAM150 matrix corresponding to the gap ('*') and amino acids 'A' and 'C' (again, they are not in the same order as shown in Figure 14-8 because dictionaries are not ordered).

Input Gap Penalties

The three options when performing an alignment are to align two matching residues, to allow for a mismatch that corresponds to a substitution, or to insert a gap that corresponds to an insertion or deletion. Although any gap scoring function can be used, two common ones, linear and affine, are supported by the algorithm shown in Figure 14-10. If the user specifies a linear gap type, then all gaps have the same penalty, and so, the GapOpen penalty is set to 0 and the GapExtend penalty is input from the user. If an affine gap penalty is used, the user specifies both the GapOpen and the GapExtend penalties. Recall, when using the affine gap penalty function, that the gap opening penalty is usually relatively large and the gap extension penalty is relatively small to avoid overly penalizing long insertions or deletions.

Suite of Pairwise Sequence Alignment Algorithms

SPA supports a suite of alignment algorithms including modified Needleman-Wunsch global alignment, ends-free global alignment, and Smith-Waterman local alignment. These algorithms will use a linear gap penalty model. Possible extensions to this basic alignment tool, including a discussion of an algorithm that supports affine gap penalties and more time- and space-efficient algorithms, are discussed at the end of the chapter.

Longest common subsequence (LCS)

The pairwise sequence alignment algorithms are similar to a classic computer science algorithm, longest common subsequence (LCS). The goal of LCS is to find the longest sequence of characters that is common among two strings. Note that in a subsequence, the characters must appear in the same order as they are in the strings but do not have to be consecutive in the strings. For example, in the string ACTGCT, some of the possible subsequences are T, TT, ACT, AGCT, and CTGCT. TCG is not a subsequence because the characters do not appear in that order in the string. We will first study the LCS algorithm and then discuss how the algorithm can be extended to implement pairwise sequence alignment algorithms.

The LCS and pairwise sequence alignment algorithms use the *dynamic programming* paradigm introduced by Richard Bellman in the mid-1950s. Dynamic programming is used in mathematics and computer science for solving problems with overlapping subproblems where the optimal solution can be formed from the optimal solutions to smaller subproblems. Figure 14-11 provides an abstract view of how dynamic programming works.

Let's assume we can use a matrix to score the optimal solution and that the optimal score can be found in the lower right corner of a matrix. Figure 14-11A–D shows four subproblems in green, blue, red, and yellow, respectively. The dark shaded cells represent the cells that contain the scores of each subproblem. Because this problem can be solved using dynamic programming, the optimal score of the yellow subproblem in Figure 14-11D, at $i = 3$ and $j = 3$, can be solved using the solutions of the green, blue, and red

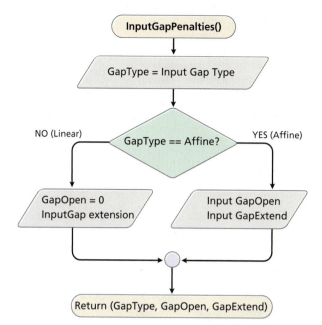

FIG. 14-10. Flow chart algorithm for InputGapPenalties.

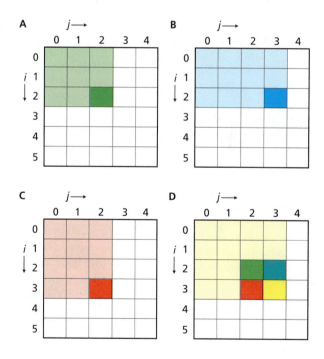

FIG. 14-11. An abstract view of how dynamic programming can be used to solve problems with overlapping subproblems. Assume that the optimal solution to each of the shaded subproblems in panels A–D is found in the lower right hand corner of the subproblem. Panel D illustrates how the optimal solution to the yellow subproblem shown in panel D can be computed using the optimal solution to the subproblem in panel A (green or *diagonal* element), the optimal solution to the subproblem in panel B (blue or *up* element), and the optimal solution to the subproblem in panel C (red or *left* element).

longest common subsequence (LCS) algorithm An algorithm that determines the longest common sequence of characters between two strings. The LCS algorithm is a starting point for the design of pairwise sequence alignment programs.

recurrence (also known as recurrence relation) An equation that defines an element in a sequence based on one or more of the earlier elements in the sequence.

traceback matrix A matrix created in some dynamic programming algorithms, such as longest common sequence (LCS) and sequence pairwise alignment (SPA), that serves to keep track of all of the possible highest scoring paths so that once scoring is complete and the highest scoring path is known, it can be used by these algorithms to create the output sequence for LCS or aligned sequences for SPA.

subproblems. The optimal score for the yellow subproblem, found at $i = 3$ and $j = 3$, can be calculated using the optimal score of the green subproblem at $i = 2$ and $j = 2$ (*the diagonal element*); the optimal score of the blue subproblem at $i = 2$ and $j = 3$ (*the up element*); and the optimal score of the red subproblem at $i = 3$ and $j = 2$ (*the left element*).

Intuitively, for LCS we can use dynamic programming to find the longest common subsequence of two strings by finding the longest common subsequence of two smaller substrings and building upon the solutions to the smaller substrings. Although LCS finds the longest subsequence of two strings, because we will be extending LCS to implement local and global pairwise sequence alignment on DNA and protein sequences, we will refer to the two input strings as sequence 1 and sequence 2 (abbreviated as *seq1* and *seq2*). To determine the longest common subsequence, a scoring matrix M is used to record the maximum number of common characters between the two strings at every location in the strings. As illustrated in Figure 14-11, the solution to the problem at indices i and j depends on the solutions to three subproblems, diagonal (at indices $i-1$ and $j-1$), left (at indices i and $j-1$), and above or up (at indices $i-1$ and j). Given two strings *seq1* and *seq2*, equation 14.1 shows the recurrence for computing the LCS scoring matrix M. The score for element $M_{i,j}$ depends on whether or not the ith character of string *seq1*, $seq1_i$, matches the jth character of string *seq2*, $seq2_j$. If they are the same, the longest common subsequence is increased by 1. That is, the maximum score for element $M_{i,j}$ will be the score of the diagonal $M_{i-1,j-1}$ plus 1 (in LCS, a match always yields the maximum score). If there is not a match, then the score of element $M_{i,j}$ is the maximum score of $M_{i-1,j}$ and $M_{i,j-1}$.

$$M_{i,j} = max \begin{cases} M_{i-1,j-1} + 1 & if\ seq1_i == seq2_j & (TB_{i,j} = Diagonal) \\ M_{i-1,j} & & (TB_{i,j} = Up) \\ M_{i,j-1} & & (TB_{i,j} = Left) \end{cases} \quad (14.1)$$

As the maximum score for each cell in M is recorded, the location of the cell used to compute the maximum score is also recorded in the traceback matrix, TB. After the two strings have been fully compared, the traceback matrix is used to print the longest common subsequence. If the score at $M_{i,j}$ was derived from $M_{i-1,j-1}$, that corresponds to the previous row and the previous column, so a *Diagonal* will be recorded in $TB_{i,j}$. If the score at $M_{i,j}$ was derived from $M_{i-1,j}$, that corresponds to the previous row and the same column, so an *Up* is recorded in $TB_{i,j}$. Likewise, if the score at $M_{i,j}$ was derived from $M_{i,j-1}$, that corresponds to the same row and the previous column, so a *Left* is recorded in $TB_{i,j}$.

Note that if both the left and up scores are the same—that is, $M_{i,j-1}$ is equal to $M_{i-1,j}$—then there may be more than one longest common subsequence. In this case, you can arbitrarily find one of the longest common subsequences by always picking either *Left* or *Up* when there is a tie. Or you can record both *Left* and *Up* when there is a tie. In this case, a more complex algorithm would be required to print all possible longest common subsequences.

The first step in the program design process involves understanding the problem, identifying inputs and outputs, and working out the problem by hand to create a test case with sample inputs and output. Let's work out an LCS example. The inputs are the two strings, which, as mentioned previously, we will refer to as sequence 1 and sequence 2 (abbreviated as *seq1* and *seq2*). The output is the longest common subsequence *LCS*. Consider the two sequences below. Let's see how we can create the scoring and traceback matrices, M and TB, to derive *LCS*.

Input sequences

Sequence 1 = TACGC	($n = 5$)
Sequence 2 = ATCAGC	($m = 6$)

Output

$$LCS = TAGC$$

First, notice that *Sequence 1* has five characters, so it has a length *n* equal to 5, and *Sequence 2* has six characters, so it has a length *m* equal to 6. We will create the scoring matrix *M* and the traceback matrix *TB* with *n*+1 rows and *m*+1 columns as shown in Figure 14-12. For ease of reference, the LCS recurrence is shown in Figure 14-12A and the *M* and *TB* matrices are shown separately in Figures 14-12B and 14-12C. Typically it is easier to understand what is happening if the *M* and *TB* matrices are shown together, as in Figure 14-12D, where *M* is in red and *TB* is in blue. *Sequence 1* is shown beside the matrix because the rows in *M* and *TB* correspond to the characters in *Sequence 1*. Likewise, *Sequence 2* is shown on top of the matrix because the columns in *M* and *TB* correspond to the characters in *Sequence 2*. The extra row and column are added because the two sequences may not match at the beginning (or at all). To handle this condition, the first row and column of *M* will be initialized to zero. For LCS, the first row and column of *TB* do not need to be initialized to anything, as we will see shortly.

Now let's use the recurrence from equation 14.1 to compute *M* and *TB* to find the longest common subsequence. Figure *14-12* shows how *M* and *TB* are being filled in, one row at a time. Recall that if there is not a match, then we will take the maximum of the up or left score. If they are the same, we need to pick either up or left. For our algorithm, we will always select up when there is a tie. So, in this example, if there is not a match and the up score is greater than or equal to the left score, we will choose up. Otherwise, we will choose left. Consider $M_{1,1}$, because $seq1_i \neq seq2_j$ (T $\neq$ A) and the up score, $M_{0,1}$, and left score, $M_{1,0}$, are both zero, we will choose up. That is, $M_{1,1} = M_{0,1} = 0$ and $TB_{1,1}$ is *Up*. Consider $M_{2,4}$, because $seq1_i == seq2_j$ (both are A), we will choose diagonal, and thus, $M_{2,4} = M_{1,3} + 1 = 1 + 1 = 2$ and $TB_{2,4}$ is *Diagonal*. Lastly, let's consider what will happen next with $M_{2,5}$, because $seq1_i \neq seq2_j$ (A $\neq$ G) and the up score, $M_{1,5}$ (1), is less than the left score $M_{2,4}$ (2), we will choose left. That is $M_{2,5} = M_{2,4} = 2$ and $TB_{2,5}$ is *Left*. Try to fill out the rest of *M* and *TB* on your own and compare your answer to Figure 14-13.

Looking at Figure 14-13, we can find the maximum number of characters the two sequences have in common in the lower right corner of the *M* matrix. We refer to this as the LCS score. In this case, the LCS score is 4. To find out the four characters that comprise the longest common subsequence, we can use the traceback matrix. During traceback, the arrows provide a path along which the longest common subsequence can be found. Whenever a *Diagonal* is encountered, it corresponds to a common character in the subsequence. In addition, when a *Diagonal* is encountered, we know that the next cell along the traceback path will be at indices *i−1, j−1* (previous row and previous column). When a *Left* is encountered, the next cell along the traceback path will be at indices *i,j−1* (same row, previous column). Likewise, when an *Up* is encountered, the next cell along the traceback path will be at indices *i−1,j* (previous row, same column). Traceback starts at the lower right corner of the *TB* matrix, at $TB_{n,m} = TB_{5,6}$. Traceback will end when we have reached the beginning of one of the sequences—that is, when either *i* or *j* is zero. This makes sense because once we have reached the end of one sequence there cannot be any more characters in common between the two sequences.

The cells along the traceback path are outlined in bold in Figure 14-13. Starting at the lower right corner, $TB_{5,6}$, we can follow the bolded path. If we were to print out the

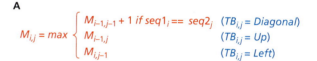

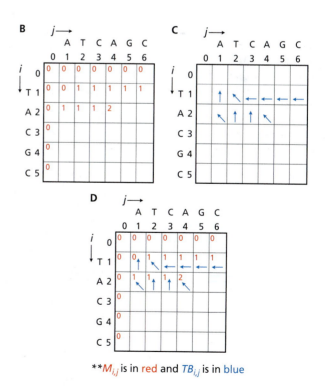

FIG. 14-12. Longest common subsequence (LCS) scoring example in which sequence 1 = TACGC and sequence 2 = ATCAGC. **A.** The LCS recurrence. **B.** Matrix *M*. **C.** Matrix *TB*. **D.** Both matrices shown together. Referring to the shaded cell, because $a_2 == b_4$ (both are 'A'), $M_{2,4} = M_{1,3} + 1 = 2$ and $TB_{2,4}$ is diagonal.

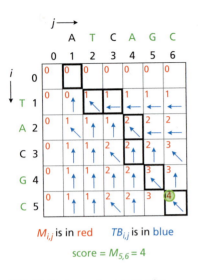

$M_{i,j}$ is in red $TB_{i,j}$ is in blue

score = $M_{5,6}$ = 4

FIG. 14-13. Longest common subsequence (LCS) traceback example in which sequence 1 = TACGC and sequence 2 = ATCAGC and the longest common subsequence is TAGC.

common characters whenever a *Diagonal* is encountered, the subsequence would be "CGAT," which is actually the reverse of the longest common subsequence. To solve this, we can trace back along the path first, and then, go backward along the path (in this example from $TB_{0,1}$ to $TB_{5,6}$) and print out the characters for every *Diagonal*. In this case, we would print the subsequence "TAGC" correctly. Look again at *Sequence 1* and *Sequence 2*, and you should be able to convince yourself that this is one of the longest common subsequences. There are no longer subsequences, but there are three others with a score of 4 ("TAGC," "ACGC," and "TCGC"). Because we always chose *Up* when there was a tie, we found only one sequence. As mentioned before, if we record a tie instead in that case, we could use a more complex algorithm to find all of the longest common subsequences.

Let's revisit the abstract concept of dynamic programming presented in Figure 14-11 to see how the LCS algorithm is in fact using the solution to smaller subproblems to solve a problem. Figure 14-14 shows the partial solution from Figure 14-12, but in this figure, we have used the color scheme from Figure 14-11 to outline the submatrices of four subproblems. For the four subproblems, the score of the subproblem is shown in italics and can be found in the lower right corner of each submatrix. For each subproblem, traceback starts in the lower right corner of each submatrix and continues back, following the bolded black arrows, until the first row or column is reached. As mentioned before, as you go backward along the traceback path, whenever a diagonal is encountered, the corresponding character from *Sequence 1* (or *Sequence 2*, because it is a match) is printed. Doing so will produce the longest common subsequence for that subproblem.

Let's consider the four subproblems highlighted in Figure 14-14.

1. Green subproblem (Figure 14-14A)—find longest common subsequence between T and ATC. The score of this solution (at $M_{1,3}$) is 1, and the longest common subsequence is T.

2. Red subproblem (Figure 14-14B)—find the longest common subsequence between TA and ATC. The score of this solution (at $M_{2,3}$) is 1, and the longest common subsequence is T.

3. Blue subproblem (Figure 14-14C) – find the longest common subsequence between T and ATCA. The score of this solution (at $M_{1,4}$) is 1, and the longest common subsequence is T.

4. Yellow subproblem (Figure 14-14D)—find the longest common subsequence between TA and ATCA. The score of this solution (at $M_{2,4}$) is 2, and the longest common subsequence is TA.

Figure 14-14D highlights how the LCS recurrence (equation 14-1) uses the solutions to the green (diagonal), red (up), and blue (left) subproblems to solve the yellow subproblem. In this case, because there was a match, the solution to the yellow subproblem builds upon the green (diagonal) subproblem. You can see how the traceback path of the yellow subproblem (Figure 14-14D) builds upon the traceback path of the green subproblem (Figure 14-14A) and, as a result, how the longest common subsequence of the yellow subproblem (TA) builds upon the longest common subsequence of the green solution (T). Note that

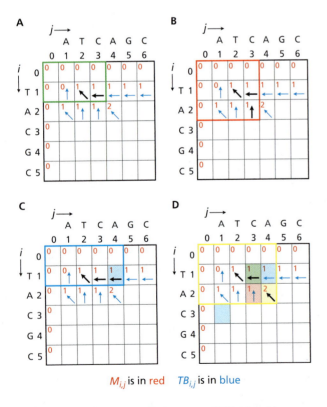

$M_{i,j}$ is in red $TB_{i,j}$ is in blue

FIG. 14-14. Longest common subsequence (LCS) subproblems illustrate how dynamic programming works.

although the longest common subsequence for each of the three subproblems is the same in this particular example (T), this is not the case in general. To verify this, explore the different subproblems of the complete solution in Figure 14-13.

Typically, matrices are traversed row by row (or column by column). In this case the LCS algorithm traverses the M and TB matrices row by row going from left to right (traversing column by column from top to bottom would also work). Convince yourself that by doing so, for any given subproblem, the solution of the three subproblems it depends upon (diagonal, up, and left) will have already been computed. Of course, the first row and column must be first be initialized (to zero for LCS).

Now that we have worked out the problem by hand, the next step in the program design process is to develop the algorithms. In this case, the pseudo-code algorithms for *Score_LCS* and *Print_LCS* are provided in Figures 14-15 and 14-16. *Score_LCS* takes the two sequences *seq1* and *seq2* as inputs and returns the *score* and the traceback matrix *TB*. After working out the example problem, the *Score_LCS* algorithm, which is based on equation 14-1, should be easy to follow. After setting n to the length of *seq1* and m to the length of *seq2* it initializes the first column and first row of M to zero. The nested *for* loops compute the values of the M and TB matrices, the outer loop going row by row, and the inner loop going through each column within a row. Recall that i indicates the current row and j indicates the current column.

The inputs of the *Print_LCS* algorithm are *seq1*, the traceback matrix *TB*, and the current value of i and j. Note that rather than using a loop to iterate over the traceback path in *TB*, *Print_LCS* uses a recursive algorithm that calls itself. Starting at the lower right corner ($i = n$ and $j = m$), we can follow the traceback path by using the traceback directions (*Diagonal* = 'D', *Up* = 'U', and *Left* = 'L') to determine the next values for i and j. If *Print_LCS* were used for the example in Figure 14-13, the first call would pass *seq1*, *TB*, and the values of n and m, specifically 5 and 6. The *Print_LCS* algorithm would then call itself six times with the following values of (i,j): (4,5), (3,4), (2,4), (1,3), (1,2), and (0,1). As mentioned previously, the traceback will stop when either i or j is zero, indicating that the end of one of the sequences has been reached, so there cannot be any other characters in common. When the *Print_LCS* function returns, it returns to the next instruction after the function call. In the *Left* and *Up* cases, there are no instructions after the function call, so it will return again, either to the previous *Print_LCS* (if it was originally a recursive call) in the call stack or to the program that originally called it. In the *Diagonal* case, it will print the corresponding character from *seq1* at index i before returning. You should hand test these algorithms using the example given in Figure 14-13.

The next step of the program design process is to code the algorithm in a programming language such as Python. This is left for you as an end-of-chapter exercise. After you implement it in Python, or some other language, you can again test your program using the example sequences from Figure 14-12. You can compare the M and TB matrices that your program generates with those shown in Figure 14-13.

Score_LCS(seq1, seq2)

```
n = length of seq1
m = length of seq2
for i = 0 to n
    M_{i,0} = 0
for j = 0 to m
    M_{0,j} = 0
for i = 1 to n
    for j = 1 to m
        if(seq1_i == seq2_j)
            M_{i,j} = M_{i-1,j-1} + 1; TB_{i,j} = 'D'    -- Diagonal
        if(M_{i-1,j} >= M_{i,j-1})
            M_{i,j} = M_{i-1,j}; TB_{i,j} = 'U'          -- Up
        else
            M_{i,j} = M_{i,j-1}; TB_{i,j} = 'L'          -- Left
    score = M_{n,m}
    return(score, TB)
```

FIG. 14-15. Score_LCS pseudo-code algorithm to compute the scoring matrix (M) and the traceback matrix (*TB*). To specify the traceback directions, 'D' is used for diagonal, 'U' for up, and 'L' for left. Note that the indices of the input sequences to the Score_LCS algorithm range from 1 to n for sequence 1 and from 1 to m for sequence 2. When coding this algorithm in Python, if you use strings to represent sequences a and b, you'll have to make adjustments to the algorithm because strings begin at index 0. One easy solution is to add a space character at the beginning of each sequence, but you'll have to take that into account when computing m and n. Because the same problem exists for Print_LCS, sequences 1 and 2 should be modified once, before calling either function.

Print_LCS(seq1,TB,i,j)

   if i == 0 *or* j == 0

      return

   if $TB_{i,j}$ == 'D'

      Print_LCS(seq1,TB,i-1,j-1)

      print $seq1_i$

   else if $TB_{i,j}$ == 'U'

      Print_LCS(seq1,TB,i-1,j)

   else

      Print_LCS(seq1,TB,i,j-1)

FIG. 14-16. Print_LCS is a recursive algorithm (pseudo-code) to print the longest common subsequence. (This algorithm assumes that sequences 1 and 2 start at index 1.)

Global alignment

Global alignment is used to align two sequences, protein or DNA, from end to end. As we know, protein and DNA sequences can be represented using strings, where each character in the string represents an amino acid residue (in proteins) or nucleotide (in DNA). A modified Needleman-Wunsch pairwise global alignment algorithm can be implemented using a similar algorithm to *LCS*. Compared to the original Needleman-Wunsch algorithm, which used fixed scoring for matches and mismatches, mismatches can be scored using a substitution matrix such as PAM or BLOSUM. If fixed scoring is desired, a simple substitution matrix can be derived from fixed match and mismatch scores. The algorithm shown for a modified Needleman-Wunsch pairwise global alignment algorithm will also allow for insertions and deletions by assigning a fixed or linear gap penalty. Given two sequences *seq1* and *seq2*, equation 14.2 shows the recurrence for computing the scoring matrix *M* for *Global Alignment*.

$$
M_{i,j} = max \begin{cases} M_{i-1,j-1} + S_{seq1_i,seq2_j} & (TB_{i,j} = Diagonal) \\ M_{i-1,j} + w & (TB_{i,j} = Up) \\ M_{i,j-1} + w & (TB_{i,j} = Left) \end{cases} \quad (14.2)
$$

Comparing this equation to the *LCS* equation 14.1, you can identify the modifications that you will need to make to the *LCS* algorithm. Instead of checking whether there is a match between $seq1_i$ and $seq2_j$, the substitution score, $<S_{seq1_i,seq2_j}>$, is used. Note that, unlike *LCS*, the *Diagonal* case will not necessarily be the highest score even if there is a match. Another difference between *LCS* and *Global Alignment* is that a fixed gap penalty, *w*, will be added to the left or up score. *Up* corresponds to progressing along *Sequence 1* (previous row), but staying in the same location in *Sequence 2* (same column). Thus, the *Up* case corresponds to a gap in *Sequence 2*. If we consider the evolution of these sequences, this gap in *Sequence 2* could be due to an insertion in *Sequence 1* or a deletion in *Sequence 2*. Likewise, the *Left* case corresponds to a gap in *Sequence 1*.

When developing the pairwise sequence alignment algorithms, your algorithm needs to address the following three issues (conditions):

1. How should the first row and first column of the scoring matrix *M* be initialized?

2. After computing *M* and *TB*, where can the score of the alignment be found in *M* and where should traceback start in *TB*?

3. Given *TB*, where should traceback end?

These issues need to be addressed for all types of pairwise sequence alignment including global, ends-free global, and local. The answer to the first question also depends on how gap penalties are modeled.

Let's consider the first issue for global alignment with a fixed (linear) gap penalty. A score in the first row ($i = 0$) represents a gap in *Sequence 1* being aligned with an amino acid or nucleotide in *Sequence 2*. If you are at index $(0,1)$, that means that the first residue of *Sequence 2* is aligned with a gap at the beginning of *Sequence 1* and the score should be *w*, which is the gap penalty (a negative value). If you are at index $(0,2)$, that means that the first two residues of *Sequence 2* are aligned with two gaps at the beginning of *Sequence 1* and thus the score should be *2w*. Generalizing, cells in the first row of *M* should be initialized to *w*j* where *j* ranges from 0 to *m*. Likewise, cells in the first column of *M* should be initialized to *w*i* where *i* ranges from 0 to *n*.

The second issue is where to find the score of the alignment in M and where to start traceback in TB. Because this is a global alignment that will align the sequences end to end, the score of the alignment will be found at $M_{n,m}$ and traceback will start at $TB_{n,m}$. Again, because the alignment is from end to end, the solution to the third issue is that traceback will end at $TB_{0,0}$.

Local alignment

Whereas global alignment aligns two sequences from end to end, local alignment will find the highest scoring local region of alignment between two sequences. Recall that highly conserved regions among sequences can be used to identify regions with functional or structural significance. Smith-Waterman proposed a recurrence equation similar to the one shown in equation 14.3. This equation for *Local Alignment* assumes a fixed/linear gap penalty. This recurrence is very similar to equation 14.2 for global alignment except that it also has a 0 as one of the possible scores because negative scores are not allowed. A score of 0 indicates the end of a region of similarity, and thus, traceback would stop at this point in local alignment.

$$M_{i,j} = max \begin{cases} 0 & (TB_{i,j} = Stop) \\ M_{i-1,j-1} + S_{seq1_i,seq2_j} & (TB_{i,j} = Diagonal) \\ M_{i-1,j} + w & (TB_{i,j} = Up) \\ M_{i,j-1} + w & (TB_{i,j} = Left) \end{cases} \quad (14.3)$$

Let's revisit the three issues that need to be addressed when developing a pairwise sequence alignment algorithm (see previous section). First, because movement along the first row or column corresponds to inserting a gap that has a negative score, the first row and column are initialized to the minimum score, zero. Second, the score of the best possible local alignment will be the maximum value in M, called $M_{x,y}$. To find this score, you need to search M for the maximum score. Traceback will start at $TB_{x,y}$. Third, traceback will stop when a *Stop* direction is encountered, which corresponds to a score of 0 in M.

Output Alignment

After the sequences have been scored, the aligned sequences must be printed. Because *LCS* finds the longest *common* subsequence between two sequences, *Print_LCS* prints only one string. For pairwise sequence alignment, *Print_LCS* must be adapted to print two sequences. Protein and DNA sequences tend to be very long. Therefore, instead of printing the sequences directly, it is better to adapt *Print_LCS* to first form an alignment string for each aligned sequence containing both residues and gaps. This modified algorithm would more aptly be called *Gen_Align* for generate alignments. After the strings are formed another algorithm, called *Print_Align*, can be developed to print the alignment. The *Print_Align* algorithm can follow the EMBOSS alignment format below.

```
195 NYTSPSATPRPPAPGPPQSRGT-----SPLQPGSYPEYQASGADSWPPAA 239
    ..:.|..  ||.| ||||...|.     .|.:||        :..||..
212 GASGPMG-PRGP-PGPPGKNGDDGEAGKPGRPG---------ERGPPGP 249


240 ENSFPGANFGVPPAEPEPIPKGSRPGGSPRGVSFQFPFPALHGASTKPFP 289
    :    ||. |:|.....|..||.|           .|..|.||.....|
250 Q----GAR-GLPGTAGLPGMKGHR-----------GFSGLDGAKGDAGP 282
```

Notice that alignment wraps over several lines and only 50 characters are printed per line per sequence. At the beginning and end of each line are the numbers of the first and last residues of each sequence segment. Thus, 'N' is residue 195 of the top sequence and 'G' is residue 212 of the bottom sequence. Due to gaps, these residue numbers are different. EMBOSS uses the hyphen ('−') to indicate a gap. In between each sequence segment is a markup line that shows where sequences are mismatched, gapped, identical, or similar. EMBOSS generally uses a space for a mismatch or a gap. A period ('.') is used when there is little or no similarity (a score less than 1.0), a colon (':') is used for a similarity which scores greater than or equal to 1.0, and a vertical bar ('|') is used to indicate a match regardless of the score.

A header block should also be included before the aligned sequences. As discussed earlier, header blocks contain information about the alignment input arguments. Following the header section there is often a data section. Reporting the inputs used for the alignment allows the alignment to be reproduced and verified by others. Headers typically start with a pound/hash character ('#'). At a minimum, your alignment should include an output data section before the alignment that contains the score of the alignment. For example, the output data section may contain a line similar to this:

$$\text{\# Score: 729.5}$$

The example local alignment shown in Figure 14-1 illustrates other useful information that can be included in your header and data sections.

14.6 ALGORITHM COMPLEXITY

algorithm complexity Quantifies the amount of memory space (space complexity) or time (time complexity) that an algorithm requires as a function of the size of the input.

Now that we have finished the introduction to our simple pairwise sequence alignment tool, SPA, let's look at possible extensions. First, though, we need to introduce an important concept called *algorithm complexity*, which refers to the amount of time and memory space that the algorithm requires.

There are many different algorithms that can be developed to solve the same problem. Once an algorithm is proven correct, you can evaluate its efficiency to determine if it is the best algorithm to use. This is referred to as computing the algorithm complexity. The complexity of an algorithm can be computed for time and for space. The time complexity provides an approximation of how long it would take to execute the algorithm, and the space complexity provides an approximation of how much memory is needed to execute the algorithm. Complexity is determined relative to the size of the problem, N.

When evaluating algorithm complexity, we are interested in the worst-case scenario. That is, what is the longest possible execution time of the algorithm for a given problem size N? For sequence alignment, N is the size of the longest sequence being aligned. Furthermore, we are interested in the efficiency of the algorithm as the problem size grows, which is defined by the asymptotic behavior of the algorithm. For example, if the worst-case execution time for a given algorithm has been determined to be $4N^2 + 6N + 8$, the asymptotic behavior is dominated by the highest term, $4N^2$. Ignoring the constant 4, we would say that the algorithm complexity of this algorithm is on the *order* of N^2, which can be written as $O(N^2)$. This notation is referred to as big-O notation in algorithm complexity analysis.

Let's take a look at the algorithm complexity of the original Needleman-Wunsch and Smith-Waterman algorithms that were discussed in Chapter 5. As illustrated in Figure 14-17A, the original Needleman-Wunsch algorithm computes the best alignment at each location in the scoring matrix by finding the maximum score in the previous row and column of the submatrix. In the worst-case scenario, the two sequences being aligned have the same length, which we will call N. For an

$N \times N$ matrix, N^2 scores must be computed. Finding the maximum score of a row and column worst case takes $2N$ comparisons. Thus, to compute the alignment score requires a total of $N^2 \times 2N$ or $2N^3$ computations. Thus, the algorithm complexity of the original Needleman-Wunsch scoring algorithm is $O(N^3)$.

Now let's consider the Smith-Waterman scoring algorithm illustrated in Figure 14-17B. In this case, the best alignment score at each location in the scoring matrix is determined by using the scores of the cell directly above, directly diagonal, and directly to the left. Because there are N^2 cells in the scoring matrix, to compute the overall scoring matrix requires $3N^2$ computations. Thus, the algorithm complexity of the Smith-Waterman scoring algorithm is $O(N^2)$. Searching for the maximum score in the scoring matrix also has $O(N^2)$ complexity. Finally, traceback has $O(N)$ complexity because it needs to traverse only one path through the matrix. Thus, the algorithm complexity for the entire Smith-Waterman local alignment algorithm is $O(N^2)$. Similarly, the algorithm complexity of the modified Needleman-Wunsch algorithm covered in this chapter is also $O(N^2)$. To align two proteins with 400 amino acid residues each, the original Needleman-Wunsch algorithm would take on the order of 400^3, or 64 million computations. On the other hand, the Smith-Waterman algorithm and the modified Needleman-Wunsch algorithm would take on the order of 400^2, or 160,000 computations. For this reason, we used the modified Needleman-Wunsch algorithm (see equation 14.2 above) for global alignments.

Although the Smith-Waterman or modified Needleman-Wunsch algorithm is acceptable for comparing a pair of protein sequences, the computational cost to compare a protein sequence against a large database of sequences using these algorithms is prohibitive. These algorithms will guarantee the best possible alignment for a set of specified input arguments, but other *heuristic*-based algorithms such as BLAST have been developed to perform sequence alignment with much less computational time (implementing BLAST is beyond the scope of this chapter).

In addition to computational time, the memory space required for storing the scoring and traceback matrices is $O(N^2)$. Because the Smith-Waterman algorithm uses only the immediate neighboring cells in the previous row and previous column, at any given time only the two rows are needed to compute the score (this is also true for the modified Needleman-Wunsch algorithm). Thus, it is fairly simple to modify the scoring algorithm to use two rows, which requires $2N$ or $O(N)$ space for the scoring matrix M. Recall that the purpose of the scoring matrix is to determine the optimal alignment that is derived from the information in the traceback matrix, which is still $O(N^2)$. Thus, the overall space requirement is still $O(N^2)$. Hirschberg developed a space-efficient alignment algorithm that uses a divide-and-conquer approach, which still requires $O(N^2)$ computational time but only $O(N)$ space.

A pioneer in algorithms and algorithm complexity, Richard Karp, is most famously known for identifying combinatorial problems whose solutions are nondeterministic in polynomial time (these problems are referred to as NP-complete problems). Because no known solution to the problem in polynomial time exists, the time to solve these problems grows very quickly as the size of the problem grows (some could take billions of years to solve). Instead of using an exact solution for these types of problems, computer scientists use heuristics to generate an approximate solution. Karp, a trained mathematician and early pioneer in the field of computer science, has focused the latter part of his career in the field of bioinformatics and understands the importance of the intersectionality of the various disciplines, mathematics, computer science, and biology to the field of bioinformatics (see Box 14.1).

A Example of computing the original Needleman-Wunsch global alignment scoring algorithm.

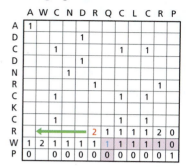

B Example of computing Smith-Waterman local alignment scoring algorithm.

		A	W	C	N
		0	-4	-8	-12	-16
A		-4	4	0	-4	-8
D		-8	0	0	-3	-3
C		-12	-4	-2	9	
D		-16				
N		-20				

FIG. 14-17 Examples of original Needleman-Wunsch and Smith-Waterman scoring algorithms taken from Chapter 5 pairwise alignment examples. A. Example of computing the original Needleman-Wunscsh global alignment scoring algorithm. B. Example of computing Smith-Waterman local alignment scoring algorithm.

Richard Karp

RICHARD M. KARP (born 1935) is Director of the Simons Institute for the Theory of Computing at the University of California at Berkeley. He received his bachelor's, master's, and doctoral degrees in applied mathematics from Harvard, and has been at UC Berkeley since 1968 (apart from a four-year period as a professor at the University of Washington). Widely known for his work on algorithms and algorithm complexity, most notably for his contributions to the concept of NP-completeness, Karp received the highest recognition in the field of computer science, the Turing Award, in 1985. He was also awarded the U.S. National Medal of Science, the Benjamin Franklin Medal in Computer and Cognitive Science, the Kyoto Prize, Fulkerson Prize, Harvey Prize (Technion), Centennial Medal (Harvard), Lanchester Prize, Von Neumann Theory Prize, Von Neumann Lectureship, and the Babbage Prize.

Richard Karp has also been conducting bioinformatics research since 1991. His recent work focuses on transcriptional regulation of genes, discovery of conserved regulatory pathways, and analysis of genetic variations in humans. He is interested in finding the genetic basis of complex diseases so that more effective modes of treatment can be developed. Karp understands the multifaceted and complex nature of bioinformatics. He states, "Solving biological problems requires far more than clever algorithms: it involves a creative partnership between biologists and mathematical scientists to arrive at an appropriate mathematical model, the acquisition and use of diverse sources of data, and statistical methods to show that the biological patterns and regularities that we discover could not be due to chance."

REFERENCES

Karp, R. M. 1972. "Reducibility among Combinatorial Problems." In *Complexity of Computer Computations*, edited by R. E. Miller and J. W. Thatcher, 85–103. New York: Plenum.

14.7 EXTENSIONS TO SIMPLE PAIRWISE ALIGNMENT TOOL

There are several extensions that can be added to the basic pairwise sequence alignment tool presented in this chapter. Brief summaries of these extensions are provided here, but the details are beyond the scope of this chapter. The extensions are listed in order of implementation complexity: from simpler to more complex.

Print all local alignments greater than a threshold. As mentioned previously, highly conserved regions typically have functional or structural significance. There may be several highly conserved regions. Local alignment can be extended to find any regions with a score greater than some threshold (this will include local alignments that have the same optimal score).

Print all possible optimal alignments. Local and global alignment, as presented in this chapter, will always have a bias toward inserting a gap in *Sequence 2* (because it selects *up* if there is a tie between the left and up scores). Global and local alignment can be modified to print out all possible optimal alignments in order to let the biologists decide which one is the most logical.

Add a graphical user interface (GUI) to your alignment tool. In this chapter we recommend starting with a simple, text-based, menu-driven interface because substance is more important than style. That is, the ability to align sequences is more significant to a biologist than being able to select options using drop-down menus, for instance. Time permitting, though, a GUI typically provides a friendlier user interface. Python has a huge number of GUI frameworks that you are encouraged to explore.

Add a database interface to your alignment tool. Oftentimes biologists want to be able to compare a given sequence against a database of sequences. A simple implementation can consist of creating a flat file of sequences and a query sequence. The query sequence can be aligned against each of the sequences in the flat file, and the alignments that have a score greater than some threshold can be displayed. Rather than using a simple score, the algorithm can compute the statistical significance of each alignment, as discussed in Chapter 12. A more complex algorithm is to interface your alignment tool to an online database. Further extensions can be made to speed up the queries using a BLAST or FASTA heuristic algorithm.

Implement a space-efficient alignment algorithm. As mentioned, it is relatively simple to reduce the space required for the scoring matrix M because when computing a row in M, only the current and previous rows are needed. However, the entire traceback matrix TB is still required for printing the alignment. Dan Hirschberg proposed a divide-and-conquer algorithm that does not require a traceback matrix. This algorithm requires $O(N)$ space, where N is the length of the longer sequence being aligned. More advanced programmers are encouraged to research and implement Hirschberg's space-efficient algorithm.

Implement an alignment algorithm that supports affine gap penalties. The simple alignment tool presented in this chapter used a linear gap penalty. Every gap has a fixed penalty w, and gaps of length L will have a linear gap penalty of $w*L$. This may overly penalize long gaps. Rather, the affine gap penalty model penalizes the insertion of a gap more than the length of the gap. A gap of length L will have an affine gap penalty of $w_{open} + w_{extend}*L$. Osamu Gotoh proposed a distance-based global alignment algorithm that uses the affine gap penalty model. Gotoh's algorithm uses three scoring matrices, one to keep track of a gap opening in *Sequence 1*, one to keep track of a gap opening in *Sequence 2*, and the main scoring matrix that keeps track of matches and mismatches. The intuition behind his algorithm is that when a gap is opened, it may not have the lowest score, but as the gap is extended, it may ultimately have been better to have opened a gap. The extra two matrices allow that option, because they keep track of what would happen if a gap had been opened. Advanced programmers are encouraged to research and implement Gotoh's alignment algorithm with affine gap penalties.

SUMMARY

As this book has demonstrated, many bioinformatics tools have been developed to help biologists understand the vast field of biology and many more are needed as the field continues to evolve. In this chapter we have explored how to implement a classic bioinformatics tool: pairwise sequence alignment. The goal of the chapter was to give you the basics of designing a tool so that you can create your own tool if desired. Understanding how to develop tools can also make you a more adept tool user by demystifying how computer software works. As a developer, you realize the potential power as well as the limitations of software applications. For example, as discussed in this chapter, you can develop a tool that allows you to find all possible statistically significant local alignments between two sequences. However, it requires a biologist to select a substitution matrix and gap penalty model that will yield a good local alignment that captures the conserved regions of the proteins or DNA sequences.

EXERCISES

These exercises are designed to help you develop the knowledge and skills needed to successfully implement the simple pairwise alignment tool SPA.

1. The first step in program development is to understand the problem, its inputs, and its outputs. A good way to do that is to generate a solution by hand. In addition to helping you understand how to solve the problem, the solution can also be used to test your program. In this problem you will generate the scoring matrix (*M*), the traceback matrix (*TB*), the alignment score, and the pairwise alignment for the following problems:

    a. Global alignment using the PAM150 substitution matrix and a linear gap penalty with $w = -4$. Use the global alignment recurrence equation 14.2.
    b. Local alignment using the BLOSUM62 substitution matrix and a linear gap penalty with $w = -4$. Use the local alignment recurrence equation 14.3.

    Both problems will use the same input sequences. Note that these short sequences were created for testing purposes, so their alignments do not necessarily have any biological significance.

    > Sequence 1 = PRPPATP
    > Sequence 2 = PQAP

    Remember to consider the following issues:

    i. How should the first row and first column of the scoring matrix *M* be initialized?
    ii. After computing *M* and *TB*, where can the score of the alignment be found in *M* and where should traceback start in *TB*?
    iii. Given *TB*, where should traceback end?

2. The EMBOSS sequence alignment format includes the residue numbers for each sequence at the beginning and end of each segment of the alignment as follows:

```
195 NYTSPSATPRPPAPGPPQSRGT-----SPLQPGSYPEYQASGADSWPPAA 239
    ..:.|.. ||.| ||||...|.     .|.:|| 		  :..||..
212 GASGPMG-PRGP-PGPPGKNGDDGEAGKPGRPG----------ERGPPGP 249

240 ENSFPGANFGVPPAEPEPIPKGSRPGGSPRGVSFQFPFPALHGASTKPFP 289
    :   ||. |:|.....|..||.|       .|.|.||.....|
250 Q----GAR-GLPGTAGLPGMKGHR-----------GFSGLDGAKGDAGP 282
```

    Assuming you are given the starting residue number for each sequence, the alignment string (containing residues and gaps) for each sequence, and the number of residues to print per line, write an algorithm for displaying the aligned sequence in EMBOSS format. Also, describe how you can determine the starting residue numbers for local alignment (for global, the starting residue numbers are 1).

3. Implement the *InputSubstitutionMatrix* algorithm shown in Figure 14-9 as a Python function. The function should return the substitution matrix, which will be a dictionary of dictionaries. To test your *InputSubstitutionMatrix* function, write a test script that calls the function to create the substitution matrix. Use the PAM150 substitution matrix shown in Figure 14-8. This matrix and others can be downloaded from ftp://ftp.ncbi.nih.gov/blast/matrices/. Because Python dictionaries are not ordered, it will be difficult to verify the output matrix by simply printing it. Instead, your test script should include a series of dictionary lookups that you can verify against the PAM150 matrix shown in Figure 14-8.

4. Implement the LCS in Python.

    a. Implement the function *Score_LCS(seq1, seq2)* shown in Figure 14-15, where *seq1* and *seq2* are strings. The function should return the *score*, the length of the common subsequence, and *TB*, the traceback matrix. Hint: Append a dummy character at the beginning of each sequence so that the indices of *M* and *TB* match the sequence indices. To test your *Score_LCS* function, create a test script using the sample data from the example shown in Figure 14-13 (*seq1* = TACGC and *seq2* = ATCAGC). Verify that your *TB* matrix is the same as shown in Figure 14-13 and that your score is 4.
    b. Implement the function *Print_LCS(seq1, TB, i, j)* shown in Figure 14-16, where *seq1* is a sequence, *TB* is the trace back matrix, and *i*, and *j* are indices that indicate the current location in *TB*. Note that *i* refers to the row of *TB* and *j* refers to the current column of *TB*. The *Print_LCS* function should print the longest common subsequence. To test your *Print_LCS* function, create a test script using the sample sequence 1 and *TB* matrix shown in Figure 14-13. Notice, if you have already completed part a, you can use the *TB* matrix created by *Score_LCS*. Otherwise, you can type in the data shown in Figure 14-13 into a list of lists. Note that the initial values for *i* and *j* are *n*, the length of *Sequence 1*, and *m*, the length of *Sequence 2*.
    c. To practice using Python modules, create a Python module (file) called *score.py* for your *Score_LCS* function and another Python module (file) called *printLCS.py* for your *Print_LCS* function. Create a third Python module (file) called *lcs.py* for the main script. The main script should import *score.py* and *printLCS.py*. The main script should prompt the user for two input sequences (strings), input the two sequences, and call the *Score_LCS* and *Print_LCS*

functions to to compute and display the score and the longest common subsequence. Again, test the main script in *lcs.py* on the sample data from Figure 14-13. In addition, create your own test sequences (strings) for verifying that the program works.

5. Modify the *Print_LCS* function to append characters to a string called *lcs* instead of directly printing the characters. The *Print_LCS* function should return the string *lcs* and then the test script can print *lcs*. Verify the correctness of *Print_LCS* as discussed in problem 4b.

6. Any problem that can be solved recursively can also be solved iteratively. Write an iterative algorithm for *Print_LCS(seq1, TB, i, j)*. Code your algorithm in Python and verify its correctness as discussed in problem 4b.

7. **Ends-free global alignment (also known as semi-global alignment) performs end-to-end global alignment, but it does not penalize gaps at the beginning or end of the aligned sequences. Define the three necessary conditions for extending LCS to implement ends-free global alignment.**
   i. How should the first row and first column of the scoring matrix *M* be initialized?
   ii. After computing *M* and *TB*, where can the score of the alignment be found in *M* and where should traceback start in *TB*?
   iii. Given *TB*, where should traceback end?

8. **Given your solution to problem 7, repeat problem 1 for ends-free global alignment using the PAM150 substitution matrix and a linear gap penalty with *w* = −4. Use the global alignment recurrence equation 14.2.**

PROJECT

Implement SPA, the simple pairwise sequence alignment tool shown in Figure 14-4. Note that you are encouraged to customize your tool's name.

1. **Base project—global pairwise sequence alignment:**

   Extend the LCS algorithm to implement global alignment using dynamic programming. Use PAM and BLOSUM substitution matrices and assume a fixed gap penalty. You can either prompt the user for the gap penalty (preferred) or get it from the substitution matrices (value in last row/column).

   The program should prompt the user to enter the substitution matrix filename and two sequences. The program should display the aligned sequences, showing gaps (-) in each sequence and the matches (|) between the sequences, and the alignment score.

2. **Extensions (select at least one of the following):**
   i. Extend your program to support local alignment algorithm.

   ii. Extend your program to support ends-free global alignment.
   iii. Modify your program to work with a query sequence and a database. In this case, instead of prompting the user for two sequences, the program should prompt the user for the query sequence and the database flat file. The format *of each entry in* the database flat file is:

   >sequence identifying information
   {sequence without spaces or newline characters}

   The program should also prompt the user for a score threshold. All sequences that exceed that score will be displayed (ideally in order from highest score to lowest score).

   iv. Extend local alignment algorithm to print out all alignments with maximum score.
   v. Extend your program to support affine gap penalties. Use the gap penalty from the scoring matrix for your gap opening penalty.

   Note: A solution is not provided for the project.

ANSWERS TO THOUGHT QUESTIONS

14-1.

a. The first amino acid ('K') in the top sequence of this snippet is residue 291 of the human p53 protein. How do we know that? The slice of the alignment starts at residue 288, and moving left we count each amino acid (ignoring gaps). Similarly, the first

amino acid ('K') in the bottom sequence of this snippet is residue 321 of human p73.

b. The snippet below shows the alignment with the corresponding score for each residue pair using the BLOSUM62 substitution matrix and for each gap using a gap open penalty of −10 and a gap extend penalty of −0.5.

```
K K G    E      P    H    H    E    L    P    P  G S T K R A
|  .  |                                              .  :  :  |  |  |
K N G    -      -    -    -    -    -    -    -  A A S K R A
5  0  6  −10.5  −0.5 −0.5 −0.5 −0.5 −0.5 −0.5 −0.5 0 1 1 5 5 4
```

The alignment score is: 5 + 6 + −10.5 + 7*(−0.5) + 0 + 1 + 1 + 5 + 5 + 4 = 13.[8] Notice that be-

---

[8] Note that alignment programs do not need to compute the score this way. The score is actually computed and used by the algorithm to determine the optimal alignment.

cause the affine gap penalty is being used, when a gap is opened, both the gap open penalty of −10 and the gap extend penalty of −0.5 are applied. After the gap is opened, each subsequent gap is only penalized by the gap extend penalty.

c. The length of this snippet is 17.

d. The identity of the snippet is 5/17 (29.4%).

e. The percent similarity of the snippet, the pairs of residues with scores greater than or equal to one, is 7/17 (41.2%).

f. The number of gaps in the snippet is 8/17 (47.1%).

REFERENCES

Dasgupta, S., C. H. Papadimitriou, and U. V. Vazirani. 2008. *Algorithms*. Boston: McGraw-Hill Higher Education.

EMBOSS. 2012. "Alignment Formats." Accessed November 12. http://emboss.sourceforge.net/docs/themes/AlignFormats .html.

Gotoh, O. 1982. "An Improved Algorithm for Matching Biological Sequences." *Journal of Molecular Biology* 162: 705–708.

Hetland, M. L. 2008. *Beginning Python from Novice to Professional*, 2nd ed. Berkeley, CA: Apress.

Hirschberg, D. S. 1974. "A Linear Space Algorithm for Computing Maximal Common Subsequences." *Communications of the ACM* 18: 341–343.

Jones, N. C., and P. A. Pevzner. 2004. *An Introduction to Bioinformatics Algorithms*. Cambridge, MA: MIT Press.

Karp, R. M. 1972. "Reducibility among Combinatorial Problems." In *Complexity of Computer Computations*, edited by R. E. Miller and J. W. Thatcher, 85–103. New York: Plenum.

National Center for Biotechnology Information. 2012. "Web BLAST Page Options." Accessed November 9. http://www.ncbi.nlm.nih.gov/blast/blastcgihelp.shtml.

Needleman, S. B., and C. D. Wunsch. 1970. "A General Method Applicable to the Search for Similarities in Amino Acid Sequence of Two Proteins." *Journal of Molecular Biology* 48: 443–453.

Smith, T. F., and M. S. Waterman. 1981. "Identification of Common Molecular Subsequences." *Journal of Molecular Biology* 147: 195–197.

GLOSSARY

2D gel electrophoresis A method used to separate proteins by size and isoelectric point. The method is used in proteome studies because many proteins can be visualized at one time. In addition, relative levels of proteins can be easily measured and the proteins can be identified by subsequent techniques. (10)

addition principle The principle that when outcomes can be divided into non-overlapping cases, the total number of outcomes is the sum of the number of outcomes in each case. (11)

advantageous mutation A mutation that increases the propensity of an organism to create viable progeny. (3)

algorithm A set of steps to accomplish a task. (13, 14)

algorithm complexity Quantifies the amount of memory space (space complexity) or time (time complexity) that an algorithm requires as a function of the size of the input. (14)

alignment score threshold A similarity score used by the BLAST program to determine whether that score is to be reported to the user. (6)

allele Alternative sequence variant that occurs at a particular locus in a species. (3)

alternative splicing A process that produces different messenger RNAs (mRNAs) from a single transcript. The different mRNAs are called alternatively spliced variants. (2)

amino acid A molecule composed of an alpha carbon, a carboxyl group, an amino group, and a side chain. Amino acids can be polymerized to form proteins with distinct molecular functions. Each side chain gives a unique chemical property to its amino acid. (1)

amplification (also known as DNA amplification) DNA copy number is increased several-fold. This can occur in some parts of a chromosome in cancers. Amplification is also used to describe the process of polymerase chain reaction (PCR). (3, 9)

ancestral gene The precursor gene from an extinct species that gave rise to at least one gene in a living species. (3)

assembly (also known as genome assembly) Ordering of sequenced DNA segments. (9)

B-factor (B-value, Debye-Waller factor, temperature factor) A measure of the deviation of an atom from its average location. The B-factor usually ranges from 20 to 80 Å^2. (7)

Bayes' law A rule to compute a conditional probability that is at the heart of Bayesian inference. (11)

Bayesian inference A method to update the likelihood of a hypothesis based on observed data. For example, the likelihood of a particular genetic makeup of parents can be computed based on the genetic makeup of the children. (11)

Bernoulli random variable The simplest discrete random variable, with just two possible values, namely 0 and 1. (11)

Bernoulli trial An experiment that has only two possible outcomes. (11)

binomial random variable A discrete random variable that counts the number of successes in a fixed number of Bernoulli trials that all have the same success probability. (11)

bit score A measure of the similarity between sequences that does not depend on the size of the database that was searched. (12)

BLAST (Basic Local Alignment Sequence Tool) A popular heuristic program that rapidly compares a query sequence to a subject sequence. (6)

BLOSUM substitution matrix A general term of a set of amino acid substitution matrices derived from amino acids changes observed in multiply aligned sequences found in motifs and domains. (4)

bootstrapping A statistical method that uses rearranged character data to measure the robustness of the topology of a phylogenetic tree. Each clade of a tree may be assigned a bootstrap value, which is based on the percentage of bootstrapping pseudoreplicates that match the given clade topology in the phylogenetic tree. (8)

bridge amplification A PCR amplification method that uses a DNA strand as a template. The DNA strand is hybridized to a primer bound to a surface. The second strand synthesis creates a free end that is complementary to a second primer bound to a surface. It also creates a bridge such that one end is bound to the surface and the second end is hybridized to a surface-bound primer. Successive rounds of PCR amplify the DNA. (10)

catalytic site Region within an enzyme that binds to a substrate, assists in the conversion of the substrate into the product, and releases the product. (3)

cDNA Complementary DNA synthesized from RNA using reverse transcriptase. The final product consists of two strands. (2)

CDS The protein coding sequence within a nucleotide sequence. (2)

cell cycle A sequence of events that lead to cell division. The phases are G1 (gap 1), S (DNA synthesis), G2 (gap 2), and M (mitosis). Another phase is G0, a quiescent phase where cells exit the cell cycle for long periods of time. (10)

central dogma A term that explains the relationship between DNA, RNA, and proteins. Briefly, DNA replicates and serves as the template for its transcription into RNA. RNA serves as the template for its translation into protein. RNA can also be reverse transcribed into DNA. (1)

centromere A protein-DNA complex located on a chromosome. The centromere divides the chromosome into two unequal segments. The longer segment is the q-arm and the short segment is the p-arm. During mitosis spindle fibers attach to the centromere and the two chromatids of the chromosome are pulled away from each other. (9)

Chou-Fasman method A secondary structure prediction program created in 1974 that uses a sliding window. Experimental solved crystal structures are used to predict the amino acid secondary structures from primary sequence information. (7)

chromosome A segment of the genome tightly wound and combined with protein. The segments range in size. In diploid organisms, such as humans, there are two sets of almost identical segments (with the exception of the XY set in males). After genome replication each replicated segment is called a chromatid (a pair of identical chromatids are called sister chromatids). Chromatids are held together by a protein structure called a centromere near the center of the segment. The ends of the chromatids are called telomeres. (1)

clade A group of species connected to a common node. (8)

Clustal Omega A multiple sequence alignment program that uses unweighted pair group method with arithmetic mean (UPGMA) to cluster sequences and hidden Markov models to align sequences. (6)

ClustalW A popular multiple sequence alignment program that creates a rooted tree from a distance matrix. The rooted tree, called a guide tree, is used to determine the order of progressive pairwise alignments. The sequence at the tip of the longest branch is added last to the alignment. Through clustering, an intermediate unrooted tree is created. Sequences are associated with different weights depending on the length of the branches that connect them to the guide tree. (6)

clustering Grouping of similar things. Clustering is used in a number of bioinformatics software programs. In BLOSUM substitution matrices and ClustalW, it is used to increase (i.e., add more weight) the contribution of unique sequences in the scoring. (4)

coding strand The DNA strand that is identical in sequence to the RNA transcript with the exception that DNA has T's and the RNA has U's. (2)

codon Three nucleotides that code for an amino acid or a signal to terminate protein translation. (1)

complement of an event The set of outcomes that are in the sample space but not in the event. (11)

complementary strand The DNA strand that binds to the written strand through hydrogen bonds formed by pairs of nucleotide bases. (1)

conditional probability The probability of an event given that another event has occurred. (11)

conditional statements (if-tests) Statements in Python that perform different operations depending on whether the condition is true. (13)

configuration A particular arrangement of all atoms in space in a system at a particular time. (7)

conserved region A segment of a sequence that, after multiple alignments with other ortholog or paralog sequences, has a higher fraction of identical or similar aligned residues than other segments of a sequence. (3)

continuous random variable A random variable with values that lie in a continuous interval or collection of intervals. (11)

copy number variation (CNV) A DNA segment that is present at a number that is different from the number found in a reference genome with a usual copy number. For humans and other diploid organisms the usual copy number of the reference genome is two. (1, 9)

coverage In RNA-seq it is the number of fragment reads per kilobase per million fragments mapped. Coverage gives the relative frequency a particular region of the genome is detected by RNA-seq. (10)

CpG islands Regions in the genome enriched for the CpG dinucleotide sequence. CpG islands often exist near promoters. Methylation of cytosines within CpG islands correlates with gene silencing. (9)

data (in computer science) Information that is stored in computer memory and manipulated by computer operations. (13)

data type Defines the kind of a data value and determines how the value can be used in a programming language. Some data types are mutable, and some are immutable. Examples of data types are integer, floating point, and string. (13)

debug Process of identifying and fixing errors (bugs) in a computer program. (13)

deleterious mutation A mutation that decreases the propensity of an organism to create viable progeny. (3)

dependence and independence Two events are independent if the occurrence of one event does not change the probability that the other one occurs. If the events are not independent, then they are called dependent. (11)

dictionaries (keys and values) A Python structure for organizing data *values* that can be looked up (accessed) using *keys*. (13)

dideoxy sequencing (also known as Sanger sequencing or chain termination sequencing) A method of DNA sequencing that uses a DNA polymerase to incorporate unnatural dideoxyribose-containing nucleotides (ddNTPs) to terminate DNA synthesis. The ddNTPs have fluorescent tags that allow them to be detected by a laser-fluorescence detector. This was the dominant method of DNA sequencing from the early 1980s to 2010. (9)

discrete random variable A random variable with values that form a finite or countable set. (11) *Also, see related glossary terms* **Bernoulli random variable, binomial random variable,** *and* **Poisson random variable.**

distribution The description of a random variable, consisting of the possible values and their probabilities. In the case of a discrete random variable, this description takes the form of a probability mass function; in the case of a continuous random variable, a probability density function. (11) *Also, see related glossary terms* **extreme value distribution** *and* **normal (Gaussian) distribution.**

disulfide bond A covalent bond created between two sulfur atoms. Some cysteine amino acid residues form disulfide bonds to stabilize the tertiary and quaternary structures of proteins. (3)

divergence time (time of divergence) Period of time in which two species last shared a common ancestor. (8)

DNA (deoxyribonucleic acid) A chain of nucleotides where each nucleotide is of one of four bases, adenine, guanine, cytosine, thymine and a deoxyribose sugar and a single phosphate within each nucleotide. Most genomes are composed of DNA. (1)

dot plot A plot created by placing dots in a matrix to create a main diagonal when two sequences are similar. (5)

Dotter A sliding window program that compares the similarity of two proteins by producing a dot plot. (5)

E-value For a given similarity score s, the expected number of HSPs with at least score s, assuming random sequences. Used as a measure to assess whether similarity score s could have been obtained by random chance. Note, sometimes s is capitalized to S. (6, 12)

elements (also known as entries) Individual items in a matrix. (4)

enhancer A specialized response element located more than 200 nucleotides away from a gene promoter. The enhancer can be altered in its location and in its orientation relative to the promoter and still affect transcriptional activation. (10)

event A set of outcomes of an experiment to which a probability can be assigned. (11)

evolutionary distance The estimated number of substitutions that have occurred since two species shared a common ancestor genome. (8)

exome The portion of the genome that codes for protein and "functional" RNAs such as tRNAs and rRNAs. In humans, the exome constitutes less than 2% of the entire genome. (9)

exon A segment of a gene that is transcribed as part of the initial transcript. The initial transcript undergoes splicing to keep the RNA encoded by the exon. (2) *Also see related glossary term* **intron.**

exon shuffling A process in which the exon of a gene is duplicated in the same gene, or copied and moved to another gene. Exon shuffling is thought to contribute to the modular nature of proteins (3)

expected value An alternative name for the mean of a distribution. (11)

extreme value distribution A distribution of a random variable that is itself the maximum of a set of random variables, used in bioinformatics to assess whether an alignment score indicates biological significance of the alignment. (12)

FASTA format Also known as Pearson format, formatted data in which the first line begins with ">" to signify that it is the header line. The following lines contain a nucleotide or amino acid sequence. (2)

fitness The ability of the organism to survive to the age where it reproduces and creates viable offspring. (3)

floating point number A format used to represent real numbers in computer systems. (13)

fluorescence in situ hybridization (FISH) A method of staining chromosomes with oligonucleotides attached to molecules that fluoresce when exposed to light of a particular wavelength. (9)

for loop A Python statement used to iterate over an ordered sequence such as a list, string, or tuple. As it iterates over the sequence, each value in the sequence can be operated upon. (13)

function A named encapsulated sequence of instructions that performs a specific task. Functions can be called by name, passed input data, and return results after performing the task. (13)

gamma distance model A nucleotide substitution model that considers a variable rate of substitution due to relative frequencies of nucleotides in genomes. (8)

gap penalty A value that is subtracted from the similarity score in alignment programs. The penalty may be linear or affine. A linear gap penalty means that the same value is subtracted for each gap added to optimize sequence alignment. An affine gap penalty, which is more commonly used in sequence alignment programs, is composed of two components: a gap opening penalty and a linear gap penalty. (5)

GenBank An annotated database that contains nucleotide sequences derived from DNA or RNA sources. Each record in the database consists of three sections: header, feature keys section, and nucleotide sequence. (2)

gene A segment of the genome that produces a protein or a functional RNA, such as transfer RNA, small nuclear RNA, ribosomal RNA, and so on. (1)

genetic code A series of codons that cause amino acids to be placed in a specific order within a protein. All organisms use the same genetic code. (1)

genome The DNA found in the organism. For some cells a separate genome exists in some organelles. In such cases, we distinguish the two genomes by saying there is a nuclear genome and an organelle genome (mitochondrial genome, chloroplast genome, etc.). In rare instances the genome can be composed of RNA, but this is limited to RNA viruses. (1)

genotype The set of alleles contained within the genome. (9)

global alignment Optimal pairing of two sequences over the entire lengths of the sequences. (3, 5)

global variable A variable defined outside of any function and can be used by any function. (13)

GOR method A secondary structure prediction program that uses a sliding window. It predicts the secondary structure of each amino acid residue by considering each amino acid in a 17-residue window. The window slides in increments of one residue until the secondary structure of each amino acid is predicted. (7)

haplotype A set of linked genetic markers located on the same chromosome. (9)

HapMap database A collection of SNPs and tag SNP information from different ethnic populations worldwide. (9)

heatmap A display that shows color gradations that correlate to level of gene expression. (10)

hidden Markov model (HMM) A probabilistic model that assigns likelihoods to all possible combinations of gaps, matches, and mismatches to determine the most likely multiple sequence alignment (MSA) or set of possible MSAs. (12)

high-scoring segment pair Two amino acid or nucleotide sequences that, upon alignment, produce a similarity score s above some alignment threshold A. (6)

homolog A gene related to other genes by evolutionary descent from a common ancestral DNA sequence. A homolog may be an ortholog or a paralog. (3) *Also, see related glossary terms* **ortholog** *and* **paralog**.

homology modeling A protein structure prediction method that uses a protein template of known structure to build a structural model of a sequence. This method works well when the percent sequence identity is ≥50%. (7)

horizontal gene transfer (also known as lateral gene transfer) Transfer of DNA from one cell to another or from one organism to another without sexual or asexual reproduction. An example is when a virus captures a host gene from one cell and transfers the host gene to another cell. (3)

hydropathy plot A plot created from an output of a sliding window program that shows the hydrophobic areas of proteins. (5)

identifier (in computer science) Variable name. It is important when coding to create meaningful identifiers that make it easy to follow the data flow. (13)

identity (also known as percent identity) In two optimally aligned sequences, the number of identical residues divided by the number of residues plus gaps in the alignment multiplied by 100. (3)

identity score Sum of matched residues in two optimally aligned sequences. (4)

immutable (in computer science) Used to describe a type of data in Python that cannot be altered without creating a new location (variable). Immutable data types include integers, floats, strings, and tuples. For example, assume an integer variable x has been assigned and holds the value 5; the statement $x = x + 1$ actually creates a new variable called x in a new location and assigns it the value 6. (13)

indel A mutation that results in the insertion or deletion of nucleotides into the genome. (3)

integer A type of data in Python for representing whole numbers that can be positive, negative, or zero. (13)

intersection of events The set of outcomes that are in every one of the events referred to. (11)

intron A segment of a gene that is transcribed as part of the initial transcript. The initial transcript undergoes splicing to remove the RNA encoded by the intron. (2) *Also see related glossary term* **exon**.

inverse Fourier transform A mathematical treatment of a Fourier transform (diffraction pattern) used in X-ray crystallography to depict electron densities of atoms. (7)

inversion (also known as DNA inversion) The reversal of a DNA segment in one genome relative to another genome. (9)

isoelectric focusing (IEF) A method used to separate proteins by isoelectric point. A strip of gel contains ampholytes that create a pH gradient. Proteins migrate by electrophoresis until they settle at a position where their net charges are neutral. (10)

isoelectric point The pH of a solution where a dissolved molecule has a net neutral charge. (7)

Jukes-Cantor model A model of evolutionary distance that predicts the actual number of nucleotide (amino acid residue) replacements based on the observed fraction of sites that differ between two gene sequences. (8, 12)

k-permutations of an n-element set The number of different ways one can select k items from n possible choices, when the order of selection matters. (11)

Kimura two-parameter model A nucleotide substitution model that explicitly considers that transversions are less frequent than transitions. (8)

lac operon A bacterial gene regulatory system responsible for regulated import and hydrolysis of lactose. (10)

law of total probability The rule that the total probability of an event can be computed as a weighted average of conditional probabilities. (11)

Li-Fraumeni syndrome A rare familial cancer exhibiting autosomal dominant inheritance and early onset of tumors, multiple tumors within an individual, and multiple affected family members. The most common cancer types are soft tissue sarcomas, osteosarcomas, breast cancer, brain tumors, leukemia, and adrenocortical carcinoma. Most Li-Fraumeni syndrome patients inherit one mutant *TP53* allele. The second *TP53* allele is commonly observed to be mutated in the cancer tissue, but not normal tissue. (2)

liquid chromatography mass spectrometry (LC-MS) A two-step method that first separates molecules by column chromatography. This step usually uses an HPLC. Eluant from the column is ionized and sent through a mass spectrometer. The charge-to-mass ratio of each molecule is determined with high precision. (10)

list A comma-separated sequence of elements delimited by square brackets. Elements within a list can be of any data type, though usually they are the same data type. (13)

local alignment Optimal pairing of two subsequences within two sequences such that the similarity score remains above a set threshold. (3, 5)

local gapped alignment A type of BLAST program that bridges two pairs of aligned sequences that are separated by relatively few gaps. (6)

local variable A variable with a local scope is visible only within the block of code in which it is defined. For example, variables defined within a function by default can be used only within the function. (13)

locus A specific sequence on a chromosome that is experimentally detectable. (3)

long interspersed elements (LINEs) DNA repeats derived from class I transposon retrotransposition. LINEs code, or did at one time code, for enzymes with reverse transcriptase activity and other proteins. (9)

long terminal repeat (LTR) transposons DNA that is similar to retrovirus reverse transcriptase and retrovirus repeats. (9)

longest common subsequence (LCS) algorithm An algorithm that determines the longest common sequence of characters between two strings. The LCS algorithm is a starting point for the design of pairwise sequence alignment programs. (14)

low complexity region A sequence with a repeated pattern of amino acids or nucleotides. (6)

main diagonal Elements located in the longest diagonal of a matrix, starting at the upper left corner. (4)

Markov chain A discrete-time model that represents transitions of an object from one condition (state) to another without memory. That is, the probability of a state depends only on the prior state and not on the sequence of states that preceded it. (12)

masking A feature of BLAST that allows the user to exclude a region of the query sequence from contributing to the alignment. It is used when there is a region in the query sequence that is very common to many sequences in the database. The user is more interested in regions of similarity found between the unmasked areas of the query and the database. (6)

matrix A rectangular array of numbers, symbols or expressions. (4)

matrix (in computer science) A two-dimensional array or list. Can be formed in Python using list of lists— that is, a list of elements in which each element is another list. (14)

maximum similarity score The highest similarity score created by extending the hit by the BLAST program. (6)

MDM2 A ubiquitin ligase that places multiple ubiquitin units onto p53, which marks p53 for destruction. MDM2 also binds to the transactivation domain of p53 and escorts p53 from the nucleus to the cytoplasm. (2)

mean of a distribution The mean of all the possible values of the random variable described by the distribution, weighted by how likely they are. (11) *Also, see related glossary term* **sample mean**.

methods (in computer science) A function that belongs to a class. Example *upper()* is a method for the string class. Given the object *str*, an instance of the string class that has been assigned a string value, the expression *str.upper()* will invoke the method *upper()* to convert every lowercase character in the string to an uppercase character. (13)

methylome The location of all of the methylated CpG sequences in the genome. (9)

microarray technique A method to specify the level of all or a subset of transcripts in the organism. Usually, the method requires reverse transcription of RNA into cDNA, labeling the cDNA and hybridization of the cDNA to complementary single strand DNA molecules bound to a surface. (10)

microRNA (miRNA) RNA molecules approximately 20 nucleotides in length that silence gene expression by hybridizing to mRNAs. The targeted mRNAs are degraded by the RNA-induced silencing complex. (10)

microsatellites Segments of DNA, known as simple repeats, that have a range of 1–3 nucleotides that are repeated in tandem in blocks of up to 200 nucleotides in length. The lengths of these blocks often vary from individual to individual and are used for DNA profiling. (3, 9)

misfolding Macromolecules, especially proteins, may not correctly fold to their native states due to thermal or chemical stress. Such proteins either must be degraded or refolded to their native states. Excess buildup of misfolded proteins, also known as protein aggregation, can lead to disease including Alzheimer's and Parkinson's. (7)

module A file containing Python definitions and statements. The file name is the module name with the suffix .py appended. (13)

molecular clock (molecular evolutionary clock) A theory that states that the rate of observable sequence change is linearly proportional to divergence times. (8)

molecular dynamics (or molecular dynamics simulation) A method of protein structure prediction that uses heat energy and molecular mechanics to predict atom positions. (7)

molecular mechanics (MM) A method of modeling structure and movement of molecules using Newtonian (or classical) physics principles. Free energy of the molecule is estimated from a force field composed of the sum of separate expressions that describe the potential energy of a single macromolecular conformation. (7)

molecular viewer Software program that uses PDB files as input and displays structures from these PDB files. (7)

MS/MS spectrum A spectrum that plots relative abundance of molecules versus mass/charge (m/z) ratio. To generate the MS/MS spectrum, peptides are fragmented by a neutral gas. The masses of the fragments and the original peptide can be used to deduce the peptide sequence. (10)

multiregional hypothesis An explanation of the origin of modern *Homo sapiens* that supposes that more than one common ancestor was located in different geographical locations. (8)

multiple sequence alignment (MSA) Alignment of more than two protein or nucleotide sequences. MSA is useful for detection of conserved regions of proteins that may be of functional and structural significance. (6)

multiplication principle The principle that if there are *a* ways of doing something and *b* ways of doing another thing, then there are $a \cdot b$ ways of doing both things. (11)

mutable (in computer science) Used to describe a type of data in Python that can be altered. Mutable data types include lists and dictionaries. (13)

mutation An alteration in DNA sequence that produces a sequence that is different from normal DNA and is passed on to daughter cells. (1, 3) *See related glossary term* **neutral mutation**.

native state Natural state of macromolecule. It is thought that the native state is the structure of the macromolecule in nature at which it is fully functional. (7)

natural selection A process by which biological traits become more or less common in a population over a period of time. The traits have a beneficial or harmful effect on the production of progeny and the progeny's survival. (3)

neighbor-joining method An evolutionary distance matrix-based clustering algorithm that creates a phylogenetic tree using the shortest possible evolutionary time. Pairs of OTUs are identified that minimize total branch length within the tree. At each stage of the clustering process the total branch length is minimized. (8)

neighborhood words A list of words that are similar to or identical to the query words. The words in the list have, upon alignment to the query words, similarity scores above a word threshold set by the user or the BLAST program. (6)

neutral mutation A mutation that does not alter the ability of the organism to produce viable progeny (i.e., does not alter the fitness of the organism). (1, 3)

Newick tree format (also known as Newick notation or New Hampshire tree format) A list of rules and distances used by phylogenetic tree drawing programs to create phylogenetic trees. (8)

next-gen sequencing Post-dideoxy sequencing methods that quickly sequence DNA samples. Many of these methods use the principle of sequencing by synthesizing DNA strands one nucleotide at a time in parallel. (9)

nonsynonymous mutation A mutation in the coding region of a gene that alters the protein sequence. (3) *Also, see related glossary term* **synonymous mutation**.

normal (Gaussian) distribution A very important probability distribution in statistics. Physical quantities (such as measurement errors) that result from the actions of many independent processes are typically normally distributed. (11)

Northern blotting An experimental method used to determine relative steady-state level and length of specific transcripts. (10)

nuclear magnetic resonance (NMR) spectroscopy The perturbation of atom nuclei by the application of a magnetic field to gain information about the location of neighboring atoms. An NMR spectrum is produced by chemical shifts produced by nuclei resonance frequencies relative to the resonance frequency of a reference molecule. (7)

nuclear run-on An experimental method used to determine the rate of transcription. As DNA is transcribed, labeled nucleotides are incorporated into the transcripts. The transcripts are hybridized to oligonucleotides, and unhybridized molecules are washed away. From the amount of label incorporated into the transcripts within a specified time period the transcription rate can be calculated. (10)

nucleotide A molecule composed of a nitrogenous base, a sugar, and a minimum of one phosphate. (1)

object-oriented programming (OOP) language A programming language that provides a set of rules for creating and managing objects. Examples of OOP languages are Python, C++, and Java. (13)

oncogene A mutated form of a gene whose normal function is to promote cell growth or cell survival. The mutation may cause abnormally high levels of oncogene product or abnormally high levels of oncogene function. (3) *Also, see related glossary terms* **proto-oncogene** *and* **tumor suppressor gene**.

open reading frame A segment of DNA that has the potential to code for a protein because it begins with a start codon (codes for Met) and there are no stop codons for a relatively long stretch of DNA. (9)

operational taxonomic unit (OTU) A species or group of species at the external node of a phylogenetic tree. (8)

operators Symbols used to indicate the arithmetic or logical operation to perform on data values. Overloaded operators have different meanings depending on the data type. Examples of operators are + (add or concatenate), * (multiply or replicated), & (bitwise AND), && (logical AND), and ** (power). (13)

organelle Membrane-bound entities in cells, which perform specialized functions in eukaryotic cells. Examples include mitochondria, chloroplast, Golgi apparatus, nucleus, and nucleolus. (1)

origin of replication The location within the genome or plasmid where DNA replication starts. (9)

ortholog One gene of a set of genes that descended from a single gene in a common ancestor. The set of genes diverged from one another due to the evolution of a new species. (3) *Also, see related glossary terms* **homolog** *and* **paralog**.

outgroup A species or group of species approximately equally unrelated to other species in a rooted tree. (8)

P-value The probability of seeing a value of a random variable that is as large as or larger than a given value. (12)

p53 A tumor suppressor protein (canonical length in humans is 393 residues) responsible for initiating cell cycle arrest, apoptosis, or DNA repair in response to cell stress, such as DNA damage, ribosome denaturation, and oncogene activation. (1)

PAM substitution matrix A general term for a set of amino acid substitution matrices based on the principle of missense mutations accepted by natural selection. These were the first amino acid substitution matrices that were widely used to calculate similarity scores in aligned sequences. (4)

PAM1 mutational probability matrix A matrix where an average of 1% of the amino acids have changed during evolution. (4)

paper chromatography Experimental method that separates molecules on the basis of charge and hydrophobicity. (1)

paralog One gene of a set of genes that that underwent a duplication event in a common ancestor. The set of genes diverged from one another due to evolution of

a new species. (3) *Also, see related glossary terms* **homolog** *and* **ortholog**.

peptide mass fingerprint Masses of peptides created from proteins by protease digestion are measured by mass spectrometry. Knowledge of the masses and the protease is used to identify the proteins. (10)

phage (also known as bacteriophage) A DNA or RNA virus that attacks bacteria. (9)

phenotype The physical appearance of an organism that is largely dictated by the genotype. (9)

phylogenetic tree (phylogram) A diagram that portrays evolutionary relationships in which the branch lengths are proportional to evolutionary time. (8)

phylogenetics A type of phylogeny that uses character data (observable traits) to describe evolutionary relationships. One type of character data that can be used to describe evolutionary relationships is sequence information. When sequence is extensively used, it is called molecular phylogenetics. (8)

phylogeny The description of evolutionary relationships of organisms, or groups of organisms. (8)

Poisson correction A method of generating a rate of sequence substitution that accounts for multiple substitutions. (8)

Poisson random variable A discrete random variable that counts the number of events occurring in a fixed interval of time if these events occur with a fixed average rate and independently of each other. A Poisson random variable is used in the Jukes-Cantor model to count nucleotide changes occurring over time. (11)

polycistronic RNA A transcript that codes for more than one protein. Polycistronic RNAs have more than one start codon and more than one stop codon. (10)

polymerase chain reaction (PCR) A method to amplify the level of DNA in an exponential manner. The method is used extensively in molecular biology and in other fields. (9)

polymorphism A DNA sequence alteration observed in more than 1% of the population of the species. The most common polymorphism is a single nucleotide polymorphism (SNP). (3)

prior and posterior probabilities In Bayesian inference, the probabilities before the observation of new data and after the observed data have been used to reassess the probabilities, respectively. (11)

probability density function A function that describes the relative likelihood of a continuous random variable to take on a given value. (11)

probability mass function A function that gives the probabilities that a discrete random variable is exactly equal to each of its possible values. (11)

probability of an event A measure of the likelihood that the event will occur, that is, that any one of the outcomes in the event occurs. (11)

program A set of instructions that tell a computer what to do in order to perform a task. (13)

protein One or more chains of amino acid residues where each chain is a minimum of 50 residues long. Proteins perform most of the biochemical functions in the organism. (1)

protein data bank A database that contains Cartesian coordinates of atoms of biomolecules. The majority of the coordinates are derived from X-ray crystallography experiments. Molecular viewers can be used to display the atoms on a computer screen. (2, 7)

protein half-life The time it takes for 50% of a population of newly synthesized proteins to be degraded. (10)

proteome The identity and level of proteins expressed from the genome in a cell. (10)

proto-oncogene A gene whose normal function is to promote cell growth or cell survival. This is the wild-type version of the oncogene. (3) *Also, see related glossary terms* **oncogene** *and* **tumor suppressor gene**.

pseudo-code An informal way of writing an algorithm or program. Pseudo-code represents the program's functions and data flow, but does not require strict syntax. (13, 14)

pseudogene A gene that does not produce a protein or functional RNA such as tRNA or rRNA. Some pseudogenes express RNAs that influence protein expression by binding to microRNAs. (9)

PSI-BLAST (Position-Specific Iterated BLAST) A software program that creates a position-specific substitution matrix (PSSM) from the top hits of a BLAST run. The PSSM is used as a substitution matrix for more rounds of BLAST searching. PSI-BLAST is useful for detecting distant homologs of proteins. (6)

PSIPRED A secondary structure prediction program that uses a neural network and PSI-BLAST. (7)

Q_3 An equation used to measure the secondary structure prediction accuracy of software programs. (7)

query words Short segments of a query sequence used to create a hit in BLAST. (6)

R-factor (reliability factor, residual factor, R-value, R_{work}) A measure of how well the final crystal structure model predicts the observed data used to create the model. An R-factor value in the 0.15-0.25 range is considered satisfactory. (7)

random variable A variable that can take on a set of possible different values each with an associated probability. (11) *Also, see related glossary terms* **continuous random variable** *and* **discrete random variable**.

read (also known as sequencing read) An experimentally determined sequence obtained from a single sequencing run on a segment of DNA. Reads can be 25 to 1,000 nucleotides in length depending on the sequencing method. (9)

recent African origin hypothesis An explanation of the origin of modern *Homo sapiens* that supposes that a single common ancestor arose in sub-Saharan Africa. (8)

recombination (also known as crossing over) A process that occurs during meiosis where two parental chromosomes exchange DNA. This results in offspring with genotypes that differ from those of their parents. (9)

recurrence (also known as recurrence relation) An equation that defines an element in a sequence based on one or more of the earlier elements in the sequence. (14)

RefSeq A secondary database derived from GenBank that contains wild-type sequences. (2)

renaturation The process of folding an unfolded or misfolded protein into its native state. (7)

response element (also known as responsive element) Segment of DNA that binds to protein activators and repressors. Upon binding, the DNA-protein complex influences gene transcription rates, often by binding to RNA polymerase II. (10)

retrotransposon A transposon that requires an RNA intermediate. Also known as class I transposon. (3)

retrovirus A virus containing two copies of a RNA genome and reverse transcriptase enzyme. In part of the virus life cycle its genome is reverse transcribed into double strand viral DNA and this incorporates into the host cell genome. (3)

RNA (ribonucleic acid) A chain of nucleotides where each nucleotide is one of four bases, adenine, guanine, cytosine, uracil, and a ribose sugar and a single phosphate within each nucleotide. RNA is transcribed from DNA and performs many functions including the coding of proteins. (1)

root mean square deviation (RMSD) A method used to quantify the difference of locations of atoms in two structures of the same atom composition. RMSD is often used to calculate the differences between a predicted protein structure and an experimentally derived protein structure. (7)

root node A bifurcation point within a phylogenetic tree. The single branch that bifurcates connects to the Tree of Life. (8)

rooted tree A phylogenetic tree that specifies the order of all speciation events. Rooted trees contain an outgroup, which is required to order the speciation events. (8)

sample mean The arithmetic mean of a set of observed values of a random variable, typically used as an estimate of the mean of the distribution of that random variable. (12)

sample space The set (collection) of all possible outcomes or results of an experiment in probability theory. (11)

sample standard deviation A value computed from observations of a random variable, used as an estimate of the standard deviation of the distribution of the random variable. This value is approximately equal to the standard deviation of the observed values. (12)

satellites Segments of 4–200 nucleotides that are repeated in tandem up to a total of hundreds of kilobases. They are called satellites because when mammalian DNA is centrifuged through a CsCl density column, satellite DNA migrates to a position in the column that is different from the major DNA fraction. (9)

script A program or sequence of instructions that is interpreted by another program (a run-time environment) that will carry out the execution of the program. In contrast, *compiled* programs are executed directly by the computer hardware. (13)

sequence (in computer science) An ordered collection of objects or elements. There are three types of sequences in Python: lists, strings, and tuples. Sequences can be indexed to access specific elements and sliced to create subsequences. (13)

sequence alignment Optimized pairwise matching of nucleotide or amino acid sequences. (1)

short interspersed elements (SINEs) DNA that was reverse transcribed from RNA. The RNAs, originally generated by RNA polymerase III, were 5S rRNA, tRNA, and others. (9)

sickle cell anemia Disease caused by a mutation that results in an abnormal sickle shape of the erythrocyte. The abnormal erythrocytes stick to endothelial cells of the blood capillaries and prevent blood flow. (1)

significant decay A decrease in the similarity score created by extending the hit further than the maximum similarity score. (6)

similarity (also known as percent similarity) In two optimally aligned sequences, the number of similar residues divided by the number of residues plus gaps in the alignment multiplied by 100. (3)

similarity score In two optimally aligned sequences, the sum of scores of residue matches and mismatches. Mismatches include residue mismatches and residues aligned with gaps. (4, 6)

single nucleotide polymorphism (SNP) A single base difference between one individual and another at the same position in the genome. In humans, this occurs in approximately 1 nucleotide in every 1,200 nucleotides. (9)

sliding window program A sliding window is a segment that partitions information. A sliding window program performs a calculation on information in the segment and then the segment moves incrementally. At each increment the program repeats the calculation. The output from each calculation can be displayed in a graphical format. (5)

Sov (fractional overlap of segments) An equation used to measure the secondary structure prediction accuracy of software programs. This equation takes the order of secondary structures into account and is thought to be superior to Q_3 as a measure of structure prediction accuracy. (7)

standard deviation of a distribution The square root of a variance, used to express how spread out the corresponding values are in the units of those values. (11) *Also, see related glossary term* **sample standard deviation.**

state diagram A graph that shows the transitions between states of a Markov chain and the probabilities of each transition. (12)

steric hindrance Electrostatic repulsion (also known as van der Waals repulsion) caused by close proximity of electron clouds of two or more atoms. Steric hindrance can restrict atom movement. (7)

string A sequence of characters enclosed by quotes. A built-in data type in Python. (13)

synonymous mutation A mutation in the coding region of a gene that does not alter the protein sequence. This is also known as a silent mutation. (3) *Also, see related glossary term* **nonsynonymous mutation**.

syntax A set of rules that describe how to organize symbols to create correctly structured code as defined by a specific programming language. The code must have proper syntax in order for it to be correctly interpreted or compiled. (13)

synteny The similar arrangement of genes in the genomes of two species that share a common ancestor. (3, 9)

Tajima relative rate test (Tajima test) A calculation used to determine if two sequences are undergoing similar rates of substitution. In the calculation, substitution frequencies are computed by comparing sequences to an outgroup sequence. (8)

telomere The end of a linear chromosome composed of repeated nucleotide sequences. (9)

template strand In replication and in transcription, the strand that is being used as the basis for synthesis of the new nucleotide strand. The template strand is complementary to the newly synthesized strand. (2)

threading A protein structure prediction method that uses the CATH database to predict the structure of each residue of a protein sequence. Each residue is tested for optimal compatibility with each fold in the CATH database. (7)

TP53 The gene that codes for p53, located on chromosome location 17p13.1, approximately in the base pair range 7,571,720-7,590,868. The gene codes for multiple alternatively spliced forms of mRNA. (1)

traceback matrix A matrix created in some dynamic programming algorithms, such as longest common sequence (LCS) and sequence pairwise alignment (SPA), that serves to keep track of all of the possible highest scoring paths so that once scoring is complete and the highest scoring path is known, it can be used by these algorithms to create the output sequence for LCS or aligned sequences for SPA. (14)

transcription The process of polymerizing nucleotides to produce RNA from DNA. (1)

transcriptome The identity, length, and relative level of all transcripts expressed in a cell. (10)

transition A mutation that results in a substitution of one purine for another or one pyrimidine for another. (3)

transition matrix A matrix containing all the transition probabilities of a Markov chain. (12)

transition probabilities The probabilities that describe how likely it is for each state in a Markov chain to transition to each other possible subsequent state in one time step. (12)

translation The process of polymerizing amino acids in an order dictated by messenger RNA. (1)

translocation Mutation where one part of a chromosome fuses to another chromosome. (3)

transmembrane protein Proteins that traverse a biological membrane. The Kyte-Doolittle sliding window program was the first program to successfully predict which regions of a transmembrane protein traverse a biological membrane. (13)

transposon (also known as transposable elements, mobile elements, and mobile genetic elements) Segment of DNA moved from one location to another in the genome. There are two mechanisms for this movement: direct transfer and indirect transfer. (3)

transversion A mutation that results in the substitution of one purine for one pyrimidine or vice versa. (3)

trimming Shortening of aligned sequences of extended hits until the maximum similarity score is obtained. (6)

tumor suppressor gene A gene whose normal function is to suppress cancer formation by restraining cell growth or causing apoptosis. In cancers, tumor suppressors are inactivated. (3) *Also, see related glossary terms* **oncogene** *and* **proto-oncogene**.

tuple An immutable sequence of elements of any data type. Tuples are delimited by parentheses, and elements are separated by commas. Useful for returning multiple values from a function, for assigning multiple values simultaneously, and for defining a constant set of values. (13)

union of events The set of outcomes that are in at least one of the events referred to. (11)

unit cell A small three-dimensional segment that contains atoms. The environment of the atoms at the vertices formed by the intersection of the unit edges must be identical. The unit can be repeated to create a three-dimensional structure of the molecule. (7)

unit evolutionary period Time required for a protein sequence to change by 1%. (8)

unrooted tree A phylogenetic tree that has no root and therefore does not show an order of speciation. (8)

UPGMA (unweighted pair group method with arithmetic mean) A clustering method that uses a distance matrix to create a phylogenetic tree. UPGMA assumes a constant molecular clock for all species in the tree. It creates an ultrametric tree. (8)

UTR Untranslated region at either of the two extreme ends of the gene. This is a segment of RNA that is not translated into protein and does not contain the stop codon. The 5′ UTR is a UTR that is at the 5′ end of the gene, and the 3′ UTR is located at the 3′ end of the gene. (2)

variable A symbolic name associated with a data storage location within a program. The value of a variable can be changed as the program executes. (13)

variance of a distribution The variance of all the possible values of the random variable described by the distribution, weighted by how likely they are. This quantity is a measure of the spread of the values of the random variable around the mean of the distribution. (11) *Also, see related glossary term* **standard deviation of a distribution**

Venn diagram A graphical tool for visualizing the union of events, the intersection of events, and the complement of an event. (11)

vertical gene transfer Transfer of DNA through sexual or asexual reproduction (i.e., from parent to offspring or parent cell to daughter cell). (3)

while loop Used for repeating a block of code until a defined condition is met. (13)

whole genome shotgun sequencing (WGSS) A sequencing method that includes random fragmentation of many copies of a genome. The fragments are sequenced and assembled to produce the entire genome sequence. (9)

X-ray crystallography A method that uses X-ray bombardment of crystallized molecules to create a diffraction pattern. The diffraction pattern is used to create a three-dimensional arrangement of the atoms in the molecules. (7)

z-score Expresses how many standard deviations the value of a random variable lies above or below the mean of that variable. (11)

CREDITS

Chapter 1

Box 1-1 Image and article reprinted with permission from the News Syndication. From "Bride with rare cancer gene opts for mastectomy after wedding" by Russell Jenkins –The Times (London) July 25, 2007.

Figure 1-12 Adapted from M.W. Davidson and The Florida State University Research Foundation, "Molecular Expressions, Cell Biology and Microscopy Structure and Function of Cells & Viruses," http://micro.magnet.fsu.edu/cells/ (accessed January 18, 2016).

Figure 1-13 Adapted from National Human Genome Research Institute, https://www.genome.gov (accessed January 18, 2016).

Figure 1-14 Part A image adapted from Public Domain Images, http://www.public-domain-image.com/free-images/science/microscopy-images/comparative-ultrastructural-morphology-between-normal-red-blood-cells-rbcs-and-a-sickle-cell-rbc (accessed January 18, 2016). Part B and C images adapted from National Heart, Lung, and Blood Institute; National Institutes of Health; U.S. Department of Health and Human Services http://www.nhlbi.nih.gov/health/health-topics/topics/sca (accessed January

Figure 1-16 Images reprinted by permission from Macmillan Publishers Ltd: Nature; V. M. Ingram, "A Specific Chemical Difference Between Globins of Normal and Sickle-Cell Anæmia Hæmoglobins," *Nature* 178 (1956): 792–794.

Chapter 2

Box 2-1 Article courtesy of Los Alamos Science, Los Alamos National Laboratory. Image from "Genbank" by W. B. Goad in Los Alamos Science No. 9, Fall 1983 p52–63.

Figure 2-2 Courtesy of the National Library of Medicine.

Figure 2-3 Courtesy of the National Library of Medicine.

Chapter 3

Box 3-1 Acc. 90–105 - Science Service, Records, 1920s-1970s Smithsonian Institution Archives.

Figure 3-2 Image from https://commons.wikimedia.org/wiki/File:Trichoplax_adhaerens_photograph.png. Creative Commons Attribution 4.0 International license. Source: Eitel M, Osigus H-J, DeSalle R, Schierwater B (2013) Global Diversity of the Placozoa. PLoS ONE 8(4):e57131.

Figure 3-3 The results here are in whole or part based upon data generated by the TCGA Research Network: http://cancergenome.nih.gov/." Credit for data analysis and image: Robyn Latimer.

Figure 3-4 Program: EMBOSS Needle. Website: http://www.ebi.ac.uk/Tools/psa/emboss_needle/ Reference: McWilliam H, Li W, Uludag M, Squizzato S, Park YM, Buso N, Cowley AP, Lopez R. Analysis Tool Web Services from the EMBL-EBI. Nucleic Acids Res. 2013 Jul;41(Web Server issue):W597-600. doi: 10.1093/nar/gkt376. Epub 2013 May 13. PubMed PMID: 23671338; PubMed Central PMCID: PMC3692137

Figure 3-12 Program: EMBOSS Needle. Website: http://www.ebi.ac.uk/Tools/psa/emboss_needle/ Reference: McWilliam H, Li W, Uludag M, Squizzato S, Park YM, Buso N, Cowley AP, Lopez R. Analysis Tool Web Services from the EMBL-EBI. Nucleic Acids Res. 2013 Jul;41(Web Server issue):W597-600. doi: 10.1093/nar/gkt376. Epub 2013 May 13. PubMed PMID: 23671338; PubMed Central PMCID: PMC3692137.

Figure 3-13 Program: EMBOSS Needle. Website: http://www.ebi.ac.uk/Tools/psa/emboss_needle/ Reference: McWilliam H, Li W, Uludag M, Squizzato S, Park YM, Buso N, Cowley AP, Lopez R. Analysis Tool Web Services from the EMBL-EBI. Nucleic Acids Res. 2013 Jul;41(Web Server issue):W597-600. doi: 10.1093/nar/gkt376. Epub 2013 May 13. PubMed PMID: 23671338; PubMed Central PMCID: PMC3692137.

Figure 3-14 Adapted from J. V. Moran, R. J. DeBerardinis, and H. H. Kazazian, Jr. "Exon shuffling by L1 retrotransposition," *Science* 283 (1999): 1530–1534.

Exercise 3-6 Image adapted from J. J. Yunis and O. Prakash, "The Origin of Man: a Chromosomal Pictorial Legacy," *Science* 215 (1982): 1525–1529.

Chapter 4

Box 4-1 Image of Ruth E. Dayhoff, M.D.; U.S. National Library of Medicine.

Figure 4-3 Adapted from M. O. Dayhoff, R. M. Schwartz, and B. C. Orcutt, "A Model of Evolutionary Change in Proteins," in *Atlas of Protein Sequence and Structure*, Vol. 5, Suppl. 3 (Washington, DC: National Biomedical Research Foundation, 1978).

Figure 4-4 Adapted from M. O. Dayhoff, R. M. Schwartz, and B. C. Orcutt, "A Model of Evolutionary Change in Proteins," in *Atlas of Protein Sequence and Structure*, Vol. 5, Suppl. 3 (Washington, DC: National Biomedical Research Foundation, 1978).

Figure 4-5 Adapted from D. T. Jones, W. R. Taylor, and J. M. Thornton, "The Rapid Generation of Mutation Data Matrices from Protein Sequences," *Computer Applications in the Biosciences* 8 (1992): 275–282.

Figure 4-6 Modified from M. O. Dayhoff, R. M. Schwartz, and B. C. Orcutt, "A Model of Evolutionary Change in Proteins," in *Atlas of Protein Sequence and Structure*, Vol. 5, Suppl. 3 (Washington, DC: National Biomedical Research Foundation, 1978).

Figure 4-7 Modified from M. O. Dayhoff, R. M. Schwartz, and B. C. Orcutt, "A Model of Evolutionary Change in Proteins," in *Atlas of Protein Sequence and Structure*, Vol. 5, Suppl. 3 (Washington, DC: National Biomedical Research Foundation, 1978).

Figure 4-8 Modified from M. O. Dayhoff, R. M. Schwartz, and B. C. Orcutt, "A Model of Evolutionary Change in Proteins," in *Atlas of Protein Sequence and Structure*, Vol. 5, Suppl. 3 (Washington, DC: National Biomedical Research Foundation, 1978).

Figure 4-12 Adapted from P. G. Higgs and T. K. Attwood, *Bioinformatics and Molecular Evolution* (Malden, MA: Blackwell Publishing, 2005).

Chapter 5

Figure 5-2 From J. Kyte and R. F. Doolittle, "A Simple Method for Displaying the Hydropathic Character of a Protein," *Journal of Molecular Biology* 157 (1982): 105–132.

Figure 5-4 Parts B and C adapted from A. D. Baxevanis and B. F. F. Ouellette, eds., *Bioinformatics: A Practical Guide to the Analysis of Genes and Proteins*, 2nd ed. (New York: John Wiley and Sons, 2002).

Figure 5-7 Adapted from S. B. Needleman and C. D. Wunsch, "A General Method Applicable to the Search for Similarities in the Amino Acid Sequence of Proteins," *Journal of Molecular Biology* 48 (1970): 443–453.

Chapter 6

Box 6-1 Modified from the Partnership for Public Service Monday, February 9, 2009; 12:00 AM *The Washington Post.*

Figure 6-2 Adapted from J. Pevsner, *Bioinformatics and Functional Genomics*, 2nd ed. (Hoboken, NJ: Wiley-Blackwell, 2009).

Figure 6-6 National Library of Medicine.

Figure 6-7 Part A from the Center for Disease Control, Dr. Mae Melvin, http://phil.cdc.gov/phil/home.asp (accessed March 1, 2016). Part B from Center for Disease Control, http://phil.cdc.gov/phil/home.aspk (accessed March 1, 2016).

Figure 6-8 Adapted from J. D. Thompson, D. G. Higgins, and T. J. Gibson, "CLUSTAL W: Improving the Sensitivity of Progressive Multiple Sequence Alignment Through Sequence Weighting, Position-Specific Gap Penalties and Weight Matrix Choice," *Nucleic Acids Research* 22 (1994): 4673–4680.

Figure 6-9 Adapted from J. D. Thompson, D. G. Higgins, and T. J. Gibson, "CLUSTAL W: Improving the Sensitivity of Progressive Multiple Sequence Alignment Through Sequence Weighting, Position-Specific Gap Penalties and Weight Matrix Choice," *Nucleic Acids Research* 22 (1994): 4673–4680.

Figure 6-10 Adapted from J. D. Thompson, D. G. Higgins, and T. J. Gibson, "CLUSTAL W: Improving the Sensitivity of Progressive Multiple Sequence Alignment Through Sequence Weighting, Position-Specific Gap Penalties and Weight Matrix Choice," *Nucleic Acids Research* 22 (1994): 4673–4680.

Figure 6-12 Software program courtesy of EMBL-EBI.

Chapter 7

Box 7-1 Figure 1 Adapted from Protein Data Bank file 1TSR.

Box 7-2 Figure 2 Adapted from R. Melero, S. Rajagopalan, M. Lazaro, A. C. Joerger, T. Brandt, D. B. Veprintsev, G. Lasso, D. Gil, S. H. Scheres, J. M. Carazo, A. R. Fersht, and M. Valle, "Electron Microscopy Studies on the Quaternary Structure of p53 Reveal Different Binding Modes for p53 Tetramers in Complex with DNA," *Proceedings of the National Academy of Sciences USA* 108 (2011): 557–562.

Figure 7-1 Adapted from "Protein Folding" sponsored by Wikipedia, http://en.wikipedia.org/wiki/Protein_folding/ (accessed January 18, 2016).

Figure 7-3 Image from David Leibly, UCLA-DOE Institute of Genomics and Proteomics. Protein Data Bank file number 4W6J.

Figure 7-5 From A. H. Kwan, M. Mobli, P. R. Gooley, G. F. King, and J. P. Mackay, "Macromolecular NMR for the Non-spectroscopist," *FEBS Journal* 278 (2011): 687-703.

Figure 7-6 Part A is an image of unit cells from PDB file 1STP (Protein Data Bank) (accessed July 16, 2014). Part B is an image of a cross-section of a protein created using Chimera software (version 1.4.1). The plane was clipped to expose the core of the protein. This image was taken from R. Sanchez, "Exploration of New Parameters That Could Play a Role in the Accuracy of the Cysteine Oxidation Prediction Algorithm (COPA)," MS thesis, California State University, Los Angeles (Publication No. AAI 1515168).

Figure 7-8 The image in part A is from "CPK Coloring," Wikipedia, http://en.wikipedia.org/wiki/CPK_coloring (accessed June 24, 2012). The images in part B–D were created with Chimera molecular viewer (version 1.6.1) using PDB file 1LW6.

Figure 7-9 From "Protein Folding," Wikipedia, http://en.wikipedia.org/wiki/Protein_folding (accessed June 24, 2012).

Figure 7-12 Reprinted by permission from Macmillan Publishers Ltd: S. Cooper, F. Khatib, A. Treuille, J. Barbero, J. Lee, M. Beenen, A. Leaver-Fay, D. Baker, Z. Popovic, and F. Players, "Predicting Protein Structures with a Multiplayer Online Game," *Nature* 466 (2010): 756–760, doi:10.1038/nature09304.

Figure 7-13 Data for PDB 1RV1 from L. T. Vassilev et al. "In vivo activation of the p53 pathway by small-molecule antagonists of MDM2," Science 303 (2004): 844–848.

Figure 7-14 Adapted from M. Zvelebil and J. O. Baum, *Understanding Bioinformatics* (New York: Garland Science, Taylor & Francis Group, 2008), 418.

Figure 7-15 From M. Zvelebil and J. O. Baum, *Understanding Bioinformatics* (New York: Garland Science, Taylor & Francis Group, 2008), 425.

Figure 7-17 Adapted from A. M. Lesk, *Introduction to Bioinformatics*, 3rd ed. (Oxford, UK: Oxford University Press, 2008).

Figure 7-18 Image is a modified version of a figure from D. T. Jones, "Protein Secondary Structure Prediction Based on Position-Specific Scoring Matrices," *Journal of Molecular Biology* 292 (1999): 195-202.

Chapter 8

Box 8-1 Image of a child Neanderthal with head reconstructed from fossil skull (image credit: © P.PLailly, E.Daynes/Lookat Sciences).

Box 8-2 Photo credit: Gitschier, J. 2008. "Imagine: An Interview with Svante Pääbo." PLoS Genetics 4(3): e1000035.

Figure 8-1 Image of Leonardo Da Vinci's The Vitruvian Man from Science Source. Image of fingernail was taken from http://www.normanallan.com/Med/askdr/finger.html (accessed August 26, 2012).

Figure 8-2 From "Tree of Life (Biology)," *Wikipedia*, http://en.wikipedia.org/wiki/Tree_of_life (biology) (accessed July 12, 2012).

Figure 8-3 From "Tree of Life (Biology)," *Wikipedia,* http://en.wikipedia.org/wiki/Tree_of_life (biology) (accessed July 12, 2012).

Figure 8-5 Adapted from D. Voet and J.G. Voet, *Biochemistry*, 4th Edition (Hoboken, NJ: John Wiley and Sons, 2011), and from R.E. Dickerson, "The structures of cytochrome C and the rates of evolution," *J. Mol. Evolution* 1 (1971): 26–45.

Figure 8-6 The image for part B was taken from the collaborative project of the University of California Museum of Paleontology and the National Center for Science Education, http://www.evolution.berkeley.edu/evosite/evo101/IIE1cMolecular clocks.shtml (accessed July 26, 2012).

Figure 8-7 Modified from Andrew Rambaut's presentation "Molecular Clocks and Dating," http://www.molecularevolution.org/molevolfiles/presentations/Molecular%20Clocks.pdf (accessed July 26, 2012).

Figure 8-8 Adapted from V. M. Sarich and A. C. Wilson, "Immunological Time Scale for Hominid Evolution," *Science* 158 (1967): 1200–1203.

Figure 8-10 Courtesy of Itai Yanai.

Figure 8-12 Adapted from P. G. Higgs and T. K. Attwood, *Bioinformatics and Molecular Evolution* (Malden, MA: Blackwell Publishing, 2005).

Figure 8-14 The program used to create this image was Clustal Omega, from http://www.ebi.ac.uk/Tools/msa/ (accessed July 31, 2012).

Figure 8-16 Modified from P. S. Soltis and D. E. Soltis, "Applying the Bootstrap in Phylogeny Reconstruction," *Statistical Science* 18 (2003): 256–267.

Figure 8-20 Modified from M. Ingman, H. Kaessmann, S. Pääbo, and U. Gyllensten, "Mitochondrial Genome Variation and the Origin of Modern Humans," *Nature* 408 (2000): 708–713. Used by permission of the Nature Publishing Group.

Figure 8-21 Modified from the Recent African Origins of Modern Humans website http://en.wikipedia.org/wiki/Recent_African_origin_of_modern_humans (credit: NordNordWest (File: Spreading homo sapiens ru.svg by Urutseg) [Public domain], via Wikimedia Commons).

Figure 8-22 From "Phylogenetic tree" by Eric Gaba (Sting - fr:Sting) - NASA Astrobiology Institute, found in an article. Licensed under Public Domain via Commons https://commons.wikimedia.org/wiki/File:Phylogenetic_tree.svg#/media/File:Phylogenetic_tree.svg.

Figure 8-23 Adapted from V. A. Belyi, P. Ak, E. Markert, H. Wang, W. Hu, A. Puzio-Kuter, and A. J. Levine, "The Origins and Evolution of the p53 Family of Genes," *Cold Spring Harbor Perspectives in Biology* 2 (2010): a001198.

Chapter 9

Box 9-2 Image modified from Sheryl Ozinsky, "Can DNA Profiling Be the Answer to Reduced Crime?" OH WATCH, January 9, 2010, http://www.ohwatch.co.za/oh-watch-blog/2010/01/09/can-dna-profiling-be-the-answer-toreduced-crime/.

Figure 9-1 Image for part D from http://missinglink.ucsf.edu/lm/molecularmethods/images/sequencing.htm (accessed August 11, 2014).

Figure 9-5 Adapted from https://www.nanoporetech.com/ (accessed August 2,2014).

Figure 9-6 Image for part B is from Jean-Yves Sgro, http://www.virology.wisc.edu/virusworld/images/174_phix.jpg (accessed January 31, 2013).

Figure 9-8 Modified from Fleishmann et al., "Whole Genome Random Sequencing and Assembly of *Haemophilus influenzae* Rd," *Science* 269 (1995): 496–512. Used by permission of AAAS.

Figure 9-9 From Fleishmann et al., "Whole Genome Random Sequencing "Assembly of" (add space) Haemophilus influenzae Rd," *Science* 269 (1995): 496–512. Used by permission of AAAS.

Figure 9-11 Image generated by SynMap software program, part of a suite of programs collectively known as CoGe, Accelerating Comparative Genomics, http://genomevolution.org/CoGe/ (accessed February 16, 2013).

Figure 9-12 Giemsa-stained image from "Human Genome," Wikipedia, https://en.wikipedia.org/wiki/Human_genome (accessed February 16, 2013). CFTR gene chromosomal position image from https://ghr.nlm.nih.gov/primer/howgeneswork/genelocation (accessed August 3, 2014). FISH-stained image from http://www.genomenewsnetwork.org/resources/whats_a_genome/Chp1_2_1.shtml (credit the J. Craig Venter Institute) (accessed July 22, 2015).

Figure 9-13 Modified from https://genomics.soe.ucsc.edu/research/comp_genomics/human_chimp_mouse (accessed July 22, 2015).

Figure 9-14 Image generated by SynMap software program, part of a suite of programs collectively known as CoGe, Accelerating Comparative Genomics, http://genomevolution.org/CoGe/ (accessed February 24, 2013).

Figure 9-15 Data from L. I. Patrushev and I. G. Minkevich, "The Problem of the Eukaryotic Genome Size," *Biochemistry* (Moscow) 73 (2008): 1519-1552.

Figure 9-16 Adapted from P. Papasaikas and J. Valcarcel, "Evolution. Splicing in 4D," *Science* 338 (2012): 1547-1548.

Figure 9-17 Adapted from "Copy Number Variation," Wikipedia, http://en.wikipedia.org/wiki/Copy-number_variation#media viewer/File:Geneduplication.png (accessed July 31, 2014).

Figure 9-18 From the UCSC Genome browser (accessed April 3, 2013).W. J. Kent, C. W. Sugnet, T. S. Furey, K. M. Roskin, T. H. Pringle, A. M. Zahler, and D. Haussler, "The Human Genome Browser at UCSC," *Genome Research* 12 (2002): 996–1006.

Figure 9-19 From "Hypersensitive Site," Wikipedia, https://en.wikipedia.org/wiki/Hypersensitive_site (accessed September 20, 2013).

Figure 9-20 From the UCSC Genome browser (accessed April 3, 2013). W. J. Kent, C. W. Sugnet, T. S. Furey, K. M. Roskin, T. H. Pringle, A. M. Zahler, and D. Haussler, "The Human Genome Browser at UCSC," *Genome Research* 12 (2002): 996–1006.

Figure 9-22 Adapted from the International HapMap Project website sponsored by the National Center for Biotechnology Information (http://hapmap.ncbi.nlm.nih.gov/originhaplotype.html.en)

Figure 9-23 From https://www.23andme.com/.

Chapter 10

Box 10-1 Bio adapted from the Howard Hughes Medical Institute website, http://www.hhmi.org/research/investigators/brown_bio.html (accessed April 28, 2013).

Figure 10-1 Image of results from colorimeter experiment was taken from W. F. Loomis Jr. and B. Magasanik, "Glucose-

Lactose Diauxie in Escherichia coli," *Journal of Bacteriology* 93(1967): 1397–1401, and modified for clarity.

Figure 10-4 Part B adapted by permission from Nature Publishing group: H. L. Hinds, C. T. Ashley, J. S. Sutcliffe, D. L. Nelson, S. T. Warren, D. E. Housman, M. Schalling, "Tissue specific expression of FMR-1 provides evidence for a functional role in fragile X syndrome," *Nat Genet*. 3 (1993): 36–43.

Figure 10-6 Part B: Microarray image was modified from M. B. Eisen, P. T. Spellman, P. O. Brown, and D. Botstein, "Cluster Analysis and Display of Genome-Wide Expression Patterns," *Proceedings of the National Academy of Sciences* USA 95 (1998): 14863–14868.

Figure 10-7 Adapted from webinar by Christopher Mason, "RNA Sequencing: The Right Choice for Any Lab," October 5, 2010, http://www.illumina.com/events.ilmn#webinars (accessed April 19, 2013).

Figure 10-9 Modified from http://seq.molbiol.ru/sch_clon_ampl.html (accessed April 19, 2013).

Figure 10-10 Part A image modified from B. J. Hass and M. C. Zody, "Advancing RNA-Seq Analysis," *Nature Biotechnology* 28 (2010): 421–423. Part B image modified from J. M. Toung, M. Morley, M. Li, and V. G. Cheung, "RNA-Sequence Analysis of Human B-Cells," *Genome Research* 21 (2011): 991-998.

Figure 10-11 Adapted from "Proteomics/Protein Separations-Electrophoresis/Two Dimensional Polyacrylamide Gel Electrophoresis(2D-PAGE)," http://en.wikibooks.org/wiki/Proteomics/Protein_Separations_Electrophoresis/Two_Dimensional_Polyacrylamide_Gel_Electrophoresis(2D-PAGE) (accessed April 13, 2013).

Figure 10-12 From SWISS-2DPAGE, http://world-2dpage.expasy.org/swiss-2dpage/ (accessed April 14, 2013).

Figure 10-14 Adapted from A. M. Barsotti and C. Prives, "Noncoding RNAs: The Missing 'Linc' in p53-Mediated Repression," *Cell* 142 (2010): 358–360.

Exercise 10-6 Image adapted from Alan Wilson, former head of the Schistosomiasis Research Group at University of York.

Chapter 13

Box 13-1 Photo supplied by Russell Doolittle.

Figure 13-1 From J. Kyte and R. F. Doolittle, "A Simple Method for Displaying the Hydropathic Character of a Protein," *Journal of Molecular Biology* 157 (1982): 105–132.

Chapter 14

Box 14-1 Photo from https://simons.berkeley.edu/people/richard-karp (accessed May 25, 2016). Permission granted by Richard Karp.

INDEX